The Big Picture of Climate Change on Planet Earth

A "big picture" system-level perspective of the dynamic astrophysical, geophysical, natural, and human factors that affect the regional surface temperatures, climatic anomalies, adverse weather events, natural catastrophes, and rising ocean levels

by

Herbert A. Hutchinson

Scientist, System Safety Engineer, Genealogist, Inventor

Durham, North Carolina
2025

VERITY PUBLICATIONS

Published in the United States of America

ISBN 979-8-89395-178-3 (SC)
ISBN 979-8-89395-179-0 (HC)
ISBN 979-8-89395-180-6 (Ebook)

Library of Congress Control Number: 2025914565

Verity Publications
145 W Broadway
Long Beach, CA 90802, USA
www.stellarliterary.com

Order Information and Rights Permission:

Quantity sales. Special discounts might be available on quantity purchases by corporations, associations, and others. For details, contact the publisher at the address above.

For Book Rights Adaptation and other Rights Permission.
Call us at toll-free 1-888-945-8513 or send us an email at admin@stellarliterary.com.

Disclaimer

The science profession has earned a well-deserved reputation for advancing the knowledge of the natural and physical structure of the solar system and of the environment in which the human inhabitants of the planet Earth conduct their lives. The title of this book and factual contents herein illustrates the scope and complexity of the problems resulting from human/natural environmental interactions. When any scientific observation is made, there is always a possibility of varying interpretation as to the importance and completeness of the variables in the observation. The terminology "fallacious" is not intended to point a finger of blame for a fallacious observation. Instead, constructive criticism is used to show that an original fallacious interpretation of a database of information that is continually being updated (e.g., with advanced technology sensors on satellites) is not an error but requires more complete and more accurate measurements to be made defining the outdated knowns and previously unknowns in that database.

Examples of how the fallacies of the Earth - being a flat circular disc floating on a vast ocean and with the Sun orbiting about the flat disk - were updated by legendary philosophers and mathematicians (and also the first "naysayers) such as Nicolas Copernicus, Galileo de Galilei, and Johannes Kepler in the 1600s are legendary in science history. Updates by more recent scientists, physicists, and mathematicians as Svante Arrhenius, Milutin Milankovic, Charles Keeling, and NASA's Edwin Hubble and James Webb in the twentieth century are equal in significance to the knowledge base of Man.

The learning process in science is a cumulative one that ideally is based on astute observations, common sense logic, accurate and complete measurement of all significant factors, unbiased forensic-like analysis, and with the purest of ethics in interpretation and presentation to peers and the public. However, scientists are also humans, and humans occasionally commit misjudgments in moments of distraction, carelessness, or being unintentionally inconsiderate of the consequences when generating a piece of scientific information that may affect the safety, health, comfort or economics of a segment of the population affected by that information. The interpretation(s) of the myriad informational database(s) in this book – scientific or non-scientific – is/are entirely those of an unbiased qualified and experienced professional scientist.

The Author earned those qualifications and gained the experience to analyze complex databases and to write this book (thanks to those listed in the Dedication section of this book). The Author is also a member (now inactive) of, but not as a representative for, the not-for-profit Mensa International Society. The Author pledges the integrity of the purpose for and the contents of this book in keeping with the Mensa Society Constitution: *"to identify and foster human intelligence for the benefit of humanity."* It should not go unnoticed that the US Constitution also starts with "*We, the People*, . . . ". As the note on the cover of this Author's previous book "The Facts and Fallacies of Climate Change on Our Planet" indicates, *"It is far more about people than about just the environment".* This book "The Big Picture of Climate Change on Planet Earth" updates that previous conclusion to add the admonition "*Climate change is a regional non-uniform problem not just an "average" global problem.*"

The book has also been written with the intent to avoid further unrealistic and wasteful expenditure of resources to achieve the then-optimistic – but now outdated - goals (global temperature limiting and net zero carbon emissions) set in the middle of the twentieth century. The more recent assessment is aimed at setting more-achievable goals and realistic priorities of technologies for needed public attention and cooperation in pursuing better (more efficient and with less polluting by-products) methods and technologies for the harnessing and application of natural and geophysical sources of clean and reliable energy.

Further, the Author has taken all measures to assure the information in this book contains no data, information, or less-than-professional interpretation of the environmental database that is believed to have been misleadingly retouched from its published form in reputable scientific journals and reports. Similarly, this book has not "cherry-picked" such that only a portion of the database selected that is forced to agree with a scientific hypothesis that has been incompletely analyzed or presented in a form intended to mislead the public and governmental environmental protection regulators. The seriousness and urgency for resolution of the safety and health "crisis" due to climate change, is much exaggerated by a certain segment of the environmental science profession focused only on protecting the environment. The urgency for corrective actions is clarified herein and presented as motivation rather than as a threat of an apocalyptic nature to the human population if the unrealistic goals are not met.

Most of the photographs, diagrams, and database information related to climate change, global warming, and ocean levels rising were obtained by the Author from the databases of information in the public domain from the several government organizations identified in the Acknowledgment section of this book. Inasmuch as those organizations were and are being funded by Government grants (whose source is taxpayers' tax monies) and the Author being a taxpayer, the Author is using this book to disseminate factually updated information items – as a public service – that government organizations have already disseminated to the public. Just as that was not done by any government-funded organization to make a profit, the Author also is disseminating that information already in the public domain not for profit but also to add a different interpretation or perspective that results in a more factually based alternate interpretation. Where known, the Author assures that credit for the photograph, diagram, tabulated database, etc. has been given below the item to its creator. Any small infraction on a copyright restriction that may be contained anywhere in this book is purely unintended and for which the Author pre-apologizes to the creator who "used the sweat of his or her brow" to create that item of information (photograph, diagram, analysis document, or statement).

The Author has striven to avoid and dispel any criticism that the Author (long-retired) has published this book in collusion with any person or organization labeled as "nay-sayers" of the environmentalists' claim that Man burning fossil fuels is creating crisis-level emissions having carbon-based content that ominously traps heat in the atmosphere near Earth's surface. This book expressly avoids placing unwarranted blame on humans for the observed and measured temperatures on planet Earth becoming elevated and rising since the start of the "Industrial Revolution" of the late 19th century. The Author commends the science and engineering professions that have brought the developed countries of the World into a safer, healthier, more comfortable, and more prosperous state of existence for the human population. More needs to be done by the more developed countries to humanely assist those in lesser-developed countries who cannot do that in their present natural environment and status of economy.

The Author chooses not to accept any form of personal financial gain from publication of this book which was written as a public service. Any and all royalties or profits gained from the sale of this book "The Big Picture of Climate Change on Planet Earth" to better inform and enlist the cooperation of the public (domestic and elsewhere) will be donated to the Salvation Army International and ear-marked to specifically go to their Emergency Services sector. That organization has provided emergency aid to the surviving victims of major natural catastrophes around the world and seek no front-page news coverage for having done that. In the same vein that this book has followed, it is "for the benefit of humanity."

Fact Sheet

Based on an independent, unbiased analysis of valid databases of scientific information conducted “for the benefit of humanity”, the facts and fallacies about global warming and climate change are summarized in this Fact Sheet as follows:

1) There were no discernible anthropogenic greenhouse gas effects found in the several scientific databases when assessed in a professionally ethical “big picture” system engineering analysis approach by the unbiased Author. Environmental scientists disregarding the complex factors causing climate change and “cherry-picking” the total scientific databases and, in collusion with the sensationalistic news media, have intentionally misinformed and alarmed the public to believe there is an imminent catastrophic environmental crisis.

2) The surface temperatures of the planet (land and water) are rising but are found to be caused by combined astrophysical, geophysical, and natural factors beyond human control, NOT by “anthropogenic greenhouse gas” emissions. Mathematical calculations by Dr. Milutin Milankovic confirm the cyclical frequency at which geophysical parametric variables appear in meteorological databases.

3) The combined heat-trapping natural (geophysical) greenhouse gas emissions (measured in Vostok Antarctica pre-Industrial Revolution ice core sample analysis to be 295 ppmv parts per million by volume (0.0295% concentration) and water vapor (at a global average of 1.4%) total to approximately 1.43% average concentration of natural heat-trapping gases in Earth’s atmosphere affecting different regions to different extent depending on their latitude, and not in a “one-size-fits-all” average global manner.

4) Concentrated analysis of the Vostok ice core sample database (by ice research scientist Dr. Nicolas Caillon of France) found that it is temperature that drives natural carbon dioxide in the atmosphere and not the converse. The lag of carbon dioxide peak behind the temperature peak (and with the absence of human influence) was determined to be approximately 800 years.

5) The amount of anthropogenic (human and industrial generated) greenhouse gases measured by NOAA at Mauna Loa are currently (April 28, 2025) at 135 ppmv (430 ppmv total – 295 ppmv natural) or 0.0135% concentration in the atmosphere. The small concentration of anthropogenic emissions is an insignificant factor to global warming or climate change.

6) The ratio of natural (atmospheric chemistry + water vapor) greenhouse gas concentration to that of anthropogenic (human and industrial) greenhouse gas concentration is 1.43% to 0.0135% or 106 to 1 (as shown in climate scientist Dr. S. Fred Singer’ “A Closer Look at the Numbers” analysis). Environmental scientists disregard the effect of water vapor which contradicts their fallacious hypothesis that anthropogenic greenhouse gases dominate the global warming effect on climate change.

7) The statistical approach used by environmental scientists of “averaging” and “linearizing” meteorological data masks the factual seasonal temperature extremes of hot and cold at regional topographical elevations to which all life forms in a latitudinal region (not merely globally) must adapt. A better method (considering latitude, elevation, and ocean proximity effects) for estimating average annual temperature was developed using World Meteorological Organization database statistics. (LEO)

8) The technology of coal burning in the USA must be reduced because coal ash residue contains toxins and carcinogens which, when disposed in landfills factually pollute underground aquafers endangering life (human, animate and agricultural). Combustion of fossil fuels – oil, petroleum fuels, and natural gas - generates no emissions affecting regional or global surface temperatures, climate, or ocean levels.

It is scientifically unrealistic – not “naysaying” - to expect that reducing anthropogenic greenhouse gas emissions by eliminating combustion of all fossil fuels (coal, oil, and natural gas) will have a significant effect on retarding or avoiding global warming, climate changes, or ocean levels rising. Technologies for mitigating the effects of the inevitable surface temperature rises, climate changes, and ocean surface levels rising and the urgent need to better control anthropogenic pollution (especially plastic materials and leakages/spillages of oil and natural gas) must be escalated to a high priority – national and international - as the projected growing population will exacerbate human safety, health, sanitation, and/or environmental pollution hazards.

Dedication

This book is quinarily dedicated to:

This Author's father Karl – who taught his son two basic rules of life: 1) things do not just happen, they are caused to happen; and 2) some things can be prevented from happening, and some things cannot.

This Author's mother Mildred – who taught her son that the more important things in life are not things but are the people in every person's life who serve as role models, teachers, advisors – and friends.

The Brooklyn Technical High School – where this Author was taught early in life how to think about scientific problems that need to be solved with a logical thought process and the use of the basic tools of scientific knowledge (especially physics and mathematics).

The Georgia Institute of Technology – where this Author was taught the basics of science, technology, engineering, and mathematics (STEM) and the systems engineering processes for applying that knowledge to the development of new technologies and systems for the benefit of Mankind.

USAF Lieutenant General James T. Stewart (Commander, Aeronautical Systems Division) – who enabled this Author (as a Chief System Engineer in Federal Civil Service) to apply everything the above four dedicatees taught this Author in the integration of a set of advanced technologies into a complex system that became, and is still in use (since 1975) in the military aircraft inventory of 28 nations for the defense of the free-World. The most meaningful reward given to this Author is described in an extract of a letter of commendation (October 28, 1975) from Lt. Gen. Stewart:

Mr. Herbert A. Hutchinson
Deputy for Engineering
Aeronautical Systems Division
Wright-Patterson AFB OH 45433

Dear Mr. Hutchinson

It is a pleasure to award you the Air Force Systems Command Certificate of Merit. Your performance as Chief Systems Engineer, Deputy for the Air Combat Fighter/F-16 System Program Office from July 1974 to July 1975 is deserving of this recognition.

As a result of your dedicated efforts, the Air Combat Fighter made a successful transition from the prototype program, through source selection, into a rigorous full scale development program. The F-16 Air Combat Fighter has introduced many unique technological innovations into weapon systems development.

I commend you for this outstanding performance and wish you continued success.

(This Author subsequently was merit-promoted to Federal Civil Service Grade GS-15 and selected to be the Chief of the USAF Flight Technology Division, 1979 to 1982.)

Acknowledgement

There is a large and constantly growing database of information for the topics of global warming, climate change, and ocean levels rising describing their causes and effects on the land, water, and atmospheric resources and the human population of planet Earth. The total database is compiled from three major scientific observation and measurement activities and a major mathematical analysis to understand and define the factors that are the "causative" variables and those that are the "effect" variables affected by the multiple, complex factors associated with global warming, climate change, and the ocean levels rising. There are many organizations within the government and independent scientific, technology, engineering, and mathematical (STEM) research academies that deserve acknowledgment for the enormous applications of human analytical efforts, administrative labor, and funds to observe, photograph, measure, document, compile, and interpret the large complex databases. Listing the myriad of participants would be analogous to this book appearing to be a telephone directory.

The four primary types of databases and the dominant contributing and originating organizations and their staffs given special acknowledgment for this book are:

1) The Astrophysical and Geophysical factors database (1900-Present)
Special acknowledgement is given to:
The Milutin Milankovic Astrophysical Calculation Team, the original Russian and subsequent International (Russian, United States, British, Scandinavian, and French) Ice Core Research teams at the Vostok Research Station (Antarctica), the US Geological Survey, the Smithsonian Institute for Science, the National Geographic Society, the National Science Foundation, the Encyclopedia Brittanica, and the International Union of Geological Sciences.

2) The Natural factors database (1880-Present)
Special acknowledgement is given to:
The National Oceanic and Atmospheric Administration (including the National Weather Service, the National Hurricane Center, the National Tornado Center, the National Snow and Ice Data Center, and the Keeling Atmospheric Observatory at Mauna Loa), and the US Geological Survey Agency (Earthquake Hazards Reduction Program; Volcano Hazards Program).

3) The Weather and Climatic factors database (1900-Present)
Special acknowledgement is given to:
The National Weather Service, and the World Meteorological Organization.

4) The Earth Research factors database from land-based meteorological stations and space-based satellites and International Space Stations (1960-Present)
Special acknowledgement is given to:
The National Oceanic and Atmospheric Administration, the National Aeronautics and Space Administration, the National Academy of Science, the National Science Foundation, the Smithsonian Society for Science, and the National Geographic Society

Special acknowledgment is also given to the United Nations Population Division and Intergovernmental Panel on Climate Change (IPCC), the US Department of Energy, the US Department of Commerce, and to the many other scientific organizations and scientific research and analysis groups world-wide (governmental and institutional) supporting the activities to protect the human population and the environment.

Preface

Before delving into the very complex topics of global warming, climate change, and oceans' surface levels rising, it should be a question in the minds of readers of this book why the Author, whose background is detailed in the About the Author section of this book, has no formal education or experience in those branches of science to be considered credible. The principles of science, technology, engineering, and mathematics (STEM) are applied in a process known as integration – which in less technical terminology translates to the bringing together (integrating – bringing several into one) the multiple principles of the separate domains of science, technology, engineering, and mathematics. The product of integration is a system intended to perform a useful function. An analogy is that of a conductor of an orchestra or symphony harmonizing the music and the timing of the several instruments. The Author's education and experience gained during a highly successful professional career of over forty years is summarized in the About the Author at the end of the book.

This Author chose to specialize in the aspect of engineering termed as System Engineering with the emphasis on human safety while piloting or being transported in an aircraft system in the atmosphere of the planet. From the investigation of several aircraft crashes and accidents, the process of searching the system (in a "big picture" manner) for a root cause of the crash/accident – being a technical design flaw, some factor in the manner in which the system was operated outside of its design "envelope" (limits of speed at a specific altitude/density/temperature combination), or simply human error. For the factors of global warming and climate change, temperature is a common variable in the effect elevated temperature has on the environment, the population, or on mechanical systems.

Specifically of high interest to aeronautical engineers is the "hot day take-off" condition which increases the distance on the runway the aircraft uses to accelerate to a rotation and lift-off speed. Commercial aircraft operators sometimes on extremely hot days must delay the take-off of an aircraft (especially at hot airports such as Phoenix or Las Vegas) because there is insufficient runway length for safe take-off. Similarly, military aircraft operating on forward airbases with the shortest possible runway lengths are designed to have an afterburner on the jet engine to overcome the hot day take-off limitation to accomplish the mission.

This book later develops a temperature envelope of the extreme temperatures – at the extremes for that latitude - that everything on the planet's surface, animate or inanimate and developed and in development must endure and adapt to or will cease to exist. A preview of that temperature envelope – for the extreme hot and extreme cold latitudes of the United States – is shown as follows: Regions at other latitudes must develop the temperature envelope that is appropriate for those latitudes defining the region.

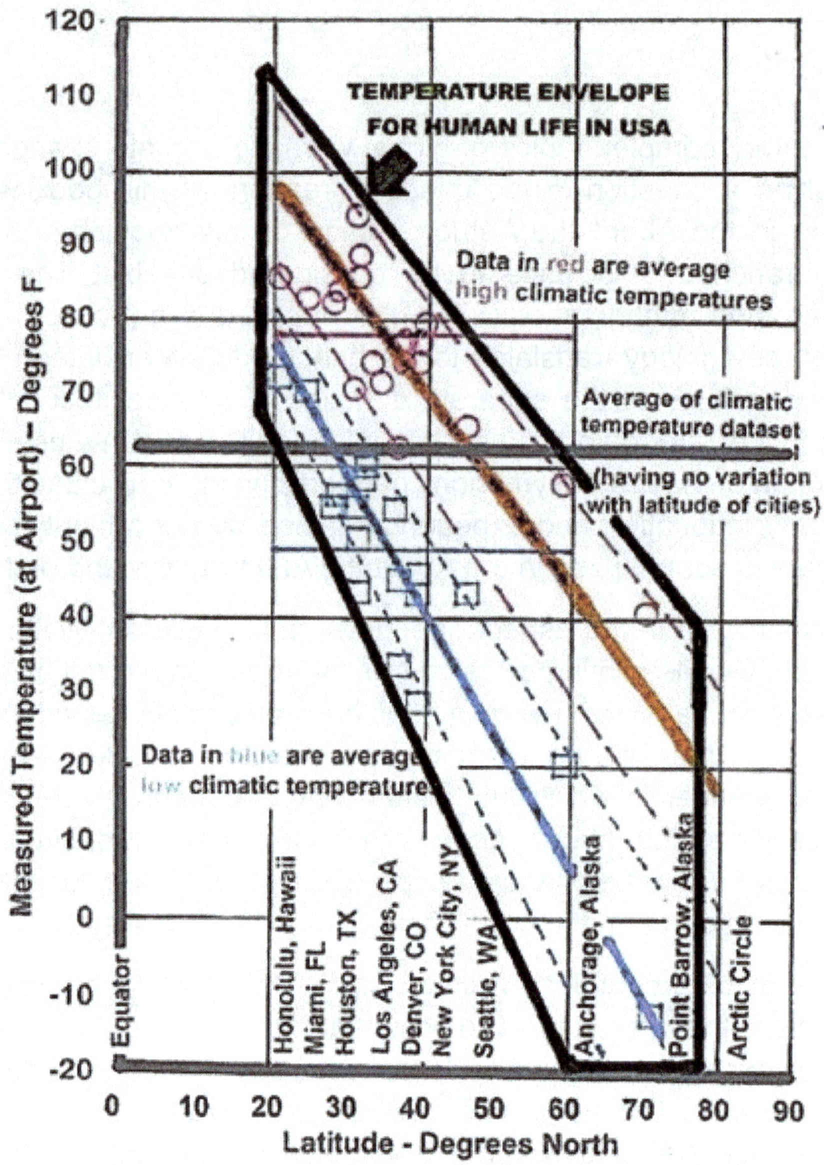

(Source: World Meteorological Organization WMO database in graphic form (by Author)

The increase in surface temperature of the planet is therefore a very important multi-discipline mandating the correct interpretation for what part(s) of the system of the planet – the environment, industrial machinery function, the human and animal population, the transportation infrastructure, and to the agronomy variables (forestry, food plants, botanicals, etc.), oceanography. Temperature literally touches everything on the surface of the planet. This book contains some of the more important details of the "big picture" of what temperature change means to system analysts and attempts to place the findings into proper perspective. What is observed in a test tube may be scientifically correct but may be of little consequence to other professionals applying scientific observations to the real-world scale of the planet.

Two statues "Past" (on left) and "Future" (on right), in front of the US National Archives building in Washington, D.C., have the philosophies of "Study the past" and "What is past is prologue" engraved on their base pedestals.

In the age of the philosophers and astronomers many centuries ago those words of wisdom were very sagely advice. Based on his education and experience in development of new technologies for airplanes, which entailed crash investigations when things did not happen quite as planned, the Author added "and present" to the "past" of the National Archives engravings philosophies. That set of information perhaps explains why a system safety engineer has analyzed the scientific topics of global warming and climate change. In the solving of any complex problem, it is also sagely advice to get "a second opinion", or in science and engineering "listen and learn" from another "perspective". Mountain climbers gain one perspective if they are at the base of a mountain looking up, different from at the top of the same mountain looking down. The mountain hasn't changed – only the perspective.

This book shows in the perspective of environmental scientists who (circa 1898) found carbon dioxide gas in a laboratory beaker will retain heat when irradiated with a heat lamp. The science of transfer of heat energy is valid in a laboratory but is shown to behave differently in the humongous non-homogenous atmosphere of the planet. Carbon dioxide absorbs heat, but if the laboratory experiment were conducted with carbon dioxide being mixed with other gases in the beaker and has a lower "concentration", there will be a lesser amount of heat absorbed. As the front cover of this book reveals, this book explains the big picture perspective of climate change – caused by the carbon dioxide in different concentrations in the atmosphere of the planet. The environmental scientists' "grass roots" perspective is proven in this book to be fallacious (inaccurate) because they did not consider other "big picture" perspectives and rushed to convince the public that Man burning fossil fuels is the dominant cause for global surface temperatures that have been increasing since the start of the Industrial Revolution 200 years ago.

Environmental protection scientists have developed a hypothesis that Man burning fossil fuels since the onset of the "Industrial Revolution" has created a heat-trapping "greenhouse gas effect" in the lower atmosphere. Using inferential logic, a "settled science" has concluded the Man-made carbon dioxide in the atmosphere is the dominant cause for the average global temperature of the planet to be rising at a "crisis" rate. Peer scientists are in general agreement that the average global temperature of the planet is on a rising trend but that it is not at a crisis rate and is probably caused by other factors that are not Man-related. The purpose of this book was/is to independently research and analyze the existing databases of information pertinent to the differences within the scientific community and to clarify the facts that the public should know. As an indicator of how the public is being misinformed about Man and industry having used coal to become developed nations is reflected in a news article recently published:

Coal is the global climate scourge that just won't go away

BY MARK GONGLOFF
Bloomberg Opinion

It's nearly impossible to get every country in the world to agree on anything. But at the latest United Nations climate talks, leaders representing all the people of Earth held hands and sang in perfect harmony of their shared belief that coal is the absolute worst.

And yet too many of them still can't seem to get enough of the one fossil fuel that makes oil look like a slacker when it comes to ruining the atmosphere for human habitation. It's a grim reminder that transitioning the global economy to clean energy will be hard, and exorbitantly expensive, and that wealthier nations must share much more of the burden than they have so far.

It took decades of climate talks for the world to publicly admit, at the COP28 confab in Dubai, that fossil fuels generally are the main culprit behind the greenhouse-gas emissions heating the planet. But a consensus of scorn for coal formed a bit earlier: The world vowed to do away with "unabated" coal power at COP26 in Glasgow in 2021, a pledge it repeated in Dubai without debate. (More on that "unabated" qualifier in a minute.)

That's not working out so great. Global coal consumption has surged more than 5% since that Glasgow agreement, based on International Energy Agency estimates.

The IEA this month optimistically suggested global coal demand peaked this year at a record 8.5 gigatons and would gently ebb to 8.3 gigatons by 2026. If this sounds like marginal progress, I guess you could argue it's still better than nothing.

But a year ago, the IEA suggested coal demand had already essentially peaked at around 8 gigatons and would be barely higher in 2025. In its latest report, it revised its estimate of 2022 demand up significantly, from 8 gigatons to 8.4 gigatons. Last year's peak-and-plateau turned out to be a false summit. The IEA's new claim of marginal progress masks what has actually been a lurch in the wrong direction.

Coal is humanity's biggest greenhouse-gas polluter, churning out 40% of all energy-related emissions. After dipping during the pandemic, coal emissions surged to a record high of 15.5 gigatons in 2022. That was more than the entire energy emissions of China and three times as much as the U.S. Not only is this terrible for the climate in the long term, but coal pollution has immediate effects on public health. We won't be able to stop global heating without first stopping the burning of coal.

Coal is in rapid, welcome decline in the U.S. and the EU, tumbling 21% and 23%, respectively, in just the past year, according to the IEA – and this despite the war in Ukraine triggering worries that Europe would be forced to make up for a lack of natural gas by burning the dirtier stuff. In the U.S., renewables now produce more electricity than coal. Even Japan, which has been much more leisurely about ditching coal, burned 8% less this year.

The story is much different beyond the developed world. China, India and fast-growing Southeast Asian economies have all been ramping up coal use lately. All have joined the global pledges to phase out coal – or at least the "unabated" kind, meaning that which doesn't rely on dodgy carbon-capture technology to negate its greenhouse-gas effects. But it's still a cheap, reliable source of power when renewables aren't available, a common problem for developing economies.

The best solution is to bolster renewable capacity to take its place. But that takes money – lots and lots of money. China has been throwing wads of cash at clean energy, and coal's demise seems inevitable there, my Bloomberg Opinion colleague David Fickling has written. But poorer developing nations will need a hand. In 2021, the Group of Seven leading industrial nations set up the Just Energy Transition Partnership to help some of the world's most dedicated burners of coal to switch to cleaner energy. The results have been underwhelming.

At COP28, even as they finally vowed to ditch fossil fuels, global leaders conveniently omitted the details of who will pay for that monumental change. The world's struggle to kick its coal habit is a striking example of why it behooves well-to-do countries to get much more serious about helping their poorer cousins hurry their transitions. Whatever the means, the key is to remember that the coal burned in Vietnam profoundly affects the climate in Vermont and Vienna.

(Source: Bloomberg Opinion in public domain news media)

Throughout this book, references will be made to the values and pitfalls in the use of inferential logic. An example of such logic is: Truth is a dominant factor in life. Humans seek the truth. Scientists are humans. By inferential logic, scientists seek and speak the truth. Any scientist (or human) who seeks the truth should not be ostracized as a "naysayer" or "contrary scientist" for doing that. A caution is given that there are some in science who wish to "warp" the truth to gain recognition (and to continue receiving grant monies from sources biased toward protecting the environment instead of protecting the safety of the human population).

Historical records reflect that at least three times before (in the Ancient Chinese, Greek, Roman, and Egyptian periods and again later in the "Age of Discovery"), philosophers and scientists made then-valid observations, but which were interpreted incorrectly using fallacious logic and were later changed – with improved telescopes - in interpretation.

The first was the belief – derived from observation - by a group of poets and several philosophers (circa 4th and 5th centuries B.C.) that the Earth was a flat disk, or a "plane" made up of *terra firma* – land,

and floating on a vast *la mar - ocean*). Science, by that name, had not then been identified so the group was not referred to as "scientists". Nevertheless, that misinformed group had a few reappearances down through centuries of scientific history and were referred to as "Flat Earthers". In 1956, a contemporary group in England registered their name as the International Flat Earth Society (shortened to the Flat Earth Society), but which was soon after disbanded when NASA satellite photos became available in the mid twentieth century which factually proved the Earth to be spherical. (The Author adds that from the perspective of a misinformed land-based observer, when standing on the Earth it can seem to be a flat circular disk as many members of the public agreed with the Flat Earthers until the NASA database convinced them otherwise. It is emphasized throughout this book that the sum of two half-truths does not lead to a whole truth – only to the better definition of the two half-truths.

The second was the 'observation' by scientists in the "Age of Discovery" (circa 1428 – 1620) that the Sun revolved about the Earth because it was "obvious" (a word derived from the two roots 'ob' – 'near', and 'vis' – 'see') that every day the Sun was observed to "rise" above the Eastern horizon in the morning, travel across the sky for about ten to twelve hours and then observed to "settle" below the Western horizon for another ten or twelve hours.

The first scientifically-defined concept of the planetary system is attributed to the Greek philosopher and astronomer Claudius Ptolemy (c.100 – c.170 AD). However, the Ptolemy concept of which planets were orbiting what center body was greatly influenced by powerful theologians who believed that everything in the universe (including the Sun then regarded as another planet) centered on – and revolved around - the planet Earth which was made by the Master Creator of all life. The Ptolemy diagram shown below was the first documented form of the astrophysical database of the planetary system. Ptolemy's diagram shows the center planet to be Erde (a Germanic name later anglicized to Earth) in a western hemisphere and an eastern hemisphere and the other then-known planets in circular orbits with Earth at their center:

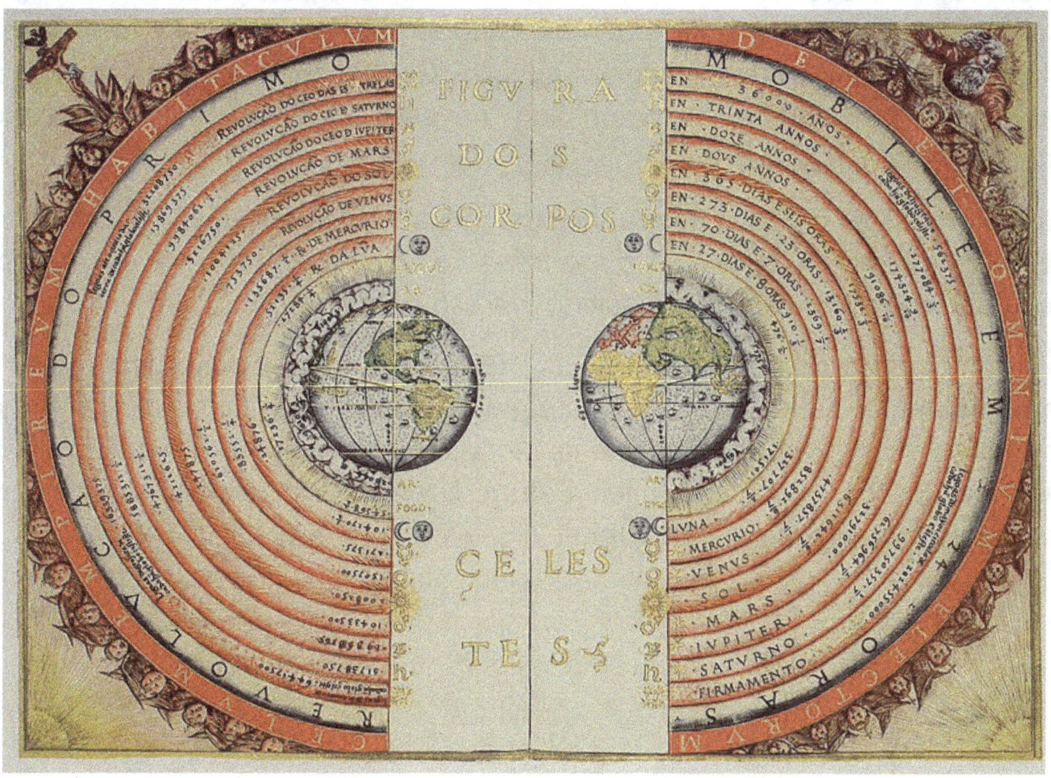

(Courtesy of Encyclopedia Brittanica)

In keeping with the Greek philosophy at that time, the known planets were all given the names of Greek Gods – Mercury, Venus, Mars, Saturn, Jupiter, Uranus, and Neptune. The Sun was known as Sol (shown in the diagram above to be the fourth planet in orbit of the planet Earth), and the Moon (correctly shown as orbiting the planet Earth) was originally named Luna. (An interesting item of trivia is that persons who act strangely when there is a full Moon are named as being "lunatics".)

That cycle of events was visually timed using "sun dials" tracing the shadow of the Sun's movement. That also gave rise to the terminologies of 'sunrise' and 'sunset' and which the public of that period accepted the "obvious" interpretation by reputable scientists as fact. Humans believe what they want to believe – and if repeatedly given fallacious misinformation by those who represent science.

Once pointed out to them, the fallacious interpretation of the sunrise and sunset was noted by the public and the terms 'day" (or "daylight") and 'night' have become more common in public use. After all, "seeing is believing" (but only if that which was seen is interpreted correctly). The physical fact is that the Sun was/is not moving at all, and it was/is the planet that was moving (rotating and translating) relative to the Sun. There are many photographs and diagrams in this book to assure scientifically correct interpretation of observations for their illusional "message". Many meteorologists continue the use of "sunrise" and "sunset". A rainbow is a beautiful sight, but when its cause is explained in scientific terminology that should not diminish the beauty of the rainbow.

The Polish philosopher and astronomer Nicolaus Copernicus (1473 – 1543) was the first known scientifically-qualified critic of Ptolemy's concept of the planetary system and was the first known theoretician to describe the system of planets to revolve around the Sun at its center. That became the theory of "heliocentrism". However, the powerful theologians of that time period banished any further research that conflicted with that of Ptolemy. It was Galileo Galilei (1564 – 1642) who was the first scientist to more scientifically (with improved telescopic instruments) and (with the mathematical genius Johannes Kepler (1571 – 1630) of Germany as a peer) mathematically confirmed Copernicus' theory of heliocentrism of the solar system and how the various planets elliptically orbited the Sun not the planet Earth. It was then that the identification of anyone (Galileo and Keppler included) who disagreed with the theologians to be heretics in religion and "naysayers" of the poets and philosophers who continued to maintain that Earth was the center of the Universe.

Kepler made a statement which gave this Author much inspiration for writing this book: *"The diversity of the phenomena of nature is so great, and the treasures hidden in the heavens so rich, precisely in order, that the human mind shall never be lacking in fresh nourishment."*

Based on the combined works of Copernicus, Galileo, and Keppler (and other lesser-known astronomers and mathematicians), the solar system was generated with the Sun at the center and the four "inner" planets (Mercury, Venus, Earth and its Moon, and Mars in their unique orbits about the Sun: There were also four "outer " planets (Jupiter, Saturn, Uranus, and Neptune) observed and they are also shown on the diagram in their unique orbits:

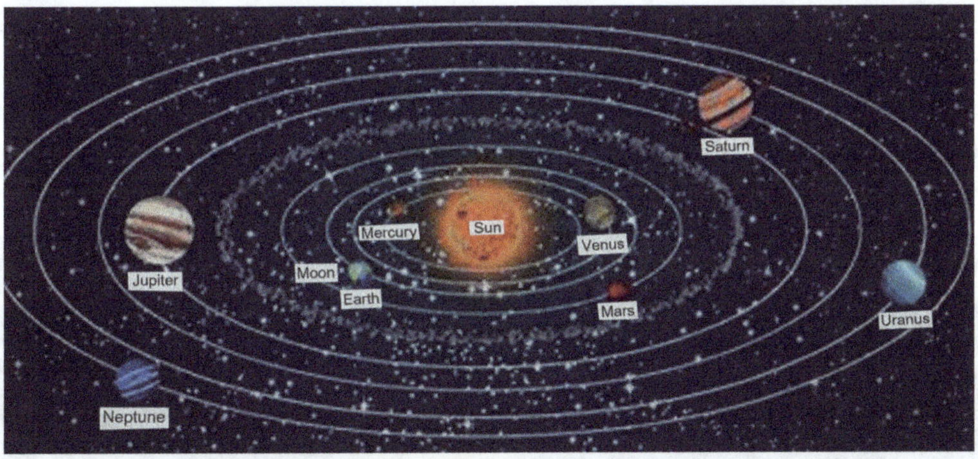

(Source: Encyclopedia Britanica with Author's annotations)

In the "Age of Reason" which rapidly followed the "Age of Discovery", observations combined with mathematical theorems (principally those of Johannes Keppler (in 1620) and Rene' Descartes (in 1644), and later by Sir Isaac Newton (in 1687) describing the "obvious" interpretations with numerical models proved that the Copernican interpretation of "heliocentrism" (the Sun being at the center of our galaxy) was correct. What was known as fact then as the "big picture" of the solar system was improved continuously as observation technology improved.

The complexity of our solar system makes the "big picture" at one point in time becoming only like one grain of sand on a beach. The NASA Edwin Hubbell Space Telescope and the current NASA James Webb Space Telescope represent the ultimate in telescope technology. Those telescopes has allowed astronomy to see the full "big picture" of what comprises the universe and the solar system within our "Milky Way" galaxy. A summarized knowledge of the solar system should assist Readers in understanding the complexity and the enormity of the geophysics involved.

The preceding "big picture" perspective of the solar system adds much understanding of the several complex interacting dynamic factors that affect the temperatures and climatic conditions on our planet Earth. As shown, the Earth is in a large elliptically shaped orbit about the Sun and makes one full orbit about the Sun in about 365.24 days. The distances reveal why the Author has chosen the terminology "big picture" as an appropriate descriptor of the process of how the Earth's surface temperature is caused to vary.

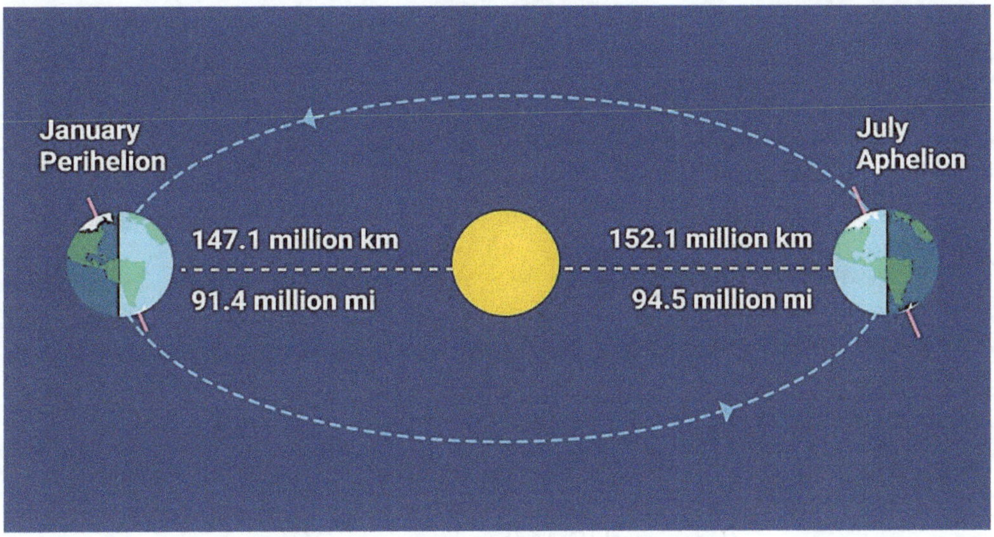

(Source: Encyclopedia Britanica)

The Earth itself is also rotating about its own axis of rotation and makes one rotation every 24 hours – with those 24 hours (a day) divided into almost 12 hours in the direct path of the Sun's radiation and the remainder of the 24 hours in darkness (the hemisphere not receiving the Sun's radiation). To further compound the complexity of those motions, the axis of the Earth is tilted (about 23.5 degrees relative to the Sun) and is also wobbling like a toy top in "precession".

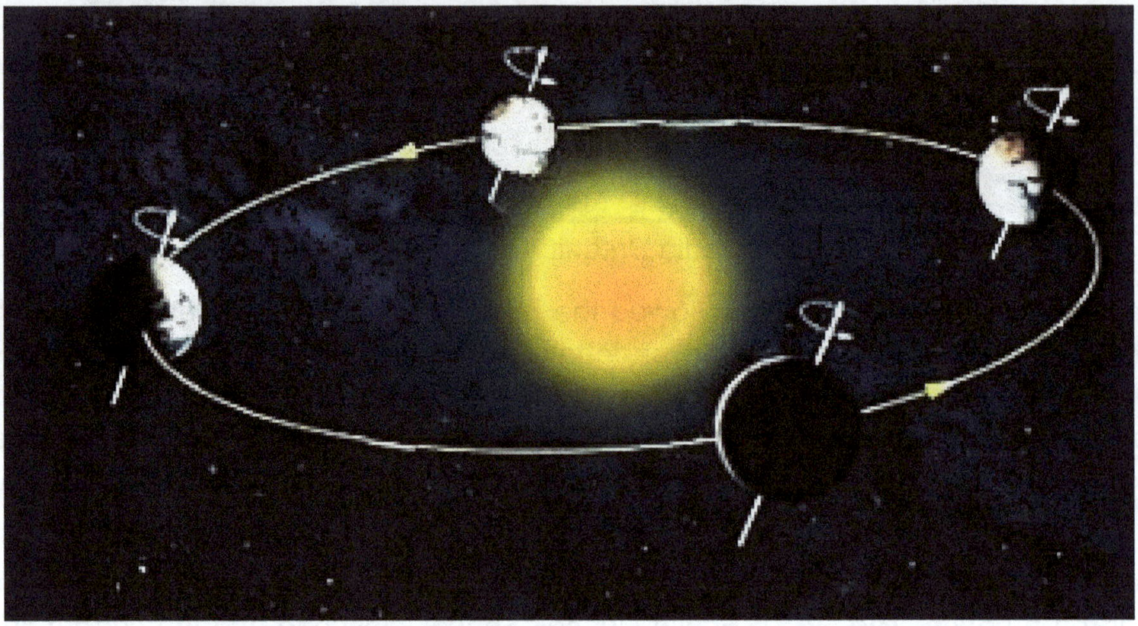

(Credit to unknown graphic artist)   The Solar System

The tilt angle of Earth's axis of rotation is the determining factor why the surface temperature of the planet has a widely varying surface temperature depending on its "seasonal" location within its annual orbit. For the latitudes of the United States, the extremes of the hottest the temperature will attain (in the mid-summer season) and the coldest temperature (in the mid-winter season) are graphically shown as follows:

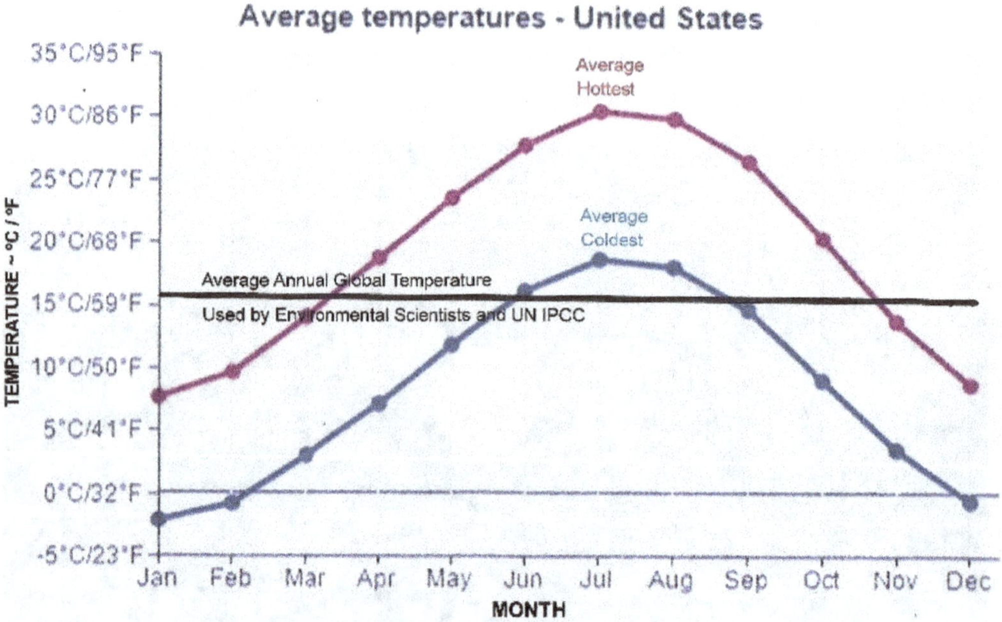

(Source: US Meteorological Office

Those geophysical variables of planet Earth have been and are disregarded by environmental scientists since those dynamic geophysical factors dispel their fallacious hypothesis that Man burning fossil fuels is the dominant cause of global warming and climate change. It is incomprehensible (to this Author and thousands of other scientists and engineers) how environmental scientists have arrived at the conclusion that anthropogenic causes for the Earth's surface temperature dominate the amount of anthropogenic "greenhouse gases" resulting from the combustion of the fossil fuels – coal, oil, and natural gas. The goal set by the United Nations Intergovernmental Panel on Climate Change (whose membership is predominantly environmental scientists) have set a temperature limit of 2 °C (3.6 °F) rise in average global annual temperature by 2050 to be attained by the elimination of combusting fossil fuels. When an increment of 2 °C (3.6 °F) is compared to the difference in the extremes of hot and cold temperature in the USA being more than 30 °C (54 °F) apart, the question that must be asked (and which caused this Author to research the entire subject of global warming and climate change in a "big picture" perspective) is: "Does a reduction of 2 °C (3.6 °F) make a significant difference to the population of the USA or of any other country?"

The process by which that far reaching conclusion was reached is proven later in this book when researching and analyzing several other databases of pertinent scientific information to be fallacious.

The Earth and its only Moon are in orbit of the Sun. It is a basic fact (proven mathematically by the mathematician Johannes Kepler based on the observations of "naysayer" astronomer Galileo di Galilei) that the amount of heat energy from the Sun changes – increases and decreases - with the distance each of the four inner planets and their moons is from the Sun that opens the path to understanding the primary causes for temperature variations on Earth 's surface. with time - seasonal, annual, decadal, and centurial.

Within the major factor of distance from the Sun, there are additional dynamic factors which are not evident from an illustration of the Earth in static positions at four different points in its orbit. Those geophysical factors are a result of the Earth rotating about a tilted, wobbling axis of rotation and the tilt angle also processing like a toy top. Adding to the complexity of the temperature on the Earth is the fact that the orbit Earth follows cycles between a shape which is elliptical but with varying amounts of ellipticity – i.e., cycling between being less elliptical to more elliptical. That variation is referred to as ellipticity. The major factors – which do not vary from the effects of atmosphere on the planet's surface – are illustrated in the following diagram:

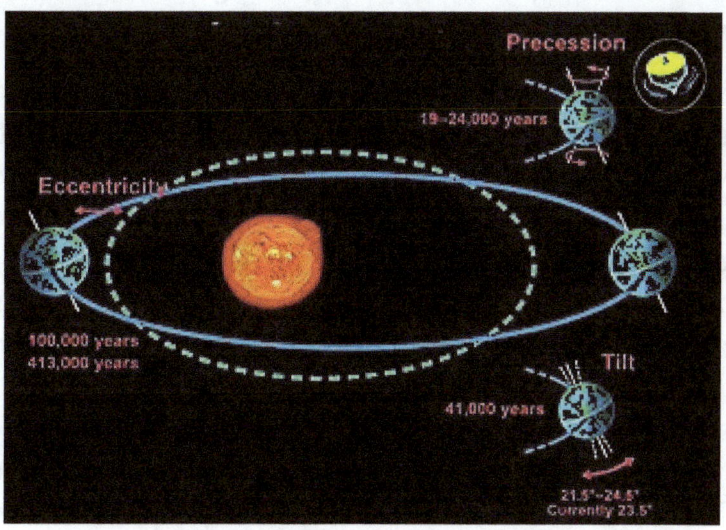

(Source: Smithsonian Institute for Science)

Brilliant mathematicians such as Johannes Kepler (1571-1630) of Germany and more recently by Milutin Milankovic (1879-1958) of Serbia-Hungary have determined the number of years in their respective cyclical periods (numerically estimated as shown in the diagram) of Eccentricity, Tilt angle, and Precession.

It is these dynamic factors in combination which alter the angle at which the Sun's radiated heat energy impinges on the Earth's surface and creates the term known as Obliquity. To assist Readers in understanding why the tilt angle is vitally important in the understanding of the causes for global warming to vary at different points on Earth's surface, the following diagram may be of assistance:

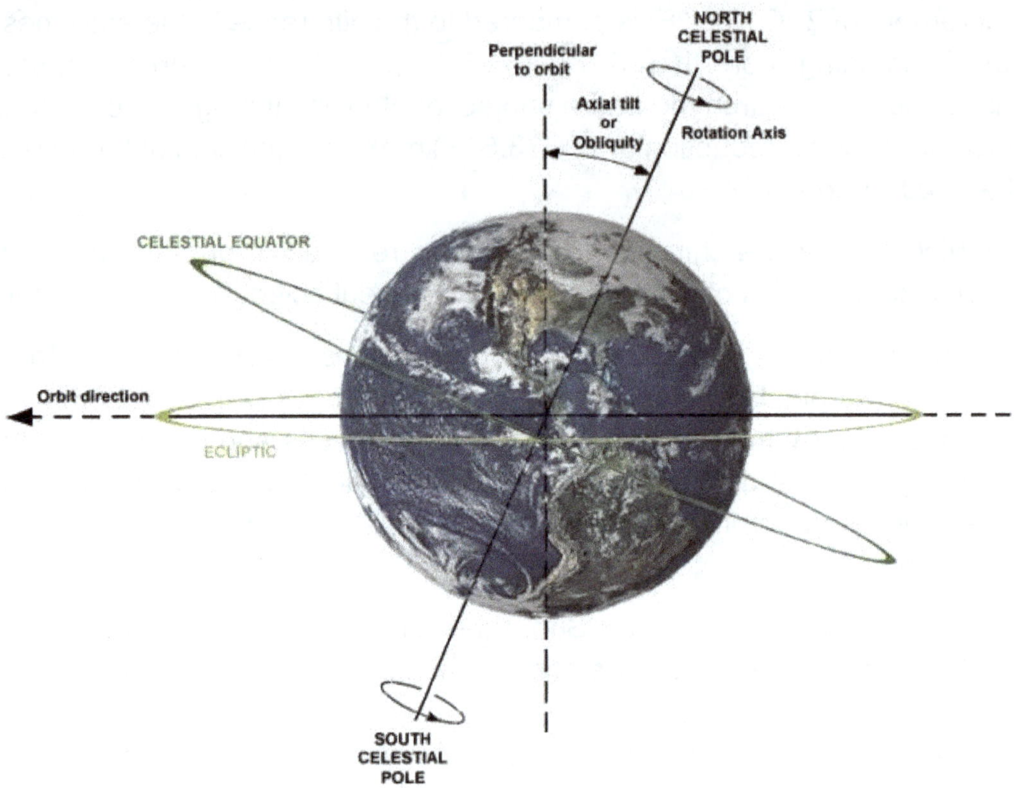

(Source: Smithsonian Institute for Science)

Further compounding the dynamics of the system of rotational, lateral, and changes in angle of obliquity, the planets each have a unique atmosphere through which the Sun's radiated heat energy must pass before impinging on Earth's surface.

A tabulation – compiled from various NASA descriptions for each planet individually - is shown next that summarizes what the known temperature changes (from measurements by satellites and space vehicles with an impressive variety of sensors having extremely high accuracy are for the planets with the distances from the Sun and with indication of the amount of carbon dioxide in their atmospheres. Because, the temperature varies with the distance the planet is from the Sun and each planet having a vastly different orbital path around the Sun, NASA chose to make the comparisons with the mean (similar to average) distance the planet was from the Sun. The mean temperatures of each planet are presented in the three most common forms of temperature Centigrade (which is used commonly in the metric system of measurement), Fahrenheit (used in the Imperial system of measurement – in the USA), and Kelvin (which is the most useful scientific tool when considering the near absolute 0 temperature known to scientists). The thickness of the atmosphere is less accurate (atmospheres on planets are not homogenous) but is as best as can be estimated by NASA sensors.

Distance,Temperature, & Atmosphere of Inner Planets

	Mean Dist. from Sun		Mean Temperature			Percent CO_2 in and Thickness of Atmosphere
Planet	Miles x 10^6	Km x 10^6	°C (H / L)	°F (H / L)	K (H / L)	
Mercury	36	58	167 (430 -180)	333 (800 -290)	440 (703 93)	0% None
Venus	67.2	108	464	867	737	96% Thin
Earth	93	150	15	59	288	0.04% Thick (with 3% Water Vapor)
Moon	93	150	-5 (129 -130)	21 (250 -208)	270 (400 140)	0% None
Mars	141.6	228	-65 (-1 -125)	-85 (30 -193)	208 (272 148)	95% Thin

Note: There was no reliable data found for the average global hot (H) and average global cold (L) temperature for Venus and Earth.

(Source: NASA database; Tabulation by Author)

A more detailed analysis is included in the appropriate later chapter of the book, but for this Preface suffice to say that the presence of a high percent concentration of carbon dioxide in a thin or no atmosphere does not have any effect on the expected Kepler distance - temperature relationship. Conversely a large percentage concentration of carbon dioxide in a thick atmosphere (as on planet Venus) has a profound effect on the expected Kepler estimated surface temperature. Planet Earth has a very low concentration of carbon dioxide in its thick atmosphere and as shown in the following relationship of distance and temperature for the four inner planets and the Moon, only Venus shows a large effect of carbon dioxide on the temperature of Venus.

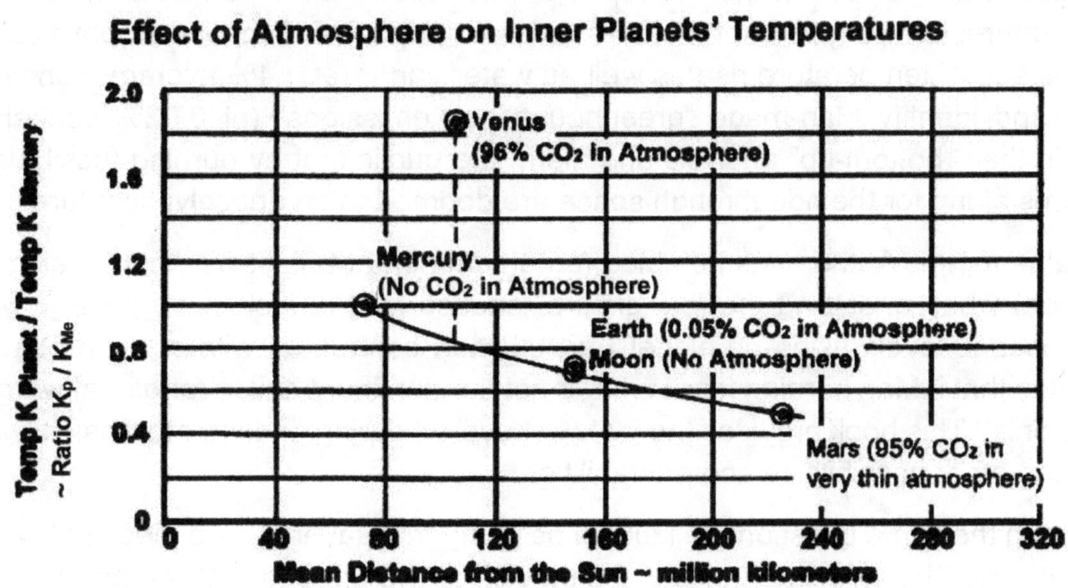

(Source: Author's analysis of NASA database for inner planets)

There is no industry or presence of Man on either Venus or Mercury and that fact raises two questions: 1) What else besides carbon dioxide at high concentration (over 90%) in the atmosphere causes the temperature of the planets to be what they are?; and 2) By what process have environmental scientists been led to conclude that carbon dioxide at a total concentration of less than 0.05% (of which 0.02% is Man-made) in the atmosphere of Earth is the dominant cause for the surface temperature on Earth to be rising and at a rate that environmental scientists have declared to be a "crisis"?

Despite the many astrophysical, geophysical and natural possible causes for temperature increases in regions of the planet and for climate change, environmental scientists, the Environmental Protection Agency, and the news media in the United States have conspired to influence the public and government leaders in many countries to point the finger of blame at the "big" industries (oil, steel, coal, automotive, etc.) as being the dominant contributors to the creation of "greenhouse gas" emissions. After considering the scientific knowns of the solar system shown above, any experienced system safety engineer would question (as this Author does) if and how the Industrial Revolution could possibly be the dominant cause for the planet Earth to be in a Man-caused temperature safety "crisis" when the Man-made carbon dioxide emissions are less than 0.02% of the atmosphere of the planet.

The opening sections of this book contain a combined scientific and system engineering analysis aided by mathematical forms of statistical analysis of the several databases of scientific data (i.e., STEM) to address the first question with a top-down ("big picture") analysis approach. Following that, this book contains a reconstruction of the historical trail (much like a genealogy specialist conducts research for family histories) of the information and data that have led current environmental scientists to fallaciously conclude that the 0.02% of Man-made carbon dioxide in the atmosphere is the dominant cause of the temperature and climate changes on planet Earth. Readers should note that when two different perspectives are used and consider the same "mountain" of information and data, a different conclusion can be, and has been, derived depending on a "top-down" (commonly referred to as a "big picture") perspective or "bottom-up" (commonly referred to as a "grassroots") perspective of that same mountain.

Just as most everything in the solar system varies in some form of exponential, there is little reason to not believe that the concentration of "greenhouse gasses" – which must include the most powerful greenhouse gas water vapor - in the atmosphere also follows an exponential effect on the planets' surface temperature. Ironically, environmental scientists ignore the effect of astrophysical, geophysical, and natural causes of temperature rise as well as water vapor (at 1.4% average concentration in the atmosphere). and identify Man-made "greenhouse gas emissions" (at 0.02% concentration in the atmosphere) in their "bottom-up" analysis and alarm the public that by burning fossil fuels, the planet and all life forms along for the ride through space are doomed to an apocalyptic future.

On a personal note, this Author was humbled (as a common ant must be when seeing the entire ant colony in action) when analyzing the "big picture" causes for the factors of global warming, climate change, and oceans levels rising on planet Earth. When confronted with a low probability, scientific curiosity dictates that if Man burning fossil fuels is not the dominant cause for global warming on Earth, then what are or is? This book provides the facts uncovered during and after forensically analyzing and evaluating those other possibilities and probabilities.

Others pondering the same question are labeled as being "naysayers". As a piece of trivia – historically based – the Catholic Church in the Period of the Inquisition (13th through 16th century) labeled scientists who disbelieved that the Earth was made by God and created the Earth to be at the center of motion of other celestial bodies' to be "naysayers". Galileo was the most famous "naysayer" but there were many others – that is the basis for science to identify possibilities and dismiss impossibilities. The laws of society in the judicial sciences apply that "examination of evidence" in courtroom scenarios every day in all parts of the developed world with the same purpose. While physical sciences use the precision of mathematics to determine the "most probable" – above 99% probability - cause for a problem, the judicial science is based on a "beyond reasonable doubt" determination of the cause for an anti-societal act.

Rulings made recently (2024) by the US Supreme Court against the US Environmental Protection Administration (EPA), which attempted to identify Man-made carbon dioxide gas as being an atmospheric pollutant, are based on the fact that with precise scientific measurements of total carbon dioxide in the atmosphere showing that of a measurement of 415 parts per million by volume (0.0415%) total carbon dioxide being in the atmosphere, 280 parts per million by volume (0.0280%) are of pre-Industrial Revolution natural origin – based on the Vostok Ice Core Samples database (analyzed later) that are scientifically/historically valid for the past 420,000 years. It is beyond reasonable doubt that, at most, 135 parts per million by volume (0.0135%) of carbon dioxide in the atmosphere is Man-made. In essence, the Supreme Court "beyond reasonable doubt" ruling is a wise judgment that if Nature does not pollute the atmosphere by itself with natural carbon dioxide gas, then neither can/does Man pollute the atmosphere with 48% less (135/280) carbon dioxide gas than that of Nature. The Supreme Court did not strike down other forms of Man-made pollution that are not made by Nature thereby allowing – with valid reasoning - true pollutants (containing toxins, carcinogens and biologic hazardous components) of air, land, and water to be lawfully controlled by stringent EPA anti-pollution regulations.

It is hastened to be added that the EPA case submittal to the Supreme Court was not an attempt by the EPA to obtain the Supreme Court recognition that Man-generated "greenhouse gas emissions" are a major cause for global warming and/or climate change, it was aimed only to seek Supreme Court approval to permit more stringent control by the EPA on Man-generated pollutants and higher penalties EPA could lawfully levy on offenders polluting the environment. It is also added that the ash residue left after coal is combusted and needing disposal does contain toxins and carcinogens that are finding their way after disposal in landfills into the underground water aquafers and into agricultural products serving the local public. This book supports discontinuing the combustion of coal in the contiguous "lower 48" states of the USA because of its true polluting properties (toxins and carcinogens) but not because of its effect (if any) on the atmosphere causing global warming and/or climate change. Whether or not to continue the use of coal in other countries is a matter for the environmental protection organization in each country or region to decide. It is not a mandatory recommendation but if continued anywhere, the coal ash residue must not be disposed of such that it pollutes the land or water of the region and become a safety hazard for the local human population.

While the Author was in attendance at Brooklyn Technical High School in 1946, a most enlightening topic (revealed in the Earth Science curriculum) was the calculations (done circa 1938) of a Serbian geophysicist and applied mathematician Milutin Milankovic (1879 – 1958) describing how Earth travelled in an irregular elliptical orbit about the Sun and how Earth simultaneously was/is in a rotation mode with its own tilted and wobbling axis of rotation during the orbit. (It is a recollection of those Earth Science and Applied Mathematics classes in the mid-1940s that is now believed to have "planted the seed" of thought in this Author for this book.)

The Earth is known to rotate about its axis once in almost precisely 24 hours and takes 365.24 days – a year - to complete one orbit of the Sun. The distance the Earth travels in one orbit is about 940,000 kilometers, or about 1.09 billion miles. The eminent mathematician Johannes Keppler calculated the enormous numbers describing the distances Earth was from the Sun (a median of 150,000,000 kilometers – about 93,000,000 miles) and at a median speed of travel (about 30 kilometers per sec – about 68,400 miles per hour). Most humans are unaware they are moving at that speed and for that distance – but not eligible for Frequent Flyer miles or eligible to wear a Mach 2 lapel pin.

A diagram was shown earlier (and repeated here for Readers' recollection) illustrating the complex set of dynamic factors affecting solar radiation impinging at various angles on the Earth as well as explaining the seasonal effect on Earth 's surface temperature. (The Author's Earth Science teacher in high school used an analogy of a pig being roasted on a barbeque spit to attach reality to the topic of the solar system geometric variations, dynamics and heat energy characteristics.)

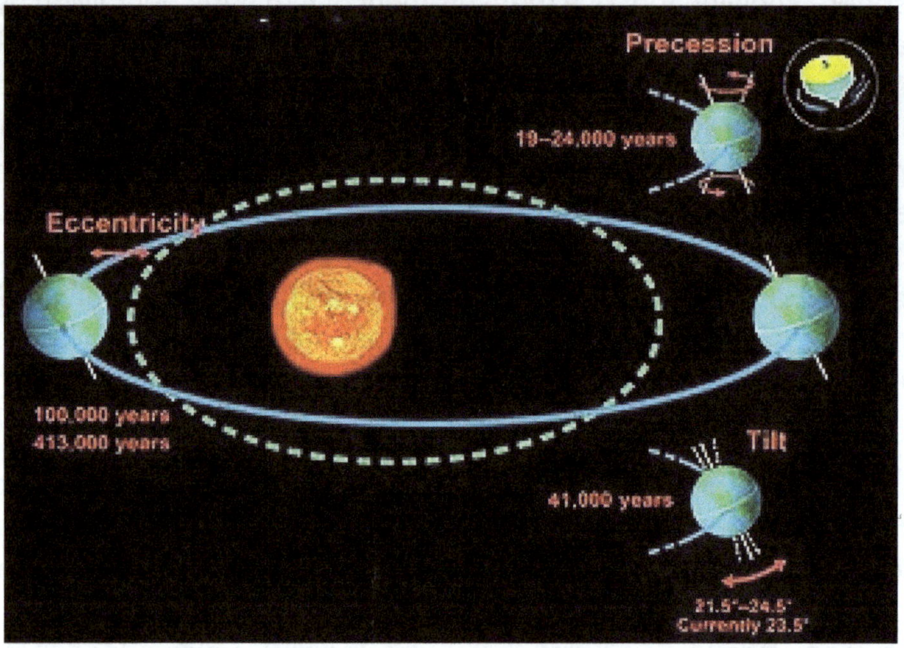

(Credit to unknown graphic artist for the Smithsonian Society)

The Milankovic calculations were described separately in the Author's Mathematics class by a different instructor and which showed graphically how a large complex mathematical database can be summarized in a manner – as Milankovic has done - that could be understood by students:

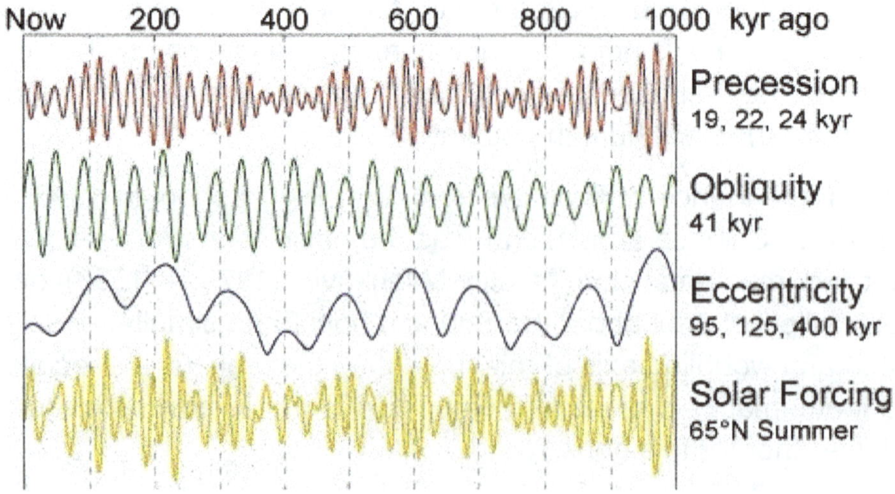

(Source: Smithsonian Institute for Science)

It should be noted by Readers that if one generation of Man at 50 years length in time is selected from the total database of Milankovic calculations of 1,000 kiloyears (1 million years), the combined effects of precession, obliquity, and eccentricity during that 50 years "slice" on solar forcing will correspond to a very different trend in the temperature on the planet. In other words, the planet temperature variation will be increasing (as it is now), decreasing, or perhaps include a peak where the trend does a near reversal.

On a smaller scale than that of Milankovic, observations were made during the ancient ages that the temperature on the surface of the Earth was on a cycle of changes that were annual – repeatedly occurring every year. When a person residing on Earth's surface wonders what is causing the annual (i.e., one year) four "seasons" of Spring, Summer, Fall (Autumn), and Winter and each season having a different range in temperature from a minimum low temperature to a maximum high temperature, the explanation is seen graphically to be due to the tilted axis of the Earth relative to the parallel rays of solar insolation from the Sun:

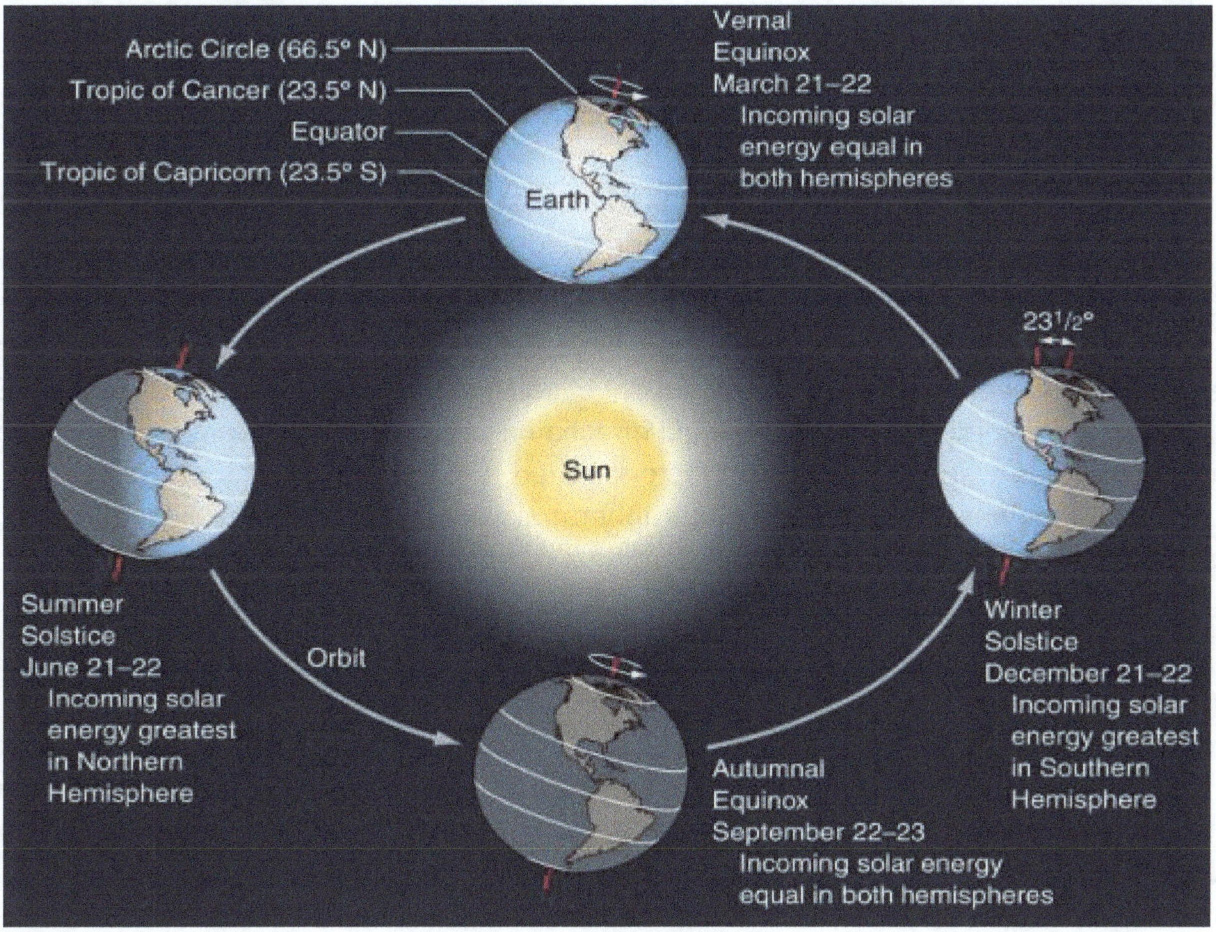

(Courtesy of NASA)

The variation in temperature of a fixed point on Earth's surface measured continuously as the year progresses is shown to vary in a regular manner. For the illustration, the average of the hottest and coldest temperatures for the average monthly temperatures of the United States is presented:

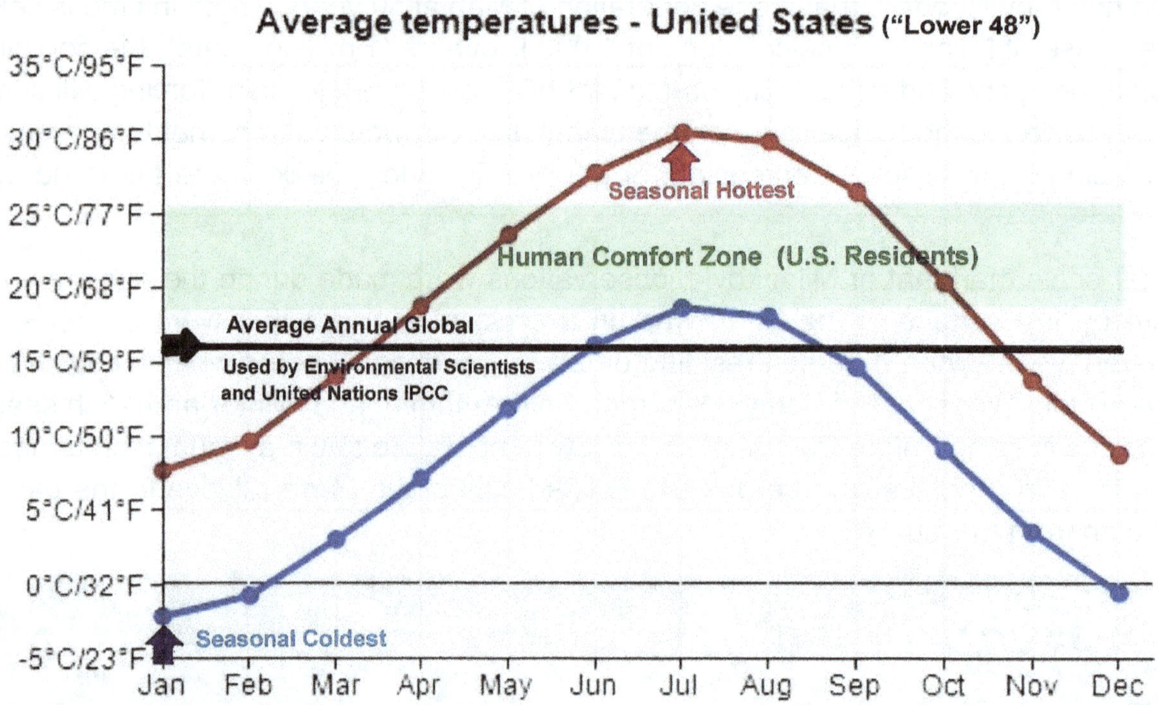

(Source: National Weather Service of the National Oceanic and Atmospheric Agency (NOAA), with Author's added descriptors)

The complex variations in surface temperature and number of daylight and darkness hours for a specific point on Earth's surface can be calculated but are beyond the scope of this book to be included herein. Early meteorologists over the years from the advent of sailing ships began analyzing the ship Captains' trip logbooks which recorded temperature, wind direction and speed, and ocean currents direction and speed and how those weather and ocean factors varied in different portions of the "seven seas" at different times of the year. Mapmakers (cartographers) also developed the definition of the continents from the ship Captains' notes and logbook recordings.

Turning attention to the 18th century to introduce the chronological history of the impact Man has had on the environment of the planet by the burning of fossil fuels, it was the invention of the practical steam engine (powered with the combustion of a pre-fossil fuel - wood) by James Watt in Scotland in 1776 that is regarded as the origin of the observation by environmental protection scientists that the by-product gasses and smoke and steam from a steam engine operation was very adverse to the environment. In the public opinion, the steam engine was "a work of the devil". The Watt steam engine was improved by the Porter-Allen high speed steam engine in 1862 and that version of the steam engine is credited as having started the Industrial Revolution – and the offensive emissions from the burning of fossil fuels (initially wood, but then soon thereafter substituted by coal as a more efficient – much higher heat energy per unit of volume and mass).

Important observations made about the atmosphere surrounding the planet and having an impact on how the solar radiation is affected (but that later were revealed as incomplete) were those made by a reputable scientist (Dr. Svante Arrhenius of Sweden in 1902) – an atmospheric chemist by specialty, and whose research earned him the Nobel Award for Chemistry in 1903. Dr. Arrhenius's observation of carbon dioxide having a "hothouse" heat-trapping characteristic initiated that scientific inquisitiveness being explored by a group of scientists in the branch of mushrooming science known as environmental science.

Regrettably, environmental scientists made a hasty and only partially valid logical interpretation of very valid scientific observations (at small scale and in a laboratory with a very closely controlled internal environment) at the start of the twentieth century. Dr. Arrhenius's scientific observation was the gas carbon dioxide when in high concentration having a heat-trapping characteristic. That heat-trapping characteristic – which is qualitatively 100% scientifically valid but which varies quantitatively with concentration – was then exaggerated into a "greenhouse gas effect" such that, when scaled up and injected into the atmosphere surrounding the planet, carbon dioxide at high concentration was concluded to be the dominant cause for the surface of the Earth to be warming and at a "crisis" rate.

Using inferential logic improperly, one fallacious illogical conclusion led to another fallacious conclusion that the only recourse for slowing/halting global warming was/is for Man to cease burning fossil fuels. The scaling up from the size of a large flask containing perhaps a volume of about 1 liter of carbon dioxide gas under carefully controlled laboratory environmental conditions into the proportion of a non-homogenous 5 miles thick atmosphere surrounding the entire planet (a sphere having a diameter of 8,000 miles and containing an irregular distribution of water vapor which is 100 times more capable of trapping heat than all other greenhouse gases) is a scientific absurdity.

The observance of carbon dioxide gas <u>when in high-concentration</u> having a heat-trapping property had been scientifically proven in very professionally conducted laboratory research and experiments – however, the correct interpretation should have included that the heat-trapping property of carbon dioxide is diminished when in a low concentration (less than 0.05%) and especially in the presence of water vapor at 1% to 3%. Nevertheless, that grossly fallacious interpretation and overexaggeration of effectiveness was made. In the absence of critical peer constructive criticism, In the absence of any naysaying – such as that of a contemporary Galileo or Copernicus - Dr. Arrhenius was awarded the Nobel Prize for Atmospheric Chemistry in 1902.

There were two basic scientific findings from the valid but misinterpreted observations and analysis work of Dr. Svante Arrhenius in the late 1890s; one of near-future implications and the other of distant-future implications. Dr. Arrhenius credibly concluded that in the near-term, the warming of the atmosphere by "hothouse" gas emissions in low concentrations from the burning of coal (by humans and by industries) had a potentially beneficial effect of warming perma-frost regions in northern Scandinavian regions and could provide substantially more arable acreage for the exponentially increasing human population in Europe. The far-term (10,000 to 25,000 years in the future) conclusion was that <u>a cumulation of high-concentration "hothouse" emissions in the atmosphere</u> could have adverse temperature and climate changes and melting of the glacial ice in Greenland and in Antarctica that could be disastrous to both the human population and the environment.

The scientific conclusions of Dr. Arrhenius are as valid – and as fallacious (for not including other factors) - now as they were in the early 1900s. Current environmental scientists have chosen to ignore any near-term beneficial warming of permafrost land effect from "greenhouse gasses", and have, without convincing scientific justification, moved the Arrhenius long-term catastrophic effects from global warming to be realized much sooner in the future so as to constitute being a "crisis" now. Many ethical scientists disagree with the environmental science hypothesis that Man burning fossil fuels – of any type (coal, oil, or natural gas) – is the dominant cause for global warming and/or climate change. Those who disagree are modern-day "naysayers" and "contrary scientists" and have no credible voice in environmental science.

That early chemistry laboratory experimentation observation (carbon dioxide being a "hothouse gas") was scientifically noteworthy but rightfully was not considered as a "crisis" in 1903. The recent activities and concerns of environmental scientists over global warming, climate change, and ocean surface levels rising 150 years later are factually proven in this book to still not be in a "crisis" level. The declaration of any problem to be in a "crisis" state when it factually is not one is indicative of the tactics being used by the environmental science faction and "green" activists to create the "fear" impression about the urgency of the problem in the minds of the public. History is rife with the practice of "fearmongering". Students in the primary grades of education are taught the moral of "Do not call 'Wolf" when there is none". In later life, it is punishable as a public safety offense to call out "FIRE" in a public place when there is no fire.

The news media is largely responsible for over-exaggerating the urgency of the climate change problem – it is sensationalism that sells newspapers. Regrettably, the news media published the following excerpt in the United States news media in 1902:

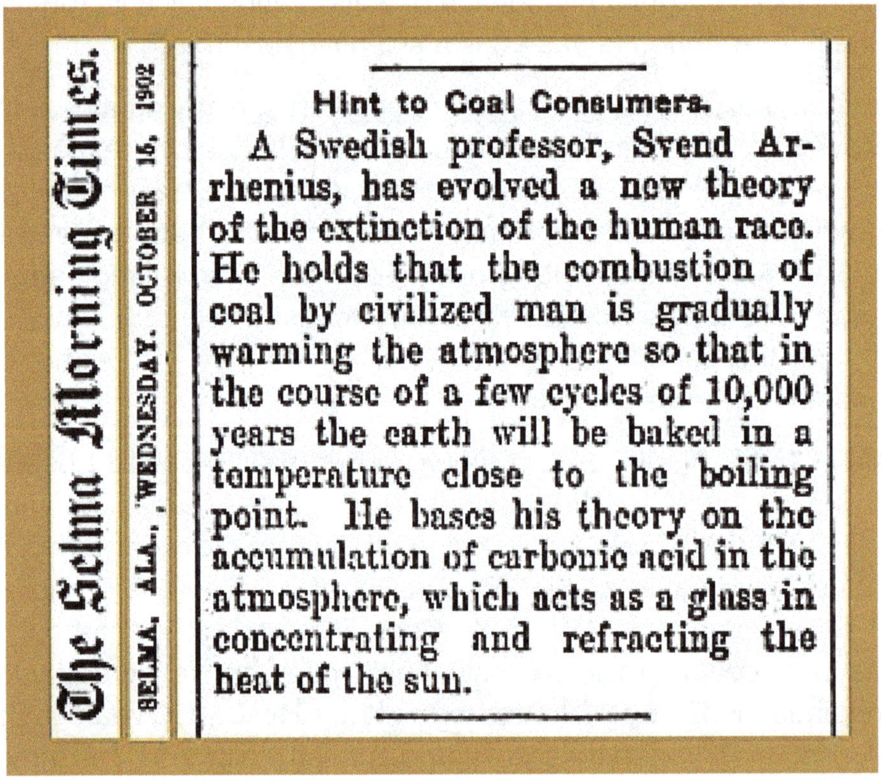

The Selma Morning Times.

SELMA, ALA., WEDNESDAY, OCTOBER 15, 1902

Hint to Coal Consumers.

A Swedish professor, Svend Arrhenius, has evolved a new theory of the extinction of the human race. He holds that the combustion of coal by civilized man is gradually warming the atmosphere so that in the course of a few cycles of 10,000 years the earth will be baked in a temperature close to the boiling point. He bases his theory on the accumulation of carbonic acid in the atmosphere, which acts as a glass in concentrating and refracting the heat of the sun.

(Source: The US National Archives, Washington, D.C.)

It was not coincidental that the reporter for the Selma Morning Times omitted the potentially beneficial near-term conclusions of Dr. Arrhenius as that would reduce the sensationalism of there being a "crisis" of apocalyptic proportion.

The fact that Dr. Arrhenius (in the late 19th century) and the more contemporary (in the twentieth and early twenty-first centuries) environmental scientists neglected to consider the several other factors (astrophysical, geophysical, and natural) that are each adding to the planet to be warming and its climate to be changing accordingly is statistically and analytically a major error of oversight and misjudgment. This book describes the "big picture" top-down analysis and dispels the results of the "bottom up" grassroot overexaggeration that Man burning fossil fuels is the dominant cause – or even a major cause – of global warming and climate change.

Following the basic learning curve process about anything and almost everything, this book has been written to serve three purposes: 1) identification of the factually most significant factors affecting the triune problem of global warming, climate change, and ocean surface levels rising; 2) scoping the magnitude of the triune problem from a "big picture" perspective not from the myopic perspective that is valid in small scale and in high concentration in a laboratory, but is largely invalid in the full scale real world at a very low concentration; and 3) establishing a priority for resolution of the climate change problem relative to other related problems visually obvious in the real world and adversely affecting the environment and the human population. The Table of Contents for this book essentially lists the steps in the analysis, resolution, and technology candidates prioritization process.

The difference in how a scientist addresses a "problem" and how an engineer addresses the same "problem" is well known to any member of a STEM (science, technology, engineering, or mathematics) profession. The Author considers STEM to be missing a letter – "A" – representing "Applied" and in a system manner, not as individual stand-alone activities). The Author requests Readers to not "shoot the messenger", or to brand a scarlet "N" in his forehead, or burn him at the stake - for revealing the "big picture" factual causes of global warming, climate change, and the ocean levels rising being different from that of the "settled science" group of environmental scientists. That revelation is preferred to be interpreted as the Author performing a public service "for the benefit of Mankind" (as stated in the Constitution of the Mensa organization) putting the problem in correct perspective. A simple "Naysayer" naming – of the kind awarded to Copernicus, Galileo, Keppler, Descartes, and Newton - is acceptable but not necessary.

For public understanding, scientists typically use "grassroots" observations in a laboratory environment using small scale experiments to identify the chemistry and basic physics of the problem and typically will examine only the obvious elements of the problem at that small scale conducted in a set of well-controlled environmental conditions. That approach to problem-solving is referred to in the STEM professions as "grassroots" or "bottom-up" form of problem resolution. (A tongue-in-cheek expression of "grassroots" is "that's where the weeds are". Conversely, a "top down" approach starts with observing factors that nurture the colorful flower and worries about eliminating the weeds later.) Almost everything visible to the human eye is interpretive, depending on the perspective. As an analogy, from a distant perspective, mountains may not appear to be a challenge to climb, but from a perspective at the base of the mountain and looking up, the mountain may appear to be insurmountable (pun intended). The height of the mountain did not change, only the perspective.

Another significant hazard of using a "bottom-up" approach in problem-solving is that it can cause major peripheral factors affecting the problem to be overlooked. (Horse-race owners intentionally often use "blinders" on racehorses to avoid distractions of the owner's racehorse by other competing horses.) The observation that a horse which has another horse's rear-end in front of it will not win the race is considered a temporary valid interpretation – becoming invalid if the horses switch positions before reaching the "wire".

Development of technologies is analogous to a horse race – a technology can "look good" at the start but its later performance could lag behind other technologies in the same "race". The use of wind power is an example of a technology appearing valid in theory under "average" conditions, but not so valid in real world application. Theory often follows the use of "averaging" the variables comprising the technology but overlooks the "not average" conditions – such as wind, cloud cover, temperature extremes (hot or cold), unpredictable natural phenomena (earthquakes, volcanic eruptions, lightning strikes, lightning-ignited wildfires, flooding, etc.) which vary in the region where the technology is to be applied after full-scale development. Later chapters of this book provide data (numerical and photographic) describing – "seeing is believing" – that "not average" conditions are above the normal for one region and below for another region at a different latitude.

One of the first databases discovered early in the research phase of this book was the climate modeling done by the eminently qualified climatologists Dr. Wladimir Köppen (1846-1940) of Russia, Germany, and Austria and Dr. Rudolf Geiger (1894-1981) of Germany. The first systematically developed climatology method was published initially in 1884 by Dr. Köppen and is seen below to be more of a global climate identification by latitudinal zones, most probably based on Dr. Köppen's observations of the meteorological notations in the maritime records from world ocean travelers in sailing ships traversing the oceans of the planet. The absence of the outlines of the continents and of the oceans reflects that the color coding chosen by Dr. Köppen was not related to the climate being different on the land masses of the different continents:

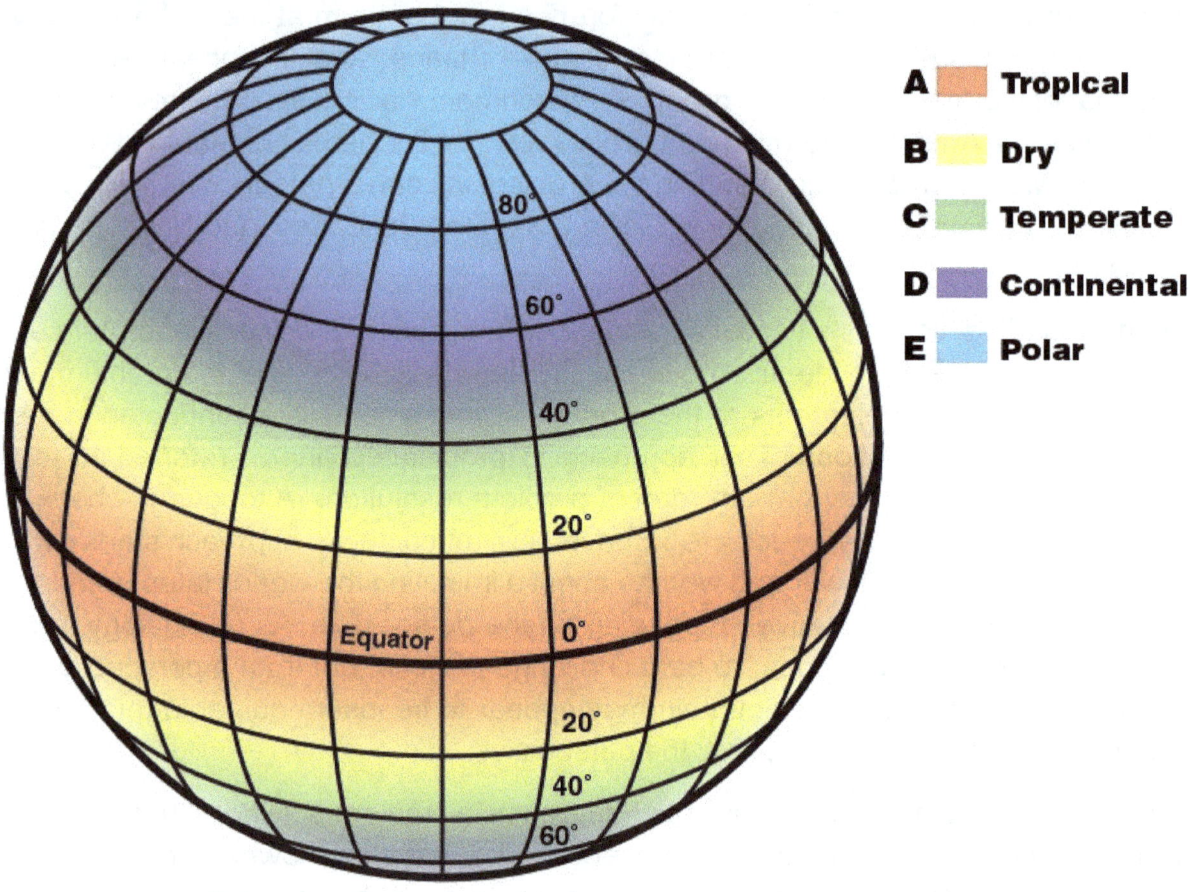

(Source: Encyclopedia Brittanica)

The climate in various regions changed during the next six decades and the original Köppen climatology model was updated in 1954 and again in 1964 by Dr. Geiger. When the second update by Dr. Geiger was published in 1964, the atmospheric chemistry works of Dr. Svante Arrhenius (1894-1940) and his "hothouse" identification of Industrial Revolution emissions being a cause of the planet surface temperatures and noted in the early 1900s to be rising in the northern latitudes of Scandinavia, were revived in scientific journals. In the cursory review of the databases of Doctors Köppen and Arrhenius, the Author found no reference by Dr. Köppen to Dr. Arrhenius's atmospheric chemistry hypothesis of Man burning coal with "hothouse" emissions into the atmosphere as being a major climatology factor.

In 1964 when the second update of the Köppen-Geiger climate model was published, the Author was in the employ of the US Air Force Aeronautical Systems Division (at Wright-Patterson Air Force Base near Dayton, Ohio) as a flight technology developer. Observing that changes in the temperature of the atmosphere had surfaced in scientific journals led to the Author being tasked (by the Commander of the USAF Aeronautical Systems Division) to determine if those changes in atmospheric temperature (caused by any factor – natural or Man-made) were of significance to the flight capability or to the shock waves from supersonic military aircraft and are perceived by the human ear as "sonic booms".

Shock waves propagating through the atmosphere are sensitive to temperature and if/when they impinge on the ground as an "N"-shaped pressure wave. Sonic booms can cause glass window breakage if they are originated by an aircraft at supersonic speed and below about 45,000 feet altitude. With or without physical damage to properties on the ground, the sonic booms are a form of noise pollution. They are of a high level of concern to the national transportation system as well as to military operations. Ultimately, sonic boom mitigation measures for public toleration reasons caused the prohibition of supersonic flight over densely populated areas which led to the economic demise of the British/French Supersonic Transport (*Concorde*) program. Military aircraft can fly at supersonic speeds only within unpopulated flight operations ranges.

The Author quickly became highly familiar with physical properties of the atmosphere (its gaseous content, its pressure profile with altitude, its temperature profile with altitude, its precipitation and lightning content, and the direction and velocity of winds at altitude) which are key factors affecting flight of an aircraft. It is that atmosphere through which the flight vehicles (military and commercial) fly in transporting humans and high-priority cargoes large distances between Region A and Region B.

Also, in 1964 when the second update of the Köppen-Geiger climate model was published, during the ensuing research addressing the USAF interest in atmospheric changes, the atmospheric gasses measurements by Dr. Charles D. Keeling (1928-2005) at the Mauna Loa Atmospheric Research Observatory in Hawaii became known and were added into the Author's "pile of stones needing to be looked under" on his desk.

Dr. Keeling scientifically developed a capability to capture samples of the atmosphere and to analyze that atmospheric sample to scientifically isolate the amount of the different gases in the sample. The book will later describe these atmospheric samples in greater detail but for this Preface the sampling was done in a real environment (i.e., not in a well-controlled laboratory environment). The environment was that of the atmospheric Research facility located at a high elevation (about 10,000 feet above sea level) on the long-dormant Mauna Loa volcano in Hawaii. Once again though, the scaling up of professionally valid scientific measurements conducted at a location in the middle of the Pacific Ocean cannot be accepted as applying everywhere else on the surface of the planet.

It is hastened to be added that this criticism is not intended in any manner to besmirch the validity of Dr. Keeling's very valid scientific measurements – for which he received the prestigious American Meteorological Society medal in 1981 and the National Medal of Science in 2002. There is justifiable question that the measurements of carbon dioxide gas (and other "greenhouse gases" in the atmosphere) are "corrected" to remove the amount of water vapor in the atmosphere. For Readers' interim information, a very recent graph, illustrating the Mauna Loa Atmospheric Observatory carbon dioxide measurement data being collected, appears as follows:

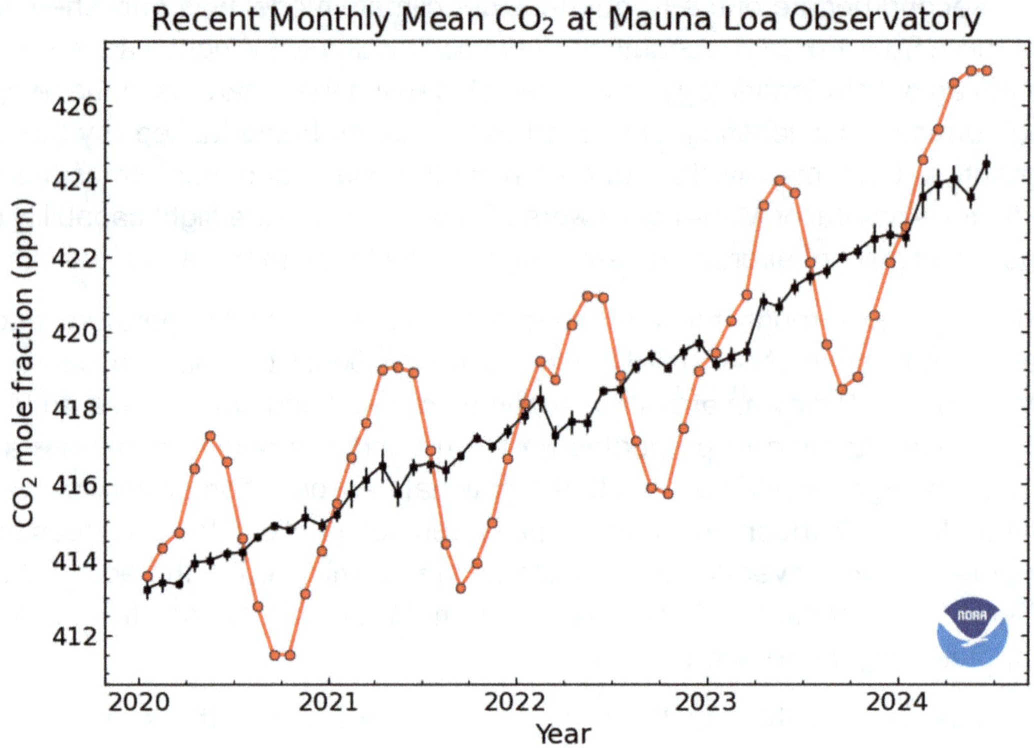

(Source: NOAA)

The significance of these “Keeling Curve” data is described in detail in later chapters. It was the rate of carbon dioxide increase reflected in the “Keeling Curve” however that was intentionally warped in interpretation by environmental scientists wanting/needing public support in eliminating the practice of Man burning fossil fuels causing carbon dioxide “greenhouse gas” atmospheric pollution and at a “crisis” rate.

<u>The data in red are the “raw” measurements and those in black are the data “adjusted” by environmental scientists to remove the water vapor in the atmosphere</u>. The high significance of the “adjustment” to remove water vapor is also discussed in greater detail within the book. Water vapor is a natural heat trapper and is more than 60 times higher in volumetric concentration in the atmosphere than are the Man-made greenhouse gasses. A second criticism is that seldom if ever do the environmental scientists volunteer the fact that the Keeling measurements are of total carbon dioxide in the atmosphere and inherently include the carbon dioxide that is produced by Nature totally independent of any Industrial Revolution or other anthropogenic means. From equally valid scientific measures – of the ice core samples collected and analyzed by ice research scientists at the International Ice Research Laboratory at Vostok, Antarctica, prove that there were/are about 285 or more parts per million of carbon dioxide in the atmosphere <u>before</u> the Industrial Revolution. That fact alone indicates that only about 130 ppmv of anthropogenic carbon dioxide was measured in 2020 and had increased to about 140 ppmv in 2025. When expressed in percentage format, the anthropogenic carbon dioxide in the atmosphere was measured to be a concentration of 0.013% in 2020 and 0.014% in 2025. In 1980 when environmental scientists began the claim that the amount (0.008%) of greenhouse gas emissions in the atmosphere was approaching a “crisis” level, there was in fact no basis for that fear-mongering tactic to eliminate the use of all fossil fuels if a “crisis” was to be avoided. Analysis of the Vostok ice core samples by the French Ice Research scientists (Dr. Caillon et al) proved that it is not carbon dioxide driving the global surface temperature but that the carbon dioxide peak

lagged the temperature peak by about 800 years. <u>Factually, 0.015% of anthropogenic carbon dioxide in the atmosphere is neither the dominant nor even a significant cause of global warming.</u>

Without water vapor in the atmosphere, the climate in every region would be classified as being "arid" since all precipitation on Earth's surface is originated from water vapor. Climate scientists use the fallacious logic: *"By eliminating everything else, Man burning fossil fuels is the dominant cause of global warming."* That fallacious hypothesis by environmental scientists is a fallacy echoed throughout this book.

A most important "observation" made during the research in this book for the factual causes of global warming and climate change revealed that environmental scientists also overlooked the importance of latitude on the temperature of a region and used a one-size-fits-all "average global temperature" as the measure of global temperature increasing. The change of any variable in a multivariable analysis is based on the difference from a "baseline" condition of the variable changing.

Temperature is a common variable – certainly not the only variable – in global warming, climate change, and ocean surface levels rising. The Author explored the effect that latitude has on the average annual temperature of 40 cities worldwide (20 in the Northern hemisphere and 20 in the Southern hemisphere), and using meteorological databases that are in the public domain (from national and international weather tracking organizations) to derive an eye-opener about the effect of latitude.

The latitude effect has been found to seldom if ever considered by environmental scientists in their pursuit of their fallacious hypothesis that Man burning fossil fuels is the dominant cause of global warming and climate change. When viewed in the "big picture" (shown in the next diagram – of the Northern hemisphere where over 85% of the human population resides), the environmental scientists' hypothesis that carbon dioxide is the "dominant" cause of surface temperature changes is once again proven fallacious.

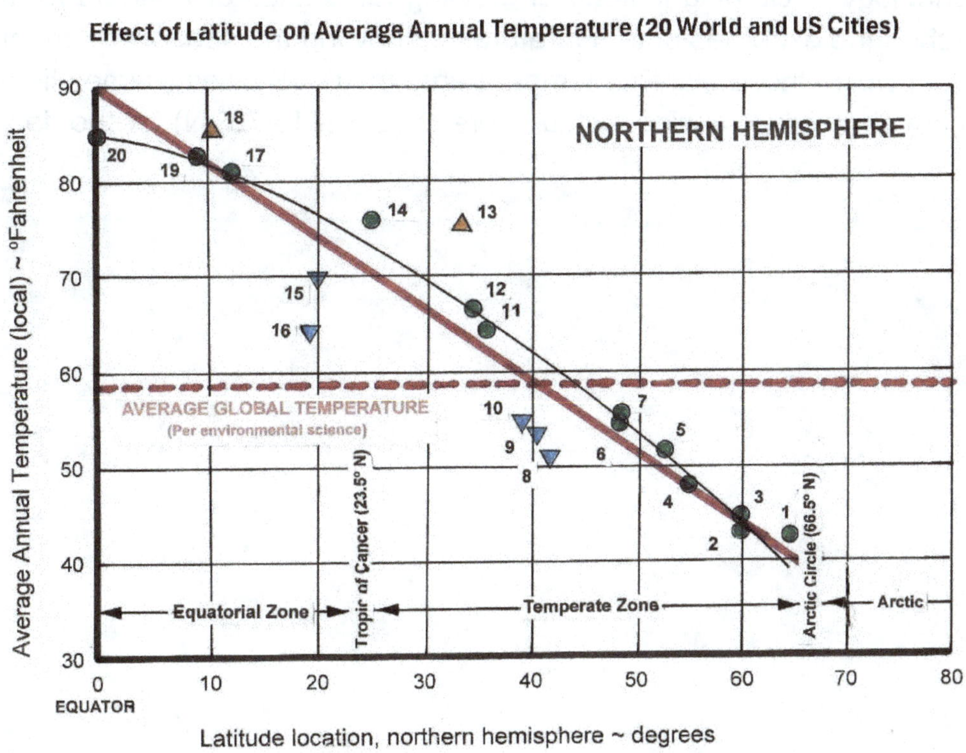

(Source: Graphic by Author from World Meteorological Organization database)

The "average global temperature" used by environmental scientists as the "baseline" from which global warming temperature is claimed to be changing due to the "greenhouse gas effect", is shown as the horizontal dashed red line (being the reference baseline) – at about 59 °F (15 °C). The circular and triangular (point up and point down) temperature data points are numbered 1-19 corresponding to cities located between latitude 0° N (the Equator, numbered as 20) and 64 ° N (Reykjavik, Iceland, just below the Arctic Circle at latitude 63.5° N). The solid red line is an approximation (a linear slope) of the effect that latitude has on average annual temperature of a location depending on its physical latitude.

The claim by environmental scientists that a 2 °C (3.6 °F) rise in average global temperature by 2050 constitutes a global temperature crisis has no basis in fact, when it is seen that the average annual temperature of cities having a substantial difference in latitude below latitude 40° N are already 5 °F (2.8 °C) to 30 °F (16.7 °C) warmer than the 2 °C (3.6 °F) used as the threshold temperature rise creating a global warming "crisis". The human populists in those cities are not in a temperature crisis as they have adapted to living in cities having average annual hot temperatures as shown on the illustration. Raising their heat index temperature by 2 °C due to climate change will require some modification to their current means of adaptation to heat to remain below a life-threatening temperature crisis level. The availability of water (for human and agricultural uses) being decreadecreasing in some regionssed may cause some regions to increase drilling of wells to access water aquafers and development of desalination technology (if near an ocean or gulf).

It is also factual that the use of average annual temperatures – global or regional or municipal - is not representative of the thermal conditions within which humans at any latitude on the planet must exist. Humans must exist and function "normally" at the extremes of temperatures surrounding their skin. Animals have adapted to that as shown by the fact that animals (land, aquatic, and avian) grow fur, scales, or feathers to maintain control of their body temperature. The type and extent of those external skin coverages is determined by the extremes of their temperature environment. Humans have developed the technology of clothing instead of growing fur, scales or feathers for the same reason. The Author researched the extremes of temperatures – showing the seasonal temperature extremes - for 17 US cities, from which those extreme temperatures in 10 US cities (names listed near bottom of illustration) were chosen to cover a wide latitude spread (20° N to 70° N) for the database graphically presented next:

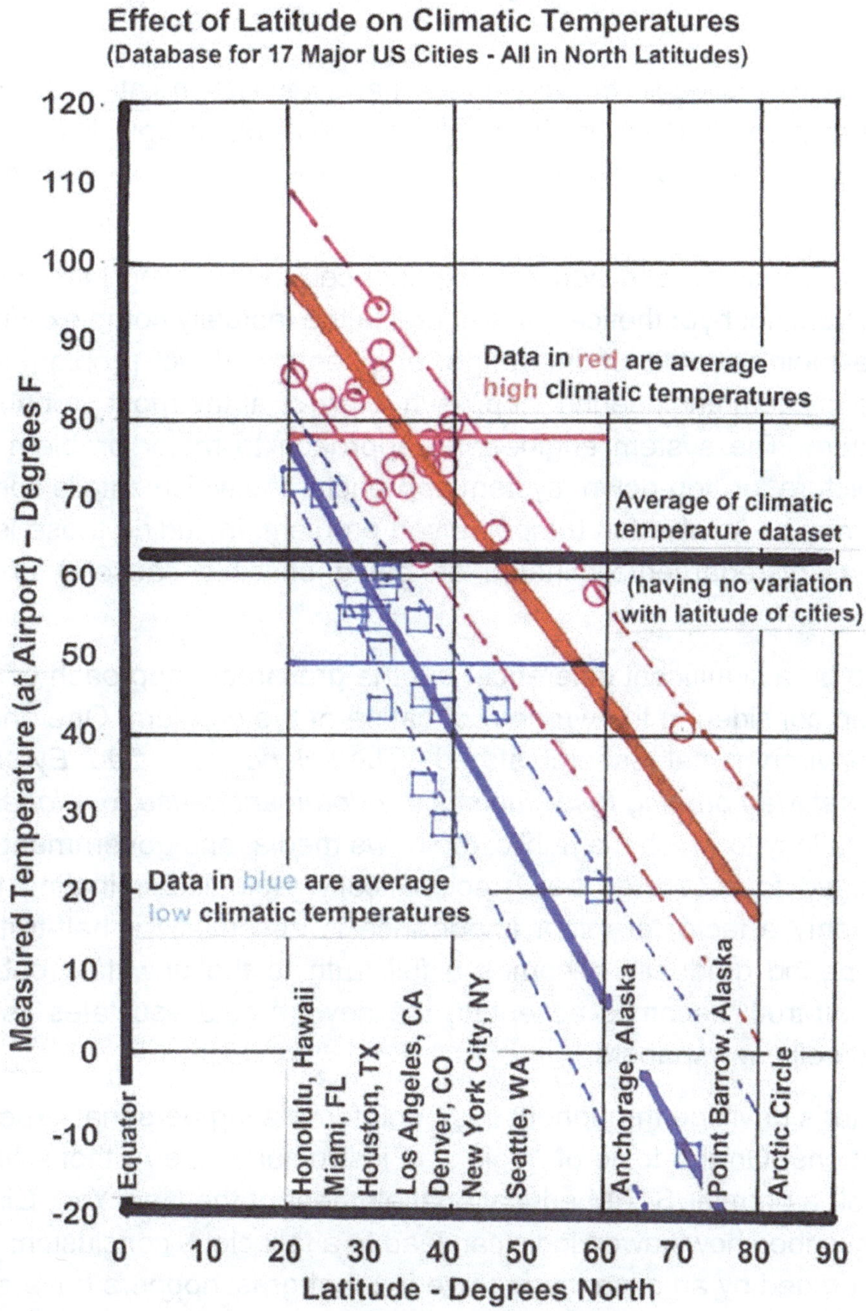

(Source: Graphic by Author from Meteorological Database in Public Domain)

The solid sloped lines (red for hot and blue for cold) represent the average annual extremes of seasonal hot and seasonal cold temperatures representing the latitude effect also on those average extreme temperatures. The dashed red and blue lines represent the average record high hot temperature (in summer) and record low cold temperature (in winter) of the cities shown. It is therefore the highest dashed red line representing the hottest temperature on record within which humans, animals, and plants/trees must survive. Correspondingly, the lowest dashed blue line representing the coldest temperature on record within which the humans, animals, and plants/trees must also survive. Some animals and some birds migrate to less hostile temperatures (e.g., birds at higher northern latitudes migrate south in the winter as their food supply is covered by snow as well as the outside temperature being too cold for their feathers to protect their skin). Humans "tough it out" and don heavier (heat retaining) clothing in the cold season and wear less and lighter clothing in the hot season. Humans in the desert regions wear much loose clothing and headwear for skin protection from intense solar radiance.

For reference purposes in this book, the environmental scientists maintain the use of an unrealistic "average global temperature" (shown as the solid horizontal black bold line near the center of the illustration) to be the baseline above which 2 °C (3.6 °F) is fallaciously considered as a "crisis" to humans.

Engineers are educated, trained, and become experienced in resolving large scale problems in the "real-world"; i.e., of factual not hypothetical nature, and in the factually complex environment that exists on the planet. Unlike their scientific counterparts, engineers conduct problem-solving and accident investigations using a "system engineering" approach to arrive at the most probable "root" cause(s) for the accident or problem. The system engineering approach to major problem solving is generally referred to as a "big picture" or "top-down" system approach, and which entails considering more of the possible factors that may be involved in the perceived problem, including those isolated factors in the problem that scientists disregarded/eliminated as being possible cause(s) of the initial observed problem.

That may not seem to be a significant difference, but the grassroots approach can – and often does – lead to a good solution considering the wrong root cause of the problem. One environmental scientist at a major center of environmental research stated: *"The science is settled. By eliminating everything else, it is concluded that Man burning fossil fuels is the dominant cause for global warming."* That is a prime example of the fallacy to which the public, the news media, and government regulators of human practices have been led to accept as being scientifically factual. Eliminating the factors of major importance to prove only a factor of minor importance is essentially a half-truth. A half-truth when repeated for a long period gradually becomes a full truth to the unwitting public and government leadership. Once a half-truth becomes accepted, the news media escalates its importance to be a "crisis" as that is what sells newspapers.

The Author uses Author's privilege throughout this book for relating personal experiences to add to his professional observations. On the topic of "logic", an instructor in the Author's high school (Brooklyn Technical High School, a premier STEM educational element of the New York City education system) science laboratory described how flawed logic can lead to a fallacious conclusion. The instructor stated a hypothesis (to be pursued by an experiment to verify) that grasshoppers have ears in their rear legs that cause them to jump from point A to point B. He described the steps in the experimental steps in the process of proving the hypothesis: 1) Capture a grasshopper and 2) place it on a table. 3) Slam a book on the table near but not on the grasshopper. 4) Observe how the grasshopper jumps in reaction to the slamming noise of the book. 5) Next, remove the rear legs of the grasshopper and 6) again place the grasshopper without rear legs in the same location on the table. 7) Using the same book, slam the book on the table at the same distance from the grasshopper. 8) Observe that the grasshopper does not react to the same noise from the slamming of the book. 9) *Q.E.D. (*the abbreviation of Latin *"quod erat demonstradum"* meaning *"Having demonstrated that which was to be demonstrated")*, the grasshopper is proven to have its ears in its rear legs.

The fallacious inferred logic of the grasshopper having ears in its rear legs applies the same fallacious logic as that used by the environmental scientist (whose name is respectfully not revealed as a professional courtesy to a scientific peer) who stated: *"When everything else is eliminated, Man burning fossil fuels is proven to be the dominant cause of global warming and climate change."*

This book has been created from the findings of research (the observations correctly interpreted) done by an engineer using a top-down logical approach in analyzing the large number of known databases of factual information providing clues to the tri-faceted problem of global warming, climate change, and ocean levels rising. One finding of major importance is that the fossil fuel technologies – that of coal, of oil, and of natural gas - each have a set of pros (good) and cons (bad) that were learned only by using those technologies extensively in the development of the planet to become a safe, healthy, comfortable and prosperous place on which humans can live.

In the research of the factual causes for global warming, the contention that global warming is at a "crisis" level of existence was also researched to determine if that is a factual contention or merely a fear-mongering tactic that scientists sometimes employ to overemphasize the importance of their scientific research efforts. Valid scientific research deserves to receive funding support from the government within which the research may aid in its development. Unfortunately, but understandably from an economic perspective, some potentially highly useful research that is of importance to the advancement of science does not warrant expending taxpayers' monies (received in the form of grants) to be pursued if the taxpayers do not reap some benefit. With global warming being declared a "crisis", the question – which this book addresses – is in two parts: 1) to what/whom/where/when is that "crisis" impacting; and 2) what are the potential consequences if the hypothetical cause(s) for the crisis – global warming and climate change of planet Earth – are not resolved in a timely manner to avert the crisis?

This book "The Big Picture of Global Warming on Planet Earth" has been written to address the first part of the two-part question – identifying the factually proven root causes for global warming and whether those causes are at a "crisis" level of significance. Of equal importance is the second part of the question – to what extent and at what rate is global warming threatening life (animate and inanimate, organic and inorganic) on the planet? A separate book by this same Author "Dealing with the Real Climate Change on Our Planet" is being published contemporarily to address the second part of the major question.

There was much technology development information uncovered during the basic research of global warming identifying the several candidate technological activities already underway to some degree of success in addressing global warming – but which have been aimed at the fallacious reduction/elimination of anthropogenic "greenhouse gas emissions" being the cause for global warming. The separate book "Dealing with the Real Climate Change on Our Planet" prioritizes 1) the technologies underway or are emerging which should be re-vectored to apply to the factual causes and effects of global warming, and 2) identifying the emerging candidate technologies having the potential to realistically mitigate the factually proven adverse effects of global warming and climate change.

In both books by this Author, an impassioned suggestion is that the young students in science classes at the high school stage of education and later in their college education be taught the factual scientific causes for the surface conditions – natural and anthropogenic - on their planet within which their future will unfold. An equally impassioned suggestion is that far more attention be given to the reduction/elimination of anthropogenic pollution. The reduction/elimination of the use of the fossil fuel coal is well underway to avoid its toxic and carcinogenic ash residue polluting the aquafers below landfills and trash dumps and polluting the oceans. Other fossil fuels (oil and natural gas) not having that ash residue need not be eliminated for the purpose of reducing "greenhouse gas emissions"; however, the use of any fuel in liquid or gas state presents pollution problems with leakages and spills that must also be reduced/eliminated.

Contents

Introduction

Among the most important factors affecting global warming, climate change, and the ocean levels rising on planet Earth are those in the "big picture" astrophysics of the solar system and the relationship of planet Earth with respect to the Sun. Scientists in the Age of Discovery several centuries ago established the unrefutable fact that planet Earth is in a large non-circular orbit about the Sun, and the planet is rotating about its unique tilted axis of rotation while moving at incredible speed (averaging 67,100 mph or 30 kilometers per second) along an irregular elliptically-shaped orbital path. That travel speed is not sensed by a human statically-seated watching a televised football game (perhaps the Georgia Tech Yellow Jackets defeating the University of Georgia Bulldogs?) because there are no air molecules in space and therefore no dynamic pressure being imposed - on the football or on the spectators. Not seen or felt is the ever-present physical "force" (factually an acceleration, not a force) of gravity on everything that has mass on Earth's surface. Without the presence of gravity, everything not physically attached to the planet would be centrifugally spun off into space.

The combination of those primary astrophysical factors is the dominant cause (by varying the distance of a point on the planet from the source of all heat energy, the Sun) of the incoming solar radiation and the temperature that is generated on Earth's surface.

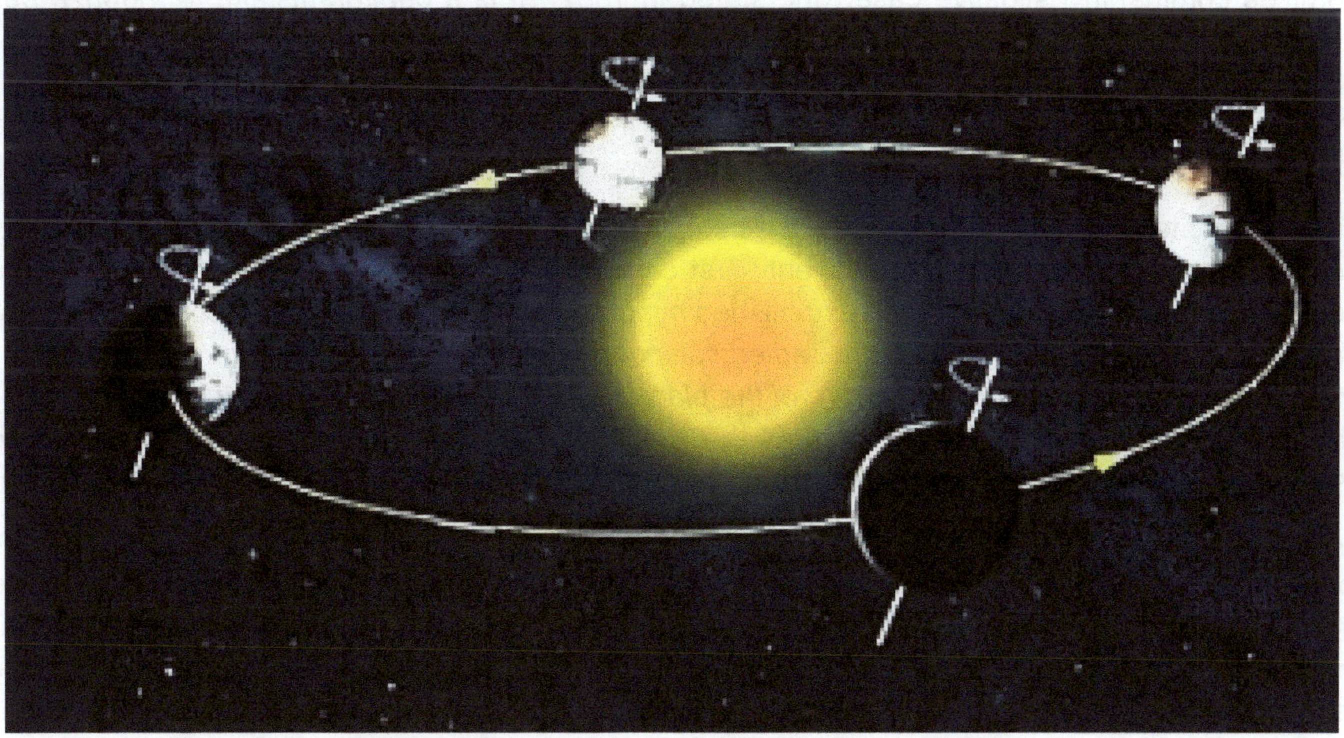

(Source: The Smithsonian Institute for Science)

Scientists have compressed the terminology of "incoming solar radiation" to the singular term "insolation" (not to be confused with "insulation").

There are only some portions of the planet's surface that are exposed to the Sun's solar heat that are being illuminated at the same moments in time (as shown above). Humans have referred to those conditions as "daylight" and occurring between "sunrise" and "sunset". "Sunrise" and "sunset" are scientific misnomers because the Sun does not rise or set. As the Earth rotates, the constant rays from the Sun begin to appear on the eastern horizon and approximately 10 hours later will disappear below the western horizon. The terminologies of "light" and "dark" are more descriptive of the daily phenomenon.

Astrophysical based factors are not affected by anything that has been or is being done by Man on the planet (with the exception of Man if engaged in a global nuclear war). Depending on where a person is situated latitudinally on Earth's surface, that person will experience a change in temperature daily, seasonally, and to the extremes that Nature creates in regions where the land surface of the planet interfaces with the ocean water surface. The extremes of those astrophysically-caused temperature changes – hot and cold – have occasional peaks in temperature of multiples of tens of degrees (on any scale). On any given day, all forms of life on planet's surface may be exposed to about a 40 degrees temperature difference from the pre-dawn time period to the early afternoon period. A change of one or two degrees – from any cause - between those two hot and cold extremes 40 degrees apart is of little significance to humans (but also cannot be simply ignored if important to sensitive natural life-forms).

Scientists generate scientific theories of how the natural world functions, whereas engineers develop the technologies for applying those scientific theories into the factual world environment. Extensive testing, analysis, and modification of every technology to perform efficiently and safely in the "real" world of astrophysical, geophysical, natural, and anthropogenic factors is mandatory. The British writer Sir Arthur Conan Doyle (1859-1930), whose world famous fictional character Sherlock Holmes epitomizes criminal investigative casework, stated in 1887; *"It is a capital mistake to theorize before one has substantiating data. Insensibly one begins to twist facts to suit theories instead of theories to suit facts."* Sir Arthur wrote in one of the fictional Sherlock Holmes series (Sherlock advising his famous crime investigation colleague Dr. Watson): *"When all the impossibilities are removed, all that remains is truth."*

A highly-educated scientist at a major US scientific laboratory in Houston, Texas, investigating the causes of global warming in the early twenty-first century, applied the Sherlock Holmes investigative philosophy in generating the now well-established hypothesis of the environmental science profession: "*After eliminating everything else, it is a settled science conclusion that Man burning fossil fuels is the dominant cause of the rise in temperature of the planet's surface."* This book contains the data and the Author's analysis of that data that illustrates the practice of "eliminating everything else" is not valid reasoning if that "everything else" contains well-established scientific facts that dwarf the fallacious conclusion that Man burning fossil fuels is the dominant cause of global warming.

Another contrast in approach is that scientists use "averaging" and "linearizing" so as to be able to generate models of observed phenomena for predictive applications. Experienced engineers have learned that when some element in a system fails it is because an extreme condition (temperature, pressure, vibration, electromagnetic interference, mechanical force or stress, or combination of those conditions) was imposed on the system for which one or more of the subsystems within the system had not been designed. Engineers conduct "qualification tests" in test facilities on every safety-critical subsystem at the extremes of those conditions and before approving the system to be mass-produced. Scientists are not bound by any safety regulations in development of theories.

The image shown earlier in the Preface (colored planet mean temperature with the USA outlined in white) begs the question: "How can environmental scientists realistically expect that an "average global temperature" sufficiently represents the surface temperature equally for all regions of the planet?" A related question is "Aren't the humans populating the planet at all latitudes north and south of the

Equator exposed to different extremes of hot and cold that are experienced at different levels depending on the latitude of the region containing that population?" To explore those questions, and to analyze the problem which is global in nature, it was found useful to start the analysis by using the known facts about the annual average temperatures that exist in the "lower 48" states comprising the USA.

In the "big picture" mode of analysis, the illustration in graphic form of the average annual temperature is most revealing:

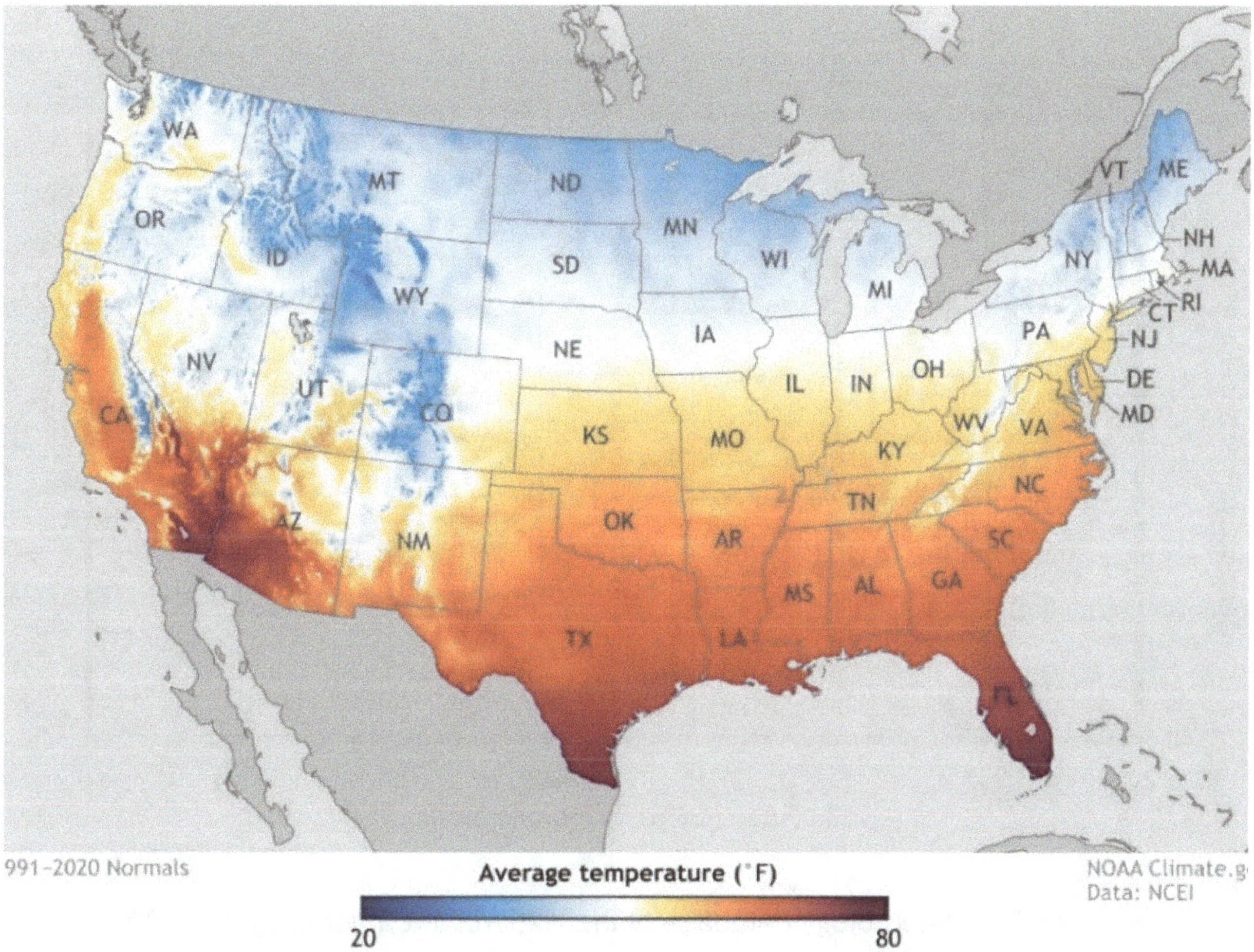

(Source: The National Weather Service, NOAA)

With the exception of persons who have the unfortunate optical affliction of being color-blind, there can be little doubt that the hottest average annual temperatures are in the states nearest the southern borders of the USA (i.e., in the lower latitudes within the Northern hemisphere). The coldest average annual temperatures are in two swaths – those states that are in the mountainous regions and those states that are on the northern border with Canada. Those visual clues to what are the factors causing that distribution of average annual temperatures to not be consistent with the environmental scientists contention that it is the atmospheric "greenhouse gas emissions" that is causing global warming – and use an "average annual global temperature" as their metric of measure. If the atmosphere was a liquid and the United States (48) were shaped like a bowl that might be a valid scientific conclusion that the heavier components in the liquid were settling in the bottom of the bowl. That is a fallacious conclusion since the atmosphere is not a liquid but is a gas, and the topography of the United States is not shaped like a bowl.

There are other geophysical factors involved in the average temperature distribution – the latitude being the most common. A chart was created showing graphically the extremes in temperature (i.e., the average hottest (red) and average coldest (blue) temperatures) to which the population in eight major cities in the United States "lower 48" are subjected:

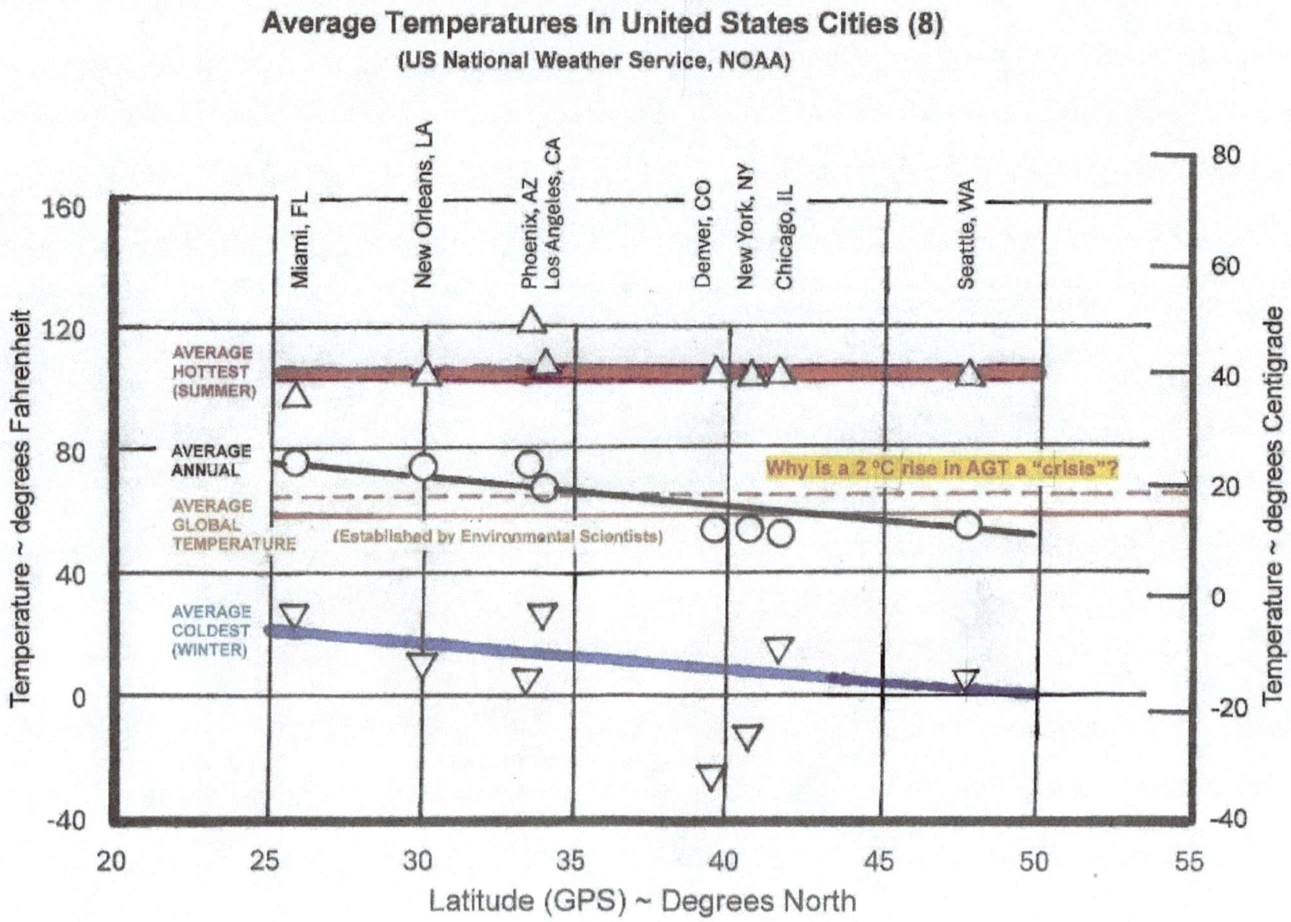

(Source: The Meteorology database of the National Weather Service, NOAA)

When the Average Annual Temperature data points were added, it was a surprize to the Author to note how closely the data correlates with the latitude change known for the eight cities chosen to be analyzed. The significance of a 2 °C rise in "average global temperature" (the horizontal thin red solid and dashed lines) is made less significant when examining the above chart for the city of Seattle, Washington. Seattle is at a latitude of 47° 48' N in the Northwest corner of the contiguous United States. The extreme average hottest temperature obtained for Seattle from meteorological records was 105 °F (40 °C) – during summer months, and the extreme average coldest temperature (also from meteorological records) was 5 °F (-15 °C) during winter months. The difference in temperature between the hottest extreme and the coldest extreme is shown to be 110 °F (55 °C). By coincidence, the average annual temperature of Seattle is 55 °F (12.8 °C) and only 4 °F (2.2 °C) cooler than the environmental scientist's "near-crisis" 60.8 °F (16 °C). The 2 °C rise above 16 °C is considered by environmental scientists as being a "crisis" or "tipping point" temperature, but there are typically no facts presented as to factually what will happen and to what or whom if the "crisis" temperature is exceeded.

There are an average of five annual deaths in Seattle from hypothermia (extreme cold) and two from hyperthermia (extreme hot) – mostly to "homeless" humans. The Author has no medical training or education, but it is common knowledge that the human body world-wide has an average "normal" body temperature of 37 °C (98.6 °F). The human body is on average about 20 °C (36 °F) above the environmental scientist's "crisis" level 18 °C (64.4 °F). However, the normal body temperature of humans is only 3 °C (5.4 °F) below the average extreme hottest temperature in Seattle while being 52 °C (93.6 °F) above the average extreme coldest temperature in Seattle. If the human body were a machine, a reliability engineer would caution that the further a machine operates from its normal operating temperature, it is more likely to fail in acceptable function. There is much difference between human blood and body water with that of the hydraulic fluid and oils in machines, but more machines fail when close to their minimum cold temperature limit (where fluids become more viscous and do not flow as well to lubricate the machinery) than fail when close to their maximum hot temperature limit.

Water in the human blood and in the human body cell structure – or anywhere else - freezes at 0 °C (32.2 °F) and thermodynamically exposing a human body to colder than 0 °C (32.2 °F) for a substantial length of time should cause hypothermia to eventually set in. (This is a well known fact to the Author who came very close to losing both ears from frostbite to the -30 °F (-34 °C) temperature just below latitude 38° N while serving in the US Army in Korea during the winter of 1953. An unusually strong cold wave had moved down from Manchuria causing a wave of frostbite injuries – mostly to toes - to Army personnel.) The Seattle fatality statistics of 5 deaths from hypothermia vs. 2 deaths from hyperthermia are consistent with what the laws of thermodynamics dictate. This simplistic analysis does suggest that the environmental scientist's hypothesis that Man burning fossil fuels is creating a "crisis" in atmospheric temperature is a fallacious claim if based on medical injury to humans. Whether or not the anemones on coral reefs are hypersensitive to a 2 °C (3.6 °F) temperature change in their water environment may be regarded as a "crisis" for marine biologists, but it is not a valid basis for concern of that small change in atmospheric and land surface temperature posing a "crisis" to human life.

Deaths from any cause always stir human emotions – and are fodder for the sensationalistic news media - but a statistically meaningful, non-emotional fact is that a 2 °C rise in temperature for cities having a latitude of greater than 40° (N or S) can have a net, albeit small, benefit – on the human population and probably on the ecology. That is consistent with the claim of the atmospheric chemist Dr. Svante Arrhenius of Stockholm, Sweden, in 1902 and who received a Nobel Prize in Chemistry in 1904 for his observations about the "hothouse" effect of varying concentrations of carbon dioxide in the atmosphere above Scandinavia.

A more extensive analysis of the effect of latitude (and elevation) on forty world cities is contained in a later chapter. The scientific database has been explored sufficiently to more completely address the hypothetical question: "Why does a 2 °C rise in "average global temperature" constitute a crisis?" Added to the question must be: "to whom, to what, how, when, and why?" In a similar manner, the Author found during the research for the problem that each factor explored generated other factors to be explored. An amusing comment made years ago by an engineer under the supervision of the Author described it as "Every stone I turned over to see what was under it only revealed more stones to be overturned."

In the "big picture" mode of analysis, careful analysis identifies the factors that significantly influence the system being analyzed and developed. The system can be a machine, a collection of technologies, or a human. For this book, the environmental science faction of the STEM (Science, Technology, Engineering, and Mathematics) process has in the past few decades stated (in various papers and editorials in scientific magazines and in the news media): *"The science is settled. By eliminating everything else, it is concluded that Man burning fossil fuels is the dominant cause of global warming and climate change."*

Anyone experienced in developing technologies views the statement "The science is settled . . ." to be a wish rather than a factual claim, as the science of a new process, a new technology, a new invention, a new discovery with the development of better observation and inspection instruments, is almost never a "settled" science because advancements are continually being made in science and off-spring technologies. The basic principle of knowledge being a stepping-stone process – with the stones (commonly referred to as "progress milestones") not regularly spaced in a linear pattern with time - is shown in a simplified "Learning Curve" diagram:

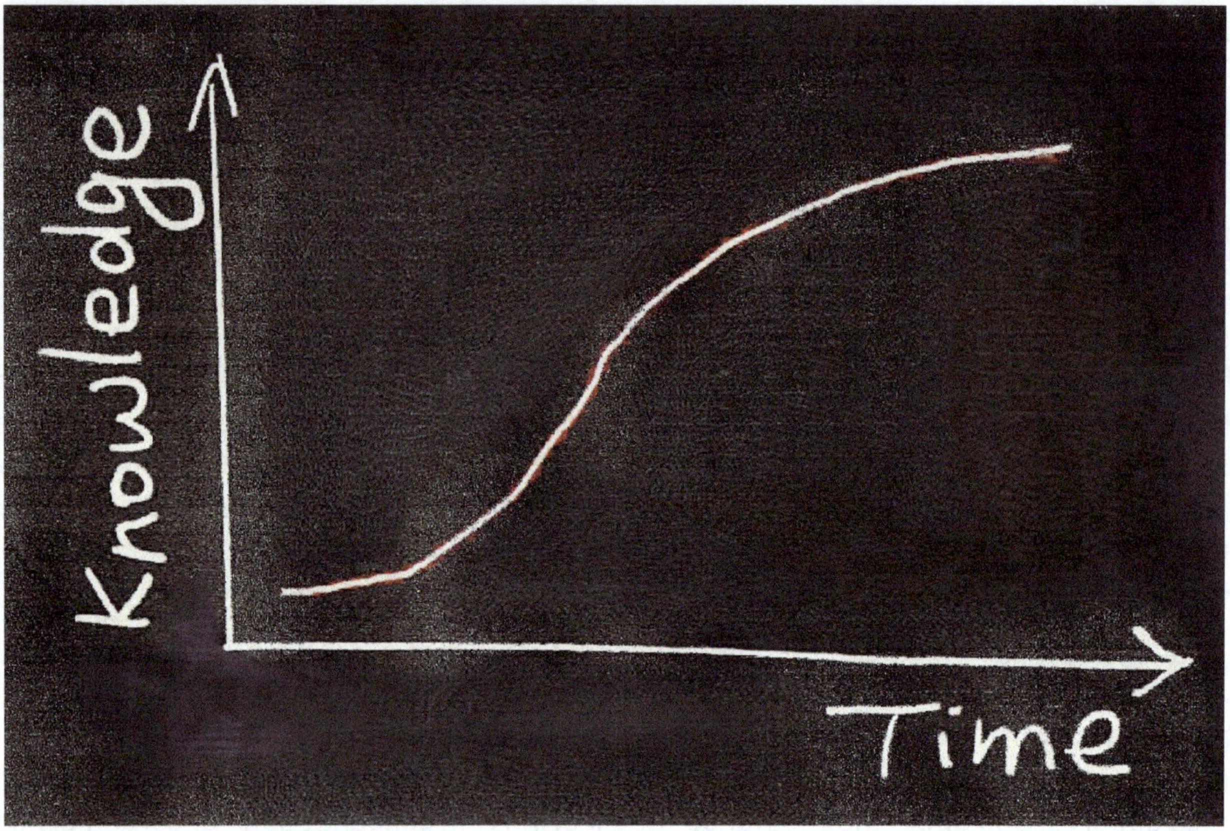

This curve applies in much the same way when learning anything about anything. Babies learn how to crawl, walk, and run in a "learning curve" rate of storing knowledge in memory. Being a student in high school is the age at which the most knowledge is acquired in the shortest time. Attending college is the next stepping stone, followed by being employed and applying all that stored basic knowledge to accomplish specific task assigned by a supervisor. The Author worked with an associate who when asked: "What gave you the inspiration to think "outside the box"? His reply was: "What box?"

The portion of the environmental scientist's hypothesis: *"When everything else is eliminated . . "* is a red flag that the person or group may have eliminated factors that are in fact more important than the singular factor claimed to be the cause for global warming (such as: *"Man burning fossil fuels is the dominant cause. . ."*). That elimination process is in effect "cherry-picking" the full database of information that exists to prove the wished-for end result. Ignoring a factor because little is factually known about the factor does not justify that factor to summarily be "eliminated" from further analysis.

The Author has long been researching these topics – global warming, climate change, and ocean surface levels rising - as a matter of personal interest and as a public service. It was a personal goal to identify and prioritize the importance of the many factors that influence the temperature of the planet's surface where humans and many other life forms live. The Author's "learning curve" for the three-pronged "problem" was resumed in the late 1990s and about at the Time = 0 date on the preceding graphic of a typical learning curve. One of the earliest charts found that captured the Author's interest was that shown in a report of the National Oceanic and Atmospheric Administration:

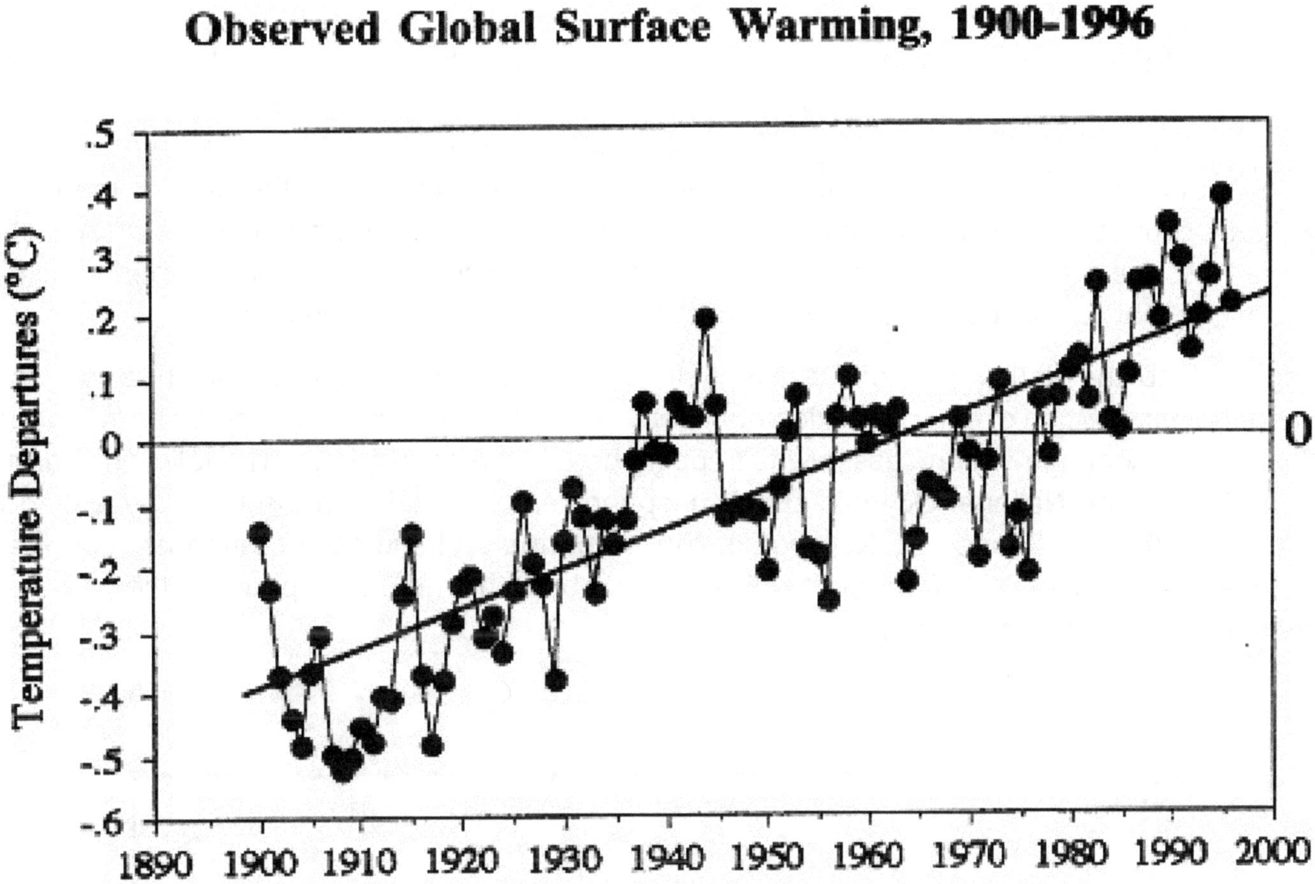

(Source: NOAA)

In undergraduate school at Georgia Tech (in the previous millennium), the Author was taught the discipline of how to present information on a chart such as the one above. In the temperature scale on the vertical axis (that for the "Temperature Departures"), two basic observations generated the Author's questions: 1) "temperature departures" are departures from what reference condition?; and 2) on closer investigation, the scale of the temperature departures were at first glance interpreted to be from 6 °C below the reference condition and to 5 °C above the reference condition. On closer look, it was seen that the generator of the chart had omitted the "0" before the faintly printed decimal point on all of the scalar temperatures (on the small vertical scale). Those "nit-picks" may seem unimportant but it changes the misinterpretation of the temperature departure to be from -5 °C in the lowest point of data to 4 °C in the highest point of the data (an increase of 9 °C in the ninety-six years from 1900 to 1996). The correct mathematics is the "departure" being -0.5 °C to 0.4 °C (a difference of 0.9 °C in the ninety-six years from 1900 to 1996). An actual rate of increase in temperature of 0.1 °C per decade is far less than the incorrect 1 °C per decade – but neither rate is an obvious "crisis" rate.

A more troubling observation made by the Author however was the sloped solid black line constructed (by an environmental scientist in the "settled science" faction) to represent an "average" of the "peaky" nature of the data. The data "spread" is seen to be a temperature variation of + / – 0.4 °C which is "averaged out" by the environmental scientists' sloped solid black line. A professional statistician would be disturbed by that "averaging out" process instead of analyzing what it was that was causing the large deviations ("scatter") in the data. Oblivious to the unprofessional analysis being done, the environmental science faction "sold" their claim that the global waming was at a convincing rate of 0.1 degrees per decade and that a "crisis with dire consequences" was looming. The fallacious solution derived by environmental scientists was to cause Man to cease combusting all fossil fuels for satisfying his demand for heat energy. Environmental scientists failed – and have persuaded the public to accept their failure as being a scientific conclusion - to consider that the small cyclic variation of the measured data points were due to small scale temperature changes caused by astrophysical and geophysical effects that varied in that cyclical manner and caused the slight increase in the data. If the increase was factually caused by Man's introduction of "greenhouse gasses into the atmosphere" at a steady increasing rate, the datapoints would not exhibit a cyclic varying trend with time. Sherlock Holmes would consider that as "tampering of the evidence" to convict a not gulty perpetrator.

Stepping back to the earlier part of the twentieth century, the environmental science faction had presented the observation of carbon dioxide being emitted by humans (and moreso by heavy industrial activities) in their coal-burning processes was trapping heat in the atmosphere. The following chart was issued in 1996 by the National Oceanic and Atmospheric Administration in the USA presenting valid measurements (by Dr. Charles D. Keeling) of the purportedly offending "greenhouse gas" carbon dioxide (chemical symbol CO_2) from the 1958 period to about the 1996 time period: (the portion after 1996 was added later as an update)

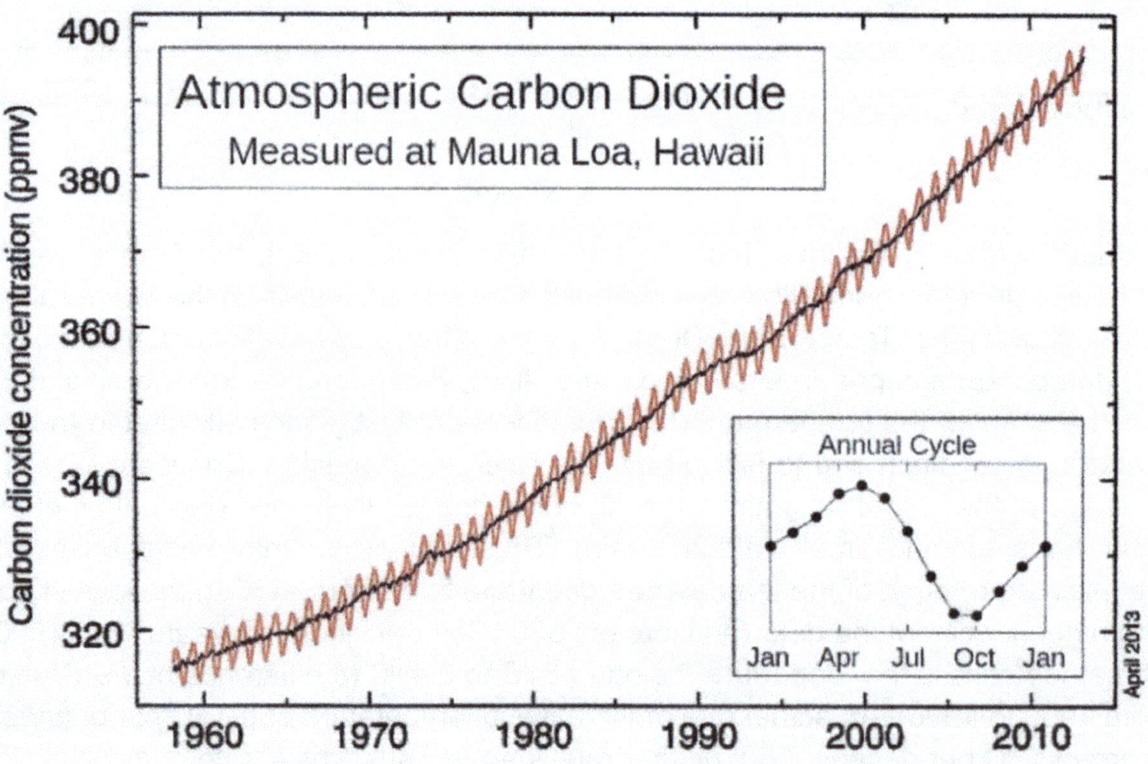

(Source: The National Oceanic and Atmospheric Administration, NOAA)

Dr. Charles Keeling (1928 – 2005) made the measurements of carbon dioxide in the atmosphere at his Mauna Loa Atmospheric Research Laboratory, and there is no question of the validity of his measurements. He standardized the measurements to be for dry atmospheric conditions so as not to confuse the issue of how much of the "greenhouse gas effect" was due to carbon dioxide and how much was due to water vapor. That "standardization" adjustment (to exclude water vapor) to the measure field data is not criticized, but does affect the interpretation of the data. Environmental scientists made the correlation that there was an increase in the "average global temperature" (from the previous chart) and reasoned that the Keeling-measured carbon dioxide increases (which contain both natural causes and Man-made causes) in the atmosphere were a result of the Industrial Revolution burning coal starting in the late 1890s.

Very few people (scientists or the general public) inquired what the cause was for the red wavy line around the curvilinear black median line (the carbon dioxide content in the atmosphere on which all attention was fixed). The small inset in the lower right was added by Dr. Keeling to scientifically indicate that there probably was a seasonal effect at which industries and the human population emit carbon dioxide emissions, but it is now known that it was the seasonal effect of temperature on the natural factors of the environment that vary seasonally in processing of total (natural and Man-made) carbon dioxide in the atmosphere. The ice core samples collected and analyzed by scientists at the Vostok Ice Core Research Station showing that carbon dioxide in the atmosphere lags behind the temperature but varies in the same cyclical pattern as the temperature brought environmental scientists to realize that the seasonal temperature changes were annually causing one cycle of one year period of the red line data on the Keeling curve but that there was a climate change in temperature due to other astrophysical factors (described in the mathematical calculations of a team of mathematicians applying the factors of obliquity, Earth's tilted axis wobbling (precessing), and the irregular eccentricity of Earth's orbit around the Sun – which combine to cause the seasonal changes measured by Dr. Keeling. That gain in knowledge did not support the environmental scientists' hypothesis (Man burning fossil fuels being the dominant cause of global warming and climate change) and was squelched from being revealed to the public. This book is formatted in much the same manner of gaining knowledge about atmospheric temperature change and climate change as time progresses on the "learning curve".

Once the fallacious correlation was introduced to the public by the environmental scientists, their hypothesis of "global warming being caused by Man burning fossil fuels in developed countries", the global warming "crisis" was "off and running". The United Nations formed an Intergovernmental Panel on Climate Change (IPCC) to pursue what needed to be done to control the greenhouse gasses by all nations burning coal to satisfy their energy needs. The news media escalated the word "control" to mean "combat" of the climate change.

NASA satellite data began appearing in the atmospheric database (the learning curve) in the mid 1970s and NASA generated the following portrayal of global warming from the NASA "big picture" perspective for the period 1979 to 1997:

Global Satellite Temperatures, 1979-1997

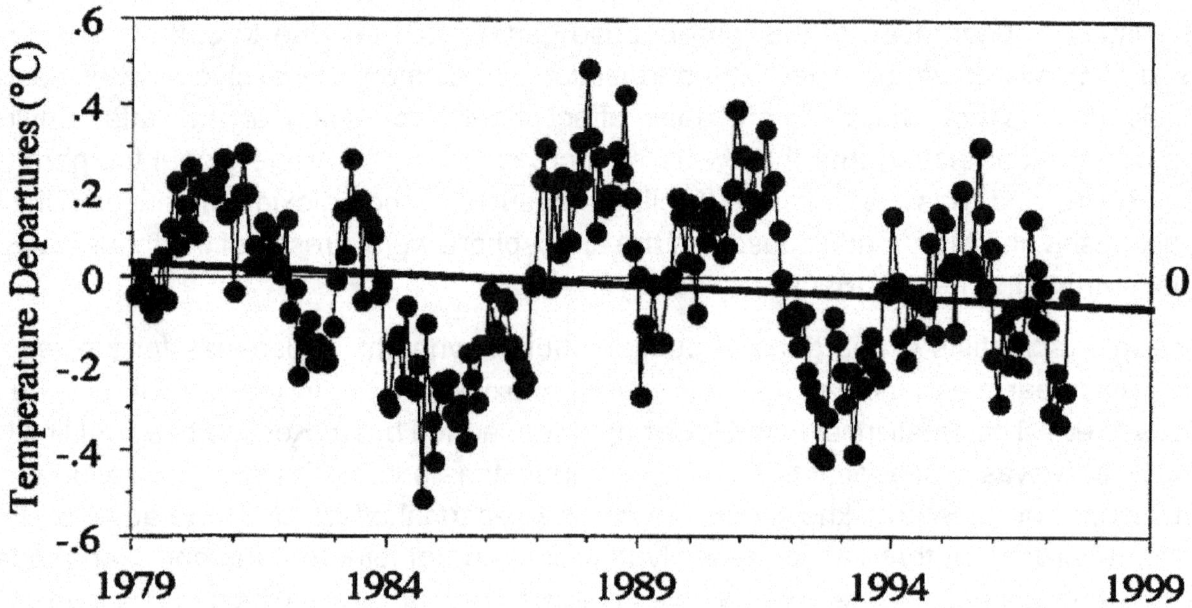

(Source: NASA satellite database provided to NOAA)

It appears that the scientist (or scientific organization in environmental science) who generated the above "Global Satellite Temperatures, 1979-1997" chart was the same creator of the Observed Global Surface Warming, 1900-1996" chart shown earlier in this Preface. The Global Satellite Temperature reflected the same + or – extremes ("scatter") in the Temperature Departures data points and a linear approximation was made as shown (the straight black line through the data). However, for the period 1979-1997, the more recent chart suggested that there was no, or only a slightly insignificant, decrease of global warming for that period. That observation of temperature departure decrease conflicted with the US environmental scientists' hypothesis as the combustion of coal had not decreased correspondingly for that same time period. The result was that the conflicting data was apparently "cherry-picked" from the database being used by the UN IPCC in setting goals for the reduction of Man-made "greenhouse gasses" and thereby limiting the rise in average global temperature (using the original meteorological database and not considering the satellite measurement database). It must once again be stated that the humans exposed to the temperatures shown by the blackened circular data points on the graph must survive those factual temperatures (no matter how much different from the "average" normal), not just the simplified linear thin black line "averaging" the database for simplified scientific modeling purposes.

A most important database started to become available in the late 1970s which seems not to be acknowledged by the environmental science profession – not surprizing as this database reveals that there has been a varying amount of carbon dioxide in the atmosphere long before Man began burning fossil fuels. Shortly after World War II had ended (September 1945), scientists of Russia (then-known as the Union of Soviet Socialistic Republics, USSR) began drilling through the thick glacial ice covering the eastern portion of the Antarctic Continent. Their objective was to obtain core samples of the ice that hopefully would have "recorded" the constituents of the atmosphere entrapped in the air bubbles at the center of the core samples and which by very sensitive spectrographic and other forensic instrumentation and processing techniques might provide valuable insight into the past history of the planet. A photo of that ice core sampling/measurement/analysis process is shown:

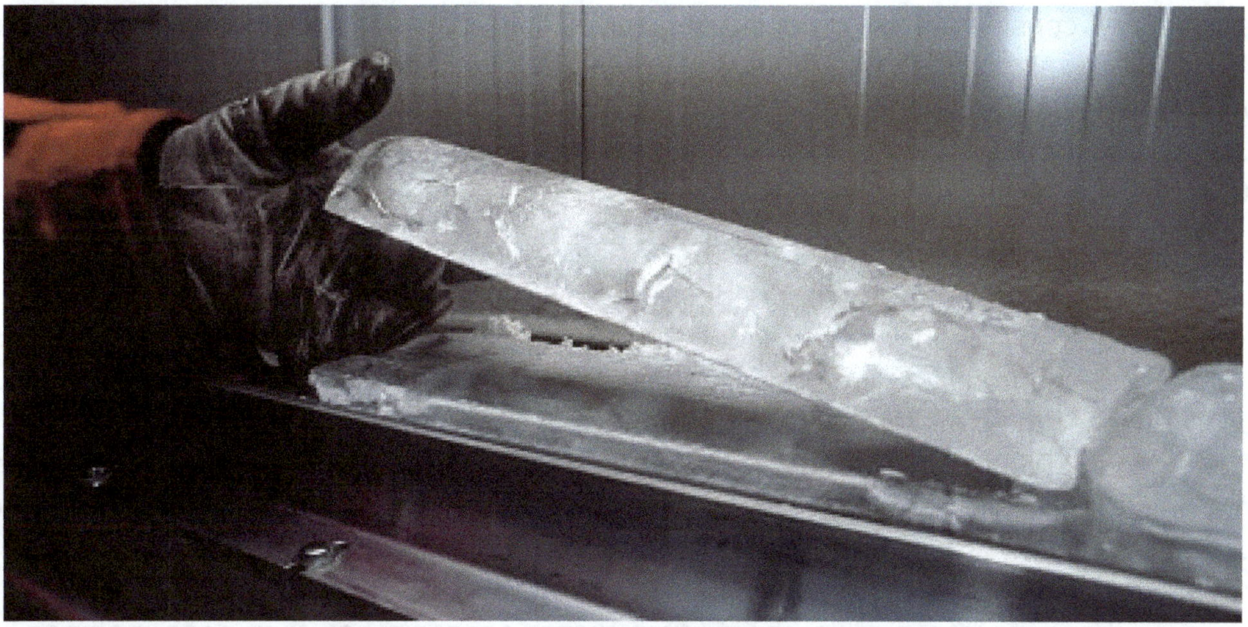

(Source: Smithsonian Institute for Science)

The mesmerizing details of that scientific venture at the Vostok Ice Research Station (at latitude about 78° S) are analyzed in more detail later within this book. After several years of drilling and analyzing, a summary of the results of that venture are shown next, revealing the history of the planet's atmosphere for the past 420,000 years:

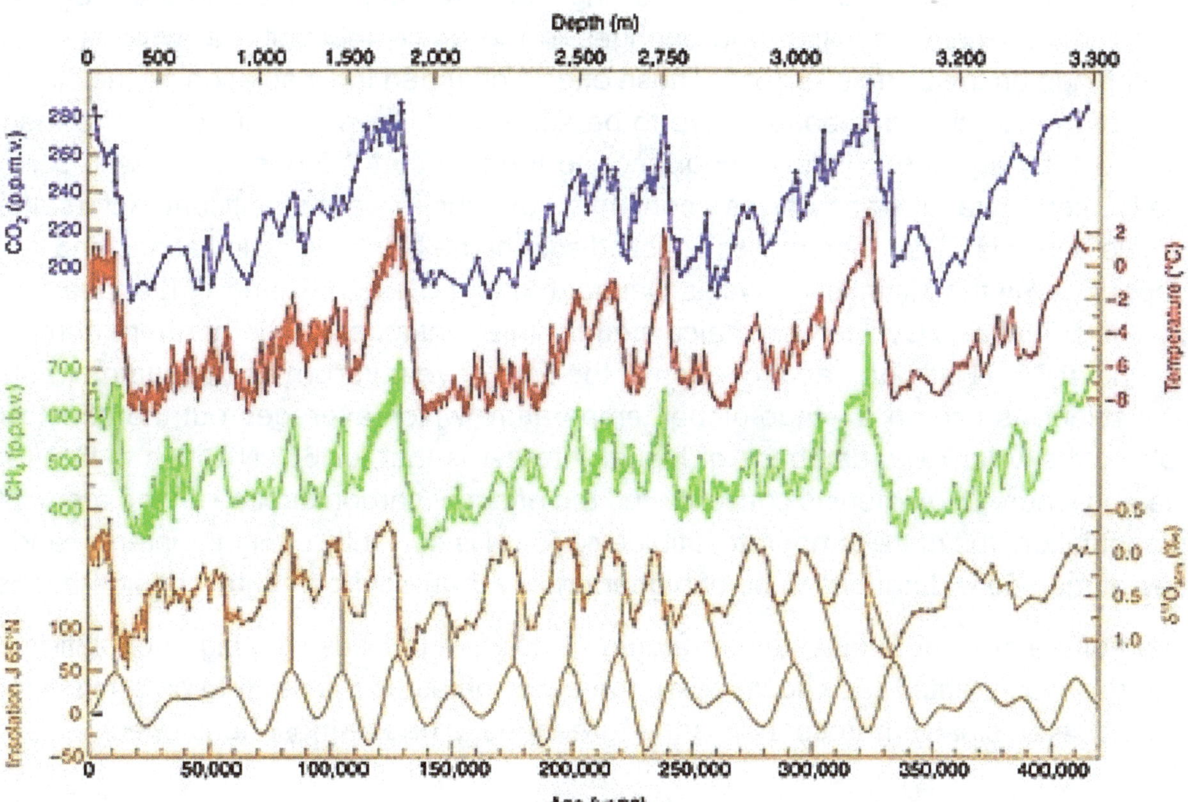

(Source: The Smithsonian Institute for Science)

Readers' attention is drawn to the top three "traces"; in blue (carbon dioxide concentration, measured in parts per million by volume, ppmv), in red (temperature departure from "normal", in °Celsius), and in green (methane concentration, measured in parts per billion by volume, ppbv). Both carbon dioxide and methane are considered as "greenhouse gasses". To be especially noted is that there is an amazingly good correlation in those traces – all three seem to "peak" in a spike pointing upward or in a valley pointing downward at about the same time. The scientific "big picture" significance is that these three variables occur in a series of four cycles between maximum high and minimum low with about 120,000 years between the extreme peaks.

The temperature (red) trace indicates (from the ice core analysis) that whatever is causing the temperature increases and decreases, the planet is now near the peak hot temperature at the end of the fourth cycle and has been for the past 10,000 years. The significance of that observation is that without any emissions from Man's or industrial activities, the planet has been at the temperature that very closely matches the peak temperature of "now", about 120,000 years before "now", 240,000 years before "now", and 320,000 years before "now". The traces for carbon dioxide and methane gases show the same peaks and valleys of their concentration in general as does the temperature peaks and valleys. To a scientist, it arouses the probability that one of those three variables is driving the other two variables. From the data alone, that question is narrowed by observing that of the two gases carbon dioxide and methane, carbon dioxide has a much higher concentration than does methane and it is inferred that the two dominant factors in the database are temperature and carbon dioxide

The two traces below the ice core sample traces are those generated by a mathematical calculation team headed by a brilliant astrophysicist and mathematician Dr. Milutin Milankovic (1879 – 1958) of Serbia and the region known today as Slovakia. Dr. Milankovic applied the then-known factors of astrophysics affecting the planet's motion through space in orbit about the Sun and with the planet's tilted axis of rotation that was precessing ("wobbling") and varying tilt angle in the irregular (varying ellipticity) orbit of the Sun. In addition, the changes in the angle (obliquity) at which the Sun's heat energy rays impinge on the Earth's surface. Those effects changed the insolation on the planet as seen at the bottom of the ice core database above to be somewhat "sawtooth" in form. The magnitude of changes in insolation are of secondary importance to the fact that the ups and downs correlate very well with the ice core data "ups and downs", confirming that those were significant factors affecting the ice core sample analysis. Environmental scientists discount the Milnkovic calculations and have issued scientific papers: "Why the Milankovic Cycles Cannot Explain Climate Change". It appears that that is a half-truth. The Milankovic Cycles were calculated for three latitudes in the northern hemisphere - at 55° , at 60°, and at 65° N latitude - and for each at the time of year known as the summer solstice. The environmetal scientists use an average global temperature which averages out the effect of latitude. The example of the Milankovic database of higher interest is that for 65° N which is that of northern Scandinavia and where Dr. Arrhenius concentrated the effect of carbon dioxide on the atmosphere. For the human population, the more temperate latitude of 55° N is currently of higher interest and this book which places human considerations to be of higher priority than environmental considerations.

When a graph of the average annual temperatures of world wide cities with high population density at different latitudes was created (as shown next), the effect of latitude is seen to be a highly correlative temperature indicator, clearly moreso than is the "greenhouse gas" atmospheric effect:

Effect of Latitude on Average Annual Temperature (20 World and US Cities)

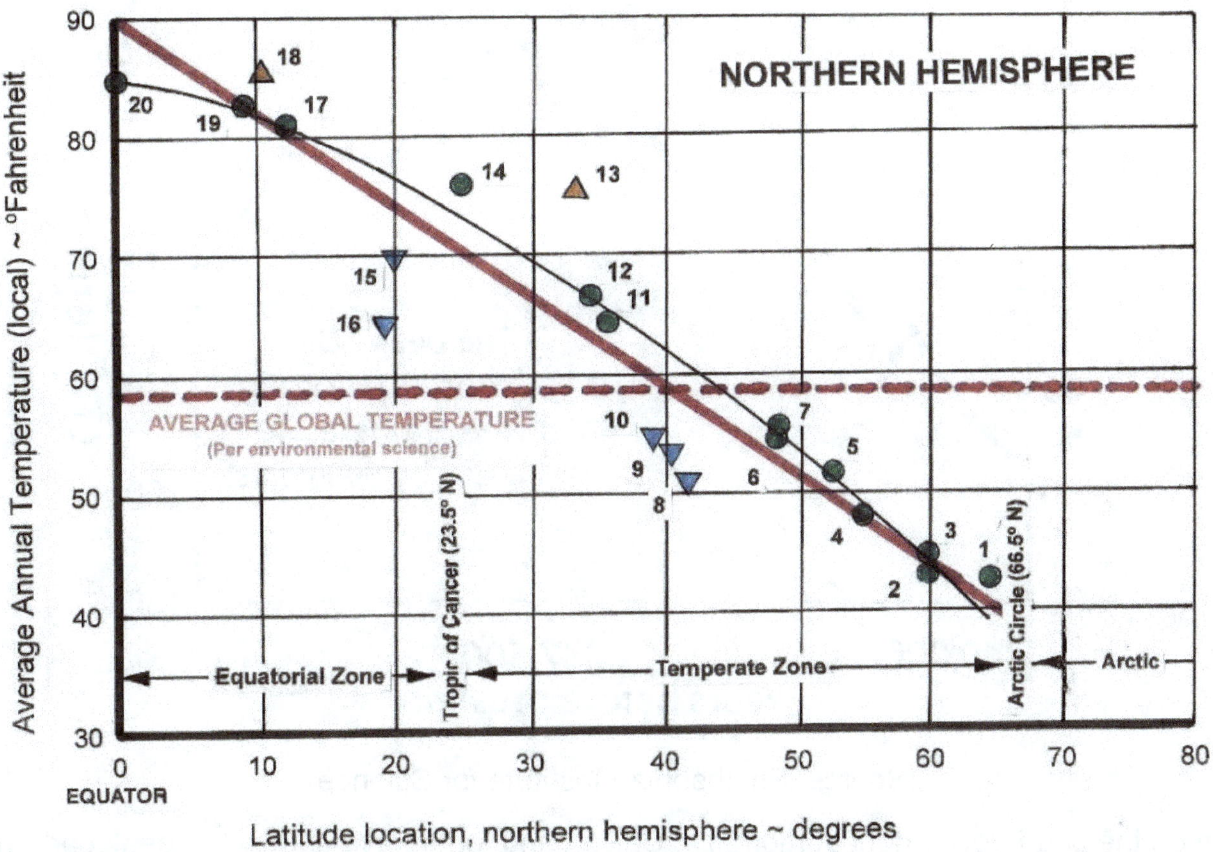

(Source: World Meteorological Organization Database, Chart by Author)

When correction of temperature is made for the average elevation of each city, the agreement of the corrected temperature (for both latitude and elevation effect), the estimation of each city's annual average temperature recorded in meteorological records correlates extremely well (within 3 or 4 degrees F) with the red linear "averaging" line above. (That comparison of adjusted temperature for latitude and elevation effects is included in a later chapter). The environmental scientists' one-size-fits-all "average global temperature" (shown on the chart) is seen to be a very sophomoric approximation to surface temperature reality of what is "normal".

The database from the Russian Vostok Ice Research Station attracted the interest of scientists from other nations, and a French scientist re-analyzed the Russian interpretation with emphasis on answering the question: "Does temperature drive the carbon dioxide and methane levels, or is it that carbon dioxide and methane drive the temperature levels?"

The answer (analytical finding) to that important question is seen in the next chart created by the French analysis team (scientist Caillon, et al.) analyzing the Vostok Ice Core database in closer detail:

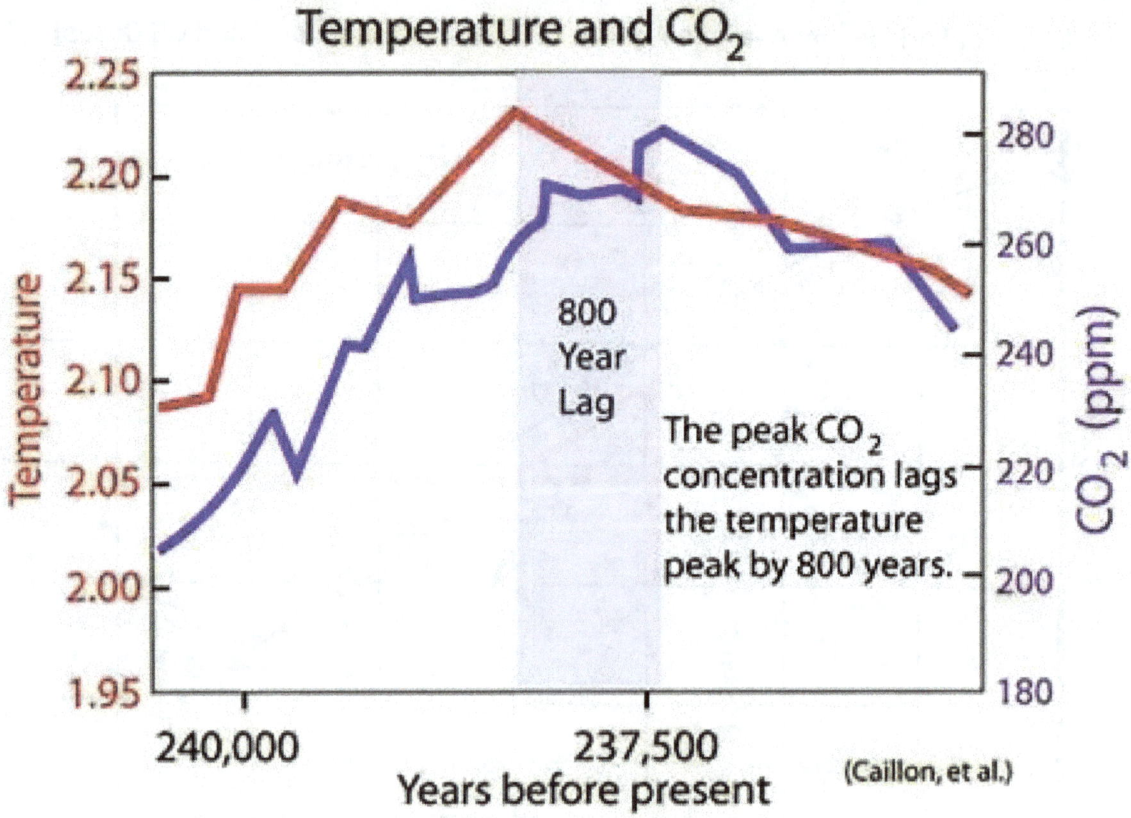

(Source: Smithsonian Institute for Science)

As shown on the chart (lower right portion), Dr. Caillon and team have shown the peak carbon dioxide concentration to lag the temperature peak by about 800 years - which signifies that the surface temperature of the location of the Vostok Ice Research Station (at latitude 78° S) is not driven by the concentration of the total carbon dioxide in the atmosphere. The Vostok database has been given further analysis by the Author in the later chapters of this book and once again points out that there is yet to be made known any database that confirms that Man burning fossil fuels – and creating heat-trapping "greenhouse gas emissions" - is a major cause for the global warming and climate change increases which environmental scientists are claiming are at "crisis" levels. Further, (underlined for its extremely important implications to the global warming problem) there is no indication that Man ceasing the burning of all fossil fuels in order to limit the rise in "average global temperature" will have any significant effect on that temperature.

In the early years of the twenty-first (current) century, the Author began serious "Big Picture" thinking as to the causes for global warming and climate change. This thought-provoking activity, by an independent data analyst performing the analysis as a public service with no affiliation to any organization (especially "big oil", "big steel", or "big automotive"), was spurred by the Author's penchant for problem-solving. After researching other databases that had been "eliminated" by environmental science, the Author came to the realistic conclusion that Man burning fossil fuels was neither the dominant cause nor even a major cause for global warming or climate change or ocean levels rising.

The research showed indications that several factors other than Man burning fossil fuels were probably more effective in both increasing and decreasing the surface temperature on regions of the planet. A summary of some of the factors – including the primary factor of solar insolation from the Sun - which environmental scientists should not (but apparently did) "eliminate" for scientific convenience are shown in the complex diagram:

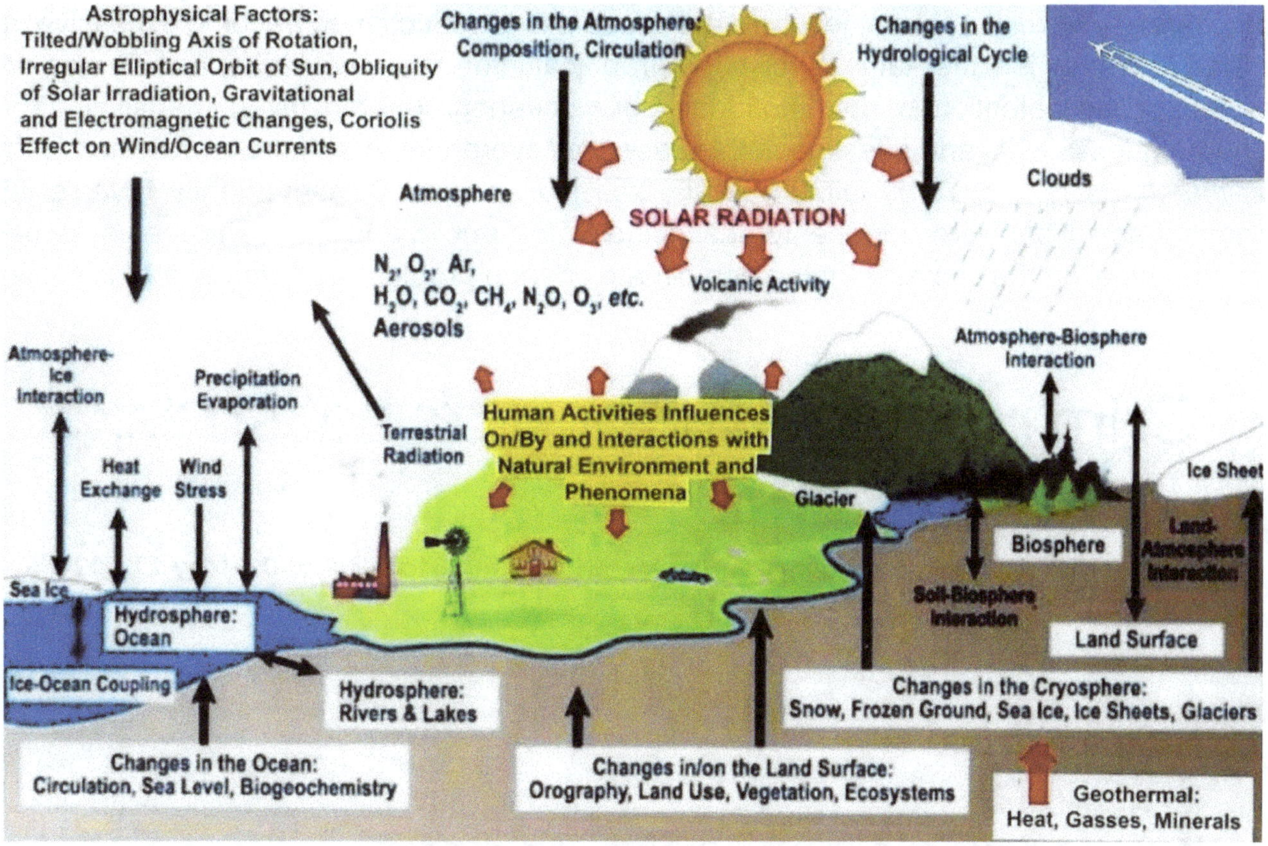

(Source: Factors discovered in database research, with annotations by Author)

The Author enjoyed traveling by automobile to locations in the entire 50 United States (to Hawaii by airplane, and to Alaska by both airplane and cruise ship) where the local cultural history of the region's residents were/are taught in Social Studies to high school students, including the Author's offspring. Those travels gave the Author and his children an opportunity to see (in the big picture) the magnificent beauty – as well as experience the adversities in the weather - of Nature in the land, water, and atmospheric environments that daily steers human lives.

The Author resided in the coastal region of New York City for 23 years during his early life and later for 34 years with his large family in Southern California (Thousand Oaks area of Ventura County adjacent to Los Angeles County). This gave the opportunity to first-hand witness and experience the factors that influence living 24/7/365 for a lengthy period near an ocean (the Atlantic Ocean off the densely-populated New York, and the Pacific Ocean off the beautiful coast of Southern California). Swimming with friends as a teenager in the warm Gulf Stream of the Atlantic Ocean at beaches on the East Coast and fishing in the cooler waters of the Pacific Ocean in the Aleutian Current on the West Coast were lessons not taught in schools about the importance of the oceans on human lives. The influence of the weather conditions (temperature, winds, precipitation, humidity, cloudiness, etc.) also became evident. The weather in Los Angeles is greatly influenced by "off-shore" winds (dubbed "the Santa Anas") from the hot, arid desert southwest region (Arizona and Nevada) or "on-shore" winds from the cooler Pacific Ocean. Similarly, the "on-shore" winds bringing high humidity and higher temperatures into New York City during "hot as Hell" summers added to the many factors – and challenges - experienced in everyday living in a wide variety of environments.

Only recently (as this book was on the verge of being readied for submittal to the publisher) did a piece of most informative information come to the attention of the Author. The Author has long been trying to establish the basis of the environmental scientists' claim that the surface temperature of the world –

land and oceans – were at a “crisis” level. To not cause this Introduction to become excessive in length – more than it already is – the Author shows the graphical information which apparently was prepared by a very eager, well-intentioned, but emotional young oceanographer involved in analyzing the effect of Man burning fossil fuels and affecting in this case the “average sea surface temperature”. The sub-title *“Off-the-charts temperatures impact corals, sea ice and more*” inadvertently reveals what the environmental scientists were not openly describing to the public and the news media as being the basis for their labeling the global warming and clmate change problem as being at a “crisis level”:

CLIMATE

Grim results of the ocean’s hot streak

Off-the-charts temperatures impact corals, sea ice and more

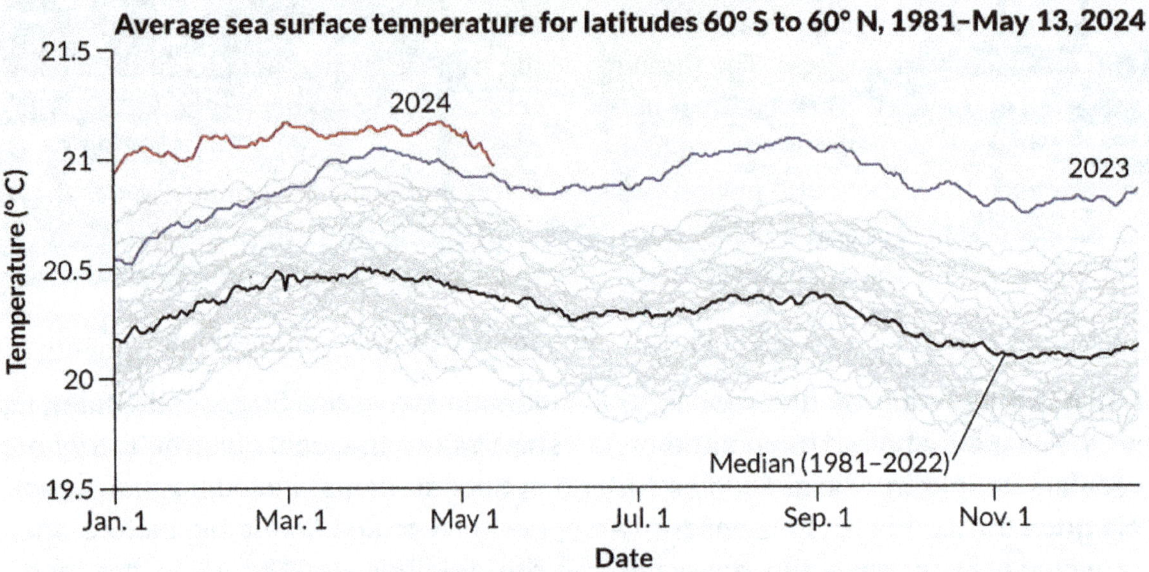

Soaring sea temps Since March 2023, the average sea surface temperature of the extra-polar ocean, at latitudes from 60° S to 60° N, has broken the daily record every single day, NOAA data show. And recent daily temperatures have shattered records that were set last spring.

SOURCE: CLIMATE REANALYZER/UNIV. OF MAINE; NOAA

(Source: Science Magazine)

The Author admires the enthusiasm of the young scientist who created the chart, but in parallel admonishes those in the environmental science (the “settled science” faction) for white-washing the true-reason and genuine concern over coral bleaching in order not to create doubt of their fallacious hypothesis that it is Man burning fossil fuels who is the biggest enemy of the anemones on the ocean floor. That same chart above appeared in the proceedings of the UN Intergovernmental Panel on Climate Change as:

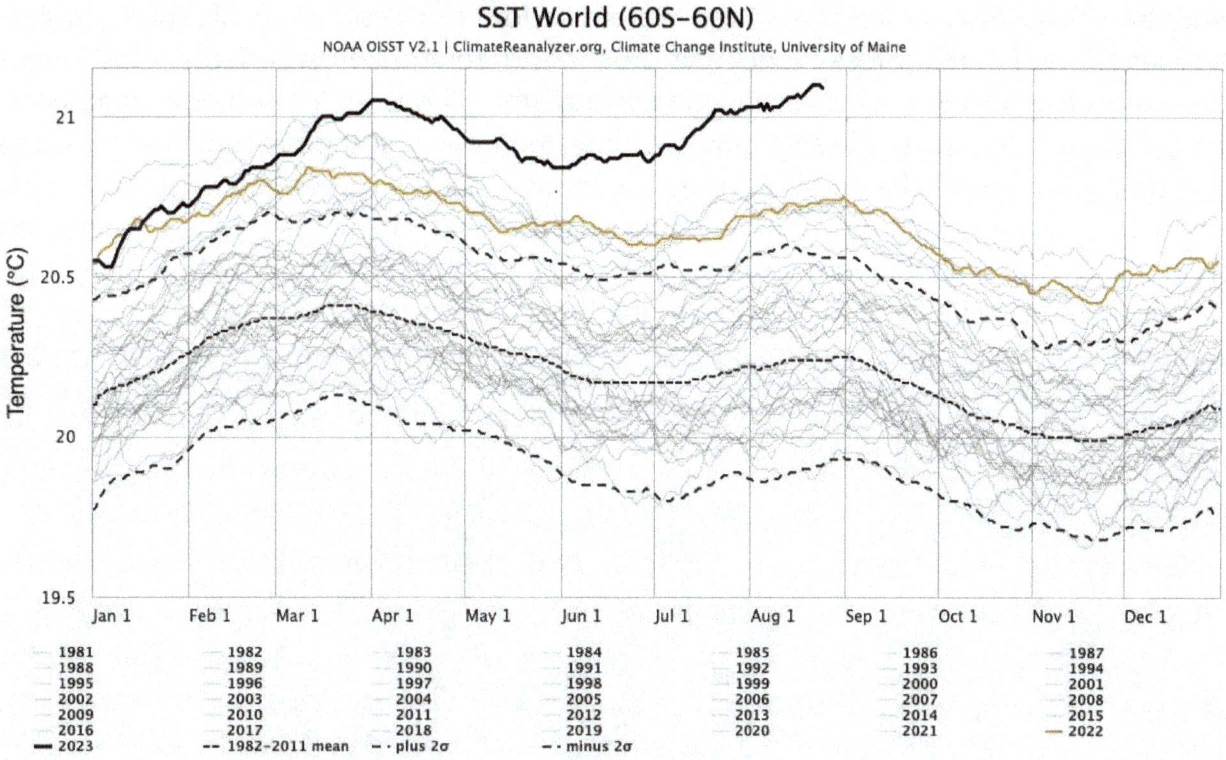

(Source; UN Climate Assessment Report)

The emotional distress that the environmental science profession was lamenting, ostensibly because the Great Barrier Reef of Australia was being devastated by Man burning fossil fuels, prompted the Author to generate the following diagram:

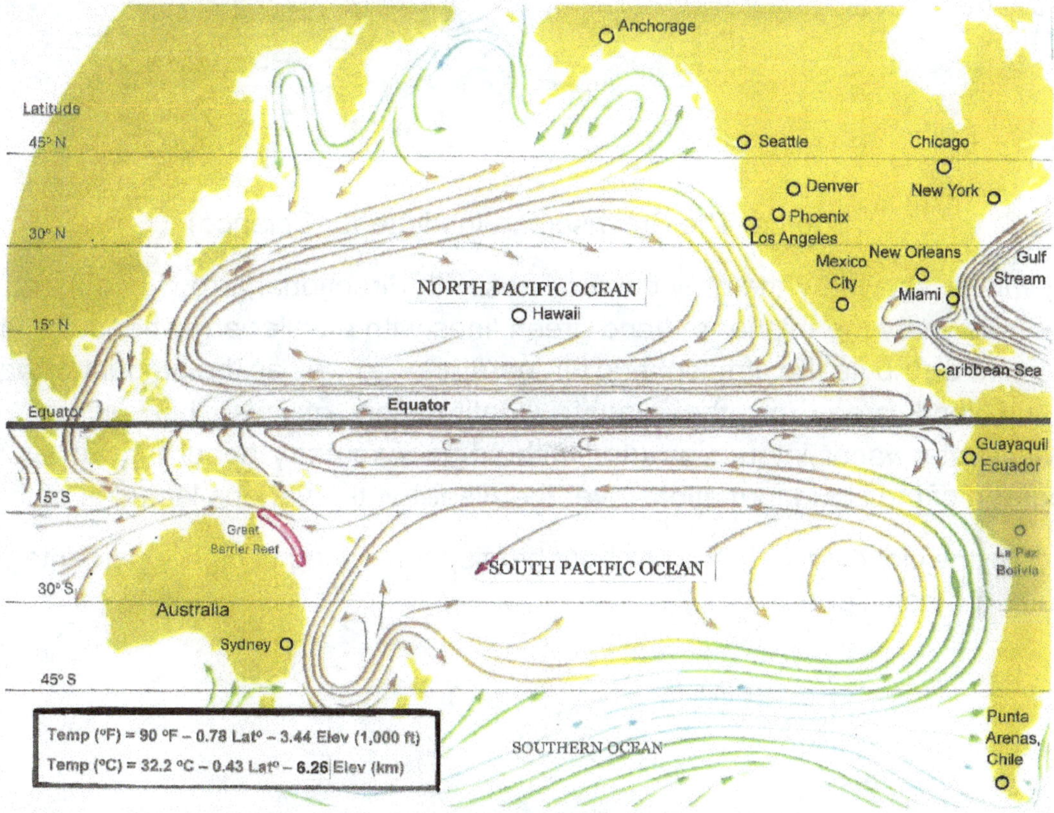

(Source: Background Ocean Currents are from Rand McNally "Atlas of the World")

Environmental scientists have long assigned the cause for coral reefs to be bleaching white to be the adverse result of Man burning fossil fuels and creating "greenhouse gas effects" on the atmosphere. The ocean currents in the South Pacific Ocean flowing along the Equator and becoming hotter as they approach the Great Barrier Reef in Australia are depicted clearly in the preceding diagram as being a far more important cause for the coral reef thermal stress – not the "greenhouse gas emissions effect". The mathematical equations in the lower left corner of the chart accounts for the effect of latitude and elevation on the average annual temperature of any city, but the temperature obtained mathematically (for a specific latitude and elevation – shown in the inset block at the lower left of the illustration) must also be adjusted for the cooling or warming efect of the ocean currents and prevailing wind currents.

The winds (and ocean currents flow) on the Earth's surface have been characterized by atmospheric scientists to generally be as seen next:

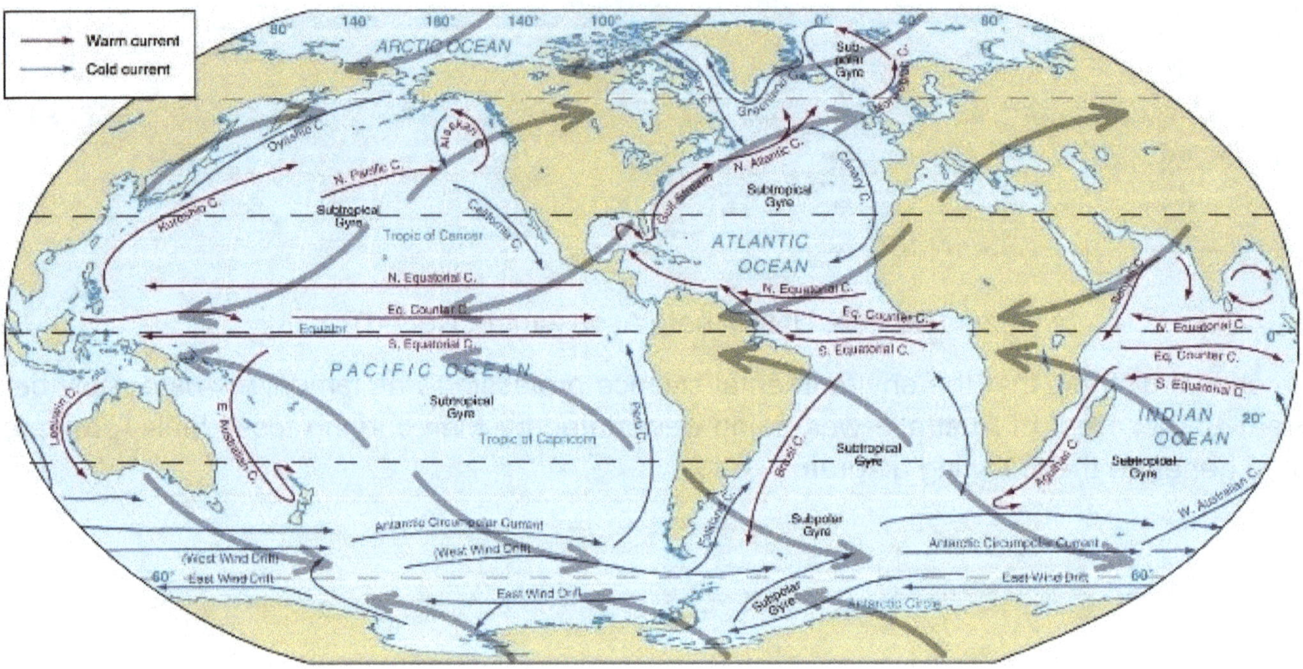

(Source: The Smithsonian Institute for Science)

The prevailing winds (shown by the gray curved lines with directional arrowheads) and the prevailing ocean currents (shown by the thinner red and green lines with smaller arrowheads) unmistakingly are seen to interact and distribute the atmospheric temperatures as well as the amount of water vapor over the oceans and land areas on the planet. Confirmation of this "big picture" effect of the temperature/wind/water vapor interaction (and with some symmetry about the Equator) is seen in a satellite photograph of the cloud formations over the planet at the various latitudes of the planet:

To be noted also is the cloud cover in the northern hemisphere being of higher percentage than is cloud cover in the southern hemisphere.

(Source: NASA)

For scientific convenience in modeling the temperature of the atmosphere, environmental scientists disregard the presence of water vapor in the atmosphere because of its erratic distribution shown by the varied direction and temperature of ocean currents and prevailing winds. It should be obvious that precipitation is an unmistakeable result of water vapor in the atmosphere being transformed (condensed) from gaseous form to liquid form we humans enjoy as "rain". Without precipitation very few forms of life can exist on the planet.

The Author gave the environmental science the benefit of having made a misjudgment on the importance of precipitation on human living, but the environment also cannot survive without adequate water in the growing process. However, the environmental science profession also chose to ignore the professional work of an eminent atmospheric scientist and peer Dr. S. Fred Singer (1924 – 2020) who repeatedly warned the environmental scientists that water vapor in the atmosphere was a far more effective heat-trapper than was the other "greenhouse gasses" from Man burning fossil fuels. This book later shows what the difference is, but key to the water vapor issue, Dr. Singer had repeatedly shown his now-famous comparison between water vapor and the "greenhouse gasses" graphically and numerically compared in the next chart:

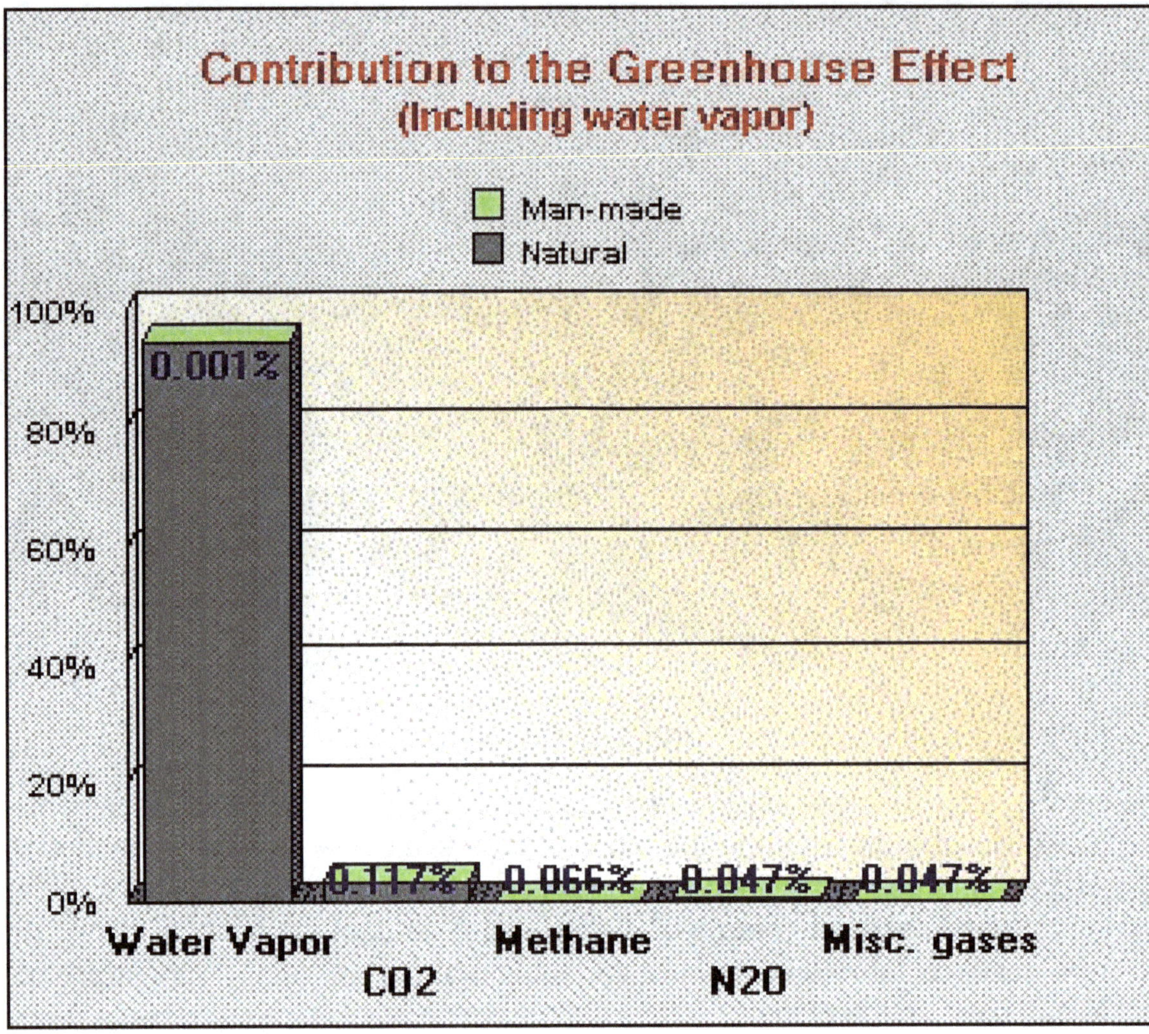

(Source: Dr. S. Fred Singer, University of Virginia, "A Closer Look at the Numbers")

Despite the importance of water vapor greatly overshadowing the significance of Man burning fossil fuels causing global warming and climate change, the environmental science profession has "circled the wagons" in defense of their fallacious hypothesis that Man burning fossil fuels is the dominant cause for global warming.

<u>The preceding set of information clearly shows those factors that scientifically are the dominant natural factors affecting global warming and climate change, not the factor of Man burning fossil fuels and the "greenhouse gas effect".</u>

The Author elected to document the results of his research efforts in an unpublished paper entitled "The Facts About Climate Change on the Planet" in about 2009. In a somewhat rebellious attitude to presenting problems "up the chain of command" (as the Author learned during his service in the US Army), it was decided to bring the results of his database research to the attention of the highest levels of US government. The Department of Energy was wisely created in October 1977 by President Jimmy Carter (1924 – 2024) with enactment of the Department of Energy Oganization Act. In 2017, the Author sent a copy of "The Facts About Climate Change on the Planet" (not published) to then-President Donald Trump with a request that his Secretary of Energy (the Honorable Rick Perry) consider the findings and recommendations made by the Author. Whether the Author's paper had a direct effect on President Trump's decision to retract US participation (in establishing unrealistic climate change goals and climate controls by legislation) from the UN Paris Climate Accord in 2017 is unknown.

A response to the Author's submittal of "The Facts About Climate Change on the Planet" in 2017 was received from then- Trump in a letter dated June 11, 2019, and a copy of the President's response letter appears next:

THE WHITE HOUSE

WASHINGTON

June 11, 2019

Mr. Herbert A. Hutchinson
Paso Robles, California

Dear Mr. Hutchinson,

Thank you for sharing your views on energy and environmental policy.

My Administration is strongly committed to environmental protection, including keeping our air and water clean and being good stewards of our environment. This requires all of us to do our part, and I want to thank you for your commitment to preserving our environment for future generations.

As we continue to protect our environment, we must also ensure that American workers have greater opportunities to provide for their families. That is why my Administration is also committed to protecting American workers and American companies from unnecessary regulatory burdens. Going forward, regulations must balance our stewardship of the environment with the needs of a growing economy.

Thank you again for sharing your views. As President, I am committed to unleashing America's economic potential, while also protecting the environment. I am confident that together we can preserve America's natural blessings for future generations.

Sincerely,

(Source: Author's Personal Correspondence Files)

Following receipt of the above letter from then-President Donald Trump, the Author began dedicated actions to update (with NASA satellite and other significant databases) his earlier research on the topics of global warming, climate change, and oceans surface levels rising in the paper "The Facts About Climate Change on the Planet" which the Author had sent to President Trump earlier. Those updates are included in this book.

Turning attention from the causes of global warming, climate change, and the ocean levels rising, the latter half of this book addresses what needs to be done to mitigate those effects and the more serious effects of natural catastrophes. The final reports of accident crash investigations not only contain the finding of a most probable cause for the accident, but contain recommendations that will minimize the accident from being repeated. A statement often heard at Accident Review Board hearings is "To make a mistake is human. To repeat the same mistake is unforgivable." There is a unquestionable need to revise the approach to "combat" climate change that has been fostered for the past several decades. The environmental science hypothesis that calls for the elimination of Man burning fossil fuels because of adverse effects on the environment, not only should but must be revised to consider human safety, health, comfort, and transportation as having a higher priority than the aquatic undersea life.

Humans and most living things must endure the extremes of whatever Nature throws at them, not the "average" of those extremes. Since those extremes are caused by natural factors and relegating Man-made factors to almost being insignificant, the content of this book does not support any recommendation that all fossil fuel useage should be terminated. The technology of coal burning has become outdated (and hazardous because of its polluting ash residue not because of its "greenhouse gas" emissions), but there is still the need for the burning of oil and natural gas until clean, reliable fuel technologies are developed to full operational use in the real world climatic extremes and at all latitudinal locations.

Current technologies (e. g., solar power, wind power, fossil fuels, nuclear fuels, hydrodynamic power, etc.) are all candidate technologies due for upgrading with cleaner, more reliable technologies. Newer technologies (thorium nuclear, geothermal, all-electric transportation vehicles, development of electric storage batteries – of a safer technology than being of lithium-ion technology - for diverse uses and applications , desalination of ocean saline water by distillation vice reverse osmosis, etc. are emerging but there must be a tailored approach for the different regions of the world to use the resources indigenous to the region. The application of a "one-size-fits-all" goal to eliminate fossil fuel emissions solely for limiting an "average global temperature increase to 2 °Celsius" is not a smart approach for countries and regions who only have fossil fuels (oil and/or natural gas) as their source of energy. In the United States, coal burning technology is no longer a necessary technology, and cleaner, more reliable technologies are closely reaching usefulness without harmful by-products (gaseous and solid, toxic and carcinogenic). The other fossil fuels (oil and natural gas) do not have the same harmful by-products from coal combustion and it is wasteful to place great emphasis ceasing the use of <u>all</u> fossil fuels simply because one type (coal) is technologically obsolete. That is not the same status in all countries. A major finding in this book of temperature being very dependent on latitude steers each region to control the temperature within its latitudinal extremes north and south by efficient use of the resources uniquely available within its physical boundaries and governmental jurisdictions.

The climate cannot be regulated to suit Man's desires. The climate is changing from causes beyond human control, and the climate change is different for each different region. There are no regulatory means that can be enacted to control climate change, but there are regulatory means for controlling true Man-caused pollution. Pollution is proportional to the size of the population, and with estimates that the population will grow by a factor of two or three times beyond the 8+ billion humans as of 2024 on the planet, the probability is very high that the amount of pollution will also likely occur. The finding that the United Nations "one-size-fits-all" global goal of net zero carbon emissions by 2050 calls for the goals to be regionally tailored to allow countries or groups of countries to develop the technologies for meeting the energy demands of the population within the region and with its indigenous resources. It is the United Nations approach that needs to be changed, not the climate change factors. That is evident with the "combat" of climate change currently embedded in the United Nations thinking:

UN chief urges countries not to surrender on climate fight

BY ARITZ PARRA
AND FRANK JORDANS
Associated Press

U.N. Secretary-General Antonio Guterres urged countries Monday not to lose hope in the fight against climate change, as representatives from nearly 200 countries gathered in Madrid for a two-week meeting on tackling global warming.

In his opening speech to delegates, Guterres cited recent scientific data showing that levels of heat-trapping gases have hit a record high, reaching levels not seen for at least 3 million years when sea levels were 33-66 feet higher than today.

Unless emissions of carbon dioxide and other greenhouse gases are sharply cut, temperatures could rise to twice the threshold set in the 2015 Paris accord by the end of the century, he warned.

"Do we really want to be remembered as the generation that buried its head in the sand, that fiddled while the planet burned?" Guterres asked.

His appeal came after Chile's environment minister, Carolina Schmidt, said the Dec. 2-13 meeting needs to lay the groundwork for moving toward carbon-neutral economies while being sensitive to the poorest and those most vulnerable to rising temperatures - something that policymakers have termed "just transition."

"Those who don't want to see it will be on the wrong side of history," said Schmidt, who is chairing the meeting. She called on governments to make more ambitious pledges to reduce emissions of greenhouse gases ahead of a deadline to do so next year.

The summit, which moved to the Spanish capital after Chile had to pull out amid anti-government protests, aims to put the finishing touches to the rules governing the 2015 Paris accord.

That involves creating a functioning international emissions-trading system and compensating poor countries for losses they suffer from rising sea levels and other consequences of climate change.

"We have a common challenge but with differentiated needs and urgencies, which we can only overcome if we work together," said Schmidt.

Countries agreed in Paris four years ago to limit global warming to well below 3.6 Fahrenheit, ideally 2.7F by the end of the century compared with pre-industrial times. Already, average temperatures have increased by about 1C, leaving little room for the more ambitious target to be met.

Guterres called out big greenhouse gas emitters that are still building coal-fired power plants, saying that unless the world stops burning coal "all our efforts to tackle climate change will be doomed."

He noted that had countries started cutting their emissions drastically a decade ago, reaching the Paris goal would have been much easier.

"Ten years ago, if countries had acted on the science, they would have needed to reduce emissions by 3.3% each year," he said. "Today, we need to reduce emissions by 7.6% each year."

(Source: Associated Press)

The need for better United Nations understanding of the causes for global warming and climate change requires coordination – not direction – of all countries in mitigation of the effects of global surface temperature changes and climate changes in all regions. That change in approach will become more important as the world population doubles and triples its population size after about 2100. The Antarctic ice core samples indicates the planet will at some point in the not-too-distant future reverse the global warming trend and be entering into another "Little Ice Age" trend where the heat energy demands for humans will be greatly different from what they are at present. When that reversal in global temperature and climate occurs it is anticipated that an assessment of the technologies needed for colder temperatures and climates will be accomplished similar to the manner in which the technologies have been assessed for this book. The assessments must consider both the effect on the human population as well as on the environment – it is not an either/or choice.

Realistically, this book will scientifically prove that coal is the offender on pollution, not the dominant cause of climate change (as the public has been led to believe). As an obsolete technology, the time has come for coal technology to be replaced. The article shown next (representing many like it in the news media) well-summarizes the attitude the public holds about coal:

Coal is the global climate scourge that just won't go away

BY MARK GONGLOFF
Bloomberg Opinion

It's nearly impossible to get every country in the world to agree on anything. But at the latest United Nations climate talks, leaders representing all the people of Earth held hands and sang in perfect harmony of their shared belief that coal is the absolute worst.

And yet too many of them still can't seem to get enough of the one fossil fuel that makes oil look like a slacker when it comes to ruining the atmosphere for human habitation. It's a grim reminder that transitioning the global economy to clean energy will be hard, and exorbitantly expensive, and that wealthier nations must share much more of the burden than they have so far.

It took decades of climate talks for the world to publicly admit, at the COP28 confab in Dubai, that fossil fuels generally are the main culprit behind the greenhouse-gas emissions heating the planet. But a consensus of scorn for coal formed a bit earlier: The world vowed to do away with "unabated" coal power at COP26 in Glasgow in 2021, a pledge it repeated in Dubai without debate. (More on that "unabated" qualifier in a minute.)

That's not working out so great. Global coal consumption has surged more than 5% since that Glasgow agreement, based on International Energy Agency estimates.

The IEA this month optimistically suggested global coal demand peaked this year at a record 8.5 gigatons and would gently ebb to 8.3 gigatons by 2026. If this sounds like marginal progress, I guess you could argue it's still better than nothing.

But a year ago, the IEA suggested coal demand had already essentially peaked at around 8 gigatons and would be barely higher in 2025. In its latest report, it revised its estimate of 2022 demand up significantly, from 8 gigatons to 8.4 gigatons. Last year's peak-and-plateau turned out to be a false summit. The IEA's new claim of marginal progress masks what has actually been a lurch in the wrong direction.

Coal is humanity's biggest greenhouse-gas polluter, churning out 40% of all energy-related emissions. After dipping during the pandemic, coal emissions surged to a record high of 15.5 gigatons in 2022. That was more than the entire energy emissions of China and three times as much as the U.S. Not only is this terrible for the climate in the long term, but coal pollution has immediate effects on public health. We won't be able to stop global heating without first stopping the burning of coal.

Coal is in rapid, welcome decline in the U.S. and the EU, tumbling 21% and 23%, respectively, in just the past year, according to the IEA – and this despite the war in Ukraine triggering worries that Europe would be forced to make up for a lack of natural gas by burning the dirtier stuff. In the U.S., renewables now produce more electricity than coal. Even Japan, which has been much more leisurely about ditching coal, burned 8% less this year.

The story is much different beyond the developed world. China, India and fast-growing Southeast Asian economies have all been ramping up coal use lately. All have joined the global pledges to phase out coal – or at least the "unabated" kind, meaning that which doesn't rely on dodgy carbon-capture technology to negate its greenhouse-gas effects. But it's still a cheap, reliable source of power when renewables aren't available, a common problem for developing economies.

The best solution is to bolster renewable capacity to take its place. But that takes money – lots and lots of money. China has been throwing wads of cash at clean energy, and coal's demise seems inevitable there, my Bloomberg Opinion colleague David Fickling has written. But poorer developing nations will need a hand. In 2021, the Group of Seven leading industrial nations set up the Just Energy Transition Partnership to help some of the world's most dedicated burners of coal to switch to cleaner energy. The results have been underwhelming.

At COP28, even as they finally vowed to ditch fossil fuels, global leaders conveniently omitted the details of who will pay for that monumental change. The world's struggle to kick its coal habit is a striking example of why it behooves well-to-do countries to get much more serious about helping their poorer cousins hurry their transitions. Whatever the means, the key is to remember that the coal burned in Vietnam profoundly affects the climate in Vermont and Vienna.

There is a quote (unknown source) that "Two half-truths do not add to be a whole truth" and the above article contains more than three half-truths. The above article fallacious states: *"The coal burned in Vietnam profoundly effects the climate in Vermont and Vienna."* One of the greatest challenges will be for converting those fallacious beligerant beliefs in the public domain into factual-based actions – realistically funded -for the benefit of Mankind such as reducing pollution. The climate will change to whatever Nature will cause it to become. Man can only mitigate those effects, not control them with such bravado as "Combat climate change – to the last man standing".

While the time has come for coal technology to be eliminated as a source of energy in the United States, developing countries remain dependent on coal because of its availability and economic value compared to use of other fuel technologies. The work of Dr. Fred Singer has scientifically shown that water vapor traps over 60 times larger heat in the atmosphere (the "greenhouse gas effect") than does carbon dioxide yet the environmental science activists disregard that truth and maintain that carbon dioxide is the dominant cause for global warming and climate change. Readers are shown in the chapters of this book that the factors of astrophysics, geophysics, and natural events are far more influential on the planet's surface temperature than the combined effect of all "greenhouse gas effect" emissions.

A critically important factor environmental scientists have not considered fully – apparently by intention – is the effect the protection of the environment causes severe impacts economically and legislatively to the human population. Every region has a different size and distribution of its population as indicated on the following population density maps issued by the UN World Population Division for the population of the World (upper left), the population of Europe (upper right), the population of Asia (lower left), and the population of the United States by the US Census Bureau (lower right):

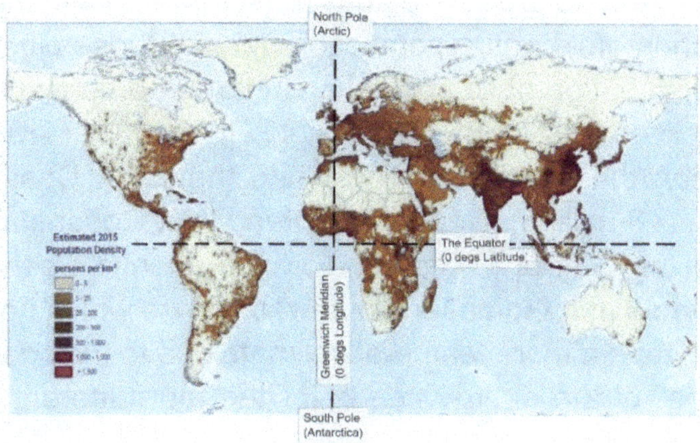

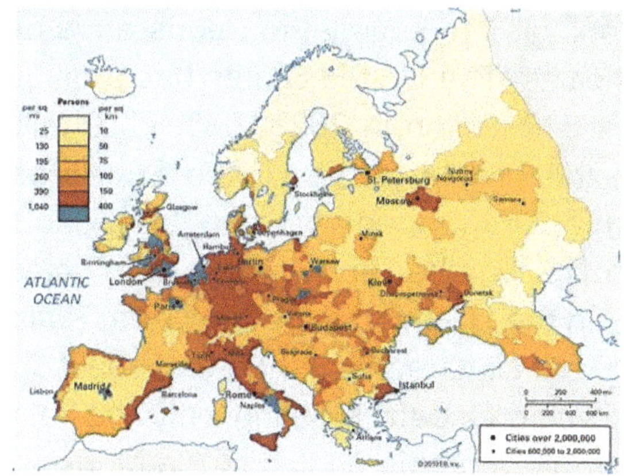

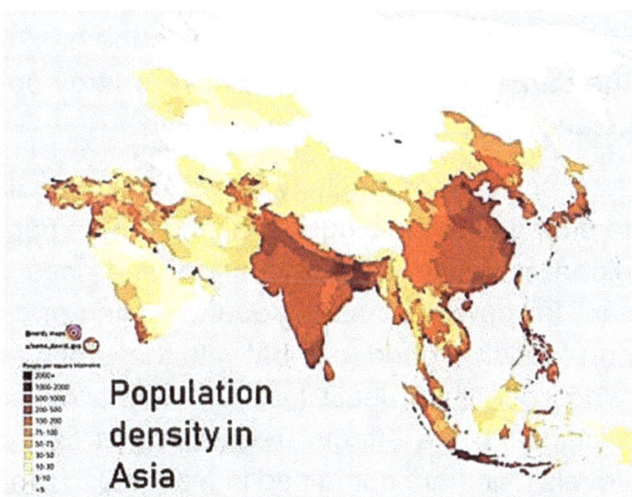

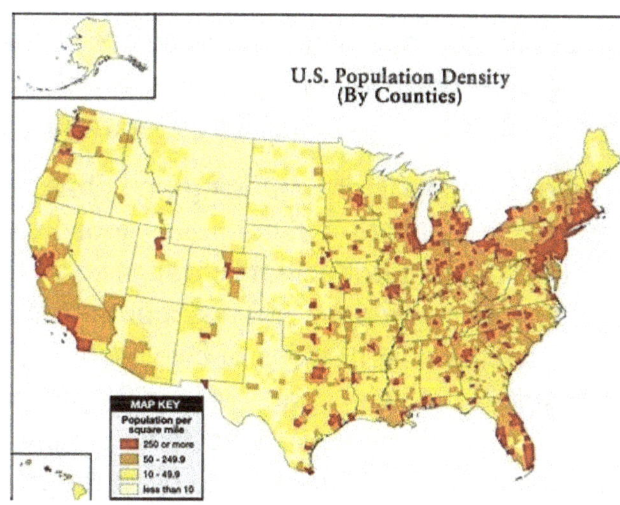

The premise and use of an "average global temperature" might be of scientific interest to environmental scientists in studying the environment, but the process of averaging masks the realism of the environment being so differently configured regionally as shown. This book confirms that the surface temperature of the planet is strongly affected by the latitude of the different regions and averaging masks that effect. If that same practice of averaging were to be applied to the realism of the distribution of the human population being so randomly distributed on the exposed land of the planet, an "average human population" as the descriptive variable would be equally statistically invalid. The surface temperature varying with latitude should first be noticed by the distribution of the color (representing temperature) on the image of the planet seen on the front cover of this book. That is reality not hypothesis.

The UN World Population Division also estimates that the human population of the World will double (to 16 billion) or triple (to 24 billion) by 2300 from the total population of the planet as of 2024 (over 8 billion). If the energy being produced at present is stretching to its maximum rate of production from all energy sources (including coal) with the demands and needs of a population of 8 billion, how will the energy production facilities be able to satisfy a population of 16 or 24 billion people? That is the "big picture" problem that looms as the "crisis" needing to be planned for starting now. The door should be left open to start using coal once again regardless of its "greenhouse gas effect" characteristic if in fact the future temperature of the planet reverses toward another "Little Ice Age". Most important, if there are world pollution problems with an 8 billion population, what is being planned to cope with the pollution from a population double or triple that size?

For larger countries such as China and India, the distribution of electrical energy from any source is a challenge to serve the large population residing in remote areas. Regions must resort to a combination of technologies to satisfy the needs ands demands of its entire population not just those clustered in coastal cities. The United States has a unique distribution of its population in three portions: Of a total of about 340 million in 2023, 53% (179 million) lived east of the Mississippi River, 33% (113 million) lived in the middle of the United States between the Mississippi River and the West Coast states (California, Oregon, and Washington), and 14% (48 million) lived in the three West Coast states.The Eastern US weather and climate is affected by the Gulf of Mexico and by the Atlantic Ocean warm current. In the US middle are the Great Plains states and the desert southwest states which have their two different climates, and the West coast states have a more temperate climate due to the coolcurrent of the Pacific Ocean. Meeting only the "average" of the energy needs for the four different regions would create a problem in all four regions.

The use of oil and natural gas in all regions of the world should be continued if that is the best option for that region – not eliminating their use because they are a fossil fuel. Alternative energy sources depend on resources that differ for all regions on the planet.

The climate is undoubtedly changing and change in how science, technology, government regulators, and the public mitigates – not combats - those climate changes is the bigger challenge. This book has provided the "big picture" scientific definition of the global warming, climate change, and oceans sea level rising problem– it does not describe a "battle plan" for how to combat global warming or climate change. (From the observations of a veteran of a war on foreign ground, combat with the wrong enemy seldom results in a "win". The "winner" is the one who suffers the least losses.). The prospects of emerging and yet-to-be-developed technologies for mitigating the effects (regardless of causes) of global warming, climate change, and oceans surface levels rising are contained in a sequel to this book by this Author.

An interesting and informative database became available most recently that has resulted from the environmental science hypothesis that the burning of fossil fuels was the dominant cause of global temperature rise and climate change on the planet. That fallacious hypothesis was accepted without peer review of its implications on the development of reliable and efficient technologies that would be needed to replace the burning of fossil fuels. As Readers well know the development of solar and wind energy were rapidly initiated, as was the efforts to discontinue the use of all fossil fuels. However, while the use of coal has been reduced rapidly in the United States since about 1997, most coal burning facilities transitioned to the burning of oil or natural gas – the other two main fossil fuels found in abundance in the world. The Department of Energy issued the following summary of the trends in energy technologies since 2005:

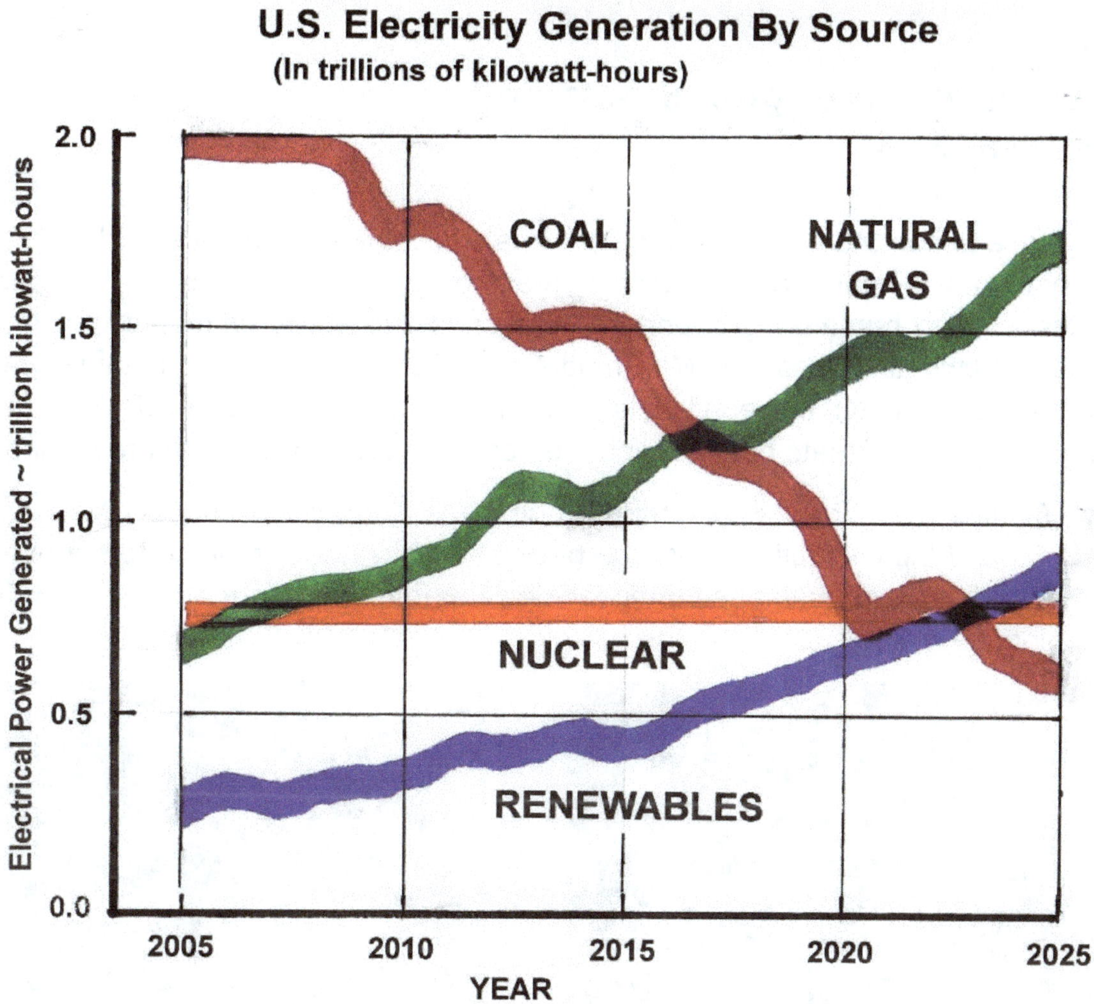

(Source: US Department of Energy DOE)

Information later in this book and in the sequel to this book "The Big Picture of Climate Change on Planet Earth" by this Author explains the reasons why electricity generating technologists have developed the energy generating technologies shown above.

Chapter 1
Databases important to global warming and climate change

The planet's surface temperature is changing, the climate is changing, and oceans levels are changing, and cyclically not simply linearly from steadily increasing "greenhouse gasses". The rates of change are not at crisis level to humanity, but perhaps are critical to some elements of the ecology. The Introduction has shown that it is <u>not</u> Man-made greenhouse gasses dominantly causing global warming, climate change, and the ocean levels rising. That raises the question of "What then <u>is</u> causing regional warming, regional climate change, and the ocean levels and temperatures rising?

There are two statues – of "Past" and "Future" - near the entrance of the US National Archives in Washington, D.C., and the inscriptions near the base of both statues combine to serve as a reminder to all researchers perusing an immense multi-faceted database of information: "Study the past" and "The past is prologue".

(Source: Photographs made by Author at US National Archives, Washington, D.C.)

Many factors in the environment (geology, weather, climate, ocean levels, atmosphere, life forms, agronomy, etc.) on the planet have continually been changing from the time (over 4 billion years ago) when the planet's surface was almost in a totally molten state with no water anywhere. Once that violence stabilized to "normal", most changes in some of those factors seem to be cyclical between the extremes of very hot and very cold. Later chapters in this book show how those changes have been captured in the geology of the planet's crust and uncovered by geologists and archaeologists seeking an understanding of how the Earth was formed, what its weather and climate was, how and why it changed and continues to change.

Water was placed on Earth in abundance – on more than 70 percent of its surface area. Yet there is no standing water on any other planets. All forms of life have evolved on planet Earth being in the "habitable zone" of the universe, but when and how Man came into being is still only known as an approximation where and when. There exist still many unanswered questions about planet Earth, and predicting its future is fraught with many uncertainties about its past.

Whenever some set of conditions and circumstances exhibit a change(s) from their state of "normalcy" or equilibrium, humans affected by those changes try to understand what factors are causing the change(s), to what level of importance can be attached to each of the factors, and what may be the consequences of disregarding those changes. For this book, it is the changes in the climate that prompts the attention of scientists and the world population, and the setting of realistic and appropriate goals for control of the undesirable products of human progress, Unfortunately, some goals set – e.g., to reduce the heat-trapping emissions created and introduced into the atmosphere by Man burning fossil fuels to satisfy his energy needs for heat energy to power the machinery of most technology developments – are of a "one-size-fits-all" type and aimed at the "average" driver condition of global temperature. The earth is a complex system with many variables that are varying in a non-linear, non-average manner. Man trying to control natural events is similar to being the manager at a rattlesnake ranch. "Linearizing", "averaging", and "cherry-picking" data from databases may be necessary simplifications to create predictive models, but those models are at best approximations to factual non-linearities and dynamics.

Humans must survive in the hot and cold extremes that Nature creates, not just at the average of those extremes. Furthermore, humans reside in many different regions (at different latitudes) of the planet and where the extremes (of temperature, amount of precipitation, prevailing winds, seasonal weather changes, proximity to a large body of water, flooding or drought, etc.) vary from region to region. The present scientific focus on global warming and global climate change is an early step toward understanding regional changes in the planet's surface temperature and climate, but it is one rung at the bottom of the ladder of knowledge. The planet surface is made of tectonic plates which are frequently moving relative to each other, and the entire set of tectonic plates on the not exactly spherical planet are rotating about a tilting, wobbling axis of rotation and in an irregular shaped elliptical path in orbit of the Sun – steady-state mathematics in two dimensions are not applicable to dynamic three-dimensional variations.

To avoid the need for every region to repeat an investigation of the causes for unwanted changes and incur a lot of duplicated effort and waste of resources expended - perhaps even human lives needlessly lost - a bank of information known as a database is gradually developed for each region. That bank of information in the database is valuable only to the extent of its factual validity. A quote of a relatively unknown member of the US public (Bobby V. – described only anonymously as "a Jersey City tenement dweller") is stated very astutely despite him admitting to having a low level of STEM education: "Data beats assumption and intuition; unbiased, relevant and accurate interpretation of a set of data beats data that is not." To that quote should perhaps be added that: "A database needs also to have complete and pertinent data within its boundaries, obtained by accurate measurement, analyzed by a spectrum of talents and interests and continually being upgraded as new data becomes known".

The purpose of this chapter is to identify and scientifically analyze the databases of known, factual information in a few of the more important databases pertaining to the surface temperatures and precipitation on the surface of the planet as they affect the climate, and the state of oceans on planet Earth. The scope of this analysis contains factually significant factors that primarily may affect the safety, health, and comfort of humans either beneficially or adversely. The analysis also considers the effect of temperature changes and climate changes on the environment and ecology but only to the extent that the effect is a factor interacting with human safety, health, or comfort. Steps and activities that may be undertaken by scientists in all branches of science – not singly the environmental protection branch - and/or the public to control, reduce, or eliminate the most significant hazardous (to humans, animals, and the environment). The multi-faceted factors are identified and discussed as appropriate in later chapters.

To best understand the problems brought about by weather extremes and climate extremes, the public must first understand the difference between weather and climate. For starters, weather is comprised of "short-lived" variables (quantitative measurements of ambient temperature, barometric pressure, wind velocity and directional vector, dew point and humidity, precipitation (in liquid form such as rain or in quasi-solid form such as snow, hail, or graupel) in the short time period of 6-8 hours, 3-5 days and 1-2 weeks (but not longer than a few weeks) in a locality and region and on the surface of that locality – a municipality, a county, or a state. (That sentence is one of the typical compound sentences in this book because at this time there is still no one scientist or scientific organization that is capable of understanding the true significance of all the factors "in someone else's back yard".)

Climate is the long-term variable (with no specified number of years but generally accepted in the climate science profession to be not less than twenty or more years) and involves assessing "qualitative average" trends of the two primary variables of ambient temperature and amount of precipitation that were imposed on the surface of a larger surface area (a county, state, or group of states or of an entire small country) during the selected period of not less than twenty years.

The word "changes" also is perceived differently by different persons of different backgrounds in education, social standing, culture, economic situation, etc. A change is from one condition to another – and perhaps not enough is known about the ends of that rope to know which end is the more important to be changed. The term "climate change" has become a trigger word that is polarizing environmental scientists from the general public, and indeed within the science profession as well. Very few people – scientists or not – can argue that the climate is changing because in fact the average climate is changing. The climate change arguments center on what is factually causing the climate to be changing, not whether or not it is changing.

To the average member of the public even that "average" is debatable, as the extremes in public behavior are equally as disturbing as are the extremes in the weather and the climate changes. Most interest is in how the weather and changes are affecting the near-term lives of the public. Changes in their surrounding environment that affect their food supply, choice of location where to live and can be afforded, or activities that are important to their comfort and sense of safety, are important to the public both near-term and far-term. Scientists who have developed a focus on but one factor affecting the public hold that isolated factor to be more important than other factors, and tend to lose a common-sense "big picture" calibration of the relative importance of that factor to other factors outside that branch of science.

It is regrettable that some unethical scientists and the sensationalistic news media have intentionally misinformed the gullible public that climate change – due to Man-created "greenhouse gasses" is causing severe weather-caused phenomena (tornadoes and hurricanes which are seasonal-related, and storms and storm surges causing local floods and landslides - which are unpredictably caused by natural local wind currents and out-of-equilibrium heat energy states in combination with amount of water vapor in the atmosphere - and increased lightning strikes causing massive and deadly wildfires). There is yet to be presented a credible scientific database to support any fallacious claim that climate change is causing those weather-related events and catastrophes to be more extreme and/or more frequent, as later chapters of this book describes.

It is much the same as adding a column of numbers (e.g., the "average global temperature" for the planet calculated and tabulated for a period of twenty years) and reporting that the "average global climate" of the planet – i.e., the sum of the "average global temperature" for each of the twenty years and dividing that sum by twenty – is changing because the twenty-year period average is causing one or more of the individual years to have a temperature which is above the average. That is termed as "circuitous reasoning" and is fallacious. Mathematically (unless environmental scientists have developed a "new math" for analysis of a database) the "global temperature" for each of the twenty years tabulated is an independent variable (and as this book later indicates should be "regional" rather than "global"). The "average global climate" for those same twenty years is a dependent variable. The average of any sum of twenty numbers depends on the sum obtained by adding those twenty numbers. Dependent variables in any mathematical analysis do not affect the independent variables. The quip "The tail does not wag the dog" is a reminder of how the climate change process would be better pursued.

Factually, contrary to the claims of some environmental scientists, the climate (the dependent variable) does not affect the weather (the independent variable) of the scientific claims, but it is factually correct that the weather can, and is, directly affecting the climate and indirectly affecting the human population who are exposed to the effects of regional and local weather variations. It is the regional climate which determines the total contents of a person's closet and dresser drawers, not the weather (which changes daily, seasonally, and regionally in duration). Humans who live in the land of "Midnight Sun" have different lifestyles than those who must "make hay while the Sun shines".

The news media is rife with fallacious statements by environmental scientists that the severity and frequency of tornadoes and hurricanes are increasing because the climate is changing. They then attach a statement that "the frequency and intensity is due to Man burning fossil fuels. Determining a trend (by linearizing and averaging a set of datapoints that is fluctuating between extreme "peaks" and "valleys") may be of scientific value in modeling the "average" conditions of a variable in the near term, but a "locus" method of analyzing the set of data varying between extremes is a better way of predicting what the high and low extremes will likely be in the far term.

An example – which is more completely analyzed later in this chapter - of the value of the locus method versus the simplistic averaging method is shown for the database collected from the ice core samples obtained by drilling into the Antarctic ice sheet and analyzing the gasses trapped in the bubbles and carbon-dating the solid "stuff".

Five Million Years of Global Temperature Change

(Based on Sediment Cores at Vostok, Antarctica)

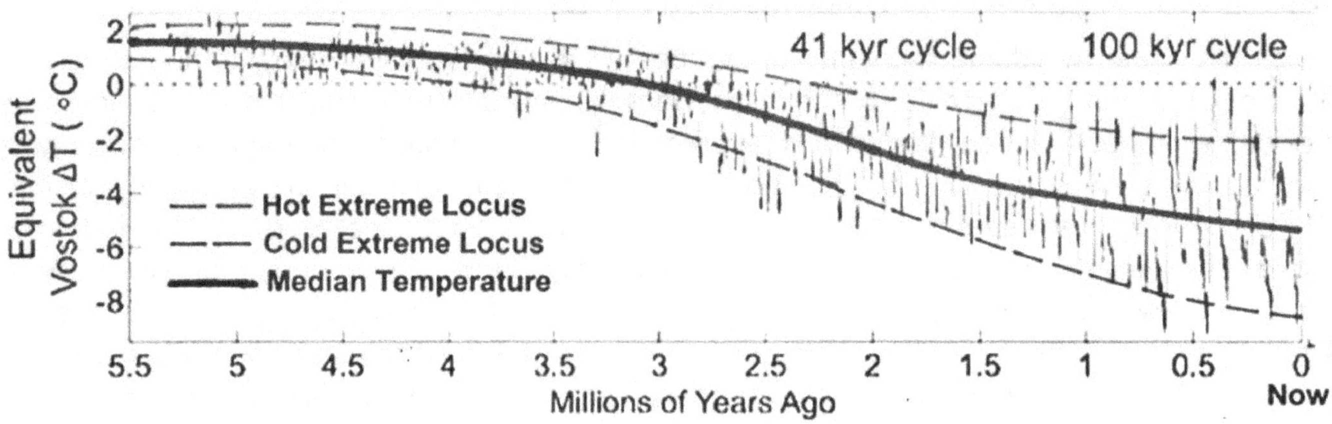

(Source: Analysis by Author of Antarctic Ice Core Database Extended by Milankovic Calculations)

Of highest interest is the time marked as “Now”. The red (“hottest”) locus line indicates a temperature anomaly (i.e., difference from “normal” or average) of about -2 degrees Celsius (Centigrade), and the blue (“coldest”) locus line indicates a temperature anomaly of about -9 degrees C anomaly. The solid black median line indicates a temperature anomaly of about -5.5 degrees C. If any one of the “squiggles” between those hot and cold locus extremes were to be fully occurring in the near past (100 years ago) and the near future (100 years from now) it could be interpreted that Earth is experiencing global warming to be on the rise from what is considered as normal and to be increasing toward the hot extreme locus. The “normal” is the median line; therefore, a rise in temperature from normal (the ice core no anomaly is indexed to -5.5 degrees as being the 0 degrees anomaly “normal”) to -2 degrees (at the hottest locus) is a temperature increase of 3.5 degrees Celsius above “normal”.

For all purposes, the locus lines represent the extremes that have historically caused without the presence of Man on the planet. Historically, when the planet temperature peaked at the red line due to natural causes, the planet’s “thermostat” (whatever that includes) activated and reversed the temperature toward the blue locus line. The “mirror image” of that is that when the planet’s temperature peaked at the blue line due to natural causes, the planet’s thermostat” activated and reversed the temperature back to the heating increase at the red locus line once again. If all living things (animate or inanimate) were not capable of surviving an increase in “global” temperature of 3.5 degrees hotter than “normal” and/or a decrease in “global” temperature of 3.5 degrees colder than “normal”, it should come as no surprise that living things on the planet did not or will not survive.

The inability to explain something does not mean it does not exist or can be ignored simply because it cannot be fully explained. Life would be boring if we knew exactly what lies ahead in the future. There is nothing Man can do to avoid that pattern of natural-caused global heating and cooling between 3.5 degrees hotter and 3.5 degrees colder than normal. It is hastened to be added that Man should do nothing that adds to the inevitable 3.5 degrees hotter increase in temperature. This book addresses the aspects of what Man is doing and should and should not do to avoid making a bad situation worse than it already is and is likely to worsen for reasons beyond control.

Perhaps one of the moves scientists and government planners could do would be to put fossil fuel burning facilities in “moth-balls” status during segments when on a temperature rise above the median line, and re-activate them when on a temperature decline below the median line? Fossil fuels stored in underground geothermally-heated compartments would be available all year round as they are now, whereas alternative energy fuels stored or operating above ground (solar panels encased in ice and wind and turbines frozen from rotating are of no use when encased in ice and much harder to be accessed if beneath a glacier). Nuclear facilities that are powered by uranium require a water-cooling system circulating around the reactor but if the cooling system is frozen "how now brown cow?" The development of reliable alternative fuels for use in the next 200 years is addressed later in this book.

The point to be made in considering any database is that the database in its entirety must be considered and not "cherry-picked" for the effect that confirms the applicability of only a small portion of the database. Another point to be made for a database that is historically valid in its content, any new data that emerges does not replace the previous historical data in that database, it only broadens and extends the database. The emergence of Man on the planet does not and cannot undo the natural astrophysical and geophysical factors that are in the records, it can only affect the future aspects of the planet – on its human, animal, and ecological passengers.

The topics addressed in the next few chapters of this book are to identify and attempt to quantify how much of an effect on Earth's surface temperature and climate are those other factors (astrophysical, geophysical, and natural) have had. To those findings then will be added the anthropogenic (Man caused) factors having been observed by scientists and explained with valid scientific measurements and analysis since the Age of Discovery.

Seafarers long ago began documenting (in their ship's logs) a mariners' database, cataloging the normal wind and currents of the ocean waters that affected sailing ships travelling between the continents. The goal for this mariners' database was primarily to be of assistance in route-planning by other future mariners for safety of travel and of efficiency to minimize sailing time between port A and port B. By sharing those records, many ships and their crews were saved from being lost to treacherous ocean wind and water currents by being trapped in the doldrums (lack of wind to fill the sails as the only power source of sailing ships) or from treacherous currents that exist "normally". Treacherous currents caused the loss of many ships and their crews sailing around Cape Horn on the southern tip of South America where the currents of the Atlantic Ocean and the Pacific Ocean intersect and tumultuously co-mingle. However, those original maritime records were not systematically kept or analyzed in an organized scientific manner – and earth-monitoring weather and navigation satellites have obviated the need to do so.

Fortunately, scientists and inventors have devised and invented far more accurate and extensive equipment to measure virtually all the many complex factors that weather encompasses and it is the interpretation of the newer more complete and more accurate database that climate is being measured and calculated/analyzed. The importance of having a database for weather and climate variables was also beginning to be noted in countries on most of the continents that were most heavily populated because the weather had a large effect on the growth of farmed products that were being grown to feed the growing population of those regions.

The old cliché' "Necessity is the mother of invention" began to emerge in the real world. Based on farmers observations and "hands-on" experience (no pun intended), various types of instruments were developed to measure the key factors that were observed to affect how well different types of crops would grow and in what season – thermometers for fluctuations in temperature, barometers for measuring atmospheric pressure that signal approaching weather conditions, gauges for the amount of precipitation, and anemometers for measuring and reporting wind direction and velocity. Those development activities gave birth to the measurement technology area named in science as metrology and later to meteorology to be more applicable to the forecasting of weather using constantly-improving measurements of the variables in the meteorology databases.

For the purposes of this book, the timeline history of meteorology will commence in the early 1840s when the Smithsonian Institution established an observation network across the United States in which 150 observers (using the relatively new technology of telegraphy (introduced in 1837 by Samuel Morse, and transmitting the coded dashes and dots of electrical impulses which quickly became known as the "Morse Code") regularly reported the weather conditions at those 150 weather stations. By 1849, 500 U.S. telegraph stations were reporting weather observations to the Smithsonian Institution in Washington, D.C..

In 1869, the New York Meteorology Observatory opened, and a systematically documented database of land-based weather conditions began to be recorded. In 1870, the U.S. Weather Bureau was founded, and in 1873 an International Meteorological Organization (IMO) was formed in Vienna, Austria. By 1880, the International Meteorological Organization, later to become the World Meteorological Organization (WMO), had formed the well-documented and analyzed collection of the international database of meteorological data measurements. While the meteorological stations in the WMO are predominantly land-based, the WMO also has numerous ocean-buoys providing measured data to the WMO database.

It is important to be known by Readers that "weather" as measured by the land-based meteorological stations (international and national) is not defined the same as "climate". To the public, "weather" contains the more important factors (temperature ranges, precipitation types and amounts, winds, cloudiness, humidity, etc.) that, depending on the numbers associated with each element of the weather, affects their daily, weekly, and seasonal activities in the small locality of where they reside, attend places of education, or conduct their employment tasks. "Climate" is in general of more interest and importance to commercial and business activities in a larger region (e.g., a business district or perhaps a large group of medical buildings and clinics) due to climate factors (especially precipitation and seasonal adverse weather events) affecting longer range pre-planning for transportation of manufactured/chemical items and travel plans of employees.

Within this book, it is shown that environmental scientists are somewhat co-mingling both weather and climate being affected directly or indirectly by Man burning fossil fuels and creating "greenhouse gasses" that affect the "average global temperature" and the anomalistic difference in temperature – both in weather events and in climate trends in all regions of the world. The concept of an "average global temperature" is a misleading terminology for the planet where the surface temperature (normal and the difference from normal – colder or warmer) varies with the latitude of the regions where high-density population are resident or noticeably absent. The scientific fallacy of "average global temperature" being used as a "one-size-fits-all" goal to be obtained by reducing or eliminating the emissions of carbon dioxide from Man burning fossil fuels is discussed – and dismissed as being unrealistic - throughout this book.

This chapter identifies and summarizes a few of the more important databases created since about the start of the Industrial Revolution (1890) and to the present (2024). The intent is to inform the public of those facts that are of both reliable or of questionable validity (intentionally or unintentionally fallacious) found while assessing the environmental science claims (credible in quality but unsupported by quantitative findings) that human and industrial activities are having a major detrimental affect on the environment and on the climate. There definitely is a correlation but the causes for climate change and environmental conditions are unquestionably due more to astrophysical, geophysical, and natural factors over which humans have no control. It is important to understand that an important difference exist between denying a troublesome condition exists versus a finding (by analyzing its pertinent database of factual information) that there are, or may be, alternate more scientifically valid causes for that troublesome condition to exist.

As mentioned in the Preface, environmental scientists with "blinders-on" (an outcome from the sport of horse-racing to disrupting the concentration in the racehorse being distracted by peripheral vision of other nearby horses challenging the front-running racehorses) are making observations and drawing major "tunnel-vision" conclusions from outdated (prior to 1960) and/or incomplete databases of information. Those conclusions – regardless of being intentionally or unintentionally fallaciously derived – are unnecessarily creating angst (fear) in the public who once very much respected and trusted the scientific profession for having professional integrity and truthfulness. When the additional huge costs of scientists enlisting government officials to "combat" climate change are incurred and passed on to the public as an increase in their taxes, the public trust is lost and angst turns into public rage. Like falling dominoes, the emotional extremes lead from one to another and protests are held (some by students who would be spending their time more wisely learning in classrooms rather than by violent means.)

For the news media to habitually "cherry-pick" the scientists' findings and "tunnel vision" conclusions and report only the picked-out sensational, over-exaggerated highlights with the intent to cause public unrest is reprehensible at best. That was shown to be the case with the Dr. Arrhenius "hothouse gas" database observations and conclusions that lacked specificity in the early 1900s. There is no stigma of scientific invalidity ("denial") attached to Dr. Arrhenius' database, conclusions, and projections – his interpretation of them were valid scientifically for trending purposes but not so well-founded as to cause public alarm. (As youngsters in grade school are taught, it is not proper to "cry wolf" if there is no wolf present.)

It was/is a standard practice for news media to publish incomplete and over-exaggerated information that is presented to the public – in Europe, Asia, and in the United States. It cannot be determined what the public would have believed and what actions would have been taken if World War I had not occurred to divert the public attention away from a potential major problem with the climate and to the more immediate concern at hand – that of the first World War in Europe. World War I mercifully was short, starting in about 1914 and ending in late1918, but both sides suffered horrendous and irretrievable losses – human, materiel, and economic. The seeds of concern however were planted in the public mind and in the mind of many scientists and environmental protectionist organizations in the world that ominous clouds (no pun intended) were gathering in the science of climatology.

The "Milankovic Cycle" database calculations, briefly identified in the Introduction section, were revealed in scientific journals in the late 1930s, indicating that the factors in astrophysics in the solar system had a significant and unusual effect on the solar "insolation" (incoming solar radiation expressed in units of energy (Watts) per square surface area (meters squared) of the planet Earth. Without any consideration of the atmosphere having any effect on the surface insolation, the magnitude of solar insolation was considered to be effected by a number of astrophysical factors: 1) due to the Earth rotating about a tilted axis of rotation that was "precessing" (wobbling like a toy top on a tabletop); 2) due to that tilted rotation axis causing the angle of the approaching solar insolation (heat energy from the Sun) to have an “obliquity” that caused the solar insolation to be partially absorbed and partially reflected with that ratio of absorbed to reflected insolation varying with the tilt angle and with the latitude of the planet; and 3) due to the planet having a dynamic path about the Sun in a somewhat elliptical shaped orbital path that also varied in ellipticity with the shape of the orbital path having an “eccentricity”.

The angular tilt of Earth’s rotational axis has been scientifically measured to vary in a parabolic (wave-shaped) manner between 22.1 degrees and 24.5 degrees and on a cycle (from lowest to highest and return to lowest) during a period of about 41,000 years. At present (2023), Earth’s rotational axis tilt angle is at 23.4 degrees and decreasing toward the 22.1 degrees minimum tilt angle. A diagram of the variables that Milankovic envisioned needed to be calculated is shown next:

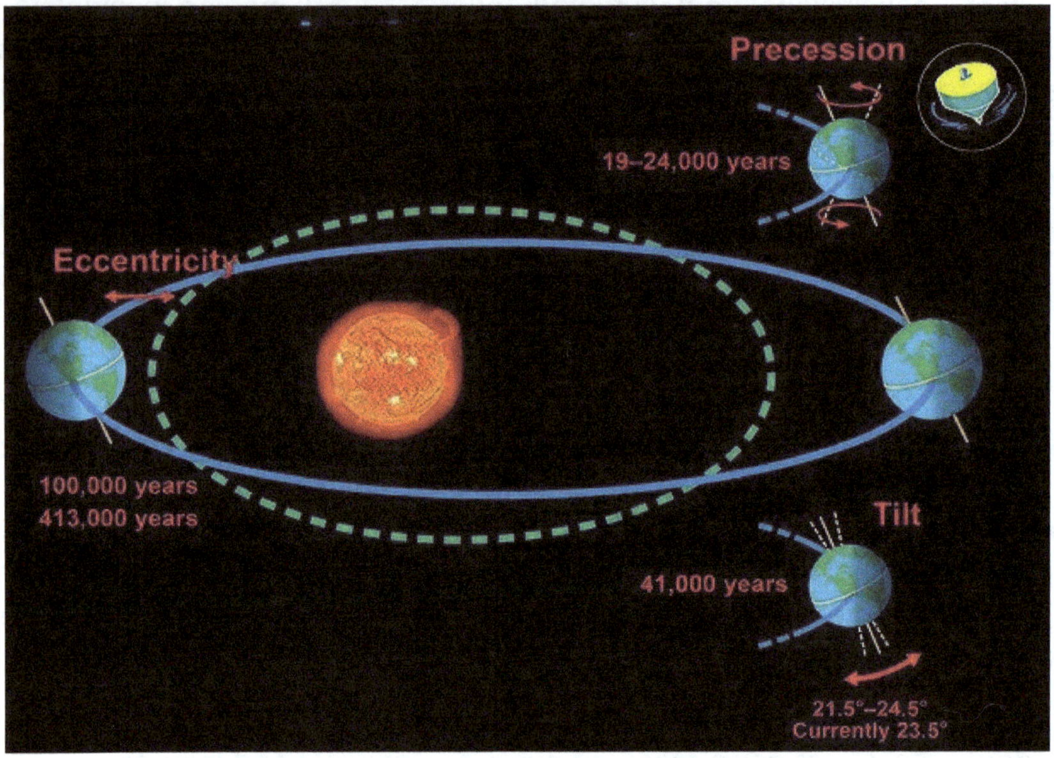

(Source: National Science Foundation)

The period covered by the Milankovic Cycles database was a large series of continuously-overlapping calculations in large time increments (50 years to correspond to significant changes in climate?) between 800,000 years in the past to 800,000 years in the future. The combined effect of the several astrophysical factors on the solar forcing function – which “drives” the magnitude of the solar energy radiance impinging on Earth’s surface – was calculated by the Milankovic mathematical team as an approximation as to how the insolation would be expected to vary due to the set of variables input at the start of the calculation. Combining the effect of small period cyclical variations in each of the three independent astrophysical variables (precession, tilt angle variation, and obliquity), the resulting total effect of the three independent variables on the solar forcing dependent variable (i.e., altering the amount of solar heat energy impinging on Earth’s surface) were calculated by the Milankovic (his Serbian name often is Anglicized (more phonetic) to be spelled Milankovitch) mathematical team and the calculated cycles - due to the combinations of Earth’s tilt angle relative to the Sun, obliquity of the Sun’s radiation waves on Earth’s surface, and tilt axis precession on solar irradiation - graphically summarized as shown next:

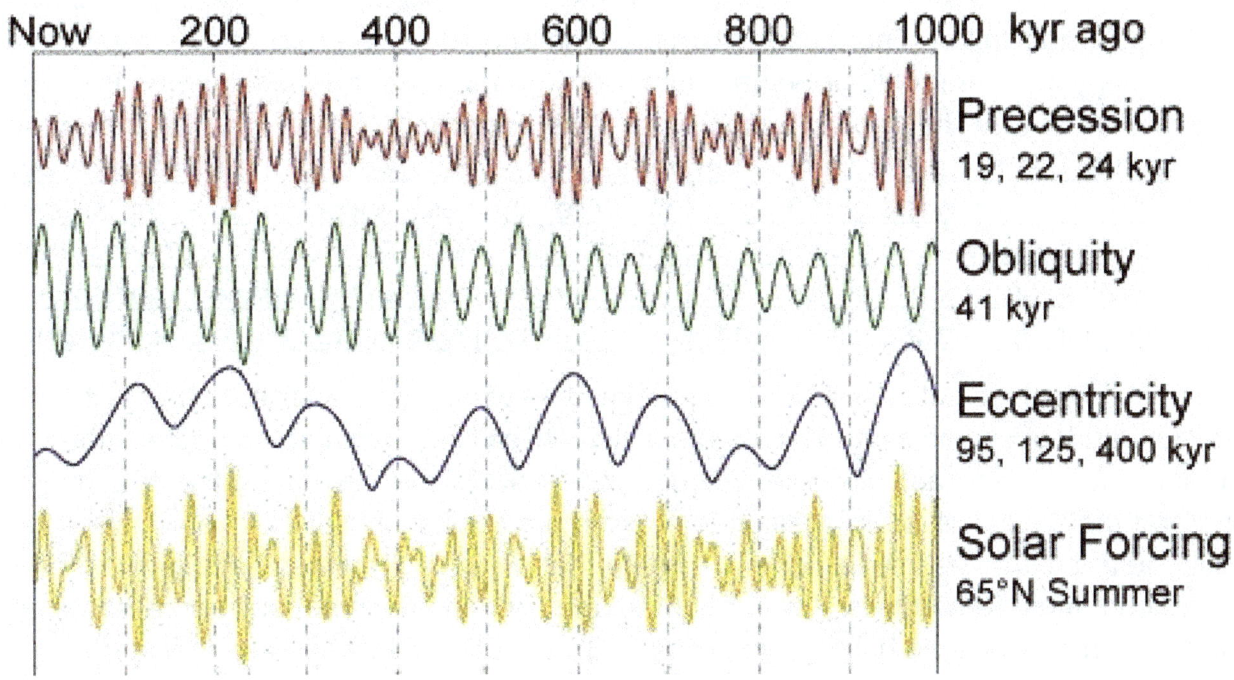

(Courtesy of the Society for Science)

The cyclic variations of the three astrophysics factors that were calculated individually by the Milankovic calculation team are seen in the upper three wavy "squiggles" to establish the cyclical effect – but not the factual magnitude - each "driver" factor was calculated to have on the solar heat energy being radiated from the Sun on Earth's surface at the latitude of 65 degrees North and in the summer solstice location of Earth on its orbital path. The combined effect of the three astrophysical factors (precessiojn, obliquity, and eccentricity) on solar insolation is shown to vary in an irregular cyclical manner as the lowest "squiggle" on the summary chart above.

It must be cautioned that the atmospheric conditions, especially irregular amounts and distribution of water vapor and clouds in the atmosphere that the solar irradiation must pass through before causing any temperature change on the planet's surface temperature are not incorporated in the Milankovic mathematical team calculations. Those atmospheric irregularities, however, do not alter the validity of the cyclic nature of the basic calculations.

Similarly, the time period shown on the horizontal axis above is in an incremental scale of 1,000 kiloyears (i.e., there are 1 million years elapsing from the "Now" to "1,000 kyr ago" limits of the diagram). The cyclic period for each of the three factors (precession has three shorter cyclical periods of 19, 22, and 24 kyrs; eccentricity has three longer periods of 95, 125, and 400 kyrs; but obliquity has a regular intermediary cyclic variation of 41 kyrs). The 3 cyclic periods for the precession are not the same as those of eccentricity and the single cycle for obliquity is different from the 6 cyclic periods of precession and eccentricity. What that means is that superimposing the combined effect of the 3 independent variables on the solar forcing dependent variable (in yellow) causes the solar forcing variation to have a total of 7 cycles – with periods of 19, 22, 24, 41, 95, 125, and 400 kyrs. The simplest explanation is that the astrophysical effects causes the solar forcing variable to be more "squiggly" than the other three variables that are "forcing" solar forcing to be so erratic in period and amplitude.

One most important first observation that possibly is not obvious to the public is that at the time period designated as t = 0 (Now) in the above diagram, the solar insolation – after "crunching" all the astrophysical effects numbers – mathematically shows that the planet should be approaching or be at a peak in the solar insolation. That observation – which should not be misconstrued by the public to mean that astrophysics is the only or primary factor driving the present "global warming" – provides a partial explanation of why the "global temperature" has been on a steady rise for the previous 200 years. That factual rise in Earth's surface temperature – as estimated by the Milankovic Cycles and confirmed by scientific analysis of the Antarctic ice core samples - is continuing at the present (2024) and largely explains why environmental scientists and the public are recognizing that factual temperature increase as "global warming". When that rapid rise in Earth's surface temperature will peak and reverse direction (toward a cooling trend) should be soon (within the next few centuries).

This book cannot and does not provide a more precise date when the temperature "big picture" variation will reverse its trend, but when all the databases of scientifically accurate and complete information are analyzed and interpreted factually, the science profession, the news media, the governmental regulators, and the public will be in position to make more realistic plans and approaches for abating and adapting to the extremes in weather and climate that are being manifested during the generations of near future centuries. Regardless of whether Man does or does not meet any unrealistic temperature goals and limits on greenhouse gas emissions by 2037 or 2050 or 2100, in a few centuries a peak hot temperature from the astrophysical effects alone will be reached and reverse its trend to be a cooling surface temperature trend.

Ironically, the regions that are suffering the extreme hot temperatures now may then become the regions where Man must/will migrate to so as not to succumb – individually, nationally, or world regionally to survive the hazards of hypothermia, frozen water, frozen land (much beneath glacial ice) preventing growth of agricultural food supplies for Man's survival, and snow blizzards of epic proportion replacing hurricanes. The two polar regions may stop melting of their glacial ice and there will probably be another "Little Ice Age" that has previously destroyed entire species of life-form. Additional comparison of the Milankovic team calculations with the Vostok ice core sample that cover the most recent one million years will show additional data as to Earth presently having a "global warming" trend reversing to be a "global cooling" trend.

The coincidence that the Industrial Revolution (using fossil fuels) began within the past 200 years with a peak from astrophysical effects may be historically interesting but the facts in the database suggest that when the Industrial Revolution was in its infancy the astrophysical factors had already started the global warming (temperature rising) trend identified by environmental scientists. (It will be shown later in this database analysis chapter that the Vostok Ice Core Database data also shows the planet surface temperature rising as well as the concentration of carbon dioxide in the atmosphere increasing before the Industrial Revolution had come into existence.)

Because the period of years for the factors of precession, obliquity, and eccentricity to complete some of the complex-shaped astrophysical cycles are large (some more than for just a few hundred years), it may be useful to show how the annual four-seasonal effect occurs in one orbit of Earth. The Earth is in a large irregular shaped orbit around the Sun and at enormous distance from the Sun at its nearest and furthest point in that orbit. A simplistic diagram (not to scale) is shown next to illustrate that the ever-changing distance from the Sun is one of the primary causes for Earth's surface temperature to be changing:

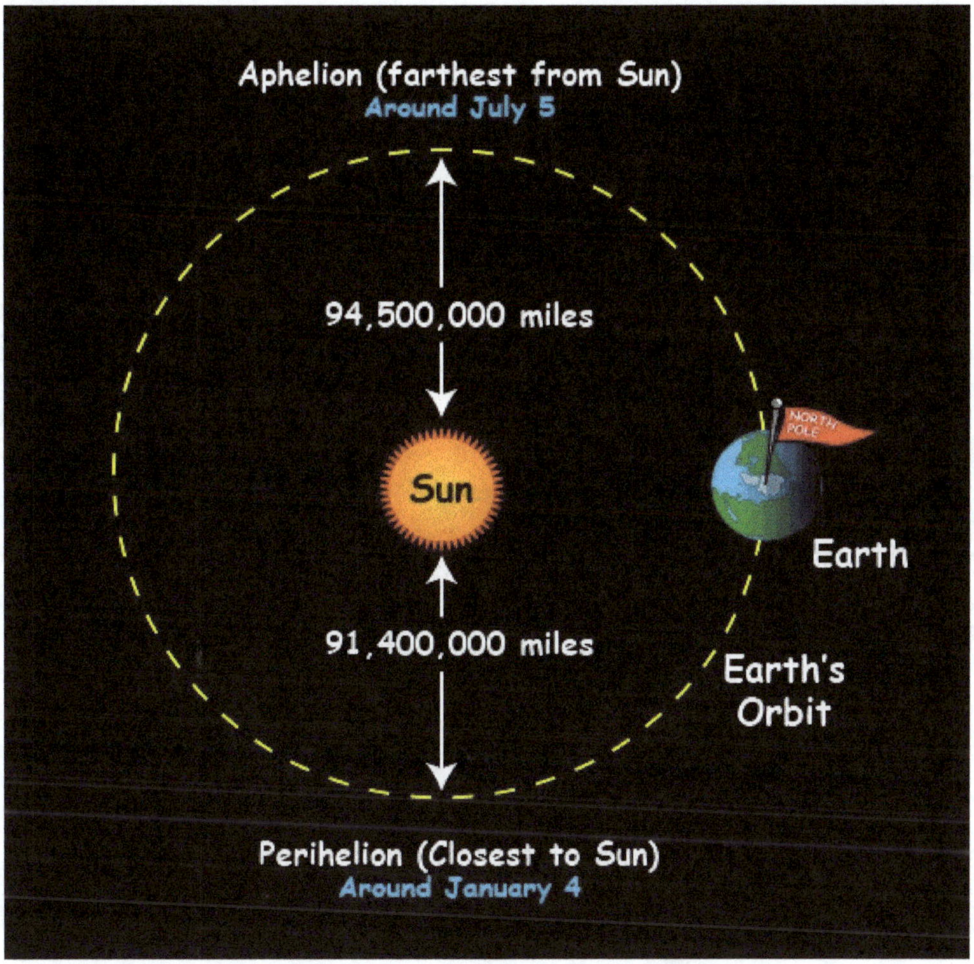

(Source: National Science Foundation)

Earth's surface temperature also changes substantially depending upon the "season" of each year (Spring, Summer, Autumn, and Winter). The labor-intensive Milankovic team calculations were reduced by 75% by calculating only the relative position of the planet Earth in its "Summer solstice" location of the orbits. The "hot-house effect" that Dr. Arrhenius had earlier (1902) identified (due to a concentration of carbon dioxide in the atmosphere surrounding Earth's spherical (3-dimensional) mass) would have been of most significance during the hotter summer period when Earth is nearest to the sun (perihelion). Accordingly, the Milankovic calculation team was justified in calculating the astrophysical effects that were manifested only at the summer solstice conditions of the planet when the Earth's surface temperature is at its hottest state. From a human perspective, global warming is the largest problem when the base surface temperature is at its hottest state due only to its distance from the Sun. The effect the atmosphere (with whatever gasses are in its composition) has on altering that surface temperature is explained later.

Realistically, the human population on Earth is exposed to a "seasonal (Spring, Summer, Autumn, and Winter) effect" unique to all latitudinal locations on Earth and accordingly exposed to the cyclically varying average temperatures at that latitude between hot and cold and back to hot each year. The seasonal variation is due to the tilt angle of the planet (the same as used in the Milankovic Cycles mathematical team's analysis) exposing different portions of the planet to the solar radiance in each of the four quadrants of the planet's elliptically-shaped orbital path. The "seasons" and their four points of temperature inflection (two at opposing equinoxes and two at opposing solstices) are graphically shown next:

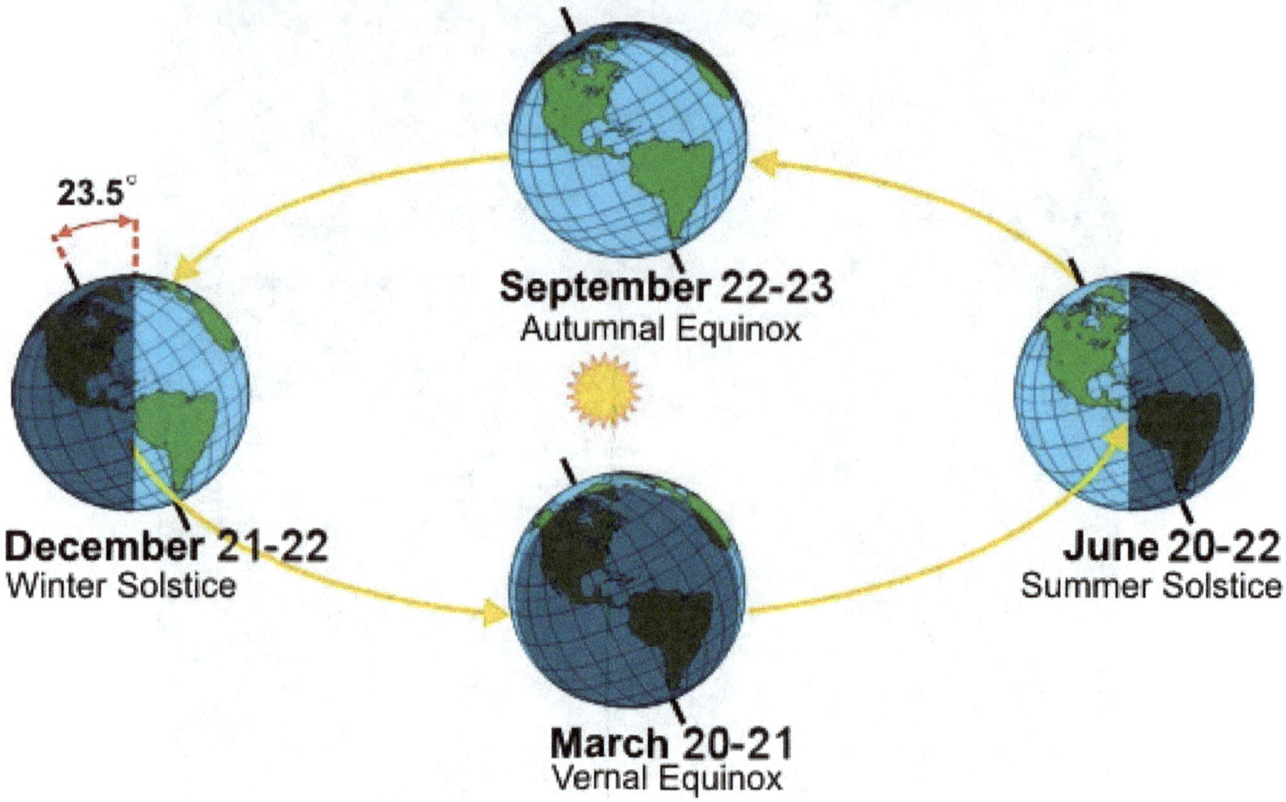

Note: The Sun (small yellow sphere at center) is shown for its relative position in Earth's orbital path, but the Sun's diameter is in fact more than 109 times larger than the diameter of Earth (about 7,900 miles or 12,750 kilometers).

(Source: US National Science Foundation)

One elliptical orbit of Earth around the Sun is accomplished in 365.248 days, giving rise to the present calendar need for adding a day (February 29th) to the month of February in a Leap Year every four years to account for the 0.248 decimal portion of the Earth's orbit occurring once every 365.248 days. Earth travels at a speed of about 66,620 miles per hour averaged along its 584 million miles length orbital path during those 365.248 days. Similar to nature, astrophysics does not follow nicely-formed geometrical paths of motion or movement between points A and B. A most important point being made is that of the environmental scientists using an "average global temperature" as the measure of extent and rate of global warming of the planet. Humans must survive the variation in surface temperature each day – varying a range of temperatures of forty to fifty degrees from the coolest during night time to the warmest in early afternoon. The "average temperature" during a day lasts for perhaps one hour out of the 24 hours per day, but humans must wear layers of clothing on each for the extremes in temperature expected for that singular day. As the season of the year changes – as it does four seasons annually – that daily "average temperature" will change as the average for each of the four seasons Spring, Summer, Autumn (Fall), and Winter. The lowest "average" (or "normal") temperature of a location in the Winter season can be as much as eighty degrees different than the warmest temperature in the mid-Summer season at that same location.

Inasmuch as the environmental protection scientists have a charter to protect the environment – not the human population – the setting of goals to control the increase in "average global temperature" by setting a "one size fits all" average temperature goal not exceeding more than two degrees "average global temperature" increase on a distant target date (2050) may have scientific meaning to protection of ocean floor life forms but is of little significance to humans in public exposed to daily temperature extremes of forty or fifty degrees and seasonal extremes of eighty or ninety degrees. It seems apparent that in order to get the public to be concerned over efforts to curtail their generation of undesirable gasses and residues from the burning of fossil fuels, by declaring a "crisis" to exist and suggest that there are "tipping points" in the "average global temperature" beyond which they may be disastrous results. But why are environmental protection scientists not identifying exactly what are those apocalyptic consequences that will be realized by the human population in all locations of the planet if the "average global temperature" exceeds two degrees in temperature in 2050?

For those Readers not familiar with the descriptors 'equinox' and "solstice", these names come from old maritime records (sailing ship Captains' logs) four to five hundred years ago who spent many months on ships traversing the major oceans of planet Earth and made the following observations about how the number of hours of the day (when it was light between sunrise and sunset) and number of hours of the night (when it was dark) varied in length, and that it was different between sailing in the northern hemisphere or in the southern hemisphere. In the previous diagram describing the "seasons", approximate dates are given and the quadrants of the orbit given "seasonal" names with which most humans are familiar:

1) The Spring season is seen to start on March 20-21 when the Earth is at the Vernal equinox and the hours of daylight will be equal to the number of hours of darkness. Spring season lasts approximately three months (one quarter of a 12 months year) and the average surface temperature from March 20-21 through June 20-22 is normally "comfortable to warm".

2) The Summer season is seen to start on June 20-22 when the Earth is at the Summer solstice, and the length of day and night are in an opposite ratio depending if the observer is in the northern or southern hemisphere. If north of the Equator, at the Summer solstice will show the longest daylight hours of the day, and the night will be the shortest darkness hours of the day. However, if south of the Equator, at the Summer solstice will show the longest darkness hours of the day, and the shortest daylight hours of the day. The Summer again lasts approximately three months, and the average surface temperature from June 20-22 through September 22-23 is normally "warm to hot".

3) The Autumn (Fall is often used as the name for this season due to the leaves on deciduous trees falling in late September or early October) season is seen to start on September 22-23 when the Earth is at the Autumnal equinox – and again the number of hours of daylight will equal the number of hours of darkness. The Autumn season lasts approximately three months and the average surface temperature from September 22-23 to December 21-22 is normally "warm to cold".

4) The Winter season is seen to start on December 21-22 when the Earth is at the Winter solstice, and the length of day and night are again in an opposite ratio depending if the observer is in the northern or southern hemisphere. If north of the Equator, at the Winter solstice will show the shortest daylight hours, and the longest darkness hours of the day. If south of the Equator, at the Winter solstice will show the longest daylight hours of the day and shortest hours of darkness of the day. The Winter lasts approximately three months, and the average temperature from December 21-22 through March 20-21 is normally "cold to cool".

The effect of latitude on Earth's climate (with "climate" defined as being a combination of the surface temperature and "normal" amount of precipitation averaged over a period of not less than 20 years) is well-defined from the "Age of Discovery" about 500 years ago, and the climate is illustrated to vary with latitudinal zones of the spherical planet as shown next in the color graphic indicating the approximate "bandwidth" of the different latitudinal climate zones:

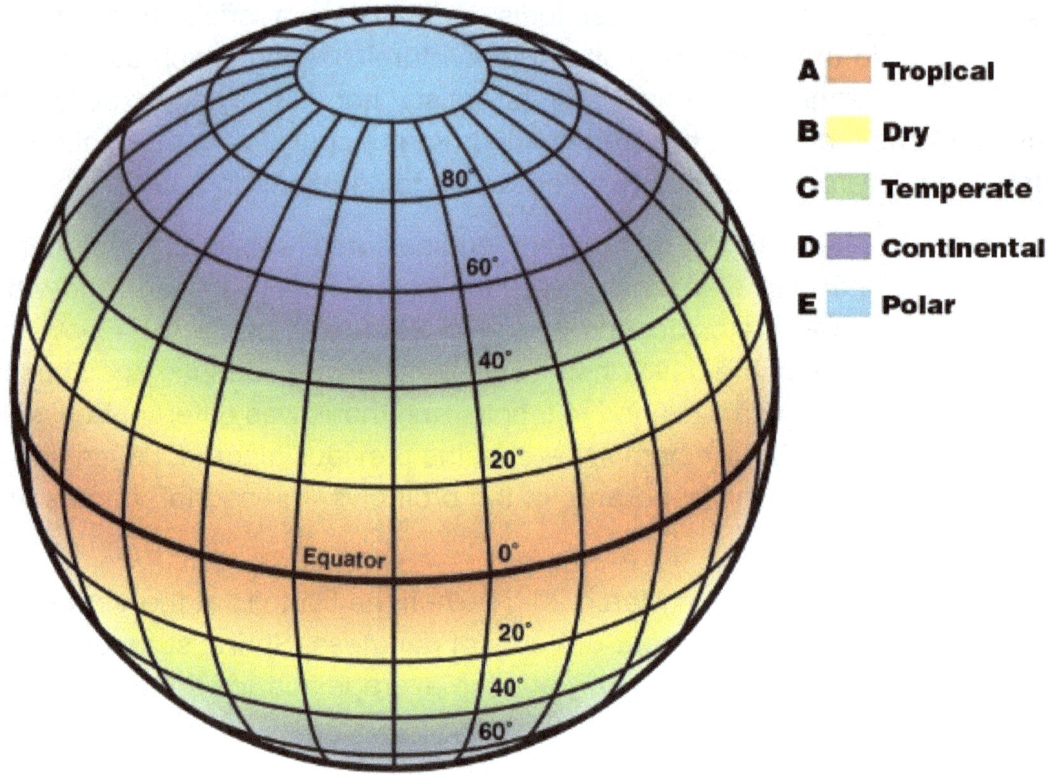

(Source: NASA Earth Resources database)

The five descriptive terms Tropical, Dry, Temperate, Continental, and Polar were not precisely defined numericaly by a specific temperature range or with an associated numerical range of amount of preciptation, but were more descriptive as to the comfort level to humans on Earth's surface in that range of latitudes and to the ability of farmers to grow agricultural crops in that "climatic zone".

Although the above illustration is a graphical summary of the outdated Age of Discovery (16th and 17th century) database, the more recent NASA satellite-based image shown on the front cover of this book is more informative about the latitudinal effect on the planet's surface temperature using the numerically-quantified NASA satellite database information. The cover illustration also shows (in different shades of color such as those in the same sequence as those in a rainbow – red, orange, yellow, green, blue, indigo, violet) the latitudinal extremes of each temperature zone for the surface areas of the several continents on Earth's surface. The temperature distribution on the planet's surface illustrated on the front cover does not have a temperature scale (numerical/colored) attached to it, and a version that does have the temperature color-coded scale numerically defined is shown next:

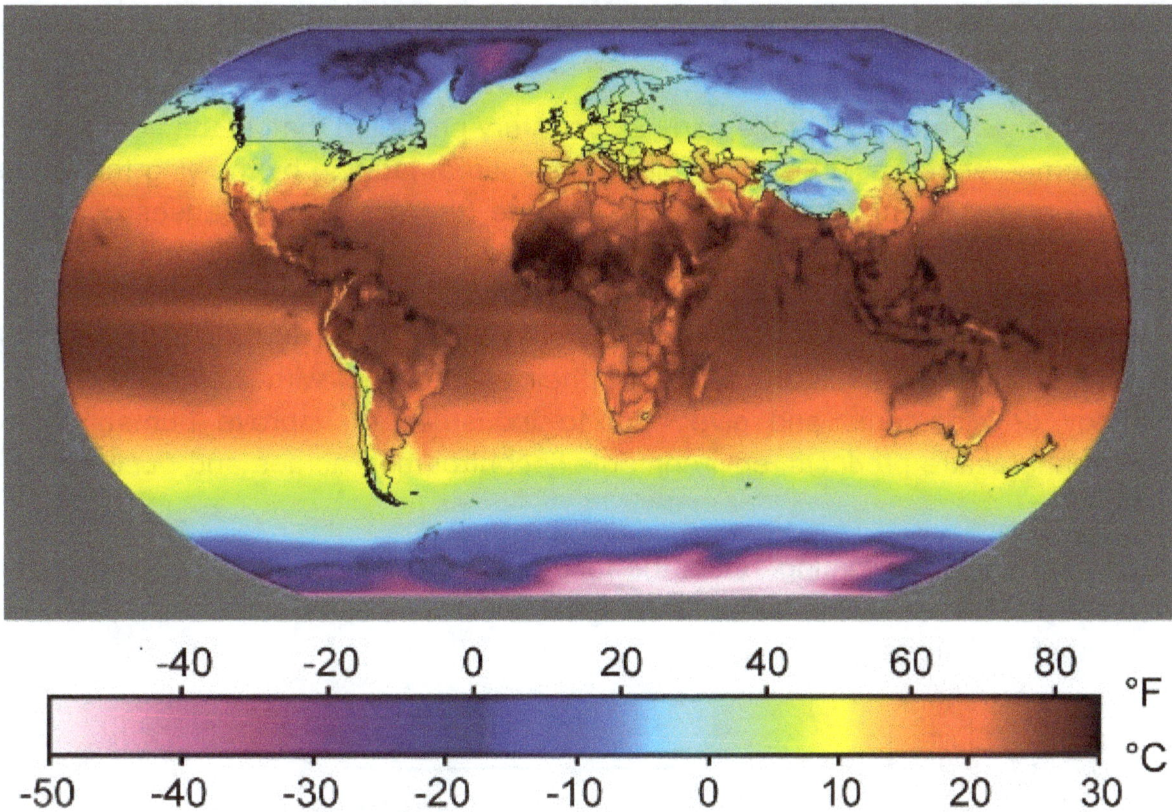

Annual Mean Temperature

(Source: NASA Earth Resources database)

Anyone having any level of scientific education might immediately question "How can environmental scientists use a mathematically-calculated "average global temperature" of any specific number to represent that entire rainbow-like splash of colors in the NASA satellite-based diagram?" If a person were to be fortunate enough to observe a spectacular rainbow in the sky and had a son who was an environmental scientist, would the environmental science son explain that a rainbow has an average color of green? The annual mean temperatures that are not in the same color (albeit in different shades of that same color) but there is no "average red", "average yellow", "average green", "average blue", or "average violet". Humans reside/work/go to school in one of those five different temperature/color zones all over the planet's surface. Would a person living in Uganda (where the NASA annual mean temperature is measured by satellites to be about 21.2 degrees C / 70.2 degrees F) expect to be affected by a two degrees rise in the United Nations calulated "average global temperature" the same as another person living in northern Scotland (where the NASA annual mean temperature measured by satellites is 8.1 degrees C / 46.6 degrees F)?

It should also be obvious that there is no indication for the distribution of precipitation in the above illustration as was alluded to in the descriptors in the Age of Discovery global illustration shown on a previous page; therefore, the above illustration applies only to the global warming portion of the trifecta problem of global warming, climate change, and ocean surface levels rising. Clearly, the NASA satellite-based average global database shows the variation of that annual mean temperature with latitudinal zone is the more realistic database on which to base any environmental scientific goals.

However, some environmental scientists could not abandon the thought that Dr. Svante Arrhenius (1859 – 1927) of Sweden had brought to the attention of the world from before 1941- regarding the

Industrial Revolution and "hothouse gas" emissions from the burning of fossil fuels that were causing the atmosphere to warm.

The Arrhenius database generated in the early 20th century projected that, in the near-term, the rise in Earth's surface temperature could be of benefit to agricultural farming by thawing frozen earth in the extreme latitudes above about 55 degrees north latitude (the northern expanses of Russia, China, Northern Europe/Scandinavia, Canada, and Alaska being in that "too cold for most agricultural and meat protein production" region of the planet) sparked the interest in several other scientists and mathematicians in Europe and in the United States to expand and further develop the Arrhenius postulation of: "If and how humans are damaging their environment with "hothouse gasses" as they make technology developments and give birth to industries to improve human safety, health, transportation, manufacture products, environmentally-control their homes and businesses, and their food/water supply".

In the 1902 Selma, Alabama Morning Times the following news article appeared and contained only Arrhenius' long-term adverse effects of "hothouse gas" in the atmosphere, and was totally mute about the near-term beneficial effects.

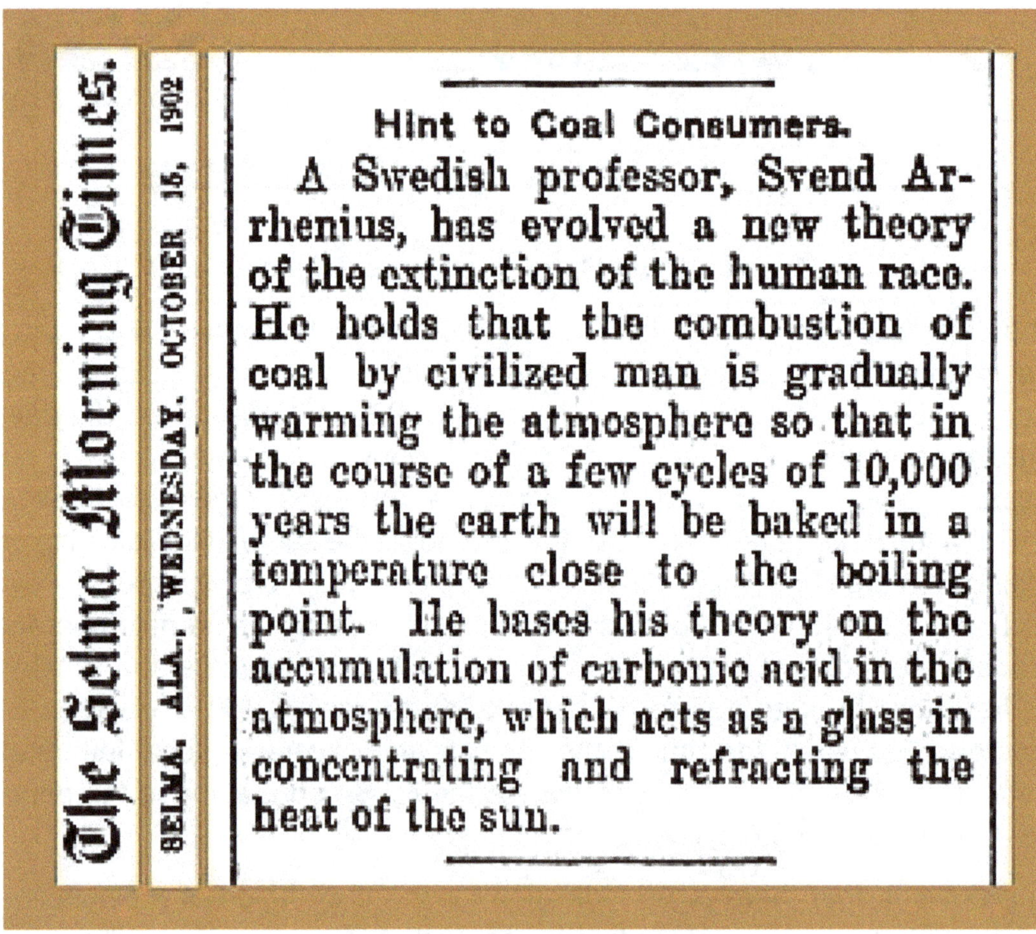

The Selma Morning Times.

SELMA, ALA., WEDNESDAY, OCTOBER 15, 1902

Hint to Coal Consumers.

A Swedish professor, Svend Arrhenius, has evolved a new theory of the extinction of the human race. He holds that the combustion of coal by civilized man is gradually warming the atmosphere so that in the course of a few cycles of 10,000 years the earth will be baked in a temperature close to the boiling point. He bases his theory on the accumulation of carbonic acid in the atmosphere, which acts as a glass in concentrating and refracting the heat of the sun.

(Source: US National Archives)

Dr. Arrhenius was awarded the Nobel Prize for Chemistry in 1903. However, the military-political unrest in Europe in the first few decades of the twentieth century put almost all of Dr. Arrhenius' research findings "on the back burner".

The analysis of the primary factors in databases affecting the weather, the climate changing, and the ocean levels rising, references most commonly found were to the "Keeling Curve" (the direct measurement and recording of carbon dioxide in the atmosphere) and to the "UN Intergovernmental Panel on Climate Change" (calling for limiting the "average global temperature" to two degrees Celsius (Centigrade) by 2050 by controlling human-caused "greenhouse gasses" emissions from the combustion of fossil fuels).

In the group of environmental scientists exploring environmental problems rather than technological opportunities (such as the use of nuclear energy from depleted radioactive materials) was Dr. Charles David Keeling (1928 to 2005). Charles Keeling was born and lived his early development years in the western Pennsylvania coal-mine and steel mills area near Wilkes-Barre, and lived with the local atmospheric disturbances aggravated, if not directly caused, by those unsafe, and unhealthy environmental conditions. There had not yet been highly effective environmental protection controls imposed on those Man-caused operations and processes, and Keeling dedicated the rest of his professional life toward improving the environment. He obtained an undergraduate degree in general chemistry, a graduate degree in atmospheric chemistry, and earned a Doctorate degree in environmental chemistry from Northwestern University in 1953. In 1956-1957 while employed at the Scripps Institution of Oceanography near San Diego, California, he developed the first scientifical instrument to accurately measure carbon dioxide in atmospheric samples. His developmental years in an unhealthy industrial environment led him to seek employment in collecting atmospheric measurements (of carbon dioxide in the atmosphere) intended to better understand and to enhance control over the quality of the atmosphere.

In 1958, the International Geophysical Year (IGY) organization funded Dr. Keeling and appointed him to develop an atmospheric research laboratory at the Mauna Loa Observatory in Hawaii (at latitude 19.5 degrees North latitude and at an elevation of 3,400 meters (11,134 feet) near the summit of the Mauna Loa inactive volcano). A photograph of the Mauna Loa Observatory is shown next:

(Source: National Oceanic and Atmospheric Administration)

The Mauna Loa Observatory was chosen to be at high elevation and in a location without any significant plants or trees that exchange carbon dioxide with the atmosphere that could influence the carbon dioxide measurements (and several other gaseous components of the atmosphere being measured).

It should be noted that water vapor in the atmosphere – its presence shown in the form of co-altitude clouds in the photo background – is inherently collected in the data. It should also be noted that humans exhale carbon dioxide with every exhalation. Dr. Keeling purposefully located the sensor and atmospheric collection instrumentation outside (seen in the unpainted – to preclude paint vapors - wooden structure to the right of the main building where processing of the data into meteorological parameters is done) to assure that exhalations of the human staff did not enter into the measurement process.

A simplistic "pie-chart" of the measurements collecting atmospheric data is shown next to give Readers a picture of the relative proportions of gasses in the atmosphere that were/are measured at Mauna Loa.

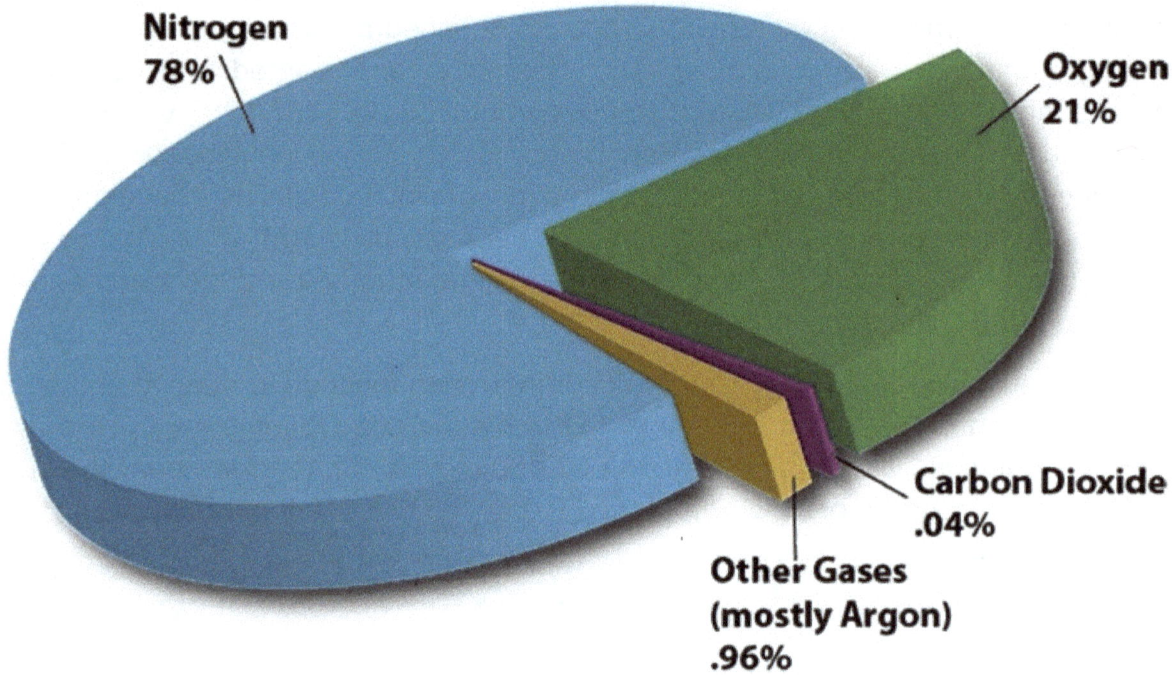

While the simplistic pie-chart is mostly accurate in the "big picture", it is not representative of the actual atmosphere which contains water vapor in varying concentration and distribution in the atmosphere.

The planet's surface area is covered with more than 75 percent water. When water is warmed, the water surface evaporates into water vapor. The concentration of water vapor varies from being near zero over desert areas and up to about 4% when over ocean waters. A portion of the water vapor rises in altitude until it reaches a cool temperature of the atmosphere that condenses the water vapor into the common cloud. As the next photograph shows, that cloud cover indicates that the atmosphere below the clouds had high water vapor content. There are seldom any clouds over the arid desert "belt" of North Africa, the Mid-East, and southern Asia which exist on over 6.5% of land area of the planet.

(Source: NASA)

This photograph and a Department of Energy tabulation, shown next, that disregards water vapor as a variable in global warming and climate change, spurred further research by the Author into the effect water vapor may have on the planet's surface temperature. It is well-known in the science profession that water vapor has a higher heat-trapping effect on the Earth's surface temperature than does the total of other Man-made "greenhouse gasses" shown in the table below:

TABLE 1.
The Important Greenhouse Gases (except water vapor)
U.S. Department of Energy. (October, 2000) (1)

(all concentrations expressed in parts per billion)	Pre-industrial baseline	Natural additions	Man-made additions	Total (ppb) Concentration	Percent of Total
Carbon Dioxide (CO2)	288,000	68,520	11,880 (2)	368,400	99.438%
Methane (CH4)	848	577	320	1,745	0.471%
Nitrous Oxide (N2O)	285	12	15	312	0.084%
Misc. gases (CFC's, etc.)	25	0	2	27	0.007%
Total	289,158	69,109	12,217	370,484	100.00%

(Source: Department of Energy Report, 2000)

The amount of water vapor in the atmosphere varies from virtually none to a maximum of 3.2% depending mostly on the temperature of the body of water from which it is vaporized into the atmosphere. The evaporated water ("water vapor") rises to an altitude where cooler temperatures condense that water vapor into cloud formations. Scientists have developed that evaporative process summarized graphically as:

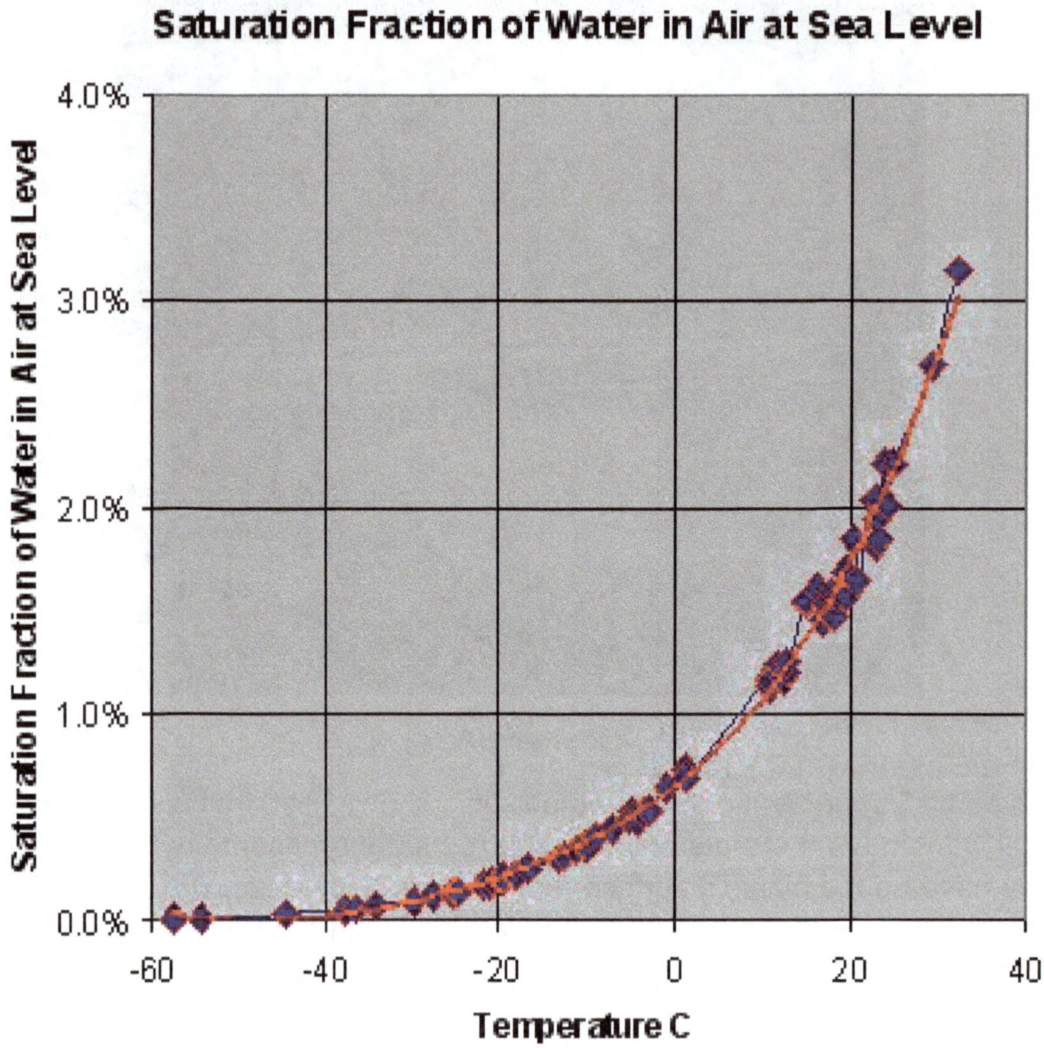

(Source: Standard Textbooks of Atmospheric Chemistry)

The maximum saturation of water vapor in the air of 3.2% is seen to occur over water that is 32 degrees Celsius (about 90 degrees Fahrenheit). The Polish-born physicist Daniel Fahrenheit (1686-1736) developed that temperature measurement scale named after him for his temperature measurement instrumentation research in the Netherlands in the 18th century. In Germany, nouns can be converted to adjectives by adding the suffice "heit", and when heit is added to the noun Fahren – which translates to "travel" in English. Figuratively Fahrenheit is the scale by which temperature travels between water with high salt contentment that is frozen (changes state from liquid to solid ice at 30 °F) and boiling (changes state from liquid water to gaseous steam) at 212 °F).

The American climatic scientist and physicist Dr. Charles Keeling recognized that the varying amount of water vapor in the atmosphere at the Mauna Loa Observatory was masking the effects of carbon dioxide and the other "greenhouse" gasses as shown in another form of the pie chart which includes everything in the atmosphere having a heat-trapping effect (the natural water vapor in blue and all other natural and Man-made gasses in various colors):

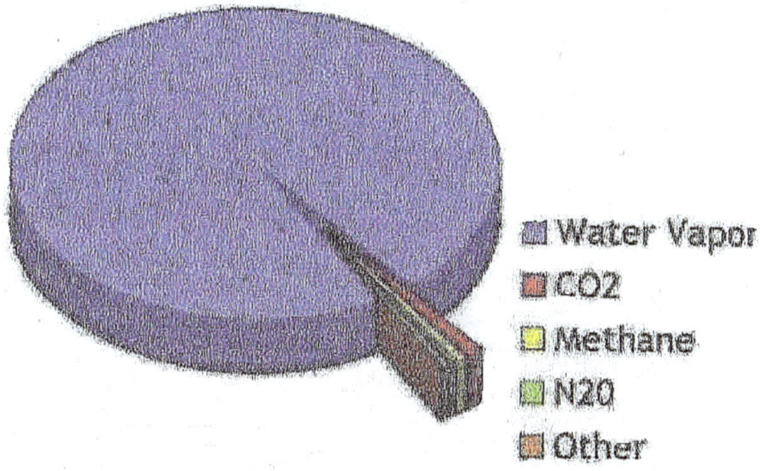

(Source: Standard Textbooks of Atmospheric Chemistry)

The carbon dioxide slice (colored red) of the pie is to be trisected for that portion which is of natural origin and the remaining portion which is the offensive Man-made portion.

It is not known whether or not Dr. Keeling advised the Scientific Advisory Committee (SAC) that his measurements at Mauna Loa served as confirmation that it was carbon dioxide alone (without water vapor consideration) being the primary cause of Earth's surface temperature increase. It is more probable that climatologist and dendrochronologist Michael E. Mann (1965 -), using the measurements from varves (sediments in lake beds) and the spacing between tree rings, did so in the early years of the twentieth century with the now famous "hockey-stick" curve (which appeared in the Third Scientific Assessment Report (AR3) of the UN Intergovernmental Panel on Climate Change (IPCC) published in 2001. Based on the "proxy" data and information from the varves of lakebeds and tree rings, Dr. Mann and Associates Raymond Bradley and Malcolm Hughes had produced in 1998 a paper that led to creation of the "hockey stick graph. A photograph of an example of the tree rings is shown next followed by a photograph of a varve site in Canada.

(Source: The British National Museum, London)

(Source: Encyclopedia Britannica)

The science of dendrochronology (counting the number and spacing between tree rings and correlating that with climate change) and the similar paleogeology chronology (analyzing the spacing and patterns of sediments known as varves in lake beds) is a tribute to the science of climatology and to advanced methods of statistical data analysis. However, the inferred content of the atmosphere from Dr. Mann's database was likely considered by fellow meteorologists to not be quite as accurate as direct measurements of carbon dioxide in the atmosphere as done at the Keeling Mauna Loa Observatory. but the information derived as to the climate (temperature and precipitation) are vital indicators inferring the carbon dioxide concentration in the atmosphere and its change with time. A graphical summary of their dendrochronological and varve chronology research in what became known to climatologists as the "hockey stick" curve is shown next:

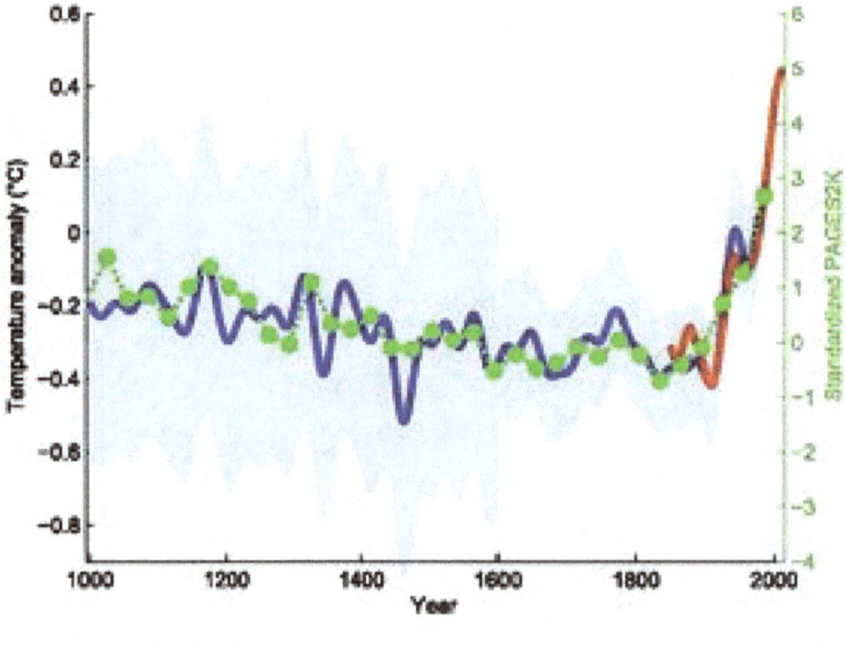

(Source: UN IPCC Third Assessment Report AR3 of 2001)

In essence, the dark blue line and dotted green line are information mostly from dendrochronological research and the red solid line is mostly from varves research. The sharp upswing in temperature anomaly at about 1850 coincides with the start of the Industrial Revolution. It is the shape of the curve. A fellow climatologist associate (Jerry Mahlman) of Drs. Mann, Bradley and Hughes dubbed the shape of the graph (resembling an ice hockey stick with the handle horizontal and the blade on a vertical angle) as being "like a hockey-stick".

There are three main considerations with using the Keeling carbon dioxide database and the Mann temperature anomaly database affecting the scientific value to the understanding of climate change: 1) The starting point for the two different sets of data is highly important to the environmental scientists' hypothesis that it is Man burning fossil fuels that is causing global warming and climate change. The Keeling Mauna Loa database starts in 1958 well after the start of the Industrial Revolution whereas the Mann hockey stick database begins at about the year 1000. The Mann hockey-stick database well brackets the pre-Industrial Revolution and the post-Industrial Revolution rise in surface temperature (where the trees grow), but that does not prove the "greenhouse gas effect" that has climatologists frothing at the mouth. The Keeling direct measurement of the total observed rapid increase of carbon dioxide in the atmosphere as shown in the Keeling database is more convincing of the effect of Man burning fossil fuels (the contention of environmental scientists claiming to be "settled science") and better supports the UN IPCC global warming hypothesis on temperature changes being a result of the Industrial Revolution production of "greenhouse gasses" especially carbon dioxide. For that fact of time span applicability (from before the Industrial Revolution to well after the Industrial Revolution), the Mann database is considered to be of greater continuity value on the emissions from the Industrial Revolution (which started in the mid-1800s). The Keeling database did not measure the carbon dioxide in the atmosphere until after 1958. The carbon dioxide measurements by Svante Arrhenius in Sweden in the early 1900s begin to show the rise in carbon dioxide from the Industrial Revolution but are not nearly as accurate as are the Keeling database measurements and are not adjusted to dry air atmospheric conditions for database standardization.

2) After 2000, the Keeling carbon dioxide measurements are likely the better scientific measurement database because the direct and very accurate measurement of carbon dioxide in the atmosphere is probably of higher significance to climatologists than is the Mann/Bradley/Hughes database obtained from trees and lakebeds (and therefore is representative of regional climate, depending on the latitude and climate of the region where the tree growth and lakebeds are located. and factually are regional not global). The Mann database produces temperature and climatic temperature data from which the level of carbon dioxide in the atmosphere is inferred and not directly measured.

3) After the advent circa 1970 of NASA satellites with sensors to measure the actual surface temperatures and the variation in temperatures very precisely being different in different regions, the conclusion throughout the climatology science profession is that the more complete "story" of climate change is now beginning to be much better known than it was known by Dr. Svante Arrhenius in 1902, by Dr. Charles Keeling from 1958 to 2001, and by Dr. Michael Mann since 2001.

The National Academy of Sciences awarded Dr. Charles Keeling the National Medal of Science for Physical Science in 2001. The Author includes Doctor Keeling in the Dedication section near the start of this book as deserving of high recognition for professionalism in not over-exaggerating (as some of his environmental science peers and the news media did and are doing) his measurements of carbon dioxide in the atmosphere as being anything but scientifically accurate and of cautionary importance. His measurements would possibly have been more valuable to understanding the scientific effect of carbon dioxide on the surface temperature of the planet if they had been made at a lower elevation (below 2,000 feet) closer to the surface of the ocean at sea level where the greenhouse gasses trap the heat energy sufficient to cause the ocean water to increase in temperature (and evaporate as the water warms), and at a latitude closer to the latitude of the peak distribution of the human population on the planet (i.e., at about 40 or 45 degrees North latitude rather than at 19.5 degrees North and in mid-Pacific Ocean) where the Man-made greenhouse gasses are not at their peak concentration from both human activities and industrial activities.

There can be little better way than to show the type of data being collected daily and weekly and annually (for the year 2014) at Dr. Keeling's Mauna Loa Atmospheric Research Laboratory to produce the most accurate database that no one in the scientific profession could have done better. From a statistical data analytical perspective, a question that remains is whether there should be other laboratories sampling the atmosphere as professionally as being done at the Mauna Loa facility. Does this Keeling Curve carbon dioxide data truly represent the atmospheric condition at/above regions of the planet where there is far more significant population and Industrial development activities?

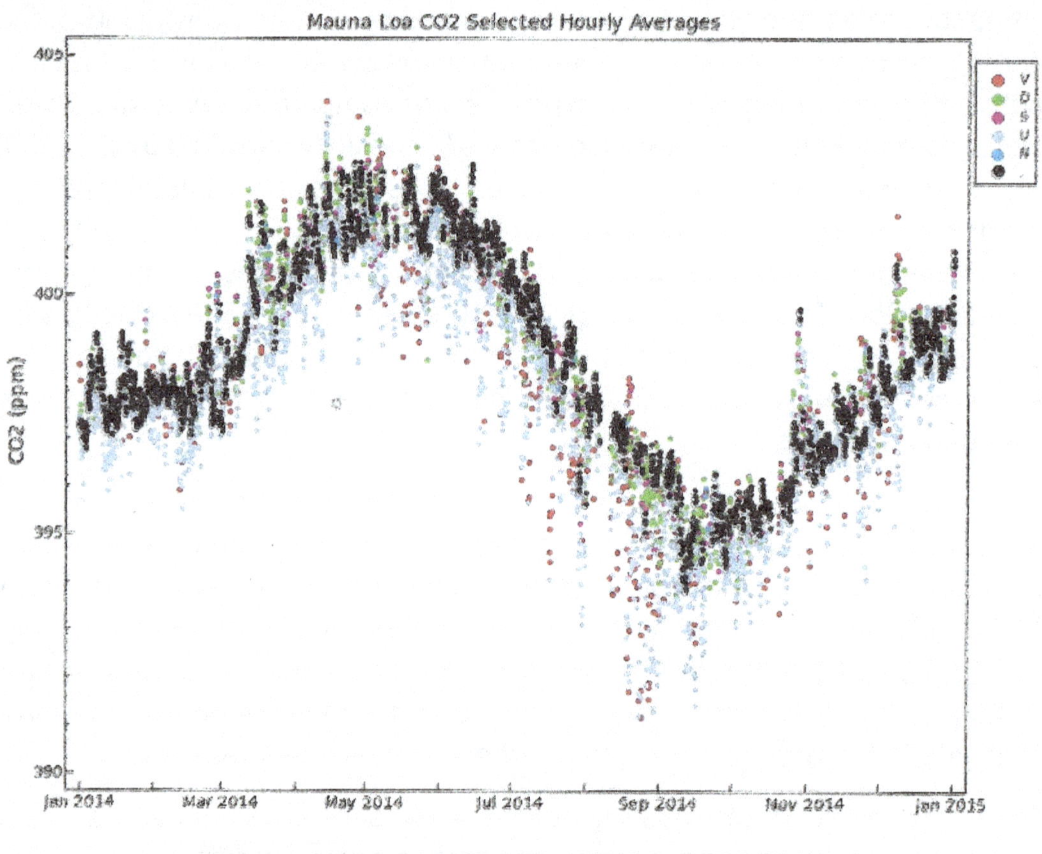

Figure 3. Plot of data with the different selection flags for a typical year (2014).

(Source: NOAA)

The rainbow dotted coloration (resembling a speckled trout) represents the raw "trapped" data during the year 2014 with adjustments to bring the measured data into a standardized dry air level of scientific presentation and for historical and modeling purposes. It is the single cycle above that is shown on the next chart as an inset.

The Author (who has a strong background in collecting large complex sets of test data in wind tunnels, the analysis of test data, and data adjustment process) commends the Keeling staff for their painstaking attention to details in issuing such important data pertinent to the global warming issue. It must be stated however, that one of the adjustments made to the test data is to remove the water vapor in the atmospheric samples. As seen in the photograph earlier of the Mauna Loa Observatory facility on top of the Mauna Loa volcano there are cloud formations at about the same elevation of the facility at 3,400 meters (11,134 feet) – indicating a water vapor of 3-4% in the atmosphere being sampled. That water vapor in the atmosphere is a very important factor that is present everywhere on the planet surface (except over arid deserts) as the Earth's surface area is over 70% water that is evaporating into the atmosphere.

By scientifically "adjusting" the Mauna Loa atmospheric test samples of the atmosphere to be "dry air" for scientific analysis purposes, the Keeling curve does not factually represent the actual amount of carbon dioxide in the atmosphere at all locations on the Earth's surface where water vapor is present. The magnitude of the correction is such that a sample test reading of 400.6 parts per million by volume of carbon dioxide in an atmosphere containing 3% (300 ppmv) water vapor is adjusted to cause the 400.6 parts of measured CO2 to be reported as being 413 parts per million (0.0413%).

Scientific peers (at the National Oceanic and Atmospheric Administration and US Environmental Protection Agency) of Dr. Keeling recognized the importance of the Keeling Curve and Mann "hockey stick" Curve "greenhouse gas" database and much like a tsunami after an earthquake flooded the scientific community of the finding – confirmed by test data – that Man burning fossil fuels is the dominant cause of global warming. Once a tsunami makes a move it is impossible to redirect it in the right direction where it will do the least harm and to control the spread of its breadth. The news media soon picked up on the sensational aspects of the misinformation and magnified the significance of both the Keeling Curve and Mann Hockey Stick Curve. Dr. Keeling only maintained the scientific fact showing that carbon dioxide in the atmosphere possibly has some effect on the environment and on atmospheric-caused "greenhouse gas" effects (just as Dr. Arrhenius had done in Sweden the early 1900s.)

It is a point not to be overlooked that Dr. Arrhenius referred to the "hot-house effect" in the atmosphere (from the emissions created by industrial practices) as being due to "carbonic acid" – which is a product of carbon dioxide and water - and not carbon dioxide in isolation. Dr. Arrhenius apparently had not made any attempt to isolate water vapor in the atmosphere as did Dr. Keeling decades later. Suffice it to say that the UN IPCC Working Groups have been struggling since the late twentieth century to arrive at a valid – concurred with by the entire scientific profession - of a "cause and effect" explanation for the true causes of global warming, climate change, and ocean levels rising. The most valuable NASA "big picture" satellite data has not caused the UN IPCC to alter its "Man burning fossil fuels being the cause of everything that is not behaving according to the UN IPCC "average global temperature" and "one-size-fits-all" goals of holding surface temperature to a 1.5 degrees Celsius rise in 2050. The associated UN IPCC goal to control Man-made carbon dioxide emissions to "net zero" by 2050 is far off-track to be successfully achieved.

Fast forwarding a bit, the scientifically accurate measurements of carbon dioxide in the atmosphere contained in the original Keeling Curve database from 1958 to 2005 (measurements made under Dr. Charles David Keeling's direction at Mauna Loa – until his death in 2005) and 2005 to 2013 (measurements made by Dr. Charles Keeling's son Ralph Keeling and the staff at Mauna Loa) – continue as the primary scientific database for the UN IPCC. The NASA satellite database and the other databases (Vostok Antarctica Ice Core Samples and Milankovic mathematical calculations of astrophysical and geophysical effects) are apparently regarded as distractions by the UN IPCC.

The most current version of the Keeling Curve is shown next:

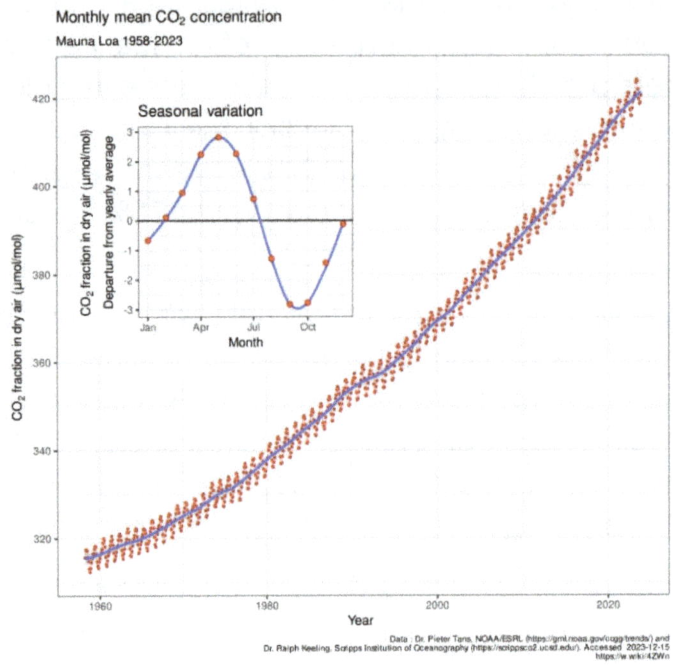

(Source: NOAA)

The serpentine data in the red color reflects the seasonal effect on carbon dioxide in the atmosphere as measured and corrected to show only the carbon dioxide in the dry atmosphere at that latitude and altitude. The black line is the median of the serpentine seasonal data as it varied from 1958 to 2023.

A review of the process used at the Keeling Mauna Loa Observatory revealed that the mole fraction (parts per million, ppmv) for the carbon dioxide in the atmosphere has been adjusted to be a larger concentration in the atmosphere by the removal of water vapor – i.e., a notation on the Keeling Curve should have – but is not shown – stating that the conditions are for a "dry atmosphere" without water vapor. A review of the most recent data on the carbon dioxide collected at the Mauna Loa Observatory from 2020 to 2024 – by Dr. Keeling's scientist son Dr. David Keeling after Dr. Keeling passed away in 2012 – still does not note that the data is for the idealistic "dry air":

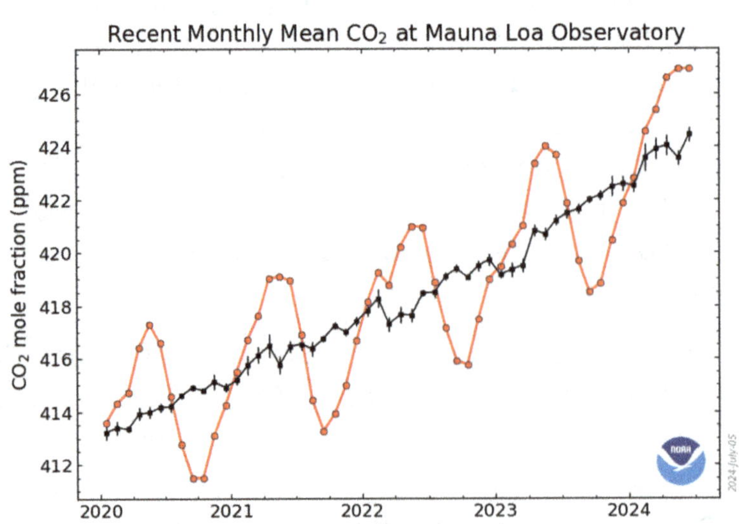

(Source: NOAA)

There are some factors associated with the Keeling Curve database that are mostly unknown to the public and to Governmental regulators who are not environmental scientists:

1) although not well noted for public and news media understanding, the pre-Industrial Revolution database on carbon dioxide in the atmosphere was estimated to be nearly 300 parts per million by volume (ppmv), and the measured rise into the 320-360 ppmv carbon dioxide concentration bracket in the thirty years period between 1960 and 1990 was claimed (by environmental protection activists and not by Dr. Keeling) to be due entirely to industrial activities and to human activities that generated carbon emissions;

2) the Keeling measurements shows a seasonal variation (in red on the curves which indicates to most scientists that the database accurately measures both natural (certainly seasonal) as well as Man-made effects on carbon dioxide – such that future efforts to reduce "greenhouse gas emissions" to "net zero" by a certain future date were/are predestined to not be met (per the more recent of the 28 annual reports of the UN Intergovernmental Panel on Climate Change).

3) the presentation of carbon dioxide measurements in mole fraction (parts per million by volume – ppmv) are made for a dry atmosphere – all water vapor removed. Environmental scientists fail to inform the public and government regulators that reduction of all Man-made carbon dioxide emissions from the atmosphere is nearly inconsequential on the "greenhouse gas effect" driving the unrealistic "net zero emissions" goals set by the UN IPCC.

The above revelations are not intended to besmirch Drs. Charles or David Keeling or their accurate measurements in any way, nor the scientific efforts of climatologist Dr. Michael E. Mann. The review of the major database information does indicate that some environmental scientists however, are not being truthful – regardless of being unintentional or intentionally over-exaggerative - with the public as to the meaning of all atmospheric measurements of "greenhouse gasses". With water vapor at 3% (300,000 ppmv), carbon dioxide at far smaller total concentration of 0.0422% (422 ppmv – 280 ppmv / 0.028% natural made, and 142 ppmv / 0.014% Man-made) in the atmosphere as the primary "greenhouse gasses", the reduction of all Man-made emissions cannot and will not have significant effect on "average global temperature" or on climate change in a global perspective. There could be some effect of reducing carbon dioxide emissions in regions where there is almost no water vapor in the atmosphere (such as over desert areas and in Antarctica where it is too cold for the evaporative process to be in effect). Except for desert nomads and the fewer than 2,000 scientists on Antarctica, the costly benefits of reducing Man-made emissions to halt global warming and climate change are and will continue to be fruitless.

In May 2009, the Author was in telephonic communication with Dr. Siegfried Fred Singer (who was then a Professor Emeritus in Climate Science at the University of Virginia) and conversed with him on his "A Closer Look at the Numbers" series of papers he had created regarding the environmental scientists ignoring the presence of water vapor in the real atmosphere. Dr. Singer was, until his death in 2020, very vocal about fellow climate scientists not acknowledging the overpowering effect of water vapor on the rising surface temperatures on the planet's surface – land and water. Two charts from presentations given by Dr. Singer at many dozens of lectures and presentations summarize graphically what the environmental scientists were hypothesizing and which he (Dr. Singer and many other ethical scientists) was pointing out in rebuttal. The first chart is the environmentalists' version without water vapor:

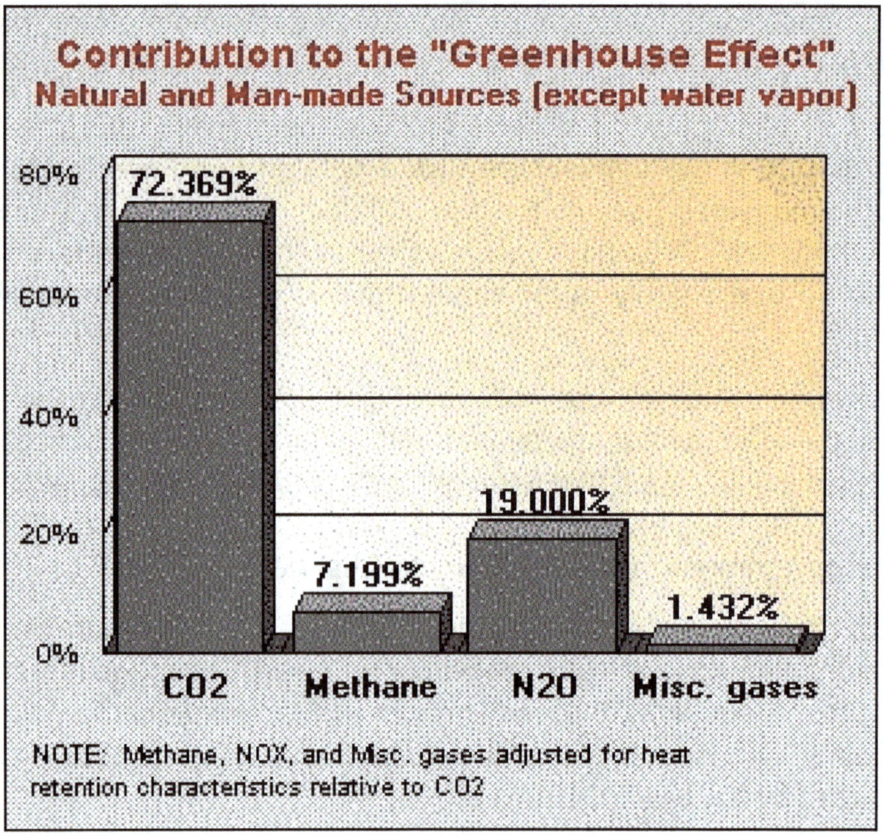

(Source, Dr. S. Fred Singer, "A Closer Look at the Numbers")

The second chart, prepared by Dr. Singer circa 2005, includes the effect of water vapor as having the same heat-trapping "greenhouse gas effect" as the other greenhouse gasses on the environmental scientists' list, is shown next:

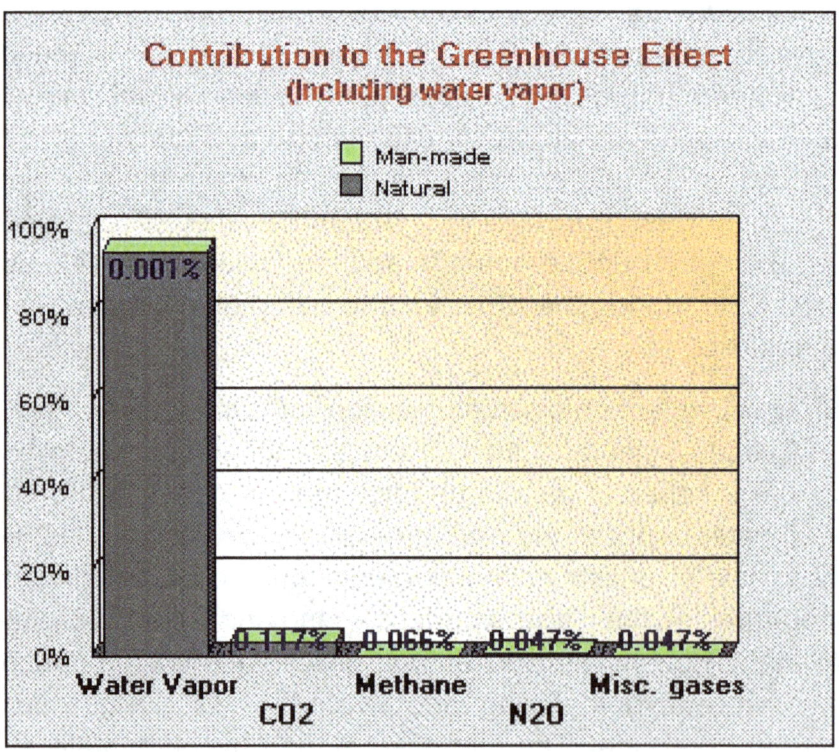

(Source: Dr. S. Fred Singer, "A Closer Look at the Numbers")

A later portion of this book suggests that the "global warming" (temperature anomaly above a normal average temperature) be tracked and reported for a "regional zone" (by its latitude) status not a fictitious "average global" status which is non-factual. Supporting this observation by many non-environment affiliated scientists (including the Author) is a database based on NASA Earth Resources satellite measurements of how carbon dioxide concentration (measured in parts per million by volume, ppmv) varies with the latitude of the climate zones of the planet. The latitudinal variable satellite-based database is graphically shown in the next chart:

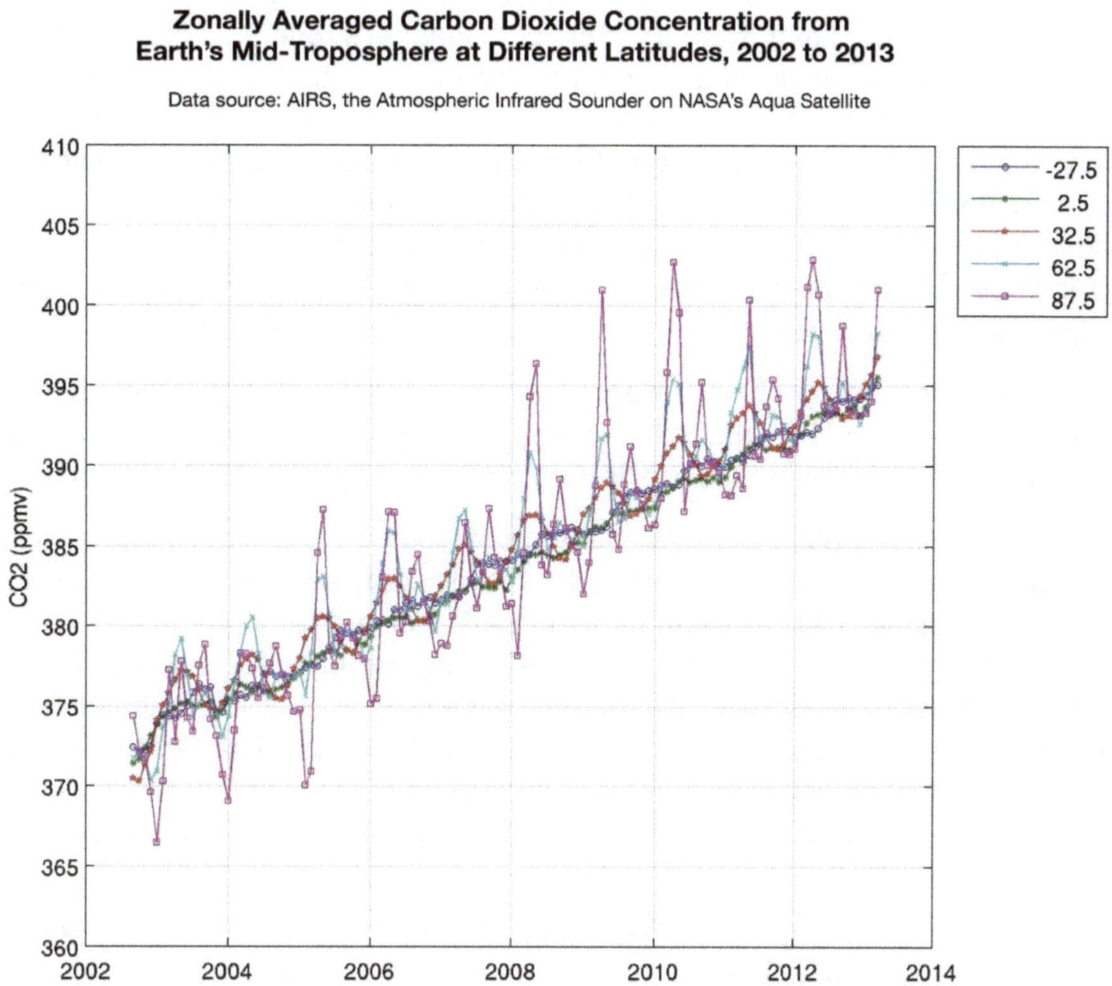

(Source: NASA Earth Resources database)

The "average carbon dioxide" concentration measured at the different latitudes shown in the symbol and scale box (at the lower right corner of the chart) may initially seem to the average public eye to just be a "scatter" in the data. To an experienced database analyst, the measured data points are interpreted to indicate that for latitudes within about 30 degrees north or south of the Equator (at 0 degrees latitude), the latitude effect on carbon dioxide concentration in Earth's mid-troposphere is small; however as the latitude increases in magnitude away from the Equatorial zone, the "anomaly" (i.e., the difference from "normal"), from that which could be represented by a scientifically more-convenient but factually less representative "global average" of carbon dioxide concentration, becomes more of significance. This is probably due to the effects of the "latitudinal zone" effect of "obliquity" – the angle at which heat energy radiation from the Sun impinges on the surface of the planet – and those degrees of obliquity determine the interaction between the surface temperature and the amount of natural-caused carbon dioxide.

The latitude data is described in degrees in the summary table insert shown on the graph above, but those degrees are not the same as the degrees in the obliquity angle. The mesurements by the satellites already incorporate the effect of the tilt angle of the planet, whereas any mathematical calculations using the latitude on the planet adjusted to account for the tilt of the planet's axis of rotation.

Geometrically, the analysis for the magnitude of the latitude anomaly can be estimated by a trigonometric (cosine) variation of the obliquity angle between the insolation vector where it impinges on a small area of the planet's surface and the tangent to the planet's surface at that latitude. Because the tilt angle of the planet's rotational axis is currently at about 23.5 degrees, the obliquity angle will be the latitude (in angular degrees) minus the tilt angle of the planet's rotational axis For the graph shown above, the tilt axis was at an angle of 23.5 degrees when the carbon dioxide measurements were collected by the satellites at the latitudes shown. This correction for the tilt of the planet's axis of rotation on the obliquity at a latitude on the planet's surface will be discussed further in a later chapter which discusses the astrophysics factors that affect the surface temperature and the interaction between the surface temperature and the naturally-caused "greenhouse gasses". There is apparently no significant correlation of latitude and the planet's axis of rotation being tilted on the anthropogenic portion of "greenhouse gasses" measured to be in the atmosphere at very low concentration.

The primary conclusion to be drawn from the above databases of Dr. Charles Keeling (and the Keeling Curve) and that of Michael E. Mann (and the "Hockey Stick" graph) is that the satellite measurements show that latitude is a very significant factor affecting the amount of carbon dioxide in the atmosphere.

The use by environmental scientists of an "average global temperature" which has some effect on the planet's surface temperature cannot be scientifically "simplified by averaging" of the combined effect of latitude and angle of obliquity (due to Earth's rotation axis being at a tilt angle of 23.5 degrees). The human population endures temperature swings of 25-35 degrees daily and must survive the extremes in temperature (hot and cold) seasonally (with temperature swings of 50-60 degrees between hot and cold extremes) not some fallacious one-size-fits-all "average global temperature" that is used for scientific convenience to represent all latitudes and all regions of the planet. Few residences above the Arctic Circle need/have air conditioning units and few residences in the Equatorial zone need/have heater units.

The IPCC goal as established at the Paris Agreement (primarily using the "hockey stick curve" of Dr. Michael Mann et al, has no latitudinal zone associated with that arbitrarily-selected goal. The National Oceanic and Atmospheric Administration (NOAA) most assuredly had a large influence on setting that two degrees Celsius limit goal, but it is believed that specific level of temperature change chosen was not for the effect it would have on the human population but more importantly on the ocean aquatic life-forms and/or atmospheric life and agricultural life-forms. The NOAA database was compiled for the observed (from about 1,000 land-based meteorological atmospheric measuring stations worldwide – mostly in the United States - increasing in number with time). The "global surface warming temperature" shortened to the more common "global warming" terminology compiled from those many measurement stations and averaged annually each year (regardless of latitude for the 2,000+ stations network) from 1900 to 1980 is shown next:

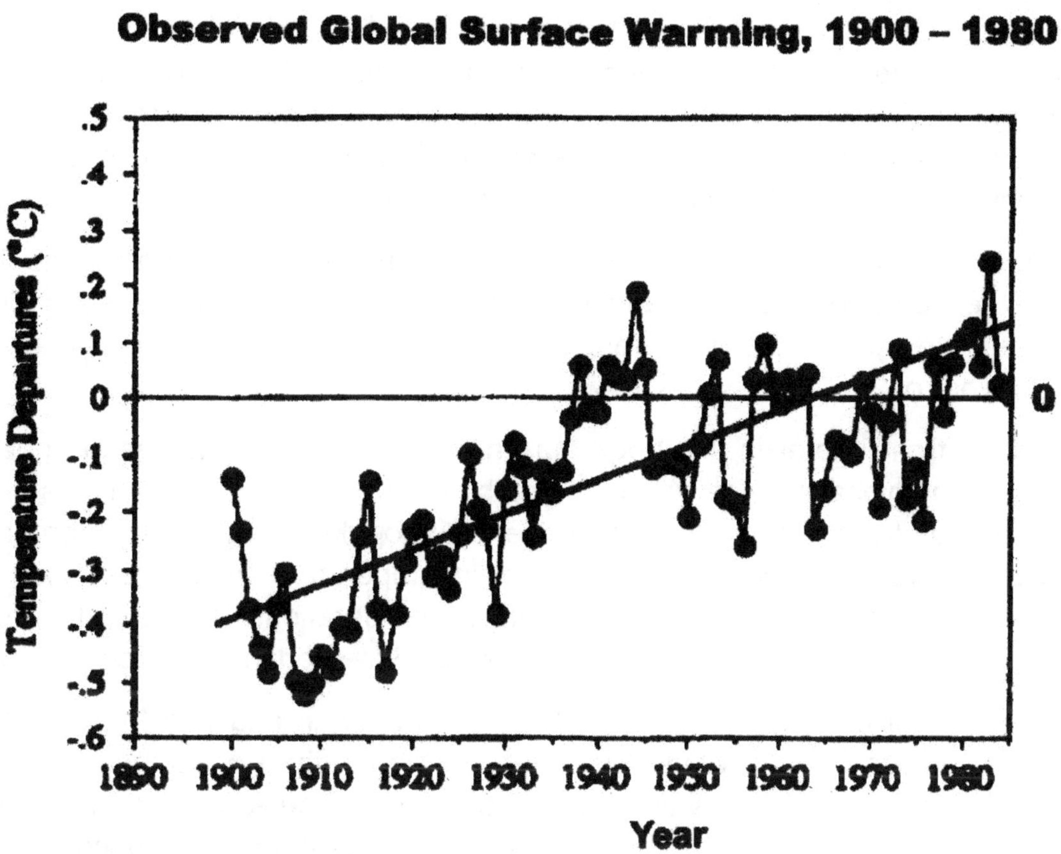

(Source: National Oceanic and Atmospheric Administration, NOAA)

The trend for the averaged global surface warming temperature was obtained by first "linearizing" the data – shown as the straight line with a positive slope indicating the trend to be a warming of the averaged, linearized global warming temperature. Although the basic data has a cyclic nature similar to that calculated by the Milankovic Cycles mathematical team, there was not found (by this Author) any comparative analysis to corellate these two databases without a bias. There was found a documented analysis explaining "Why the Milankovic Cycles Cannot Explain Global Warming" but as the title infers it was a biased attempt to dismiss any other database that conflicted with the environmental scientists' "party line" of "settled science" that Man burning fossil fuels is the dominant cause for global warming.

Any professional statistician analyzing the database of information shown above would most likely state three observations about the basic principle of how the data was collected and analyzed to represent the average temperature on the spherical planet's irregular surface (75% of that surface area covered by water and not populated):

1) the database was collected from a statistically significant number of measurement stations but those stations were not uniformly distributed in any regular pattern and mostly (almost totally) located in the northern hemisphere (i.e., how can the data be considered as being "global" when only the northern half of the globe is measured?);

2) the scale of the measured temperature (on the left vertical axis) is set at 0.1 degrees Centigrade increments which is a finer scale than is the accuracy of the measured data that has been masked by the averaging and linearization process;

3) the absence of a preceding 0 on the vertical axis scale (e.g., showing -.3 for a temperature of -0.3) is not in keeping with statistical data graphical presentation standards as it is easy for the analyst with less that perfect eyesight to miss the small decimal point and read -.3 as -3; (Author's Note: Regarding the eyesight of data analysts needing to be perfect, colleagues of the Author who analyze statistical databases for their professional career path have stated: "A good data analyst doesn't have to have perfect 20/20 vision; it only has to be good enough to pick fly crap out of ground black pepper.")

4) the cyclic nature of the basic raw data (not averaged or linearized) considered, the linear slope of the data between the eighty years of 1900 and 1980 cannot be linearly extrapolated to forecast reliably and accurately what the slope of the temperature departure/anomaly level and gradiant will be eighty years in the future year of 2050.

It will be suggested strongly later in this book that the UN IPCC "one-size-fits-all" goal based on a fallacious early "average global surface warming" database be revised to a goal that is more prioritized to protecting human safety, health, and comfort in densely populated regions than on protecting coral reef life-forms that are hypersensitive to small changes in temperature and in areas where there is little or no human population. (Author's Note: The economic considerations of unrealistic goal-setting and pursuing activities to meet those unrealistic goals are also addressed in a later chapter of this book.)

World War II raged in Europe from 1938 to 1941, but when Japan attacked the USA Naval Fleet at Pearl Harbor on December 7, 1941, World War II was started in the Pacific theater as well as in the European theater – World War II was a two-theater global war. The Arrhenius "hothouse gas effect" (carbon dioxide and carbonic acid) database from four decades earlier was once again put in a file folder in the bottom drawer of a scientific file cabinet. To bring an end to the horrific World War II, the United States developed two atomic weapons (one based on Uranium-235 and another based on Plutonium-239) and which caused the unconditional surrender of the Japanese Empire when the atomic bombs were dropped on Hiroshima and Nagasaki in Japan a few days later that ended the shooting war in the Pacific Theater. The Atomic Age – later renamed as the Nuclear Energy Age – instantly became the center of focus for almost the entire scientific profession.

(Author's Note: The Great Depression in the early 1930s had a highly-motivating effect on the public – to develop only that which would improve their prospects for a happier, safer, more prosperous future. The Author's parents, residing in the Queens borough of New York City in 1939, took the Author, age 7, and his older brother, age 9, to the 1939 New York World's Fair in Flushing Meadows, Queens, not far from the family home. The slogan of that International Fair was "The Dawn of a New Day" (as the world emerged from the Great Depression), and most of the international pavilions featured a presentation of the "world of tomorrow" envisioned for the 20th Century in all regions of the developed world. There was seen wonders such as the first household television set in the RCA Pavilion and the first automatic washing machine in the General Electric Pavilion. The Westinghouse Pavilion featured "Elektro the Moto-Man" – a 7-foot tall robot that talked, differentiated colors, and even "smoked" cigarettes. Of course, the Author and his brother most enjoyed the Heinz Food Pavilion with its 57 varieties of food-seasoning products – samples of which were freely given to visitors – and served as being a smorgasbord meal not soon to be forgotten. Eager to forget the deprivations of the Great Depression, the thought that another World War was brewing in Europe and geopolitical unrest in Asia just was not believed by the public to exist - until the Day of Infamy December 7, 1941, that took the United States into World War II.)

From the perspective of what humans can do to their environment on the planet, World War II brought more losses of human lives – estimates recalled to be more than 12 million humans (including the 4 million lost in the Holocaust atrocities), the devastation of entire cities in Europe and Asia by war actions, countless human losses due to starvation, etc. – than did the sum of all weather and climate-related catastrophes for the same period from 1938 to 1945. Compared to those Man-caused losses, the hypothesis of Man burning fossil fuels 80 years later causing the planet to be headed toward a "tipping point" at a "crisis" rate borders on the world being in another unimaginatively costly war – on the climate. After World War II (1941-1945) and the Korean War (1950-1953), a growing number of environmental scientists in the United States and Europe added two additional global concerns to the environmental problem stated in earlier sections of this book attributed to Dr. Svante Arrhenius:

1) *The global climate is being caused by humans and fossil-fuel-burning industries at a "crisis" change rate of trend to the endangerment of the environment for near-future human generations*; and,

2) *Ocean levels, directly and indirectly rising caused by climate change, are rising also at an accelerating rate due to human activities and posing danger in the near-future to humans and businesses located in areas less than five feet above current sea level.*

Except for the obvious problems brought about by ocean levels rising to inundate low-lying populated and heavily commercialized and industrialized areas, there is a distinct lack of specifics as to who, what, where, and when will suffer apocalyptic events should the dreaded "tipping point" be exceeded. Those in the generation who lived through the projected (by other scientists) calamity that would happen when the "Y2K clock" introduced the new millennium – but nothing significant happened – need have no fear of consequences of exceeding the projected "tipping point".

Acid rain formed from atmospheric gasses that combine with industrial emissions, and pollution are not drivers of the surface temperature, but are a concern in the long term for humans to reduce and control. A simple diagram showing the acid rain problem from industrial activities is shown next for Readers to distinguish the difference:

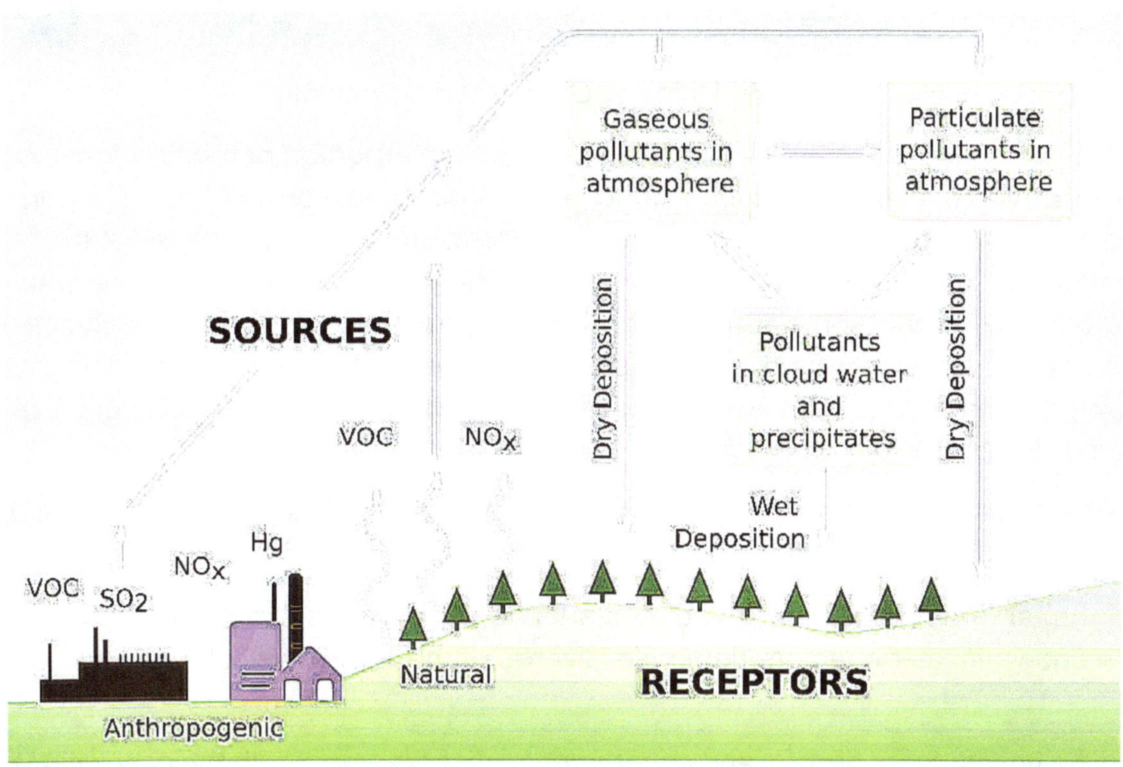

(Source: Credit to unknown graphic artist creating illustration for public domain)

As shown in the illustration, sulfur dioxide, nitrous oxide, and mercury (Hg) from anthropogenic sources as well as particulate matter (from industrial sources and from volcanic activity) are the primary polluters and "acid-rain" generators but have little if any heat-trapping properties as "greenhouse gasses". Over-exaggerating environmental activists regard carbon dioxide from the Industrial Revolution activities as being the dominant cause for global warming and seldom miss an opportunity to show a photograph of a carbon-fuel burning facility with a prolific plume of white smoke emanating from the chimney(ies).

(Source: Union of Concerned Scientists)

All fossil fuel burning facilities in the United States have a large portion of the heat generation process controlled with Environmental Protection Agency emissions reduction features. Carbon dioxide is a colorless, odorless, odorless gas and any amount of carbon dioxide being emitted in the "white smoke" of the above facility cannot be seen by the human un-aided eye. Almost all of the white "smoke" is steam vapor, but may contain some coal fly ash in a small particulate form. That particulate matter does fall out of the steam eventually and is a pollutant that needs stringent control. The best control of particulate air pollution at coal-burning facilities is to convert the facility to a different fuel which does not have the particulate fall-out problem.

A later review of the several databases dispel the claim of carbon dioxide being the "dominant" cause for global warming; the air and pollution concerns remain very valid.

In any case, the far larger causes for the planet surface temperatures, and thereby the climate, to be changing are shown in later chapters database reviews of the astrophysical, geophysical, and natural effects.

Despite there being residual scientific and governmental disagreement with the conclusions drawn from them, the Arrhenius atmospheric carbon dioxide database and the more recent Keeling Curve and

Mann "hockey-stick" curve database brought the problem to public and governmental attention and history is indebted to them for their contributions to the beginning of understanding of a problem of major public implications and justifiable concern.

The world public should be very careful before accepting the environmental scientists' claim that Man burning fossil fuels is the dominant cause for carbon dioxide in the atmosphere. When a carbon dioxide measurement at the Mauna Loa Atmospheric Research Laboratory is reported being 410 ppmv or 420 ppmv, it is not 100% due to humans and/or industry burning fossil fuels. Most of it (about 298 ppmv as shown in the Vostok ice cores database at the peak near the end of the fourth cycle and for the NOW point in time) is caused by astrophysical and natural/geophysical factors over which Man has no control. No amount of reduction in carbon dioxide emissions generated by humans burning fossil fuels can reduce the amount of carbon dioxide in the atmosphere to be less than that due to astrophysics and/or geophysics or natural causes. Practices by Man and nature such as deforestation or lightning-ignited wildfires can indirectly increase the carbon dioxide in the atmosphere by reducing the number of trees that naturally absorb total carbon dioxide (generated by Man or by other causes) from the atmosphere.

It is hastened to be stated that the reduction of carbon dioxide emissions by reducing Man's dependency on fossil fuels is not a bad idea – only that it should be done as a gradual transition in the associated technologies and not as a human "crisis" to achieve that reduction in emissions. The introduction of any type material (solid, liquid, or gas) that infiltrates as a pollutant into the air, land, or water resources on the planet must be reduced if not eliminated. Man should and must continue to develop alternative energy sources to reduce the dependency on fossil fuels for energy reliability and economic benefits.

Mining underground for coal is inherently unsafe. It is also important for the public to recognize that after coal is combusted there is a residue (ashes) containing both toxic (arsenic) and non-toxic (silicon carbide) materials that must be disposed of safely in order that the coal ashes not find their way to pollute the water and land areas near the coal-combusting facility. Anti-pollution measures have been installed at all coal-burning industrial facilities in the United States and Europe - with compliments to the Environmental Protection Agencies anywhere that has caused those anti-pollution measures to be put into effect. Unfortunately, most underdeveloped countries have not done so or plan to do so.

Both of the highly-reputed scientists (Dr. Arrhenius in his laboratory in Sweden and Dr. Keeling at the Mauna Loa Atmospheric Research Station in Hawaii) and with subsequent experiments and accurate measurements in the controlled laboratory environment and small scale of specimens (solid, liquid, or gaseous), led other environmental scientists in Europe and the United States (including Dr. Michael Mann and his associates in dendrochronology)) to believe – without considering the Vostok Ice Core samples and Milankovic astrophysical calculation databases - that it is the amount of fossil fuels (coal and oil) being burned by industrialists and by the exponentially-growing human population as the cause for global warming, climate change, and ocean levels rising.

The world population is now over 8 billion mostly in the more industrially-developed regions of the northern hemisphere. In past and recent years, with abundant fossil fuels available to satisfy their needs for heat energy, the world resorted to combusting fossil fuels that are at most minimally causing the average global temperature and "greenhouse gas" emissions to rise and indirectly causing the planet's surface temperature to be warming everywhere and trending to a dangerous level.

The news media and many governmental environmental control organizations need to update and re-plan many atmospheric control activities based on the more globally-applicable satellite measurements that became increasingly available after about 1960. The goals that were based on the outdated "average global temperature" and "applicable to all latitudes" database (i.e., those that were established before the more factual satellite measurements and advanced technology sensors came into use (post 1960) need to be restructured to build on only the valid portion of their database (e.g., the need to wean totally away from coal and transition to oil and natural gas; to develop clean "green" fuels; and to develop of alternate sources of energy that do not produce "greenhouse gas effects"). The one-size-fits-all "average global temperature" reference should/must be replaced with unique regional (not global) satellite-measured temperature and climate databases.

In 1965, President Lyndon Johnson appointed a Science Advisory Committee (SAC) to assess the importance of Dr. Keeling's measurements of carbon dioxide in the atmosphere and acting as a "greenhouse gas" in trapping heat energy that is causing the global temperature to rise at a disturbing rate and to ominous level. Dr. Keeling reported to the Science Advisory Committee that the measurements he made at the Mauna Loa Laboratory were accurate – there was a measurement made of 315 parts per million by volume (ppmv) in 1958 and 321 ppmv in 1965. Dr. Keeling pointed out to the SAC that the red line (in the "hockey-stick" measurement chart) reflected the seasonal (i.e., temperature change) effects on the measured carbon dioxide data but that the mean (black line) carbon dioxide measurements showed a rapid increase (6 ppmv in seven years) in the mean carbon dioxide in the atmosphere which is most likely due to Man-made (human and industrial) "greenhouse gas" emissions causing the rapid rise. Dr. Keeling neither drove nor actively supported the SAC report to President Johnson's staff scientists to claim that it was Man burning fossil fuels that was the primary cause for any increase in global temperature as there was a combination of Man-made carbon dioxide and naturally-generated (at least seasonal if not by other causes) carbon dioxide in the rise noted in the measurement database.

In researching (to a very limited extent) Dr. Keeling's scientific papers and reports to the International Geophysical Year scientific staff, Dr. Keeling apparently also advised that the measurements of the atmosphere in the middle troposphere (the Mauna Loa facility being more than 11,000 feet (3,400 meters) in elevation above the surface level of the warm Pacific Ocean waters surrounding all of the Hawaiian Islands). The choice for that elevated location of the Keeling atmospheric data collection was apparently mostly based on the environmental scientists' postulation that heat energy being re-radiated from the ocean would be trapped in the lower portion of the troposphere (i.e., below about 5,000 meters – about 15,000 feet) and that the heat energy trapped below and within the "greenhouse gas" layer would result in a feedback warming of the ocean waters. (It is left to those more highly educated in the thermodynamics of gasses – higher than the basic thermodynamics of gasses ("gas dynamics") taught to the Author in Aeronautical Engineering graduate school at Georgia Tech – to explain how a gas (carbon dioxide or other "heat trapper") can be "polarized" to significantly trap heat energy beneath it but simultaneously not preventing the heat energy from above it (from the Sun causing global warming) from passing through the gas and warming the ocean water below. The effect is understandable in principle but the magnitude is quite suspect in comparison to that of water vapor.)

There certainly is a wavelength difference between radiated energy from the solar source (the Sun) and the reflected energy from the Earth's surface, but that wavelength difference is not obviously so profound as to cause a significant "polarization" effect of carbon dioxide gas. It also seems probable that the reflected heat energy is very dependent on the surface texture, color, state (liquid or solid), heat absorptivity, and a myriad of other factors such as winds, water vapor content in the air, cloud layers, etc. Those variables do not exist in controlled scientific laboratory environments. Once again, what is happening in the real world is typically what is not happening during laboratory measurements and observations. The panes of glass in a “greenhouse” ceiling allow radiated heat energy to pass both ways but prevents heat from below not being convected through the glass into the atmosphere above the glass. That terminology “greenhouse effect” is largely a misnomer intended to mystify the gullible public.

When the Arrhenius database was presented to the scientific world in the early decades of the 1900s, and indicated that there could be a beneficial near-term warming effect on frozen land to improve the food supply for the growing population in the northern hemisphere (especially in Scandinavian countries), Russian environmental scientists took great notice given the fact that Russia has so much land (almost all of Siberia) that is frozen and not very arable. In 1957, Russia established the Russian Vostok Ice Research Facility on Antarctica. The Vostok Research Station is at latitude -78.47 degrees South and elevation of 3,488 meters (11,444 feet) in approximately the center of the East Antarctica Ice Sheet. The Vostok (translated from Russian into English, Vostok meaning “East”) Research Station was intended to drill ice core samples (to be analyzed for gaseous content and particles to be carbon-dated to establish age) and to record magnetometry parameters.

In the 1970s, the Soviet Union drilled a set of cores 500-952 meters (1,640 -3,123 feet) deep into the ice sheet below the Vostok Ice Station on Antarctica. In 1984, a second set of ice core drilling reached a depth of 2,202 meters (7,223 feet). In 1990, a third set of ice core drilling reached a depth of 2,546 meters (8,351 feet), and in 1993 to 1996 and with scientists from other nations than Russia, sets of ice core drilling reached a depth of 3,623 meters (11,883 feet). Drilling was halted in 1996 at that depth when space-based radar (aboard the European satellite ERS-1) revealed that the drilling to obtain ice core samples was approaching (130 meters – 426 feet) from the surface of a subglacial body of fresh water, one of the largest lakes in the world. Subsequently, the radar systems in satellites identified this lake (Vostok Lake) to be one, perhaps the largest, of 140 subglacial lakes beneath the East Antarctica Ice Sheet – none of which can be factually scientifically attributed (as an “inconvenient truth”) to recent global warming trends – natural or Man-caused. The drilling was suspended for a few years to assess the possibility of introducing any non-natural contaminants into Lake Vostok, but since the depth of the usable portion of the final ice core sample had corresponded to 420,000 years of past environmental history, it is that extent of the ice core samples that serve very factually as a most credible database for the environmental scientific community (and for this book).

Accurate analysis by the Vostok Research Station scientists of the ice core samples revealed the data shown next for the temperature and carbon dioxide:

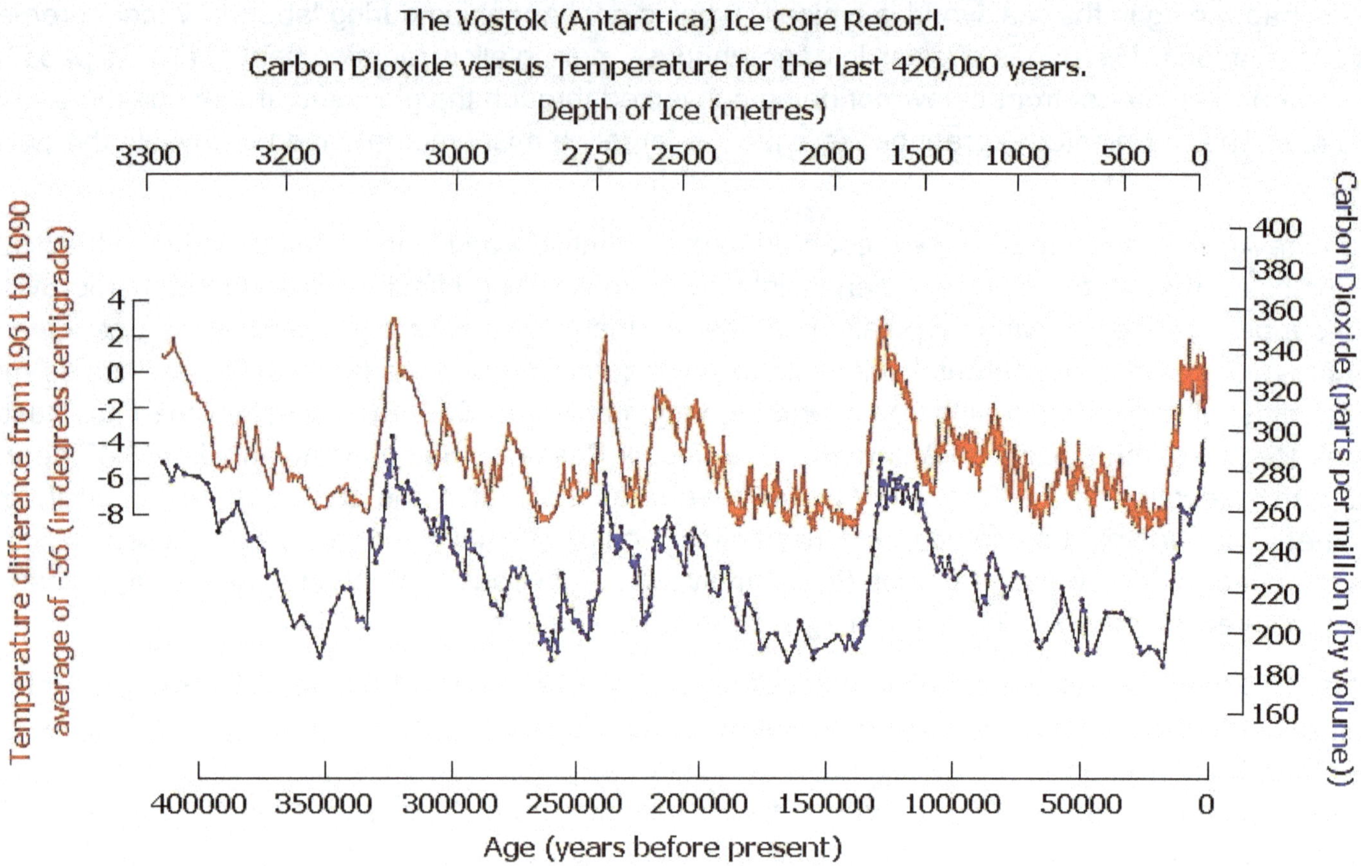

(Courtesy of the International Ice Core Research Station, Antarctica)

(Author's Note: The above chart was found in other sources, but was noted to include carbon dioxide data from the Keeling Atmospheric Research Observatory (NOAA) in Mauna Loa, Hawaii, collected as recently as October 2023. That "tack-on" data apparently was intended to show that the Industrial Revolution produced carbon dioxide emissions that were captured in the ice core samples in Antarctica – but factually were not. (The "0" time shown on the chart is accurate only up to about 1900 as the ice in the 10 inches nearest the surface had undergone melting and reforming that did not trap air bubbles containing atmospheric gasses having sufficient scientific accuracy.) Later in this book it will be shown that the wind and ocean currents are such that there is little exchange of any gasses produced by activities (natural and/or anthropogenic) that are transported from the upper latitudes in the Northern hemisphere (where the bulk of industrial and human greenhouse gas emitters are located) across the Equatorial region and to the extremely lower latitudes of the Southern hemisphere. The more scientifically authentic databases were used by this Author without comingling of one database to another.)

The Author analyzed the ice core database further to assess if any regularity in the data existed and the following chart graphically summarizes the Author's analysis:

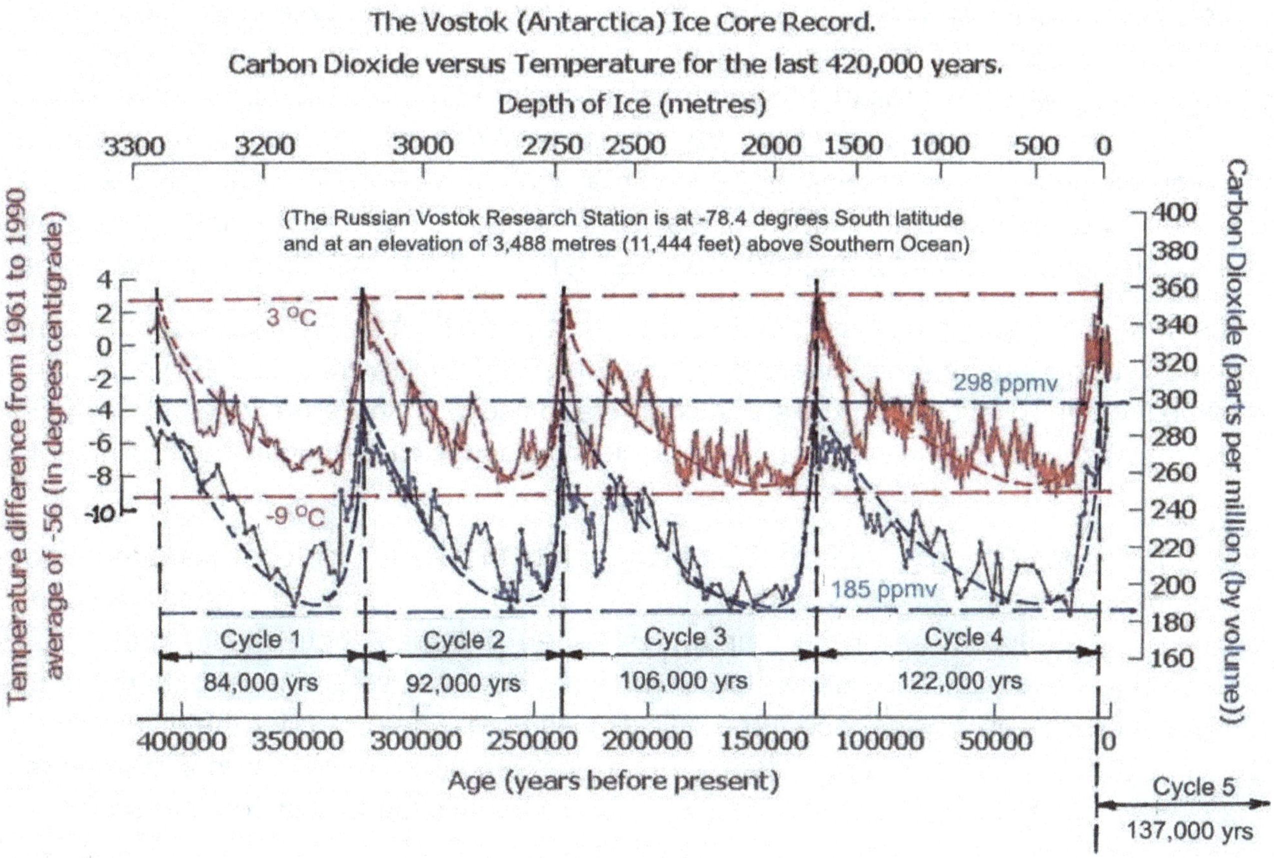

(Chart with markings from the Author's statistical analysis of the database)

A few characteristics of the database indicated repetition in both magnitude of the extremes of temperature and of the carbon dioxide concentration and the dashed colored (red for temperature and blue for carbon dioxide) horizontal lines added are the limit lines of those extremes. The temperature limits resulting from the scientists' analysis of the ice core samples show a maximum temperature difference from the 1961 to 1990 average of -56 degrees Centigrade/Celsius at Vostok Research Station of about 3 degrees warmer (-53 degrees Centigrade) in the upper red colored dashed line, and about 9 degrees colder (-65 degrees Centigrade) in the lower red colored dashed line. The same observation was made in the analysis of the carbon dioxide concentration database (shown in blue color).

The peaks in carbon dioxide were at a maximum of about 298 ppmv and at a minimum of about185 ppmv, and with the observation that the extreme peaks (warm temperature and high carbon dioxide, and cold temperature and low carbon dioxide) occurred at about the same point in time and the general shape of each curve between the peak extremes were of the same approximate shape for four cycles of different period. The period for the four cycles was measured and found to be as noted near the lower part of the database (Cycle 1 having a period of about 84,000 years, Cycle 2 having a period of about 92,000 years, Cycle 3 having a period of about 106,000 years, and Cycle 4 having a period of about 122,000 years).

An important trend was noted in the four cycles such that if the same trend is not largely distorted by the small amount of carbon dioxide produced by human activities, a fifth cycle is estimated to start within the next 200-300 years and will last about 137,000 years from now. Inasmuch as the first four cycles are due exclusively to astrophysical, geophysical, and natural factors (as will be proven in the next chapters of this book), and assuming that the alternative clean energy technologies are developed successfully (i.e., no effect of Man-made "greenhouse gasses"), early in that 5th cycle, the planet surface temperature will soon peak at a maximum temperature and head toward another glacial period similar to that of the initial four cycles. Ironically, the generations existing in that future time period of glaciation (approximately 100,000 years from now) may wonder why the current generation expended so much effort and resources to reduce global surface temperature increases while they will be developing technologies to counter the much-reduced temperatures associated with glaciation.

A major observation also needing to be pointed out is that at the time t = 0 (Now), both the temperature and the carbon dioxide in the atmosphere had been on a rise for several hundred years, long before the start of the Industrial Revolution. The surprisingly steep rise in temperature long before the Industrial Revolution was observed also in the Milankovic Cycles astrophysical calculations shown earlier in this chapter.

The Author constructed "smooth" variations on the graph of the Vostok ice core database (the red line for temperature, and the blue line for carbon dioxide) of the rather erratic ("jagged") measured data which consistently indicates that in all four cycles: when declining, both the temperature and the associated carbon dioxide declined more slowly with time; however, when rising, both the temperature and the associated carbon dioxide rose more quickly with time. The cause for the difference in rate of decrease with time and rate of increase with time for both temperature and carbon dioxide is beyond the scope of this book to be analyzed, but certainly should be investigated by the International ice core research team and the results shared with the United Nations Intergovernmental Panel on Climate Change (IPCC) to use in revising the fallacious "greenhouse gas emissions" that are proven in this book to not be the dominant cause for global surface temperatures presently rising.

A working graph was constructed – not shown here – and from that working graph it was estimated that a fifth cycle (Cycle 5, noted at the lower right corner of the chart, starting with the time about now (2023)), would likely repeat the same temperature and carbon dioxide limits of high and low extremes. However, Cycle 5 would have a still longer period of about 137,000 years based on the progressively increasingly lengthy periods shown in Cycles 1 through 4. An important caveat added is that Cycles 1, 2, and 3 were all without the presence of Man and Cycle 4 had the closing 20,000 years with the presence of Man and the final 200 years with the presence of a greater number of humans and much Industrial Revolution machinery burning fossil fuels for energy needs. The latter caveat includes that there will be a "greenhouse gas effect" of fossil fuels emissions that this book is presenting in as meaningful a manner as possible, but which is a lesser problem than that as forecast by environmental scientists using the pre-1950 incomplete databases and by their "averaging" and "linearizations" of data.

A major conclusion drawn from the above chart however, is that there is an unusually close relationship between temperatures and carbon dioxide measured from the ice core samples that varied in a synchronous manner – almost every peak and valley within each cycle seem to be 100% synchronous in time of occurrence. The carbon dioxide had to be all produced naturally, and the temperature variations were being caused by a combination of astrophysical, geophysical, and natural factors since there was no presence of Man burning fossil fuels. As a reminder to Readers, the Milankovic calculations did not include any effect of carbon dioxide presence in varying concentrations in the atmosphere at that latitude of the Vostok station as the ice cores contain and do reveal.

The Vostok ice core-indicated carbon dioxide and temperature levels variations with time were magnified by Vostok International ice research scientists with a most important qualitative observation – a lag of the carbon dioxide behind the temperature was analyzed and was/is quantatatively confirmed.

In a paper written circa 2014 by geologist Euan Mearns "The Vostok Ice Core: Temperature, CO2, and CH4", reporting on the results of an analysis of the ice core samples corresponding to the time interval 137,383 to 102,052 years ago, geologist Mearns stated:

"Many in the climate establishment will not like what was found. The Eemian warm period between 130,000 and 115,000 years ago was the last warm period before the current warm period, the Holocene. Graphs presented in 2008 by geologist Jo Nova showing the relationship between CO2 and the temperature from the Vostok Ice Cores were analyzed (by Jo Nova) and it was found that the average lag of CO2 behind the temperature, in their respective modes of increasing, peaking, and then falling, was about 800 years. This lag indicates that CO2 could not be the cause of the rise or fall of temperature.

This finding is yet another nail in the coffin of climate alarmism, and is all the more reason for the demise of the United Nations Framework Convention on Climate Change, which became effective in 1994, and its offspring the Paris climate accord (also known as the Paris Agreement) which entered into force in 2016 ."

Graphically, the 800 years lag of carbon dioxide behind the temperature is shown next:

CO_2 lags temperature rises in ice core data

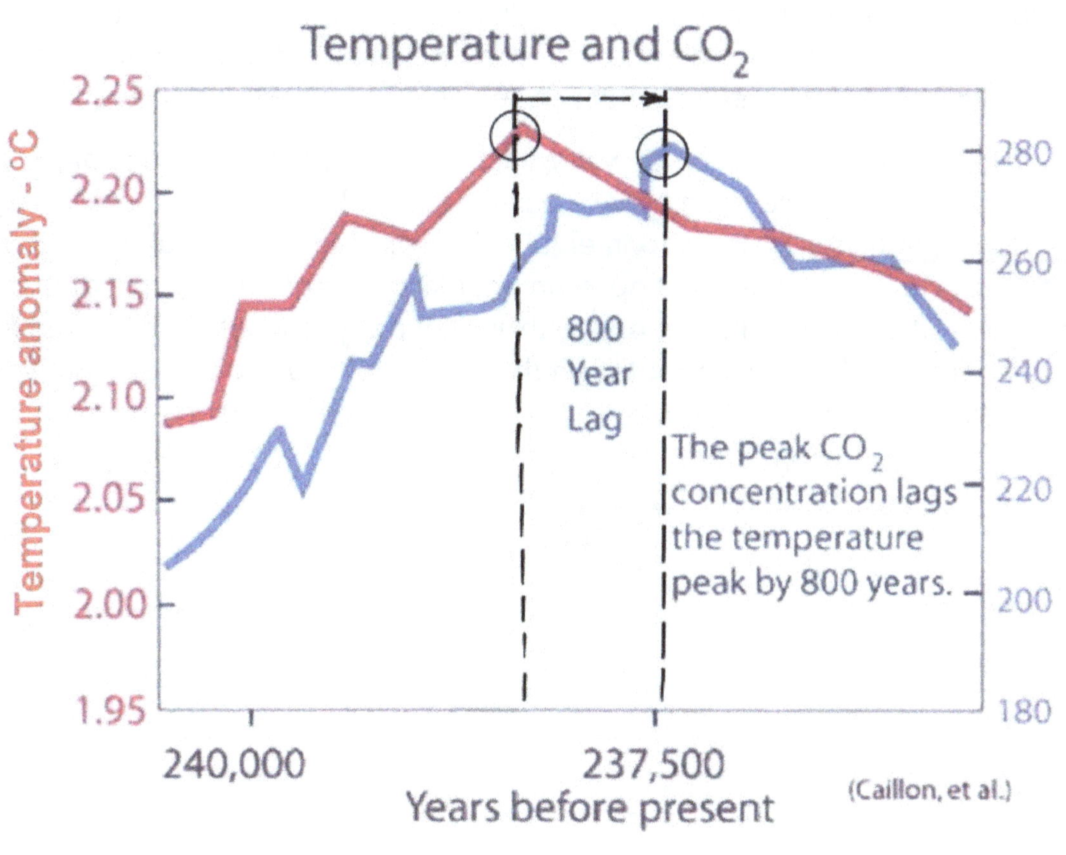

(Courtesy of the International Ice Core Research Station, Antarctica)

These two preceding graphs of the ice core database jn two different time scales showed the variations in carbon dioxide levels – before the known presence of human life and the Man-made (human and industrial) "greenhouse gasses" – to "lag" behind the temperature variations by an average of about 800 years.

A well-conducted review of the phenomenon of carbon dioxide lagging behind temperature was conducted (and updated in 2007) by astute geoscientists at the Scripps Institution of Oceanography (University of California, San Diego) which refined the initial 800 years lag conclusion. Based on their analysis of the ice core samples for the long term (5000 years) cycle, it is a high probability that although the temperature drives the carbon dioxide in the first 800 years of a 5,000 years heating/cooling cycle, the Scripps Institution refinement indicates that carbon dioxide could have had a major "feedback" effect on global warming during the last 4,200 years of the 5,000 years trend. Based on the two perspectives of "lagging" and "driving", the environmental scientists are now apparently conceding that Man-made carbon dioxide shown in the Keeling hockey stick curve is not entirely due to the Industrial Revolution, and the effect of Man-made carbon dioxide on global warming may be amplifying the natural carbon dioxide effect on global warming.

From a "big picture" system perspective, the issue of whether temperature drives carbon dioxide or carbon dioxide drives temperature is a minor issue remaining to be settled. The Antarctica ice core samples indicates that the planet is nearing the end of the fourth historical cycle of heating and cooling and return to heating shown in the past 420,000 years. All indications are that the fifth cycle is imminent and in the near future (within the next 800 years), the current temperature increases will reverse to a cooling trend (and already may have globally if not entirely regionally).

An alternate possibility (and strongly the higher probability) is a different "forcing" factor (astrophysical, geophysical, and/or natural, singlularly or in combination) exists that is driving both the temperature and the low concentration of carbon dioxide (measured in the Vostok ice core samples to vary between a minimum of 185 ppmv (0.0185%) and maximum of 298 ppmv (0.0298%)) as dependent variables.

There is little question there is a presence of water vapor in the atmosphere factually contributing to global warming, climate change, and ocean levels rising. It should be obvious that Man burning fossil fuels is NOT the dominant factor that environmental scientists and the sensationalistic news media have been propagandizing to the public and government officials. A good example of that gross over-exaggeration is shown in the next chart – and with its source purposefully not identified to avoid embarrassment of a fellow professional scientist (at the NASA Jet Propulsion Laboratory) who stated: *"by eliminating everything else, scientists have proven that humans creating greenhouse gas emissions is the dominant cause for global warming":*

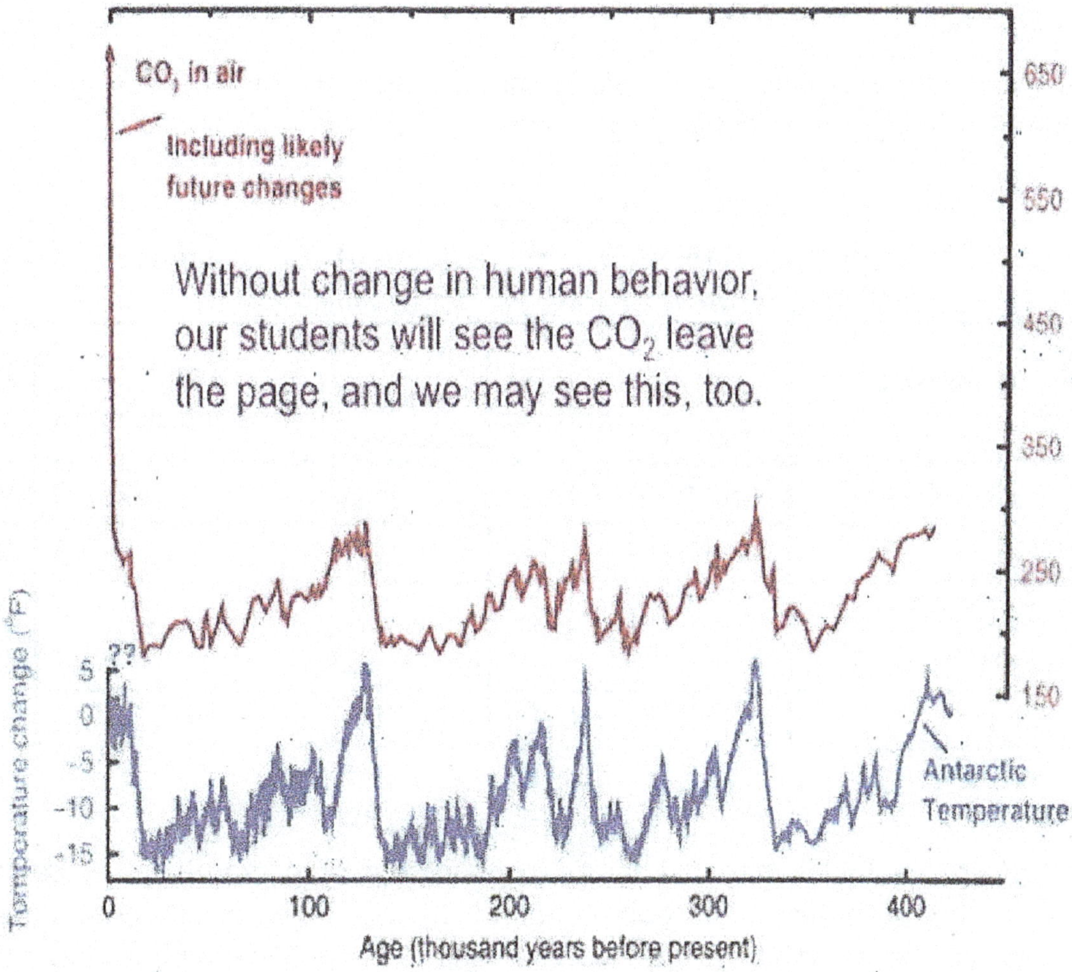

(Source intentionally not identified as a professional courtesy)

If the data does not fit on the page, the scale for the data should be reduced in the size of the increments. Any set of data can be overexaggerated by expanding the scale in which the data is presented. It is unprofessional to compare "apple" data with "orange" data with the intention to alarm the public with overexaggerated data. The temperature scale in degrees is a measure of magnitude, whereas the CO2 scale is actually a ratio (usually concentrations of one element in another element is referred to as a concentration being expressed in parts per million. The scale for CO2 shown in the illustration above is in 100 ppmv increments – but if presented in percent format, the scale equivalent to 100 ppmv increments in percent format is in 0.01% increments. It is shown in a few places in this book that the public well understands the percent notation (a penny is 1% of a dollar). A 0.01% of a penny is meaningless. Scientists use the ppmv scale with an objective of developing meaningful trends observed in the set of scientific data containing "trace" values of components rather than using the percent scale.

The recent (November 2023) measured carbon dioxide concentration in the atmosphere by the Keeling Atmospheric Research team at Mauna Loa was 423 parts per million by volume (ppmv) and not that as shown above exceeding 650 ppmv - with a notation of "Including likely future changes". The inference of using the same color and density of a line as the valid data and adding that the "likely future changes" in carbon dioxide in the air (to a level of above 650 ppmv) as an extension to the valid and professional Vostok ice core data is not only professionally unethical it is an embarrassment to the entire scientific profession. Valid databases should not mix speculative data in the same illustration as accurately measured data.

A good example of scientists magnifying the scale of data to delude the public into a state of alarm over the size of the numbers being presented is shown next:

TABLE 1.
The Important Greenhouse Gases (except water vapor)
U.S. Department of Energy. (October. 2000) (1)

(all concentrations expressed in parts per billion)	Pre-industrial baseline	Natural additions	Man-made additions	Total (ppb) Concentration	Percent of Total
Carbon Dioxide (CO2)	288,000	68,520	11,880 (2)	368,400	**99.438%**
Methane (CH4)	848	577	320	1,745	**0.471%**
Nitrous Oxide (N2O)	285	12	15	312	**0.084%**
Misc. gases (CFCs, etc.)	25	0	2	27	**0.007%**
Total	289,158	69,109	12,217	370,484	100.00%

(Source: Department of Energy)

In this table, the scale of the data is in parts per billion ppb instead of the same data in parts per million ppm. The data in the Total Concentration (ppb) column indicates the Keeling measurement of 368.4 ppmv in 2000 as 368,400 ppmb. Which is the most alarming to the public – 0.03684%, 368.4 ppmv, or 368,400 ppmb? They are all, in fact, the same measurement. What is also a deceptive practice is to make a simple note: "(except water vapor)". Environmental scientists at the Mauna Loa Observatory where the 368.4 ppmv concentration of carbon dioxide was originally measured have adjusted the measured Greenhouse Data to be "without water vapor" – so that the concentration of "Important Greenhouse Gasses" takes on more importance. Water vapor in the atmosphere varies between 0% to 4% depending on the meteorological temperature, and relative humidity of the atmosphere, and scientifically water vapor is a far more effective (by a factor of about 9 to 1) "heat-trapper" than are the "Important Greenhouse Gasses". As an average, there is 1.4% of water vapor in the atmosphere over most populated regions - and anyone who has resided in an East Coast city knows when the "humidew" ("street language" for humidity) is high. 1.4% water vapor is equivalent to 14,000 ppmv and 14,000,000 ppmb. It should not take much interpretation by the public that a concentration of 1.4% / 14,000 ppmv /14,000,000 ppmb of a powerful heat-trapping component of the atmosphere is far more important to global warming than is a 0.03684% / 368.4 ppmv / 368,400 ppmb concentration of carbon dioxide (of which only 0.015% / 156 ppmv / 156,000 ppmb is of anthropogenic source). In ratio form, there is almost 100 to 1 more water vapor than anthropogenic carbon dioxide in the atmosphere.

The news media (and some unethical environmental scientists) simplifies the numerical data to the sensationalistic reporting to the public: "Man-made greenhouse gas emissions are destroying the planet". Man-made greenhouse gas emissions – as well as all other gasses emitted by any process that have toxic or carcinogenic properties - need to be controlled. That is not solely for improving the environment or "combatting" climate change as a "crisis"– that is a societal problem for any developed country to protect its population (safety, health, comfort, mobility, economy, etc.) as new technologies evolve. Eventually, common sense will prevail. "If you want scrambled eggs, you have to break some shells."

As a lead-in to the next important database to be discussed, the analysis charts of the Vostok Ice Core database have the appearance of the electrocardiograms (EKGs) done on human cardiac health patients. A natural question that must be asked then is: "What is causing such "squiggly" variations noted in the ice core samples?" And a corollary question is: "Are those "squiggles" related to those that were noted earlier in the Milutin Milankovic astrophysical calculations in the pre-1950 database?"

The Milankovic calculations of the early years of the 1930s, which were introduced earlier in this chapter (as a pre-1950s database), reflect the complex changes due to astrophysical factor variations in solar irradiation on Earth's surface. When the Milankovic calculations are compared with the Vostok measurement database from the ice-core drilling in Antarctica in the mid-twentieth century, there are a few most interesting and important observations that can be and are made as shown next:

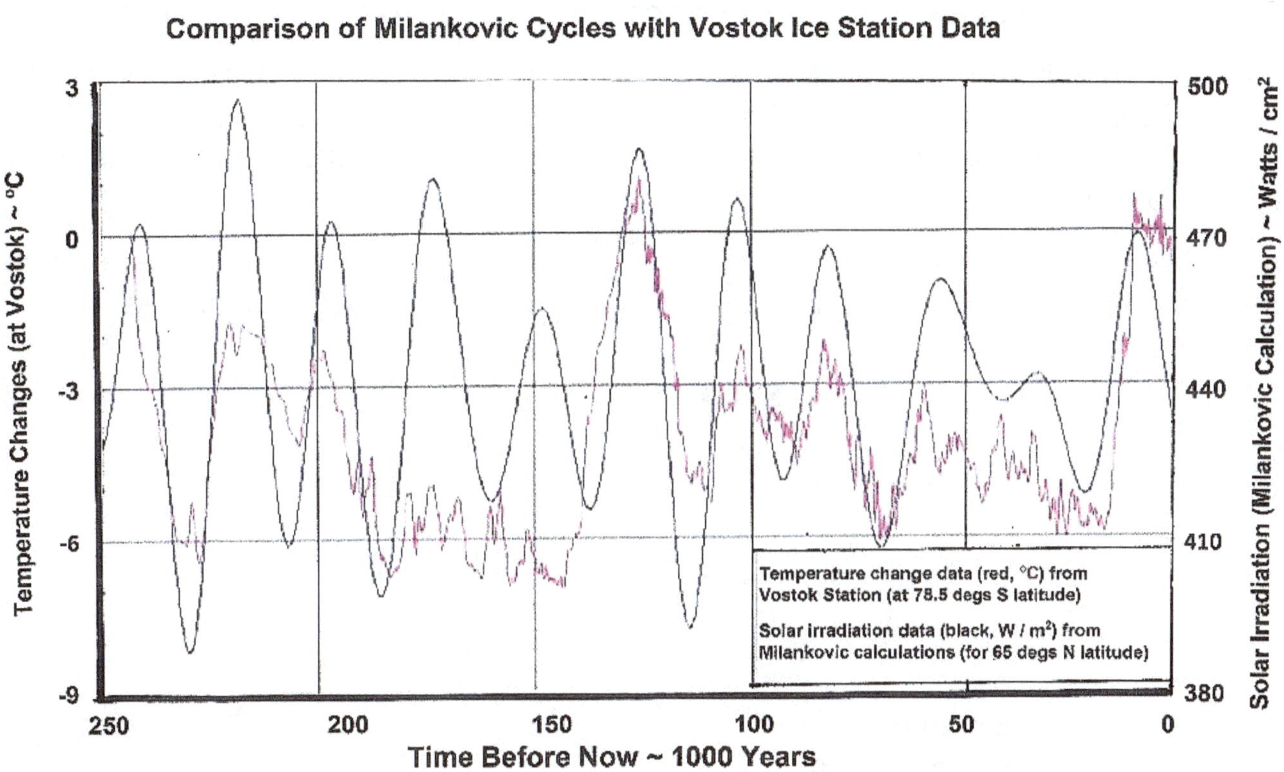

(Courtesy of US Geological Service, Dept of Interior)

To a professional data analyst, this comparison would be rated statistically as being an excellent correlation with the timing of the cyclic variations of astrophysical nature on the temperature derived from the Vostok ice cores (shown in red) physically collected compared to the Milankovic calculations (shown in black) of those same astrophysical factors and their effect on solar irradiation impinging on Earth's surface. The Vostok ice core samples inherently include the effect of all factors that affect Earth's atmosphere. Whereas, the Milankovic calculations do not consider any atmospheric or natural/geophysical effects on Earth's surface temperature – only astrophysical effects. Pursuing that thought then, the scientific conclusion could easily be reached that it is that difference in atmospheric effect on temperature evidenced in the Vostok ice core measurement database that has the same overall periodicity but resembles an “erosion” of the larger amplitude of solar insolation in the Milankovic calculated database.

There is an aspect of “the greenhouse gas” effect on open water sufaces of the oceans, but there is no standing water on the exposed surface of the Antarctica continent. Therefore, this “feedback” effect could not have a direct effect or interaction with the surface ice layers in the ice layers deeper in the ice on the Antarctic continent. Open water surfaces and ice layers absorb and reflect solar irradiation affecting the planet surface temperature differently as will be discussed later in this book.

The quality of correlation in the chart of the magnitude of the peak hot temperature anomalies and the peak low temperature anomalies between the Vostok ice core data and the Milankovic calculated data is rated as being statistically not good. It is hastened to be added that there could be a latitude effect on temperature but that effect is yet to be precisely defined. The Vostok ice core measurements data were obtained from the ice cores at latitude -78.5 degrees south whereas the Milankovic calculations were done for latitude 65 degrees north (and at the summer season equinox). Later (in the 1960-1990 time period), the NASA weather satellite database revealed a significant effect of latitude (and the associated “obliquity” factor) as discussed briefly near the beginning of this Introduction.

In each of the four cycles there are the Milankovic-type “squiggles”, characteristic of the effect of Earth’s tilted axis “precessing” on cycles of 19,000, 22,000, and 24,000 years, the “obliquity” effect cycling with about 41,000 years period, and the “eccentricity” (of Earth’s elliptical orbit) effect occurring at 95,000, 125,000, and 400,000 years cyclical periods.

The chart shown next (and which was shown at the start of this chapter) may be of interest in realizing there are more than one statistical method that can be used when analyzing databases:

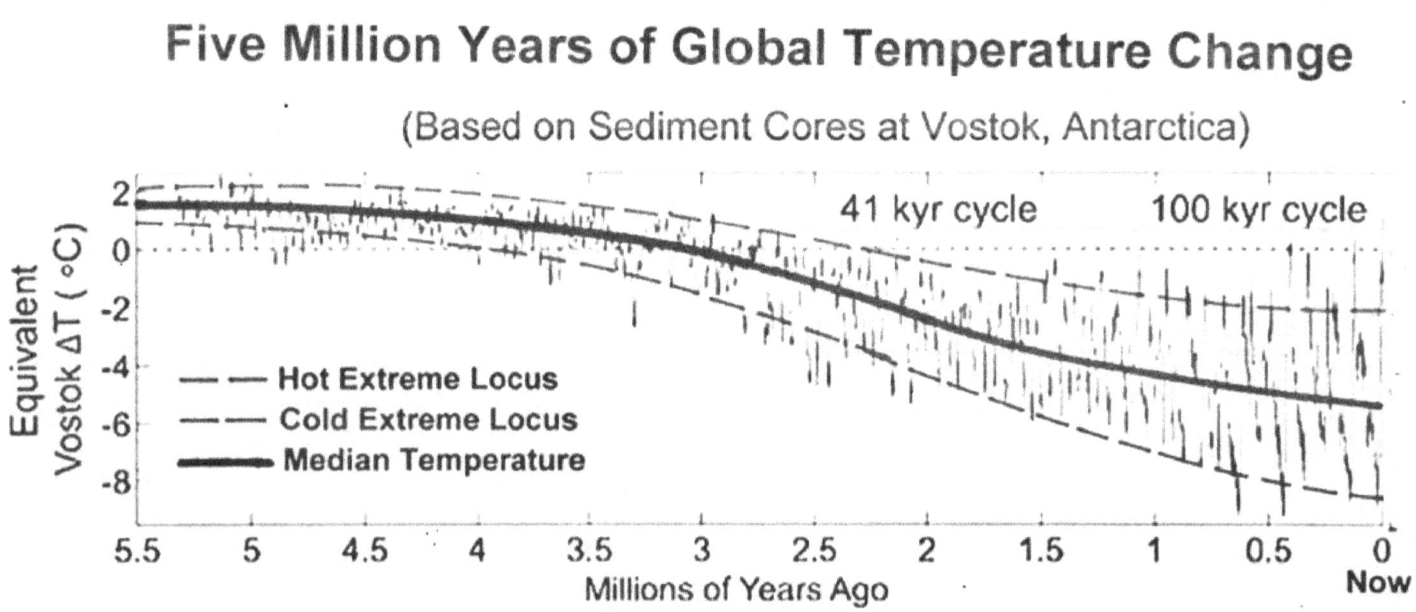

(Author’s Alternate “Big Picture” analysis of Vostok Ice Core Database)

Environmental scientists seem to be prone to “averaging” and “linearizing” cyclic type data with an objective of developing a “trend” from which a model can be created and used for predicting what the variable that is linearized and averaged portends for the future of that variable. If that was a valid statistical analysis technique (which the Author contends it not to be valid statistically), the trend of the average temperature at the Vostok station would be -5.5 degrees Centigrade cooler than "normal" for the entire database of about 400,000 years.

The above Cornucopian-shaped database being analyzed with a locus of extremes (the "mountain peaks" and the "canyon floors") and the curvilinear median half-way between those extremes illustrates again why averaging and linearization, as used by environmental scientists, and the use of only small portions of an enormous database to prove the validity of an idealistic hypothesis (to focus on the need to phase-out the practice of Man burning fossil fuels in order to alter the environment to be less hostile) misses the more critical need for Man to do whatever is necessary to survive the extremes – not the averages - that nature causes to be changing with time. It is relatively easier for Man to adapt to living on a planet that is gradually cooling as the energy balance of heating inputs (solar insolation) varies cyclically, but the steady loss of internal energy stored when the planet was born through terra firma and into the atmosphere offsets the solar insolation. The median line of the extremes of the loci (plural of locus) shows that the planet is gradually cooling with eons of time and in a Cornucopian shaped manner. There was a lower rate of cooling at first as the terra firma was being added to the planet and insulating the molten core from losing heat energy to the extremely cold atmosphere. The rate of cooling accelerated as both the terra firma and the "thickness" of the atmosphere slowed to form the planet in its near final configuration (diameter of the spherical terra firma and thickness of its surrounding atmosphere) to what they are now.

Contrary to the median temperature line having a Cornucopian (reflex curvature) general form, the extremes (the outer locus envelopes) are becoming larger (the top of the mountains are getting colder than the median and the floors in the valleys are getting hotter than the median). The time periods between the peak highs and lows are also apparently longer and may be due to the fact it takes longer for nature seeking equilibrium to go from a hotter peak to a cooler valley floor. The median line may be thought of as being the equilibrium temperature and the extremes of the peak highs and valley floors signal a need for the planet's temperature to return to the equilibrium condition. That may be the best lay description that can be criticized by scientists for being too simplistic, and possibly too "nerdy" by the public. The only simpler analogy would be to that of the human body which auto-regulates its temperature to be closer to a normal 98.6 degrees Fahrenheit (37 degrees Centigrade) if the person has a "fever" or if the person has been "out playing in the snow" too long.

It is hoped by the Author that Readers are beginning to recognize that conclusions drawn with the use of only a portion of any single database – and ignoring (intentionally or unintentionally) whether or not that database is complete or is the most current – can drive large groups of scientists to cause the news media and lobbyists to pressure their representatives in government to mandate programs and activities that are not totally wasteful but would be more valuable if a slightly different direction the "corrective action" were to take. The other "side of the coin" is that incomplete and/or outdated databases not containing the full set of knowns and unknowns about a problem also affects a valid analysis of the total problem. The databases analyzed by the Author and being presented in this chapter do not provide a clear explanation of how much effect that the human population could factually have on global warming, climate change, and/or on oceans levels rising by reducing or ideally eliminating "greenhouse gas" emissions.

To further illustrate that deciding what is the best course for humans to follow based on the known information, it would be unwise to state that the most current "big picture" database of information for what is known about global warming, climate change, and ocean levels rising is preferable than only considering outdated or incomplete databases. This chapter thus far has presented the highlights of four databases – the Arrhenius database of the early 1900s, the Milankovic Cycles database of the 1930s, the Keeling database from 1960 to the present, and the Vostok ice core database from 800,000 years in the past to the present. Thus far, the Milankovic Cycles database is known to contain only the astrophysical effects on surface temperature without considering the natural effects, atmospheric effects, or human effects on that surface temperature. The Arrhenius and Keeling databases define only the effect of carbon dioxide on the atmosphere without considering any geophysical effects and do not differentiate what effects the man-made and naturally generated carbon dioxide have on the atmosphere.

The Vostok ice core database presents the total resultant effects (astrophysical, natural, geophysical, and human when present) on surface temperature and on carbon dioxide and methane in the atmosphere, but is not useful for determining how much is due to which factors. The Author added some human presence information to the Vostok Ice Core database presence which might also add to understanding of the effect humans and their man-caused and industrial "greenhouse gas" emissions have caused to the extent of surface temperature as captured in the thick ice sheet on the Antarctic continent.

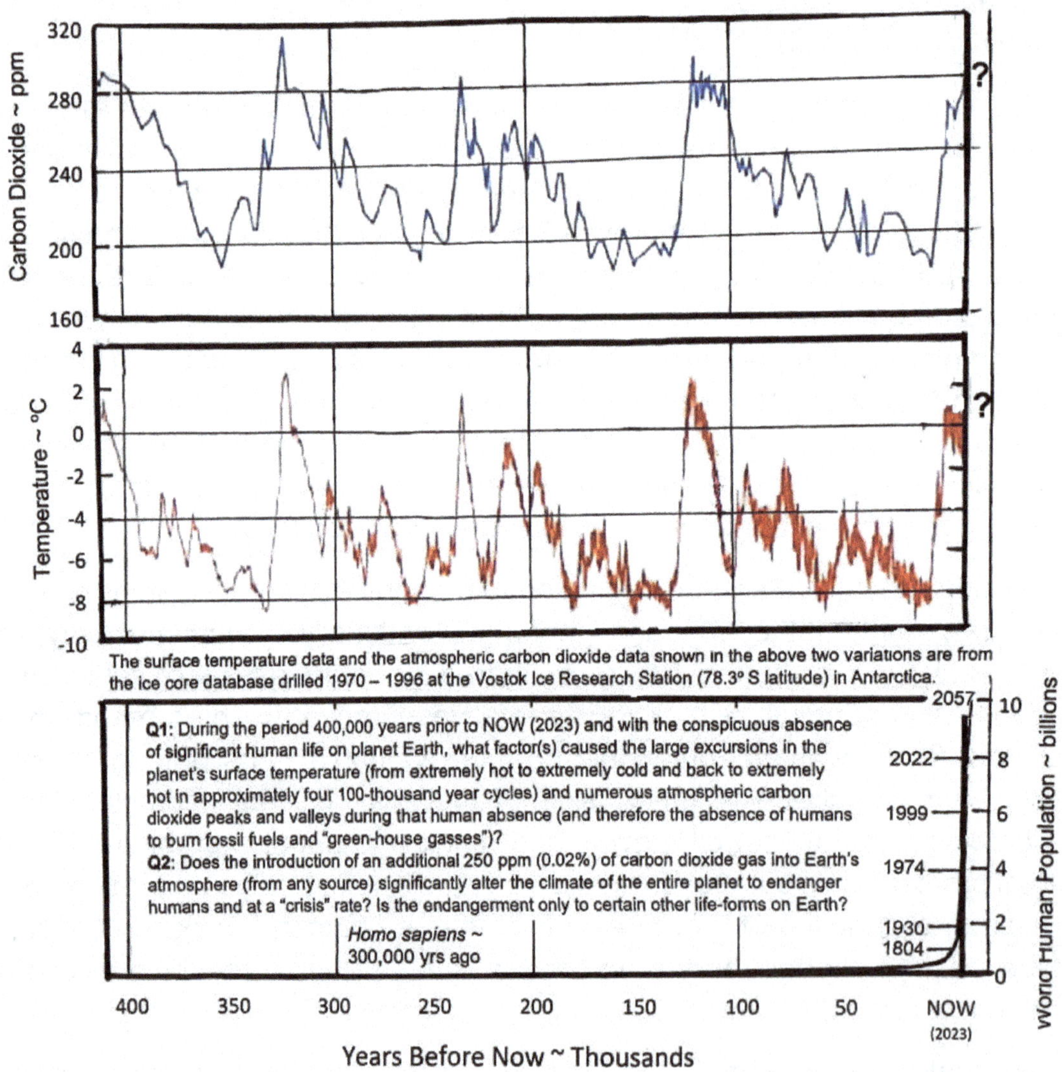

(Source: Vostok Ice Core Research Station with Author additions)

It can be seen in the above information that adding the exponentially rapid growth of human population on the planet has significantly increased only from about 15,000 years ago and first reached the 1 billion population count in 1804 when the Industrial Revolution also was in its early rapid growth and production of human-related products. If the ice core database had not become available or only had reached the depth corresponding to 50,000 years ago, the data might suggest to an analyst that the rapid rise in temperature was in fact caused by the onset of the Industrial Revolution when the world population was less than 1.5 billion humans. (The world population in 2023 was 8 billion).

From the full database of ice core data, however, it is obvious that there are three prior temperature peaks (with approximately an average of 110,000 years between those three peaks) and where the temperature level peaked out at being 2 degrees Centigrade above the average annual temperature (-55 degrees Centigrade; -67 degrees Fahrenheit)) for that latitude (-78.5 degrees South). The third of those temperature peaks occurred about 120,000 years prior to now (2023), and therefore the trend of the three prior temperature peaks suggests that the temperature peak occurring now is in keeping with that past peak temperature trend starting 400,000 years ago.

Weather tracking, navigation, and communication (and several other special purpose for the military) satellites came into being as a result of the NASA Man to the Moon thrust in space technologies announced by President Kennedy in May 1961. The NASA Man in Space program was most educational to the general public to begin to understand how vast the solar system is, and for the public to appreciate the wonderful (but not perfect) environment humans enjoy on Earth compared to what the environment on the Moon is now known to be based on scientific exploration by remotely-operated exploration vehicles and in-person by nine heroic astronauts.

(Author's note: When astronaut John Young jumped around on the Moon to show its lower gravitational acceleration field was a personally amusing event to the Author, due to John Young having attended Georgia Tech undergraduate school in Aeronautical Engineering and attended some classes with the Author. He graduated in class of 1952 and the Author graduated in the class of 1953. The total absence of water on the Moon's visible surface was most striking to this Author; however, the photographs from above the planet and looking back through the atmosphere and seeing one half of the entire surface of the planet in one "big picture" with its blue water background and with the white cloud formations was awesome. One of those photos has been in the Author's files – courtesy of John Young – and has been shown earlier. It is so awesome that (at the risk of being redundant) it is shown again next as an indicator of how valuable the NASA database is to the understanding of the "big picture" of global warming, climate change, and ocean levels rising.)

In a side-by-side presentation, the other hemisphere is shown and illustrates the vastness of the planet to have arid desert regions.

Courtesy of NASA (Astronaut John Young, L; NASA Earth Resources database, R)

A database chart from the NASA Earth Resources Program database is shown next to illustrate the conclusions that can be drawn differently from a small portion of the "big picture" and using a "linearization" technique that masks the cyclic variations in the data as shown in the Milankovic and Vostok Ice Core databases:

Global Satellite Temperatures, 1979-1997

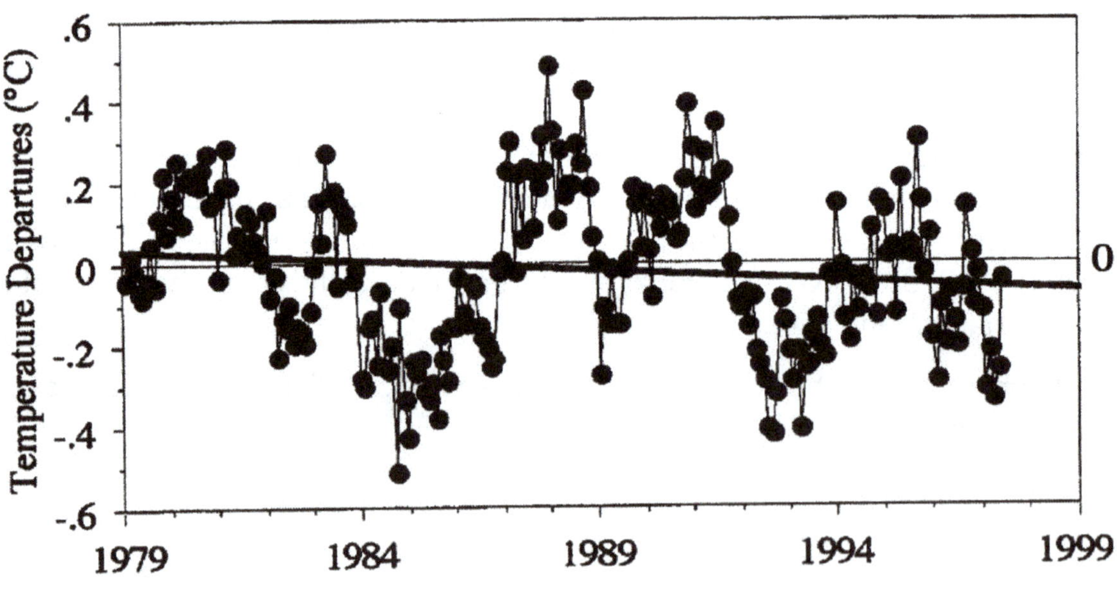

(Source: NASA Earth Resources database)

An immediate observation a credible analyst would make is that the temperature (labeled by NASA as a "departure") is apparently cyclic in nature and also the temperature anomaly is accurately measured to have less than 0.4 degree (warmer and cooler) than normal. At what latitude the data was measured is not shown and that critical piece of information is missing (intentionally or intentionally). The conclusion many analysts could draw from the above chart (after a "linearization" is made) would be based on the solid black line having a slightly decreasing slope of temperature with time, and that the global temperature anomaly based on satellite data is that a slight decreasing trend in global temperature is shown.

However, that satellite-based conclusion is not supportive of the environmental "settled science" fallacious proclamation that global temperature is rising due to Man burning fossil fuels and producing "greenhouse gas emissions" at a crisis rate. Environmental scientists have not changed their "settled science" database obtained with the now outdated temperature departure database obtained with the "one size fits all" and "linearized" database obtained with the averaging of average annual temperature from about 2,000 land-based meteorological measurements stations (based near large populated areas and many at airports) collected in almost the entire twentieth century (1900 – 1996):

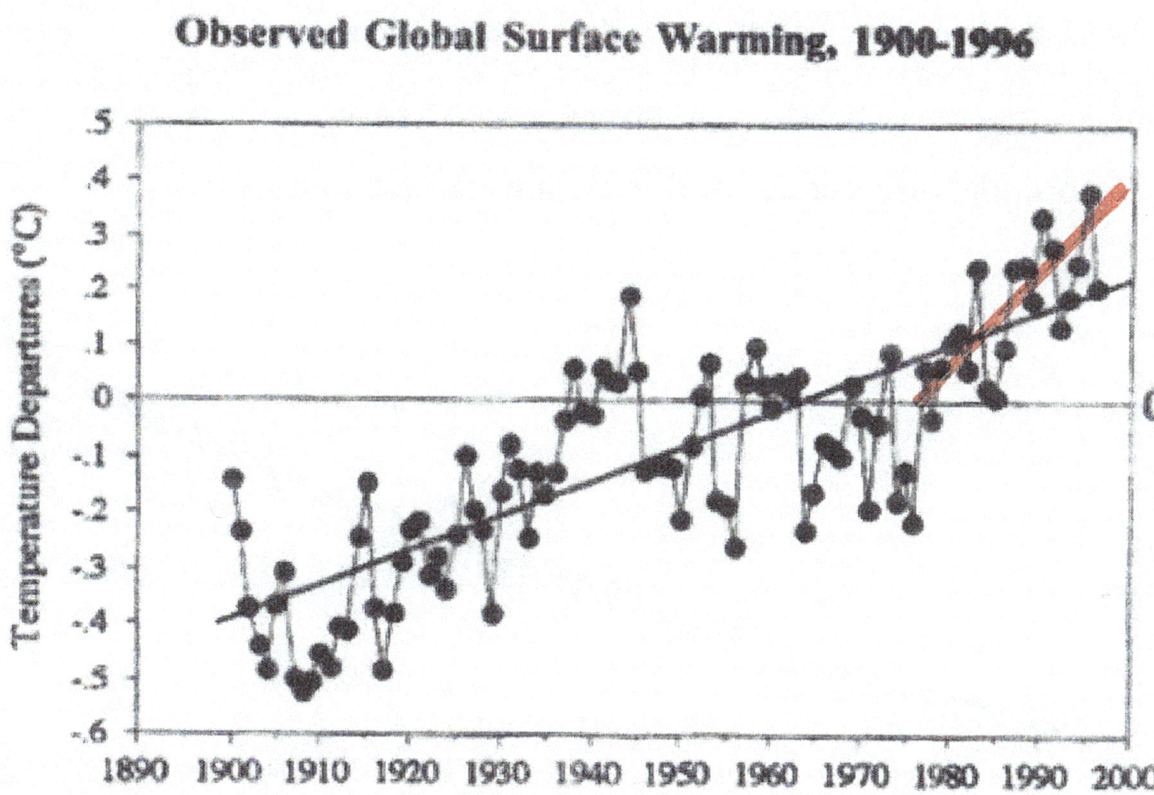

(Source: Data in black from NOAA databases, data in red by Author)

This chart may be recognized as having been shown and discussed earlier in this chapter but it was only the portion from 1900 to 1980 that was shown earlier. A point to be noted is that despite the NASA satellite global temperature shown earlier for the period of years 1979 to 1997 having a slightly cooling trend (a 0.1 degrees Centigrade decrease in 18 years), the NOAA land-based meteorological database shown above for that same period 1979 to 1997 indicates either the linearization (black line) of a warming trend (0.1 degrees Centigrade increase in the same 18 years) or an even greater warming trend (0.4 degrees Centigrade in 18 years using the red line linearization of the NOAA data for only that 18 year period from 1979 to 1997. Neither the NOAA nor the NASA satellite databases show either a warming or cooling trend that is of a magnitude to be considered at a "crisis" rate to human life. The comment made earlier is repeated that the environmental protection agencies in a government are chartered to protect the environment not the human population. Ocean floor anemones on the Great

Barrier Reef off Australia may bleach white with a two degrees rise in temperature whereas a two degrees rise in temperature over a span of 25 years is largely unnoticeable to the average member of the public.

With the reasoning that Man burning fossil fuels is a relatively constant addition of "greenhouse gasses" into the atmosphere, the cyclical variations in the data would infer that seasonal or the less rapidly-varying Milankovic cyclical astrophysical factors were partially manifested in the cyclical data in that relatively short period of 18 years. The major conclusion to be drawn professionally is that the forces of nature come into effect to restore temperature excursions from uncontrollable energy imbalances to an equilibrium status whenever it is caused to be anomalistic from "normal" (i.e., in equilibrium) by large weather variations. Since the climate of a region is based on a time period of about 50 years, the temperature cyclical changes during an 18-year time span cannot be directly attributed to climate change.

Updates were/are also continually being made to the previously discussed Keeling Curve chart and the most recent update included in this book is the September 22, 2023, measurement of the concentration of carbon dioxide in the atmosphere being 421 ppmv as shown next:

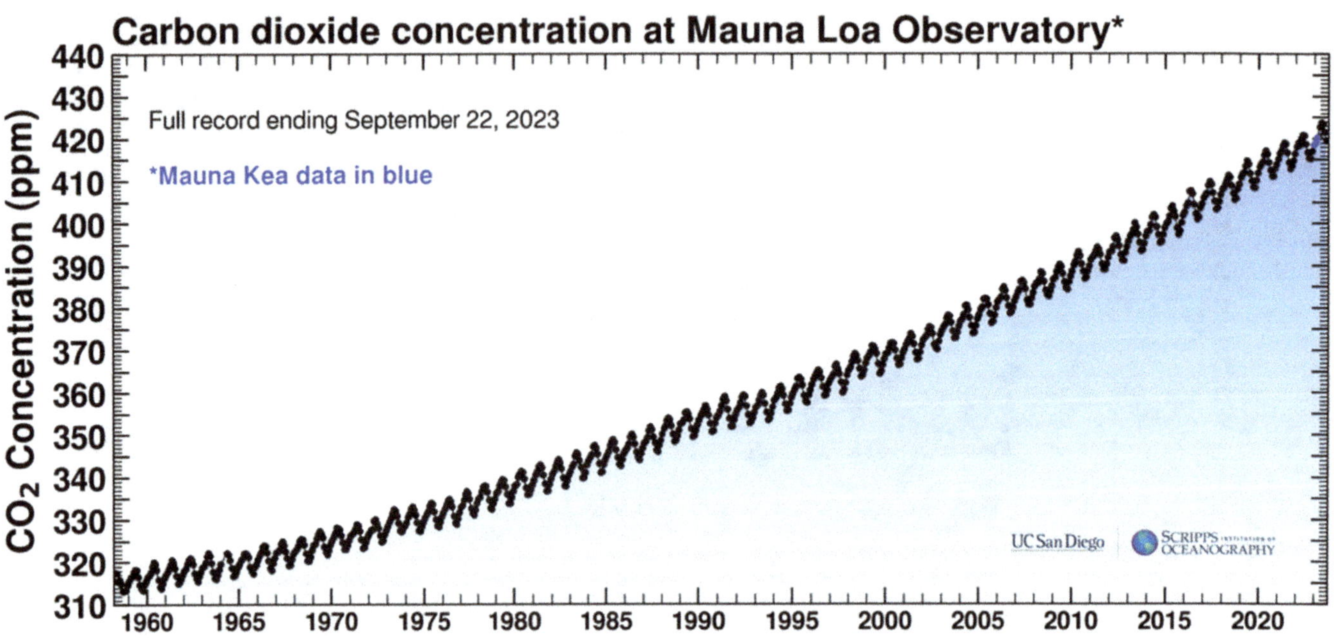

(Courtesy of NOAA - and the Scripps Institution of Oceanography?)

Readers may note that the slope of the NOAA Curve has the appearance of having decreased from the original version to this most recent extended version, but that slope change is only an optical illusion due to the scale at the left being reduced and the time scale at the bottom being increased. (That "scale setting" to emphasize or exaggerate statistics is referred to by graphical statistical analysts as "chartsmanship." In a similar manner, choosing to use a scalar dimension to emphasize or deemphasize the importance of a variable is an all-too-common statistical analysis deceptive practice. In scientific laboratory experiments when dealing with very small concentrations of a chemical or mineral, the number of parts per million is used rather than a percent. The Keeling Curve published by NOAA for the Public to see that the slope of the carbon dioxide in the atmosphere starts at 318 ppmv in 1960 is very steep and non-linear.

It was shown earlier that presenting that same ppmv measurement in a percentage of concentration (0.0318%) does not place much emphasis on the amount of carbon dioxide in the atmosphere. In 2000, the measurement of carbon dioxide in the atmosphere had risen to 368.4 ppmv (0.0368%) as seen on the Keeling Curve chart above, and environmental scientists escalated their concern about anthropogenic "greenhouse gasses" to be approaching a "tipping point" and there being a "crisis" for the future of the planet. When the same carbon dioxide measurements made at the Mauna Loa Observatory were graphically analyzed in a percent form rather than the exaggerated small scientific concentration of "parts per million by volume", the environmental scientists claim of a "crisis" and approaching a "tipping point" in the near-future lacks credibility. Two statistical methods were used in generating the extrapolation of the Mauna Loa carbon dioxide concentration in the atmosphere – one is that of "linearizing the database" which environmental scientists use in order to determine a trend, and the other being an "exponential" extension of the database to better match the slight curvature upward with time as shown in the exaggerated Mauna Loa database. That less-"chartsmanship" chart is shown next for the projection of the Mauna Loa carbon dioxide in the atmosphere database from the present (2023) to the year 2200:

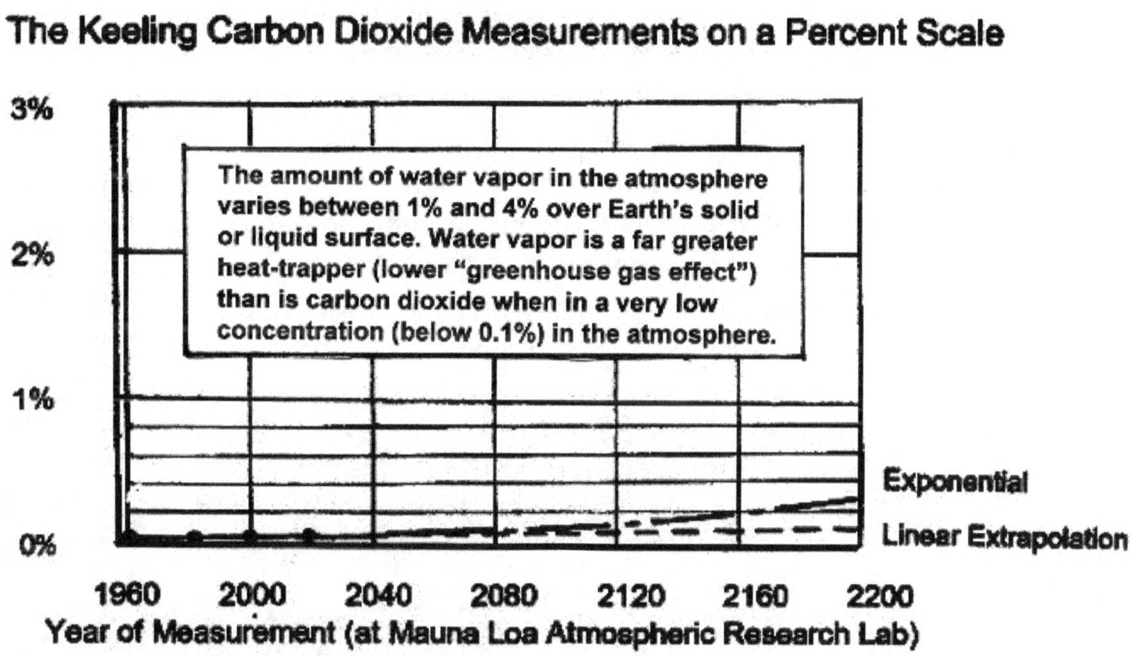

That level of projected incease of 0.28% of carbon dioxide in the atmosphere in 2200 – of which Man-made emissions would constitute 1/4th of that 0.28% (i.e., 0.07%) – seems unlikely that the small increase in carbon dioxide concentration that is Man-made is of sufficiently high significance to cause a system safety engineer to become concerned about as being an apocalyptic safety hazard to humans alive in 2200. It clearly might be considered by environmental scientists to be a disaster to land-based animal and plant life and aquatic-based ocean lifeforms in the natural environment. For it to be considered an "crisis" to humans needing to be brought under control by 2050 does not appear to warrant extreme penalties being imposed on users who convert from the total dependency on coal to a temporary dependency on oil and natural gas as fossil fuels are phased out and alternative emission-free fuels are phased in. When all databases (astrophysics past and present, Milankovic Cycles of 1902, the Vostok Ice Core database of the mid-twentieth century, the Keeling Mauna Loa atmospheric measurements from 1960 to 1990, and the preponderance of NASA satellite-based measurements of

Earth's resources coming into the combined database in the "big picture") are considered, the environmental scientists' (mostly in NOAA and some in NASA) hypothesis that "settled science indicates that Man burning fossil fuels is the dominant cause of global warming, climate change, and ocean levels rising" is no longer valid (if it ever was).

The Vostok Ice Core database indicates at the most current useable portion of the ice cores that there was at least 290 ppmv (0.029%) of carbon dioxide that is of natural source at the start of the Industrial Revolution. That indicates from the Keeling database most recent measurement of carbon dioxide to be a total of 415 ppmv (0.0415%), which further indicates that Man and Industrial Revolution carbon dioxide is less than 30% (415 – 290+) / 415) of the total 415 ppmv and the other natural and geophysical portion is greater than 70% (290+ / of the total 415 ppmv. Any "crisis" due to carbon dioxide in the atmosphere measured to be of unprecedented levels is due more to astrophysical (Milankovic factors), geophysical (volcanic eruptions), and natural (deforestation and wildfires initiated by lightning strikes) than by Man and Industrial burning of fossil fuels.

A more detailed description of the effects of water vapor will be discussed in a later chapter of this book, but suffice it for this first chapter evaluating the attributes of several valid databases to "jump right to the chase". The amount of water vapor in the atmosphere contributes directly to the meteorological database variable commonly known as "humidity". When the outside air temperature is elevated to about 80 degrees Fahrenheit (27 degrees Celsius/Centigrade) and the air has a low humidity, a human will be able to tolerate that combination in reasonable comfort. Conversely, anyone who has been in New York City or Washington, D.C., on a hot humid day will agree that when the temperature is above 85 degrees Fahrenheit and the humidity is about the same number in percent (e.g., 85%), it feels "hot as Hell". The National Weather Service has generated a "Heat Index" – a form of database - which tabulates the combination of temperature and humidity and the level of danger to human health and life at extreme combinations. That Heat Index is shown next:

NWS Heat Index — Temperature (°F)

Relative Humidity (%)	80	82	84	86	88	90	92	94	96	98	100	102	104	106	108	110
40	80	81	83	85	88	91	94	97	101	105	109	114	119	124	130	136
45	80	82	84	87	89	93	96	100	104	109	114	119	124	130	137	
50	81	83	85	88	91	95	99	103	108	113	118	124	131	137		
55	81	84	86	89	93	97	101	106	112	117	124	130	137			
60	82	84	88	91	95	100	105	110	116	123	129	137				
65	82	85	89	93	98	103	108	114	121	128	136					
70	83	86	90	95	100	105	112	119	126	134						
75	84	88	92	97	103	109	116	124	132							
80	84	89	94	100	106	113	121	129								
85	85	90	96	102	110	117	126	135								
90	86	91	98	105	113	122	131									
95	86	93	100	108	117	127										
100	87	95	103	112	121	132										

Likelihood of Heat Disorders with Prolonged Exposure or Strenuous Activity

Caution Extreme Caution Danger Extreme Danger

(Courtesy of the National Weather Service, NOAA)

The likelihood is low that global warming will ever get to "average global temperature" being hotter than average human body temperature (98.6 degrees Fahrenheit = 37 degrees Centigrade/Celsius), but there are regions of the planet (in the Equatorial climatic zone where there is commonly a high temperature and high humidity especially after a rain event has occurred that approaches dangerous health hazards to humans and animals. Not to belabor the point, but environmental scientists cannot summarily disregard the Heat Index cautions and/or water vapor being present in the atmosphere. A two-degrees rise in "average global temperature" may be important to prevent a coral reef from "bleaching out" but is insignificant to humans when compared to the danger posed when a Heat Index condition in the red and dark red color in the chart above exists.

Once again, there are regions of the planet (even in the United States) where Heat Index can be in the "danger" category but there is almost zero probability that anywhere in the polar regions will a dangerous combination of temperatures and humidity to be of concern to human inhabitants. Nevertheless, there are regions of the United States that have life-threatening conditions that frequently exist in the winter months of the seasons. It is a repeated theme of this book that humans must exist at the extremes of temperature – hot and cold. The environmental scientists have caused such an uproar in government about global warming that the public is becoming numb to the "cry wolf" tactics on hot environmental extremes and are not being prepared for the cold environmental extremes. The National Weather Service does publish the "Wind Chill Index" in the northern states on occasion but newspaper coverage in the southern states does not publish any sensationalism about hypothermia as they do hyperthermia. For completeness, the public should be aware of the hazards of the cold extremes:

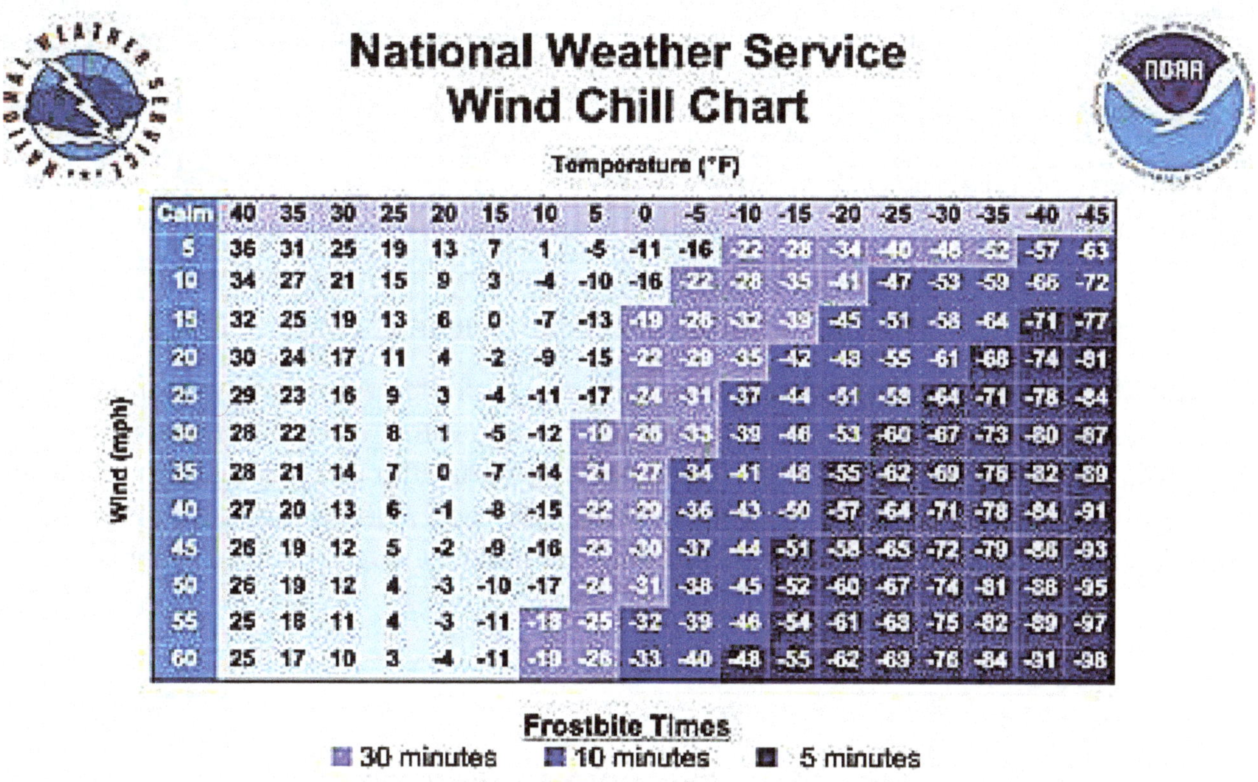
National Weather Service
Wind Chill Chart

Temperature (°F); Wind (mph)

Calm	40	35	30	25	20	15	10	5	0	-5	-10	-15	-20	-25	-30	-35	-40	-45
5	36	31	25	19	13	7	1	-5	-11	-16	-22	-28	-34	-40	-46	-52	-57	-63
10	34	27	21	15	9	3	-4	-10	-16	-22	-28	-35	-41	-47	-53	-59	-66	-72
15	32	25	19	13	6	0	-7	-13	-19	-26	-32	-39	-45	-51	-58	-64	-71	-77
20	30	24	17	11	4	-2	-9	-15	-22	-29	-35	-42	-48	-55	-61	-68	-74	-81
25	29	23	16	9	3	-4	-11	-17	-24	-31	-37	-44	-51	-58	-64	-71	-78	-84
30	28	22	15	8	1	-5	-12	-19	-26	-33	-39	-46	-53	-60	-67	-73	-80	-87
35	28	21	14	7	0	-7	-14	-21	-27	-34	-41	-48	-55	-62	-69	-76	-82	-89
40	27	20	13	6	-1	-8	-15	-22	-29	-36	-43	-50	-57	-64	-71	-78	-84	-91
45	26	19	12	5	-2	-9	-16	-23	-30	-37	-44	-51	-58	-65	-72	-79	-86	-93
50	26	19	12	4	-3	-10	-17	-24	-31	-38	-45	-52	-60	-67	-74	-81	-88	-95
55	25	18	11	4	-3	-11	-18	-25	-32	-39	-46	-54	-61	-68	-75	-82	-89	-97
60	25	17	10	3	-4	-11	-19	-26	-33	-40	-48	-55	-62	-69	-76	-84	-91	-98

(Source: National Weather Service)

Due to those known databases identified above being only partially considered or not being considered at all, current generation environmental scientists have not yet factually identified all of the life forms -

human and non-human – that are being endangered and specifically how. Nor have scientists factually enumerated <u>why</u> the changing environment observed in the past 200 years is considered to be a "crisis". The tactic of identifying any problem as a crisis when factually there is no crisis to get attention to perhaps a factual non-crisis problem is well known. Humans are taught in the first grades of elementary school not to "cry wolf" – and later not to cry out "FIRE" in a public place when there is no fire.

Environmental scientists resorting to claiming the current environmental changes as being a "crisis" without defining what the crisis could bring and to whom must not remember from their very early safety education - or is it that they (and the sensationalized news media) have since learned - that fear in the public mind is a powerful psychological weapon. When anything is repeated many times and for a long time, that "thing" is accepted as being factually correct. An article appeared (June 2022) in a very reputable news magazine published weekly to keep the public informed on what is happening in the world almost real time. It typifies what the public has accepted as being factual:

Health & Science NEWS 21

CO_2 levels the highest in human history

Humans have never lived in a world with as much carbon dioxide in the atmosphere as they are experiencing now, reports *USA Today*. New measurements by the National Oceanic and Atmospheric Administration show that the average level of the greenhouse gas is more than 50 percent higher than in pre-industrial times and the highest in at least 4 million years. Before the Industrial Revolution, CO_2 levels had hovered around 280 parts per million for thousands of years; this May, they hit a peak of 421 ppm. The cause is the burning of fossil fuels, NOAA scientists say, and the result is worsening climate change. Earth's temperature has jumped two-thirds of a degree Fahrenheit in just the past 20 years, fueling extreme weather, including longer and more severe droughts, wildfires, flooding, and tropical storms. To stave off a tipping point that could leave parts of the planet inhospitable to humans, CO_2 levels need to start falling. Instead, despite international pledges to cut emissions, the levels are rising. Humans pumped a record 36.3 billion tons of CO_2 into the atmosphere last year and are on track to best that record this year. "It's depressing that we've lacked the collective willpower to slow the relentless rise in CO_2," says geochemist Ralph Keeling of the Scripps Institution of Oceanography. "Fossil fuel use may no longer be accelerating, but we are still racing at top speed toward a global catastrophe."

Burning coal in Zhangjiakou, China

(Source: The Week magazine, June 2022)

The emissions from the chimnies ("smokestacks") shown in photos accompanying most "greenhouse gas" news media articles are from fossil-fuel burning facilities within the US but are mostly steam and water vapor with controlled lesser amounts of harmful emissions after passing through"scrubbers", heat exchangers, and filters mandated by Environmental Protection Agency (federal, state, and county) control regulations. A diagram shown next illustrates that emissions reduction apparatus:

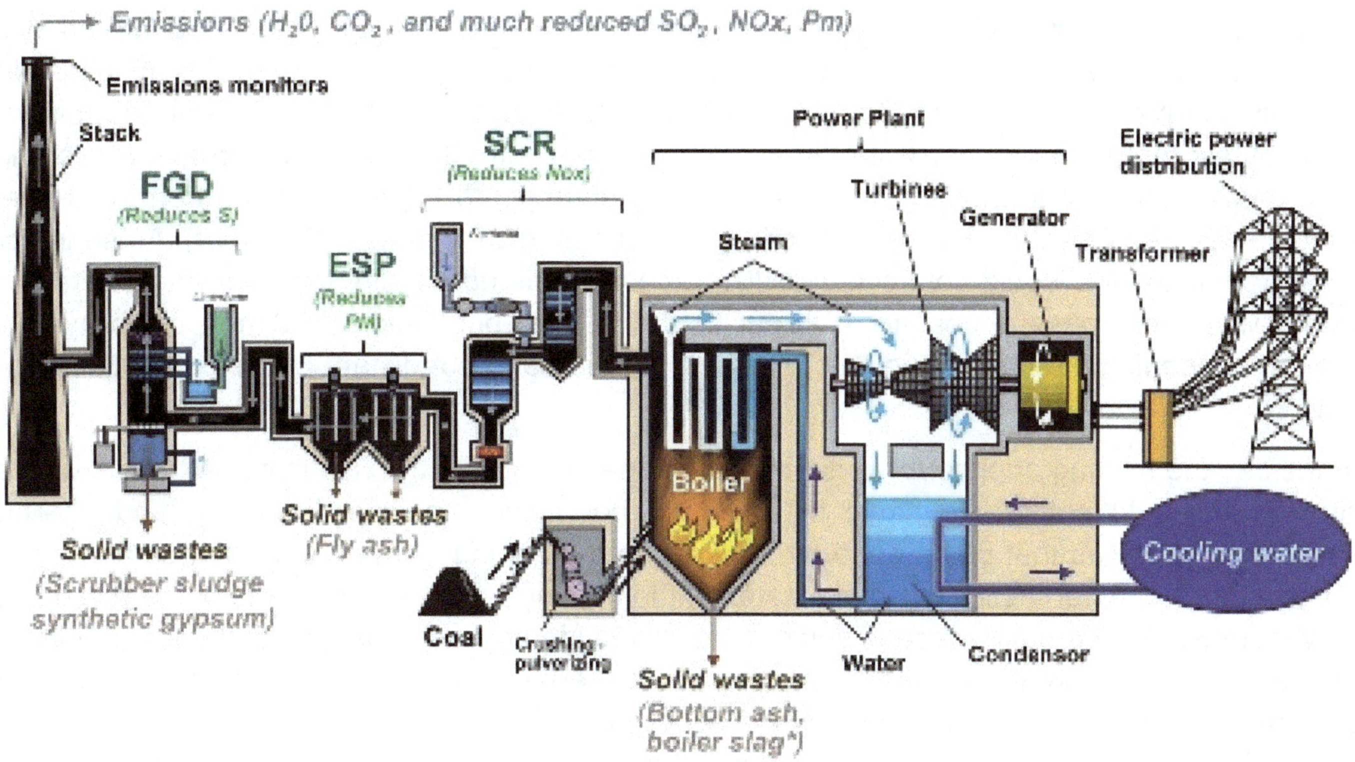

(Source: Coal Institute of America)

There is no claim that all of the heat-trapping carbon dioxide emissions are removed by the emissions control apparatus but the apparatus (when maintained in fully functional operation) meets the EPA standards for reduction of the "greenhouse gasses". There are no coal-burning facilities currently in operation in the USA that either have the above emission controls apparatus in place or have converted from coal-burning to oil or natural gas burning for producing electrical heat energy. It is worth mentioning that the Consolidated Edison Company of New York has converted their coal-burning facilities to oil-burning and also uses the steam that has passed through the turbines generating electricity for the New York City region and the adjacent Brooklyn Navy Yard to be routed through steam tunnels servicing the steam heating of over 80% of the buildings on the Manhattan Island portion of Greater New York City during the cool/cold months of the year.

The world public should be very cautious – and skeptical - when environmental scientists claim that Man burning fossil fuels is the dominant cause for carbon dioxide in the atmosphere. When a carbon dioxide measurement at the Mauna Loa Atmospheric Research Laboratory is reported being 420 ppmv or 430 ppmv, that total measured is not all due to humans and/or industry burning fossil fuels. Most of it (about 295 ppmv) is caused by astrophysical, geophysical, and natural factors over which Man has no control. No efforts to reduce the carbon dioxide emissions generated by humans burning fossil fuels can cause the amount of carbon dioxide in the atmosphere to be less than that due to astrophysics, geophysics, and/or natural factors.

The third aspect of the global warming, climate changing, ocean levels rising trifecta concern is that of the ocean levels rising. The level of the ocean surface in a long-term aspect is caused by one or both of two primary factors: 1) the addition of large amounts of water from on land; and 2) the temperature of the water increased by some means other than the known irradiation of heat energy from the Sun, and causing the ocean water to expand. Factor 1) will be discussed first.

The greatest concern of environmental scientists regarding ocean levels rising is centered on the melting of the ice in the ice sheets on Greenland and on the Antarctic Continent. According to the NASA sponsored National Snow and Ice Data Center (NSIDC) database, the Antarctic and Greenland Ice Sheets contain more than 99 percent of freshwater ice on Earth. If both of those massive ice sheets were to totally transform from solid state (ice) to liquid state (water) due to melting from heat energy, they would raise sea level on Earth's surface by an estimated 67.4 meters (223 feet). A rise in ocean level of that magnitude would not occur rapidly – and perhaps not at all if the melt rate ceases - and therefore would not pose a near term threat to the human population, but the possibility of that cataclysm happening at all bears being considered for the safety of distant future populations on the planet.

Inasmuch as the melting of either/both is not likely to be very rapid, the rise in ocean levels would be gradual and should not create panic or angst in the public. Analyses must be and are being accomplished (such as that being done by NSIDC), and planning to avert that total cataclysm must begin now to account for the intermediate states of the melting in lower increments of the ocean levels rise. What must also be done – and is being done – is to identify the cause(s) for the ice sheets to be melting at what rate, and to estimate if that rate and its causes are affecting the rising ocean surface level hazards to humans and all aspects of the ecosystem. If it is any human-caused factor adding to the otherwise natural-caused melting rate, common sense is that those human-caused factors must be controlled to reduce the effect of even near-term small rises (greater than 5 feet) in ocean surface level.

The melting of sea ice (floating ice) in the Arctic was highlighted by former Vice President Al Gore in his documentary film "An Inconvenient Truth" and as it affected the food cycle of the animal and aquatic life forms in the Arctic Ocean. Melting of ice floating on a body of water does not increase the surface level height as the laws of physics dictate that floating ice does not increase in volume as it transforms from solid state to liquid state (except for the insignificant volume of air bubbles that are trapped in the ice when it was first formed. The concern about ice in the Arctic melting is that ice in the large thick ice sheet of Greenland.

Environmentalists persist to apply their fallacious hypothesis that Man burning fossil fuels is creating the "greenhouse gas effect" trapping heat energy below the greenhouse gas and causing a rise in temperature of everything on Earth's surface – in all state forms of solid, liquid, and gas. That hypothesis will be examined further in appropriate chapters of this book. Identifying and describing the geophysical aspects of the glaciers and the geophysical and natural factors causing their temperatures and melt rate to change is contained in this chapter.

The ice sheet on Greenland is enormous and its physical description is well defined by Brittanica. The Greenland ice sheet **is** the single ice sheet or glacier covering about 80 percent of the island of Greenland and the largest ice mass in the Northern Hemisphere. In volume, it contains 12 percent of the world's glacier ice, and, if it completely melted, sea level would rise almost 24 feet (7.4 meters). A satellite view of Greenland illustrates the extent to which the Greenland Ice Sheet covers its host terra firma and the amount of sea ice (floating) in the Arctic Ocean:

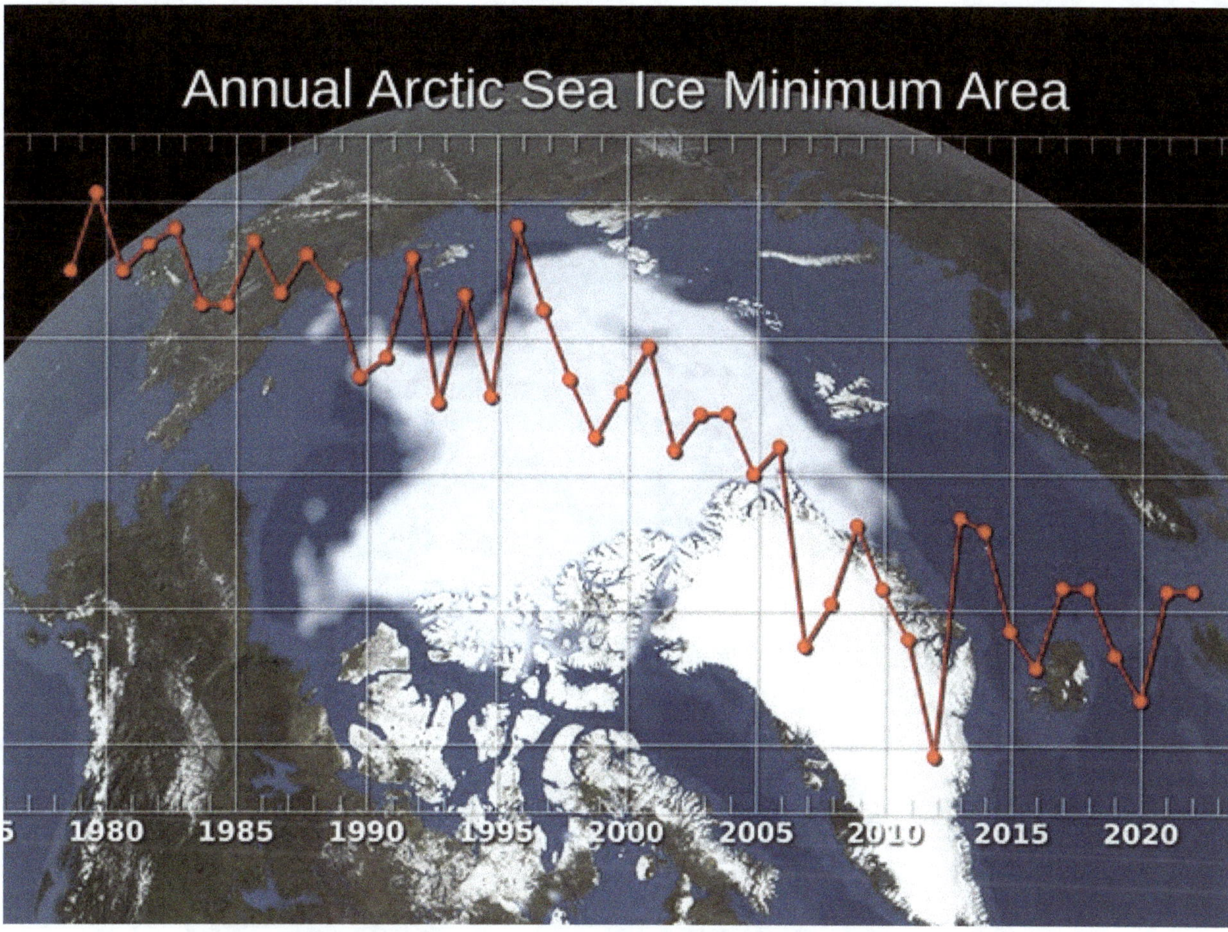

(Source: The National Snow and Ice Data Center database)

Globally, the ice sheet mass on Greenland is second in size to only the ice mass that covers Antarctica (in the Southern hemisphere). The Greenland Ice Sheet extends 1,380 miles (2,220 km) north-south, has a maximum width of 680 miles (1,094 kilometers) near its northern margin, and has an average thickness of about 4,920 feet (1,500 meters). The Greenland Ice Sheet extends from approximately 60° N to 80° N and thus is not in the polar zone, unlike the Antarctic Ice Sheet.

The Greenland Ice Sheet is protected by cold waters of mostly Arctic origin against the temperate Atlantic waters from the southeast. In volume it contains 12 percent of the world's glacier ice, and, if it completely melted, sea level would rise 24 feet (7.4 meters). The Greenland Ice Sheet is the single ice sheet or glacier covering about 80 percent of the island of Greenland and the largest ice mass in the Northern Hemisphere, globally second in size to only the ice mass that covers Antarctica.

Of much greater concern is (as it should be) the enormous Antarctic Ice Sheet of East Antarctica, and of secondary concern the ice shelves of West Antarctica.

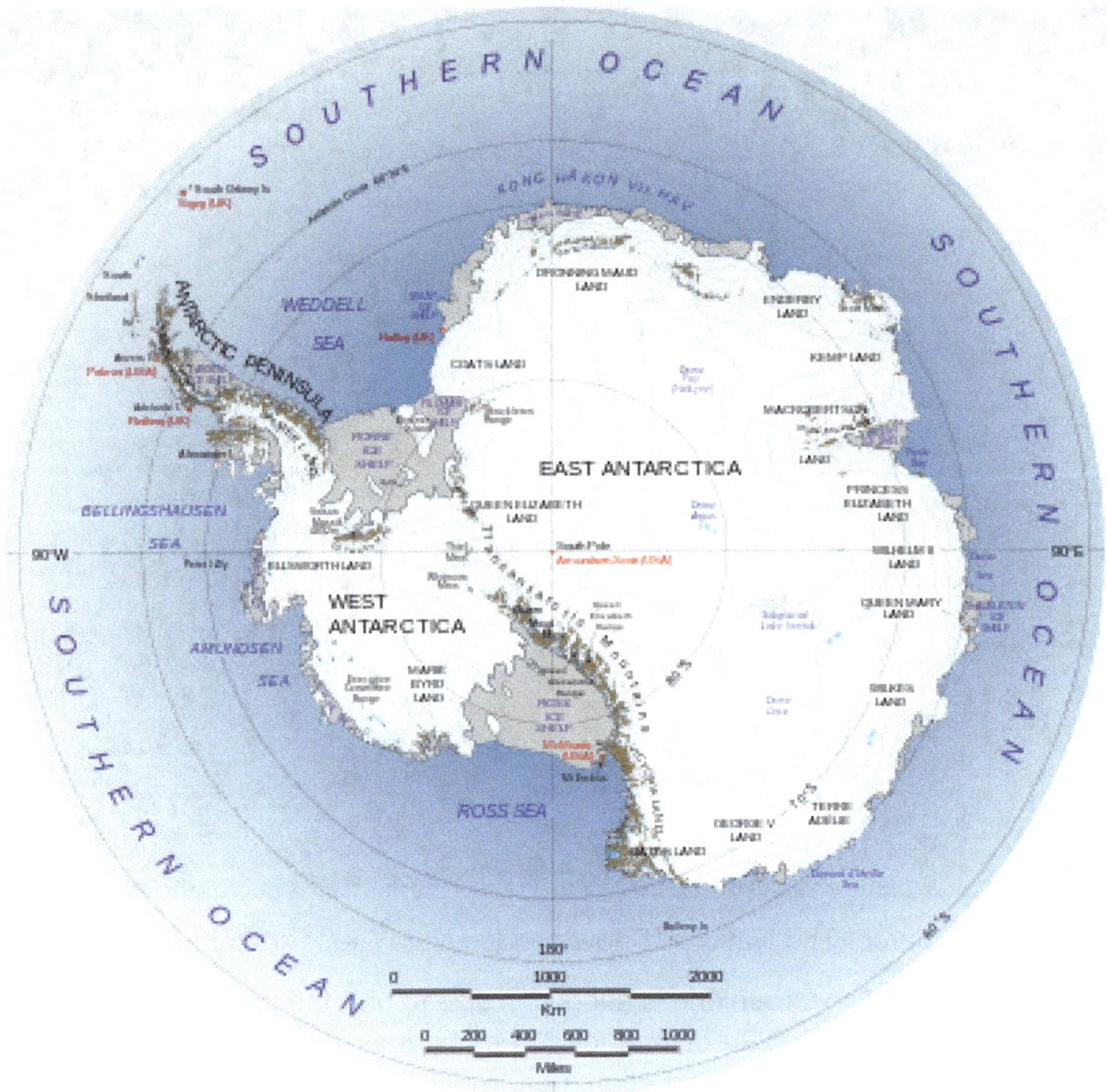

(Source: The National Geographic Society)

The next chapter provides further information on the knowns of the astrophysical factors affecting the environment on the planet and on the humans who are tenants co-sharing the planet with a myriad of other life-forms that are known to exist. The chapter shows the truly “big picture” of how the planet we stand on is actually a very small part of the solar system, and provides some indication of planet Earth not being as small (it is approximately spherical with an average diameter of 7,926 miles – 12,756 kilometers) and stationary (it rotates one revolution about a tilted wobbling axis in 24 hours and orbits the Sun on an irregular shape elliptical orbit every 365.24 days) as the public “sees” it on maps and diagrams on two-dimensional static pieces of paper.

Chapter 2
Astrophysical Factors Affecting the Planet

The Introduction and previous chapter analyzing several pertinent databases have led to the obvious conclusion that considering only Man burning fossil fuels is not the dominant or single factor affecting the weather, the climate, or the ocean surface levels to rise. The "big picture" approach is more tedious and time-consuming but is proven herein to have been well-worth the extra effort.

Many cultures and persons – scientists and a multitude of other "ists" - with many different backgrounds have "nibbled" at some small subset of the many different factors involved in analyzing all the forms of life on Earth. The more that has been and is being learned, the more is identified yet to be learned. A "tongue-in-cheek" expression is: "I turned the stone over to see what was under it, and found more stones to be overturned".

The ancient Egyptian culture (circa 3,000 BCE) came to believe that the Sun (observed to be so dominant on the Egyptian environment) was their God Ra who shined rays of light and warmth on everything on the land they lived on. At that time, humans had no concept of a solar system at which the Sun was at the center of the universe – or even that there was a universe. Egyptians recorded their observations on tablets and sculptures (colored with dyes extracted from crushed biological and mineral materials mixed with water), one of which is illustrated next indicating Rameses' observing his wife being bathed in sunrays from Ra.:

(Source: Encyclopedia Brittanica)

The human optical system is based on the human eye having a convex lens in each eye – two eyes for redundancy (in case a human loses one eye in an accident, and in that unfortunate circumstance, the redundant eye brought into use to avoid a lifetime in darkness). But it is assumed that most scientists have both eyes and have "normal" eyesight – in clarity and with depth perception - on which to make scientific observations. In street language vernacular, "What you see is what you get." The databases described in this book are all based on what humans (including scientists) have observed – with their eyes and with instruments and sensors aiding their eyes to see distant objects much farther or much closer than "normal". The state of the instruments and sensors generally improves with time and a brief summary of the optical technology and resulting observations and interpretations are included throughout the different databases described in this book.

Turning to the technologies of telescopic instruments assisting Man in identifying how planet Earth "fits" into the solar system and how its atmosphere changed with time, the earliest scientific knowledge that became known about what is in the skies surrounding the planet Earth was obtained by analyzing what was visually observed initially only with the human optical system. At one point in time, the only instrument Man had developed to estimate what time it is was the sun dial (with no moving parts) and observing how the shadow of the Sun moved on the face of the dial as time of day lapsed. On cloudy days, time lost its ability to be "measured", but that does not infer that time was not changing. It is somewhat indicative of why relying on solar energy panels being developed is less reliable because of the dependency on receiving and storing solar energy – but only when the Sun is not occluded.

Few records if any were kept of the temperature and how much precipitation on a daily, weekly, monthly or yearly basis. That too should not infer that because they were not measured that those variables were not changing and with some long-term pattern of change. The history of instruments to aid those observations made by the unaided eye of humans can only be briefly summarized for this book.

Reverting to the historic trail of optics at the time of the 2nd century *Anno Domini,* Claudius Ptolemy (c.100 AD – c. 179 AD) of ancient Greece wrote (in his work *"Optica"*) about the properties of light including reflection, refraction, and color. Collaborating with contemporary astronomers, mathematicians, and philosophers, Ptolemy generated the first known scientifically interpreted (from observations aided by primitive lenses and with mathematical calculations) diagram of the solar system:

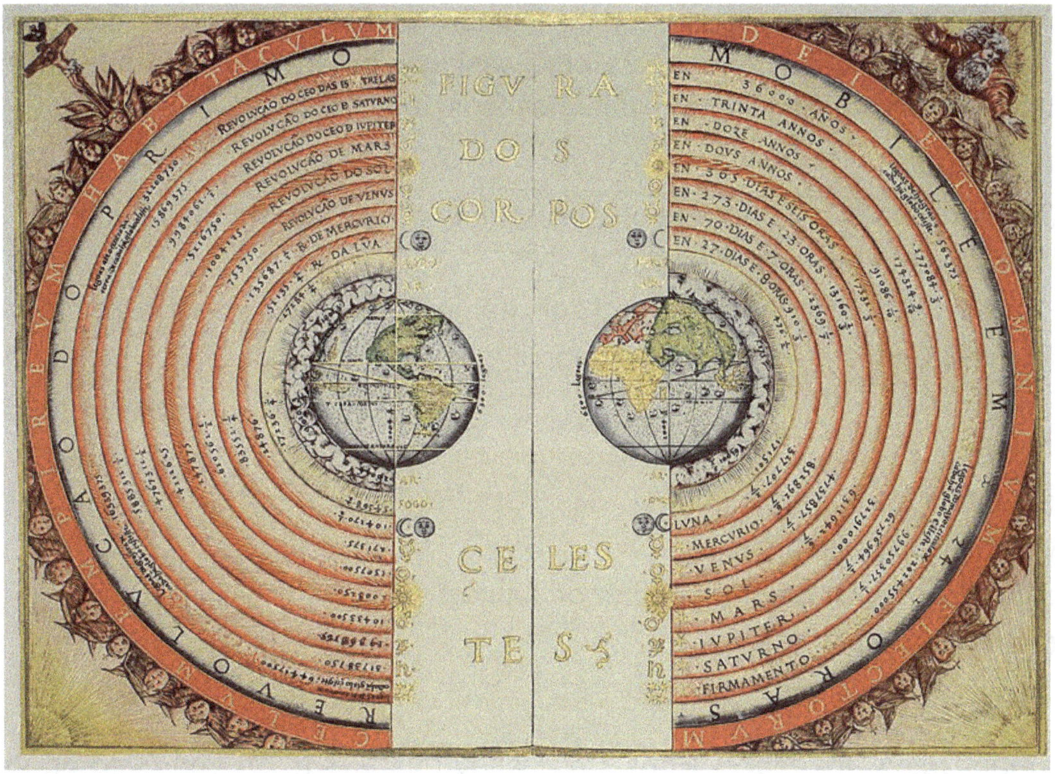

(Source: Encyclopedia Brittanica) Universe pre-heliocentrism

The fallacious interpretation by Ptolemy of the incomplete database at that point in time is seen in the depiction of planet Earth being at the center of the solar system, with the Sun (Sol) in circular orbit about Earth – not the factual converse. The Ptolemy concept was influenced largely by the religious theologians of the time proclaiming that Man is a creation of God, and therefore the only planet known to have human inhabitants must also be the center of the universe also created by God. That concept was to be challenged several hundred years later by Copernicus, Galileo et al.

The technology of optical systems to augment human visual observations improved during the next twelve or thirteen centuries and telescopic systems began to be constructed of the type shown next, attributed to its inventor Johannes Hevelius in the early 1600s:

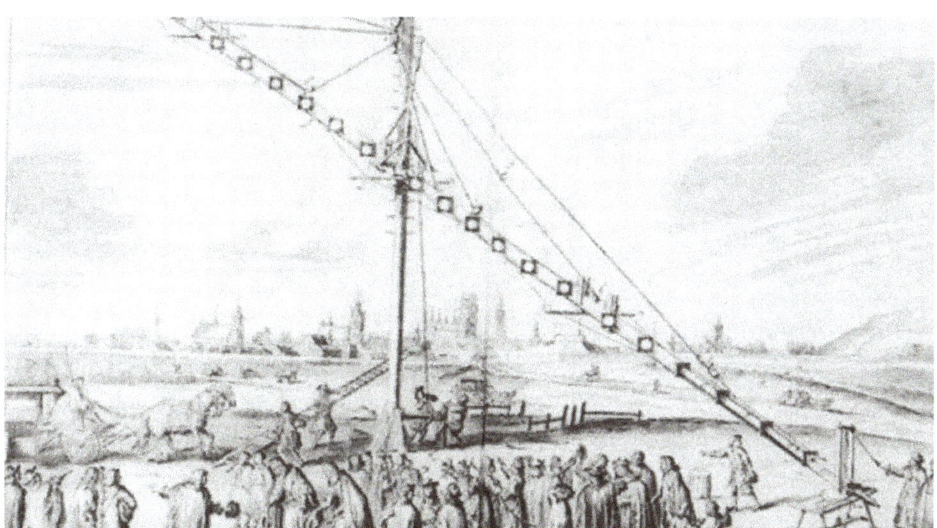

(Source: The National Science Foundation)

The names of the eight major planets are from Roman mythology and are listed below in order of distance from the Sun – although their names were assigned before their distances relative to the Sun and orbital paths were known from the earliest telescopic technology developments.

<u>Inner planets</u> (Terrestrials)

Mercury (Roman God of communication and speedy messenger of the Gods)

Venus (Roman Goddess of beauty)

Earth (Roman God Terra – ground and soil; name Earth is from Germanic erde)

Mars (Roman God of war due to red color like blood)

<u>Outer planets</u> (Galactica)

Jupiter (Roman God of sky)

Saturn (Roman God of time)

Uranus (After Roman God Saturn, father of Jupiter)

Neptune (Roman God of the sea)

In 1608, a German-Dutch lens-maker Hans Lippershey (1570 – 1619) improved on the size of what was to become known as "telescopes", and was the first known applicant for patent of a telescope. In 1609 the Italian astronomer, scientists, and engineer Galileo Galelei (1564 – 1642) made a still more improved telescope based largely on Lippershey's concept and that design made its place in optical science history:

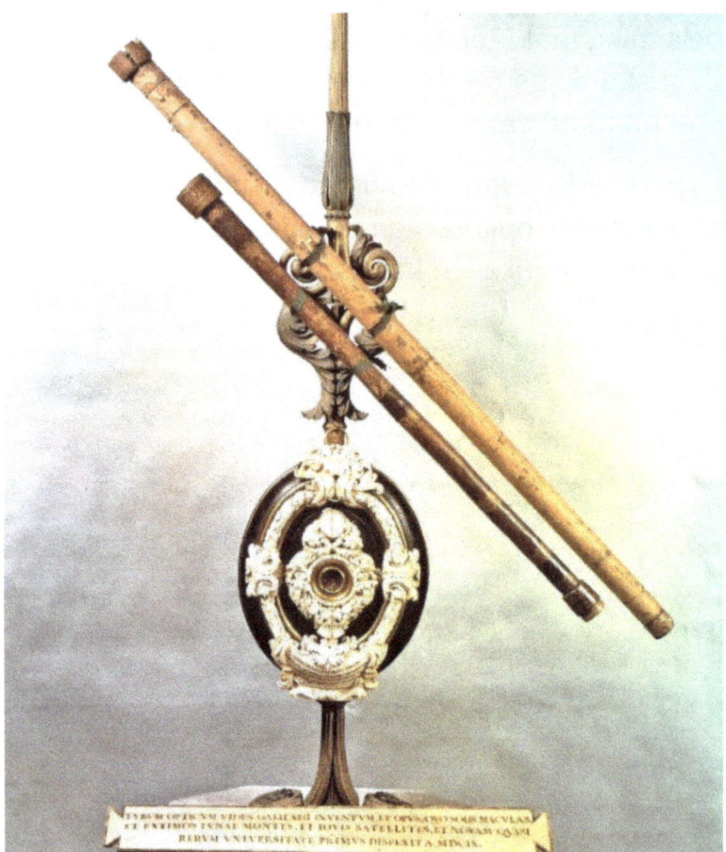

(Source: The National Science Foundation)

Credit is given to Copernicus (and other well-known astronomers such as Galileo and Keppler) for scientifically better defining the orbits of the eight major planets and scientifically correcting the major flaw in Ptolemy's observation that all the planets were in orbit around Earth. Astronomers verified that Earth's Moon (Luna) was correctly shown to have its orbit centered on the Earth. The observations correcting the Ptolemy database by Galileo were extensions of the research conducted by Nicolaus Copernicus (1473 – 1543) of Poland – first to claim that Earth and the other planets revolved around the Sun (that factor being identified as "heliocentrism"). Those combined findings - aided by history's great mathematicians such as Johannes Keppler (1571 – 1630) of Germany - corrected Ptolemy's flaw.

The solar system was shown to have the Sun at the center of the orbital paths and relative sizes of the eight major planets (all named after Greek gods) as indicated below:

(Source: The National Science Foundation)

The concept of heliocentrism (the Sun at the center of the planetary system) was hailed as a major scientific advancement in astronomy but was considered adverse to then-religious theologians' belief that Earth was the center of all that God created. Galileo was forced to suppress that which he scientifically had proven (heliocentrism), and made the following two statements (made to express disappointment in the theologians who pressured Galileo to denounce the concept of heliocentrism) quoted often since his demise in 1642:

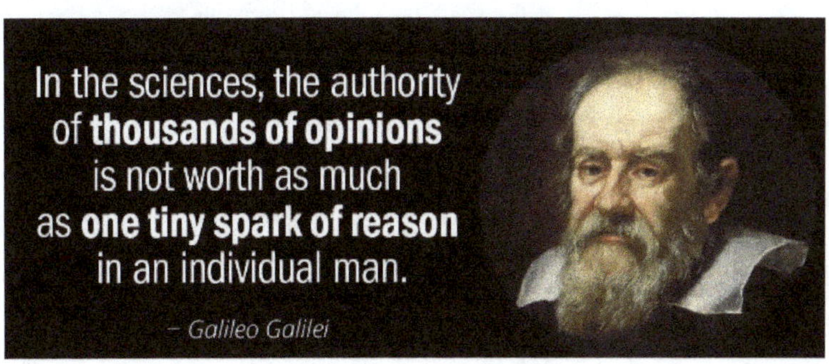

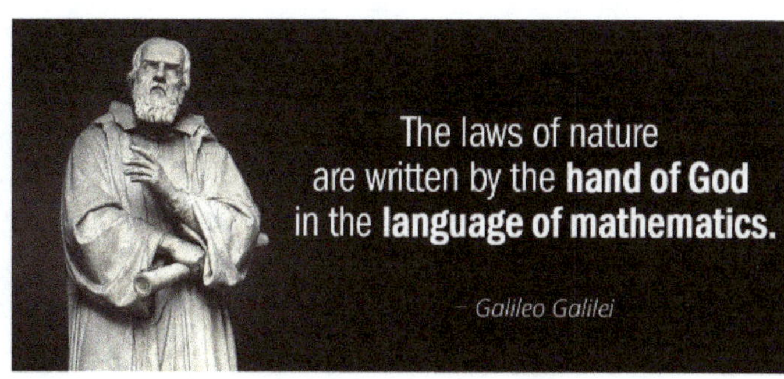

[Author's Note: Galileo is regarded as being the ultimate scientist because he maintained an open-mind when making the many observations of the ultra-complex solar system, and interpreted the database he and other astronomers were compiling with no coterie of activists influencing what they observed to be the factual truth. It is humbling to any scientist to maintain truthfulness in what they are observing and in the interpretation of the true meaning for their observations – whether made in a small, environmentally-controlled laboratory or as witnessed in the infinite reaches of space where nature has control over the events it wishes to happen and where and when to cause the events. The Author has a core belief in the Galileo quotes and that the environmental science profession will observe and apply the information that is contained in this book for the benefit of the public. During the research and analysis phases of this book, the Author maintained the Galileo philosophy of “Let the chips fall where they may” and to “Tell it like it is, not the way that many would prefer it to be.” The factual truth is the truth – convenient or inconvenient, like it or not.]

A first conclusion astronomers in the “Age of Discovery” reached was that the four inner (nearest to the Sun) planets of Mercury, Venus, Earth, and Mars had a substantially firm composition (solid and liquid matter) and these were named after the Latin *“terra firma”* as being the “Terrestrials”. The outer four planets (furthest from the Sun) of Jupiter, Saturn, Uranus, and Neptune were observed to be collections of compressed gases and some small fragments of mineral matter, and these four planets were much like the other bodies in the Milky Way galaxy. The outer four planets were named as the “Galactica”. The exterior appearance of the eight planets and the Moon of Earth were observed with the improved telescopic instruments prior to Galileo (i.e., prior to 1,500) and their relative position, their size, their color and patterns of their surfaces were visualized, and their distances from the Sun were derived from a mixture of logic, comparative observations, astronomical measurements and mathematical/trigonometric calculations. Their dynamic factors (rotation rate about their axis, the tilt of their axis, their orbital path shape and dimensions and number of hours, days, months for each planet in one orbit around the Sun) were also derived from a mixture of observations and calculations.

Based on that improved Galileo database of astrophysical observations, measurements, and calculations, the solar system was better described. It is not physically possible to represent everything known about the solar system on a small size two-dimensional page of a book, in static (steady-state status) of the three-dimensional dynamics, and in the ever-changing variable of time that each factor has uniquely and astrophysical been orchestrated in the solar system. The best this book can do is to present the “learn something new every day” information in two-dimensional illustrations, schematics, diagrams, photographs, and tables of measurements made at a specific year in time.

Shown in the next illustration is the solar system as it was observed (and measured and calculated) to exist during the “Age of Discovery” circa 1400 - 1600:

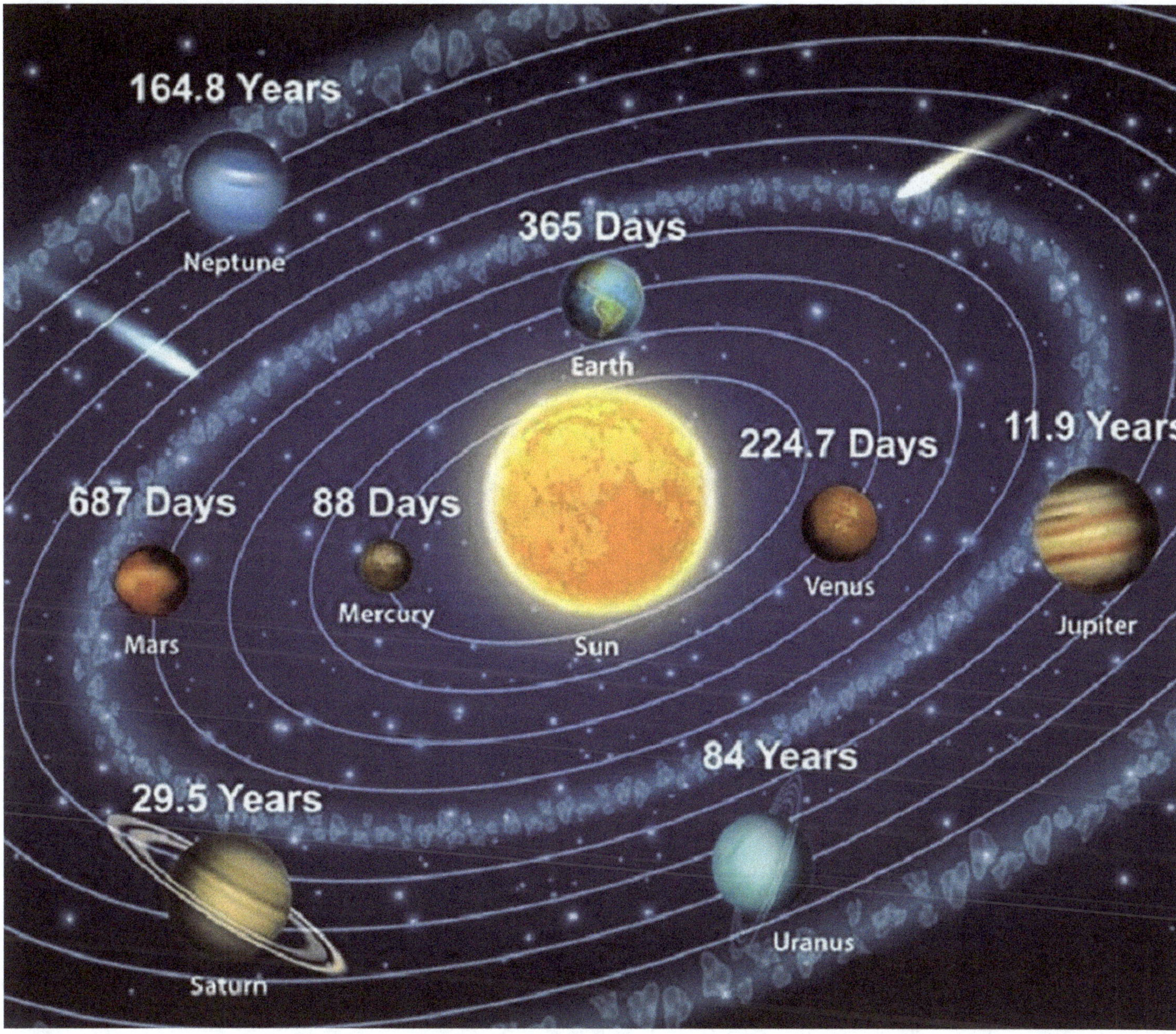

(Source: NASA, with graphical modifications by Author)

Hand-held versions of the Galileo telescope were created soon thereafter and used extensively for navigating by the stars on the sailing ships used by the mariners on all the major oceans during and centuries after the Age of Discovery. Before continuing with the further definition of Earth's surface and interior composition, the importance of astronomical measurement instruments in understanding the origin of the planet and its surface topography deserves to be highlighted to the public. Every database of information owes its value to the scientific instruments used in advancing the foundational information in that database, and to those who developed those instruments.

Singling out the planet Earth for this book, much more definition was added about the only planet known to be occupied with a myriad of life forms – not the least of which is Man. By the Age of Discover, and with the genius of scientists (Galileo) and mathematicians (Kepler) and astrophysicists, the following characteristics of the Earth rotating about a tilted axis and in an irregular orbit, the following description became accepted as factual:

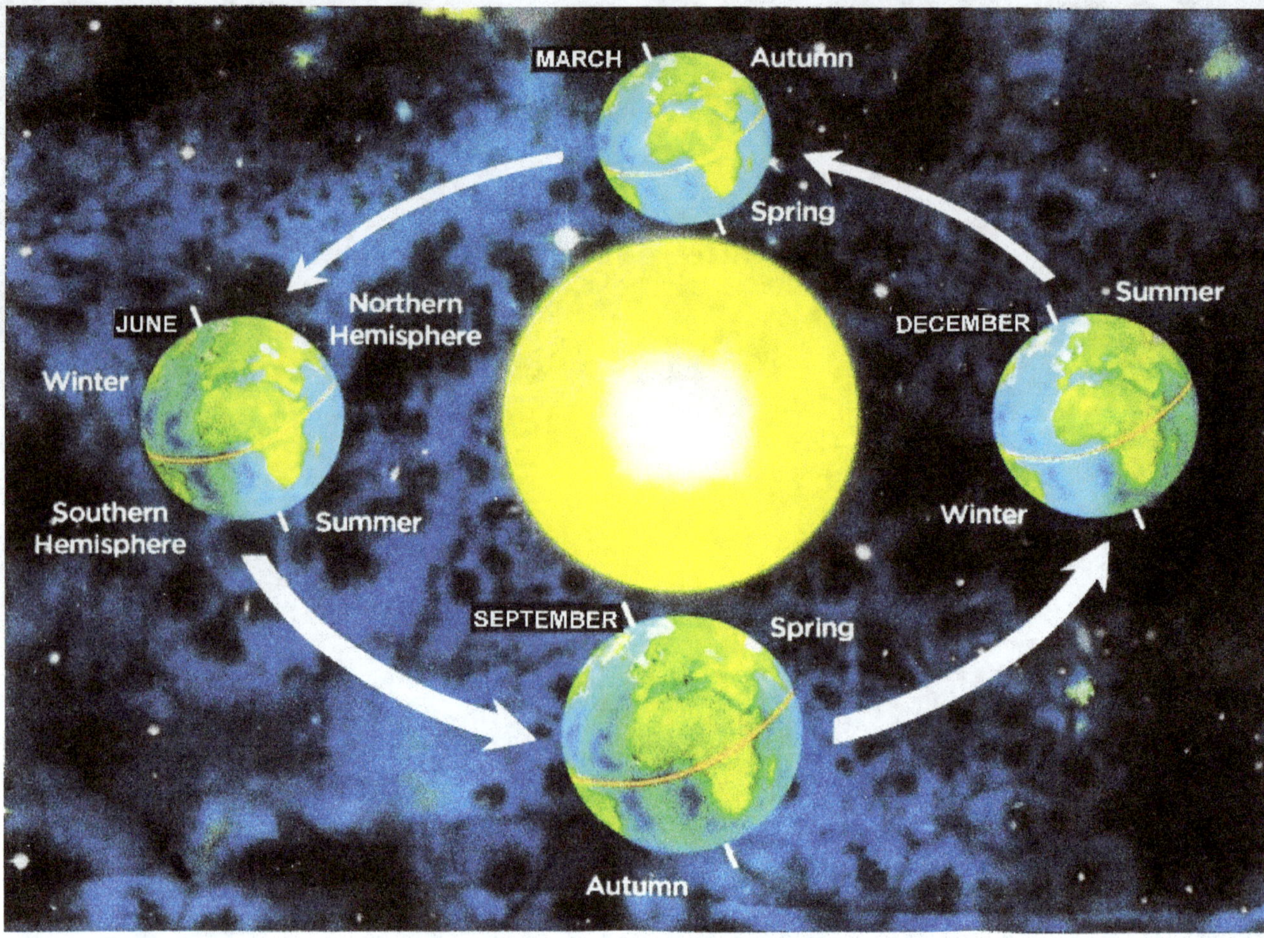

(Source: Smithsonian Institution for Science)

Every country has its pioneers in almost every technology, and one who is singled out for his brilliant contributions to the science of optical measurement and sensor technology is Dr. Edwin P. Hubble (1889 – 1953) of the United States. As with many technological developments, funding to advance technology is critical to what can and will be developed with those funds. The United States lesser-known amateur scientist, astronomer, and philanthropist John D. Hooker (1838 – 1911) was a large supporter of, and contributor to, the Carnegie Institute for Science.

John Hooker made a fortune as a producer of iron works in California, and provided initial funding for a 100-inch (2.5 meters) telescope to the Carnegie Institute for Science who established the Mt. Wilson Observatory in Pasadena, California, and George E. Hale (1868-1938) was named as the Director of the Mt. Wilson Observatory. John Hooker passed away in 1911, and regrettably never was availed the use of the 100-inch telescope he sponsored when it became operational at Mount Wilson in 1917. Hale made an offer to Dr. Edwin P. Hubble to operate and conduct astronomical research with that world's largest telescope, which Hubble immediately accepted. Dr. Hubble reported to Mt. Wilson in 1919 and a leap in astronomy history began in the United States. A photo of the Hooker-Hubble 100-inch telescope at the Mount Wilson Observatory in Los Angeles, California, is shown next:

(Source: The Mt. Wilson Observatory)

The 100-inch telescope used by Edwin Hubble generated information in the 1920-1930 decade about the Universe that astounded scientists world-wide, and demonstrated to visionaries like George Hale the need for even larger collectors. The designer of the 100-inch telescope was George W. Ritchey (1864 -1945) and the Rockefeller Foundation in 1928 opened discussions with George Hale and George Ritchey to design and construct both a new facility at Mount Palomar in Pasadena, California, designed specifically to house an unprecedented astronomical telescope having an almost unimaginable 200-inches diameter parabolic primary mirror. Construction of the observatory facilities on Mount Palomar and its new dome started in 1936; however, World War II (1938 – 1945) caused interruptions in construction activities of both the new dome and the unprecedented 200-inch telescope.

In 1948, the 200-inch telescope in the new dome at Mount Palomar was dedicated, and Dr. Edwin Hubble was again requested by George Hale to be the Director of the Mount Palomar facility with the world's largest telescope. The next pair of photographs shows the unique Mount Palomar dome (which has the rotational capability for pointing the 200-inch telescope in a full 360 degrees upper hemisphere coverage) and the 200-inch telescope within that dome:

(Source: The Carnegie Institution)

In the first few months of its use, which started on January 26, 1949, and under the direction of Dr. Edwin Hubble, slight distortions of stellar images obtained with the 200-inch parabolic mirror were detected. Corrections to the distortions were made throughout 1949, and the fully operational telescope system began collecting images distortion-free in early 1950. With the basic optics functioning as designed, the telescope was adapted with modern infrared array imagers, spectrographs, and an adaptive optics system. George Hale passed away in 1938 and never got to see the fruits of his efforts to develop the world's largest land-based telescope system.

Edwin Hubble continued operating the Mount Palomar facility and the 200-inch telescope until 1953, helping NASA to develop the Apollo Program to put the first Man-on-the-Moon, but Dr. Hubble passed away on September 28, 1953, at the too-young age of 63 and did not see that historic moment on July 20, 1969, of Neil Armstrong taking "one giant leap for Man; one giant leap for Mankind". The Mount Palomar facility is still in operation today and was inspirational to Dr. Hubble's research in proving that the universe is expanding in time. Of most importance to this book is that the 200-inch telescope with its unique array of infrared, spectrographic, and optical components has enabled the temperatures of the surfaces of the planets in the solar system and the gaseous components comprising their unique atmospheres to be sensed and measured with impressive accuracy and completeness.

A tabulated database of the distances the four inner planets (the terrestrials) are from the Sun and the temperatures on the surfaces of the four terrestrial planets and the Earth's Moon is shown next: (It was necessary to rotate the presentation of the table into a vertical, near full-page format in order to display the numerical astrophysical database in a sufficiently large size to make the highly important data therein legible)

Surface Temperatures on the Sun and Four Inner Planets & Earth's Moon

Planet	Distances from Sun ~ million km			Solar Radiation at Planet ~ watts / m²	Surface Temperatures (observed/measured) ~ degrees								
					Hottest			Mean			Coldest		
	Closest	Mean	Farthest	(Planet at Mean distance)	K	°C	°F	K	°C	°F	K	°C	°F
Sun	0	Reference	0	64×10^6	5873	5600	10112	5780	5507	9941	4273	4000	7232
Mercury	46.001	57.9	69.817	9116	703	430	806	398	125	267	93	-180	-292
Venus	107.476	108	108.943	2611	746	473	883	737	464	867	726	453	847
Earth	147.098	149.6	152.098	1366	344	71	160	288	15	59	184	-89	-128
Earth's Moon	146.714	149.6	152.482	1366	407	134	275	263	-10	-14	120	-153	-243
Mars	206.655	227.9	249.232	589	293	20	68	208	-65	-85	133	-140	-220

Notes: K (Kelvin) = C (Centigrade) + 273 F (Fahrenheit) = 9 / 5 C + 32

1) Kelvin is the absolute temperature scale; coldest temperature in science is liquid hydrogen at/near 0 K

2) Water changes state from liquid to solid ("freezes to ice") at 0° C = 32° F = 273 K

3) Water changes state from liquid to gas (boils to steam") at 100° C = 212° F = 373 K

4) Carbon dioxide (CO_2) changes state from gas to solid ("sublimation" to "dry ice") at -78.5° C = -109.3° F = 194.5 K

(Source: NASA, tabulation by Author)

Using the spectrographic sensing and wavelength discernment capabilities of the 200-inch Mount Palomar telescope to identify the type and concentration of gasses comprising the atmosphere of the four inner planets (and the Earth's Moon), and, for scientific interest, inclusion of the four outer planets (the Galactica), a tabulated numerical database for the atmospheres on the eight major planets and Earth's Moon in the solar system is shown next:

Atmospheres of the Planets (Courtesy of NASA)

Planet	Atmospheric Mass ~ Kilograms	Nitrogen	Hydrogen	Oxygen	Argon	Helium	Carbon Dioxide	Carbon Monoxide	Methane	Sodium	Other	* Water Vapor
Sun	3.0×10^{30}		71%			26%					3%	
Mercury	1,000		22%	42%		6%				22%	8%	
Venus	4.8×10^{20}		4%				96%					
Earth	1.4×10^{21}	77%		21%	1%		0.04%	0.01%			0.95%	0.1 - 4%
Moon	100,000				70%	21%				1%		
Mars	2.5×10^{16}	2.7%			1.6%		95%				0.7%	
Jupiter	1.9×10^{27}		89.8%			10.2%						
Saturn	5.4×10^{26}		96.3%			3.2%					0.5%	
Uranus	8.6×10^{25}		82.5%			15.2%			2.3%			
Neptune	1.0×10^{26}		80%			19%			1%			

Notes

* Earth is the only planet having significant water on its surface or water vapor in its atmosphere.

Earth's surface is covered by more than 75% of its area with water (in its oceans, seas, gulfs, bays, lakes, rivers, etc.).

(Source: NASA, tabulation by Author)

Using the "3 bears" relationship (too big in red, too small in yellow and red, and "just right" in green), the suitability of the 5 inner planets and Earth's Moon for human life is shown. The table shown next relates the percent concentrations of gasses and the mean temperature on the surface of Mercury, Venus, Earth, and Mars, the absence/presence of oxygen and carbon dioxide in their atmospheres, and the absence/presence of water in abundance on their surfaces and water vapor in their atmosphere as determined by the planetary scientists at NASA, NOAA, and many other science centers.

Inner Planets - Atmospheric Gas, Water, and Surface Temp. Conditions

	Atmosphere				Mean Surface Temp.		
	Gasses		Water, H_2O		Kelvin	°C	°F
Planet	Oxygen, O_2	Carbon Dioxide, CO_2	Water Vapor, H_2O	Liquid H_2O, Surface	Units	Degrees	Degrees
Mercury	42%	0%	0%	0%	398	125	267
Venus	0%	96%	0%	0%	737	464	867
Earth	21%	0.05%	0.1-4%	70%	288	15	59
Moon	0%	0%	0%	0%	263	-10	-14
Mars	0%	95%	0%	0%	208	-65	-85

(Source: Analysis of NASA and NOAA Databases by Author)

It should be noted in the tabulated database of atmospheres on the inner four planets that planet Earth is the only planet of the four major planets and the Moon to have the known presence of water – and on Earth over 75% of its surface is covered by water. Depending on the temperature of that water, there can be as much as 4% of the atmosphere containing water in vapor form. The significance of that much water vapor (1-4%) in the atmosphere is not related to factually being a scientific astrophysical factor and will be further discussed in the next chapter (Chapter 3) of this book.

It is emphasized however, and as indicated in the above table by the color coding (red, yellow, green convention), the human lifeform has evolved to be viable and sustainable only in the environment on Earth because of the combined availability of sufficient oxygen and water satisfying those basic elemental needs for human life, as well as the small amount of carbon dioxide in the atmosphere for the growing of plant lifeforms in the forestry and agricultural sector of the ecosystem. An observation – and open question that environmental scientists since Svante Arrhenius in 1902 was unable to answer is that in the atmosphere of Venus and of Mars there is 96% and 95% of carbon dioxide in the atmosphere of those two planets. If there were never any humans on either Venus or Mars, by what process has all that carbon dioxide been produced in such a high – nearly 100% - concentration on those planets which bracket Earth in their distance from the Sun? On Earth, the concentration of carbon dioxide measured at the Keeling Mauna Loa Atmospheric Research Laboratory is currently (2024) in the total measurement of 435 parts per million (0.0435%). Accepting that of that total 435 ppmv carbon dioxide in the atmosphere is an accurate scientific measurement, 295 ppmv (0.0295%) is that which existed pre-Industrial Revolution from astrophysical, geophysical, and natural causes and the remainder 140 ppmv (0.014%) is due to human activities creating carbon dioxide, by what logic does the environmental science profession reason that Man burning fossil fuels is the dominant cause for global warming and the ensuing climatic temperature increase of the planet? A parallel question is by what logic can all other scientifically identified factors that are proven herein to be factors more prominent than Man burning fossil fuels in causing carbon dioxide to exist in the atmosphere of planet Earth?

Data in tabular form (and color-coded or not) leave much room for interpretation. Using the first vertical table (database of distances and temperatures) for the planets and the Moon, the astrophysical laws of the German astronomer and astro-mathematician Johannes Kepler (1571-1630), the Kepler Law of major importance that solar heat energy radiation on a planet is inversely proportional to the distance the planet is from the Sun is graphically shown next:

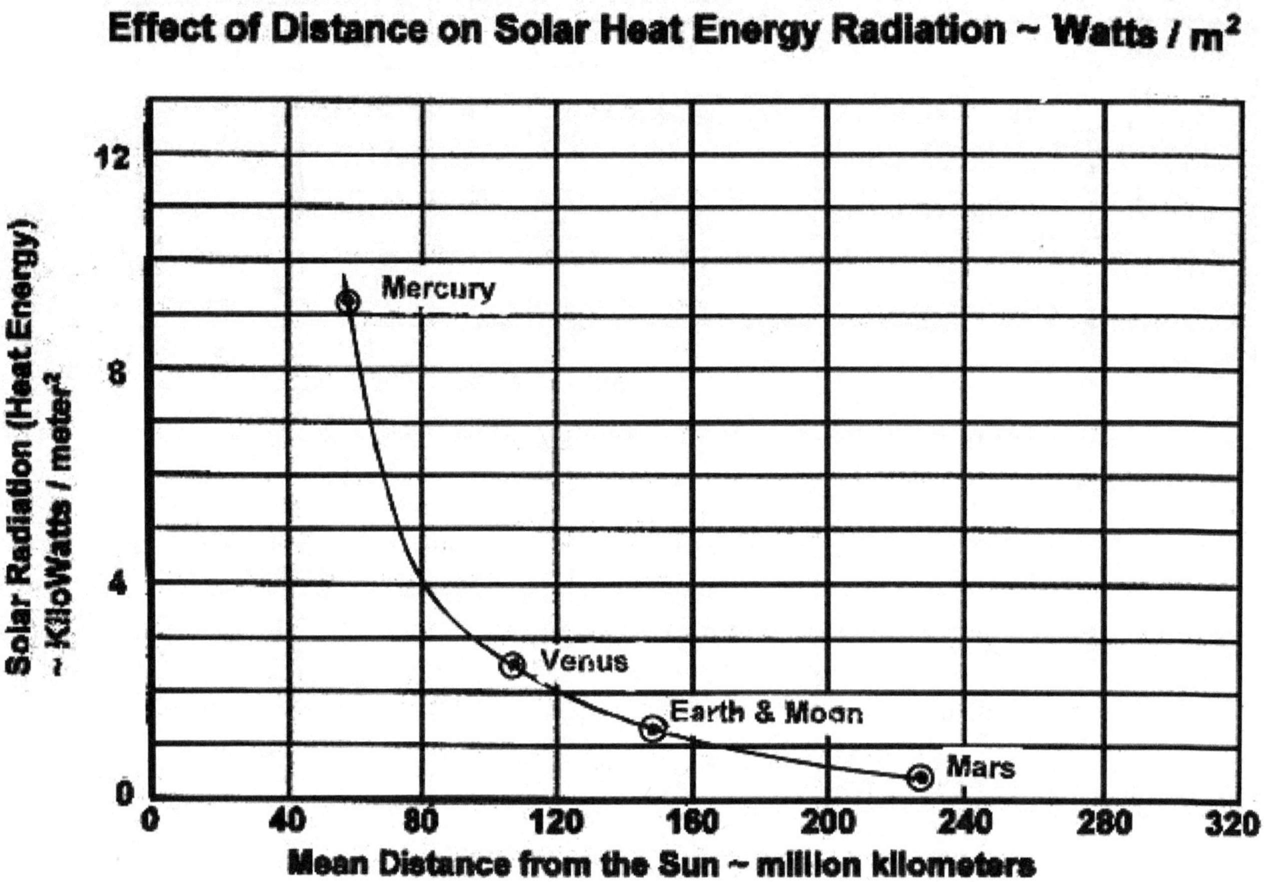

(Source: Chart created by Author from NASA tabular database)

In lay terms, the above diagram could be described as the “Moth circling a light bulb”, where the warmth attracts the moth to fly ever closer to the source of the warmth it senses. As the moth realizes too late, when the moth gets too close to the source of warmth, the moth is incinerated by the over-temperature (beyond the moth’s heat energy tolerance) to its physical structure. The large extremes in the distance from the Earth to the Sun and human innate tolerance to relatively wide temperature differences in heat extremes (by multiple causes) on planet Earth has led astronomers to refer to planet Earth as being in the “habitable zone”. The extremely hot temperatures on Mercury and Venus places those planets to be outside the “habitable zone” for humans; the extremely cold temperature of Mars causes Mars to also be located outside the “habitable zone” for humans.

While the above graph indicates a mathematically “smooth” variation in the insolation (incoming solar radiation) on the distant surfaces of the planets and the Moon, Readers must remember that the mathematical “gymnastics” of Kepler’s Laws of Astrodynamics are performed without accounting for the unique atmosphere surrounding each planet and the Moon and those laws do not consider the effect that atmosphere has on the actual surface temperature of the planet cloaked in its real and very non-homogenous atmosphere. To address the question "Does the anthropogenic (Man-made) carbon dioxide in Earth's atmosphere dominate all other factors (including astrophysics) on the level of temperature on Earth's surface?", a simple analysis was conducted with the NASA astrophysics database.

The tabulated databases (two tabulations of data shown just prior to the "moth and heat lamp" diagram) provide the measurements of the surface temperatures of the inner planets as the distance from the Sun varies. The second table provides the atmospheric gaseous content information needed to graphically illustrate an approximation of the effect of carbon dioxide in the atmosphere of each planet (having a carbon dioxide content at some concentration in its atmosphere) on the surface temperature for that planet's insolation and atmosphere.

The atmospheres on the eight major planets within the solar system vary widely in the gaseous content and the thickness of each atmosphere. Scientists have only an estimation of the likelihood of the thickness of the atmosphere on each planet and only a slightly better understanding of the content (as shown in the NASA table above). The diagram shown next was displayed in the Astrophysical Journal in 2017:

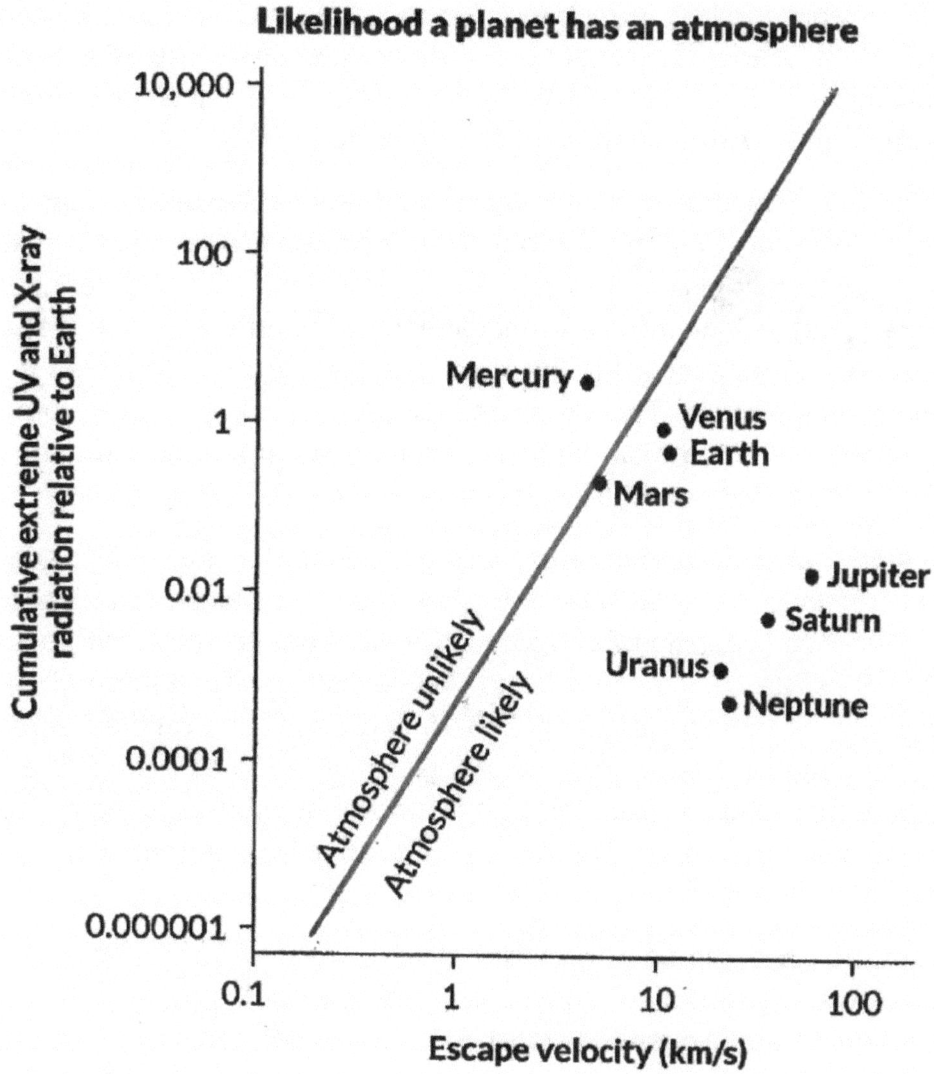

(Source: The Astrophysical Journal, 2017)

It is seen by the green line defining the likelihood of an atmosphere on the four inner planets (the "Terrestrials") that planet Mercury is not likely to have an atmosphere, Mars has a very thin atmosphere (bordering on none) and Earth (having a multi-gas atmosphere of which carbon dioxide - total of natural about 0.03% and Man-made at 0.02% but increasing - is below 0.05%). The four outer planets (the "Galactica") Jupiter, Saturn, Uranus, and Neptune have much thicker atmospheres as they are essentially without a solid core above which another thicker atmosphere could form. NASA is exploring the surface of Mars as another location for human habitation but not that of Mercury (forbiddingly hot), Venus (with 96% carbon dioxide atmosphere and also extremely hot), or the extremely cold four outer planets.

An observation of the atmospheres of Venus having a 96% concentration of carbon dioxide and Earth having a 0.03% concentration of carbon dioxide led the Swedish atmospheric chemist Svante Arrhenius (1859 – 1927) in the last decade of the nineteenth and first decade of the twentieth century to issue an atmospheric alarm to the public. Arrhenius is credited in atmospheric science history for developing a brilliant physical chemistry relationship known as the Arrhenius Equation. The research leading to the Arrhenius Equation included the conduct of laboratory experiments defining the heat-trapping properties of carbon dioxide gas. That research led to Dr. Arrhenius being awarded the Nobel Prize for Chemistry in 1904 and led to two very important observations about carbonic acid and the "hothouse" – the forerunner of the more modern terminology "greenhouse gas" - effects of carbon dioxide on the atmosphere of Earth. With Readers' indulgence and before identifying the advisory/alarm issued by Dr. Arrhenius, the following few (three) paragraphs are provided to enlighten environmental scientists of the scientific importance of Dr. Arrhenius' scientific research.

The Encyclopedia Brittanica describes the Arrhenius mathematical expression as defining the effect of temperature on the velocity of a chemical reaction, the basis of all predictive expressions used for calculating reaction-rate constants. In the Arrhenius equation, k is the reaction-rate constant, A represents the frequency at which atoms and molecules collide in a way that leads to a reaction, E is the activation energy for the reaction, R is the ideal gas constant (8.314 joules per kelvin per mole), and T is the absolute temperature. The equation is commonly given in the form of an exponential function, $k = A\exp(-E/RT)$,and it predicts that a small increase in reaction temperature will produce a marked increase in the magnitude of the reaction-rate constant.).

To understand the significance for temperature being measured in the Kelvin scale, Lord Kelvin (Sir William Thomson of Scotland, 1824-1907) established the Kelvin scale of units such that the coldest substance known to science (freezing of a theoretically ideal gas to solid form) establishes the 0 units condition for the Kelvin scale. Since there is no theoretically colder temperature than "absolute 0" (where Centigrade would be the equivalent of -273.15 degrees C and on the Fahrenheit scale -459.67 degrees F), there is no negative Kelvin units occurring or needed to be used in the science profession. On the hot end of the temperature scales, the hottest known substance is the center of the Sun where the temperature is estimated to be 15 million Kelvin units, 15 million degrees Centigrade, and 27 million degrees Fahrenheit.

The Centigrade (now more commonly termed Celsius) scale of temperature uses 100 "degrees" between water in the solid (frozen) state and water in the gaseous state resulting from "boiling" water (transformation from liquid state to gaseous state). The Fahrenheit scale is based on water (liquid state) "freezing" (i.e., transforming from liquid state into solid state – ice) at 32 degrees Fahrenheit and transforming from liquid state into gaseous state – again by boiling the liquid water into gaseous state - at 212 degrees Fahrenheit. In both the Fahrenheit and Centigrade scales of temperature measurement, there are many natural temperatures below the 32 and 0 degrees mark respectively on the Fahrenheit and Centigrade scales.

The advisory and alarm that Dr. Arrhenius is stated here as an interaction type effect between the astrophysical and the geophysical effects on planet Earth. The geophysical effect will be described in greater detail in the geophysical chapter of this book (the next chapter). The first part of the Dr. Arrhenius' two-part scientific advisory was that of near-term significance. The Arrhenius Equation, coupled with his laboratory experimentation with carbon dioxide (gaseous state) and carbonic acid (liquid state), led to his two-part assessment of the effect of Man-made carbon dioxide on the environment: 1) In the near-term, the small amount of global warming being a result of Man burning fossil fuels to support human and industrial energy needs had a beneficial effect on the environment (by opening up some land area to agricultural opportunities; and 2) in the far term, the exceedingly high concentration of carbon dioxide in the atmosphere from all causes posed a hazard to the ecology as well as to the human population of the world.

In the latter years of his scientific life, Arrhenius became profoundly passionate about mathematical concepts, data analysis and discovering their relationships and laws as applied to the natural world and environment of the Scandinavian region. Much of his brilliant scientific work was overshadowed by the events leading to World War I in Europe.

This Author humbly notes having the similarly high interest as Dr. Arrhenius in applying mathematics and data analysis to the real world of chemical and physical reactions as being a large motivating factor why the Author devoted over nine years of research leading to this book. Using the same graphical format as the previous graph for the astrophysical distance effect, by calculating a ratio of each planet's measured surface temperature in Kelvin units of temperature to that of Mercury and replacing the scale at the left of solar radiation with that Kelvin ratio referenced to Mercury, the graph shown next was created (by the Author). Adding the notation above each planet as to its concentration of carbon dioxide measured to factually be in that planet's atmosphere, an important observation can be seen:

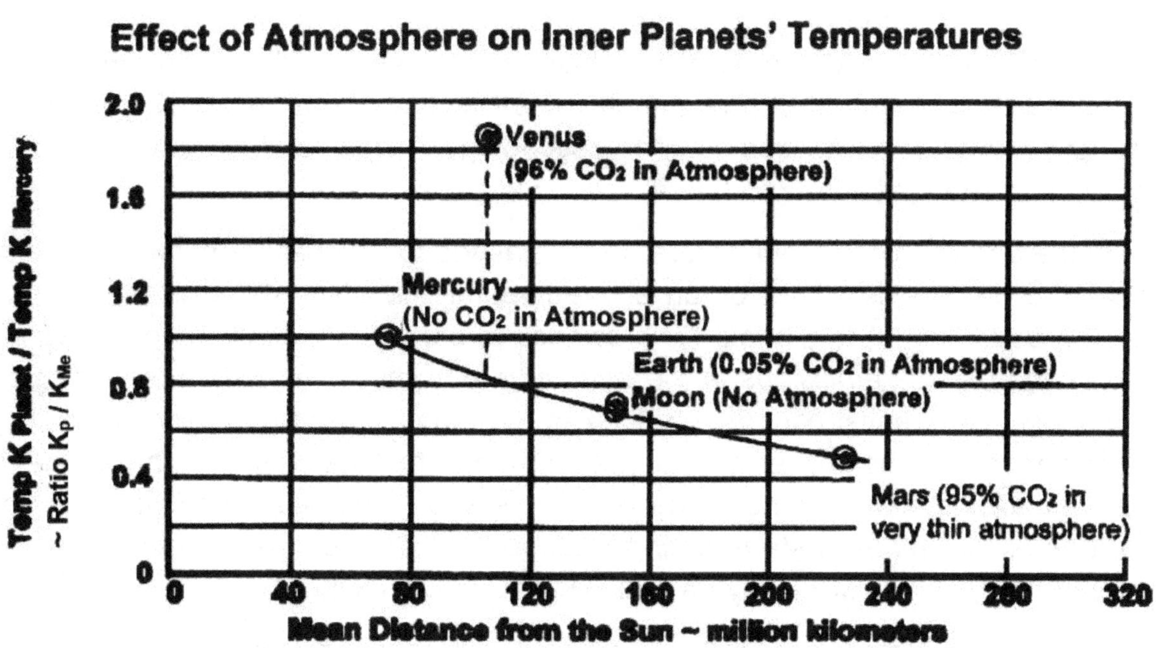

(Source: Chart generated by Author from NASA tabular databases)

The graph above draws immediate attention to the planet Venus not lying on the same "smooth" variation as the variation in surface temperature and distance effect as doe the other planets and the Moon. The planet Venus has a measured 96% concentration of carbon dioxide in its atmosphere (and insufficient other gasses that could possibly have the large effect on surface temperature seen) and is much hotter than its two neighboring bodies with no carbon dioxide in their atmospheres (Mercury and Earth's Moon).

The finding that the concentration of carbon dioxide gas in the atmosphere of planet Venus (shown earlier as being 96%) would certainly explain why its atmosphere is so incredibly hot (mean temperature of 464° C / 867° F / 737 K) and well outside the "habitable" environment for human safety and health. For Reader's interest, the mean temperature of Earth's Moon (at the same mean distance from the Sun but with no heat-trapping atmosphere) is -10° C / -14° F / 263 K, and that of planet Mars (80.8 million kilometers further from the Sun than is Earth) having a mean temperature of -65° C / -85° F / 208 K. Mars also has an atmosphere of 96% concentration of carbon dioxide but the heat trapping effect on Mars atmosphere is negligible since there is such a large reduction in solar radiation from the Sun due to Mars being over 80 million more kilometers (130 million miles) from the Sun than is Earth.

A first general conclusion is that when a planet has an atmosphere of 96% carbon dioxide concentration it will be hotter than another planet at the same distance from the Sun but having no carbon dioxide in its atmosphere. An alternate conclusion that can be drawn as a general observation is that when the carbon dioxide in a planet's atmosphere is low (Earth's atmosphere is measured at below 0.05% total – about 0.03% natural caused and about 0.02% anthropogenic caused), there may be other factors in that atmosphere that dominate the Earth's surface temperature.

Regarding the surface temperature on the planet, paleontological history from the advent of the cave Man era and the wearing of animal skins to conserve body warmth and comfort in the extremes in daily and seasonal temperature changes is sufficient proof that Man has learned how to adapt to and survive the extreme changes in the planet surface temperature at almost every latitude on planet Earth. Modern man has invented the technologies of heating, ventilating, and cooling (HVAC) machinery/systems to improve the safety and comfort of Man in Earth's widely variable temperature and water (in all three forms of state – solid, liquid, and gas) to assure his/her survivability. (The Author will not address the wardrobes that a certain segment of humans chooses to wear or not wear to attract visual attention instead of for environmental protection. Human psychology is well outside the scope of this book on weather and climate change.)

At this point in the analysis of the "big picture" of the causes for "global warming", Readers are asked to again recall the hypothesis stated by an environmental scientists stated in the "Why This Book" section near the beginning of this book: *"The science is settled. By eliminating everything else, the Man-made greenhouse gas emissions in the atmosphere is the dominant cause for global warming and climate change."* A scientist cannot be criticized for carrying what has become the battle-flag for the branch of science that has chosen to lead the "combat" on climate change, but whatever scientific group made the decision in the scientific analysis process to *"eliminate everything else"* to prove that *"greenhouse gas emissions from Man burning fossil fuels is the dominant cause for global warming and climate change"* was not sufficiently trained in statistics or in simple mathematics.

In a column of numbers, the answer for the total of those numbers depends on inclusion of the entire set of numbers in the column regardless of the erratic variations in the individual numbers. Singling out any one number from the column and eliminating all the other numbers will result in a total that bears no resemblance to what the correct answer is for the original column of multiple numbers. Similarly, the elimination of astrophysical effects on the planet's surface temperature (shown scientifically in the preceding part of this chapter) and the elimination of the effects of water – that covers over 75% of the planet's surface and is in the atmosphere at a concentration from 1% to 4% - and yet maintain the belief that Man-made "greenhouse gasses" in the atmosphere (at a concentration of 0.0295% of that atmosphere prior to the Industrial Revolution) is the dominant cause for global warming and climate change is beyond the Author's comprehension of ethical and professional science. The measured (X parts per million by volume which is the equivalent of 0.0X%) concentration of carbon dioxide in the atmosphere is shown in a later database of atmospheric measurements (commonly referred to as the "Keeling hockey-stick curve") to be calculated from there being a total of 415 parts per million (0.0415%) of carbon dioxide accurately measured in the atmosphere - of which 295 ppmv (0.0295%) is natural-created carbon dioxide and 120 ppmv (0.012%) is Man-generated carbon dioxide.

It would be a most enlightening database of information for the environmental profession to present and convince the public how that small concentration of Man-made greenhouse gasses in the vast atmosphere (0.012%) is valid reason for the public to fear for their near- and far-term existence. Equally of interest would be to learn why astrophysical effects and water concentration (1% - 4%) in the atmosphere are so unimportant to environmental scientists as to have been "eliminated" totally from further consideration by environmental scientists in the 1950 to 1990 time period. More recent databases collected by other than NOAA branches of environmental science – astrophysical, geophysical, natural, and the advent of satellite-based measurements and sensing of temperature with infrared sensors by NASA and the space agencies of other countries – are disproving that it is Man burning fossil fuels and creating "greenhouse gas effects" as being the dominant factor in global warming and climate change and removing the need for Man to discontinue burning fossil fuels as being a "crisis"-based necessity.

Authors have the option what to include – and what not to include - in their book being written. This Author would feel remiss if the awesome advancements made by NASA, in defining the many other astrophysical factors and dynamics of objects in space that may add little knowledge to the database of knowledge applicable to the trifecta problem unique to planet Earth but which are important to science in general, were not to be included (were "eliminated") for Readers' benefit.

To complete the "genealogic" history of telescopes used in past and in the astronomy technology development process, there are two developments that are more recent and of astrophysical utmost significance but which do not add to the databases for the weather and climate problems on Earth's surface being considered for this book. However, they do add to the understanding of the astrophysics of the solar system and therefore indirectly improve the quality of the databases that are of direct importance to understanding how astrophysical factors affect the trifecta problem on planet Earth. These are the Hubble Space Telescope developed in the twentieth century, and the more recent James Webb Space Telescope developments by NASA.

The Hubble Space Telescope, named for the brilliant astronomer Dr. Edwin Hubble (1889 – 1953), is a space telescope that was launched by NASA into low Earth orbit in 1990 and remains in operation. Shown next is a photograph of the Hubble Space Telescope in orbit where it explored within and beyond the solar system for thirty years:

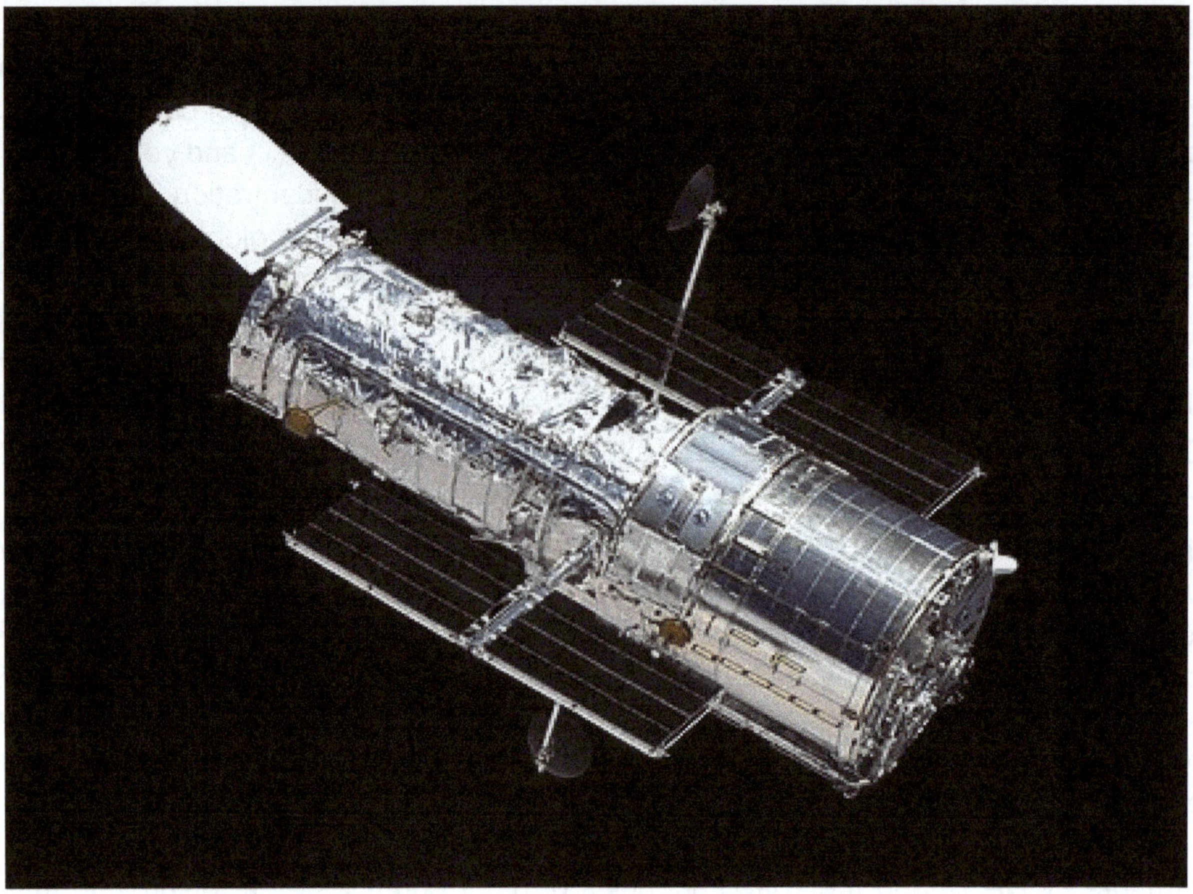

(Source: NASA)

Hubble featured a 94-inch (2.4 meters) diameter mirror and its main instruments observed in the ultraviolet, visible, and near-infrared regions of the electromagnetic spectrum. Its orbit was outside the distortion of Earth's atmosphere, allowing Hubble to capture extremely high-resolution images with substantially lower background light than ground-based telescopes. After launch, it was quickly identified that its main mirror had been ground incorrectly, but the aberrant optics were corrected to the intended quality by astronauts in an extra-curricular space mission in 1993. Hubble completed 30 years of operation in April 2020, transitioning to operation of the James Webb Space Telescope which was launched into space on December 25, 2021 (as a Christmas gift to the astronomy database by the NASA Launch Team, not by Santa's elves).

Among the huge contributions to the improved understanding of the celestial objects in our solar system from the NASA Hubble Space Telescope operations was the measurement and spectrographic analysis of the gasses comprising the atmospheres of the eight planets and Earth's Moon. A tabulated summary of those atmospheres of the eight planets and Earth's Moon was presented earlier in this chapter to highlight the amount of carbon dioxide that was sensed and measured (by spectrograph) with the Hubble Space Telescope to be in Earth's atmosphere.

The James Webb (1906 – 1992) Space Telescope with high-resolution and high-sensitivity instruments allowing it to view objects too old, too distant, or too faint for the Hubble Space Telescope. Webb's primary mirror consists of 18 hexagonal mirror segments made of collecting area of Webb is about 25 square meters, about six times that of Hubble. A photo of the James Webb Space Telescope in its fully-operational (unfolded segmented mirror) configuration in space, is shown next:

(Source: NASA)

The awesome images already obtained by the Webb Telescope are beyond words to adequately describe them. The images are given names and one of the images shows the objects many millions of light years ago shortly after being formed by the "big bang" event; The Webb Telescope images obtained are of a distance in space that has never before been seen by astronomers and is re-forming the college curricula being taught to astronomy science students. One example image obtained with the NASA Webb Space Telescope is shown next:

(Source: NASA)

The photographs with the Webb Space Telescope show that there was much dispersal of gasses and solids that apparently clustered together in different combinations to form all of the celestial objects in space, but that aspect of astrophysics is beyond the scope of this book to attempt being described.

It is important to recall that the mathematical calculations were done by Milutin Milankovic in two different time periods. The first calculations were done in the 1924 time period when mathematician Milutin Milankovic, climatologist Wladimir Köppen, and German geophysicist Alfred Wegener collaborated in studying the causes for the ice ages and the effect on the meteorology and agriculture in cold climates. That work is reported on in “Climate of the Geological Past”.

The more recent (1939) set of astrophysical calculations were done by Milankovic and his team of mathematicians who did a brilliant job of calculating the cyclic variations in the temperature for the planet. It is that set of calculations reported in “Canon of Insolation and the Ice-Age Problem” that became known as the Milankovic Cycles that are being used in this book. The calculations were done for the latitudes of 55°, 60°, and 65° North latitude because of the interest in northern Scandinavia for what global warming was doing to the regions above 65° North latitude. The only criticism of the second set of calculations is that they were done without consideration of the existence of the factual atmosphere surrounding the planet which would change the magnitude of the amplitude of the cyclic variations but not the period of the cyclic variations.

Summaries of the Milankovic calculated database indicate how the variables in the tilted wobbling axis of Earth’s rotation, the obliquity angle at which the Sun solar radiation impinges on Earth’s surface at 65 degrees North latitude and at the summer solstice, and thirdly the irregular changes in ellipticity on solar radiation – the heat energy forcing factor on Earth’s surface. Central Iceland and Northern Norway and Sweden are noted to lie at about 65 degrees North latitude on Earth’s spherical surface.

That summary of the Milankovic team calculations is shown next, and with notation of the years of the period(s) in the repetitive cyclical variation of that variable (precession, obliquity, and eccentricity):

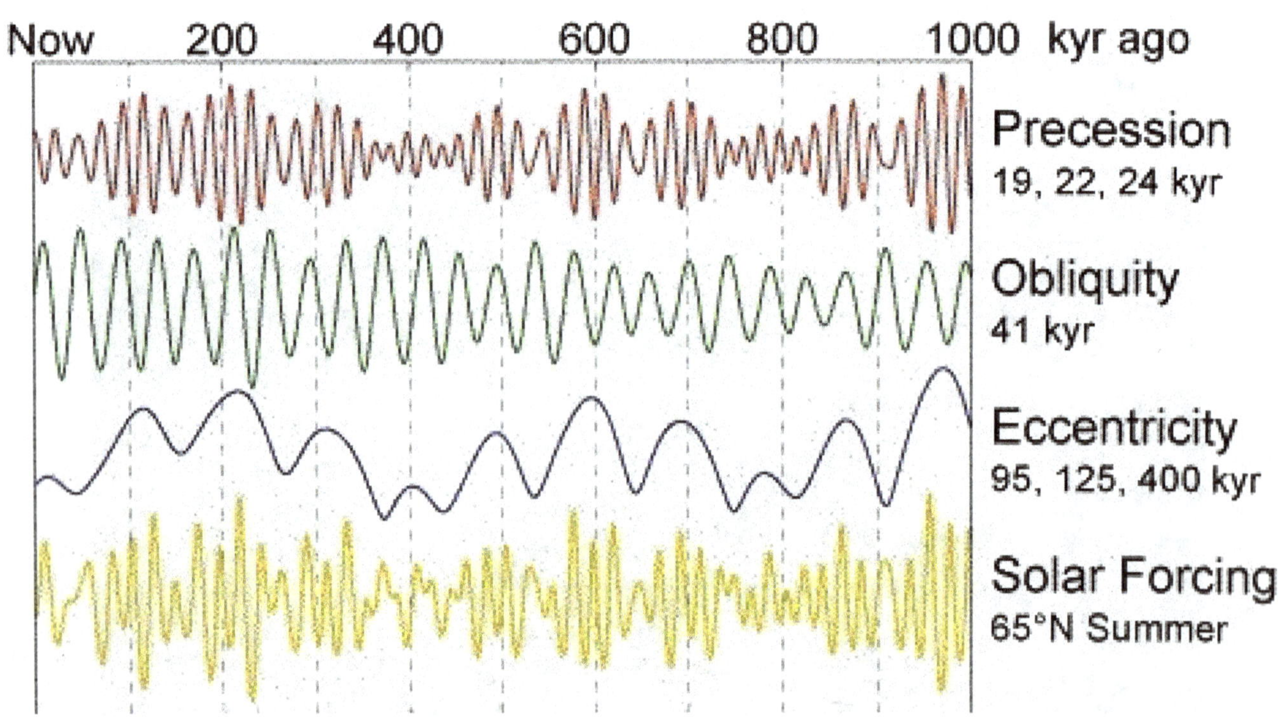

(Source: Smithsonian Institution for Science)

A question that should be asked, does the Milankovic cyclic theoretical astrophysical variations have any resemblance of correlation with the real world Vostok ice core samples? To address that question, it was shown in Chapter 1 when analyzing the known valid scientific databases for surface temperature variation that there is a correlation with the Vostok ice core database which does correlate very well with the Milankovic team calculations. For Readers convenience, that graphical correlation is repeated here:

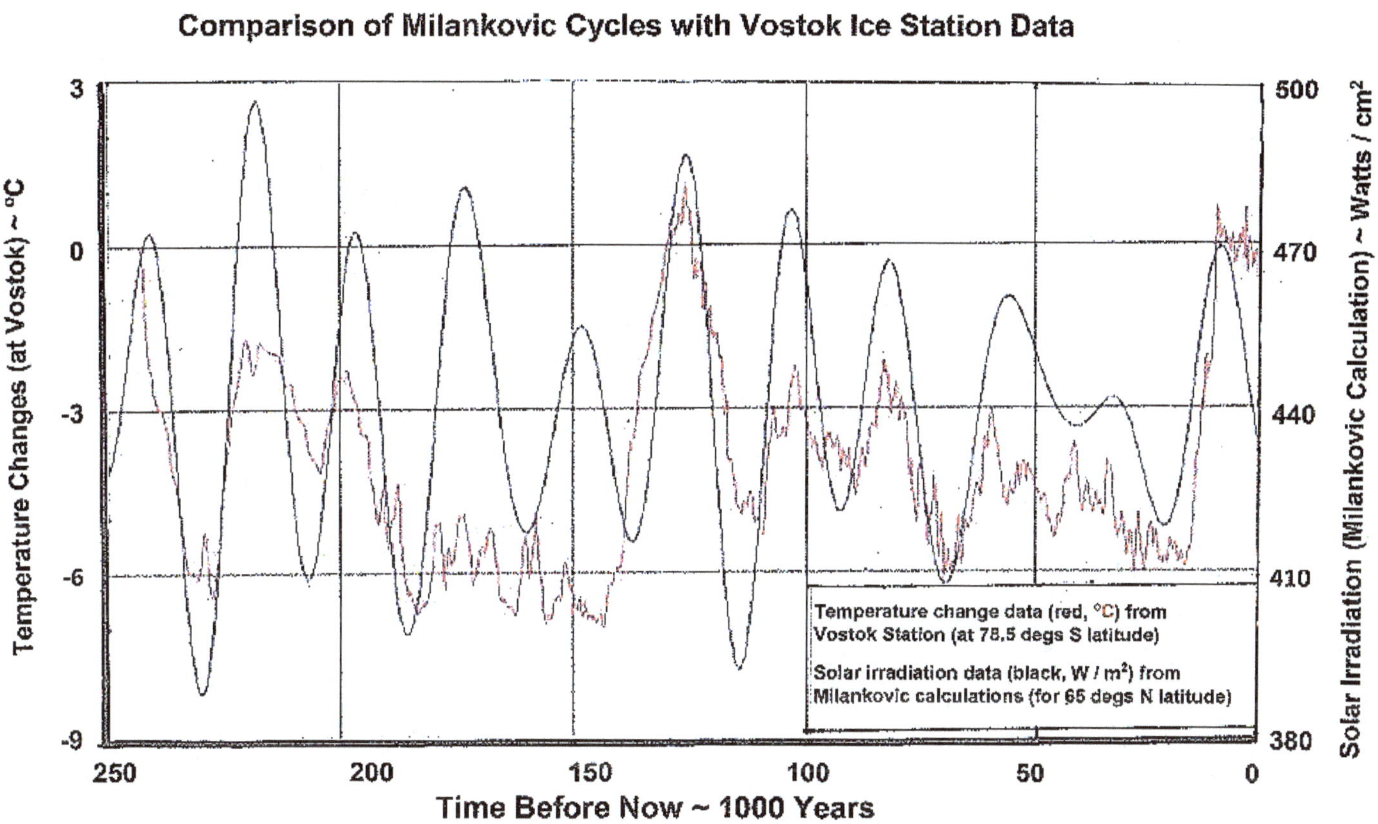

(Source: Author Correlation from Databases in Public Domain)

The Milankovic team calculations made no correction to their astrophysics calculations (in black) to account for the atmospheric changes that surely had occurred on the planet during the 250,000 years shown in the correlation. The Vostok ice cores (in red) inherently have any atmospheric changes giving the ice core data an "eroded" appearance compared to the "ideal" Milankovic calculated data. The Milankovic mathematical team also had selected 65 degrees north latitude (corresponding to approximately the latitude of Iceland and northern Finland and Sweden) for their calculations, whereas the Vostok Ice Core Research Station in Antarctica is at 78.45 degrees South latitude. That huge difference in latitude raises the question of whether the atmosphere has different characteristics in composition as well as to the different obliquity of the angle of sun's solar radiation at the greatly different latitude between 65 degrees North and 78.45 degrees South.

The NASA satellite database does show a variation in measured carbon dioxide with latitude as shown next:

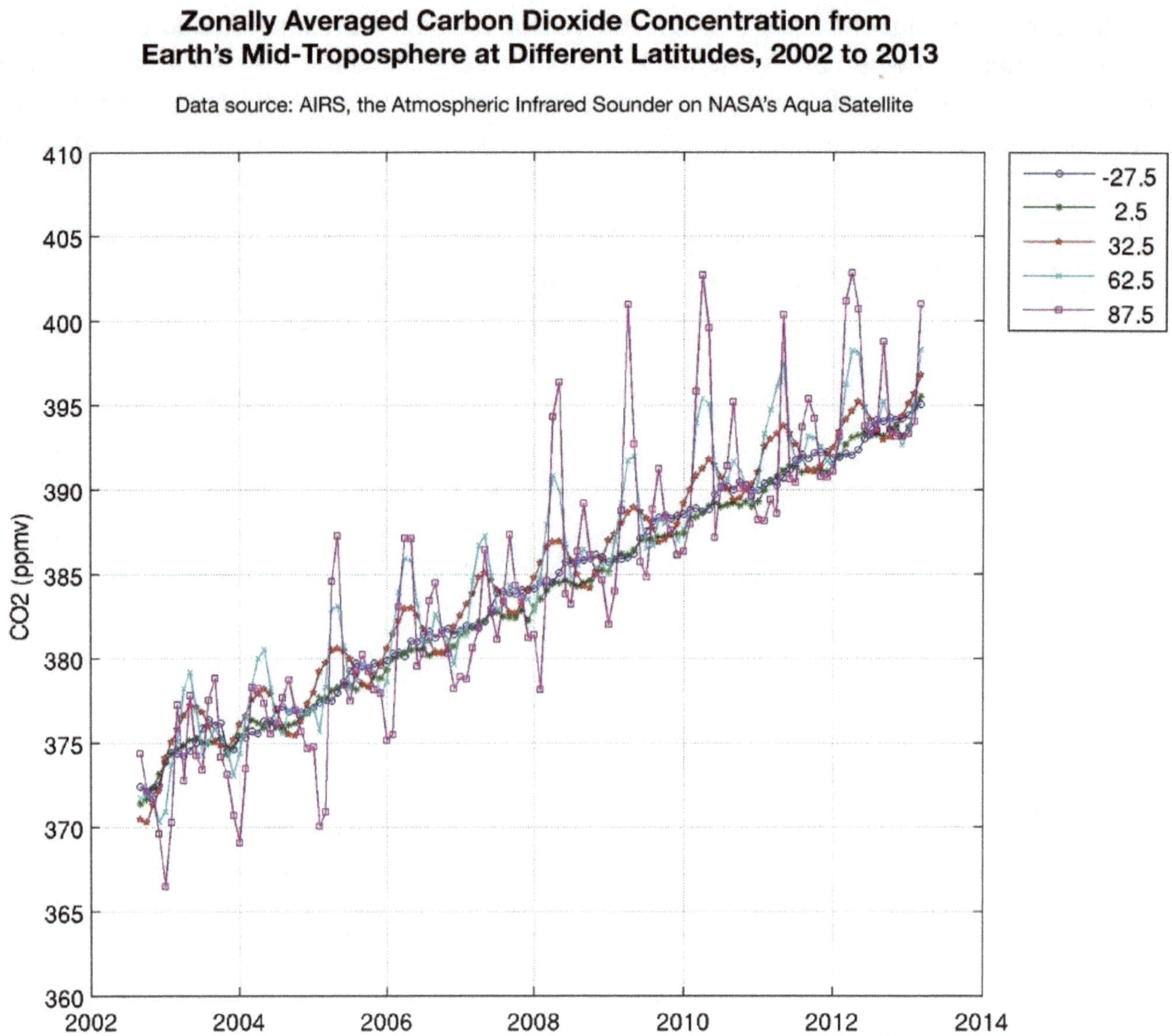

(Source: NASA)

(Author's Note: The Author has no formal education in astronomy or astrophysics and defers detailed scientific discussion of those details of science to those far more educated and experienced in those topics.)

It is from a system engineering perspective and letting the data "speak for itself" that the above "big picture" observations for the effect of latitude are made. The Author's background in data statistical analysis, however, caused the Author to analyze the meteorological data existing for worldwide cities located at various latitudes to further explore the effect that latitude may have on the surface temperature (which is not shown in the Milankovic team calculations singularly conducted for a latitude of 65° North. Because this chapter addresses only the astrophysical factors, latitude was considered to be more appropriately discussed in the geophysical aspects of the planet as contained in the next chapter. The effect of latitude was found and is shown to be one of the most profound effects on the surface temperature of the planet and Readers are strongly advised to read – and understand – the overlap in astrophysical and geophysical influences on the planet's surface temperature described further in the next chapter.

The peaks and valleys of the Milankovic theoretical astrophysical database for surface temperature and solar radiation occur at approximately the same cyclical variation with time as do the Vostok ice core samples indicate occurring. There is quite a difference in the "smoothness" between the two databases and that is attributed to the fact that local weather (temperature, pressure, and precipitation) and local climates fluctuate much depending on latitudes and everywhere with centuries of time lapses. Those meteorological variations are captured as they occurred in the ice sheet covering Antarctica – but are not accounted for in the Milankovic "theoretical" calculations. Irrespectively, the Milankovic and Vostok database lose their factual significance if those databases were to be "averaged" or "linearized" simply for scientific convenience.

A transition from the astrophysics portion of the databases, described in earlier sections and in this chapter, to the geophysical portion of the database (which is of probably higher interest to the general public (as it describes the land *terra firma* upon which they are standing) is made in the next chapter.

Chapter 3
Geophysical Factors Affecting the Planet

The category of geophysical factors affecting global warming, climate change, and/or ocean surface levels rising – due to any causal effect – is a major element of this book. This chapter is lengthy due to the many geophysical factors that have a significant causal effect on the planet's surface temperature and/or climate. Man-made factors having a causal effect on the planet's surface temperature or climate are discussed in a later chapter.

Readers were given an introduction in the preceding chapter to the astrophysical factors as the spherical planet Earth travels in its irregular elliptical orbit of the Sun – one orbit every 365.24 days - and rotates once every 24 hours about its tilted, wobbling axis as it travels. The planet can be regarded as the vehicle on which each species of life travels through time and space during the years long span of life expectancy for that species. Passengers on the planet must survive whatever environments – solid, liquid, or gas - and forces and accelerations they will be subjected to during the planet's dynamic and hazardous travel on its "programmed" never-ending travel route. The largest geophysical factors that must be considered for the effects on global warming, climate change, and oceans levels rising are those of the oceans and other large bodies of water covering over 70% of the Earth's surface, the location of humans being near those bodies of water – which serve as heating and cooling sources, the latitude of the cities relative to the Equator, and the elevation of the cities (if above 1,000 feet in average elevation).

In the Introduction of this book, it was shown that the astrodynamics (described by Milankovic) and geophysical effects of latitude, elevation, and proximity to a large flowing body of water on cities (both in the "lower 48" of the USA and in other locations) in the Northern hemisphere had a very pronounced effect on the annual average temperature of cities in that hemisphere ("top half" of the planet). The Northern hemisphere was chosen first because over 60% of the human population reside in the Northern hemisphere and it has by far the largest industrial activities on the planet in the northern hemisphere serving the population needs.

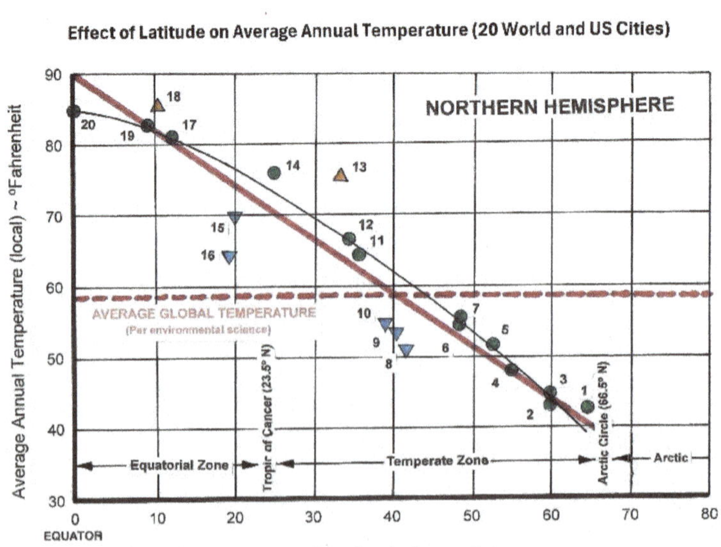

(Source: World Meteorological Organization database, chart by Author)

The Equator is numbered 1 as the reference latitude (as well as being at zero latitude and zero elevation) and the other 19 cities are numbered 2-20. As seen by the deviations from the solid red line on the graph, elevation of the city and its proximity to an ocean current – being warm or cold - are important geophysical factors that cause the data points to be either above or below the solid red line at their known latitude. A closer statistical "fit" to the data points is seen on the chart as a thin curvilinear black line; however, it is shown later in this book that the solid red line can be modeled as a simple point slop line whereas a quadratic expression is needed to describe the black line as a model.

There is no valid scientific reasoning to not believe that the latitude effect on temperature has the same effect on cities in their unique latitudes distributed over the entire planet. To verify the effect of latitude in the Southern Hemisphere is the same as seen for 20 major cities in the Northern Hemisphere and north of the Equator, a set of 20 cities were selected for the similar analysis in the Southern Hemisphere below (south of) the Equator. The next illustration in tabular form identifies the cities selected in the Southern Hemisphere for a similar analysis of their temperature variation with latitude:

World Cities Showing Latitude Effect on Temperature

(Southern Hemisphere)

	City	Elevation	Latitude	Avg. Ann. Temp.
1	Equator	0 ft (0 m)	0° 0'	85.0 °F (29.6 °C)
2	Belem, Brazil	30 ft (10 m)	1° 27'	87.4 °F (30.8 °C)
3	Guayaquil, Ecuador	16 ft (5m)	2° 11'	78.1 °F (25.6 °C)
4	Papua, New Guinea	2,188 ft (667 m)	5° 30'	81.5 °F (27.5 °C)
5	Jakarta, Indonesia	26 ft (8 m)	6° 10'	82.5 °F (28.0 °C)
6	Lagos, Nigeria	39 ft (12 m)	6° 27'	84.8 °F (29.4 °C)
7	Lima, Peru	7,956 ft (2,426 m)	12 ° 04'	63.5 °F (20.5 °C)
8	Darwin, Australia	26 ft (8 m)	12° 26'	81.3 °F (27.4 °C)
9	La Paz, Bolivia	11,975 ft (3,650 m)	16° 30'	70.0 °F (20.9 °C)
10	Rio de Janiero, Brazil	1,047 ft (319 m)	22° 54'	74.5 °F (23.6 °C)
11	Sao Paulo, Brazil	2,621 ft (799 m)	23° 33'	65.0 °F (18.0 °C)
12	Johannesburg, S. Africa	5,297 ft (1,615 m)	26° 12'	60.6 °F (15.9 °C)
13	Durban, S. Africa	272 ft (83 m)	29° 53'	77.2 °F (25.1 °C)
14	Santiago, Chile	1,887 ft (575 m)	33° 26'	61.7 °F (16.5 °C)
15	Sydney, Australia	190 ft (53 m)	33° 52'	64.5 °F (18.0 °C)
16	Cape Town, S. Africa	174 ft (53 m)	33° 55'	61.0 °F (15.9 °C)
17	Buenos Aires, Argentina	82 ft (25 m)	34° 36'	60.0 °F (16.0 °C)
18	Auckland, New Zealand	105 ft (32 m)	36° 51'	64.7 °F (18.2 °C)
19	Melbourne, Australia	607 ft (185 m)	37° 49'	59.5 °F (15.3 °C)
	"Average Global Temperature"(pre-IR)	0 ft (0 m)	Everywhere	59 °F (15 °C)
20	Punta Arenas, Chile	112 ft (34 m)	53° 10'	44.5 °F (6.9 °C)

Note: The "average global temperature" calculated by WMO in 1990 became 16 °C (60.8 °F)

(Source: World and US Geodetic Organizations; World and US Meteorological Organizations. Compilation and analysis by Author)

As done for the cities in the Northern Hemisphere, the tabulation for the cities in the Southern Hemisphere is transformed into the easier to interpret graphical form shown next:

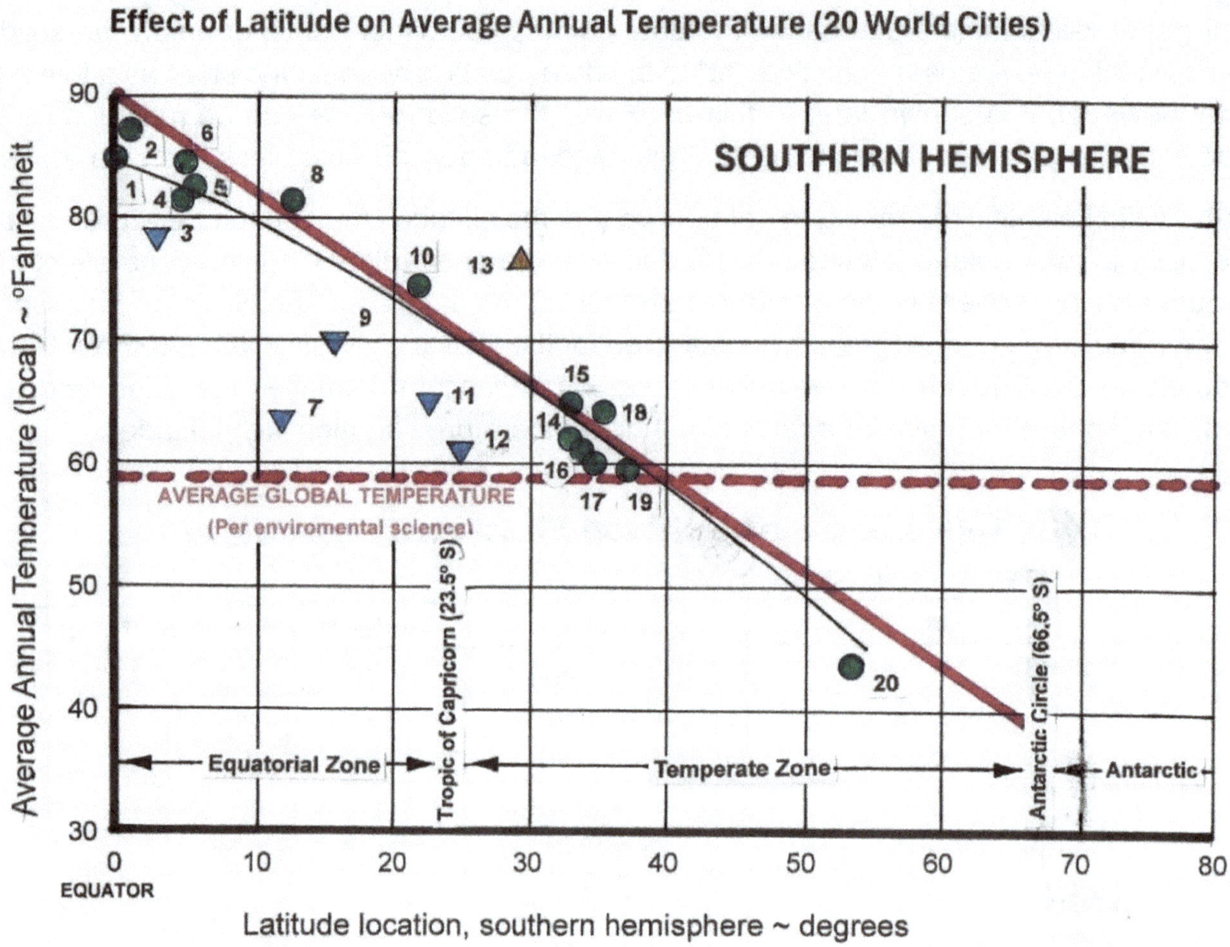

(Source: Table of Latitude Effect represented graphically, by Author)

It can easily be seen by comparing the illustrations of latitudinal effect on temperature of both the Northern Hemisphere and the Southern Hemisphere to have almost the same curvilinear variation (the thin black line) and to also have the same average linear variation (the solid red line) of temperature with latitude variation. For mathematicians or for environmental scientists wishing to model the effect of latitude on average annual temperature, the solid red line can be expressed in a simple mathematical equation for a sloped line as:

Temp (° F) = 90 °F – 0.78 L (°)

providing an estimate of the annual average temperature for cities at different latitudes in either hemisphere (North or South). In metric convention of units, the red line "model" of latitude effect on temperature is expressed as

Temp (°C) = 32.2 °C – 0.43 L (°).

Once again, the environmental scientists using the "average global temperature" approach (the horizontal dashed red line) shows no variation with latitude and is considered to be a scientific descriptor having a sophomoric level of scientific importance in describing the complexity of climatology. Mathematically, the environmental scientists' "model" of the one-size-fits-all average global surface temperature of the planet is expressed as Temp (°C) = 16 °C or in Imperial units Temp (°F) = 61 °F.

The public and much of government has been misled (primarily via the sensationalistic news media) to believe that a small rise of 2 °C (3.6 °F) above the average global temperature (the dashed red horizontal line) will cause a "crisis" to the planet being destroyed by an inferred apocalyptic set of consequences. The fallacy of that misinformation is seen graphically above – raising the horizontal dashed red line vertically upward an increment of 2 °C might change the 19 data points representing cities below latitude 40°.to be 2 °C (3.6 °F) hotter than present. If that measure of the environment being raised by 2 °C translates to the planet being thrown into hyperthermic chaos, why are those cities below 30° latitude not already in a temperature crisis state of chaos? In a later chapter, describing the importance of water on the planet, the article appearing in the Climate section of a scientific magazine was published without its emotional presentation descriptors being "fuzzed" out – and inadvertently revealed what the environmental scientists are concerned about:

CLIMATE

Grim results of the ocean's hot streak

Off-the-charts temperatures impact corals, sea ice and more

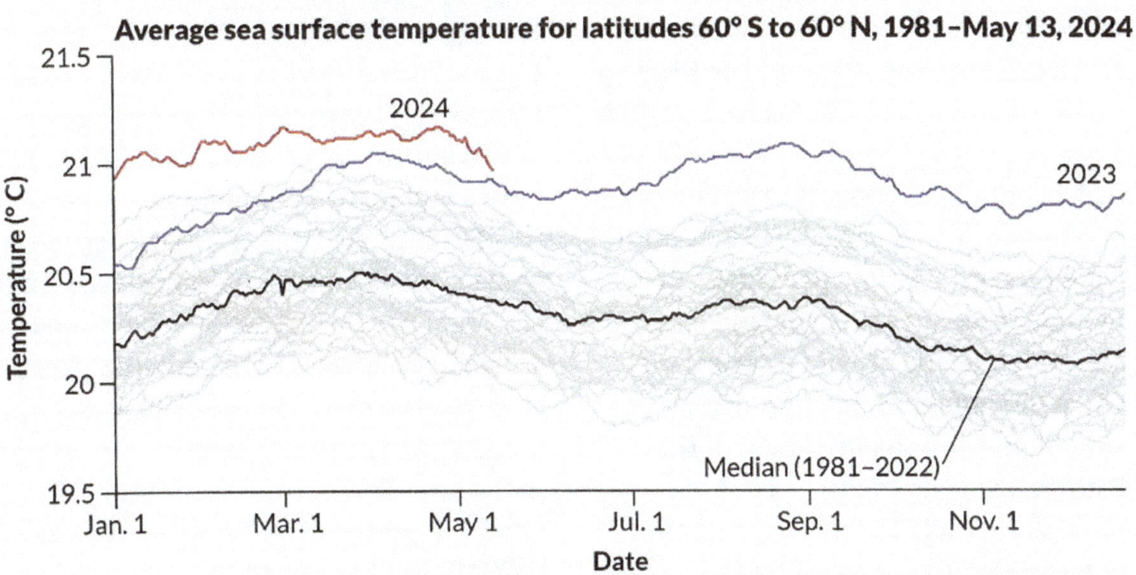

Soaring sea temps Since March 2023, the average sea surface temperature of the extra-polar ocean, at latitudes from 60° S to 60° N, has broken the daily record every single day, NOAA data show. And recent daily temperatures have shattered records that were set last spring.

SOURCE: CLIMATE REANALYZER/UNIV. OF MAINE; NOAA

(Source: Climate Reanalyzer, University of Maine, NOAA)

Readers should note several comments made to arouse public fear over Man-made "greenhouse gases" trapping heat and which is causing the "average sea surface temperatures between latitudes 60° S to 60 o N to soar". That "soar" is measured on the temperature scale (left vertical) to be about 1o C from 1981 to 2024 and is "shattering records". The largest "slip-of-the-tongue" comment is in the second line below the "Grim results of the ocean's hot streak" which reveals the concern: *"Off-the-charts temperatures impact corals, sea ice and more".* This Author wonders if the "*and more"* includes only other aquatic biological matter or perhaps includes humans? A minor wonder is exactly where are the "off the charts" data shown to be "*soaring off the chart*" – it all appears to be there?

It is also what is not on the chart that causes a wonderment. How does NOAA explain the "El Nino" effect that occurs on a greater than seasonal cyclical period and causes damaging flooding of the US Pacific Coast states - primarily Southern California? The "average sea surface temperature" method of analysis averages out the important anomalies that causes human fatalities and property damage that do not "soar off the chart". Southern Californians refer to that huge flow of water from the Hawaiian Islands region (where the massively warm "El Nino" Pacific Ocean current – a mischievous "little boy" in Spanish originates) and causes flooding in Southern California. In the "big picture", the Southern Pacific jet stream above the warm El Nino waters accelerates in a generally Easterly direction as shown in the next chart

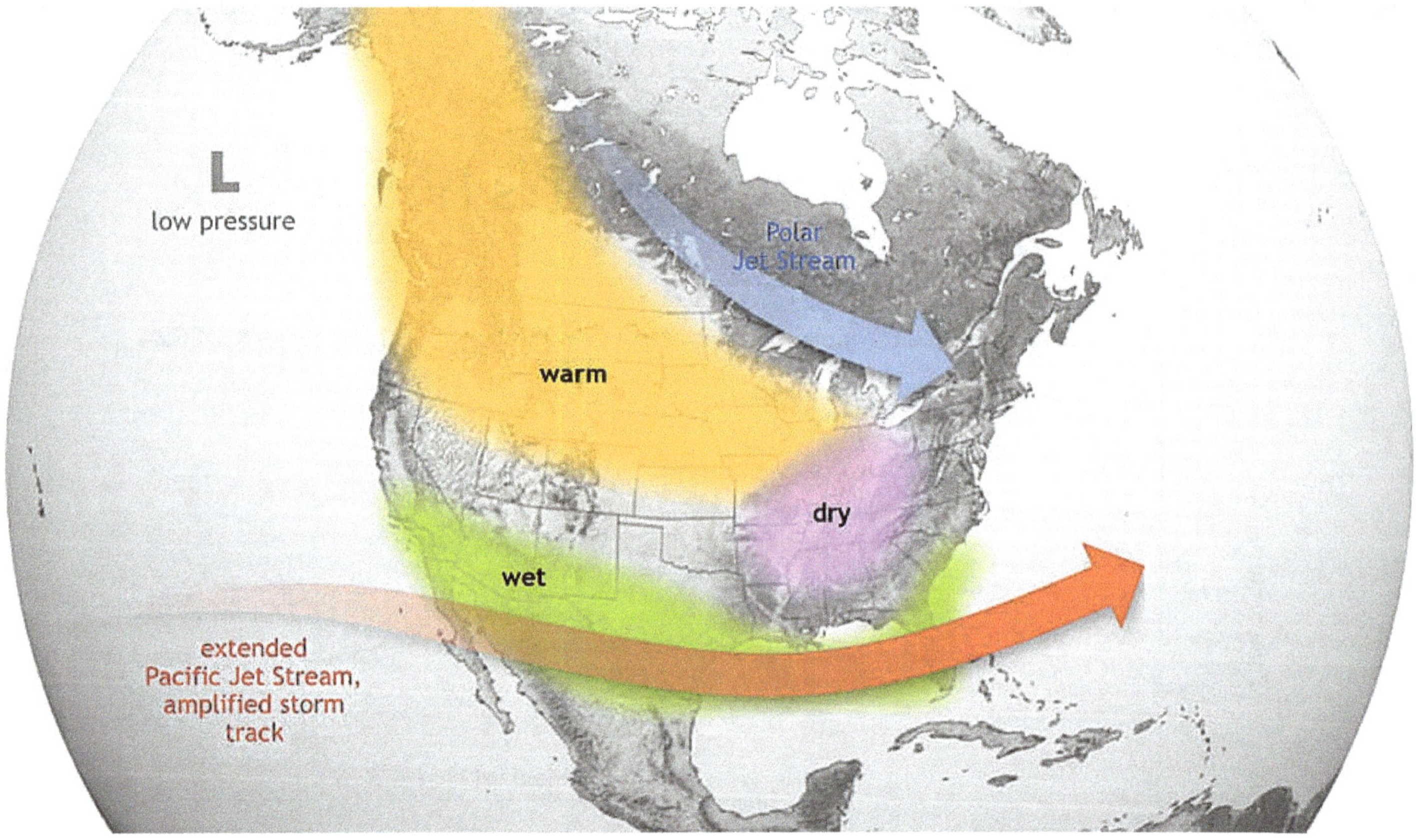

(Source: NOAA)

Californians are forewarned (by the vital National Weather Service of NOAA) of the formation of El Nino events – which occur on an irregular 2-7 year cycle - and refer to the wet, fast-flowing storm (in a swath between latitudes about 18° North and 34° North) as "The Pineapple Express". For reference purposes, the latitude of Hawaii is about 21° North, San Diego, California, is at 32° North, Mexico City is at 19° North, and Jacksonville, Florida, is at 30° North. The green areas on the diagram above get much rainfall in a very short period and flooding occurs because the massive precipitation (water) cannot seep into the ground rapidly enough and snow in the higher mountain elevations accumulates rapidly.

Another point of special interest to be noted is that the city number 20 (Punta Arenas, Chile) is the last data point - at latitude 53° 10' S on the earlier graph of latitude effect on temperature - has an average annual temperature of 44.5 °F (6.9 °C). The geophysical significance of that latitude is that Punta Arenas is at the very southern tip of South America and is the populated area in the Southern Hemisphere closest to Antarctica. Punta Arenas has an annual average temperature that is about 16 °C (29 °F) below the horizontal dashed red line representing the environmental scientists' "crisis" reference line. Environmental scientists claim that if/when the average global temperature exceeds the environmental science threshold "crisis" temperature by about 2 °C (3.6 °F), the environment and the human population will face disaster. The residents of Punta Arenas are not known to be wringing their hands in despair that the temperature of their environment will rise by 2 °C or more. There is no apparent cause to declare an environmental "crisis" in Punta Arenas over Man burning fossil fuels – the 128,000 residents of Punta Arenas have learned to wear a long-sleeve cotton shirt instead of a woolen sweater to stay comfortably warm in Punta Arenas. A much-appreciated humanitarian action would be for the Punta Arenas residents to send their no-longer needed sweaters to the 140,000 residents in Reykjavik, Iceland, or to the 5,000 residents in Point Barrow, Alaska, who then can wear the sweaters from Punta Arenas instead of their present heavy parkas).

To a professional statistician, the more important question is what are the other factors not seen that are causing the data points to not be more closely "spotted" on the solid red line – i.e., latitude is an important effect but what else other than latitude is affecting the average annual temperature of each specific city being analyzed? Two geophysical factors possibly affecting the temperatures are noted to be:1) some of the cities are near a large body of water (which is a "heat sink"); and that 2) the elevation of the city places the city above the heat sink water to which it lies in proximity.

A most important point is that it is not latitude alone that affects the average annual temperature of a location; in fact, there are several other factors and, as already mentioned, "greenhouse gas effects" are factually very low on the list for having a significant effect (if any) on the planet's temperature – contrary to environmental scientists' hypothesis of being "dominant". A good example of that is to view the earlier latitude effect on temperature graph and notice that there is a "rogue" data point number 13 (orange, point side up triangle) – representing the city of Durban in South Africa. The annual average temperature of Durban (at elevation 272 ft (83 m) is 77.2 °F (25.1 °C) and is about 10 °F warmer than the solid red line and thin solid black line at its latitude of 29° 53' S. It is also about 18 °F warmer than the environmental scientists' "average global temperature" of 59 °F (15 °C). If the environmental scientists' goal is to hold the annual global temperature rise to no more than 1.5 °C (2.7 °F) by 2050 or a "crisis"/"tipping point" will destroy human life, how is it possible that the residents of Durban, South Africa, are already 18 °F (10 °C) above that "crisis"/"tipping point" temperature and yet are still alive?

An important factor of influence is that the location of the city to a nearby ocean, with a current traversing the coast at that location, can have its climate affected by that ocean current – varying between very warm or very cold. That ocean current temperature effect on coastal cities is vividly shown in the ocean currents at the southern region of South Africa, and how those currents can cause a most significant difference in temperature for two cities apart by only 4° of latitude – Durban (at 29° 53' S latitude) and Cape Town (at 33° 55' S latitude). The straight-line distance between Cape Town and Durban is about 840 miles (1,352 kilometers), and they are influenced by two different currents as seen on the next map of the oceans and land masses surrounding South Africa:

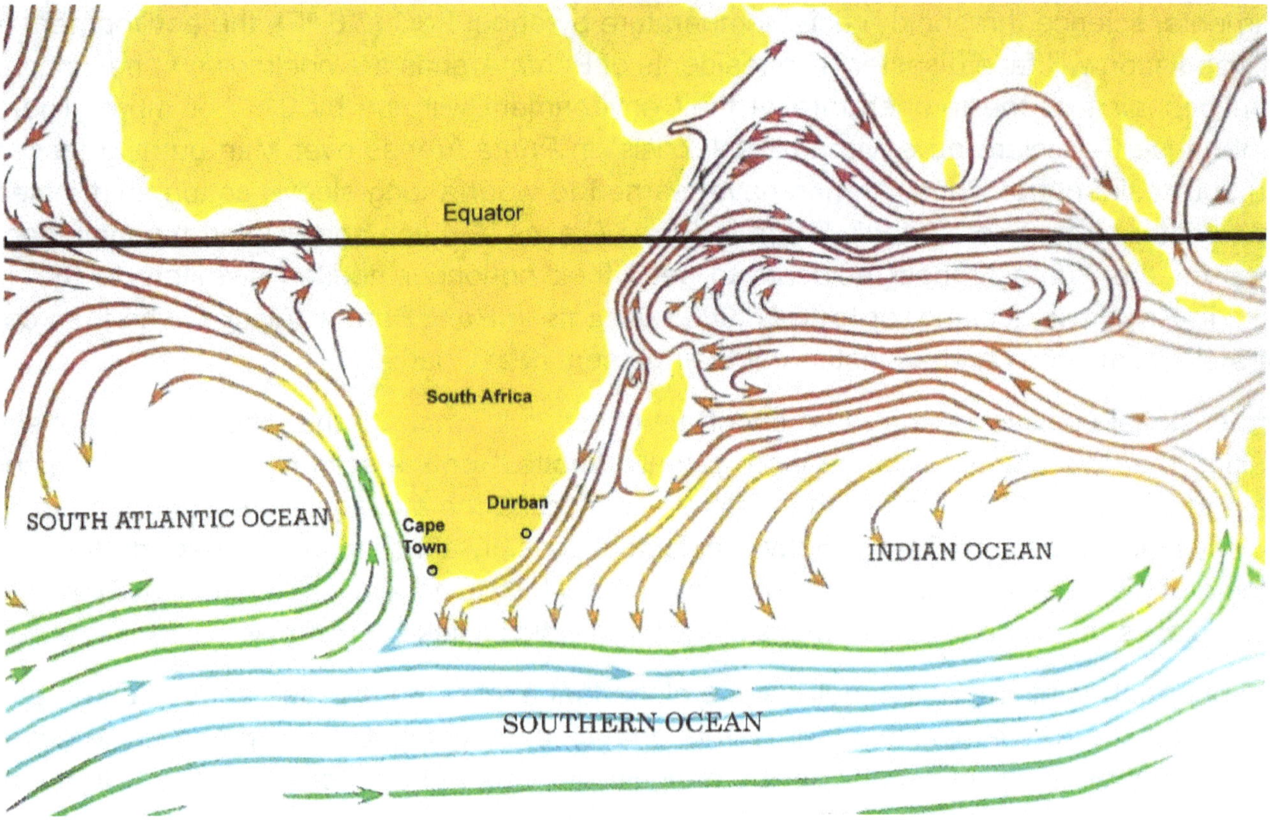

(Source: Basic map from Rand McNally Atlas of the World, with Author notations)

The average annual temperature of Durban (city number 13 on the previous Southern Hemisphere illustration) - warmed by the southwesterly current of the southern Indian Ocean - is 77.2 °F (25.1 °C). In contrast, the city of Cape Town (city number 16 on the previous Southern Hemisphere illustration) is influenced by the northeasterly branch of the current from the easterly-flowing cold Southern Ocean current and has an average annual temperature of 61.0 °F (15.9 °C). There is only a 4° latitude difference in the cities of Durban and Cape Town, but a 16 °F (9 °C) difference in their average annual temperatures. As can be seen on the Southern Hemisphere illustration, it is Durban (number 13) that is much warmer (about 10 °F) than is the solid red line estimate on the earlier chart showing latitude effect on average annual temperature for that latitude.

The elevation of a city high above the ocean surface level also can have an effect on its annual average temperature, but the elevation difference between that of Durban (at an average elevation of 170 meters (558 feet)) and that of Cape Town at sea level (0 meters and 0 feet) is not that large a difference in elevation to cause a difference in their respective annual average temperatures. A geophysical conclusion derived therefore, is that with no significant difference in latitude or elevation, the primary factor causing the 16 °F (9 °C) annual average temperature difference between Durban and Cape Town is attributed to their being exposed to the different temperatures of the two different ocean currents that flow past their coastal locations. Scientifically, there should be a dependency of every coastal city (as many thousands of them are coastal on the planet) on the temperature of the water in the large body of water flowing on their coastal boundary and the prevailing winds being a mixture of both on-shore or off-shore. Anyone familiar with the weather and climate of Southern California well understands the on-shore and off-shore wind phenomenon on the average annual temperature of the San Diego, Santa Monica, Malibu, and Zuma Beaches.

That leaves the question of the effect of elevation on the data points on the latitude effect figure for the Southern hemisphere shown on the figure on earlier pages of this chapter. It can be seen on the southern hemisphere latitude effect figure that the data points below the solid red line vary in distance (i.e., are at a colder temperature than the solid red line estimates for the latitude effect) from that line. The points numbered 4, 7, 9, 11, 12, and 14 represent the six cities among that group that have substantial elevation above the ocean surface level of the nearest ocean. For readers' convenience, the small tabular listing shown next provides the "raw" numbers that have obscure visual meaning when considered only in tabular form.

			Elevation		Average Temperature	
Point Number	**City Name and Location**	**Latitude**	**Feet**	**Meters**	**°F**	**°C**
4	Papua, New Guinea	5.5° S	2,188	667	81.5	27.5
7	Lima, Peru	12.0° S	7,956	2,426	63.5	17.5
9	La Paz, Bolivia	16.5° S	11,975	3,650	44	6.7
11	Sao Paulo, Brazil	23.5° S	2,621	799	65	18.3
12	Johannesburg, S Africa	26.2° S	5,297	1,615	60.6	15.9
14	Santiago, Chile	33.5° S	1,887	575	61.7	16.5

(Source: World Meteorological Organization)

The effect of elevation on the average annual temperature of a city is approximated by an altitude / Outside Air Temperature / ambient Pressure relationship used in the estimation of aircraft flying at different altitudes in the atmosphere. That relationship appears in Aeronautical Engineering textbooks and is established by the International Civil Aircraft Organization shown below as:

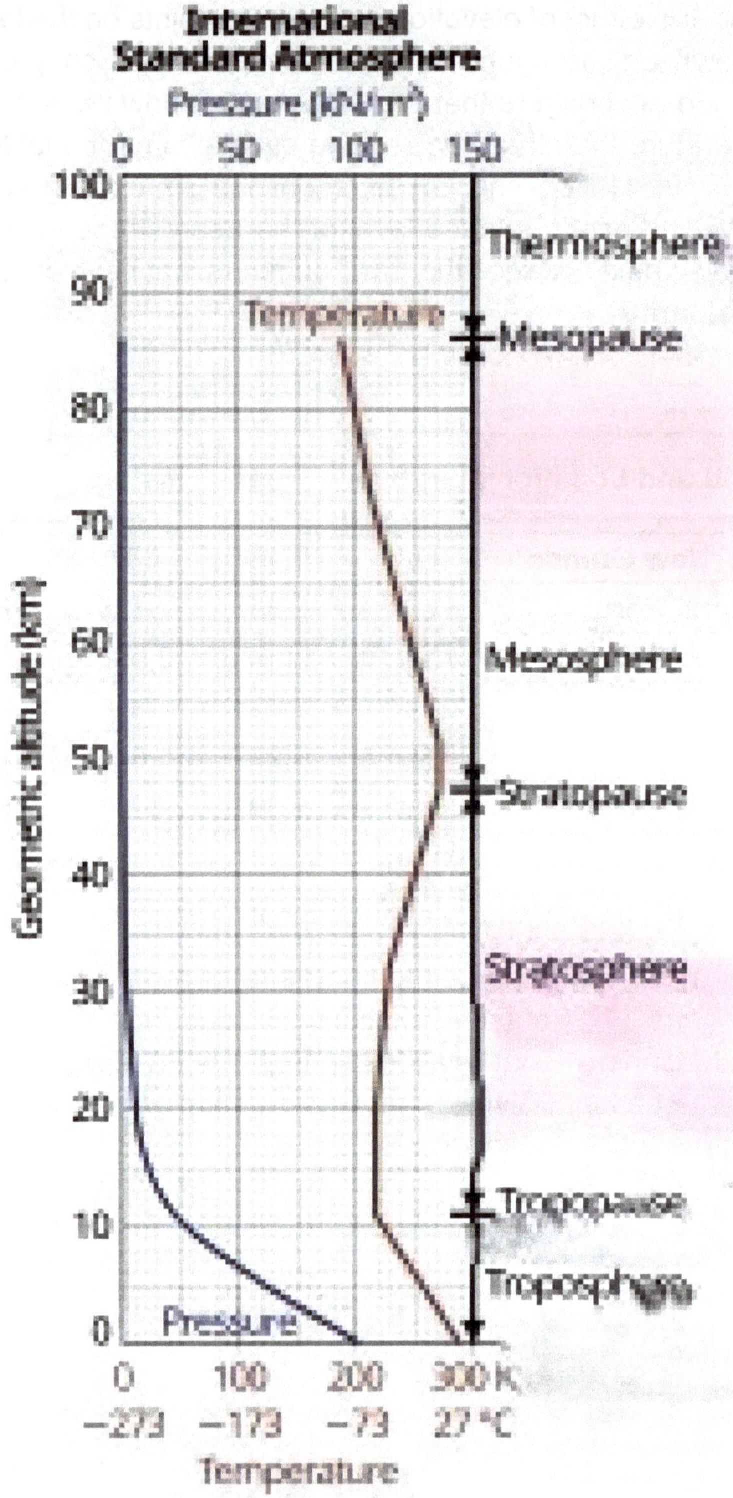

(Source: International Civil Aircraft Organization, ICAO)

The decrease in temperature for the tropopause portion of the atmosphere is referred to as lapse rate and varies linearly from 59 °F (15 °C) at Sea Level (elevation 0) to -65 °F (-54 °C) at 36,089 feet (11 kilometers) altitude/elevation. The linear red line (for temperature) is mathematically defined by the point-slope method to be:

Temp (°F) = 59 °F – 3.44 Elev (1,000 ft) in Imperial units, and

Temp (°C) =15 °C – 6.26 Elev (kilometers) in Metric units.

The "lapse rate" (i.e., the decrease in temperature as elevation increases) is the last half of the equations: -3.44 Elev (1,000 ft) Imperial and -6.26 Elev (kilometers) Metric.

Using the two lapse rates (for Imperial and for Metric) as corrections for the effect of elevation to the solid red line estimates for the effect of latitude on temperature, the correction (i.e., to bring a closer correlation with the solid red line for the six cities tabulated earlier) becomes:

(Imperial) Temp (°F) = 90 °F – 0.78 Lat° – 3.44 Elev (1,000 ft); and

(Metric) Temp (°C) = 32.2 °C – 0.43 Lat° – 6.26 Elev (km)

There was found no mathematical modeling for the effect of ocean proximity and the prevailing wind effect on the estimated temperature on a regional basis, but those can be treated as being corrections (deltas) to the above general equations modeling the combined effects of latitude and elevation. Without any basis other than making an informed estimate (observing that the oceans are warming in the latitudes where most of the population resides – and allowing that the atmosphere has in fact warmed by about 2 °C in the past century), adding an arbitrary 2.8 °C and 6 °F temperature increment to the LEO model as a first order effect of ocean temperature rise and proximity is considered sufficient in making LEO a very good model for estimating the surface temperature of the planet at any latitude, elevation, and proximity of the effects of the atmosphere and the oceans becoming warmer. Adding that temperature increment for the estimated effect of oceans' warming and their proximity to populated coastal cities, the improved LEO model becomes:

(Imperial) **Temp (°F) = 96 °F – 0.78 Lat° – 3.44 Elev (1,000 ft);** and

(Metric) **Temp (°C) = 35 °C – 0.43 Lat° – 6.26 Elev (km)**

The above models are not perfect but are infinitely more reliable for estimating the geophysical effects on the land, water, and atmosphere of the planet than is the "one-size-fits-all" average global temperature (a constant) in present use by environmental scientists.

The Author does not seek any recognition for formulating the above modeling (assigned the acronym LEO – Latitude, Elevation, and Ocean proximity), but requests only that the LEO method be used in studies and analyses being conducted by scientists, technologists, engineers, and mathematicians (i.e., the STEM community) in the United States - and in other regions of the world - when conducting their activities in developing (or improving) those technologies that may be useful in mitigating the inevitable effects of climate change.

If the WMO (World Meteorological Organization) database is ever destroyed (unlikely, but even if only remotely possible), the Latitude/Elevation/Ocean proximity (LEO) model can be of assistance in partially describing the estimated average annual temperature for any city at any latitude on the planet and with reasonable accuracy compared to what is recorded in the WMO database. A comparison was made for the six cities used in deriving the elevation effect comparing the LEO estimated temperatures with the WMO recorded temperatures to indicate the comparative accuracy of the LEO model:

Point Number	City Name and Location	Latitude	Elevation Feet	Elevation Meters	Average Temperature (WMO) °F	(LEO) °C
4	Papua, New Guinea	5.5° S	2,188	667	81.5	27.5
		Estimate by use of LEO:			84.2	28.5
				(Difference)	(+2.7)	(+1.0)
7	Lima, Peru	12.0° S	7,956	2,426	63.5	17.5
		Estimate by use of LEO:			59.3	14.7
				(Difference)	(-4.2)	(-2.8)
9	La Paz, Bolivia	16.5° S	11,975	3,650	44.0	6.7
		Estimate by use of LEO:			41.9	5.1
				(Difference)	(-2.1)	(-1.6)
11	Sao Paulo, Brazil	23.5° S	2,621	799	65.0	18.3
		Estimate by use of LEO:			68.7	19.9
				(Difference)	(+3.7)	(+1.6)
12	Johannesburg, S Africa	26.2° S	5,297	1,615	60.6	15.9
		Estimate by use of LEO:			57.3	13.6
				(Difference)	(-3.3)	(-2.3)
14	Santiago, Chile	33.5° S	1,887	575	61.7	16.5
		Estimate by use of LEO:			63.4	17.0
				(Difference)	(+1.7)	(+0.5)
		Average Net Difference:			-0.3 °F	-0.6 °C
		Maximum + Difference:			+3.7 °F	+1.6 °C
		Maximum - Difference:			-4.2 °F	-2.8 °C

(Source: Meteorological Data from World Meteorological Organization (WMO)
with use of Author Generated Latitude / Elevation / Ocean (LEO) Model)

A logical question which might be asked by Readers in the United States is “Can the LEO model be of use in understanding the factors that occur in a smaller region than that of hemispherical proportion? To pursue that question, and to satisfy the Author’s affliction with the Curse of Completeness, six large cities were selected in the United States to do a similar comparison for accuracy of the LEO model in the smaller region of the ”lower 48” (Conus) states of the United States. The six cities were shown earlier as being among the 20 in the Northern Hemisphere (and are numbered from that set of 20 cities): 14 – Miami, Florida; 13 – New Orleans, Louisiana; 11 – Los Angeles, California; 10 – Denver, Colorado; 9 – New York, New York; and 7 – Seattle, Washington. The comparison of the World Meteorological Organization (WMO) database and the Latitude / Elevation / Ocean (LEO) Model estimate for the temperature in the six selected cities (selected as having latitude differences, elevation differences, and proximity to a large body of water (ocean, gulf, or Great Lake) is shown next:

Point Number	City Name and Location	Latitude	Elevation Feet	Elevation Meters	Average Temperature (WMO) (LEO) °F	°C
14	Miami, Florida	25.8° N	6	2	76.0	24.4
		Estimate by use of LEO:			73.9	23.9
				(Difference)	-2.1	-0.5
13	New Orleans, Louisiana	30.0° N	-7	-2	75.0	23.9
		Estimate by use of LEO:			72.6	22.1
				(Difference)	-1.3	-1.8
11	Los Angeles, California	34.0° N	791	241	67.0	19.4
		Estimate by use of LEO:			66.0	18.9
				(Difference)	-1.0	-0.5
10	Denver, Colorado	39.7° N	5,525	1,684	54.0	12.2
		Estimate by use of LEO:			46.0	7.4
				(Difference)	-8.0	-4.8
9	New York, New York	40.7° N	62	19	53.0	11.7
		Estimate by use of LEO:			64.0	17.4
				(Difference)	11.0	5.7
7	Seattle, Washington	47.8° N	131	40	55.0	12.8
		Estimate by use of LEO:			58.3	14.2
				(Difference)	3.3	1.4
		Average Net Difference:			-0.3 °F	-0.6 °C
		Maximum + Difference:			+11.0 °F	+5.7 °C
		Maximum - Difference:			-8.0 °F	-4.8 °C

(Source: Meteorological Data from World Meteorological Organization (WMO)
with use of Author Generated Latitude / Elevation / Ocean (LEO) Model)

There is seen to be amazingly good estimation (better than the Author anticipated) for four of the six US cities of being less than three degrees Fahrenheit of difference for those four cities (Miami, New Orleans, Los Angeles and Seattle). The largest difference between the WMO records and the LEO estimate (warmer estimate of 11 °F / 5.7 °C) is seen in the data for New York (where the Author grew up in the first 23 years of his life). The average annual temperature method of averaging the seasonal high temperature in July with the seasonal low temperature in February masks the fact (well known to New Yorkers) that the "lake effect" of Great Lake Erie west of New York during the winter months stretches the cold weather months (the four "normal" colder "seasonal" months of November, December, January, and February) by starting in mid-October and ebbing in early March. The residents of Buffalo, New York will confirm the existence of the "lake effect" as that city records record breaking snow depths every year.

The "extra" month of the lake effect (the Great Lakes serving much like a refrigerant fluid modulating the heat energy from the surrounding and downwind areas to the East) in October and retaining that cold temperature (the "lake effect") until well into March. Much if not all of the 11 °F (5.7 °C) warmer over-estimate by the LEO model is explained as being a local anomaly. The underestimate of the Denver data point (8.0 °F (4.8 °C) – the WMO showing Denver to be warmer than estimated by LEO –

is most likely due to the opposite of the "lake effect" and being more a "desert effect". Denver is at an elevation of 5,525 feet (1,684 meters) near the ridgeline of the Rocky Mountains that stretch north to south through the desert southwest states on both flanks of the Rocky Mountains. The sands of the desert retain the heat of solar radiation by day but that heat flows upward on the sides of the mountain chain during the evening and thereby raising the temperature in the "Mile High City" on average almost the entire length of the year except for a very brutal winter months season.

The LEO model of the surface temperature on the planet cannot be considered a "giant leap by Man" to environmental science; however, compared to the environmental scientists' one-size-fits-all "average global temperature" method (as shown earlier and which considers Man-made "greenhouse gas emissions" to be the dominant cause for global warming), the LEO model is a far more credible scientific tool in better understanding the "big picture' of climate change and the factually important factors that are causing the surface temperature of regions of the planet to be warming – and cooling - differently in different regions. Averaging is too "fuzzy" a tool for the variable of temperature on which so much of importance to human and ecological lifeforms depends and to which humans must adapt.

A common thread that will be seen throughout this book is that it is not the effect that Man is having on the environment but of far greater importance is the environment – no matter how "inconvenient" – as is the driving factor determining what Man must do to survive within the limits of that environment. The Author had a most successful career developing technologies for advancing knowledge of flight vehicles moving in the atmosphere. The technologies that are useful are those that are reliable, safe to operate, and are affordable to have. The last part of this book is an evaluation of what those technologies are for Man to use within the limits of the environment of the planet. This first part is to identify what are the factors that the environment presents to humans and form the limits within which Man must survive. For an aircraft that is termed as "the flight envelope" – for humans in the environment it is the extremes of the weather and the climate that form the "envelope" within which every human conducts his/her life.

The trifecta problems of global warming, climate change, and oceans' surface levels rising are far more about their effects on all forms of life and on technologies on the planet than just their effect on the environment.

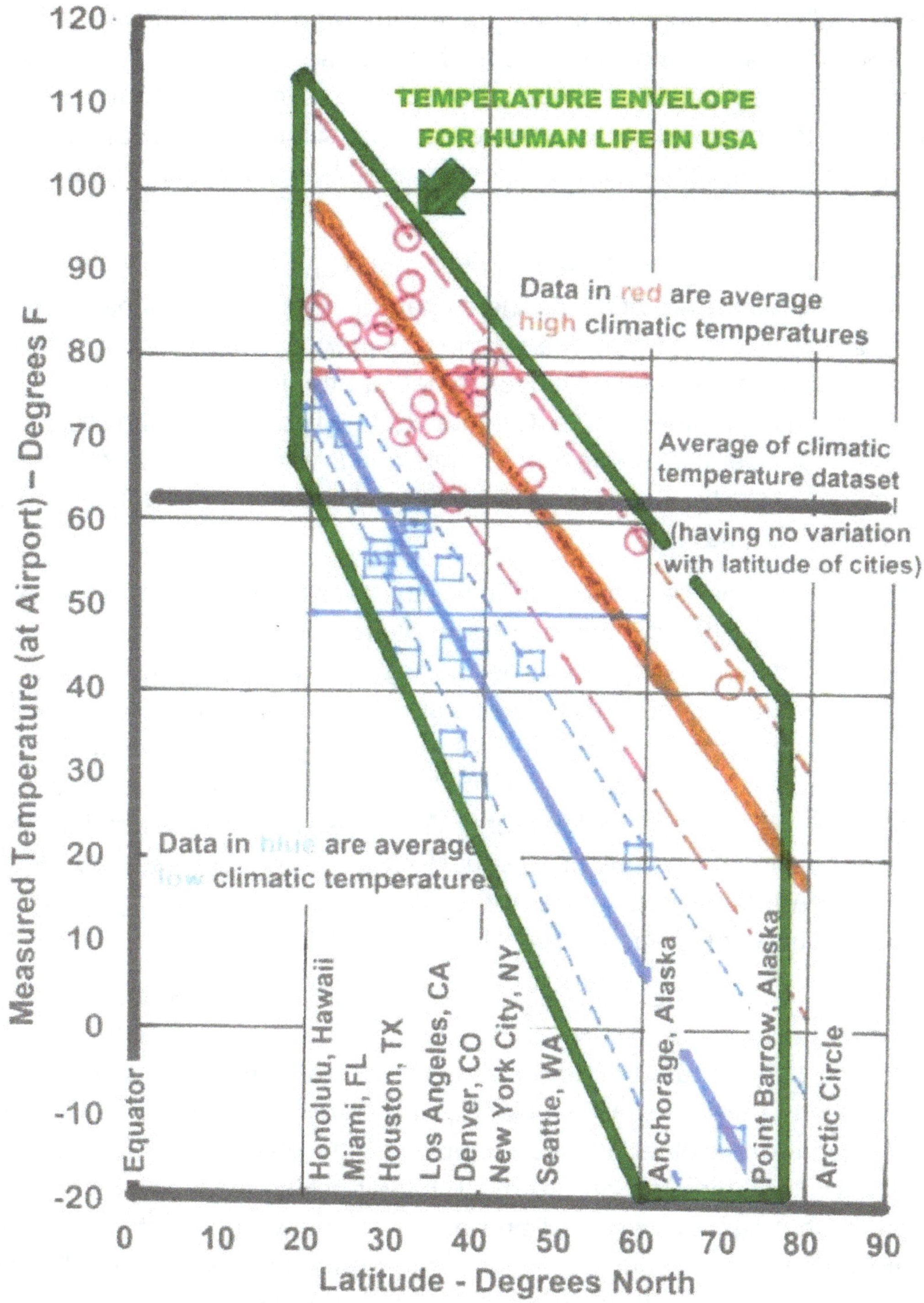

(Source: National Weather Service, NOAA) with Author Notation)

Humans within the latitudes shown above must wear the appropriate clothing and outerwear and/or control the temperature within the structure they inhabit to safely survive long term exposure to the temperatures within the temperature/latitude envelope (solid green line). Additional precautions must be taken if wind chill or heat index factors exist to reduce the limits of the temperature envelope.

Temperature is but one of the two dominant factors that affects the climate of a region – the other major variable being precipitation. The precipitation is a climatic variable and not a geophysical parameter. The precipitation aspect of climate change will be discussed in later chapters when addressing weather and climate change being more in the category of naturally-caused weather events – benign and catastrophic – and resulting in drought or flooding/mudslide/landslide events in a region. Environmental scientists have been misleading the public as to the "cause and effect" principles of science and climatology. Climate change is an "effect" not a "cause". The average annual temperature of a region is changing because of the complex combination of astrophysical, geophysical, and natural factors that are changing with time. Those changes are affecting the changes in the climate – temperature and precipitation.

Environmental scientists, eager to convince the public that Man's polluting activities (and using the ploy of *"eliminating everything else"*) are affecting the climate and creating a climate change "crisis", have perpetuated the fallacious inference that *"Man is the dominant cause of climate change"*. Man is guilty of much pollution – that does need much closer remedial attention by the entire STEM community – but Man is not guilty of changing the climate. Climate change is affecting both the environment and the human population. The scientific approach to mitigating the effects of climate change on the environment are technologically much different than what the approach should be to mitigating the effects of climate change on Man.

This chapter continues with the geophysics of the planet changing the temperature and precipitation on our land and water surface (the surface of the planet creates the extremely important interface with the atmosphere). That interface scientifically is the location where matter in solid and liquid state comprising the planet are transformed into gaseous state. Humans "see" only the visible surface of the planet and only a very limited section of that surface. An estimate is given that a six-foot tall human can see a distance to the horizon (given that the surface has no hills or structures obscuring line-of-sight to the horizon) of about 24 miles (about 39 kilometers). With those 24 miles as a radius, the area perceived by "flat-Earthers" is a maximum of about 1,810 square miles (about 4,778 square kilometers). The area of the interface between the entire planet and the atmosphere surrounding the planet is about 197 million square miles (equivalent to about 510 million square kilometers).

The ratio of the interface area between the atmosphere and the planet and the area a human can "see" is 197 million to 1.81 thousand (or in ratio form) 109,000 to 1. In a "big picture" sense, if a group of humans are concerned daily about what the weather will be like in their "What you see is what you get" in their 109,000th portion of the geophysical planet, each regional population is not overly concerned with what might happen in the other 108,999 regions of the planet in 2035. The advent of satellite weather monitoring, tropical storm tracking, and hurricane tracking technologies add to the "big picture" of climate change to be small but very real and on a regional change rather than a global average change. A very enlightening – although apparently ignored by the environmental scientists who claim that global climate change is a "settled science" – but very convincing database was presented by NASA Earth Science that conflicts with the environmental "settled science" proclamation. Using the "normal temperature" calculated over the 30-year baseline period 1951-1980, a series of illustrations for the temperature anomalies (an anomaly is the difference from "normal") was compiled from meteorological (land-based – World Meteorological Organization) and satellite (NASA space-based) records. The magnitude of temperature anomaly (between 2 °C / 3.6 °F warmer – dark red - and 2 °C / 3.6 °F colder – light blue) for the period 1932 to 2022 is indicated by the color barcode shown in the upper left corner of each image arranged as a book is read, top left to top right and proceeding from top to bottom:

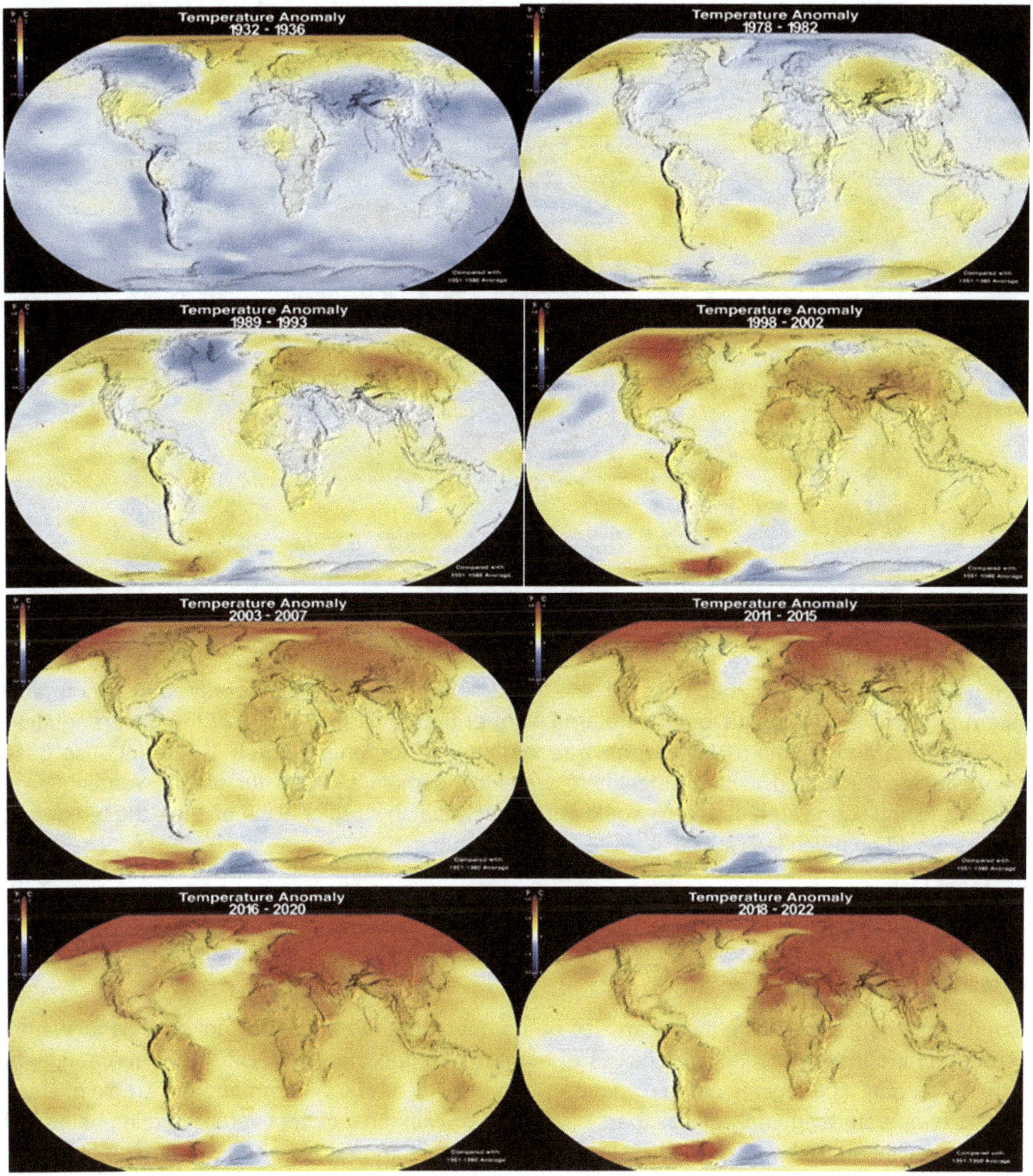

(Source: NASA Earth Resources Center and World Meteorological Organization)

There can be no doubt that in the "big picture" the planet has become warmer – but in regions not globally - since 1932 (based on the amount of red color visible in the montage of temperature anomaly illustrations). A combined "big picture" composite of land and oceans temperatures changes by regions (showing small cells within larger area regions) for 2023 indicates very obviously that some regions of the planet are warming, some regions are cooling, and some show little change independent from whatever efforts Man has been attempting to control "average global temperature".

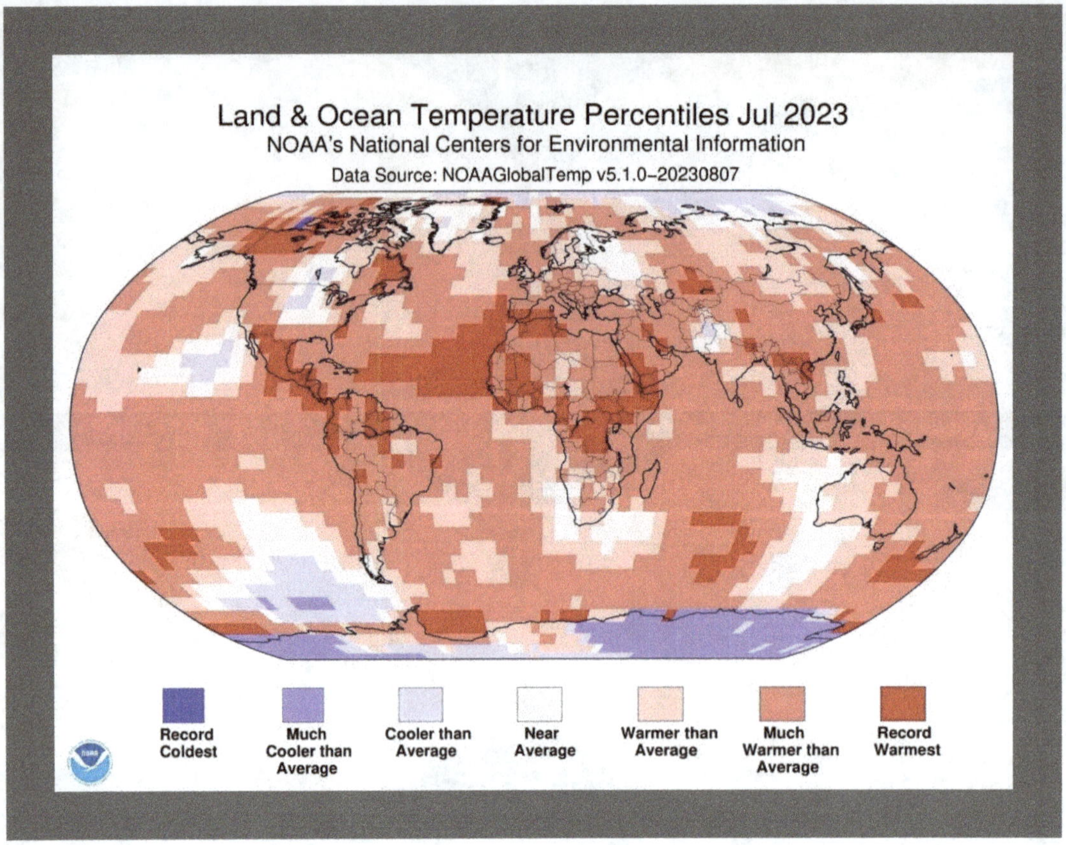

(Source: National Oceanic and Atmospheric Administration, NOAA)

For readers in the USA, the 2023 data indicate the central states have temperatures near normal, the eastern and western states have temperatures warmer than average, and only the desert southwest states (Arizona and southern Nevada) have record-setting hot temperatures. If "greenhouse gases" homogenously distributed in the atmosphere were the dominant cause for global warming on the planets land and ocean surfaces there would not be such "patchy" regional variations – the temperature changes would be much less subtle.

The 2023 data is the most recent and most comprehensive illustration regarding the regional effect of factors other than that of Man burning fossil fuels causing regional temperature changes – and for that reason is the illustration on the back cover of this book chosen for its lasting visual level of importance for Readers to remember.

From the montage set of data, there can also be no doubt that the largest extent of regional warming has been in the lesser population densities of northern Europe, northern Asia, Central America, and northern Canada. The regions of the planet between the Tropic of Capricorn and the Tropic of Cancer exhibit a temperature anomaly of less than 1.5 °C / 5.4 °F with most of the overall warming being in the Eastern hemisphere rather than the Western hemisphere. From the "big picture" (with the planet as a system of many subsystem components) perspective, there appears to be no valid reason for the environmental scientists of the United States to be declaring a "crisis" to the warming of the environment in the United States region. From that same perspective, the population distribution of the planet is such that over 60 percent of the world population is in the northeastern quadrant of the planet and the temperature anomaly montage above graphically shows the predominance of red color to be in that same northeastern quadrant.

Environmental scientists have created climate change to be in a "crisis" but as will be seen later, that "crisis" is not because humans are endangered by the small incremental change in temperature in some regions – the imaginary "crisis" is the "soaring rate" (about 0.75 °C per decade) in the average oceans surface temperature and resulting effect on marine life. Most of the public has no conception of the inner planet and the enormity of its content – and amount of heat.

While being concerned with a 2 °C rise in temperature of the ocean water, the level of temperatures within the planet almost boggles the mind. Readers should make note of the enormous magnitude of the temperatures in the various layers below Earth's outer crust and should inquisitively form a wonder of: "Wouldn't it be magnificent if some of that heat energy beneath the Earth's surface could be used to replace the heat energy being generated by humans and industries burning fossil fuels?" This "big picture" concept falls into the category of geothermal energy as an alternate form of energy that rivals the current approaches of wind and solar energy and much hydroelectric energy. That question will be addressed in more detail nearer the later chapters in this book.

As one of the many "fragments of the "Big Bang" entered into the infinite areas of what is simply termed "space", Earth is estimated to have been formed approximately 4.6 billion years ago and was formed initially as a ball of molten mass (principally iron in molten and plastic state). As layers of physical matter (in solid, liquid, or gaseous state) were deposited onto the planet with time, the Earth became formed in spherical layers much the same as that of a common onion. Each spherical layer acts somewhat like an insulation layer and such that the layers have a hot boundary (with the hotter layer beneath it) and a cooler boundary (with the cooler layer above it).

The next two diagrams show the spherical layers of the planet from its crust (continental land masses and ocean waters) down to its core, and the temperature increasing as the distance from the from the center of the planet decreases:

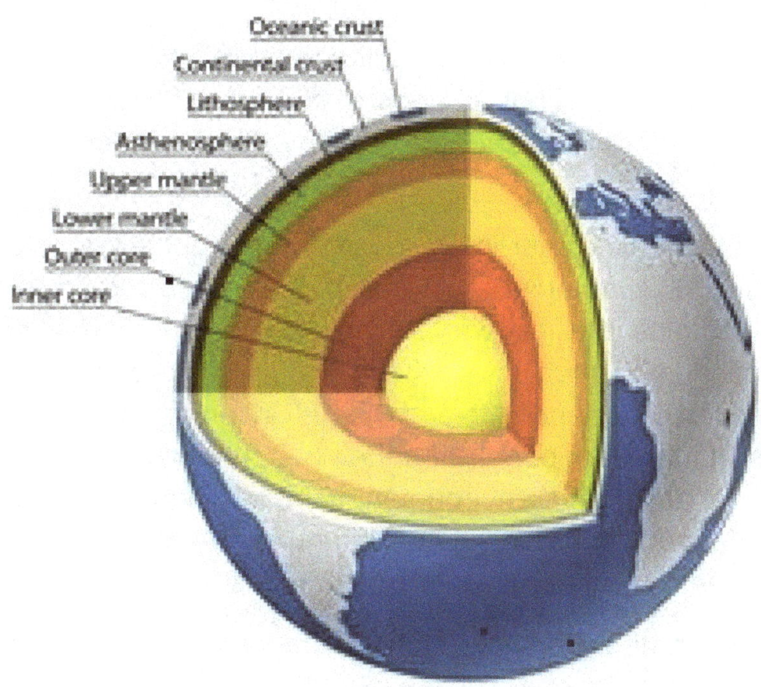

(Source: International Union of Geological Sciences)

The temperature gradients in the several "layers of the onion" are estimated to be as shown in the next diagram:

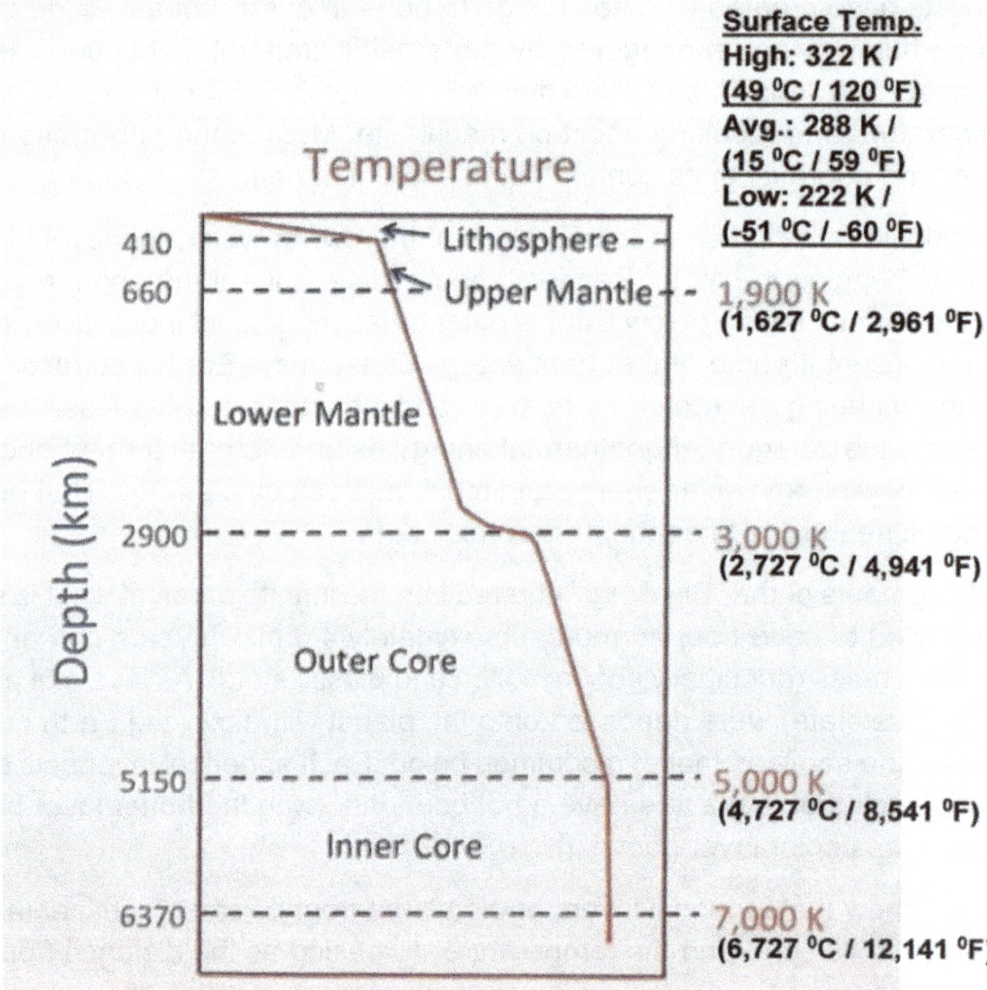

(Source: US Geological Survey with Author modifications)

The above diagrams may be a bit misleading as to the enormity of size of the planet as well as to the uniformity of its composition materials, but as each of its primary geophysical features are described there will be approximate dimensions presented in either/both tabular form, diagrammatic form, or photographic form.

The lithosphere is the most complex layer in composition and that is the layer from the surface of the Earth down (approximately 660 kilometers – 410 miles) to the upper mantle. It is the upper portion of the lithosphere (which contains the several tectonic plates in themselves of very non-uniform composition) and the land and water which are the two mediums supporting the many forms of life on the planet. The diagram has been amended to add the Kelvin temperatures shown in the more common forms of degrees Celsius (^{0}C) and degrees Fahrenheit (^{0}F). It is the surface temperature of the planet which is the least known variable in the environment to which the human population and all other life forms are subjected. The changing surface temperature and its causes and effects are the primary areas of the associated climate change being analyzed in this book.

The above diagrams of the planet and its inner temperature profile should not be interpreted as being truly representative of the inner structure of the planet as being so regularly layered in spherical geometry, because very little in the entire solar system is formed in very regular geometric patterns and move with well-ordered travel paths. Scientists have determined the surface of the planet was a much different surface – far hotter, drier, and more hazardous to volcanic activity, than what it is now. As the surface of the planet cycled between extremely hot and extremely cold conditions, and continued changing, the geophysical and environmental features on the surface of the planet also changed.

With the technological advances made in satellite sensors and imagery able to "see" within the depths of the planet, a somewhat different "big picture" illustrates how unevenly distributed the heat energy is immediately under Earth's outer crust and having an unevenly distributed thermal effect on the Earth's surface temperature.

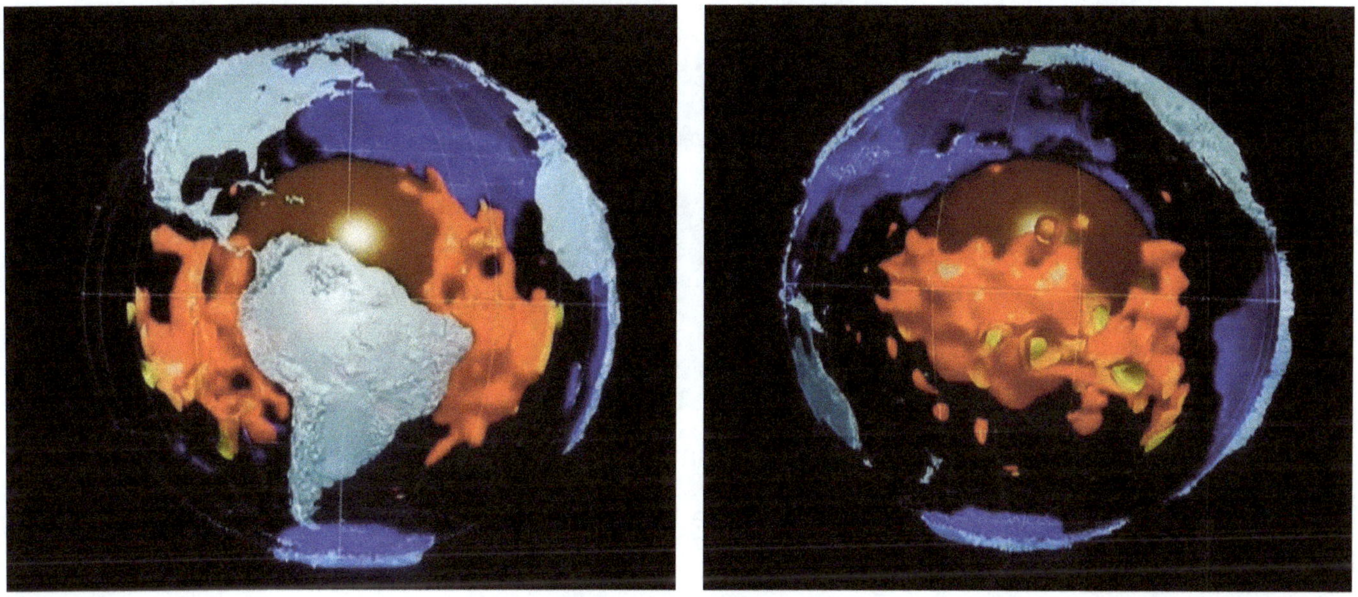

(Source: NASA)

When observing the left image above, there can be seen an immense heat source located under the eastern North Atlantic (where seasonal hurricanes are notorious for forming) and being transferred with the prevailing wind and ocean currents toward the west. Once formed, hurricanes generally take predictable westerly paths into the Caribbean Sea and Islands (Lesser Antilles and then Greater Antilles), then turning northeasterly to impact the southeastern and eastern coastal United States. Some hurricanes make the turn before the state of Florida, but some continue into the northern Gulf of Mexico before making the turn northeasterly. The US Hurricane Tracking Service (part of the National Oceanic and Atmospheric Administration, NOAA) is doing a great job, assisted by satellite imagery, in tracking the hurricanes path and intensity to give forewarning to communities that may be impacted by the approaching hurricane. There have also been scientific models developed to predict the path the hurricanes will follow, and those models are also helpful in identifying which regions should evacuate their inhabitants or to protect their property if they choose to remain.

The question is posed: "Why then are environmental scientists ignoring the above historical and visual databases and persist in distributing the fear-mongering misinformation to the public that it is the Man-made "greenhouse gas emissions" effect that is the dominant cause for the "soaring water temperatures" and the increasing frequency and severity of hurricanes and tornadoes – which are seasonal to start with?" That question is attempted to be answered with this book.

A similar question arises when observing the location of the enormous pools of heat energy under the ocean floor of the central and eastern Pacific Ocean. The islands in the state of Hawaii were formed by the movement of the tectonic Pacific Plate over the "hot spot" in the floor of that plate, allowing the large pool of magma shown in the image to erupt and form each of the major islands above the current Pacific Ocean surface level.

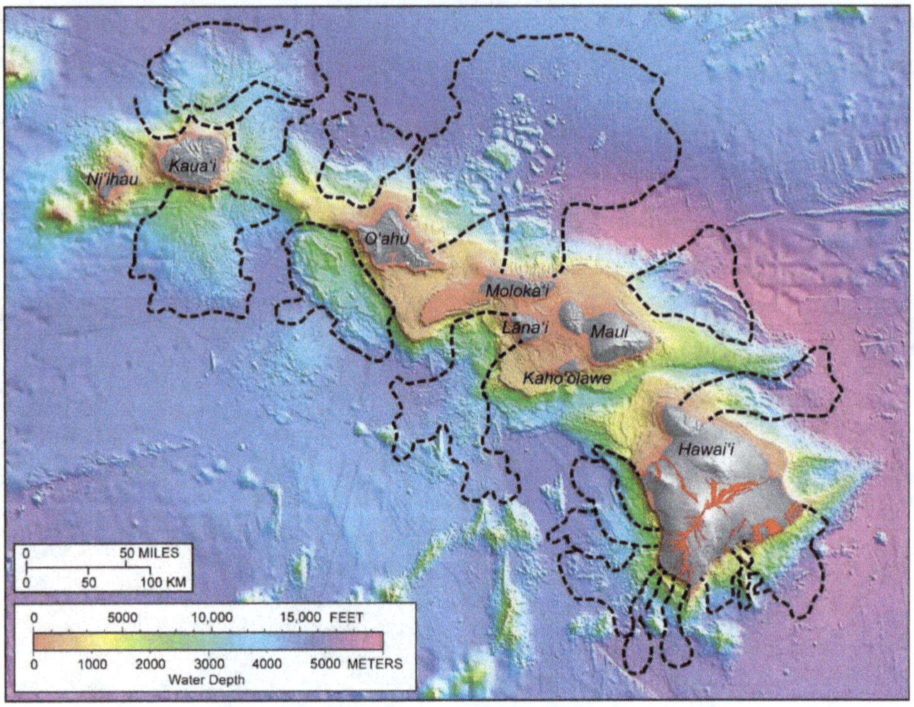

(Source: US National Geological Survey)

A simplified diagram shows the means of how the Hawaiian chain of islands were formed long ago – and are still forming:

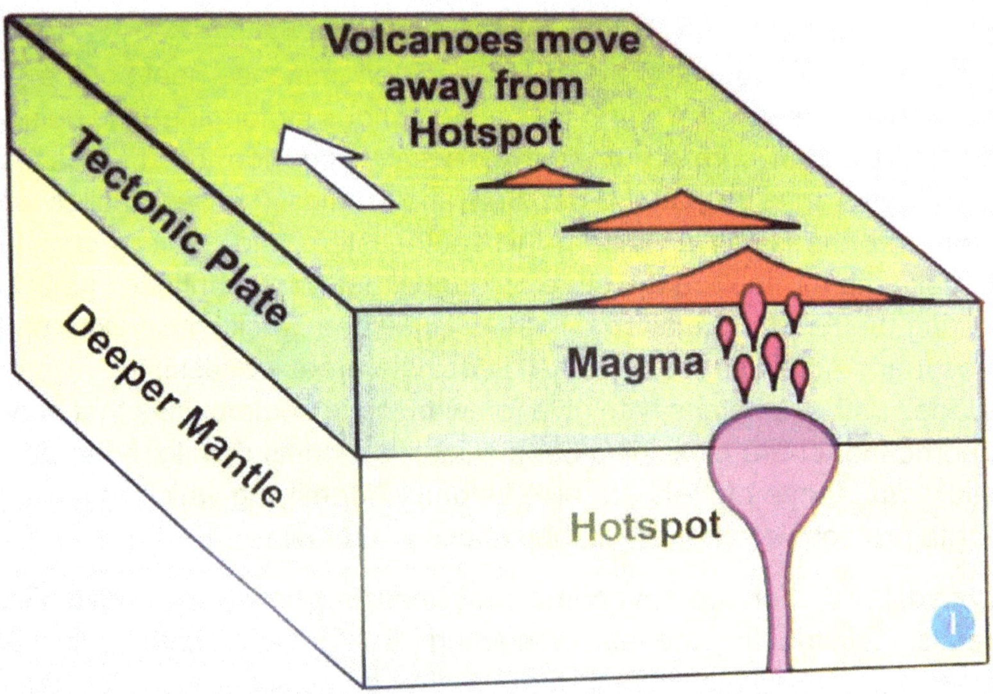

(Source: Internet Educational Website)

The volcano Kilauea on the "Big Island" of Hawaii has been erupting sporadically during the past few centuries, and did recently (June 3, 2024) have a characteristic massive lava flow – but not explosive - that caused much damage to some homes in the lava flow path (which is seen to have spared most of that community). A photograph of that scene is shown next:

(Source: US National Geological Survey)

The known locations (Central Pacific and southern portion of the North Atlantic Ocean) are also exactly where El Nino and La Nina phenomena changing the water temperature are spawned that again travel with the prevailing winds and ocean currents toward the western coasts of the United States and Mexico – the coasts of which are coincident with the Pacific "Ring of Fire" of known volcanoes. The oft referred to "Pineapple Express" is that weather phenomenon (associated with an El Nino ocean temperature and atmospheric weather pattern) when an atmospheric river brings heavy rainfall from the region of the Hawaiian islands causing flooding in the west coast states (California, Oregon, and Washington).

The persistence of the environmental scientists is admirable – and most of the environmental science has "circled the wagons" as they continue to defend their fallacious hypothesis that the Man-made "greenhouse gas effect" in the atmosphere is the dominant cause for the planet warming, the climate changing, and the ocean levels rising. Their mantra is bult on the premise that *"By eliminating everything else, settled science is proving that "Man-made greenhouse gases is the dominant cause for global warming, etc."*) The sagely old philosophers of the Age of Discovery (who burned fossil fuels to remain safely warm) saw the same effect in their contemporary scientists with the lamentations: *"There are none so blind as those who will not see."* and *"There are none so deaf as those that will not listen."*

Anyone familiar with the basic laws of physics that are taught in high school and college should be aware of the dynamic motions and forces generated when a large spherical object is in rotation about its axis and translating in a curved trajectory. (Basketball enthusiasts well understand the laws of motion in the "March Madness basketball season). Specifically, a mass not physically fixed in position on the circumference of a circular arc will tend to be moved – by a normal acceleration dependent on the rate of rotation and by the distance from the center of the sphere to the object mass - from a point nearer to the center of the rotating sphere to a location more distant on that sphere's outer surface.

The scientific name for this phenomenon is centrifugal acceleration – in layman's terminology, the capability of "slinging" a mass in an intended direction, as with a stone from a "slingshot" (*a la* David and Goliath). With the assistance of gravity and inertia and pressure forces, molten lava within the

Earth's inner layers is "slung" through the irregular cavities and tunnels within earth's inner spherical layers toward the tectonic plates that constitute Earth's surface. In the photos above, the irregular pools of molten material are seen to be beneath the floor of the oceans (Pacific and Atlantic) or seas (Caribbean or Gulf of Mexico). Volcanic eruptions occur by the ejecta being "slung" – assisted by high pressures and high temperatures that exist in the upper mantle of Earth - through a path along the edges of a tectonic plate or through a perforation ("hot spot") in the inner portion of a tectonic plate.

Many environmental scientists have the mind-set to "prove" (rather than to explore "if") it is Man burning fossil fuels that is exclusively causing the temperatures of Earth's surface to rise totally. Not all environmental scientists are scientifically myopic, but those who do, discount – most unprofessionally and ignoring basic principles of science – the proximity of geothermal heating from below rather than surface heating from the atmosphere above Earth's surface. It is that same group/organization that ignores the contribution of volcanic eruptions introducing worse hazards of particulate matter into the atmosphere that adversely affect human health and safety as opposed to the Man-made carbon dioxide in trace concentration which has little comparative effect on human health and safety.

In system engineering, the heat transfer across the interface between two subsystems (for this book, the four major subsystems are the planet, the humans occupying the surface of the planet, the ecology, and the atmosphere) such that one subsystem's functional performance might be adversely affected by an adjacent subsystem functional performance, is mandatory to be considered in a professional analysis – not "*eliminated*" from consideration simply because it is not understood.

Scientists at NOAA have long been measuring how the oceans' water temperature is varying at the depths of the ocean floors (and affecting sea floor organisms) and are theorizing the cause to be the resultant effect of "greenhouse gases" in the atmosphere causing the calculated rise in "average global temperature" and "average ocean water temperature". Life forms on the planet exist because they have adapted to the extremes of temperature in the region where they are located to which they have been and are being exposed. It is a greater probability that the enormous geothermal heat energy immediately below the oceans and seas is the true cause for the surface temperature extremes to be increasing regionally and irregularly, not in an "average global temperature" singular manner.

Scientists in some branches are able to use well-defined mathematics and various independent variables in the analysis of the dependent variables in their branch of science (astronomy, chemistry, electricity, thermodynamics, acoustics, etc.), but in the branches that are not so well defined – tectonics, meteorology ("weather"), climate, volcanism, earthquakes, hurricanes, tornadoes, and most of the environmental disturbances and anomalies – the variables are of complex combinations of factors that are difficult if not impossible to estimate if/when/where they will occur. Those unpredictable independent variables are also nearly impossible to be controlled to a different level more suitable to the safety, health, and/or comfort of Man. Scientists and inventors should be focusing and prioritizing their efforts on those developments of technologies that make the planet a safer, healthier, more comfortable place for humanity to enjoy and thrive, not simply to improve (by "reducing Man's emissions to net zero") the ecological environment.

Millions of years ago, the exterior crust portion of the planet was physically configured much differently from how it is known to currently exist (thanks to the branch of science of cartography – map making). Scientists and geologists have developed the cartographic database (much like a genealogy report for the human race) describing the changes in Earth's tectonic plate configuration. Many millions of years ago, there was only one tectonic plate named *Pangaea.* A very old French diagram in two parts estimates the “before” (*avant la separation* in French) and the “after” (*après la separation* in French) configuration of Pangaea separated as shown next:

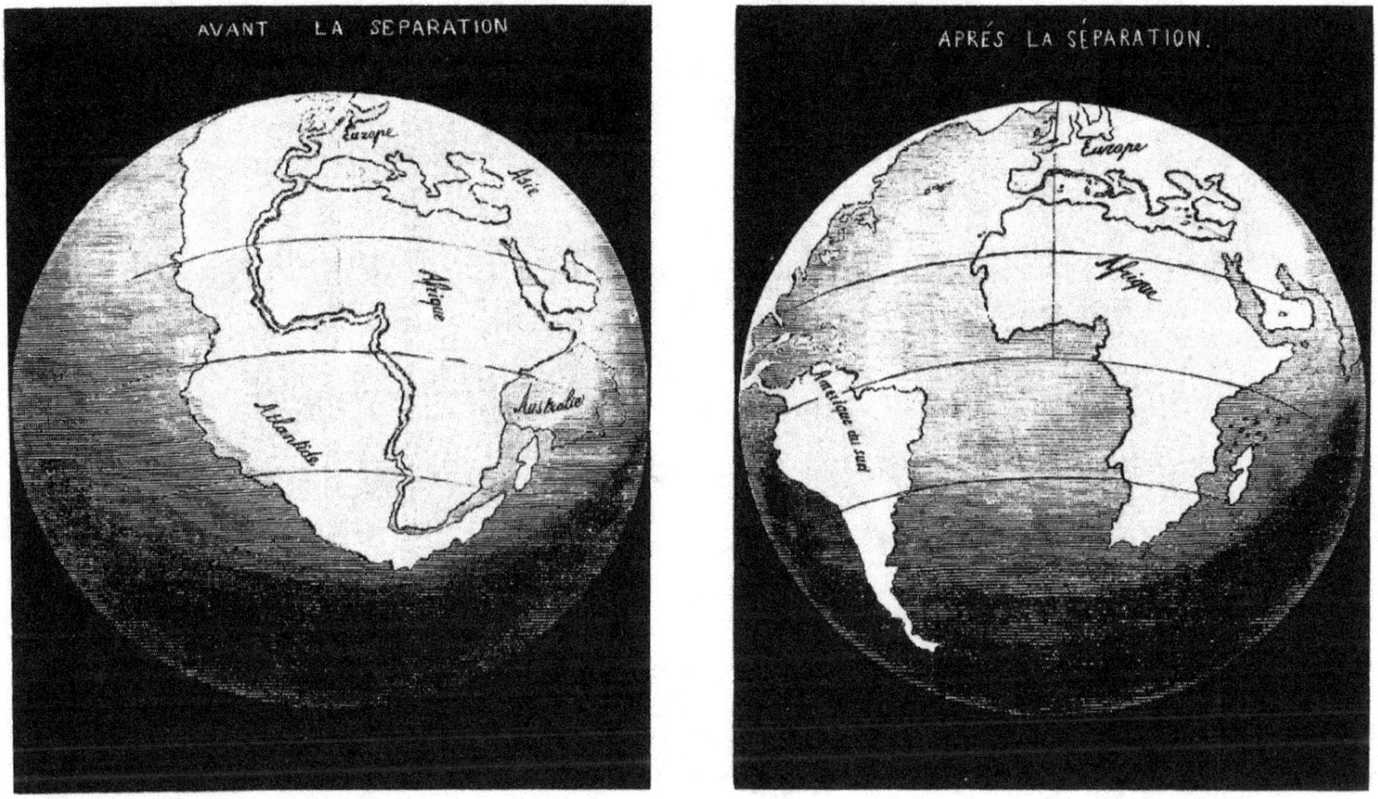

(Source: Encyclopedia Brittanica)

Geologists have developed (as “settled science”) that Pangaea fractured into smaller tectonic plates that have moved apart to eventually form the different tectonic plates as we now know them to exist. (The primary fracture boundary is shown in the left image and is seen to correlate very well with the current coastal regions of the continents in the Western hemisphere and in the Eastern hemisphere as they exist in the current database for tectonics.) Below the continents shown in the before and after diagrams above are the Earths outer crust which is fractured into tectonic plates. The time it took for the tectonic plates to move from their configuration when in the Pangaea configuration (left) to the current configuration (right) is several million years in length.

The continents as we know them today are the visible portions of the tectonic plates hidden from visibility beneath the ocean waters. If one could see the hidden full configuration of the tectonic plates, they would appear as shown next:

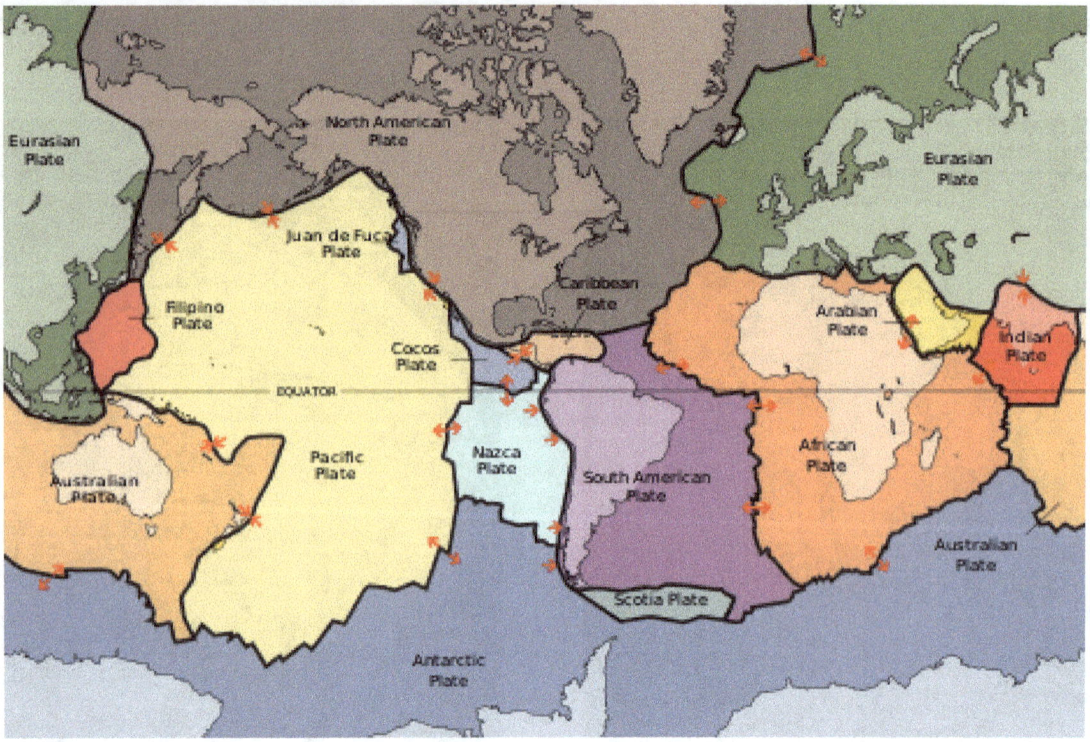

(Source: Encyclopedia Brittanica)

The plates are geologically portions of Earth's outer crust, and on the diagram of the tectonic plates configuration there are red arrows showing how the plates – and the continents (shown above in different colors) atop them – with all of them in continual motion relative to each other.

In the Norse Age (circa 800-1100), the seafarers mapped the waters in the Scandinavia (northernmost Europe) region and gave birth to the science of cartography – the making of maps.

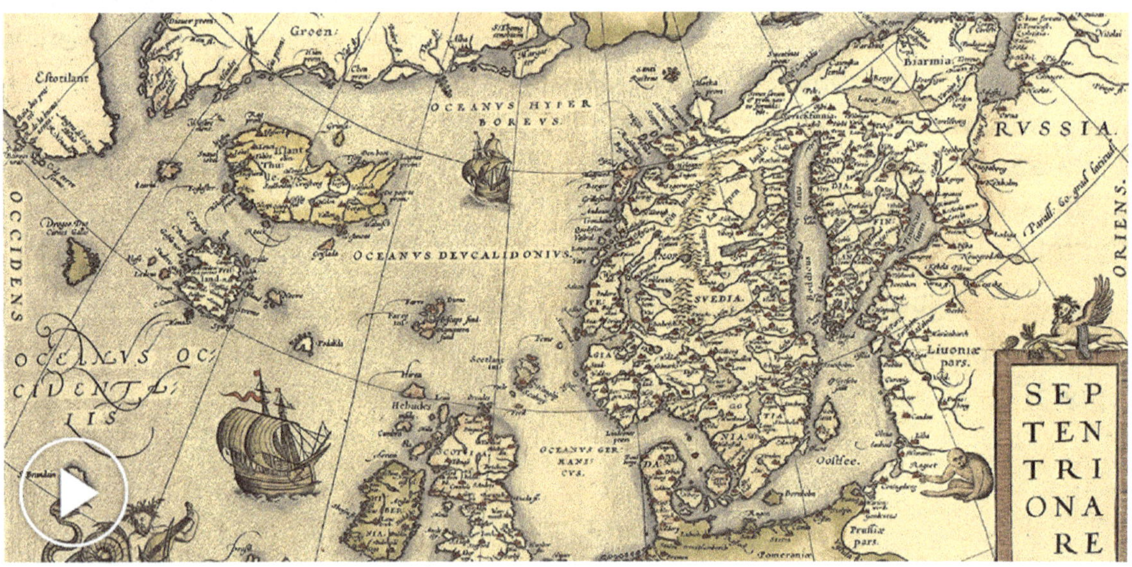

(Source: Encyclopedia Britannica)

These maps were of immense importance to the safety of the Norsemen sailing the treacherous waters of the North Sea, but were of lesser importance to the population – or lack of it – in that cold region. By the 1500s and 1600s period, the mapping (cartography) of the entire planet Earth's surface by the astronomers' and mariners' database defined the physical features (land masses and oceans) on the exposed spherical surface of the planet, as shown in the next modern map of the planet showing lines (demarking orthogonal longitude E-W and latitude N-S) in a well-defined grid pattern:

(Source: The International Union of Geological Sciences)

Various forms of aquatic life (vertebrates and invertebrates) and other forms of life formed and evolved over many eons, and it is believed that life on land (above water) evolved from different forms of aquatic species (under water) with the evolvement of lungs that extracted oxygen from the atmosphere. Vegetation and other forms of botanical growth evolved with the photosynthesis process and began to cover the Earth's surface. There was much debate over if/how Man may have evolved from some form of aquatic life (possibly from the octopus' family) and which developed lungs to breath oxygen from the air instead of through gills from the water. Other more credible beliefs are that Man (*homo erectus*) evolved from the Great Ape family of animals. Some religious theologians maintain that Man was created in the image of God. That phase of Man's presence on Earth is left to be debated by others rather than by this Author. It is a formidable challenge to the Author just to describe what "footprint" Man has left on Earth's surface long ago and is leaving now and for the predictable future.

For this book, the baseline in time of Earth's surface was chosen by the Author to be about 400,000 years before now (2024). That selection was not chosen without a basis but is a point in time where the Russian Antarctica Ice Research Station ice core sample data scientifically defines Earth's surface temperature and carbon dioxide and methane in Earth's atmosphere. Those three important factors in the total database are known to have been determined accurately and scientifically by initially Russian scientists and later by scientists from the international environmental science profession using modern-day carbon-dating and spectrographic techniques. It is to be noted that there was no presence of Man burning fossil fuels on the planet 400,000 years ago and so that all other changes that affect global surface temperature, climate, and ocean surface levels can be baselined as the reference in the "before" and "after" presence of Man on the planet.

It is important to note that the "after" in the ice core samples includes the effect of early humans but also includes continuation of the same effects – astrophysical, geophysical, and natural - that existed in the "before" period leading to the baseline. The planet continues to rotate and orbit the Sun and will be continuing that orbital movement regardless of how much fuel of any type humans may combust. Assessing what effect Man has caused by emissions from his (and Industrial) activities on the Earth's environment added to the inspiration for this book to be written for public information.

At a point in time (400,000 years ago), Earth's surface had stabilized considerably, and a map of the planet's geophysical surface had a tectonic plate configuration much like that in the illustration shown earlier and repeated in somewhat different format next:

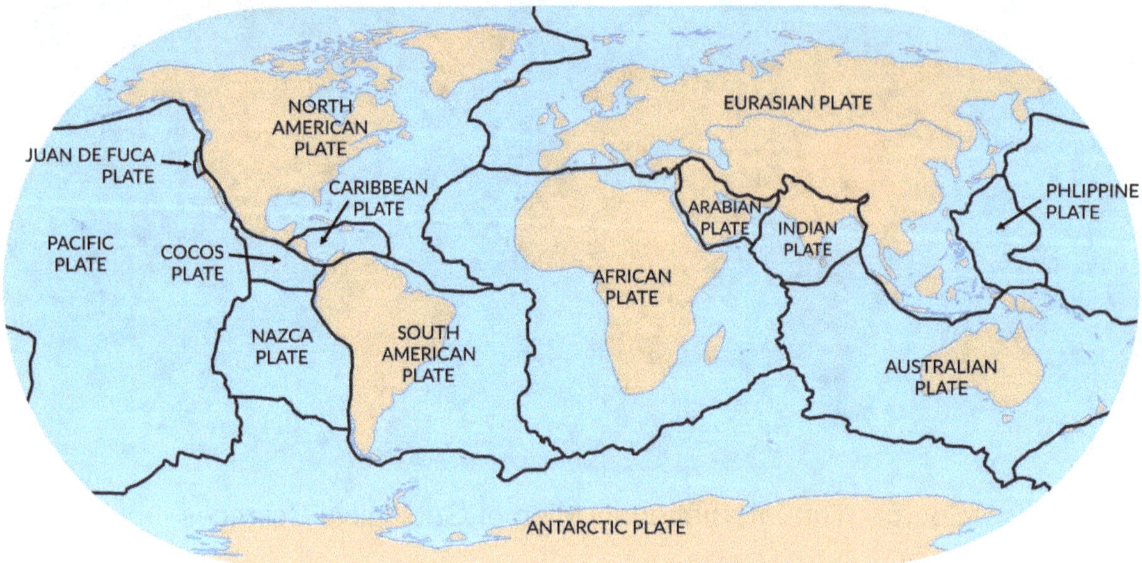

(Source: Brittanica Encyclopedia)

(Readers are asked to envision the oval shaped two-dimensional diagram above to be – in three-dimensional format – the sphere representing the planet Earth. Having a desk globe of Earth would be very useful to Readers of this book. Desk globes are typically mounted on the tilted (about 24 degrees) axis about which the planet rotates and helps to "tie together" the astrophysics and the geophysics of the planet. A larger desk globe, (preferably one that has a topographically shaped surface) is mandatory in a science classroom for students to "see" the scaled down "big picture" of the surface temperature changing daily and the longer-term climate change problem being of regional nature).

On some of the named tectonic plates (in light blue) there is an outline (in a contrasting pink color) of the "continents" as they are known in modern times world geography – North America, South America, Africa, Europe/Asia, Arabia (Mid-East), India, Australia, and Antarctica. The land areas in the pink shade colors (i.e., the continents) are presently above the current surface levels of the world's oceans, and that is the "baseline" topography used in the remainder of this book. In the oval shaped map above, there is 70-75% of the total oval area in blue designating that surface area to be water covered, and the pink colored land area (*terra firma*) is 25% of that area of the oval that is above water.

Where the edge of one tectonic plate meets the edge of an adjacent plate, there is a law of physics stating essentially "Two solid masses cannot occupy the same volume in space at the same time". Accordingly, there are three types of relative motion the plates contesting for that common space can take: 1) Plate A can slide under or over Plate B in a process known as "convergence"; 2) Plate A can move away Plate B in a process known as "divergence"; or 3) Plate A can move sideways relative to plate B in a process known as "transformance". A simplified illustration is shown next to describe what happens to the surface topography of a continent when one of those three events take place (over which Man has no control) and often results in an earthquake (and possible tsunami), a volcanic eruption, a major separation in the surface, or has created mountain ranges above a land mass or below the ocean in a long period of time, etc.

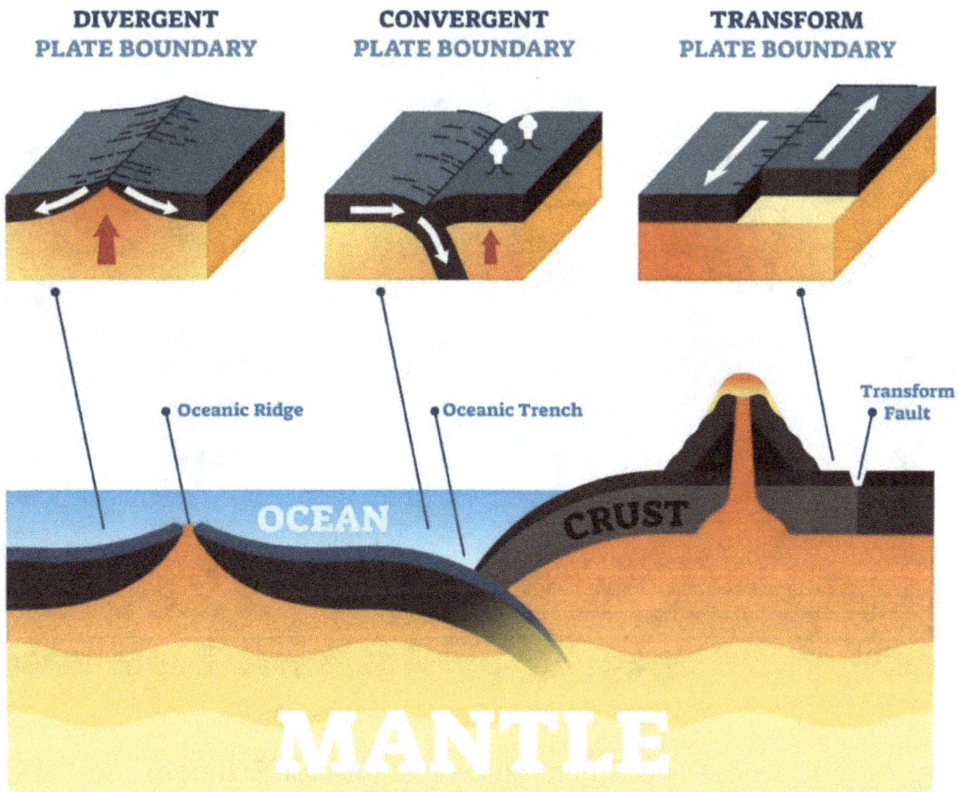

(Source: National Geological Survey)

Another phenomenon can occur along fault lines created by the relative motion of two plates. The diagram above shows the formation of oceanic ridges below the ocean floor and the classic conical shape volcanic formations on land, but there are locations where the lava emerges along a fault line in a very linear fashion – one of the rare instances where Nature kas a linear geometry. Seen below is an aerial view of such a linear eruption on Iceland:

(Credit to unknown photographer for Air & Space Quarterly Winter 2025)

Another perspective (from the Smithsonian Institute for Science) illustrates that the collision of the two plates also produces the magnificent mountain formations of the world:

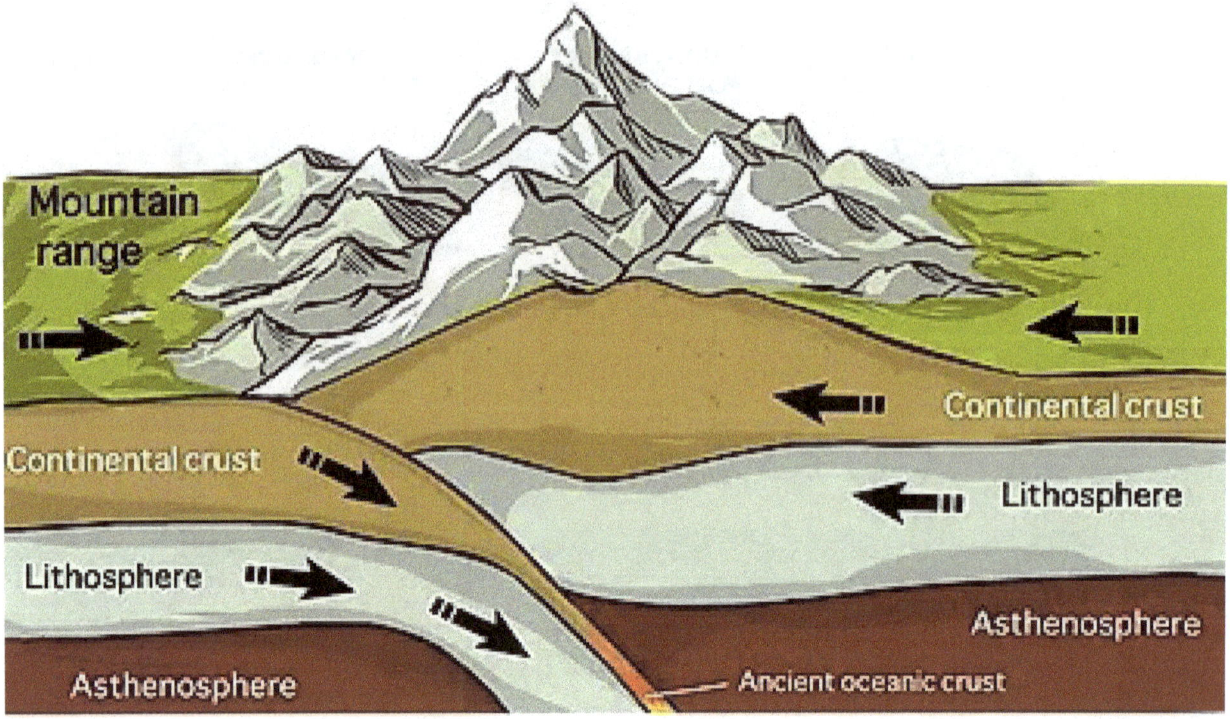

(Source: The Smithsonian Institute for Science)

The movements of the tectonic plates are imperceptible when their motion is at a slow steady rate, but when one considers how much energy it takes to move the incredibly large mass of each plate even a small fraction of an inch in any of the three axes of motion – up, down, or laterally it gives a better appreciation for how earth's surface has been changing for the past 4.65 billion years since the planet was known to have been formed. That formation is often described as being like the layers of an onion, but that is where the similarity ends.

When the plates are in divergent mode (first small insert above), the edges of the plates will cause a ridge to be formed beneath an ocean's surface or a mountain (or mountain chain) to emerge on the planet's surface. This phenomenon occurs slowly over many thousands of years and poses no urgent threat to humans but does cause some forms of the ecosystem to adapt and evolve as the mountain elevates. The "tree line" will remain at an altitude even though the mountain is thrust upward to higher elevations. In the research phase of the problem area being clarified, the statistics on the elevation of the mountains around the world and in the USA became available. The Author has tabulated an impressive database on the elevation and the number of the named mountains as shown next:

Elevation and Number of Mountains in World and in USA

Elevation at Peak			No.Named Mountains In World	No.Named Mountains In USA
Over 8,000 m (26,247 ft)			14 (All in Asia)	None
7,000 m (22,966 ft)	to	8,000 m (26247 ft)	32 (In China, India, Pakistan)	None
6,000 m (19,685 ft)	to	7,000 m (22,966 ft)	117	1 (Denali)
5,000 m (16,404 ft)	to	6,000 m (19,685 ft)	76	3
4,000 m (13,123 ft)	to	5,000 m (16,404 ft)	260	98
3,000 m (9,843 ft)	to	4,000 m (13,123 ft)	342	50+
2,000 m (6,560 ft)	to	3,000 m (9,843 ft)	305	75+
1,000 m (3,280 ft)	to	2,000 m (6,560 ft)	229	150+

Note: The elevation of Denali (formerly Mt. McKinley) in Alaska is 6,194 m (20,316 ft)
The longest mountain range is the Andes in South America at 4,500 miles.

(Source: Tabulation by Author of National Geological Survey databases)

The tallest mountain on the planet is Mount Everest in the Himalayan Mountain Range of Nepal and presently its peak is at 8,849 meters (29,032 feet) in elevation. A photograph of Mount Everest and surrounding terrain is not only visually impressive to non-scientists, but when scientists assess the mass of that amount of the planet's surface being elevated to such extreme elevations, the energy estimated to move that much mass through that distance – albeit it is at a slow rate of movement (speed much slower than the proverbial snail) - the energy from within the outer and inner mantle and outer core (molten state) had to have been mind-boggling in pounds of force and over such an expansive area.

When viewing the table above, Readers are invited to consider the number of mountains and their elevation and imagine the energy from within the planet that was expended to lift the mass of each of those mountains to their elevation indicated. It puts into perspective how small by comparison the amount of energy trapped below a Man-made "greenhouse gas" emission blanket can cause such massive energy effects when compared to that stored within the planet below its outer crust. The later sections of this book suggest tapping into that geothermal energy as a clean alternative form of energy which has thus far largely been neglected by environmental scientists. (The geothermal energy would need to be accessed with long distance drilling into the Earths outer crust and certainly at elevations much closer to sea level than are the mountain peaks tabulated above.

(Credit to Unknown Photographer)

About 800 people attempt to climb to the peak of Mount Everest every year, and as of January 2019, 5,294 have successfully reached the peak. Over 330 have died trying. The view from the peak "at the top of the world" is magnificent by most accounts, but many have reached the peak only to have a magnificent view of cloud tops (that can be seen when in a commercial airline flight at 30,000 feet cruise altitude at far lower cost and with far greater safety).

(Credit to Unknown Photographer)

There can be no doubt that the environment, weather, climate, and human habitation of a region surrounding a tall mountain is affected by the mountain and its proximity to other major features such as adjacent tall mountains or by an ocean. The wind currents, the cloud formations, and the water vapor in the atmosphere on one side of the mountain being different from the other side of the mountain a mere few miles apart.

The eight tallest mountain ranges are all in Asia (China, Pakistan, and India) whereas the United States has the Rocky Mountain/Sierra Nevada ranges near its West coast and the Appalachian range near its East coast (2,500 miles of flat prairie land between). It should be apparent that considering global warming, global climate change, and the global effects of ocean levels rising bears little common patterns within the regions of the 48 contiguous United States, and that in China, India, and Pakistan. Setting a "global emissions goal to control emissions with a "one-size-fits-=all goal of one temperature – regardless of any amount of temperature change – from that one "average global temperature" defies scientific logic. The existence of majestic mountain ranges on the planet's surface is tempered by the fact that humans choose not to live on the sides of mountains as those are prone to have landslides and rockslides far more often than risking their probability of occurrence for daily views of the placid valleys below. Beautiful orchestral music cannot match the sounds of cattle-bells on the necks of a herd of dairy cows grazing in the meadows on an alpine hillside in Switzerland. Each cow has a bell with a unique tone and the cow-herder can tell which of his herd has strayed from the main flock by the absence of that tone in the "cattle-bell symphony".

Seismologists become suspicious when the daily vibrations on a fault-line go silent as that raises the probability of a "big one" occurring soon thereafter when the tectonic plates suddenly overcome the "static friction" force retarding their transition from being at rest into constant movement.

Physicists and engineers often simplify the terminology "static friction" as "stiction", and this variable in the dynamics and forces acting to initiate a mass relative to another mass offers a partial but very basic explanation why earthquakes occur so sporadically and so "spikey" ("jerky") in motion when they do occur. A basic experiment conducted in physics laboratories in colleges placing emphasis on STEM (science, technology, engineering, and mathematics) courses illustrates static friction or stiction. The experiment invariably consists of a mass (usually a large block of wood on a metal table (a plane) that is varied in angle and to an angle where the acceleration of gravity acting on the mass gradually overcomes the static friction between the mass and the plane. Readers can conduct this simple experiment with this book being read as indicated in the diagram next:

Static Friction Example

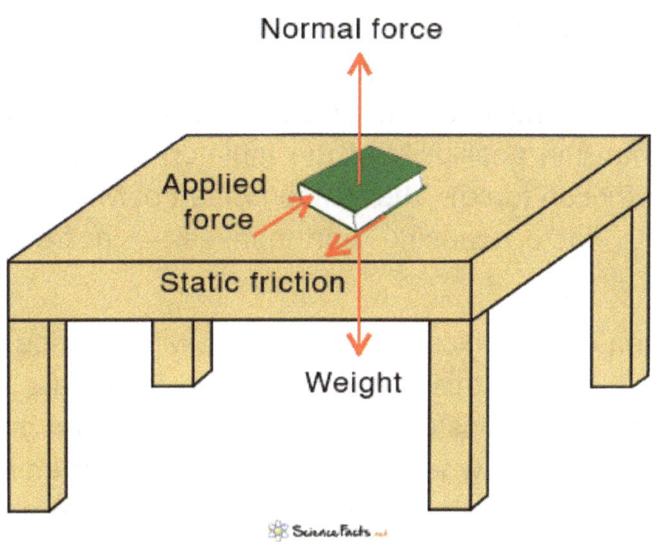

(Source: Any College Physics Textbook)

The applied force can be felt in the fingers gradually building up (to the static friction force) until the applied force is equal to the "stiction" force resisting motion of the mass. When the forces are equal, the book will start ("jump") into a sliding motion relative to the tabletop and the book will remain in motion with a smaller force being applied to the motion resisting frictional force between the two non-smooth surfaces. The weight of the book (mass) and the frictional coefficients (the smoothness of the surfaces in contact - of the bottom cover of the book and that of the tabletop - are quite different and are the variables that are unknown in nature between the two tectonic plates. That "jump" in the seismic waves is explained simplistically in the next diagram:

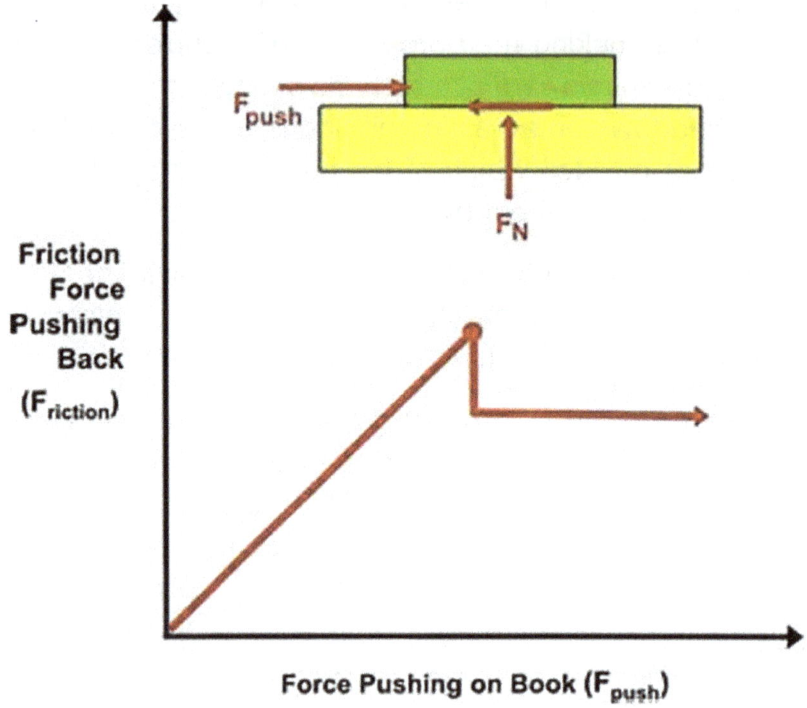

(Source: Any College Physics Textbook)

This is the same principle involved when two tectonic plates abut each other and suddenly start to move relative to each other – and the friction forces transforming into the seismic waves measured by scientists. Because the edges of the tectonic plates are not smooth and are of different materials, that static friction force causing an earthquake at the fault line is quite unpredictable until the seismic waves begin to "jump" up in a spikey pattern as the edges of the tectonic plates grind against each other. For those who may have the question in mind, the atmospheric conditions surrounding the experiment have no significant effect (other than the possibility of the tabletop being damp and the fluid acting as a lubricant to reduce the static friction force). The concentration of Man-made or natural carbon dioxide in the atmosphere is not a factor causing earthquakes – in frequency of occurrence or in severity/magnitude when occurring.

Many books have been written on tectonics and the Author will describe only the two most important aspects that relate to the topic area of this book. Those two topics are volcanoes and earthquakes, both related to the unpredictable dynamics of the tectonic plates. It can be appreciated how frustrating it must be that scientists cannot predict when a volcano will erupt, or an earthquake will occur so that warnings can be issued to nearby residents. The fatalities caused to humans and the property damage done by these two phenomena combined dwarf any estimates that are made by environmental scientists resulting from the factual causes for and effects of global warming, climate change, and ocean levels rising – in the next millennium or two. Knowing however, that most earthquakes and volcanic eruptions occur above or very close to the edges of tectonic plates, there are measurement devices and sensors that have been placed to give the earliest possible warning to the public residing in a known area where an earthquake or a volcanic eruption is most likely to occur. Only the "when" and "where" on which tectonic plate(s) will an earthquake occur or a volcano erupt is unpredictable in the long term.

It is typical that the buildup of magma (molten rock) below a volcano produces earthquake-like movements deep within the earth. There is usually at least a three- or four-day warning period (not unlike a rattlesnake shaking its tail to issue audible rattling (hence earning its name as rattlesnake) warnings to a human getting near its territory) before a major eruption. Volcanoes also may emit an increasing amount of steam or other heated gases into the atmosphere prior to an eruption and those humans who do not heed those early warning indicators are taking unnecessary risk of property damage and major injury or loss of lives (human and/or animal).

Volcanoes have been erupting on Earth for many millions of years, but their frequency has clearly decreased in recent millennia. Between 1500 and 1700 AD, there were about 600 volcanoes that were recorded (mostly in church records) to have erupted world-wide. The record since 1950 is believed to be almost complete with 2,208 eruptions recorded from 347 volcanoes. The average number of eruptions ongoing per year since 1950 is 63, with a minimum of 46 and a maximum of 85 eruptions recorded per year. In 2023, a major volcanic eruption is occurring in the Iceland region but residents near that volcano have been evacuated such that there will be few if any losses of life. Loss of property and damage to the ecosystem is not yet totaled. Conversely, Iceland is located very near the Artic Circle (at a latitude of 64 degrees north for the capital city of Reykjavik. Later in this book will be discussed the potential for using geothermal energy if there is a source for that form of clean, reliable heat energy in a region and Reykjavik is a role model city for doing that. Almost half of that city's heat energy demand is obtained from nearby geothermal activity at surface level – no drilling required.

An "eye-opener" for many Readers will be a map showing where volcanoes exist on the surface of the planet. It is hastened to be added that most are inactive and have been for a long time. But there are many that are considered active with a probability of erupting from low to high. That map is shown next:

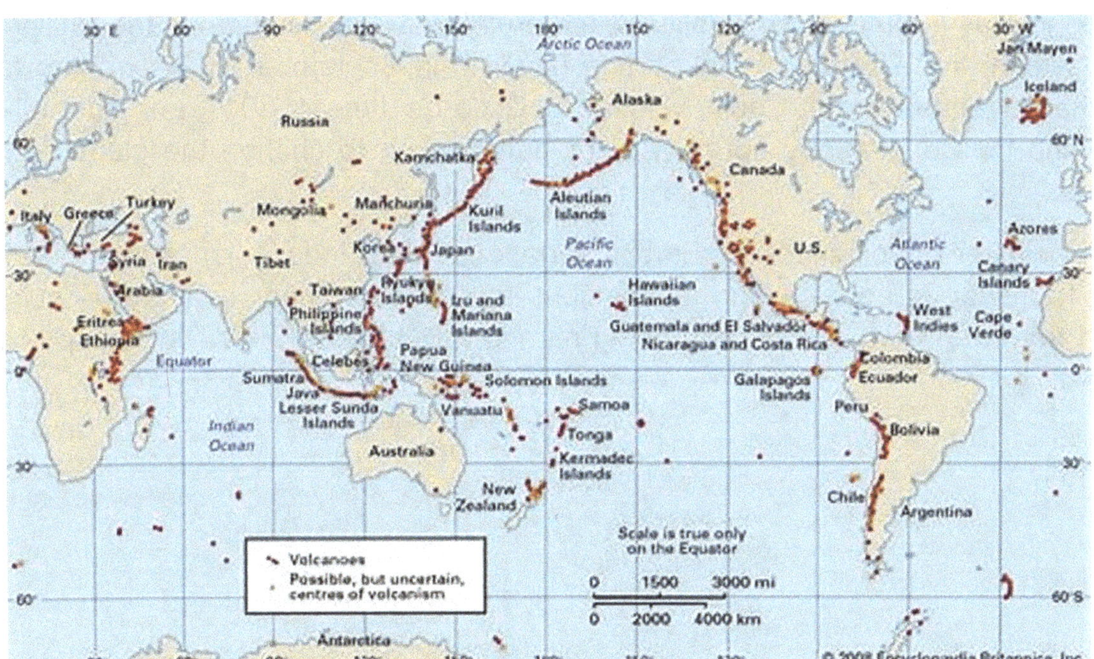

(Source: Smithsonian Institute for Science)

There is a predominance of volcanoes (each shown by a red dot in the map above) around the edges of the Pacific tectonic plate (each shown as a black dot in the lower map) and that has been dubbed as the "Ring of Fire".

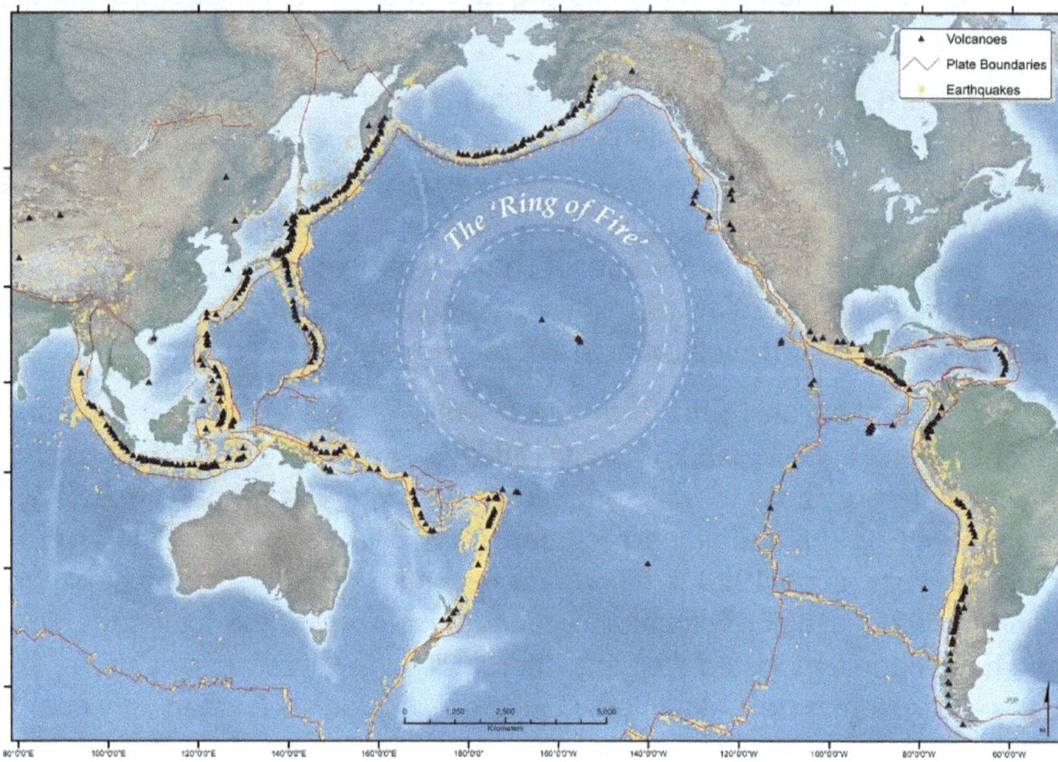

(Source: Smithsonian Institution)

What is of importance to be noticed however, is that the Hawaiian Islands chain is in the center of the Pacific Ocean and well away from the edges of the Pacific Plate. Volcanologists explain that as being a "hot spot" in the outer crust below Hawaii and each of the Hawaiian Islands is the result of a volcano formed upward from the floor of the Pacific Ocean as the tectonic Pacific Plate moved northeasterly the west coast of North and Suth America. Mount St. Helens in the US Pacific Northwest (Washington) which erupted in May 1980 (with 57 fatalities), the eruption of Mt. Pinatubo in the Philippines in 1991 (with 227 fatalities), the eruption of Mauna Loa (Hawaii) in September 2022 (with no fatalities), the eruption of the Puyehue-Cordon Caulle volcano in Chile on June 4, 2011, and currently (December 2023) in Iceland individually may not have contributed much to change the global temperature on Earth's surface.

Another major cluster of volcanoes on Earth's surface is in the Indonesian area in the Pacific Asian region. That formation is not in a ring manner but is in a more linear geometry – suggesting it be distinguished from the "Ring of Fire" and being named the "Line of Fire". That group of volcanoes is illustrated next:

(Source: National Geological Survey)

As a general comment as to the type of volcanic eruptions in the Indonesian "Line of Fire", they have historically been of more "gassy" manner of effluence than "lava" in different eruptions – and it is just the opposite in eruptions in the Hawaiian island chain. Shown below is an eruption in January 2024 on Indonesia:

(Credit to unknown photographer)

The resemblance to a nuclear bomb explosion (such as that in Japan to end World War II) is coincidental. A third historically troublesome region of volcanoes is of those that formed the West Indies in the Caribbean region of the Atlantic Ocean. Shown next is a diagram of the Caribbean Sea region and the chain of islands exposed above the current sea level formed where the Atlantic Plate abuts the Caribbean Plate:

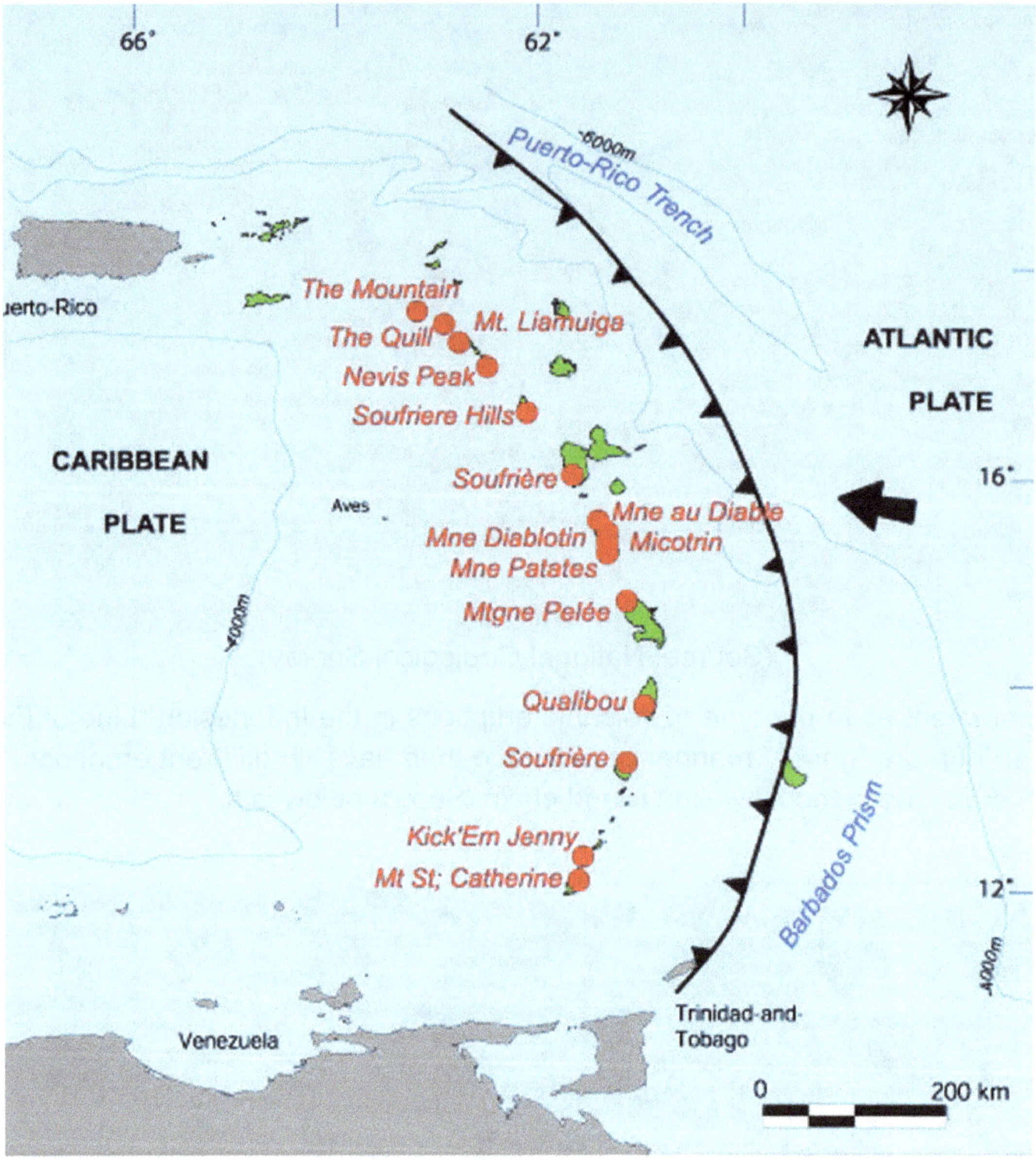

(Source: US National Geological Survey)

An interesting table of geographic and human interest regarding the islands in the Caribbean Sea is shown next, portraying the potential disaster to the over 6 million humans inhabiting the Caribbean islands and which are among the most visited places in the Northern Hemisphere seeking a relief when it is Winter in the Northern Latitudes especially those in the northeastern United States and the southern half of Canada. (The residents of the Caribbean islands refer to those millions of tourists per year as the "Snowbirds".) When volcanic activity threatens humans in larger countries there is typically an advance warning and humans evacuating to distant locations sufficiently far from where seismologists warn of an impending eruption. It can be seen in the next table that there is no distance large enough that residents on each of the volcanic Caribbean Islands can flee. Conversely, if the ocean levels do increase significantly the residents of these islands have the option of relocating further up on the side of the volcano (not a recommended action) or relocating to a larger country such as the North American or South American neighbors to their west (the better of the two options).

Caribbean Islands Most Affected by Ocean Levels Rising
(Coastal beaches and population at low height above Sea Level)

Island	Population (Census Year)	Area of Exposed Land (sq miles)	Highest Point of Island(s)(feet); (Type geologic origin)
Trinidad & Tobago	1,367,558 (2021)	1,980	3,080 (Trinidad non-volcanic; Tobago inactive volcano)
Grenada	124,610 (2021)	135	2,755 (multiple inactive volcanoes)
St. Vincent & Grenadines	110,872 (2022) 10,900 (2012)	142 17	4,049 (La Soufriere active volcano)
Barbados	277,821 (2010)	169	1,120 (upheaval, not volcanic)
St. Lucia	180,000 (2018)	238	3,120 (semi-active volcanoes)
Martinique	349,925 (2024)	436	4,583 (Mount Pelee' active volcano)
Dominica	71,293 (2011)	290	4,747 (two inactive and 3 active volcanoes)
Guadaloupe	395,925 (2024)	629	4,813 (Grand Soufriere active volcano)
Montserrat	4,390 (2022)	102	3,440 (inactive volcano)
Antigua & Barbuda	97,120 (2019)	170	1,319 (Antigua inactive volcanic; Barbuda upheaval)
St. Kitts & Nevis	54,338 (2023)	101	3,793 (Both inactive volcanic)
Anguilla	15,753 (2021)	35	240 (Of upheaval limestone)
US Virgin Islands	108,612 (2000)	134	1,555 (Mix of inactive volcanic and upheaval)
Puerto Rico (US)	3,285,874 (2020)	3,424	4,390 (inactive volcano)
Total:	6,454,991	8,002	

(Source: "Genealogy of the Hutchinson and Edghill Families of Barbados")

The highest public awareness of volcanic eruptions is of those that erupted recently in Hawaii and in Iceland. When former Vice President Al Gore produced his "An Inconvenient Truth" epic video about the loss of sea ice in the Arctic region and placing the cause to be the result of "greenhouse gas emissions" from Man burning fossil fuels, conspicuous by their absence was the volcanic activity on Iceland and the geothermal heat energy under the Greenland Ice Sheet that were not included in that "inconvenient truth".

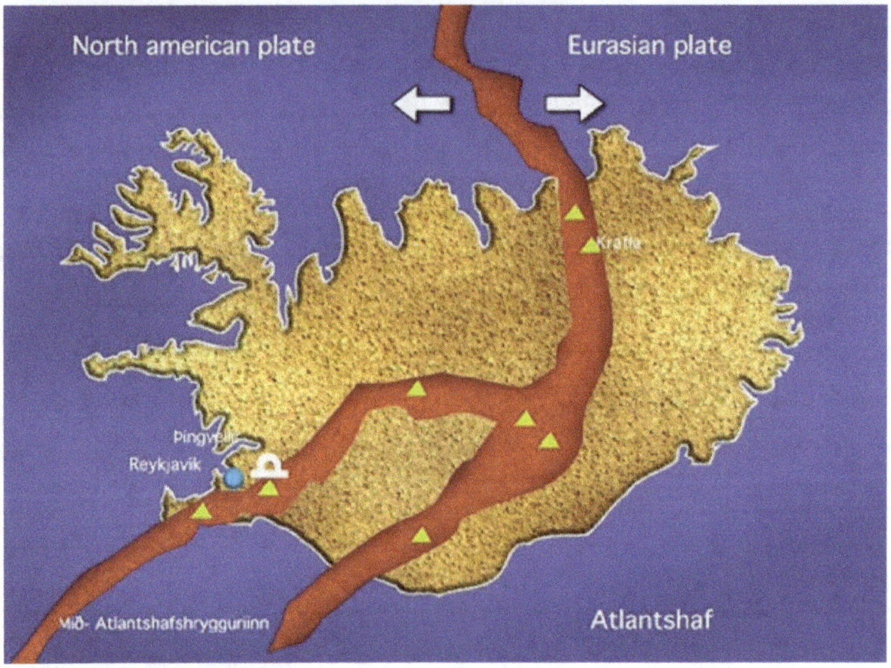

(Source: International Union of Geological Sciences)

While the volcanoes in Iceland are well-shaped in the classic conical shape, the most troublesome activity has been with the eruption of lava from the lateral subsurface tubes connecting the various volcanoes. Evidence of this is shown in the emergence of lava flowing out of a 4 miles long almost linear separation where the North American Plate and the Eurasian Plate are moving apart.

(Source: International Union of Geological Sciences)

Because the nature of the lava flow on Iceland not being a typical cataclysmic eruption but more of an oozing and horizontal flow, nearby residents were able to evacuate without the loss of any human life. It should be noted that the gases being emitted from this type of eruption are largely of natural carbon dioxide and water vapor composition, both of which are the powerful "greenhouse gases" heat trappers and bear no association with Man-made carbon dioxide from the burning of fossil fuels. It is these types of volcanic activity that produced the carbon dioxide in the atmosphere before any effect of Man even appearing on Earth's surface.

The Vostok Antarctic Research Station ice core samples "documents" that on at least four cycles in the past 420,000 years the natural carbon dioxide in the atmosphere was at a peak level of about 295 parts per million (0.0295%) and the peak hot temperature in that same cycle had peaked about 800 years prior to the peak carbon dioxide.

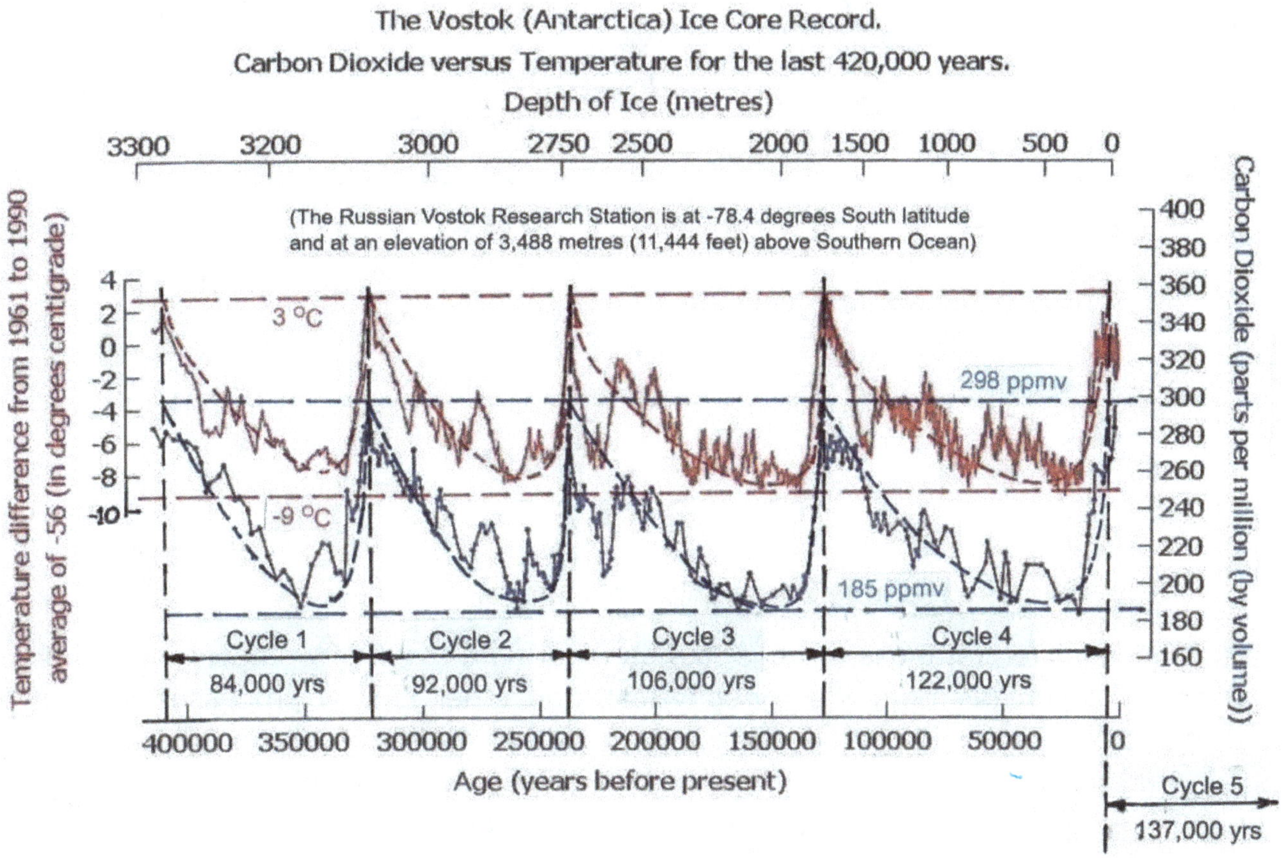

(Source: Smithsonian Institute for Science)

A closer examination of the peaks of temperature and carbon dioxide from the Vostok ice core samples in the region about 235,000 to 240,000 years ago (by an international team of scientists which included Dr. Nicolas Caillon of France) revealed an important finding regarding the "chicken or the egg" question about temperature or carbonj dioxide being the driver or the driven variable:

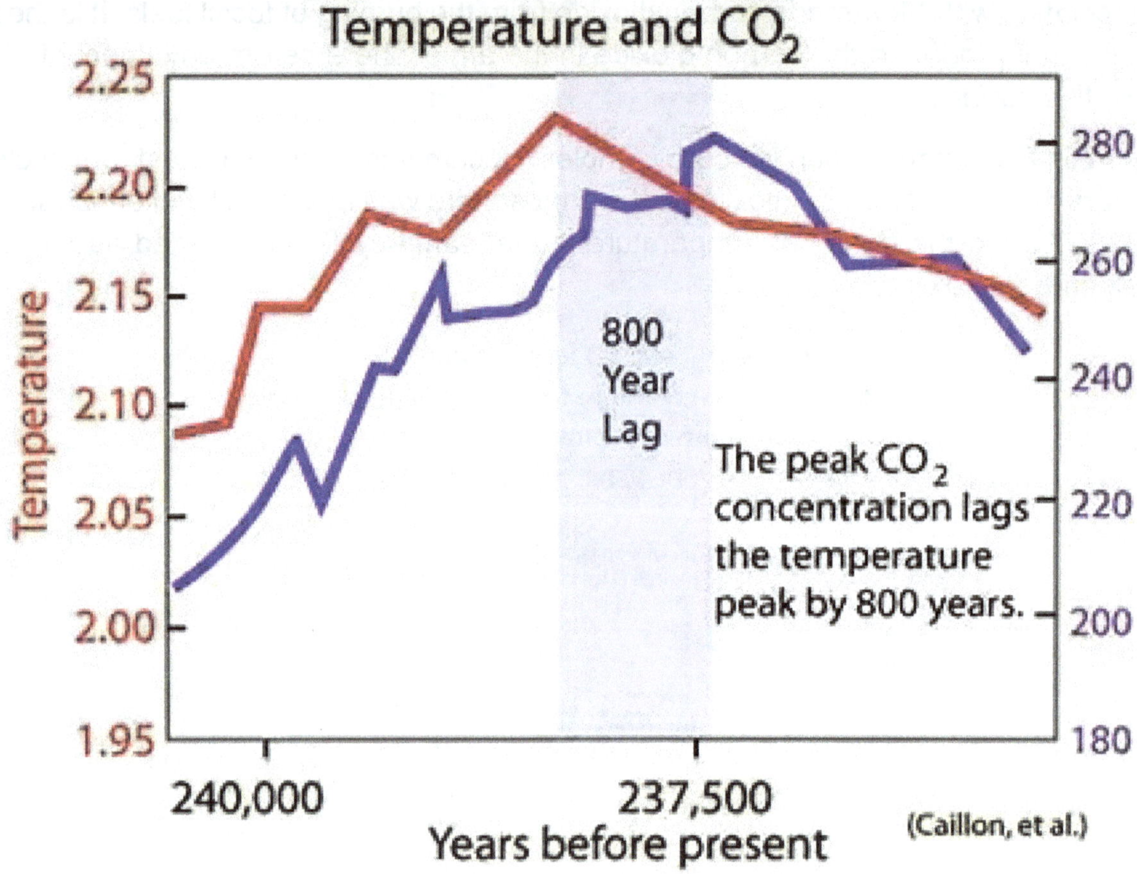

(Source: Smithsonian Institute for Science)

Reputable scientists interpret those findings to mean that as a result of whatever natural processes were and are in action (astrophysical, geophysical, and other natural processes without human influence), it is the temperature of the planet's surface that drives the carbon dioxide level in the atmosphere, and not the converse – i.e., the carbon dioxide in the atmosphere driving the surface temperature of the planet (as "settled science" fallaciously reports in NOAA publications and in the news media).

To not confuse the claim of former Vice President Al Gore of the Arctic Ocean ice melting due to Man-made "greenhouse gases", and the volcanic activity on Iceland being a primary contributor of heat causing the "Inconvenient Truth" ice in the Arctic Ocean, Iceland is geographically located immediately south of the Arctic Circle. The hyperbolic inference of the "Inconvenient Truth" is that it is the atmospheric "greenhouse gas" carbon dioxide (combination of both natural and Man-made) that is collaterally melting the massive ice sheet on Greenland – which is largely north of the Arctic Circle (but not into the northern polar cap). Before the capability of measuring the temperature profile through the ice sheet on Greenland and of the terra firma of the surface on which the Greenland ice sheet resides became available in the latter portion of the twentieth century, the "Inconvenient Truth" pointed the finger of cause to be the Man-made "greenhouse gases", and unless halted, the ice sheets of Greenland and Antarctica will melt and sea levels would rise by estimates of up to 230 feet. It is "convenient" to not mention that of that 230 feet rise, 23 feet of it would be due to the melt of the entire Greenland ice sheet and the remainder (207 feet – a ratio of about 9 to 1) would be due to the melt of the entire Antarctic ice sheet.

Mr. Gore narrated the Academy Award-winning documentary "*An Inconvenient Truth*", in 2006, and which was credited with raising international awareness about the global climate crisis. The gullible public formed many activist groups demanding the cessation of Man burning fossil fuels and governments of many countries instituted legislation that supported that reduction of Man burning fossil fuels. Mr. Gore and the United Nations Intergovernmental Panel on Climate Change (IPCC) was awarded the Nobel Peace Prize for his "Inconvenient Truth" revelation in October 2007. The Nobel Commission citation partially stated:

"On October 12, 2007, former U.S. Vice President Al Gore and the United Nation's Intergovernmental Panel on Climate Change are awarded the Nobel Peace Prize for their efforts to increase public knowledge about man-made climate change."

The former senator from Tennessee who served as President Bill Clinton's vice president from 1993 to 2001, is considered one of the first politicians to recognize (and regrettably over-exaggerate) the adverse aspects of carbon dioxide emissions, postulated as a major cause of human-induced global warming. Gore became interested in the topic of global warming during a college course he took at Harvard University. As a US Congressman, he held hearings on climate change in the late 1970s, a time when most Americans had little or no knowledge of the issue. After losing the 2000 presidential election to George W. Bush, Gore embarked on a new campaign—the fight against man-made climate change—and gave slide-show presentations around the world in an effort to "educate" the public. *An Inconvenient Truth* chronicled Gore's efforts to educate audiences with his "traveling global warming road show" – minus only the sale of the snake oil customarily sold at Wild West medicine-man shows. In the film, he details the facts and fallacies surrounding this "planetary emergency" and describes the events in his own life that led him to become an environmental crusader. It must be wondered if Mr. Gore was aware of the truth – Man burning fossil fuels is causing global warming in perhaps the same ratio of 9 to 1 as is the ratio of the amount of ice in Antarctica as there is in Greenland? Polar bears are magnificently beautiful animals to preserve their existence – but are the many millions of penguins in Antarctica not also worthy of being saved from Man's energy transgressions?

The improvement of the temperature-sensing technology aboard NASA satellites and the International Space Station in the past few decades has revealed (in the image at the top of the next page) that the surface on which the Greenland ice sheet was formed in the past few eons is a very warm surface from geothermal activity (similar to the same geothermal tectonic-geology as exists under Iceland but without the immediate presence of molten matter). In an equally revealing image, the terrain which features a large number of lakes below the ice sheet covering the Antarctica continent (and showing the extensive ice shelves of Western Antarctica that may melt long before the ice sheet above terra firma in Eastern Antarctica) is also shown on the next page:

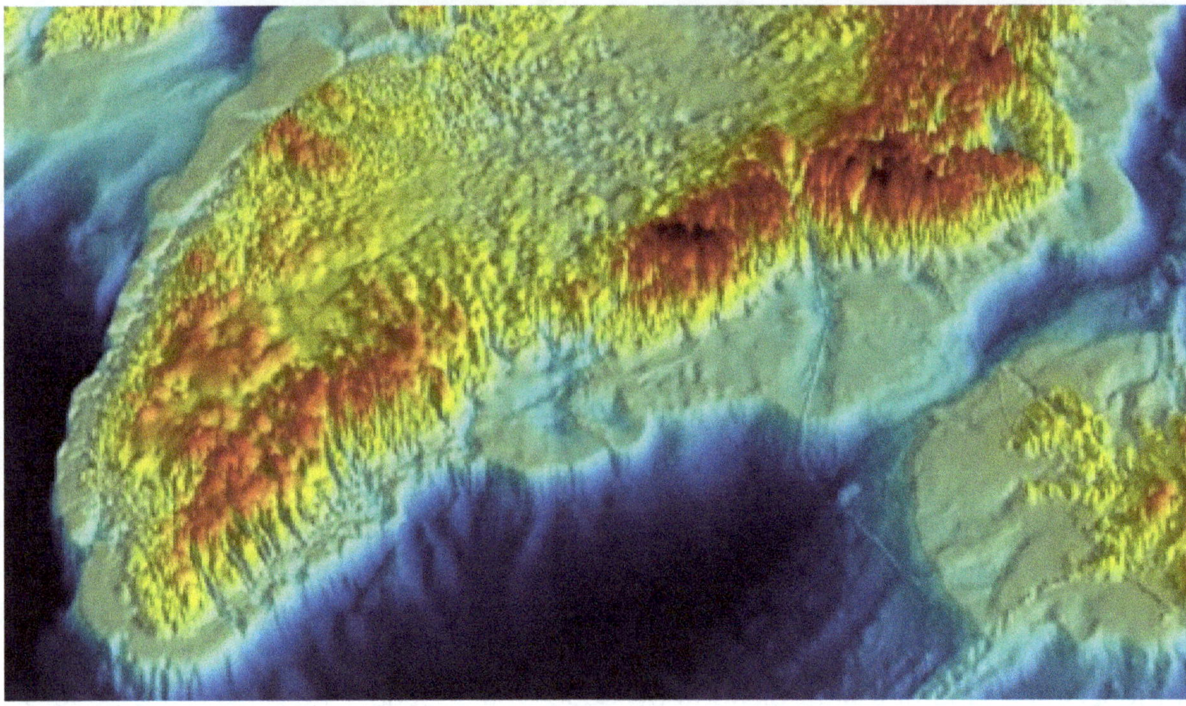

(Source: NASA) Below Greenland Ice Sheet

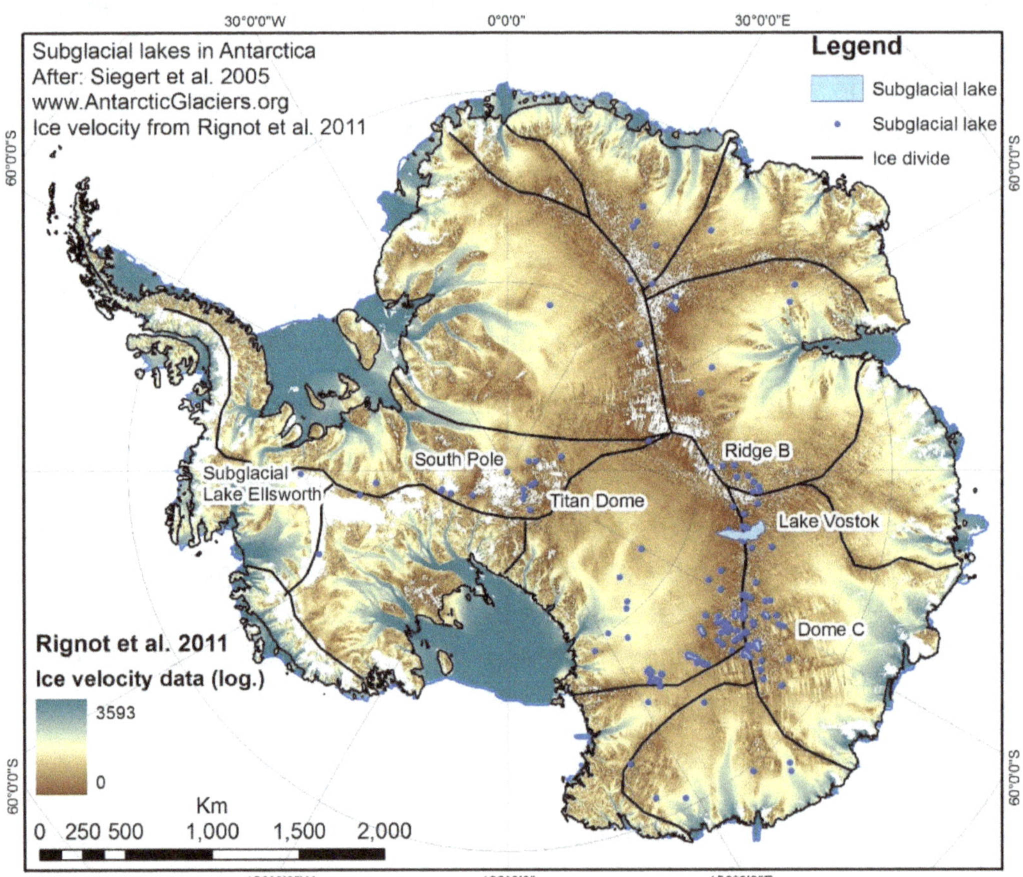

(Source: NASA) Below Antarctica Ice Sheet

A very useful "big picture" Volcanic Explosivity Index (VEI) was developed in 1982 (by Christopher G. Newhall (and an associate scientist Stephen Self) of the US Geological Survey) that reflects, in a qualitative manner since there is no physical way to measure the enormous events of a volcanic eruption, what was the volume of material (solid and/or molten) ejected onto the affected area surrounding the volcano and amount of gaseous and/or particulate material injected into the atmosphere and to what height, and how long did the continuous blast effect (if any) persist. The next chart indicates the definition used for the Volume Explosive Index (VEI ranging from VEI 0 to VEI 8+:

CRITERIA	VEI 0	VEI 1	VEI 2	VEI 3	VEI 4	VEI 5	VEI 6	VEI 7	VEI 8+
Description	non-explosive	small	moderate	moderate-large	large	very large →	→	→	→
Volume of ejecta (m^3)	$<10^4$	10^4-10^6	10^6-10^7	10^7-10^8	10^8-10^9	10^9-10^{10}	10^{10}-10^{11}	10^{11}-10^{12}	$>10^{12}$
Column height (km)	<0.1	0.1-1	1-5	3-15	10-25	>25 →	→	→	→
Qualitative description	← "gentle, effusive"	→	← "explosive"	→	← "cataclysmic, paroxysmal, colossal" →	→	→	→	→
				"severe, violent, terrific" →	→	→	→	→	→
Classification		← "Strombolian"	→		"Plinian" →	→	→	→	→
	← "Hawaiian"	→	← "Vulcanian"	→	→	"Ultraplinian" →	→	→	→
Duration of continuous blast (hours)	← <1	→	→		← >12 →	→	→	→	→
			← 1-6	→	→				
				← 6-12	→	→			
Tropospheric injection	negligible	minor	moderate	substantial →	→	→	→	→	→
Stratospheric injection	none	none	none	possible	definite	significant →	→	→	→

(Source: Newhall and Self; US National Geological Survey)

Of importance to be noted is that the Volume of ejecta (in cubic meters) – to convert a cubic meter to a cubic yard, there are 1.31 cubic yards to one cubic meter – are estimated on the VEI scale to be a range of exponents of ten (e.g., 10^4 cubic meters (10,000 cubic meters) or less is rated (combined with the other factor observations) as being of category VEI 0. The exponent numeric (4, 5, 6, …..12) is the same as the number of zero's after the 1. (As shown in a later chapter, the amount of carbon dioxide gas that environmental scientists claim has been and is being introduced into the atmosphere from Man burning fossil fuels is measured in terms of gigatons. A gigaton is 10^9 tons. Although the frequency of eruptions is only an average of 67 eruptions per year and of a gamut of VEIs, it raised the question in the Author's mind "How much carbon dioxide gas factually results from Man burning fossil fuels (especially coal) in coal-burning facilities and human residential use each year globally and how does that compare to the amount of carbon dioxide gas produced each year by volcanic action?".

To begin to understand that comparison, a statistical analysis was conducted (by the Author) to reconstruct the listed Smithsonian Institution Global Volcanism Program data into the matrix form shown in a table of Volcanic Eruptions (frequency and severity) statistics. The statistical database on volcanic eruptions from 1960 to 2023 compiled by the Smithsonian Institution Global Volcanism Program is a most interesting one – and was considered as a "red flag" needing to be examined closer by the Author.

Volcanic Eruptions Statistics by Decade, Number, and Severity Index

	Total No.	Number at Average Severity by VEI Level							
Year Span (No.Yrs)	**Total Eruptions / (Avg/Span)**	**No. at VEI 0-2**	**VEI 0-2 Yr Avg / (Span)**	**No. at VEI 3**	**VEI 3 Yr Avg / (Span)**	**No. at VEI 4**	**VEI 4 Yr Avg / (Span)**	**No. at VEI 5**	**No. at VEI 6**
1960 to 1970 (11)	649 / (59)	305	27.7 / (11)	64	5.8 / (11)	5	0.5 / (11)	1	0
1971 to 1980 (10)	627 / (63)	274	27.4 / (10)	40	5.8 / (10)	4	0.4 / (10)	1	0
1981 to 1990 (10)	626 / (63)	231	23.1 / (10)	45	4.5 / (10)	8	0.8 / (10)	1	0
1991 to 2000 (10)	636 / (64)	283	28.3 / (10)	38	3.8 / (10)	5	0.5 / (10)	1	1
2001 to 2010 (10)	778 / (78)	321	32.1 / (10)	56	5.6 / (10)	10	1.0 / (10)	0	0
2011 to 2020 (10)	705 / (71)	294	29.4 / (10)	52	5.2 / (10)	12	1.2 / (10)	1	0
2021 to 2023 (3)	240 / (80)	63 (2)	31.5 / (2)	2 (2)	1.0 / (2)	1 (2)	0.5 / (2)	1 (2)	0 (2)
(Total Span of 64 Yrs)	Avg 66.5 / Yr.								
Total 1960 - 2023	**4261 / (64)**	**1771 (63)**	**28.1 / (63)**	**297**	**4.7 / (63)**	**45**	**0.7 / (63)**	**6**	**1**

Note: There was only Total Eruptions data available for 2023; no VEI 0-2, VEI 3, VEI 4, VEI 5, or VEI 6 data.

(Source: Table generated by Author from Smithsonian Institute database)

A professional statistical analysis of the 4,261 volcanic eruptions occurring globally in approximate 10-year spans for the 64 years period from 1960 to 2023 reveals some interesting and important trends. (It is not surprising, but neither the environmental scientists nor the news media feel it to be necessary to inform the public about the role volcanic eruptions play in Earth's surface temperature changes, global climate changes, and ocean levels rising.) The information revealed about volcanic eruptions in this analysis and reported in this book hopefully are not among those *'everything's* in the fallacious "settled science" hypothesis of "*when 'everything' else is eliminated, Man burning fossil fuels is considered to be the dominant cause for global warming and climate change.*"

The Smithsonian Institution scientists evaluated each of the 4,000+ eruptions for the entire 63 years and gave each a VEI rating that seemed most appropriate all things considered. That laborious task is most valuable as it not only identified what the frequency of the different eruptions has been but includes a reasonably accurate system for estimating how much ejecta was introduced onto Earth's surface and into its atmosphere, and its oceans. The Smithsonian scientists do a far more scientific analysis than do the environmental scientists who, when they are presented with a database that does not conform with the hypothesis that "Man burning fossil fuels is the dominant cause for global warming and climate change", they simply "eliminate it" from further consideration. When presented in a tabular form (a long list of 64 lines of data), it is difficult to discern any pattern to the frequency of the volcanic eruptions or to their severity (with the VEI assessments assigned to each eruption. By graphically presenting that data, some most important observations are made and conclusions drawn.

To avoid "spotting" over 4,000 data points on the chart (giving it a "measles" appearance), the average number of volcanic eruptions occurring in the time spans, 1960 to 1970 (11 years), from 1971 to 2020 (in 6 spans of 10 years each), and from 2021 to 2023 (3 years) are presented at the mid-point of each span. Because of the large scatter – volcanoes erupt when and where they do, and except for measuring seismic activity in each volcano's vicinity, there is no long-term prediction of any singular volcano erupting.

By constructing a graph of the frequency and VEI index of severity within that frequency of volcanic eruptions (as shown earlier) provides some insight if the severity of the global eruptions is also increasing (from any cause). The graph shown next illustrates that trending of intensity effect for the differing VEI levels of the eruptions:

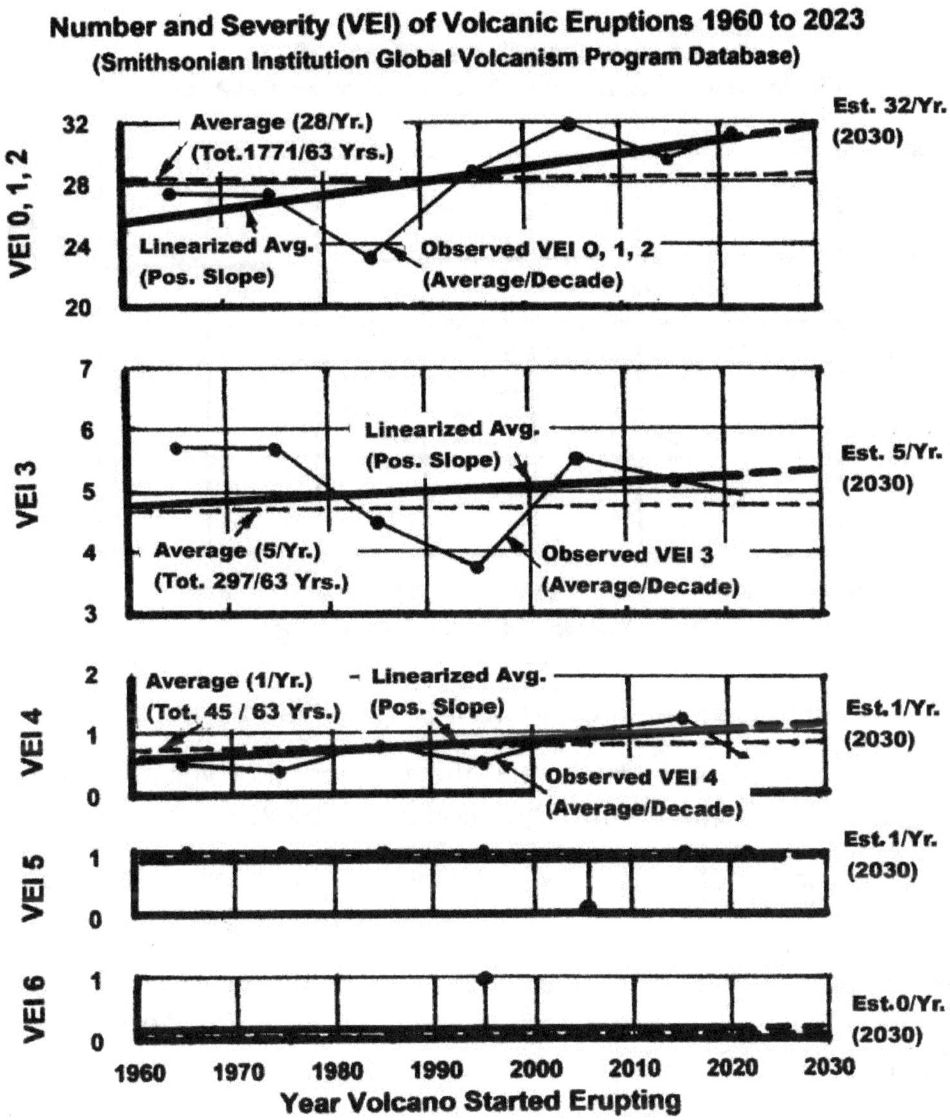

(Source: Author generated graph from Smithsonian Institution database)

The statistical interpretation of the variation in frequency is that there are several more occurrences of volcanic eruptions in the Category VEI 0, 1, and 2 per year, and there is a weak trend toward lesser frequency of the volcanic eruptions as their Category VEI level of severity increases from VEI 3 to VEI 4, 5, and 6. There should be little question whether the changes in both frequency and severity of all VEI levels of volcanic eruptions are due to unpredictable tectonic plate movements, or due to an increase in Man burning fossil fuels.

There can be an average number of all volcanic eruptions per year obtained simply by dividing the total number of all observed eruptions (4,261 below VEI 0 and from VEI 0 to VEI 6) by the 64-year time span between 1960 and 2023, but that is not a very meaningful statistic to a professional database analyst. That average number of total eruptions per year (rounded up to 67) is seen on the graph as a dashed horizontal line. It was stated earlier and restated here (and will be restated later), very little in the real-world behaves in linear form and that is especially true for volcanic activity on the planet's surface.

For the volcanic eruption data from 1960 to about 1990-2000, the majority of the events (points) are shown to be above the dashed "average" line whereas after about 2000, the majority of the points was below the "average" line as shown in the following graph:

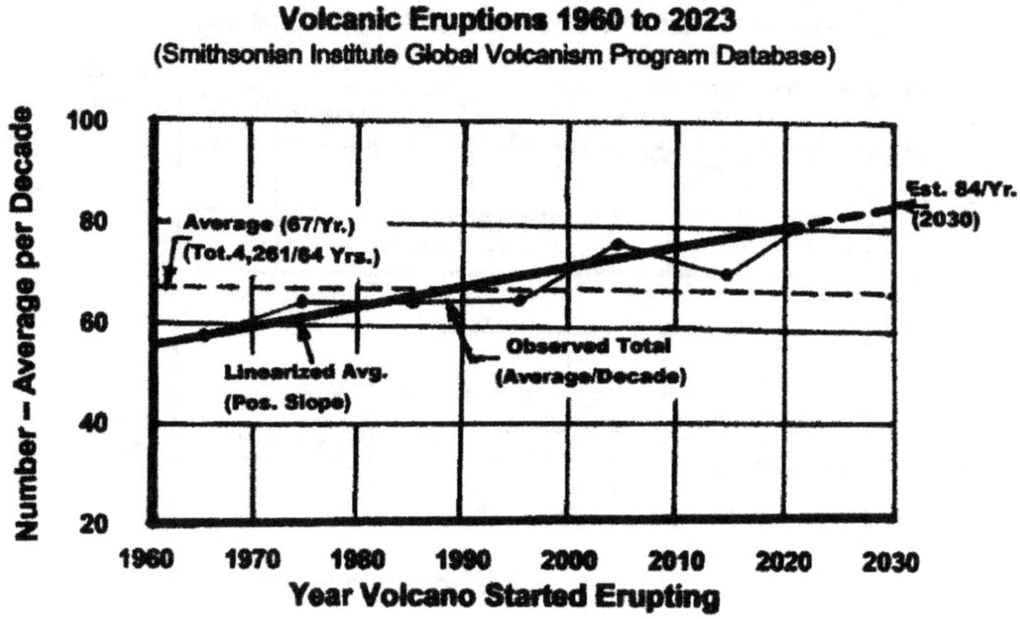

(Source: Author Generated Chart from Smithsonian Institution Database)

The overall "big picture" conclusion is made that the frequency of volcanic eruptions world-wide is increasing in number with time. A near term estimation is that by 2030, the

highest probability of the number of eruptions world-wide per year is expected to be about 84 eruptions of all VEI severity ratings.

A lengthy tabulation on a country-by-country arrangement is not feasible in this book, but a summary tabulation of the world-wide eruptions with fatalities and the region identified, during the period 1963 to 2023 is an "eye opener":

Fatalities Caused by Volcanic Eruptions on Earth 1963 to 2023

Year	Yr Total	Country & Number Fatalities
1963	1,584	Indonesia 1,584
1964	139	Indonesia 114; Chile 25
1968	87	Costa Rica 87
1969	3	Philippines 3
1971	30+	Chile 25+ & 5
1973	1	Iceland 1
1977	70	Congo Rep. 70
1979	238	Indonesia 149 & 80; Italy 9
1980	57	USA 57
1982	1,918+	Mexico 1,900+; Indonesia 18
1985	23,000+	Colombia 23,000+
1990	32	Indonesia 32
1991	890	Philippines 847; Japan 43
1992	6	Indonesia 6
1993	79	Philippines 79
1994	99	Indonesia 64 & 30; New Guinea 5
1996	3	Philippines 3
1997	19	Caribbean 19
Total Fatalities 1963-1997 was 28,255+		

Year	Yr Total	Country & Number Fatalities
2002	245	Congo Rep 245
2004	2	Indonesia 2
2005	2	El Salvador 2
2006	1	New Zealand 1
2008	1	Chile 1
2010	356	Indonesia 353; Guatemala 3
2011	35	Eritrea 31; Indonesia 4
2014	23	Indonesia 16 & 7
2017	8	Indonesia 8
2018	3,326+	Guatemala 2,800+; Indonesia 426
2019	23	New Zealand 22; Italy 1
2020	39	Philippines 39
2021	101	Indonesia 69; Congo Rep. 32
2022	1	Mexico 1
2023	10+	Colombia 10+;
Total Fatalities 2002-2023 was 4,173+		

Total Fatalities 1963-2023 was 32,428+
Fatalities 1963-2023 averaged 540+ per year

(Source: International Union of Geological Sciences database, tabulation by Author)

For unknown reason, the fatalities statistics for the global volcanic eruptions in 1998, 1999, 2000, and 2001 were not found. It is possible but not likely that no fatalities due to volcanic activities occurred during those four years.

If the average number of human fatalities from volcanic eruptions in the period 1963 to 2023 averaged 542 fatalities per year (in the tabulation of fatalities from volcanic eruptions above) and there was an average of 67 volcanic eruptions per year in that same period (1963 to 2023) (in the second table above) then using the "averaging of averages technique" (so freely used by environmental scientists as in "average global temperature"), a conclusion is that Earth loses "only" approximately 8 humans per volcanic eruption. Many of those lost were scientists making measurements and were fatally injured when the volcano unexpectedly erupted nearby.

Again, what occurs in an "average global" manner for several regions may not reflect – either too few or too many – as that common for another region. In the Indonesian region between 1963 and 2023, there were 2,962 human fatalities in 17 eruptions for an average of 174 fatalities per eruption. In the United States ("lower 48") from 1963 to 2023 there was only one eruption (Mt. St. Helens in the state of Washington) and 57 fatalities. Does that substantiate the claim of environmental "settled science" scientists': "When all other factors are considered, the conclusion of "settled science" remains that Man burning fossil fuels is the dominant threat to the human population of the World"? How many human fatalities are attributed (with scientific validity) being due to "greenhouse gas emissions" during the period 1963 to 2023? (For volcanic eruptions, that total was over 32,428.) Indonesia clearly must control better where its population resides relative to volcanic eruption hazards and have a better eruption warning system – similar to that of Hawaii in the United States and other developed countries.

As shown earlier, volcanoes erupt with a wide variety of Volcanic Eruption Indexes and with different "personalities" for the type and amount of ejecta. They also have a unique "kill/damage-mechanism" that varies significantly and with a different effect on the ecosystem. An interesting contrast was seen between the eruption of Mt. Pinatubo in the Philippines (photographs shown earlier of the vast deposit of volcanic ash at Clark Air Base and on local rooftops) compounded by the heavy rainfall from Typhoon Yunya a day after (in 1991), and the earlier eruption of Mt. St. Helens in the US state of Washington in May 1980. The eruption of Mt. St. Helens was the largest eruption in recent U.S. history and one of the best photographed eruptions as it was being well monitored by scientists before, during and after its non-typical form of eruption.

The Mt. Pinatubo eruption in the Philippines in 1991 was a vertical eruption in which there was little lava efflux, but an enormous amount of volcanic ash inserted into the atmosphere and which was carried by winds over the Luzon peninsula. Compounding the heavy ash problem to nearby locations was Typhoon Yunya which passed over the Mt. Pinatubo eruption a day after the eruption. The heavy rainfall from Yunya mixed with the thick layer of ash from Pinatubo on everything on the ground and collapsed many flat rooftops. Nearby Clark Air Base had evacuated its fleet of USAF airplanes based at Clark Air Base after a warning of Typhoon Yunya approaching. A commercial DC-10 which was unable to be flown out because of an on-board equipment malfunction became a victim to the heavy mixture of volcanic ash and typhoon rainfall:

(Credit to unknown photographer at Clark Air Base)

The Mt. St. Helens unusual eruption in the US State of Washington (on the Pacific "Rim of Fire") in early 1980 was a more lateral eruption. Based on seismic measurements being made of Mt. St. Helens which began "rumbling" for weeks before and foretelling its probable eruption, scientists and photographers began observing the bulging of the sidewall of the volcanic cone and very well documented the explosive eruption on May 18, 1980. The Mt. St. Helen eruption was especially well documented photographically as shown next:

(Source: US National Geological Survey; photo by Unknown Photographer)

The photographs of the "during" eruption from ground level (above photo) and from altitude (next photo) shows the extensive amount of mass that was ejected initially horizontally (in the initial "sidewall blow-out") and then the following plume of ejecta from within vertically to an altitude where winds in the atmosphere turned the ejecta cloud again horizontally being carried with the wind after Mt. St. Helen erupted.

(Source: US National Geological Survey; photo by Unknown Photographer)

As the ejecta plume develop and followed its wind-driven path, a GOES satellite photographically tracked (shown next) the path of the plume – providing cautionary warnings to populated areas in its path.

(Source: NASA GOES Satellite)

The Geostationary Operational Environmental Satellites (GOES) is a collaborative NOAA and NASA program providing continuous imagery and data on atmospheric conditions and solar activity (space weather). NASA builds and launches the GOES and NOAA operates them. The GOES database has provided a major portion of the updates to the environmental database for analysis in this book. What the public should "see" in the "big picture" photograph above is not only the long ejecta plume from one volcanic eruption and its shadow on the Earth's surface but notice the extensive cloud cover over the entire portion of the Earth's surface – all of which in combination dwarf the heat-trapping effect of Man-made carbon dioxide from the burning of fossil fuels.

A view several months after the Mt. St. Helen eruption returned to a near-dormant state shows the effect of the eruption of Mt. St. Helen on the surrounding environment (Spirit Lake in the foreground). The explosive blast pressure when Mt. St. Helen erupted in 1980 blew down almost every tree in the large area. The absence of forestry and vegetation with the entire region covered in grey-colored ash should be noted in the photograph. This is not typical for the state of Washington which elsewhere is abundant in green forestry and wild grasses.

(Source: National Park Service)

The acidity of the ejecta which was blown or flowed into Spirit Lake destroyed almost all the aquatic life in the lake.

An annotated photograph of the Mt. St. Helens area affected by the 1980 eruption provides some interesting geological statistics regarding the effect that the Mt. St. Helen eruption had on the environment:

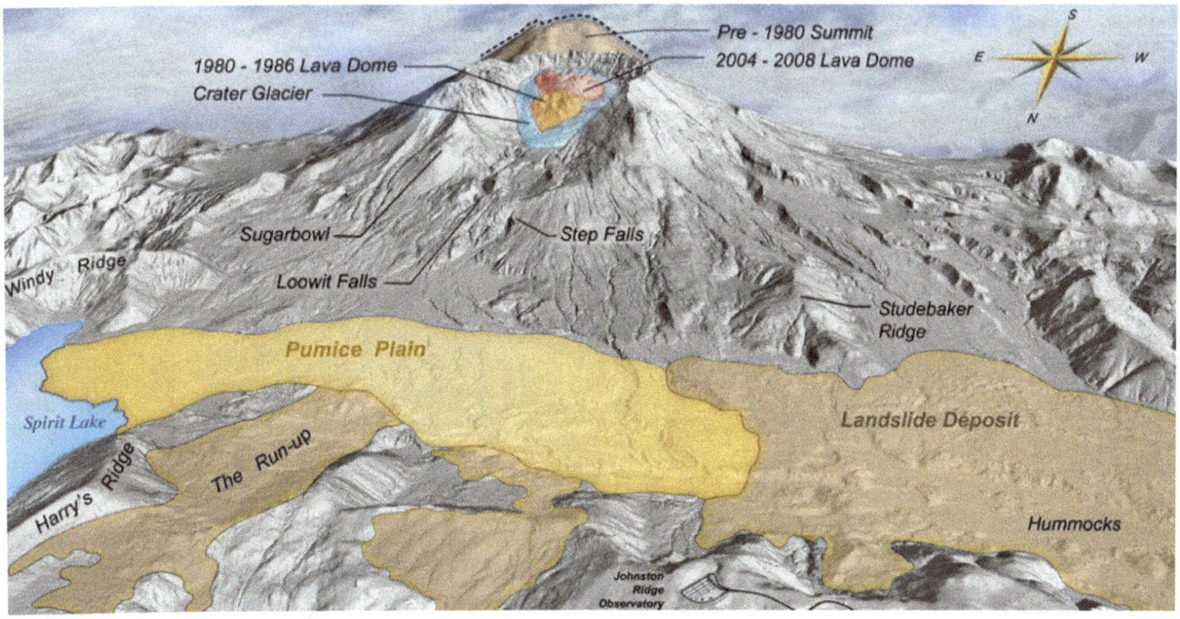

(Source: US National Geological Survey)

Readers should also note that it is natural for assaults on the environment (caused by nature itself or by inconsiderate Man) to trigger other natural forces and processes in the region to return to an equilibrium or "normal" state that existed before the disruption. A photograph of the region most affected by the Mt. St. Helens eruption in 1980 is shown next, illustrating how the region has returned (almost completely) to its natural state in 2023 (over 40 years after the eruption):

(Source: US National Park Service)

With attention turned to the ejecta from global volcanic activity having an effect on the atmosphere and the surrounding environment, an example of how environmental protection scientists are misinforming (inadvertently or intentionally) the public with over-exaggerated data to prove that a "crisis" looms ostensibly due to Man burning fossil fuels at an unusually high rate is shown next:

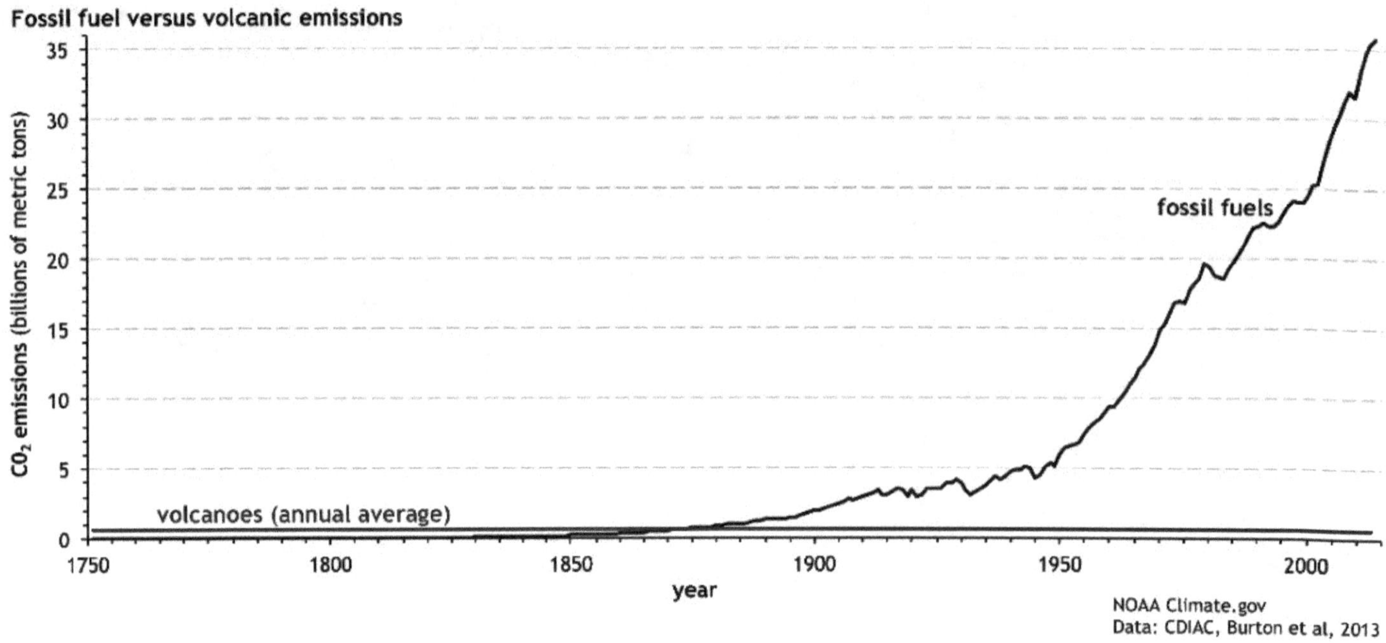

(Source: NOAA)

CAUTION: The above diagram presented to the public has several flaws, none of which may be obvious to the gullible public. Readers are cautioned to observe the intent of the chart, but not the numerical relative magnitudes of its end-effect, to illustrate that Man burning fossil fuels is the dominant cause of the global "greenhouse gas" effect.

1) The above chart graphically reflects the fallacious conclusion drawn by many environmental scientists that the human and industrial use of fossil fuels has gone exponentially out of control, but indicates comparatively that volcanic eruptions play a negligible part in the global warming and climate change "crisis";

2) The chart shows that the amount of the greenhouse gas carbon dioxide introduced into the atmosphere by World-wide volcanic eruptions is of a suspiciously low magnitude starting in 1750 and remains at that low constant level for the next 250 years (beyond 2000). The linear representation is labeled "Volcanoes (annual average)". There exists no valid database of scientific measurements (by professional volcanologists or environmentalists) of the gaseous content of all volcanic eruptions. Taking measurements of the gases would entail aircraft to fly at those altitudes into the gaseous plume and that would destroy the propulsion systems of aircraft. Therefore, the "average CO_2 in volcanic eruptions" is suspect of gross underestimation to place emphasis on the difference with "fossil fuels CO_2" measurements and overexaggerated estimates by environmental protection scientists.

3) A third question regarding the source of the Fossil Fuels data used for the above comparison and needing clarification is: "How much of the carbon dioxide in the atmosphere (shown on the chart in billions of metric tons) is actually due to Man burning fossil fuels (containing carbon) and how much is produced from natural causes? (The "Keeling Curve", provided in the previous chapter, shows the measurement of total (natural and Man-made) carbon dioxide in the atmosphere (435 parts per million by volume (ppmv) in 2023) but of that 435 ppmv, 295 ppmv of natural carbon dioxide existed in the atmosphere prior to the start of "the Industrial Revolution". Therefore only 140 ppmv of carbon dioxide in the total 435 ppmv is attributable to being Man-made. That begs the question in the above chart as to the source of the "Fossil Fuels" amount of carbon dioxide in that data shown being Man-made (from burning fossil fuels)

When coal (the major offending fossil fuel) is combusted, the combustion process draws air from the atmosphere to complete the combustion process. Air is a collection of gases with varying gas content as altitude increases. At Sea Level, atmospheric physicists use either the composition of a volume of air to be comprised of 77% nitrogen and 21% oxygen with a trace of Sulphur. If considered by mass, air is comprised of 79% nitrogen and 21% oxygen. Depending on the quality of the coal – which does vary considerably by type, (lignite, anthracite, or bituminous) – the "average" coal by mass is 83% carbon, 5% hydrogen, and 2% oxygen. Skipping over the scientific analysis (which can be found in many textbooks of Atmospheric chemistry and physics), by mass it requires 2.66 mass units of oxygen to combust 1 mass unit of coal. However, oxygen is not drawn from air only as oxygen, the mass amount of air (comprised of 79% nitrogen and 21% oxygen) drawn into the combustion process is that which provides the 2.66 mass units of oxygen to combust the 1 mass unit of air. On average, it requires a mass of air 11.28 times as much as 1 mass unit of coal (containing 83% carbon, 5% hydrogen, and 2% of oxygen.) After the thermodynamic process is done, it takes the oxygen in 11.28 mass units of air to combust 1 mass unit of coal and extract the heat of the combustion process to be used for heat energy purposes.

The bookkeeping done by environmental scientists is such that the combustion of 1 mass unit of coal introduces 12.5 mass units of carbon dioxide into the atmosphere is misleading. It is more factual that only the 1 mass unit of coal transformed from solid to gaseous state that is being added to the atmosphere – the original 11.28 mass units of air (which has the 2.66 mass units of oxygen combining with the 1 mass unit of now gaseous carbon) are also being returned after being "borrowed from the atmosphere" for the combustion process. The ratio of the unit mass of air needed to combust a unit mass of coal is approximately 12 to 1 to produce the desirable heat energy.

A law of physics is that mass can neither be created nor destroyed, regardless of any change in state (solid, liquid, or gas). When considering the Laws of Conservation of Mass, the scientific fact is that 3.66 mass units of undesirable carbon dioxide gas are being introduced into the atmosphere for each mass unit of coal, not the reported 12.5 mass units of carbon dioxide for each mass unit of coal. (Other Authors of the "nay-sayer" books in the Reference and Other Documents section at the end of this book refer to the misleading mathematics being used by the environmental scientists as "cooking the books" on coal combustion.)

Therefore, in the NOAA diagram shown earlier, of the 26 billion metric tons of total carbon dioxide generated in 2000 and introduced into the atmosphere from the combustion of fossil fuels (predominantly coal), only 2 billion metric tons of coal (the fossil fuel cited transformed from solid state to gaseous state) were introduced as the net mass change into the atmosphere – the 24 billion metric tons of air "borrowed" from and then returned" to the atmosphere should not be double-counted.

It is of high importance at this point in the research of identifying the factors adding to the understanding of global warming and climate change for Readers to note that the "big picture" approach considers the full set of pros and cons of the technology being researched. The environmental science profession was sent in the direction primarily by the observation in the late 1890s by many scientists (the most notable being Dr. Svante Arrhenius in Sweden who was awarded the Nobel Prize for Chemistry in 1903) pursuing the chemical effects of the emissions of burning coal on the atmospheric environment.

The preceding sections of this chapter address the thermodynamics of the process of combustion of coal and the gaseous and thermal emissions created in the process. The brilliant research done on the heat-trapping characteristic of the carbon dioxide generated by the combustion of the fossil fuel coal –

in a scientific laboratory and starting with only one of the products of the combustion process - diverted environmental scientists away from considering the physical residue of the combustion process that produced the carbon dioxide gas being researched in infinite detail for its heat-trapping "greenhouse gas" effect. In the "big picture" system engineering analysis (which considers the total of the pros and cons of the entire system not merely one important characteristic from scientific observation), the ash residue created from the combustion of coal is a con that has been largely overlooked. It is the disposal of that solid state coal ash residue which contains toxins and carcinogens depending on the type and quality of coal being combusted. Coal is categorized in essentially three basic types: anthracite, lignite, and bituminous, each having a different composition of carbon and other elements and impurities.

A later chapter on the fossil fuels – coal, oil, and natural gas – provides further detail on the ash residue being a pollutant of land and water entering the water and food products consumed by humans and livestock animals. In the chapter near the end of the book in which the complete set of pros and cons of 24 technology areas are assessed and prioritized (with the use of weighted trade studies), it concludes that coal technology should be reduced and eventually eliminated in the United States region for the primary reason of its ash residue being a very hazardous (toxic and carcinogenic) polluting material to human and animal health and safety. In comparative ratings, it is the pollution of land and water not that of the carbon dioxide "greenhouse gas" being emitted into the atmosphere (and having a factually small if any effect on global warming and climate change) that is of highest significance.

The NOAA comparison chart (Fossil Fuels versus Volcanic Emissions) also shows that the constant average mass units of volcanic eruption generated carbon dioxide at a total rate shown in the chart as less than 1 billion metric tons per year generated in volcanic eruptions and introduced into the atmosphere. That low rate was constant during the period from 1750 to 2000. The oxygen contained in the carbon dioxide ejected from the volcanic underground source (i.e., the oxygen in volcanic ejecta not being drawn from the atmosphere) is already combined with the carbon element both drawn from within the inner regions of the planet – such that 100% of the mass of carbon dioxide ejected during volcanic eruptions is introduced into the atmosphere as a transformation of carbon in solid state form to gaseous state form without any change in its mass during the eruption process.

It was earlier shown graphically that the volcanic eruption database for the different levels of VEI category of World-wide volcanic eruptions during the period 1960 to 2023 (after "averaging") was increasing in frequency – a higher frequency average for the lower VEI eruptions and a singularly frequency average for the highest two VEIs (VEI 5 and 6). The ejecta of each eruption varies exponentially as the VEI index increases (approximately as shown in the earlier tabulation of the Volcanic Explosivity Index developed in 1982.

The Author has made an isolated extraction of the volume of ejecta associated with the differing levels of Volcanic Explosivity Index (VEI) as shown next:

Volume of Ejecta as VEI Increases

(Newhall and Self, 1982)

Volcanic Explosivity Index	Volume of Ejecta (Observable)				
(Newhall & Self)			Other equivalences (approx)		
VEI	meters^3		kilometers^3	miles^3	yards^3
0	$< 10^2$		< 1	< 1/4	$<10^2$
1	10^2-10^4		1 to 10	1 to 3	10^2-10^4
2	10^4-10^6		10^2 - 10^3	10 - 10^3	10^4-10^6
3	10^6-10^8		10^3-10^5	10^3-10^5	10^6-10^7
4	10^8-10^9		10^5-10^6	10^5-10^6	10^7-10^8
5	10^9-10^{10}		10^6-10^7	10^6-10^7	10^8-10^9
6	10^{10}-10^{11}		10^7-10^8	10^7-10^8	10^9-10^{10}
7	10^{11}-10^{12}		10^8-10^9	10^8-10^9	10^{10}-10^{11}

Note: The ejecta in each eruption are a varying amount of solids (from particulate to boulder size), liquid (molten rock (lava, at extremely hot temperature) and various compounds of minerals/elements), and gasses (including water vapor in the form of steam.)

(Source: The National Geological Survey database reformatted by Author)

It should be noted that the small integer exponents of 10 shown in the table above reflect a multiplier of the volume of ejecta – i.e., each volcanic eruption of a VEI 5 category ejects 10^9 (1,000,000,000) to 10^{10} (10,000,000,000) cubic meters of ejecta (possibly containing only a low percent of carbon dioxide) into the atmosphere. 1 billion (10^9) cubic meters of ejecta (solid, liquid, and gas) introduced into the atmosphere by one VEI 5 volcanic eruption is far more hazardous to humans and the environment than the 13 billion tons of carbon dioxide gas (at a concentration of 435 parts per million (0.0435%) from the fossil fuels indicated in the NOAA chart (shown earlier of Fossil Fuels versus Volcanic Emissions).

Readers should take note of the decreasing frequency of volcanic eruptions as the various levels of the Volcanic Explosivity Index (VEI) increases. To be noted also is that the estimate of the volume of ejecta increases exponentially at the different VEI levels. By multiplying the number of volcanic eruption occurrences on average per year of the different levels of VEI severity by the estimated volume of ejecta associated with the different VEI categories, a simple mathematical calculation provides a high probability estimate of the total volume of ejecta that may be produced in world-wide volcanic eruptions for that year. For the year 2030, that volume is estimated with the calculation:

32 eruptions at VEI 0,1,2 at 10^4 m^3 ejecta volume = 320,000 m^3 of subtotal ejecta;

+ 5 eruptions at VEI 3 at 10^7 m^3 ejecta volume = 50,000,000 m^3 of subtotal ejecta;

+ 1 eruption at VEI 4 at 10^8 m^3 ejecta volume = 100,000,000 m^3 of subtotal ejecta;

+ 1 eruption at VEI 5 at 10^9 m^3 ejecta volume = 1,000,000,000 m^3 of subtotal ejecta

The total ejecta calculated for 2030 = 1,150,320,000 cubic meters. To attach some relative size to that large a volume, since a kilometer is 1,000 meters, a cubic kilometer is 1,000,000,000 cubic meters and the total ejecta for 2030 is estimated to be 1.15 cubic kilometers. In those countries (including the USA) where distances are measured in miles rather than kilometers, there is 0.621 miles in every kilometer. In a cubic kilometer, there will be 0.239 cubic miles. The total estimated ejecta for volcanic eruptions producing 1.15 cubic kilometers equates to an ejecta volume of 0.239 cubic miles.

Those 39 eruptions at the various VEI levels produces a high probability (over 75%) of there being a total volume of ejecta of approximately 1,150,320,000 (1.15 x 10 9) cubic meters equivalent to 1.15 billion cubic meters of high temperature, highly harmful to humans particulate, in solid, liquid/molten, and gaseous state (all heat-trapping worse than is the equivalent volume of carbon dioxide) being introduced into the atmosphere of Earth each year from 2024 to 2030. The "one-size-fits-all" human control of "greenhouse gas emissions" to net zero by 2030 is comparatively a futile and unrealistic goal with high probability of not being achieved and does not alter the volcanic eruption of ejecta estimate of 1.15 x 10 9 cubic meters per year of more ominous volcanic ejecta. The significant but not totally off-setting difference is that the ejecta heavier mass from volcanic eruptions will eventually fall to the Earth's surface whereas the gaseous carbon dioxide (natural and Man-made) remains in the atmosphere for several years.

It is those volcanic eruptions world-wide that are and will be affecting the atmosphere with heat-trapping solids/liquids/gases in total (despite their low frequency rate but considering their massive total volume of heat-trappers and blockers ejected) which actually discounts – but does not totally negate - the environmental scientists' hypothesis that Man burning fossil fuels creates greenhouse gas emissions and is the dominant cause for global warming.

It should be obvious that environmental scientists extrapolating what they observed taking place with a beaker of carbon dioxide at high concentration being heated by a heat-source in a laboratory had leapt to the fallacious conclusion that it is Man causing global warming. That is not what is happening in the very real natural world. It would be most valuable to the public if organizations such as NASA, NOAA, USNGS, and the entire environmental science profession were to pool their talents and resources and use all pieces of the many databases to identify the true scientific problem of the multiple causes of global warming, climate change, and ocean levels rising and their respective effects on the environment (atmosphere, ocean, and land) and/or on the human population. This book is a step (but not the last word) toward that objective to better understand the factual causes of global warming and climate change.

There remains another important "flaw" in the volcanic eruption data shown by NOAA, that "flaw" being that the number of average volcanic eruptions of volcanoes on the planet (reported in the Smithsonian Institute Global Volcanism Program Database and graphically shown earlier) was 57 eruptions per year average in 1960 and steadily increasing to 80 eruptions per year average in 2020. Yet, that 41% increase in eruption average rate in the period from 1960 to 2020 reported by the Smithsonian Institute Global Volcanism Program Database is not reflected on the NOAA volcanic carbon dioxide diagram shown above (which shows no increase in carbon dioxide into the atmosphere from volcanic eruptions in the period 1750 to 2000.

The originator of the chart however seems to have disregarded the fact that volcanic eruptions may not be very frequent but when a volcano erupts and expels so much ejecta into the atmosphere in the form of carbon dioxide, there is an enormous volume of other solids/liquids/gases that are introduced into the atmosphere and having far greater heat-blocking and heat-trapping properties than that of only the carbon dioxide generated by Man burning fossil fuels.

Perhaps it is not statistically valid to compare fatalities from "natural catastrophes" to fatalities from "Man-made catastrophes" – as in comparing "apples to oranges". The National Highway Transportation Safety Administration (NHTSA) reports that there are over 40,000 humans fatally injured in Man-caused automobile accidents annually in the United States alone. For the 60 years period 1963 to 2023, there were over approximately 2,400,000 humans fatally injured in Man-caused automobile crashes on the US highway network of roads. There could not be found any comparison data for the number of human fatalities directly attributed to Man-made "greenhouse gas emissions". Deaths reported to be due to heat-stroke (hyperthermia) were found to approach perhaps 1% that of the NHTSA fatality rate. According to the Center for Disease Control (CDC) records of deaths reported due to hypothermia (extreme cold) and hyperthermia (extreme heat) those due to hypothermia (extreme cold) are significantly higher – at a ratio of about nine to one for victims with otherwise normally healthy respiratory and cardiovascular systems.

Is using NHTSA highway data and Department of Energy "greenhouse gas" data a fair "apples to apples" (Man-caused to Man-caused) statistical comparison? If so, then the technology advancement made by environmental scientists to reduce motor vehicle emissions to net zero by replacing fossil-fueled vehicles with all-electric vehicles by 2050 should prove to be of utmost importance to the US and world transportation system and to the human population (minus those 2.4 million humans who lost their lives in vehicle crashes between 1963 and 2023). The oft referred to DOE chart reporting atmospheric carbon dioxide concentration is shown next:

TABLE 1.
The Important Greenhouse Gases (except water vapor)
U.S. Department of Energy. (October. 2000) (1)

(all concentrations expressed in parts per billion)	Pre-industrial baseline	Natural additions	Man-made additions	Total (ppb) Concentration	Percent of Total
Carbon Dioxide (CO2)	288,000	68,520	11,880 (2)	368,400	**99.438%**
Methane (CH4)	848	577	320	1,745	**0.471%**
Nitrous Oxide (N2O)	285	12	15	312	**0.084%**
Misc. gases (CFC's, etc.)	25	0	2	27	**0.007%**
Tota.	289.158	69.109	12.217	370.484	100.00%

(Source: US Department of Energy)

In its zeal to "educate" the public about the importance of the "greenhouse gas effect" (especially that of carbon dioxide) on the climate, environmental scientists in the DOE present the concentration of the greenhouse gases in units of parts per billion instead of the conventional NOAA units of parts per million. The general public is well familiar with the per cent (in effect parts per hundred) system of units (as in what percentage of their income is taxable), but unfamiliar with scientific significance of parts per million or parts per billion. The DOE tabulation is misleading (whether intentional or not) as the mix of per cent (showing carbon dioxide as 99.438% of the "greenhouse gas" concentration) and parts per billion (shown as 368,400 ppb for total carbon dioxide in the atmosphere) mask the reality of there being only a concentration of 0.03684% of total carbon dioxide in the atmosphere (accurately measured by NOAA). Of that total of 0.03684%, the Man-made portion of carbon dioxide in the atmosphere is 0.0073% (i.e., excluding the four times larger 0.0295% as the natural-caused (mostly from volcanic eruptions) carbon dioxide that existed in the atmosphere prior to the start of the Industrial Revolution).

Reducing the Man-caused "greenhouse gas emissions" to net zero by 2050 (set as the "one size fits all") goal by the UN Intergovernmental Panel on Climate Change IPCC) does not reduce the much larger portion of natural-caused carbon dioxide in the atmosphere. The present 2024 level of 435 ppm (0.0435%) concentration of carbon dioxide in the atmosphere, very accurately measured by NOAA can, even if the IPCC goal of "net zero by 2050" is 100% achieved, at best be reduced to the pre-Industrial Revolution level of 295 ppm (0.0295%).

A not-so-humorous question that is overheard when professional database analysts (scientists and engineers) analyze a complex set of data for root cause (by the process as done for accidents which resulted in multiple human fatalities): "Are we choking on the minnows, while the whale is swallowing us?"

Numbers and percentages are difficult to interpret and when trying to explain a composition of anything with four or more "chunks" in its composition, the use of a common "pie chart" is helpful – everyone wants a big slice of Mom's homemade apple pie. Shown next is a pie chart showing how the atmosphere (without water vapor) consists of its major gases:

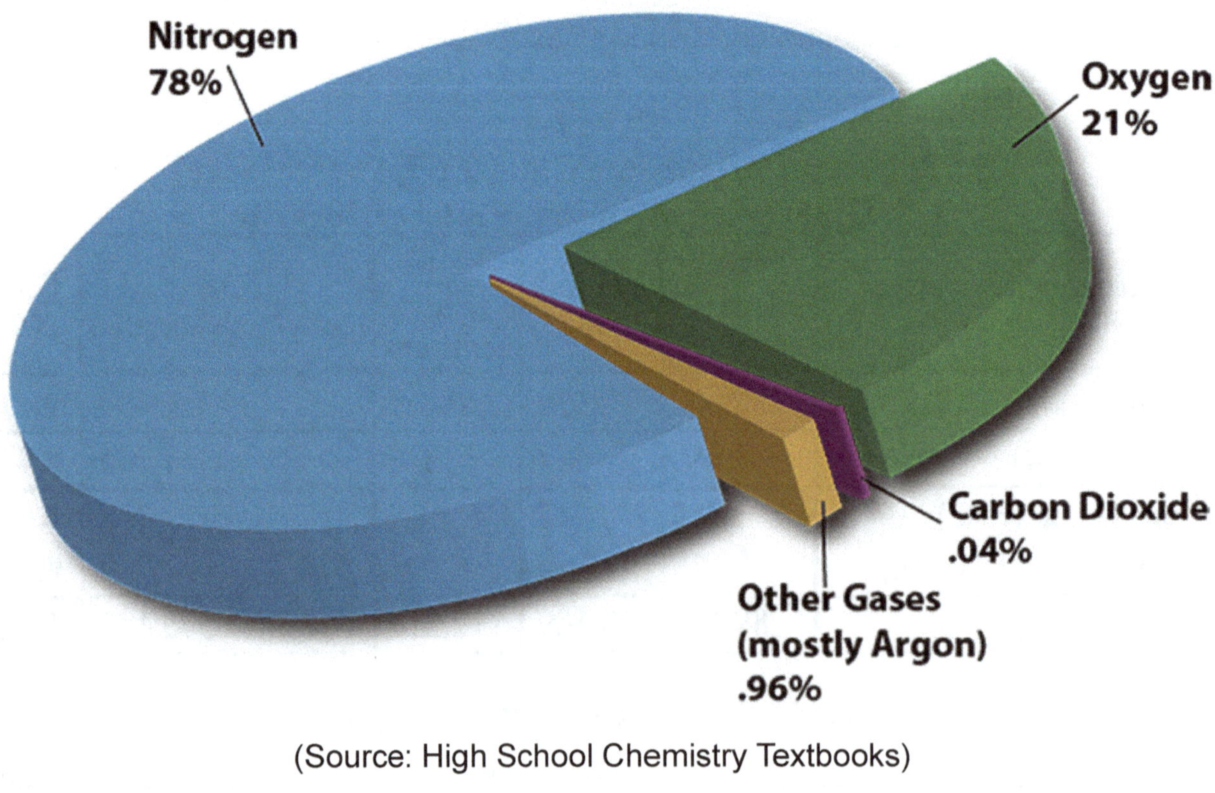

(Source: High School Chemistry Textbooks)

To be noted is the small purple slice of the pie represents the total amount of carbon dioxide in the atmosphere – shown as being 0.04% (equivalent to 400 parts per million or 40,000 parts per billion). Environmental scientists seldom show this form of presentation as it visually shows just how small the amount of carbon dioxide is measured to be as the dominant "greenhouse gas". It is explained throughout this book that of that total 0.04% (the purple sliver of the pie) of carbon dioxide in the atmosphere, there is a natural-caused portion of almost 0.03% (295 parts per million) and very slightly over 0.01% (140 parts per million) which at most is Man-caused. To put that in a better visual perspective, it equates to slicing that small sliver in purple (representing 0.04% of the 100% total pie) in half and then visualizing one of those half-slivers to be the amount of Man-made carbon dioxide in the atmosphere causing a "crisis" level of concern in the environmental science fraternity. It could be acceptable reasoning that the 0.02% of Man-made carbon dioxide was a neurotoxin which is fatal in a concentration of a few drops in a large reservoir. However, carbon dioxide is not a neurotoxin – carbon dioxide is not as effective in trapping heat as is water vapor at an equal concentration. Water vapor in the real atmosphere can be up to 4% (40,000 parts per million or 4 parts per hundred).

Each volcano also has a "personality" that defies being modeled that even the most astute environmental scientist could ever hope to achieve. The volcanic activity in Iceland – a "next door neighbor" of Greenland - tends to be more horizontal than vertical as the lava "oozes" in parallel with the surrounding topography, as opposed to the ejecta being explosively thrown to high altitude providing adequate time for residents in the area to evacuate from "harm's way". Some earthquakes are known "repeaters" (such as the Kilauea volcano on Hawaii) and while it affects the surrounding terrain and ecosystem, the lava flow oozing at a relatively slow speed poses little threat to human life. There may be some steam emitted also but relatively low toxic or climate-affecting gases in the "greenhouse gas" category.

(Source: The US National Geological Survey)

Other volcanic eruptions are with very little, if any, flow of lava, but with predominantly gas (toxic and non-toxic) and volcanic particulate matter resulting from the eruption. The most recent volcanic event that was a "gasser" was the eruptions of the Eyjafjallajökull Volcano in Iceland which lasted principally from March 20, 2010, to June 23, 2010. Volcanologists assigned a Volcanic Explosive Index of VEI 4 to the main event. Shown next is a photograph (by a land-based photographer) showing the prolific efflux of gaseous and particulate matter without any lava flow:

(Source: National Geographic Society photographic database)

Beginning on April 14, 2010, Eyjafjallajökull created an ash cloud that led to the closure of most of the European IFR (Instrument Flight Rules) airspace which caused about 20 countries to close their airspace to commercial jet traffic, and which affected approximately 10 million air travelers. The commercial jet flights were cancelled for two reasons; that of safety of flight in/through the dense gas/particulate plume, and because the particulate matter when ingested into jet engines damages the turbine blades and could lead to catastrophic results from engine shutdowns or engine damage beyond easy repair.

The Icelandic Meteorological Office (IMO) and the Institute of Earth Sciences (IES) of the University of Iceland estimate the in the ash plume, which reached to about 30,000 feet (9 kilometers) contained 250 million cubic meters (330,000,000 cubic yards) of ejected tephra. Tephra contains fine carborundum grit and glass-like particles which are abrasive to mechanical moving parts but more dangerous to the respiratory system of humans, especially those with already diseased respiratory systems. The extent of the ash plume covering all of Europe and much of the western part of Asia was mapped by the London Volcanic Ash Advisory Centre (VAAC), a part of the United Kingdom Meteorology Office is awesome in its scope and length of hazard presented to the Northern hemisphere as shown on the composite map spanning the period April 14 to 25, 2010:

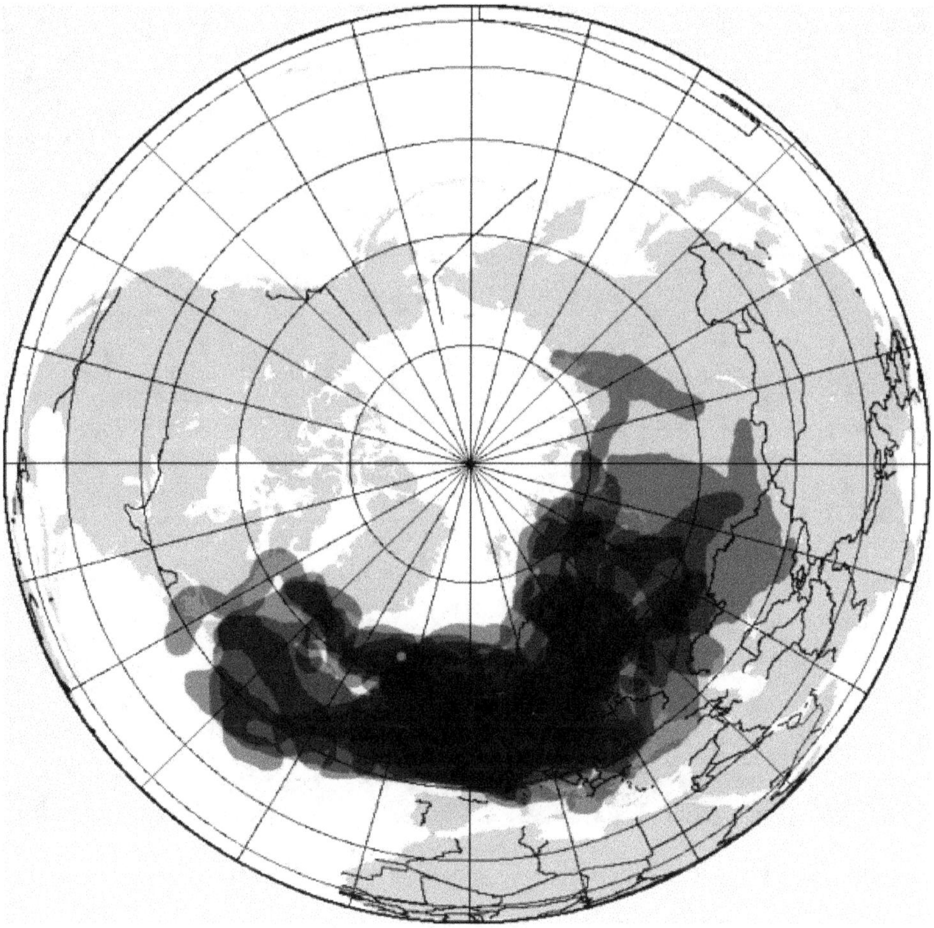

(Credit to the London Volcanic Ash Advisory Centre (VAAC), London Met Office)

Volcanic ash in the atmosphere may be disregarded by the US Environmental Protection Agency as insignificant, but apparently the London Meteorological Office in the United Kingdom has seen it to be more significant and to establish the London Volcanic Ash Advisory Centre (VAAC). The above map supports this Author's observation that weather-related and other natural catastrophic events are important to whatever they have been, are, and will be to the regions where they occur. The London Meteorological Office has the Volcanic Ash Advisory Centre whereas the US National Weather Service (part of NOAA) has the Hurricane Tracking Center and the Tornado Warning Center. They all have some effect on the global surface temperature and therefore on the global climate, yet the United Nations Intergovernmental Panel on Climate Change has representatives only from climate science universities and organizations – and "green" activists' groups.

The most recent major volcanic eruption, causing over 350 human fatalities, was of the Philippines Mount Pinatubo near Luzon in 1991. It was rated at a Volcanic Explosivity Index of 5 which includes an estimate of the ejecta to be in the range of 10^9 to 10^{10} cubic meters. An excellent photograph is shown next indicating the massive pressure that caused the Mount Pinatubo eruption to eject that much particulate (heavy) matter into the atmosphere to an elevation of about 65,000 feet (about 20 kilometers):

(Source: By Unknown Photographer at Clark AB, Philippines)

The above photo was taken from Clark Air Base 15 miles from Mt. Pinatubo and by an unknown photographer possibly on duty with the US Air Force at the vacated base (note that there are no aircraft visible on the base). The severe cumulus cloud formation seen in the photo at the base of the erupting Mt. Pinatubo were in the advancing front of Typhoon *Yunya*. It was a sheer coincidence of nature that a volcanic eruption and a typhoon striking a region would occur within a day of each other. The eruption of Mount Pinatubo deposited heavy ash up to two feet thick on the rooftops of buildings (and roofs of hangars at Clark Air Base) downwind and as far away from the base of Mt. Pinatubo as 20 miles to the northeast. There were initially no reports of damage to any structure or residences caused only by the thick but relatively low-density sponge-like ash deposits from the eruption.

However, compounding the ash deposit problem was the fact that Typhoon *Yunya* (Category 3) inundated the Luzon Peninsula the next day that Mt. Pinatubo erupted (June 15, 1991). The heavy rainfall from Typhoon *Yunya* soaked into the spongy thick ash from the Mt. Pinatubo eruption (much like a dry sponge absorbs water) on many rooftops and collapsed their roofs from the excessive weight of the "mud". 350 people were killed in the 2,013 homes and businesses that were damaged (and mostly under roofs that had collapsed).

[Author's Note: The Author was the US Air Force Chief System Engineer for development of the high technology F-16 Fighting Falcon weapon system. In October 1989 (less than two years before the Mt. Pinatubo eruption) the Author was on an official visit to Clark Air Base in the Philippines located less than 15 miles from the base of Mt. Pinatubo. A squadron of F-16s and a squadron of F-15s were based at Clark Air Base at that time. When the approaching Typhoon *Yunya* storm path initially threatened the Clark Air Base aircraft, the F-16 (and F-15) squadrons were evacuated from Clark Air Base before the typhoon was estimated to arrive and relocated to Kadena Air Base on Okinawa, Japan. The day before *Yunya* arrived, Mt. Pinatubo erupted, and Clark Air Base was impacted very heavily by the combination of rain/volcanic ash "mud" that kept gaining weight as the Typhoon *Yunya* slowly passed over. The metal roofs of many buildings (including all the F-16 and F-15 maintenance hangars) eventually collapsed almost simultaneously as shown in the next photograph:

(Source: Author's personal file; credit to unknown USAF photographer)

There are seen no aircraft (either US military operated or Philippine commercial) visible on the base as they had all been pre-evacuated safely when the alert of approaching Typhoon *Yunya* was given 4 days prior to the estimated time of *Yunya's* arrival. The USAF abandoned Clark Air Base and it remained out-of-use for more than five years until the Philippine Government cleared away all the ash and buildings debris from the Mt. Pinatubo/Typhoon *Yunya* "perfect storm" catastrophe and rebuilt new facilities. The lack of supply of reconstruction materials severely affected the rebuilding supply and caused the US Air Force to cease operations permanently at Clark Air Base. Clark Airfield was reopened (in approximately 2005) and is now used as a commercial cargo aircraft facility serving the Manilla area.]

The amount of volcanic ash from the Mt. Pinatubo eruption on nearby farmlands severely affected the local ecosystem and aggravated the homeless problem with lack of grown food and supply of clean water for years after the 1991 eruption.

The most logical recommendation that can be made for future generations to heed is to not create large communities of the population near the base of a known volcano (active or inactive). A photograph showing the hazards of doing so is shown next:

(Source: Photo by Author on Cruise Ship Approaching Port of Naples, Italy)

The volcano in the background of the above photograph is Mount Vesuvius which is legendary in volcanic eruption history. Mt. Vesuvius, (located six miles from the city of Naples in the Bay of Naples), has had eight major eruptions in the past 17,000 years. The largest Vesuvius eruption on record was in 79 A.D. when it erupted burying Herculaneum (under 66 feet of pyroclastic flow and ash) and Pompeii (under more than 20 feet of ash) – with an estimated 16,000 fatalities. Vesuvius erupted again in 1906 with about 150 casualties, and again in 1944 burying much military equipment in World War II but with no known fatalities. When the Author visited that region (to Pompeii in 2011), there was seen a small amount of wispiness (grey-white smoke) being emitted from the Mt. Vesuvius cone but that is apparently not unusual.

The most catastrophic volcanic eruption (estimated to be a magnitude 8.0) was that in 1556 of the Shaanxi volcanoes in northern China causing death to over 100,000 humans and major crushing injuries to over 730,000 others. The enormous losses and injuries resulted from the impoverished population living in caves of a soft sandstone cliff that collapsed while those killed or injured were asleep in the caves. Those living in a hazardous area do so for several differing reasons, but all too often are unaware of what lies beneath the region where they live. The opposite of that was seen by the Author while on that cruise ship which provided the opportunity to take a photograph of the Bay of Naples (shown above). As shown, the residences in the Naples area were built to withstand weather severities which are minimal in that latitude and not being on a major ocean, taking the risk that the volcano looming above them will not erupt during their lifetime. Other communities build on the rim of a volcanic caldera for the spectacular scenery in that caldera. On that same cruise and a few days after departing the Bay of Naples, passengers were treated to a sight that photographs cannot do full attribution to their unusual and awesome scenery. Shown next are three photographs of the region of the Greek islands in the Aegean Sea in the region of the Santorini (also known as Thera) volcano which erupted in circa 1600 BCE. The first is a satellite image showing the immensity of the caldera:

(Source: NASA)

For reference, the distance from Santorini to Therasia is approximately 1 mile. Cruise ships (as well as all other smaller vessels) enter the Santorini caldera at the northwestern channel, and immediately see the spectacular colors (a close second rival to those in the Grand Canyon in Arizona) of the inside walls of the caldera:

(Source: Photograph by Author, from a Princess Cruise Line ship)

To avail themselves and the many annual tourists to the island of Santorini to the entire vista of the large Santorini caldera, the local population have erected residences and commercial buildings on the top of the caldera rim as shown next:

(Source: Princess Cruise Lines Brochure given to Ship Passengers)

The all-white structures atop the Santorini caldera rim are noted to be whitewashed a gleaming white (with olive oil content to resist the weather elements) but mostly so as to not conflict with the natural varying colors of the caldera. Churches have a classic round dome painted the blue color of Greece and which matches the cobalt blue color of the water inside the caldera. The round domes of the churches and some other structures are shaped other than flat in the event of a small eruption of the Santorini volcano causing accumulation of the ash ejecta to not collapse roofs onto inhabitants.

Closer to home, since 1700 there have been 635 recorded volcanic eruptions with 215,768 major direct human injuries and fatalities and at least 62,600 indirect (from famine and disease) fatalities caused by those eruptions in Alaska (mostly in the Aleutian Islands), Hawaii, and the original "lower 48" United States. Their locations are graphically summarized next, and can be seen to be mostly on the Pacific "Rim of Fire":

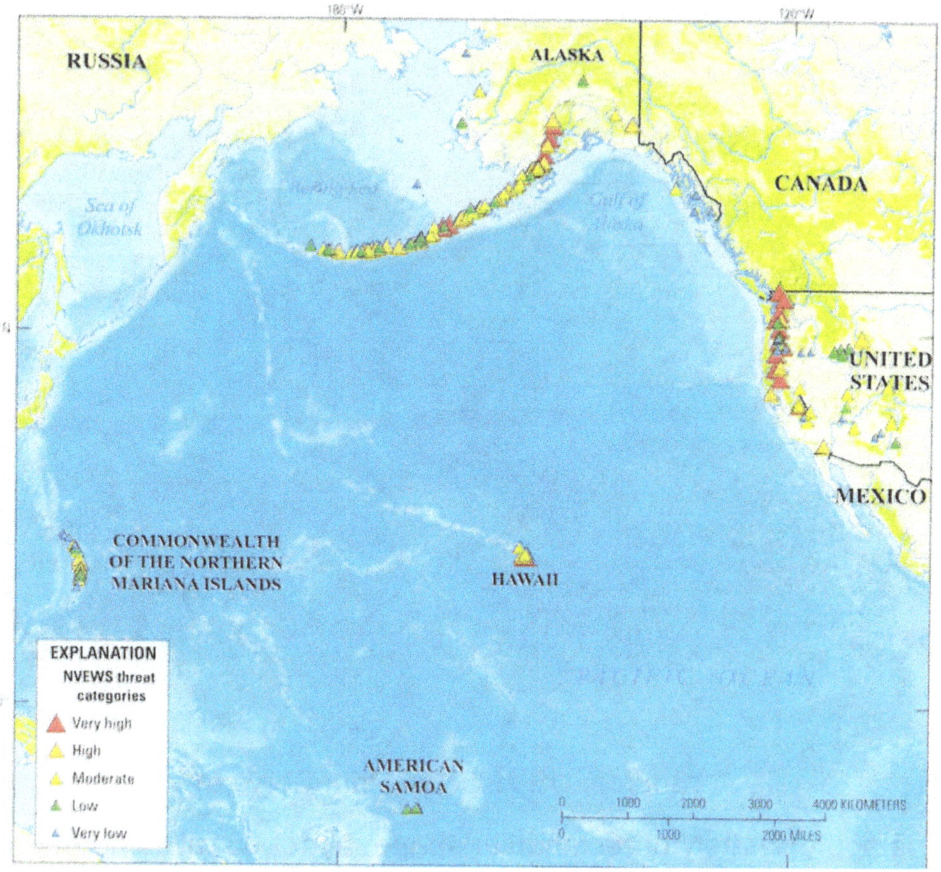

(Source: US National Geological Survey)

There are 205 named volcanoes located around the "Rim of Fire" that are of prime interest to the United States. Those 205 volcanoes are distributed (clockwise in the diagram above) with Alaska (Aleutian Islands) having 101 named volcanoes – 39 erupting since 1700; the 48 United States on the "mainland" having 161 named volcanoes – 6 erupting since 1700 (3 in Washington, 2 in California, and 1 in Oregon); and Hawaii having 25 named volcanoes with 39 eruptions since 1700. Of those eruptions, it was shown earlier that the eruption of Mt. St. Helens in 1980 directly caused 57 human fatalities, making that eruption the deadliest in the history of the USA.

The near-term outlook for volcanic eruptions being a major concern (i.e., a crisis) to residents in the populated regions of the United States is concluded to be of a cautionary level only to those residents in the states of Washington, Oregon, California, and Hawaii. The public should be aware though that there is one area in the United States where a long-term outlook for volcanic eruption of cataclysmic proportion to occur cannot, nor should, be dismissed completely. The are a wide range of estimates of the area that would be devastated physically by a major eruption of one or more of the extinct volcanoes below the surface, but the area of the United States would likely have an apocalyptic effect on ecology and population of the Western third of the United States. That large eruption would create its own weather and create havoc to the land area and atmosphere within 600-700 miles (or more depending on winds speed and direction) distance from the center of the erupting super-volcano.

That very active geothermal area is in the region of northwestern Wyoming designated as the Yellowstone National Park. Yellowstone National Park is the first National Park in the US Park System., created in 1872 by then-President Ulysses S. Grant. A diagram of the Yellowstone National Park region, and showing the major calderas contained within Yellowstone Park, is provided next:

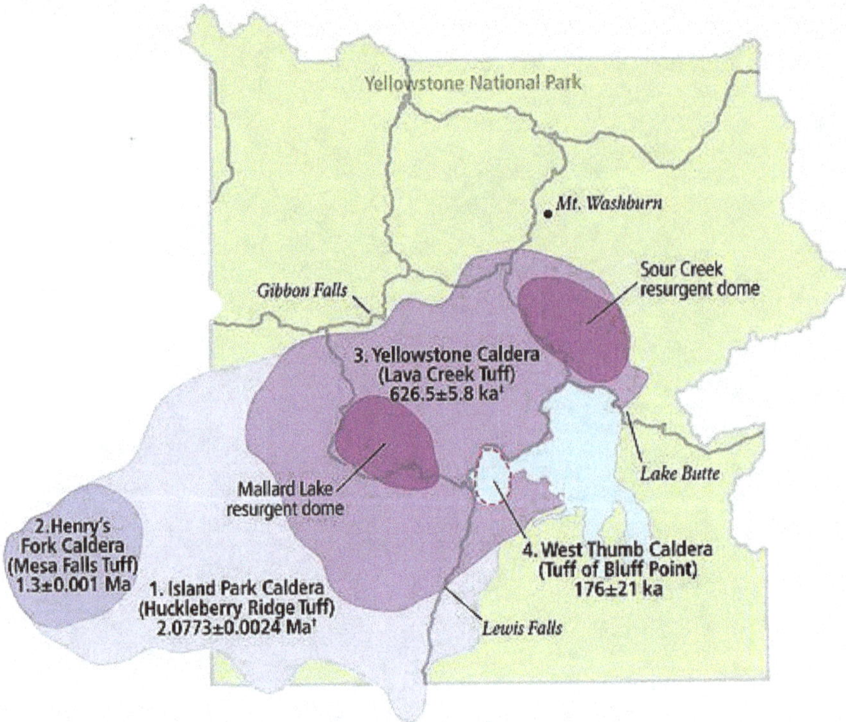

(Source: US National Park Service)

Yellowstone Park is approximately of rectangular shape, about 63 miles north to south, and 54 miles east to west. Its area is approximately 2.2 million acres (3,438 square miles). On the surface, above four major calderas, Yellowstone Park has more natural beauty than most of the US National Parks, one of which is that below showing the Yellowstone River meandering through the Hayden Meadow:

(Source: US National Park Service)

Underneath that natural scenery is however the largest known geothermal area of the world. A diagram indicating the geology below the Yellowstone Park caldera is shown next:

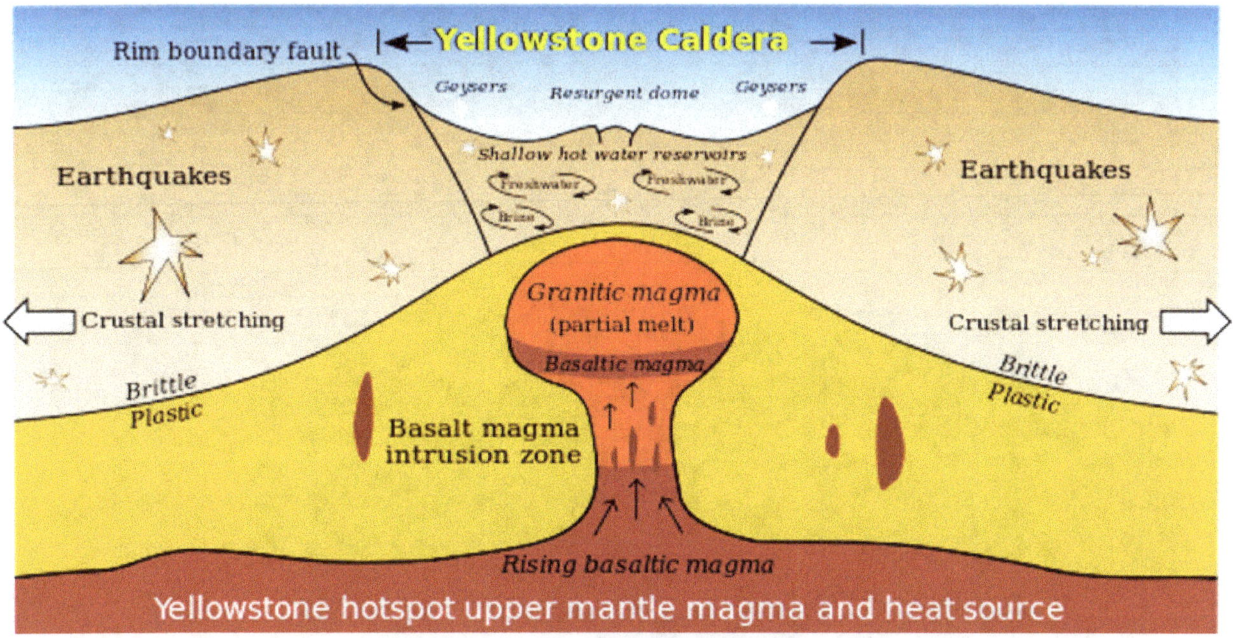

(Source: US National Park Service)

Evidence of the geothermal energy is seen at two of the most visited regions of Yellowstone Park – the first at the wondrous Old Faithful geyser which spouts water and steam into the air very regularly (approximately every 90 minutes between spouts and for a duration of about 90 seconds each spouting) as shown next):

(Source: National Park Service)

The second and by far more colorful scene is that of the Geyser Basins, one of which is shown next:

(Source: US National Park Service)

The water temperature in these geothermal features are near boiling or already in the form of steam above boiling (100°C or 212°F). As they pose no hazard to the environment (more than they obviously have/do) or to humanity, they possess the geothermal potential for alternative energy sources to the fossil fuels and to the solar and wind power sources which are cleaner than fossil fuels but are not as reliable as geothermal energy has the potential to be. The use of geothermal energy will be discussed in a later chapter dealing with fossil fuels, current natural sources, and other clean, reliable, future forms of alternative energy that should be developed as replacement of coal as the primary energy source in the twentieth century.

It is also very common that where there are volcanic and geothermal geophysical formations, there typically is seismic wave activity preceding volcanic activity. Nearby population areas are alerted to the increased probability of earthquake occurrence doing damage from the earth movement (being at the edges of two tectonic plates which spawn both volcanic eruptions and/or significant earthquakes). That leads to the second major geophysical threat primarily to humans (collapsing their residences on top of them) and with lesser consequences on the environment that results from tectonic plate movement – earthquakes. A diagram indicating the known earthquake-prone locations (along the peripheries of tectonic plates) everywhere in the world is shown next:

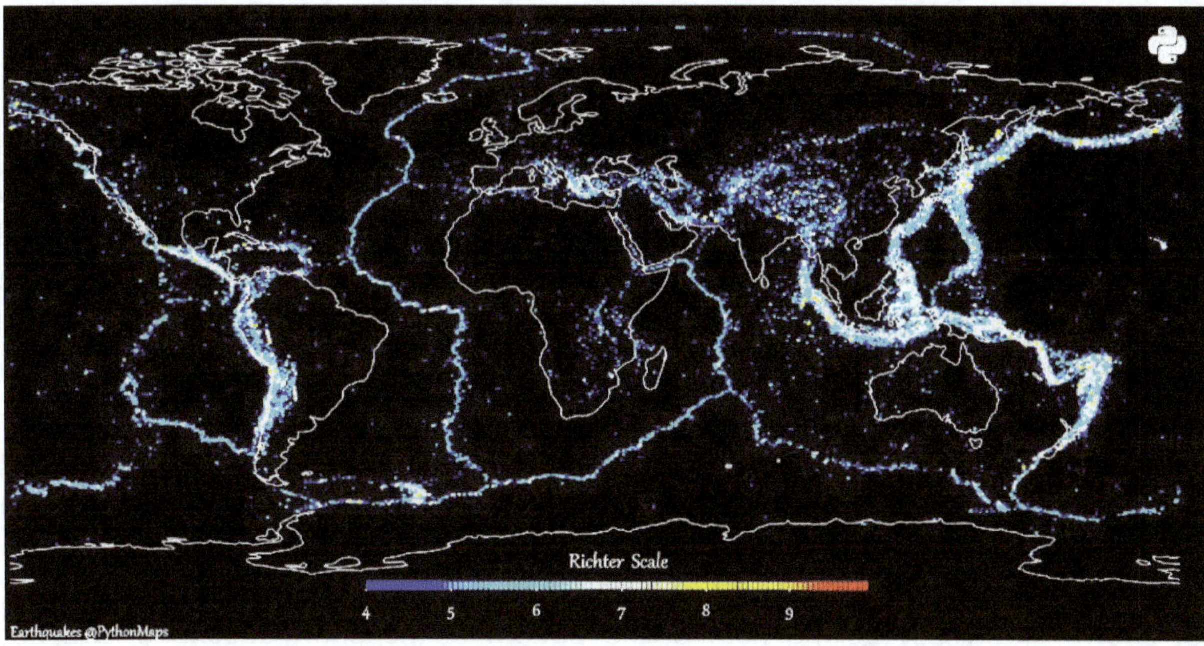

(Source: Python Maps for the US National Geological Survey)

It should be noted the similarity between volcanoes locations and earthquake epicenter locations as they are both most common to occur on the edges of tectonic plates (as shown on earlier charts).

The chart shown next has been created from the US National Geological Survey database (in listed form) of the frequency of earthquakes globally and the average occurrences per year at each of the magnitude levels of interest in this book.

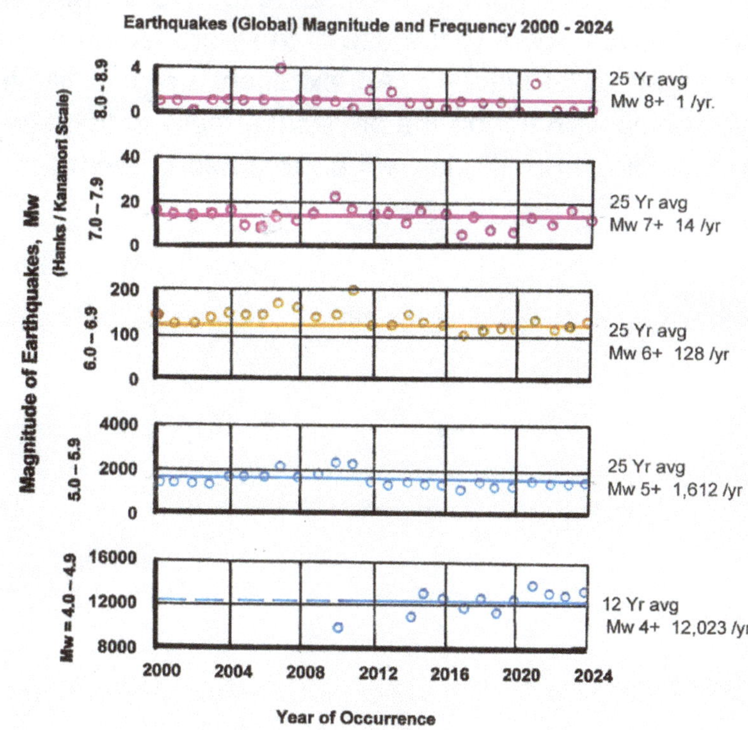

Note: Earthquakes at/above 9 on the Mw scale are too sporadic to be graphically analyzed. There is high probability that a Mw 9+ earthquake will occur every 10 – 12 years.

(Source: Graphical portrayal (by Author) of US NGS tabular database)

It is seen that there is amazingly consistent statistical data for each of the Mw momentum scale (created by Hanks and Kanamori in 1979 to replace the lesser accurate Richter Scale) levels between Mw 4.0 and 8.9. The level Mw 4+ and 5+ are colored blue as there is commonly little damage done and few lives lost at those levels. When levels above Mw 4 in a volcanic existence area are measured (by seismic sensors and recorders) to be rapidly increasing in occurrence, there is high probability that the nearest volcano has the increased potential for erupting. When an earthquake is measured to be in the Mw 6.0 to 6.9 range (colored yellow on the chart), this is considered by many to be the threshold for damage to residential and commercial structures and very dependent on the location of the epicenter relative to populated areas.

It is the Mw categories of 7.0 and above (color coded in red on the chart) that is shown in historical records to be the most damaging and most injurious to humans and animals. Somewhat mercifully, the frequency of the "really bad ones" reduces as the magnitude of Mw increases. As shown on the chart, there is on average 14 per year of the Mw 7.0 to 7.9 category earthquakes occurring globally, and for the category 8.0 to 8.9, there is an average of only 1 per year. The monstrous category Mw 9.0 and above (not shown graphically as the note at the bottom of the chart describes) are sporadic in occurrence and strike only once somewhere on the planet every 10 to 12 years.

The magnitude of an earthquake is only one measure of the damage and effect on virtually everything in its area of influence. Unlike volcanic eruptions which spew lava on the surrounding environment and ejecting enormous amounts of particulate matter and gases into the atmosphere, the mechanism of earthquakes is more physical dynamic and spreads in seismic wavelengths that causes the earth to shake violently. Another important variable is the depth at which the earthquake epicenter occurs and its location relative to a body of water above or nearly above it which creates enormous waves in the body of water named as *tsunamis*. (Tsunami is a combination word from Japanese – "*tsu*" meaning "harbor" and "*nami*" meaning "waves"; prior to the use of the word tsunami, reference was made to the earthquake-generated waves as "tidal waves"). An excellent graphic illustrating the tsunami long-distance side-effect perhaps being worse than the physical effect locally of the earthquake mechanical devastation is shown next (resulting from the largest earthquake in recorded history at Valdivia, Chile, in 1960):

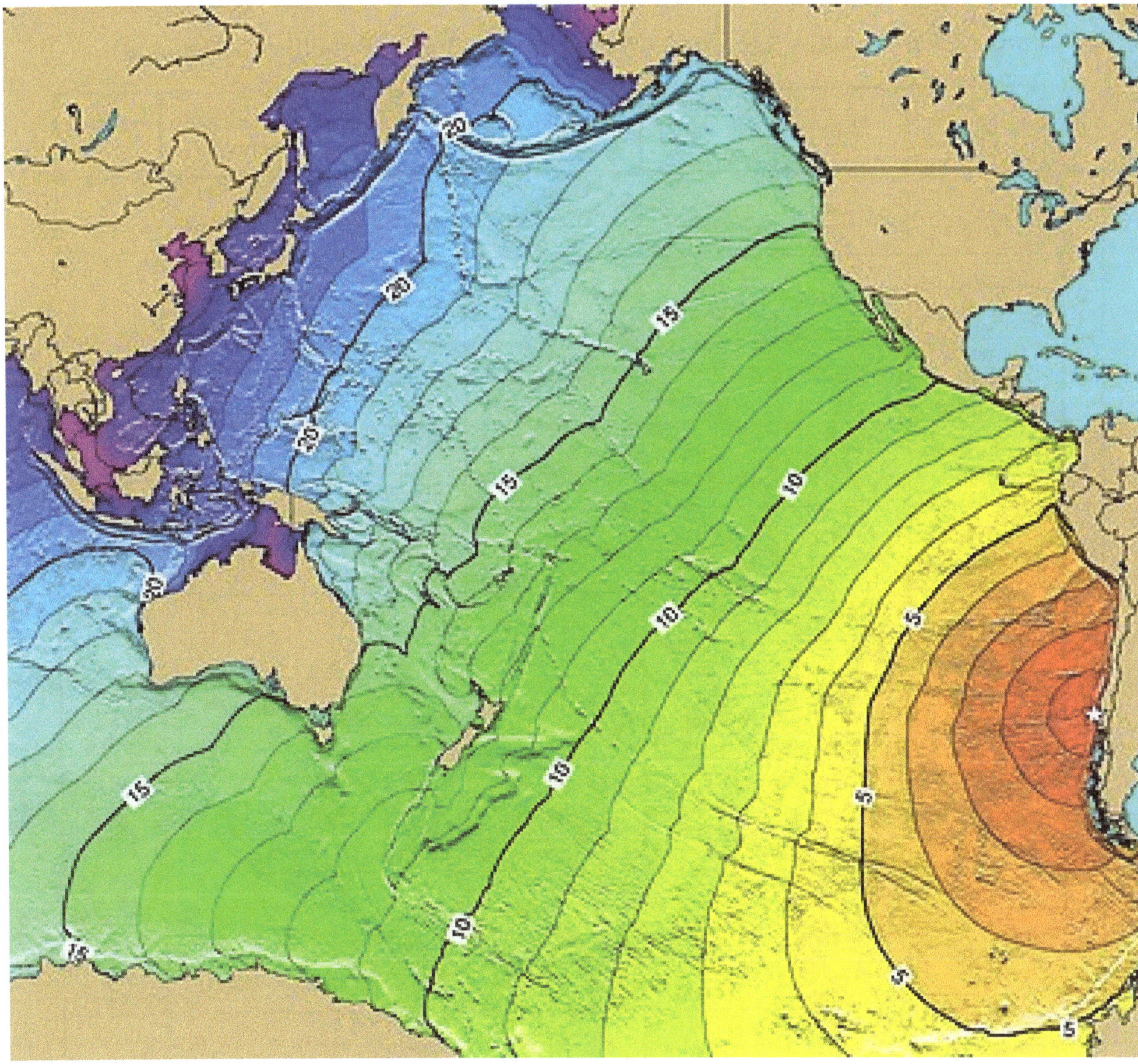

(Source: US National Geological Survey)

During earthquake occurrences globally, there is massive damage and deaths caused from the dynamic shaking and collapsing of inadequately constructed structures (especially residences) as well as by the damage and fatalities resulting on a country far away from the epicenter but impacted by the enormously high tsunami generated by the earthquake. Most importantly, it is the ability of the residents (all affected by the earthquake and tsunami and weather if made homeless) to survive the catastrophic event. A tabulation of the ten most powerful and/or destructive global earthquakes is shown next:

Largest Earthquakes Recorded (Richter or Mw Scale / Fatalities & Damage)

	Place	Date	Magnitude (Richter or Mw Scale)	Fatalities and Major Injury	Damage Cost Estimate	Generated Tsunami?	Ignited Fire?
1	Valdivia, Chile	May 1960	9.5 (Richter)	5,700 in Chile 100+ in Japan 61 in Hawaii	$700 Million	Y (82 ft. Chile; 35 ft. Hawaii; 30 ft. Japan)	N
2	Prince William Sound, Alaska	March 1964	9.2 (Richter)	128	$311 Mllion (Anchorage region)	Y (90 ft.)	N
3	Sumatra Islands, Indonesia	December 2004	9.1 (Mw scale)	300,000 est	Major	Y (100 ft.)	N
4	Tohoku, Japan	March 2011	9.1 (Mw scale)	20,000 est	$309 Billion (Nuclear Facility)	Y (132 ft.)	N
5	Port-Au-Prince, Haiti	January 2010	7.1 (Mw scale)	220,000 est	250,000+ Homes Destroyed	Y (10 ft.)	N
6	Northern Sumatra, Indonesia	April 2012	8.6 (Mw scale)	1,300+	Major	Y (3 ft.)	N
7	Unimak Island, Alaska	April 1946	8.6 (Richter)	5 in Unimak; 159 in Hawaii	Minor in Alaska; $26 M in Hawaii	115 ft. High at Hawaii	N
8	Sanriku-Oki, Japan	March 1933	8.4 (Richter)	500+ in Sanriku; 15,000+ in Honshu	Moderate in Sanriku; Massive in Honshu	100 ft. High at Honshu	N
9	San Francisco, California; USA	April 1906	7.9 (Richter)	3,000+	$524 Million (Collapses and Fires)	N	Y (Fire caused by gas line ruptures)
10	Turkey/Syria (Gazientap)	February 2023	7.8 (Mw scale)	53,537 in Turkey; 7,000+ in Syria	$149 Billion in Turkey; $14.8 Billion in Syria	N	N

Note: The Mw scale (T. Hanks and H. Kanamori, 1979) is more accurate than the Richter Scale for larger earthquakes as it estimates the total energy transformed during large earthquakes. The earlier Richter scale measures the amplitude only on a logarithmic scale.

(Source: Tabulation by Author of US National Geological Survey databases)

The photograph shown next is the damage done to a town (Sanriku) on Honshu, Japan) by the tsunami generated by the Mw 9.5 earthquake at Valdivia, Chile on May 22, 1960. The tsunami was originated at Valdivia and traveled northwesterly over 6,000 miles across the Pacific Ocean before striking Hilo, Hawaii, and then Honshu, Japan. The tsunami height decreased from 82 feet high in Chile, to 35 feet high in Hawaii, and to 30 feet high when it rolled into Sanriku, Honshu, Japan. It is incredible when speculating on how much energy was transferred from the earthquake in Chile and into the Pacific Ocean to cause that height of wave of water and transfer it at a speed of several hundred kilometers per hour and still have the energy potential to cause the flooding and damage shown in the photo next.

(Source: US National Geological Survey)

The lesser amplitude earthquake (shown at the bottom of the above tabulation as the Gaziantep, Turkey/Syria Mw 7.8 earthquake on February 2, 2023) is a contrast in the number of fatalities, damage cost, and without a tsunami contributing to those deaths/damages compared to the 1960 Valdivia earthquake. The Turkey/Syria earthquakes occurred near the intersection of the African Plate and the Arabian plate, and the initial tremor of Mw 7.8was soon followed by a Mw 7.6 tremor. An amazing statistic however was that there were 25 aftershocks of Mw 4.0 or greater within six hours of the initial tremor. Within the first 24 hours of the initial Mw 7.8 tremor there were over 570 aftershocks, and within three months after the initial tremor over 30,000 aftershocks of less than 4.3 Mw had been recorded. A diagram showing the configuration of the primary tectonic plate fault lines for that region is shown next:

(Source: International Union of Geological Sciences, IUGS)

When the 30,000 opening and aftershocks recorded seismologically were superimposed on the tectonic plate diagram shown above, there is an amazingly good correlation matching the geological database for that region.

A comparison of the number of fatalities and damage costs of the Valdivia, Chile largest earthquake ever recorded (Mw 9.5 in May 1960), with the relatively low magnitude (Mw 7.1) earthquake that occurred in Port-Au-Prince, Haiti, in January 2010 but having an unusually high number of fatalities (220,000), and that of the Turkey/Syria earthquake (Mw 7.8 in February 2023) was made. This comparison was done to better understand why there is such a wide disparity between the three earthquakes – the number of fatalities and the cost of damages and repair – not being proportional to the magnitude of the earthquake. Two variables, when considered – the population density of the region affected by the earthquake, and the quality of construction of the structures (residential and commercial) serving the population of the region – brought the correlation into better correlation. Three photographs are shown next to compare the quality of construction of the structures that were destroyed. The first photograph is that of the Valdivia, Chile area, the second photograph is that of the Port-Au-Prince, Haiti area, and the third photograph is that of the Gaziantep region in Turkey with its high damage cost and large number of fatalities.

(Source: International Union of Geological Sciences)

(Source: International Union of Geological Sciences)

(Source: International Union of Geological Sciences)

Before interpreting the statistics of a few of the most significant earthquakes in the United States (the Alaska earthquake in March 1964 and the horrific earthquake and fire in San Francisco in California in April 1906), and the massive number of fatalities and property damage they caused, it must be described that earthquakes are in two distinct modes of motion – "shakers" and "bouncers". Anyone who has lived in southern California for a long time (the Author and his family lived in Ventura County for 34 years and 4 years in San Luis Obispo County in the Central Coast region of California) can attest that it is the "bouncers" that do the most damage. A photograph of the infamous San Andreas Fault under the Southern California area is shown next:

(Source: National Geodetic Survey)

The human injuries/fatalities, damage to property (residential and commercial), and the damage done to the infrastructure (bridges, tunnels, highways, rail lines, multi-level parking structures, etc.) attributed to earthquakes deserves a book dedicated to that branch of earth science and geophysical science.

For the USA and California, the epic earthquake in San Francisco in April 1906 is the earthquake that mostly comes to mind of anyone born before 1950. It stands as the largest number of fatalities (over 3,000) and one of the costliest (over $524 million in 1906 dollars - $18.3 billion in 2024 dollars) in developed countries. Two photographs taken after and during the Richter 7.9 magnitude earthquake shown next points to the two factors that dominate the factors causing the high number of fatalities and the high amount of damage:

(Source: US National Geological – formerly Geodetic - Survey)

(Source: US National Geological – formerly Geodetic - Survey)

The upper photograph reflects that it was the initial vertical motion (a "bouncer") of the earthquake that literally bounced the otherwise structurally well-built (with steel reinforced members as a skeleton) buildings causing the non-metallic components attached to the façade and internal steel structure to fail and fall to the ground. It is seen that the skeletons of the buildings remained intact, but their facades and contents had fallen away or down. Had that earthquake of that estimated magnitude of 7.9 been a more horizontal motion (a "slider"), the buildings would have fallen when shifted off their foundation on the hilly area. Of the over 3,000 fatalities, more than 90% (over 2,700) of them were fatalities resulting from crushing type injuries. The major tremor shook the area at about 5:12 in the morning and most of those 2,700 fatalities were of humans in their beds.

The lower photo above shows that there were fires raging in the Mission District which began shortly after the initial tremor. There were again high winds in the region, and the heat from the initial fires created their own hot winds which spread throughout the Mission District. The tremors and falling debris had ruptured the gas lines in most of the residential buildings and in the main distribution pipes in the street serving the Mission District. Coincidentally, the tremor also severed the main water lines serving the Mission District such that there was no water [pressure for fire fighters to extinguish the fires. Attempts were made to explode explosives in the hope of creating firebreaks to halt the fires but those attempts also failed. Unfortunately, the explosives only served to initiate additional fires. Instead of one massive fire-front, there were over 30 fires burning at one point in time. Of the 3,000+ fatalities, the 2,700 early fatalities from crushing injuries were supplemented with over 300 fatalities of persons in the Mission District who were either also crushed instantaneously or were injured but could not be rescued because the fire had engulfed their residences. Many of those 300 succumbed to crushing and/or burn injuries if not completely incinerated by the fast-moving fires.

The fires continued to burn for four days and nights until they were extinguished by hoses from water tanker trucks that had arrived from surrounding cities of the Bay area. Over 25,000 buildings on 490 city blocks had been destroyed. The insurance companies advised the people who had lost their homes that they were not covered for damage done by earth movement but only if they had been damaged by fire. The public saw the "loophole" in that insurance coverage interpretation and some intentionally set fires to the rubble of their destroyed homes to be eligible for insurance coverage of fire damage.

At the start of the catastrophe, San Francisco had a population of over 400,000. Approximately 300,000 were made homeless and left San Francisco to be taken in by family, friends, and co-workers who lived on the other side of the Bay. A map of San Francisco reflects an outlined burn area which destroyed about 80% of the residences and commercial buildings within that outlined area. That map appears next:

(Source: San Francisco Fire Department)

Many "Lessons Learned" were derived from the San Francisco Earthquake and Fires, and a major program was started to improve the safety of buildings (commercial and residential) constructed in Los Angeles County and other surrounding Counties that straddle the many faults associated with the San Andreas Fault that runs the entire West Coast – from California to the state of Washington of the mainland (48) states. Earthquake specialists opine that it is better for the San Andreas Fault to move in several small shocks (below magnitude 7) than to move in one larger jolt of magnitude 9 or more.

In an article written in 1913 by a scientist and geologist John C. Banner made the assessment which clearly applies also to the current problems with global warming, climate change, and ocean levels rising: "The only way we know of to deal successfully with any natural phenomenon is to get acquainted with it, to find out all we can about it, and thus to meet it on its own grounds. That is the way mankind has succeeded thus far, and it is safe to conclude that it is the only way it will ever succeed."

The Author was in his residence in Thousand Oaks, California, when the Northridge Earthquake (6.7 magnitude) started at 4:30 a.m. PST on January 17, 1994, in the San Fernando Valley about 20 miles north of the City of Los Angeles. He was awakened by the sound of glass breaking. To his dismay, it was obvious that a bottle of V.S.O.R. (Very Special Old Rum) from the island of Barbados had been shaken off a shelf in the family room, and another bottle of Jim Beam had fallen on top of it with the base of the JB smashing the bottle of VSOR. Some tragedies have no limit to what gets destroyed by natural events. When the Northridge area was identified in the local news broadcasts as the approximate epicenter the number of fatalities began being broadcast. At the end, approximately 72 persons had been fatally injured directly by the damages to buildings in which they were asleep. Another three or four dozen fatalities were considered at least 50% attributable (heart attacks and a few suicides) to the Northridge earthquake. Fortunately, most if not all construction of buildings after about 1908 when building codes were amended to add earthquake-proofing based on the 1906 San Francisco disaster (earthquake and fires). A photograph of an iconic failure of an apartment complex (the upper portion meeting California earhquake construction code but the foundation apparently not) in Northridge is shown next:

(Source: The Los Angeles Times, January 18, 1994)

There were no human fatalities, and the loss of approximately 12 automobiles crushed by the collapse of the building's foundation were covered by automobile insurance policies. There were no reports known to be filed of any gas lines being ruptured and fire breaking out, but there were several reports of plumbing line breakages and leaks in water lines servicing several residences. It is largely due to strict enforcement of the earthquake-proofing provisions of California Building Codes that structural damages and fatality (72) statistics in the Northridge Earthquake of 1994 were not much larger. In comparison, the damages and fatalities in the Turkey/Syria Earthquake were high as over 70% of the structural damage was done to structures for which over 70,000 waivers for construction strength provisions in the Turkish Building Code had been granted construction waivers "for specific circumstances" (accompanied by many "under the table" cash payments.)

The prospects for earthquakes in the vicinity of the San Andreas Fault (shown earlier) are summarized well in a USGS report: The U.S. Geological Survey's most recent forecast, known as UCERF3 (Uniform California Earthquake Rupture Forecast 3), released in November 2013, estimated that an earthquake of magnitude 6.7 M or greater (i.e., equal to or greater than the 1994 Northridge earthquake) occurs about once every 6.7 years statewide. The same report also estimated there is a 7% probability that an earthquake of magnitude 8.0 or greater will occur in the next 30 years somewhere along the San Andreas Fault. A different USGS study in 2008 tried to assess the physical, social and economic consequences of a major earthquake in southern California. That study predicted that a magnitude 7.8 earthquake along the southern San Andreas Fault could cause about 1,800 deaths and $213 billion in damage. Meanwhile, in the Los Angeles County region, there are frequent earthquakes (well below a magnitude to cause significant damage or injuries to humans) occurring associated with the San Andreas Fault system.

A total of *234 earthquakes* with a magnitude of four or more have struck within 186 miles (300 km) of Los Angeles, California in the past 10 years. This comes down to a yearly average of 23 earthquakes per year, or 1 per month. On average an earthquake will hit near Los Angeles roughly every 15 days.

A relatively large number of earthquakes occurred near Los Angeles in 2019. A total of 125 earthquakes (mag 4+) were detected within 186 mi of Los Angeles that year. The strongest had a 7.1 magnitude.

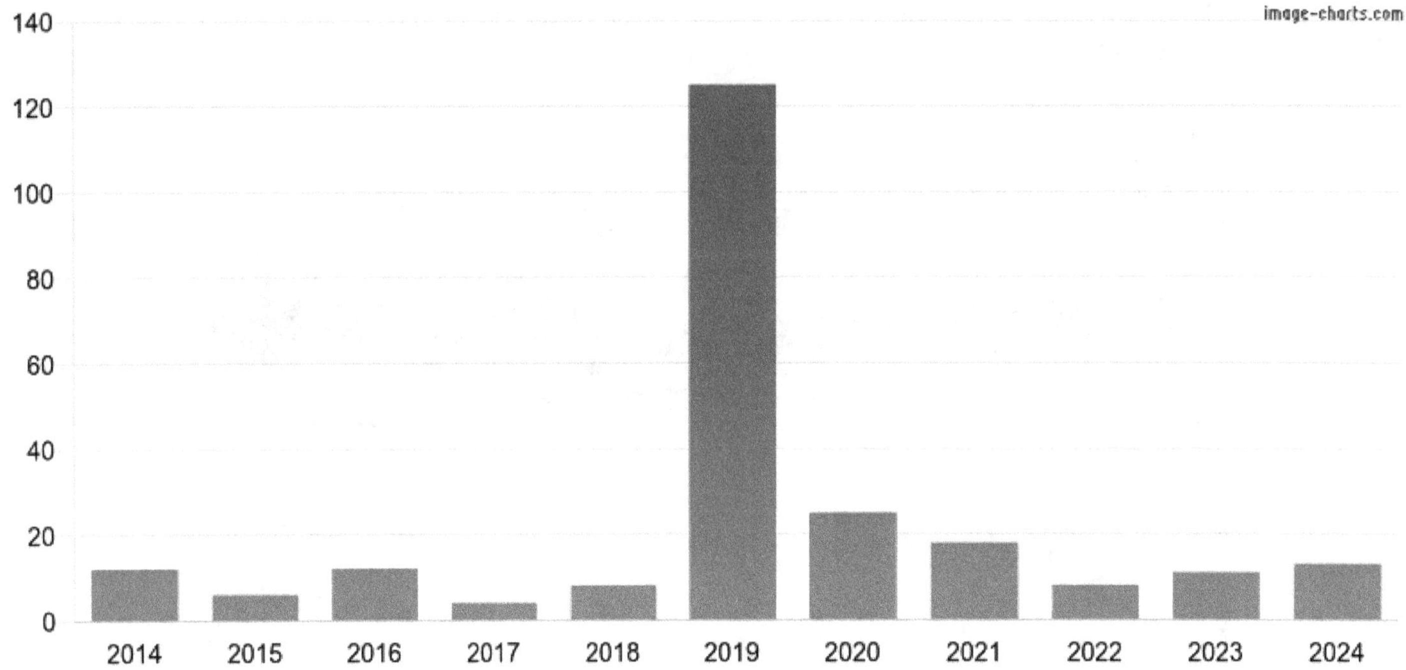

(Source: US National Geological Survey; California Institute of Technology)

The vast majority of earthquakes in the Los Angeles area are below a magnitude 4. A table shown next indicates that most are never reported as even having occurred because residents have adapted to their lifestyles so as to ignore earthquakes having a magnitude of 3 or lower. The statistics for the California region are indicated below:

Average number of earthquakes

Based on data from the past 14 years and our earthquake archive back to 1900, there are about 46,000 quakes on average per year in or near California, USA:

- Mag. 7 or higher: **0.11 quakes** per year (or 1 quake every 9.1 years)
- Mag. 6 or higher: **0.81 quakes** per year (or 1 quake every 1.2 years)
- Mag. 5 or higher: **4.6 quakes** per year
- Mag. 4 or higher: **55 quakes** per year (or 4.6 quakes per month)
- Mag. 3 or higher: **508 quakes** per year (or 42 quakes per month)
- Mag. 2 or higher: **4,200 quakes** per year (or 11.6 quakes per day)
- Mag. 1 or higher: **25,000 quakes** per year (or 68 quakes per day)

California is rated to have a "moderate" level of seismic activity. However, California has had at least 14 quakes above magnitude 7 since 1900, which suggests that larger earthquakes of this size occur infrequently, probably on average approximately every 5 to 10 years.

This chapter has shown the type and extent of damage and cause of human fatalities due to geophysical catastrophes alone and which are totally unrelated to the effect that the small concentration of carbon dioxide in the atmosphere (140 parts per million - 0.014% - of the atmosphere) may have. Environmental scientists have greatly over-exaggerated there being a "crisis" if the Man-made emissions of carbon dioxide are not reduced to net zero by 2050. Is it reasonable to expect that the failure to meet that well-intentioned goal will cause destruction and despair approaching that as shown in the foregoing database of geophysical catastrophes?

The World population – over 8 billion of us – deserve better representation to keep the important balance of control over what the environment is imposing on humans not just what humans are imposing on the environment. Taxpayers' money is well spent to prevent the human population from polluting the environment as that indirectly does protect the safety, health, and comfort of the human population to protect their food supply, the water they drink and cook with, and the air they breathe. However simply using many millions of dollars to protect marshlands and peat-covered areas that are of little value to the population is wasteful. The Environmental Protection Agencies at Federal and State level are prostituting the environmental science profession as a sample grant award of $489 of taxpayers' monies recently awarded shows:

NC lands EPA grant to cut carbon emissions naturally

BY ADAM WAGNER
awagner@newsobserver.com

To cut carbon emissions, North Carolina and three nearby states are going to look to nature.

The U.S. Environmental Protection Agency has awarded the Atlantic Conservation Coalition a $421 million Climate Pollution Reduction Grant. Along with North Carolina, the coalition includes Maryland, South Carolina and Virginia, as well as nonprofit partners like The Nature Conservancy and the N.C. Coastal Federation.

The grant is the largest the EPA has awarded for the use of natural solutions to curb the impacts of climate change.

North Carolina's portion of the award includes $50 million to the N.C. Department of Natural and Cultural Resources to execute "shovel ready" projects, with $30 million passed to the N.C. Coastal Federation to focus on projects like building living shorelines and using dredge spoils to restore threatened salt marshes. The North Carolina chapter of The Nature Conservancy will receive an additional $67.8 million to restore an estimated 33,500 acres of peatlands across North Carolina and Virginia while protecting an additional 10,500 acres.

"Protecting our natural lands for future generations is not only critical to our fight against climate change but also our state's economy and tourism industries," Gov. Roy Cooper, a Democrat, said in a statement.

North Carolina's grant projects are expected to reduce greenhouse gas emissions by more than 22.7 million metric tons by 2050, according to Michele Walker, a N.C. Department of Natural and Cultural Resources spokeswoman.

(Source: The Durham News & Observer, Monday, July 29, 2024)

Perhaps some money could be saved by allowing more carbon emissions to be cut naturally – and with EPA inspectors monitoring less frequently how nature resolves its own environmental problems without combatting itself?

Inasmuch as the preceding information in this book has proven otherwise, Man-made carbon emissions are not the dominant or even major cause of "Climate Pollution". A recent measurement of carbon dioxide in the atmosphere, as measured by NOAA (at the NOAA GML Mauna Loa Observatory on July 29, 2024) was measured to be 425.21 ppmv (parts per million). It was shown earlier that the natural carbon dioxide in the atmosphere before the start of the Industrial Revolution was 295 ppmv. The environmental scientists attribute the rise from 295 to 425.21 ppmv – a rise of 130 ppmv - of total carbon dioxide in the air to be entirely due to Man globally generating "greenhouse gases". The EPA organizations are using as a goal the "average global temperature" (which is a false scientific measure) to be held to a rise of not more than 1.5 degrees Celsius by 2050 if "net zero emissions" from Man are achieved (not likely).

The UN IPCC meeting of COP28 in December 2023 in Dubai shows that the goal (to hold "average global temperature" rise to 1.5 ^{0}C by 2050) will not be met. China and India generate over 80 percent of global Man-made emissions – and China and India are building additional fossil fuel generating facilities, not reducing that 80 percent of 130 ppmv (by 104 ppmv). Therefore, if the USA reduces their Man-made "greenhouse gas emissions" to net zero, the most reduction of Man-made CO_2 in the atmosphere would be from 130 ppmv to 104 ppmv. The marshlands and peat areas on the US East coast being protected by the EPA "cutting carbon emissions naturally" would see the total carbon dioxide in the atmosphere reduced from 425.21 ppmv as of July 2024 to not less than 399 ppmv of total CO_2. Any expectation that a realistic reduction of carbon emissions in the global atmosphere by 26 ppmv (which is in percent equivalent to a reduction of 0.0026% of CO_2 concentration in the atmosphere of the US East coast) is ludicrous at best. EPA scientists are due for a remedial course in basic mathematics. The EPA grant of $421 million dollars of taxpayers' monies to protect marshlands and peat area would be better spent building homeless shelter facilities in areas where the homeless now live without air conditioning or heat, no human waste sanitation, and having to scavenge trash cans for food.

The magnificent grandeur that Nature also has created – disguised with the identifier of this chapter as geophysics - in many places on the Earth. One such scene is one of the top favorites for the US public to visit and to appreciate why it is important to protect Man but also to protect the national treasures on its surface. This is the Yosemite Park in Northern California, and showing the Upper Falls and the Lower Falls of the Merced River through the region in its geophysical magnificence:

(Source: US National Park Service)

When viewing this scene for its natural beauty, system engineers note that gravity is the unseen factor of Nature that adds to its beauty – to wit, the magnificent falls (Yosemite Falls – upper and lower) of water. The trees grow upward despite that same gravitational acceleration opposing the trees growing away from the center of the Earth. The upheaval of that mass of granite stone to such a high elevation required an energy of magnitude that boggles the mind but again points to the natural energy below Earth's surface being almost infinite and already heated to beyond the temperature needed to generate electricity. Deep drilling technology has improved greatly in the past few decades and points to harnessing geothermal energy as being the cleanest, most abundant, most reliable and at lower cost than most current technologies to harvest that same energy above Earth's surface where everything is exposed to a naturally often hostile environment.

Those considerations are vitally – but not as a "crisis" - important when it is a certain probability that an exponentially growing population will need more area of land on Earth's surface to meet their housing, agricultural, infrastructure, and transportation needs. Much ground area could be utilized better than for use to arrange large arrays of solar collectors. Instead, the homes could be built and use those solar panels as roofing materials. This Author sees little to be gained by building more wind turbine farms off-shore and placing mechanically expensive and already low reliable wind turbine farms into the "harm's way" of natural ocean environmental challenges.

Chapter 4
Natural Factors Affecting the Planet

There is too much misinformation – intentional and unintentional - being fed to the public and government leaders (USA and International) causing them to not adequately understand the factual causes and rate of change of the weather-related catastrophes, global warming, climate change, and oceans' surface levels rising. The title of this book may be a bit misleading in that "The Big Picture of Climate Change on Our Planet" should perhaps be more correctly stated as "The Big Picture of Climate Changes and Interactions on Earth's Environment and Human Population" – but that is too long for a title.

The largest factors affecting temperature on the planet Earth's surface were shown earlier to be the astrophysical distance the Earth is from our single source of heat energy radiation the Sun, followed by geophysical (especially latitude) and natural factors. There has been found no factual database or scientifically credible factor supporting the environmental scientists' hypothesis that Man burning fossil fuels is the dominant – or even a major – cause for global warming, climate change, or ocean levels rising. Disturbingly, there was found much evidence where valid databases were partially or totally ignored if the new data (obtained with "big picture" satellite imagery) disagreed with out-of-date, incomplete ("cherry-picked"), linearized, and averaged, meteorology data. Almost nothing in Earth's astrophysical, geophysical, and natural system functions linearly or in an average manner.

Water is a most important natural feature in the "big picture" analysis of the planet's system of astrophysics, geophysical, natural, and anthropogenic (Man-made) factors being analyzed in this book for the individual and combined extent of their influence on Earth's surface temperature warming and/or climate change. It is the abundance and dynamics of water on planet Earth that distinguishes Earth from all other celestial bodies in the Universe. Water has some unique properties that cause it to be the most vital factor to all forms of life on the planet. This book prioritizes the safety and health of human life to be of primary value to the public, but in a balance with protection of the ecological system. This chapter is the largest chapter of this book because of the complexities and interactions of the many aspects of nature.

How much water exists but in the extremes of droughts and floods, how it is distributed on the Earth's surface and in its atmosphere, and the dynamics (governed by the laws of physics) associated with water movement on Earth's surface are most important factors needing to be established in understanding why and how the Earth's surface temperature and resulting climate (temperature and precipitation) are changing. Water is present on Earth in all three of the principle states of matter – solid (as ice), liquid in abundance as over 70% of Earth's surface area, and gaseous – as water vapor, humidity, or clouds in the atmosphere). Two images from a "big picture" satellite view are shown next and which are followed by a detailed tabulation of the amounts of water on the planet and where located:

The Vast Pacific Ocean — The Eastern Hemisphere

(Source: NASA)

Ocean saline water and fresh water on land covers approximately 70% of Earth's spherical surface. Numerically, the distribution of water on the planet is shown in volume totals and concentrations as water vapor in the atmosphere in the following table (not intended as an eye test chart):

Source of water	Volume of water in km³ (cu mi)	% total water	% salt water	% fresh water	% liquid surface fresh water
Oceans	1,338,000,000 (321,000,000)	96.5	99.0		
Pacific Ocean	669,880,000 (160,710,000)	48.3	49.6		
Atlantic Ocean	310,410,900 (74,471,500)	22.4	23.0		
Indian Ocean	264,000,000 (63,000,000)	19.0	19.5		
Southern Ocean	71,800,000 (17,200,000)	5.18	5.31		
Arctic Ocean	18,750,000 (4,500,000)	1.35	1.39		
Ice and snow	24,364,000 (5,845,000)	1.76		69.6	
Glaciers	24,064,000 (5,773,000)	1.74		68.7	
Antarctic ice sheet	21,600,000 (5,200,000)	1.56		61.7	
Greenland ice sheet	2,340,000 (560,000)	0.17		6.68	
Arctic islands	83,500 (20,000)	0.006		0.24	
Mountain ranges	40,600 (9,700)	0.003		0.12	
Ground ice and permafrost	300,000 (72,000)	0.022		0.86	
Groundwater	23,400,000 (5,600,000)	1.69			
Saline groundwater	12,870,000 (3,090,000)	0.93	0.95		
Fresh groundwater	10,530,000 (2,530,000)	0.76		30.1	
Soil moisture	16,500 (4,000)	0.0012		0.047	
Lakes	176,400 (42,300)	0.013			
Saline lakes	85,400 (20,500)	0.0062	0.0063		
Caspian Sea	78,200 (18,800)	0.0056	0.0058		
Other saline lakes	7,200 (1,700)	0.00052	0.00053		
Fresh water lakes	91,000 (22,000)	0.0066		0.26	87.0
African Great Lakes	30,070 (7,210)	0.0022		0.086	28.8
Lake Baikal	23,615 (5,666)	0.0017		0.067	22.6
North American Great Lakes	22,115 (5,306)	0.0016		0.063	21.1
Other fresh water lakes	15,200 (3,600)	0.0011		0.043	14.5
Atmosphere	12,900 (3,100)	0.00093		0.037	
Swamps	11,470 (2,750)	0.00083		0.033	11.0
Rivers	2,120 (510)	0.00015		0.0061	2.03
Biological water	1,120 (270)	0.000081		0.0032	

(Source: International Union of Geological Sciences database)

Readers are requested not to just visually pass over this large table of numbers but to comprehend the amount of water on Earth on and beneath its surface in the many different bodies of water (saline and fresh) and in its atmosphere.

The water vapor in the atmosphere is in a variable concentration up to 3.7% depending on the temperature of the water on the surface below. Over arid (dry) deserts, the concentration of water vapor in the atmosphere is far less and is blown by wind currents into the arid atmosphere from less arid locations. Water vapor is not visible to the human eye, but can be sensed by human skin as humidity. When warm water vapor rises to an elevation which is at a cold temperature relative to that containing the water vapor, the water vapor condenses into cloud formations, as are seen in the two photographs above.

It is also of fundamental importance to Readers of this book on climate change to understand that the variables of "global warming", "weather", "climate" and "climate change" are becoming scientifically less well-defined and the databases are being distorted in meaning (i.e., being "averaged", "misrepresented" or "cherry-picked") to create fear of a "crisis" hazarding the public rather than as factually valid analysis tools.

That is important as the environmental scientists fallaciously consider the process of Man combusting fossil fuels and creating 'greenhouse gas" emissions (in concentration totaling less than 0.02% - 130 parts per million by volume) to be the dominant cause for global warming and climate change. The 0.037% larger volume and more effective heat-trapping water vapor in the atmosphere is disregarded by environmental scientists in forming that fallacious conclusion.)

Global warming is admittedly a somewhat "fuzzy" term when calculated by environmental scientists averaging and linearizing variations in the surface temperature of the entire globe (planet) over a significant number of months and short span of years to prove that an adverse warming trend caused by Man exists. Early attempts by the Environmental Protection Agency to reduce the adverse effects – while disregarding the beneficial aspects - of fossil fuel technology, led environmental activists to induce the public to believe the use of fossil fuels is at a "crisis" level of importance, and needs to be "combatted" in order to bring the climate under control.

What are not being "seen" are the combined effects of unseen other factors (astrophysics, geophysics – gravity, Coriolis effect, natural, and a gamut of "weather" variables changing and having a near-term effect on the planet's environment and all living forms of life. The most important effect of water temperature is seen in its effect on the coastal cities that are located adjacent to the large bodies of water. An "eye-opener" for the more probable effect of water temperature on the local environment and on the local human population is in figures of the ocean currents that exist – not merely in the NOAA "average sea surface temperature" of 21 °C (69.6 °F).

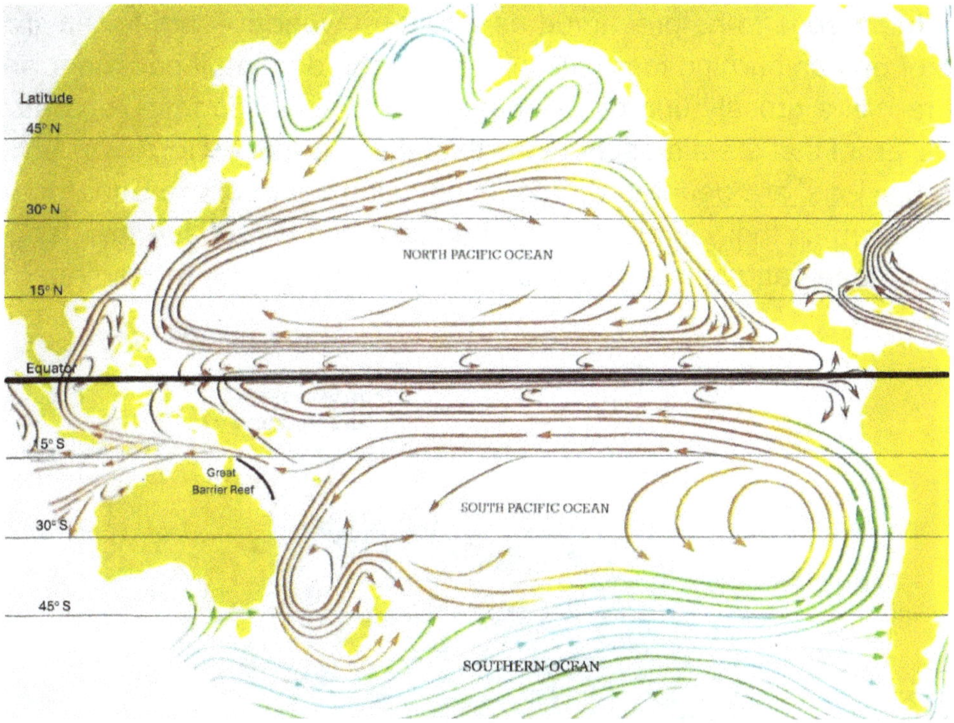

(Source: Rand McNally "Atlas of the World")

Warm and hot ocean currents (located close to the Equator depicted by the horizontal black line) are depicted with a red coloration and red arrows showing the warm current flow. Cooler and cold ocean currents are depicted with green coloration and green arrows showing the cooler current flow. For illustrative purposes, the Great Barrier Reef off the Northeastern coast of Australia (lower left on the diagram) is affected primarily by the hot Equatorial current of the South Pacific Ocean – not by the NOAA "average sea surface temperature" which is at about 21 °C (69.8 °F). The average temperature variation of the Great Barrier Reef is 24.7 °C (76.5 °F) in Winter, 26.0 °C (78.8 °F) in Spring, 28.9 °C (84.0 °F) in Summer and 27.5 °C (81.5 °F) in Autumn – all much hotter than the NOAA average SST of 21 °C (69.8 °F). Just as humans must survive the extremes of hot and cold that occur seasonally in the atmosphere, the coral animals and all aquatic life in the oceans must survive the extremes of hot and cold that occur in the ocean seasonal pattern annually. The Great Barrier Reef is a naturally beautiful scene (as shown in the next photo) to be protected, but human lives are much more so:

(Credit to unknown photographer (in a helicopter)

The Great Barrier Reef needs to be preserved as a natural wonder – but for environmental scientists to correlate the factor of Man burning fossil fuels as being the dominant cause for any bleaching that is happening on the reef is a grossly unprofessional and non-scientific fallacious attachment of cause. There is no scientific proof that the atmosphere north of the Equator (where over 95% of the Man-made "greenhouse gas emissions" are generated) affects anything south of the Equator – such as the Great Barrier Reef. That is seen (in the "big picture") in the prevailing wind patterns (generated by reputable scientists) near the surface - and in the upper atmosphere - driving the movement of weather cells and ocean currents:

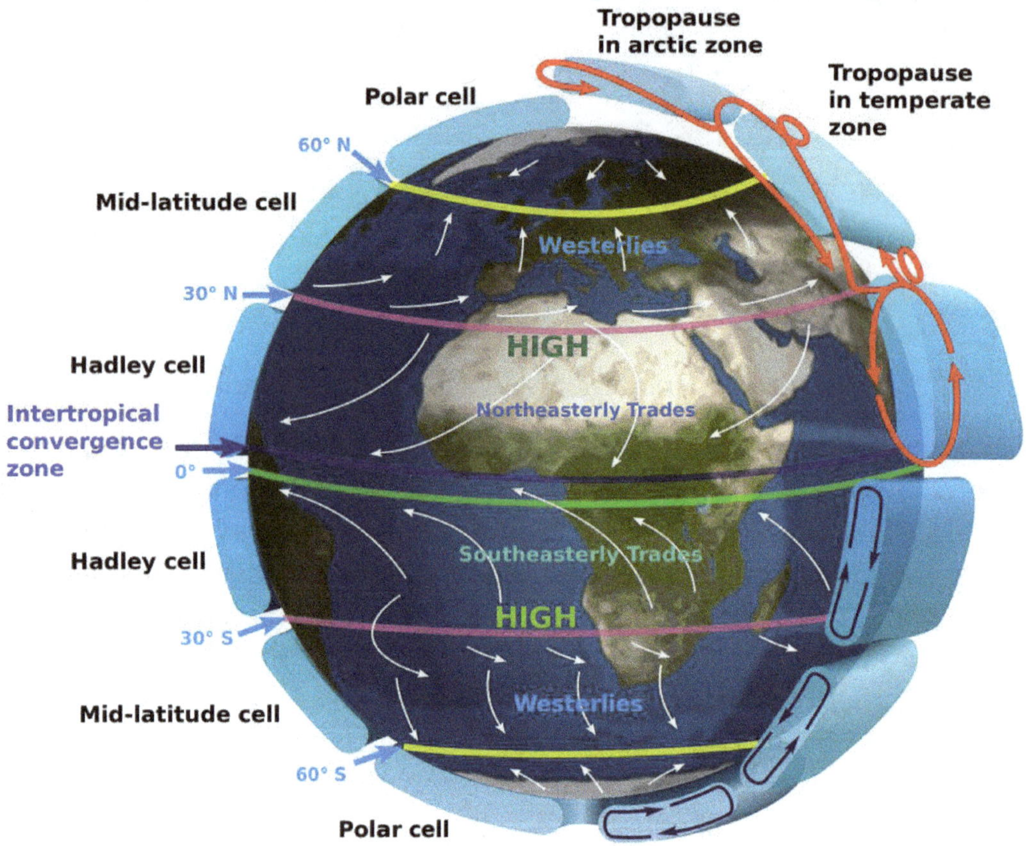

(Source: The Smithsonian Institute for Science)

The illustration of ocean currents also shows (in the upper right portion) the hot ocean currents spawning the tropical storms and seasonal hurricanes in the eastern Atlantic Ocean that devastates coastal regions of land masses interfacing with the Caribbean Sea and the Gulf of Mexico. The sea surface temperature in the Gulf Stream current is also very seasonal, averaging 73 °F in Winter, 77 °F in Spring, 84 °F in Summer, and 81 °F in Autumn. Its extreme temperatures are 70 °F in February and 87 °F in August.

The North American continent would be a much better place to live if the seasonal hurricanes were prevented from being created, but climate science has not fully understood why the tropical storms habitually are formed in the small region of the Cape Verde Islands group off the coast of West Africa. There is undoubtedly a complex interaction of the wind patterns in the atmosphere above the region and a recent satellite photograph reveals the very complex wind patterns in the Cape Verde region:

(Source: NASA and NOAA)

The Author had a career in aeronautical engineering and recognizes those wind patterns to resemble an aerodynamics phenomenon known as the "Von Karman vortex street" which also features very much like those shown above. The vortex street is seen in wind tunnel testing of cylindrical and near spherical shapes when a wind current flows over the curved surface of the body. A map of the Cape Verde region (the brown spots in the upper right quadrant of the above satellite photo are some of the Cape Verde islands group) does not cause any physical features that could act as vortex generators of the "vortex street flow" and it is surmised that it is the interaction of the colder wind patterns in the atmosphere (shown in the previous wind currents diagram) and the very hot winds flowing from the Sahara Desert region of northwestern Africa:

(Source: Rand McNally "Atlas of the World")

While it would be a most wonderful achievement if climate scientists and aerodynamics engineers could devise a “big picture” technology to control the wind patterns in the atmosphere to prevent the vortex activity that forms initially as the small vortices seen (and with low barometric pressure entering into the interaction), but that level of activity is beyond the scope of this book. The system of small vortices in the “spawning ground” coalesce into one larger vortex and that larger vortex moves westerly gaining heat energy from the ocean waters of the Atlantic Ocean and Caribbean Sea. The high temperatures that are associated with the Summer and early Fall seasons at that latitude cause the tropical storms to develop into hurricanes of different severity levels (F1 to F5) to follow a generally predictable track – headed toward the southeastern states of the United States region. The Florida peninsular shape serves as an aerodynamic guide vane for the hurricanes and their typical track westerly turns northerly to follow the Gulf Stream warm current flowing past the east coast of the United States and Nova Scotia.

Climate change is a measure of both the surface temperature of a region and the amount of precipitation (rain, snow, hail, graupel) on that region and averaged for a longer period of years. For the climate analysis in this book, the period of years for defining the climate of a region – not averaged on a global basis – is considered be no less than twenty-five years.

How the water being transformed from liquid form into water vapor that rises into the lower atmosphere and is later precipitated back to a liquid form as rain or snow and subsequently into potential other environmental problems such as flooding, mudslides, and bursting through inadequately constructed levees and dams is the topic of other chapters in this book. However, the realization that it is the distribution of massive amounts of water on Earth’s surface – too much in some regions causing flooding while too little in other regions causing droughts – that presents challenges to the environment, to the human and animal populations, and to the environment on the planet. Man is powerless to prevent the occurrence (the where, when, and how much) of temperature and precipitation extremes, but instead must develop means to channel and/or divert floodwaters from inundating populated areas and business districts. Ideally, the floodwaters in one region might be channeled into a region suffering drought conditions but each region must adopt the measures for the extreme temperatures and extremes of precipitation that apply in a regional manner not a global “one-size-fits-all” average manner.

Most Insurance companies do not cover damages done to properties by moving water (as in known flood zones), and those that do offer flood insurance charge exorbitant premiums. That should serve notice to humans to build their dream home elsewhere than in a high-risk location, but humans value the aesthetics of where to live and accept the risks of Nature – and higher cost.

(Credit to unknown photographers) “Every man’s home is his castle”

Where ocean waves are formed and to what scalar size – on a beach shore or mid-ocean – is caused by the wave energy being transformed into a mechanical form of energy. It is left to ecologists and oceanographers to explain the interactions and disturbances of wind and water affecting the environment and ecology. Informative photos of the interface between winds in the atmosphere and the water on the planet are shown as follows:

(Credit to Unknown Photographers)

The winds interfacing with the ocean surface can be either a constant velocity (from very low to very high) parallel to the water surface (laminar) or gusty (turbulent) with vorticity in small wind cells. There is no valid fact that Man burning fossil fuels has an adverse effect on the ocean waves. Less water vapor is generated with lower wind action and less of that generated is transformed into the atmosphere over placid waters than is over turbulent water surfaces. The wind, laminar or turbulent, determines whether the motion of the water will be formed into rolling waves that travel on the water surface (as in the left photo above) or become "choppy" with "white caps" in mid-ocean (as in the right photo above). The white in the frothy white caps illustrates that the temperature/warmth of the water and the wind currents play a part in evaporation in generating water vapor, and the generated water vapor being transformed from liquid to gaseous state into the atmosphere. (The explanation of "Where did all the white go?" question of children when their pale friend Mr. Snow Man melts is left to parents to explain – if they themselves know the answer.)

As in most chemical processes, including that of a liquid evaporating into gaseous form, increasing the temperature and surface area of the water (H_2O) and displacing its molecules with an air current (wind) transforms more of the water molecules into the evaporative process. The natural energy – mechanical and heat - to cause that evaporative process to occur has been shown in earlier chapters.

The presence of water on the planet's surface and water vapor in the lower atmosphere is the singular most important natural resource for several reasons. Without water, there would be no forestry, no agriculture, no botanical growths, and no agricultural foods, and no humans to exist since water is an essential for all forms of life – forestry, agriculture, and animal and human life to exist. Life in the oceans obviously would not exist if the planet became barren as have the other planets in the solar system. The human population that has fish as a main food staple would suffer loss of a main protein source. Animals and poultry (for dairy products and for protein sources) cannot exist without water in their daily diet. Water is also needed for human health, personal hygiene, and for sanitation processes.

It was only in recent years that some environmental scientists acknowledged (most have refused to do so) the fact that water vapor in the atmosphere is a more powerful trapper of heat energy than is the total of all the "greenhouse gasses" (unwanted emissions that trap heat – carbon dioxide, methane, nitrous oxide, etc.) caused by the human population on the planet burning fossil fuels for their safety, health, comfort, transportation, and industrial needs.

The following extract (from an Associated Press article by Seth Borenstein on January 31, 2022) states that acknowledgment based on then-current (2022) information: *'When it comes to measuring global warming, humidity, not just heat, matters in generating dangerous climate extremes, a new study finds. Environmental science researchers say temperature by itself is not the best way to measure climate change's weird weather and down-plays impacts of humidity in the tropics. But factoring in air moisture along with heat shows that climate change since 1980 is nearly twice as bad as previously calculated, according to their study in the January 24, 2022, Proceedings of the National Academy of Sciences."*

"The energy generated in extreme weather, such as storms, floods and rainfall is related to the amount of water in the air. So a team of scientists in the U.S. and China decided to use an obscure weather measurement called equivalent potential temperature — or theta-e — that reflects the moisture energy of the atmosphere. It is expressed in degrees, like temperature." said study co-author V. "Ram" Ramanathan, a climate scientist at the University of California San Diego's Scripps Institution of Oceanography and Cornell University. *"There are two drivers of climate change: temperature and humidity,"* Ramanathan said. *"And so far, we have measured global warming just in terms of temperature. But by adding the energy from humidity, the extremes — heat waves, rainfall and other measures of extremes — correlate much better. That is because as the world warms, the air holds more moisture, nearly 4% for every degree Fahrenheit (7% for every degree Celsius). When that moisture condenses, it releases heat or energy. That is why when it rains, now it pours,"* Ramanathan said.

"In addition, water vapor is a potent heat-trapping gas in the atmosphere that increases climate change. From 1980 to 2019, the world warmed about 1.42 degrees Fahrenheit (0.79 degrees Celsius). But taking energy from humidity into account, the world has warmed and moistened 2.66 degrees Fahrenheit (1.48 degrees Celsius), the study said. And in the tropics, the warming was as much as 7.2 degrees Fahrenheit 4 degrees Celsius). When judging by temperature alone, it looks like warming is most pronounced in North America, mid-latitudes and especially the poles — and less so in the tropics, But that is not always the case, because the high humidity in the tropics "juices up" storm activity, from regular storms to tropical cyclones and monsoons. This increase in latent energy is released in the air which leads to weather extremes: floods, storms and droughts.", Ramanathan said.

University of Illinois climate scientist Donald Wuebbles, who was not part of the study, said: *"It makes sense because water vapor is key in extreme rainfall. Both heat and humidity are important."* Environmental scientist Katharine Mach of the University of Miami, who also was not part of the study, said: *"Humidity is key in shaping the impacts of heat on human health and well-being, at present and into the future."*

Water is the second most dominant factor that controls the surface temperature of the planet and does so by the cycle of free-standing water warming to a temperature and evaporating into the atmosphere, rising (opposing the gravitational acceleration by expending the heat energy in a mass into potential energy of that mass) to an elevation where cooler temperatures exist (due to astrophysical reasons), condensing into cloud formations, and then falling back to Earth (accelerated by gravity) as precipitation (rain, snow, hail, or graupel). That process is done cyclically in accordance with the laws of energy conservation, laws of mass conservation, and in accordance with the laws of motion of a mass. This book stops short of being a textbook in atmospheric chemistry or physics – and only contains the most basic principles of fluid dynamics (i.e., liquids and molten materials flow downhill and those with high viscosity travel slower than those with low viscosity but could have more stored heat energy and more kinetic energy).

The small-scale uses of water hydrodynamic properties are illustrated later in this chapter, but its hydrodynamic power used to generate electricity on large scales will also be discussed in a later chapter of the book along with other natural sources of energy.

Logic, and a strategy of applied reasoning, was/is used extensively in this book to analyze the problems of global temperature change, climate change, and ocean levels rising, and without any bias toward selecting only those factors that are considered important by environmental scientists as the main variables "driving" the resulting effect of the factors. Using discipline and correct logic to forensically evaluate a large number of factors and set them into their correct "system" level perspective is the primary difference of this book and the flawed and biased analyses that were done by environmental scientists thirty and forty years ago - which resulted from the use and misuse of incomplete databases and/or the linearization/averaging of information. Few living and natural things perform in a "linearized" or "average" manner.

Many reputable scientists have criticized those environmental scientists and "green" activists for their flawed contention that: *"The science is settled. Humans burning fossil fuels and creating "greenhouse gas" emissions is the dominant cause for global warming and being at a "crisis" level of urgency needing to be reduced or eliminated"*

The most vocal and scientifically valid criticism of environmental scientists dismissing water vapor as the more extensive heat energy "trapper" in the atmosphere compared to the "greenhouse gasses" (carbon dioxide, methane, nitrous oxide, etc.) came from Dr. S. Fred Singer (1924 – 2020). Dr. Singer was a Professor Emeritus and was formerly Chair of the Department of Environmental Science at the University of Virginia. He did much analysis of the causes (natural and/or anthropogenic) for global warming, and climate change and reported the analytical results in a series of papers "A Closer Look at the Numbers". Perhaps the most important illustration of the importance of water vapor compared to that of natural and anthropogenic causes of the "greenhouse gasses" was shown in Dr. Singer's (and in several other reputable environmental scientists') frequent lectures and speeches as shown next:

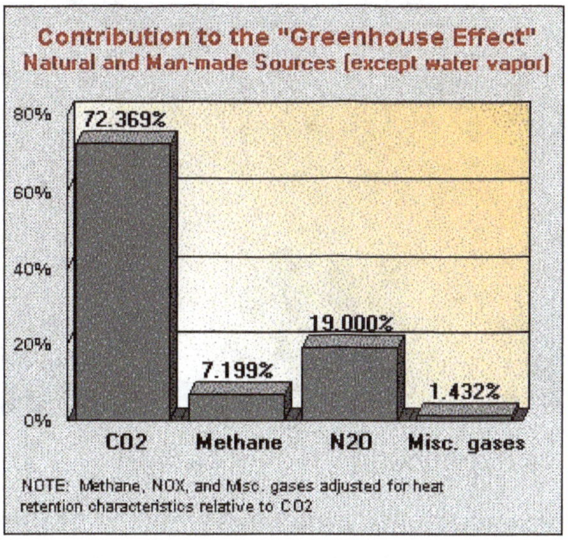

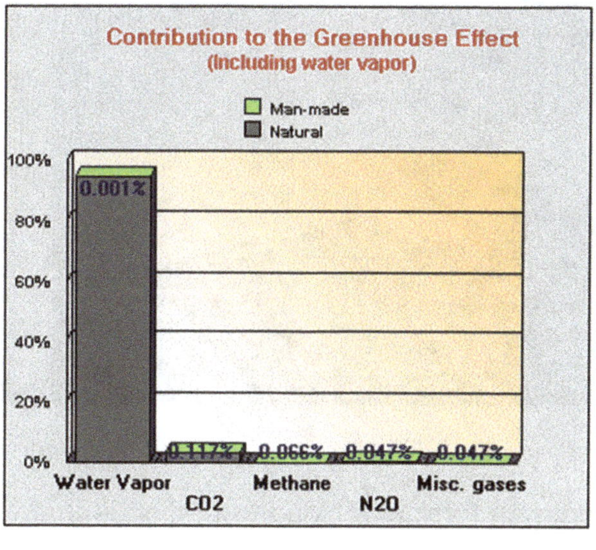

(Source: "A Closer Look at the Numbers", Dr. S. Fred Singer, formerly at U. of Virginia)

The chart on the left is based on the environmental scientists' database on "greenhouse gas effect" and not differentiating how much of the total is natural and how much is Man-made (but as the title indicates in the fine print tag-on "except water vapor"). It is obvious that the environmental scientists are concealing – probably intentionally - that water vapor is a major factor in causing the insulation effect (trapping heat) of the "Greenhouse Effect". That is tantamount to the news media asking President Abraham Lincoln's wife after his assassination: "Other than that Mrs. Lincoln, how did you enjoy the stage play?"

Readers should note that Dr. Singer indicates in the chart at the right how much of the individual gasses and water vapor is from natural cause (in grey color – and over which Man has no control) and how much is of anthropogenic (human) cause (in green color. Charts show the relative effectiveness each factor has on global warming directly. The green-colored Man-made "greenhouse gasses" (principally carbon dioxide) data are small (but not negligible) in comparison to the grey-colored natural-caused amount of the "greenhouse gas" effect; yet Man-made carbon dioxide is the main thrust of environmental scientists to eliminate or to be reduced drastically). The indirect and feedback effects of each (water vapor and Man-made greenhouse gasses) are discussed later in this chapter.

The significance of the environmental scientists claim which disregards the factual dominant effect of water vapor on the "greenhouse gas effect", coupled with the fact that there is about 80% or more of the measured total carbon dioxide in the atmosphere (0.0417% as of October 2023 compared to the 0.0350% circa 1880 pre-Industrial Revolution) that is naturally-caused, and less than 20% that is Man-caused (human and Industrial), dispels the environmental scientists' fallacious claims that Man burning fossil fuels is the dominant cause for the "greenhouse gas effect" on the planet's surface temperature. The remainder of this chapter is presented with the intent of informing the public and to generate a higher appreciation for the important role water has on their every-day life as well as their life-long dependency on water for their safety, health, and comfort.

Clouds are commonly seen above water covered surface areas, and equally common seen to be absent above desert areas – as clearly shown above. They also form over land areas in the awesomely and sometimes ominous manner as shown next:

(Credit to unknown photographers)

A very common form of a cloud formation is that of the altocumulus which is sometimes referred to as a "flock of sheep". Cloud formations – especially cumulus - are even more awesome to see in the "big picture" perspective when flying in an aircraft above them.

(Credit to unknown photographers)

The most ominous cloud formation the Author found was the one shown next:

(Credit to a very talented unknown photographer)

In any cloud formation there is a high concentration of water vapor which could fall as precipitation – rain, snow, hail, graupel, etc. No matter the shape of form or their color, it is the amount of precipitation from cloud formations and their reduction on solar insolation that contributes to the climate of a region that is of importance. Contrary to environmental scientists' misinformation to the public, the climate of the region does not have direct "feedback" causing the cloud formations to exist – but there is a cyclical "interaction" in that clouds (having an "albedo" effect) alter the temperature of the water surface which alters the evaporative rate of the water that is transformed into the atmosphere as water vapor. In common language, "the tail does not wag the dog" – the climate does not change the weather or the clouds and precipitation.

The next table of the water (ocean, seas, bays, gulfs) database numerically illustrates just how much total water in liquid and solid (floating sea ice) state there is (over 80% of Earth's surface area), and its distribution on the planet's surface: The volume of water in the five major oceans is also shown:

	Surface Area of Ocean / Sea (Est. 2021)				Volume of Oceans (Est.)	
Ocean/Sea Name	**Sq. Miles**	**Sq. kilometers**	**% of Planet**	**Rank**	**Cubic Miles**	**Cubic Km**
Pacific Ocean	65,091,238	168,723,000	33.05%	1	160,515,659	669,880,000
Atlantic Ocean	32,843,254	85,133,000	16.67%	2	74,380,203	310,410,900
Indian Ocean	27,221,172	70,560,000	13.82%	3	63,259,291	264,000,000
Southern Ocean	8,471,895	21,960,000	4.30%	4	17,204,610	71,800,000
Arctic Ocean	6,002,083	15,558,000	3.05%	5	4,490,695	18,750,000
Sub-Total Surface Area of All Oceans	**139,629,642**	**361,934,000**	**70.89%**		**319,850,458**	**1,334,840,900**
Philippine Sea	2,197,080	5,695,000	1.12%	1		
Coral Sea	1,848,308	4,791,000	0.94%	2		
Arabian Sea	1,489,912	3,862,000	0.76%	3		
South China Sea	1,185,597	3,500,000	0.69%	4		
Weddell Sea	1,080,205	2,800,000	0.55%	5		
Caribbean Sea	1,062,459	2,754,000	0.54%	6		
Mediterranean Sea	968,327	2,510,000	0.49%	7		
Gulf of Guinea	906,601	2,350,000	0.46%	8		
Tasman Sea	887,311	2,300,000	0.45%	9		
Bay of Bengal	837,931	2,172,000	0.43%	10		
Bering Sea	771,575	2,000,000	0.39%	11		
Sea of Okhotsk	610,702	1,583,000	0.31%	12		
Gulf of Mexico	597,970	1,550,000	0.30%	13		
Gulf of Alaska	591,412	1,533,000	0.30%	14		
Barents Sea	540,103	1,400,000	0.27%	15		
Norwegian Sea	533,544	1,383,000	0.27%	16		
East China Sea	481,849	1,249,000	0.24%	17		
Hudson Bay	474,519	1,230,000	0.24%	18		
Greenland Sea	464,874	1,205,000	0.24%	19		
Somov Sea	443,656	1,150,000	0.23%	20		
Sub-Total Surface Area of Seas/Gulfs/Bays	**17,973,935**	**47,017,000**	**9.22%**			
Total Surface Area of All Waters on Planet	**157,603,577**	**408,951,000**	**80.18%**			

Notes: Total surface area (land and water) of planet is 196,961,745 sq miles and 510,065,665 sq kilometers.

(Source: Tabulation by Author of International Union of Geological Service database)

In a "big picture" of water on the planet, there is seen to be about twice as much water surface area in the southern hemisphere as there is in the northern hemisphere, and over 90% of the human population lives on the land area predominant in the northern hemisphere and east of the Greenwich Meridian:

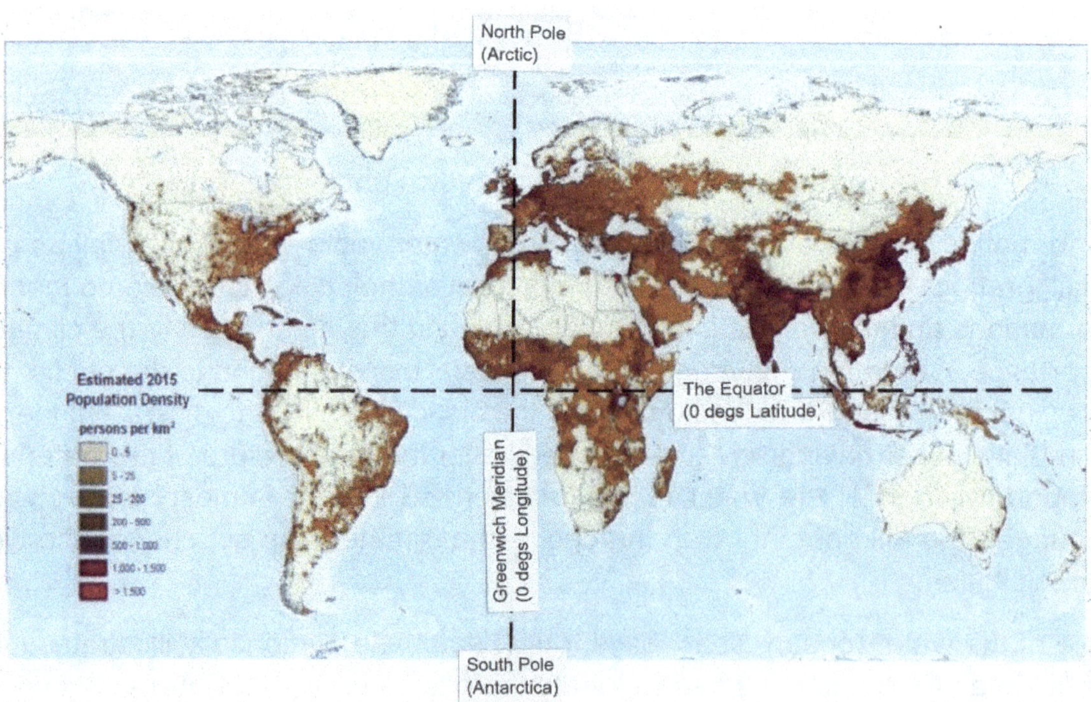

(Source: UN Population Division)

One does not need to be an environmental or climate scientist to appreciate that what the population of the North American Continent slightly does to its atmosphere and climate does not significantly affect the climate of the European and Asian Continents. The reverse of that may not be evident since the population of Europe and Asia is about 7 or 8 times that of the North American Continent. The land mass areas are at about the same ratio of 7 or 8 to 1.

A Harvard student generated an interesting form of the human population distribution of the planet by latitude and longitude as seen below:

The World's Population in 2000, by Latitude

(horizontal axis shows the sum of all population at each degree of latitude)

The World's Population in 2000, by Longitude

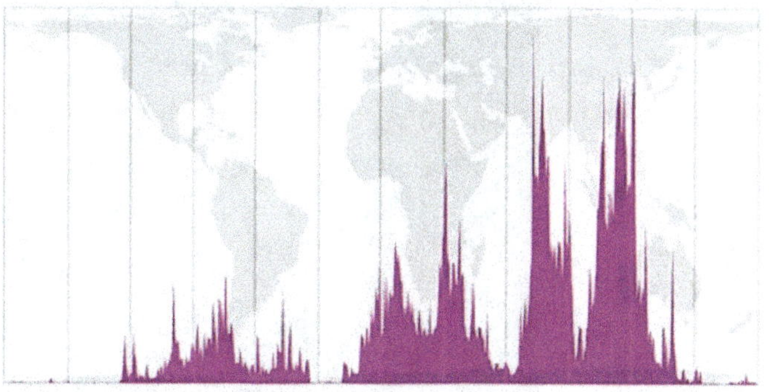

(vertical axis shows the sum of all population at each degree of longitude)

(Source: Harvard University, Unnamed graduate student)

The distribution by latitude (left illustration) is interpreted to show that the human population will choose to reside in a comfortable (temperate) climate – i.e., not uncomfortably hot or uncomfortably cold. The distribution by longitude (right illustration) is not surprising in the sense that the population chooses to live on land mass rather than on the ocean waters. The distribution showing the location by density (UN Population Division) is a clear indication of how the human population has chosen to reside at a latitude longitude climate combination that serves their needs and preferences for comfort and for the environment that supports their food supply. Deserts, jungles, and frozen land do not provide that combination.

The very large quantity of water and land area size numbers alone do not visually give the "big picture" of the magnitude of how much total water is on the planet's surface and in its atmosphere. There is also a huge amount of water in the clouds covering the planet in a somewhat random manner as can be seen in the photographs. Clouds act like a sponge – they hold a great deal of water until that water is squeezed out of them. Of great importance is the combination of the huge amount of water on its surface, the unseen water vapor in the atmosphere, and the visible clouds above it all that are among the many factors determining the temperature (seasonal and non-seasonal), the local and regional weather, and the climate of the planet

Another photograph taken by satellite shows the absence of cloud cover over North America (a relatively rare photo of a moment in meteorology where there is no cloud cover over the United States). To be noted is the dense cloud pattern off the Eastern coast of the United States which influences the weather – especially in the hot summer season - of the East Coast states (from Maine in New England to Florida in the Southeastern region). The dense cloud pattern is above the very warm Gulf Stream current from the Caribbean region and is a huge "bubble" of water vapor in the entire atmosphere from water surface level to the top of the cloud layer.

(Source: NASA)

That frequent meteorological phenomenon seen above is one of the causes for the US East coast to have very high humidity (above 80%) while the US Southwestern states (California, Arizona, and Nevada) to simultaneously have a humidity level below 15 percent and 20 degrees lower temperature. (The Author resided in New York for 23 years and in California for 38 years and can attest to the clothes "sticking" to flesh when the temperature and humidity in New York City are about the same magnitude – 85 degrees and 85 percent – while in a classroom.) That is a seasonal temperature change phenomenon, not a 10-year climate change phenomenon.

The four seasons in New York make it a most interesting location in which to reside compared to California where there is basically only one season – too warm and very dry. From 1982 to 2020, a local tongue-in-cheek expression by California "dudes" was that "It never rains in California.". The climate factor in recent decades has been changing to cycle between flooding and droughts in California, but there is no factually convincing database that the climate change is due predominantly to Man burning fossil fuels. The heavy Industrial states are in the Eastern half of the United States, but water vapor is clearly a more major "heat-trapper" over the entire "lower 48" region than is carbon dioxide emissions from heavy industries in the eastern half of the United States.

Environmental scientists have chosen to ignore water vapor (despite the concentration ratio of water vapor to all the "greenhouse gases" in the atmosphere being about 90 to 1) and environmental scientists persistently claiming that it is emissions from the Industrial Revolution that are causing global warming. That "greenhouse gas effect" – is not evidenced in "big picture" satellite photography, but the geophysical factor of latitude is clearly seen in the image on the front cover of this book. It is inconceivable how environmental scientists can use "average global temperature" based on the calculated average of measurements made at on land at over 2,000 land-based meteorological stations (most are at airports) when the planet's surface area is over 70 percent water-covered and at least another 20 percent being covered by deserts and highly mountainous regions.

Large cities are predominantly covered with concrete or asphalt-based black surfaces which retain heat. As the population grows, cities expand, airports are extended, more housing developments arise, and the extent of non-natural areas increase in size. Several book titles and film themes describe the planet well: "The Concrete Jungle". Concrete absorbs heat and radiates that heat when the surrounding air temperature is cool. Since most meteorological measurements were/are made at airports, the illusion that global warming is occurring due to Man burning fossil fuels might better be considered as the effect on "air temperature" from Man covering more of the planet's surface with concrete. As the population grows so does the amount of concrete-covered land mass.

The image on the front cover shows a very powerful effect of latitude on the surface temperature. How can meteorological measurements made on about 10% of the planet's surface area truly represent an "average global temperature" for the entire inhabitable global surface area of the planet? Mathematically it may, but realistically it does not. For environmental scientists' "convenience", a statistical "average global temperature" is a ludicrous unprofessionally deceptive practice that should not justify environmental scientists to set fallacious and unachievable carbon emission reduction goals that are regulatory imposed as US governmental and United Nations international global goals.

Earlier photos in this book (including the front cover photo) reflect the enormity of the bodies of water in liquid state and only partially portray the temperature of those bodies of water. The ice covering both the Arctic region and the Antarctic region are the solid states of water, but since they are in regions where there is little or no population of humans they are usually not considered as being of environmental interest. The tabulations shown earlier are the facts for the largest oceans, seas, gulfs, and bays, and as shown later for large inland seas, bays, and lakes.

For visual completeness of how water is physically distributed – in all three states of liquid, gas (water vapor and clouds), and solid (ice) on the planet, the two diagrams shown next are of the Arctic (North polar region) and Antarctic (South polar region). The two diagrams below of the polar regions should be of assistance to best understand the differences in amount and different type of ice at the two polar regions:

The Arctic Region (North Polar)

Antarctica (South Polar)

(Source: National Geographic Society)

A longer description of the future problem of ocean surface levels rising – and what are the causes for that hazardous eventuality - will appear in a later chapter. That later discussion will also include the three primary causes for ocean water surface levels to be rising and the adverse effects on humans and the need for upgrading of the container ports to sustain human needs.

A major difference between ice in the two polar regions is that the vast majority of ice in the Arctic region is floating "sea ice" (with the ice covering the upper 80 percent of Greenland above the Arctic Circle being on *terra firma*). By contrast, the vast majority of ice in the Antarctic region is on *terra firma* – the Antarctic continent – and only a much smaller percentage being on the ice shelves primarily on the western half of the Antarctic continent. The ratio of the total amount of ice in the Antarctic region to total ice in the Arctic region is about 9:1.

A not-so-mundane problem is that of icebergs, and the epic tragedy of the RMS *Titanic* – the "unsinkable" ship that struck an iceberg and did, most embarrassingly on its maiden voyage from England to New York. Icebergs do not have a decision-making capability as to where to go after being "calved" from their mother glacier. The winds and ocean currents determine the paths of icebergs, and the surrounding water temperature determines how long they will exist. Most science classes teach that 9/10ths of an iceberg is underwater and unseen and the expression that has become common when addressing only a small part of a bigger problem is: *"That's only the tip of the iceberg."* That is somewhat the same case the research in this book has concluded. Environmental scientists identifying Man burning fossil fuels to be the dominant cause of global warming have only identified the "*tip of the climate change iceberg*".

(Credit to unknown photographer)

A striking example of icebergs moving to where they are least expected and wanted is shown next:

(Credit to unknown photographer) Town on the coast of Greenland

Considering that icebergs are floating on water and have only about one-tenth of their mass above the surrounding water surface level, i.e., nine-tenths of their mass are unseen – the above iceberg was truly a monster. On the positive side, using the iceberg as a source of unpolluted non-saline water creates a natural resource for the residents to use as needed.

The vast Southern Ocean surrounding the Antarctic continent and the South Polar region is in a very cold but liquid state. As will be discussed in a later chapter, the thawing/melting of floating ice or ice on the ice shelves of West Antarctica does not contribute to ocean surface levels rising as they thaw/melt. It is only the ice on top of *terra firma* East Antarctica that melts and runs off into the surrounding ocean waters that pose a problem as contributory to the ocean surface levels rising in future centuries if/as the atmosphere and ocean waters elevate in temperature.

Two of the three states of water (liquid and gaseous) have a heat-trapping capability which are naturally caused to exist – and far outweigh (by a factor of about 80 or 90 to 1) the heat-trapping effects of the total concentration of natural and Man-made "greenhouse gasses" in the atmosphere. The amount of total carbon dioxide in the atmosphere was recently measured to be 420 parts per million by volume (a concentration of 0.042 percent of the atmosphere), whereas the amount of water vapor in the atmosphere is 1 – 4% depending on the temperature of the water below the location where measured). The water vapor over regions that are arid accordingly is lower than 2%, however, 80% of the Earth's surface has a watery surface emitting water vapor that does follow the following relationship:

Dewpoint (Saturation Fraction) of Water in Air at Sea Level

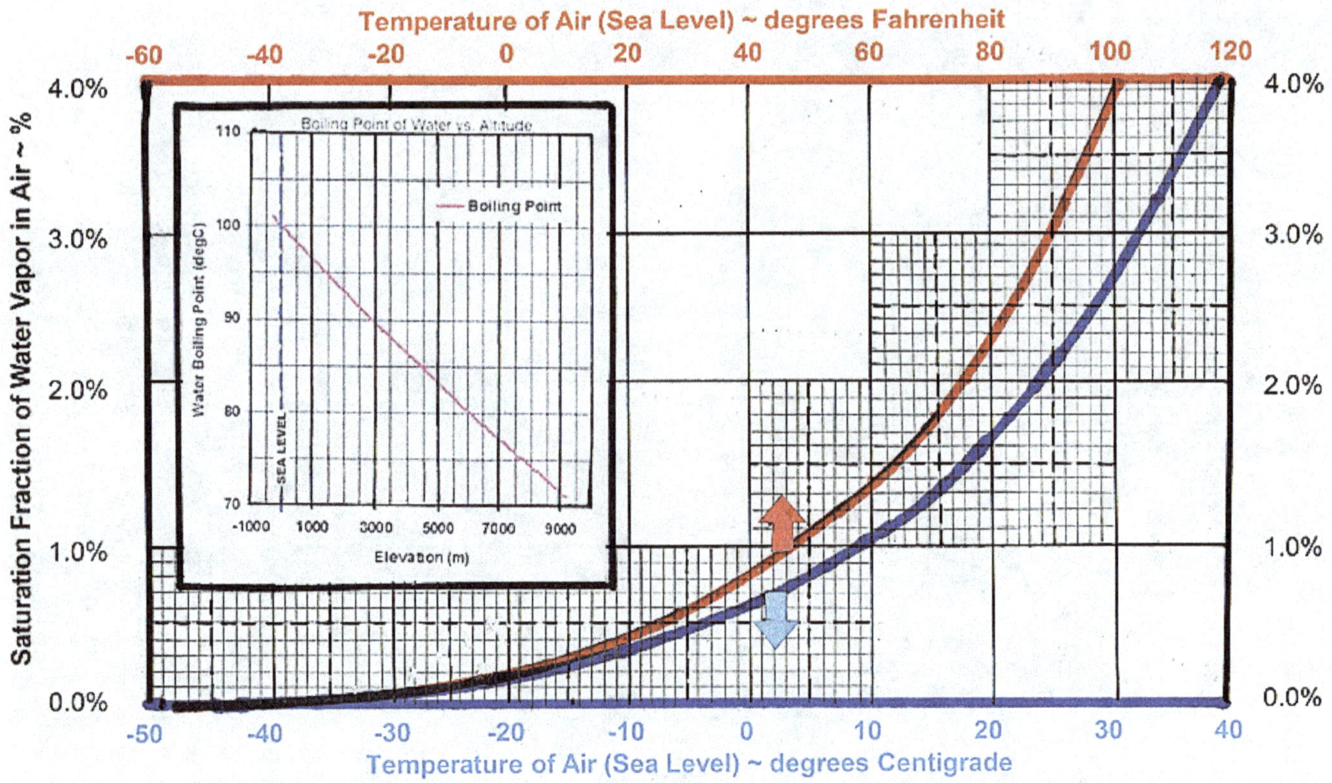

(Source: Reconstructed by Author from NWS/NOAA Database)

Anyone who has boiled a pot of water on a kitchen stove has seen the water bubbling and above the pot a small "cloud" of steam. Steam is hot water vapor with a high temperature. As shown on the small inset on the left of the above diagram, if the kitchen is at sea level elevation, the water boils at 100 degrees Centigrade (Celsius) or 212 degrees Fahrenheit. However, at an elevation higher than sea level, the temperature for boiling water reduces in magnitude such that at an elevation of Denver, Colorado, (the "Mile High City") – 5,280 feet (1,610 meters) – the water will boil at 94 degrees Centigrade (Celsius) or 201 degrees Fahrenheit. In industrial use, steam is used to generate electric power for other uses and that aspect of water will be discussed later in the "sources of energy" chapter of this book.

A simple "test of reasonableness" can be applied to the analysis of the "settled science" segment of environmental scientists, who have "circled their wagons" around their statement that "Man burning fossil fuels and thereby creating "greenhouse gas" emissions is the <u>dominant</u> cause of global warming", have yet to scientifically prove that 0.042 percent of carbon dioxide (and more than half of that 0.042 percent of carbon dioxide being introduced in the atmosphere is attributable to natural causes) that carbon dioxide in the atmosphere is the dominant cause of global warming. With only a maximum of about 0.02% of carbon dioxide in the atmosphere attributable to being Man-made, but with the higher percent (1 – 3%) of water vapor also present in the atmosphere, there is no scientific logic in the statement that Man-made carbon dioxide is the dominant cause of global warming. "Settled science" gets an F grade on the "test of reasonableness" of how such a low concentration of any gas can so hugely affect the entire atmosphere and resulting climate of the planet. The more factual assessment is that water vapor in the atmosphere is the dominant cause for global warming, in locations over 80% of the planet's surface.

For locations in the remaining 20% of the land area of the planet's surface where the highest density of human population exists, especially in cities that are over 50% concrete covered, water vapor in the form of humidity, regardless of the amount of carbon dioxide present in the atmosphere, is of great importance on the comfort of humans residing in that location. That is attested to by the Author who has spent over 20 years living in a high humidity environment in the summer seasons of New York City and of several months in Washington, DC. The previous diagram shows the saturation fraction of the atmosphere, but when there is less than that saturation fraction, the percent of that saturation amount is termed as being humidity. The relationship between humidity and dewpoint with air temperature is indicated in the diagram shown as follows:

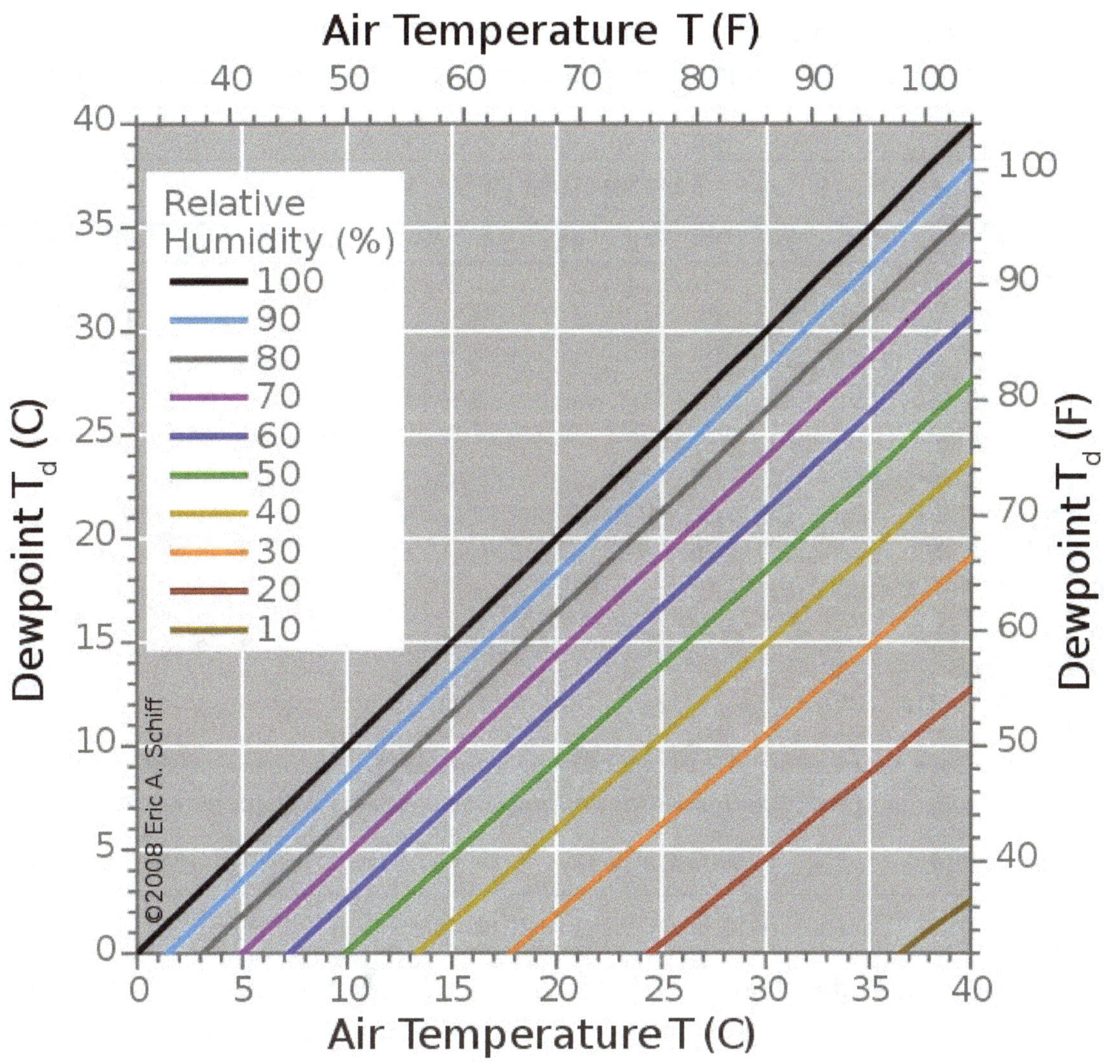

(Source: The National Weather Service, NOAA)

Anyone who has spent much time in both an extremely humid environment ("normal" for the northeast coastal region of the United States in summer months) and an arid environment (extremely low humidity but very high temperatures ("normal" for the Phoenix, Arizona, desert region of the southwest United States in summer months) can describe the difference between "wet heat" and "dry heat". Dry heat is preferred over wet heat by probably 99 humans out of 100. In Chandler (a southern suburb of Phoenix), Arizona, in 2021 and 2022 during the months of July through September, in both years there were periods of about 18 days straight with a temperature of above 100 degrees Fahrenheit but with humidity levels that were consistently lower than 20 percent. With peak temperatures of about 112 to 115 degrees F during those hot weeks, residents did not take pet animals for walks on the hot concrete sidewalks as that would cause burn injuries to the bottoms of their paws. Some residents with pets have "shoes" for their pets. Infrastructure maintenance projects were conducted during the nighttime hours for the dual purpose of least inconvenience to heavy vehicular traffic but also to avoid the excess heat on road maintenance laborers.

High humidity days coupled with high "normal" summer temperatures can create a very dangerous health hazard to humans (especially those with existing respiratory problems or those who engage in some form of exercise or labor that increases their perspiration rate). Spikes in humidity or spikes in a temperature above "normal" that last for less than twelve hours must be considered to be weather-related and not the much longer-term climate related. The National Weather Service (part of the National Oceanic and Atmospheric Administration) has made a "Heat Index" chart available to the public that warns of those combinations of humidity and temperature that are hazardous to human health:

NWS Heat Index — Temperature (°F)

Relative Humidity (%)	80	82	84	86	88	90	92	94	96	98	100	102	104	106	108	110
40	80	81	83	85	88	91	94	97	101	105	109	114	119	124	130	136
45	80	82	84	87	89	93	96	100	104	109	114	119	124	130	137	
50	81	83	85	88	91	95	99	103	108	113	118	124	131	137		
55	81	84	86	89	93	97	101	106	112	117	124	130	137			
60	82	84	88	91	95	100	105	110	116	123	129	137				
65	82	85	89	93	98	103	108	114	121	128	136					
70	83	86	90	95	100	105	112	119	126	134						
75	84	88	92	97	103	109	116	124	132							
80	84	89	94	100	106	113	121	129								
85	85	90	96	102	110	117	126	135								
90	86	91	98	105	113	122	131									
95	86	93	100	108	117	127										
100	87	95	103	112	121	132										

Likelihood of Heat Disorders with Prolonged Exposure or Strenuous Activity

Caution — Extreme Caution — Danger — Extreme Danger

(Source: The National Weather Service/NOAA)

If the "settled scientists" modify the claim to be "Total (natural and Man-made) carbon dioxide and water vapor in the atmosphere causes the "greenhouse" effect on global warming to be of urgent concern to the public, then a grade of B+ on the "test of reasonableness" could be assigned. If the "settled scientists" were to modify their claim to include astrophysical, geophysical, total carbon dioxide, and water vapor effects (such as humidity) on global warming, then a grade of A could be assigned for that test. But that IF is not likely. IF a bullfrog had wings, it would not bump its rear end so hard when it lands after each leap on land.

It is troublesome when a scientist(s), after observing a very small change in a variable in a laboratory experiment, over-exaggerates the observation and resorts to "chartsmanship" (the art of selecting scales on graphic illustrations to emphasize – and sometimes overemphasize - the visual importance of the factor) to impress his/her peers. In art, it is referred to as "artisanship". The average human temperature sensory system cannot distinguish the difference between two temperatures that are a degree or less apart. However, when a scientist records ocean water temperatures that vary seasonally as they normally do, but the "normal" for that season changes by less than one degree in a period of 30 years and reports that temperature database to show that the ocean water temperatures are "soaring" with time due to climate change – that is "chartsmanship". The "Grim results" version (the red markings are those of the Author) of a chart which was published in the Climate section in a science magazine (The Smithsonian):

CLIMATE

Grim results of the ocean's hot streak

Off-the-charts temperatures impact corals, sea ice and more

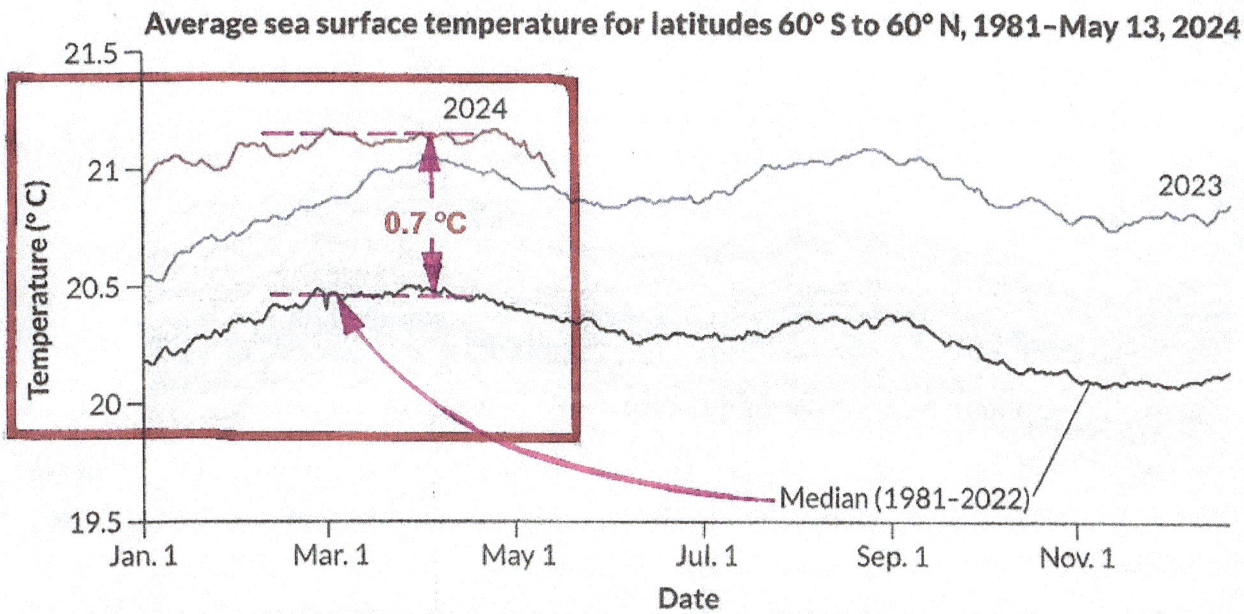

Soaring sea temps Since March 2023, the average sea surface temperature of the extra-polar ocean, at latitudes from 60° S to 60° N, has broken the daily record every single day, NOAA data show. And recent daily temperatures have shattered records that were set last spring.

SOURCE: CLIMATE REANALYZER/UNIV. OF MAINE; NOAA

Shown below is the "white-washed" version (i.e., without emotional descriptors of over-exaggerated significance) used at the UN IPCC Conference of the Parties (COP) deliberations. The COP has been described by pundits as "a group of environmental scientists all drinking their own bath water". The following diagram is shown to address that factual, but less than "angst-provoking" observation in an attempt to deceive the public by scientific "chartsmanship". The exaggerated scale on the left edge of the chart ranges from 19.5 degrees C to 21.5 degrees C which excites passionate scientists but which creates no factually proven "crisis" level of concern to the public for human safety, health, or comfort.

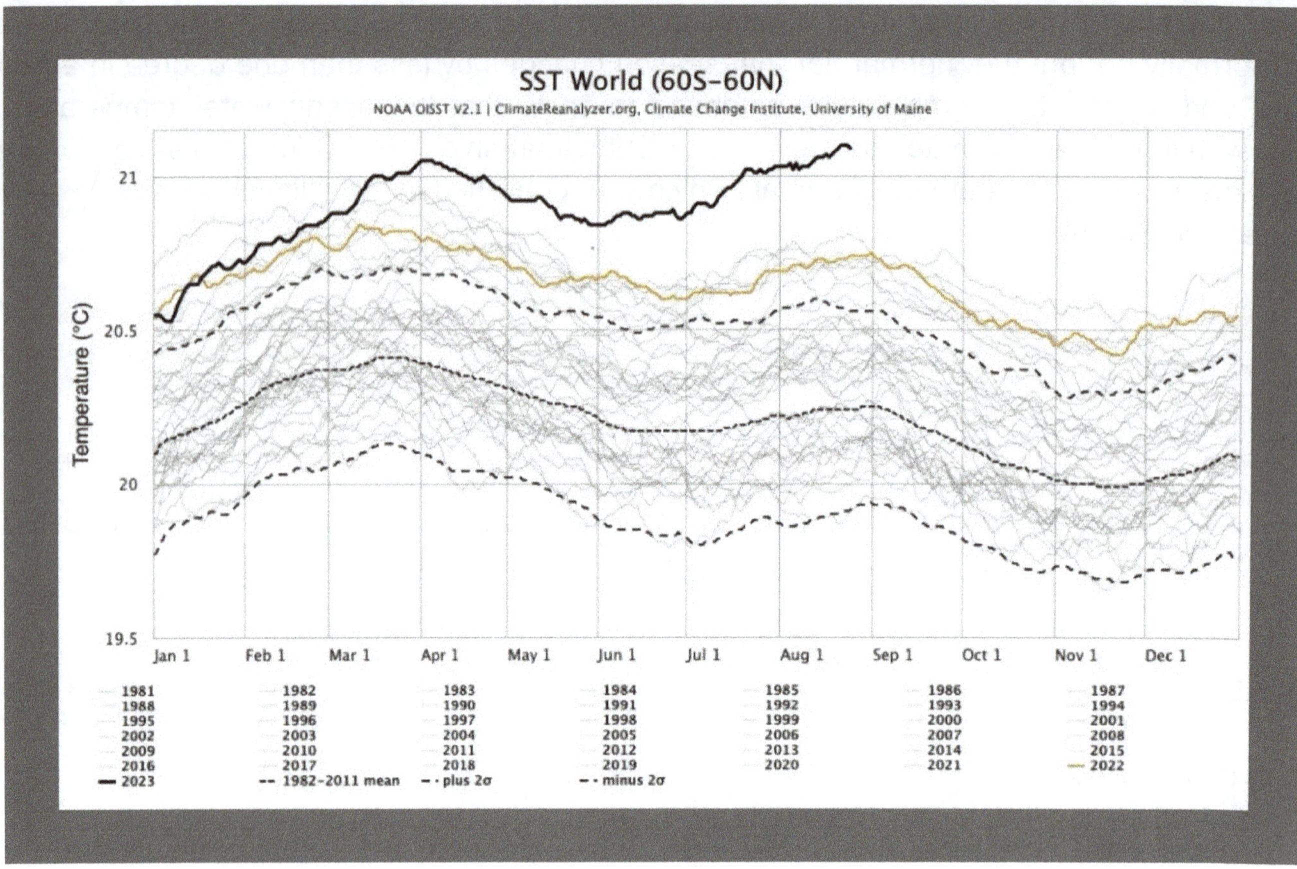

(Source: NOAA, Climate Change Institute, University of Maine)

There may be a "crisis" for some forms of ocean life, especially anemones on the Great Barrier Reef of Australia – which lose their color and "bleach white" – with less than a two degrees of water temperature change.

Readers should note that regardless of the "seasonal" variation in the World's oceans' waters temperature being less than 0.7 degree Centigrade/Celsius from the peak low and peak high temperature for each of the years 1981 to 2022, the water temperature "soaring" for partial year 2023 is apparently 0.2 degrees C above "normal". If there is reportedly an El Nino (warming ocean water temperature flowing from west to east toward the west coast of the USA) condition forming in the Pacific Ocean in 2023, that is not a Man-caused phenomenon caused by the burning of fossil fuels. El Nino originates near the Hawaiian Islands chain, all formed from volcanic eruptions eons ago, but underwater volcanoes are still forming islands that have not emerged above water surface level yet in that chain. That much underwater heating is most likely a contributor to, if not the main cause for, the El Nino warm current.

Of equal and perhaps greater significance as being a heat energy source warming the waters of the ocean floors is the fact there are possibly hundreds of thousands (if not millions) of steam vents continually flowing heated water at steam levels of temperature into the waters of ocean floors. Steam vents have been observed to be in the 15-20% of the ocean floor that has been explored (by non-alarmist oceanographic scientists such as Jacques Cousteau and family). Photographs of steam vents of two different categories (the color and content of their vented steam) are shown as follows:

(Credit to Unknown Photographers, National Geographic Society)

It has been estimated that over 80% of the ocean floor remains unexplored, and until the underwater steam vents factor is evaluated for its cause of the ocean waters being warmed – as they appear to be regionally – it is very scientifically invalid for environmental protection scientists to proclaim that the ocean levels rising are due to water temperature increases primarily traced to Man burning fossil fuels and creating the heat-trapping "greenhouse gas effect" in the atmosphere many thousands of feet distant from (i.e., far above) the ocean floors.

Supported by later information in this book, environmental scientists may be deluded with "average global temperature" on Earth's land surface area and in the "Average Sea Surface Temperature of the World". The fact remains that the surface temperature on land areas and on expansive ocean areas is dependent on the ocean currents, the wind currents, astrophysical factors, natural weather events locally, and climate change which has been repeatedly pointed out in this book to be regional in latitudinal zones on the planet. The "one-size-fits-all" singular form of "average global temperature" analysis by atmospheric scientists or ocean water scientists so desperate to develop a model for predicting the future of the planet's environment and have lost touch with the factual realities of the problem.

From another perspective of the concern of climate change and ocean water temperatures rising, when water warms it expands its volume and factually does cause rising ocean sea levels to become a human safety concern in the long-term to residences and places of business (especially large shipping ports) or cities presently only a few feet above present sea level. Because the total volume of water of the 70% of Earth's surface is so huge, a few degrees of temperature rise may represent only a very small percent of volume increase, but when the volume is so huge a small percent of temperature increase can expand a huge volume of water which has no elsewhere to expand other than up and over present coastal regions. A later chapter of this book addresses those long-term effects of ocean surface temperatures and rising levels and the need to start planning soon for infrastructure modifications in coastal areas threatened by ocean levels rising.

It is human nature to delay making a decision to relocate elsewhere until the cause for making the move is literally flooding their ankles (as shown in the next photograph):

(Credit to unknown photographer, Miami Dade County August 2023)

The importance of water on our home planet is taught in the grade schools by geography teachers who have possibly never been more than a hundred miles from home and to a beach.

Students in the pre-1950 period of attendance in public schools were required to read three literary classics – "The Iliad" by Homer, "The Rime of the Ancient Mariner" by Samuel Taylor Coleridge, and "Moby Dick" by Herman Melville. All three literary classics had the theme of ocean adventures and man interacting with his environment – real and imaginary. A rhyme from the Ancient Mariner comes to memory: "Water, water everywhere, but not a drop to drink. Water, water everywhere, and how the boards did shrink." (as the wood planks on the deck of the old sailing ships gradually dried out from the relentless solar rays (known then to the public as "sunrays"). It was not widely taught at that time in students' early development years that ocean water has a high saline content and the "not a drop to drink" lament was a subtle way of teaching that ocean water is too salty to drink. Only fresh water should be drunk – and in large quantity daily – to satisfy the human body needs for water for its internal organs to perform normally and sustain human health.

In some – but not all - high school science classes, it is taught that saline water such as that in the oceans can be made potable (safe to drink) by distilling the water and aerating the distilled water that no longer has a saline content to make it potable – for agricultural uses or for human consumption. But distillation in high school science and chemistry classes was somewhat down-played as to the "big picture" of the oceans evaporating water from their surface into the atmosphere, and then later returning the naturally distilled water vapor from the atmosphere back to Earth's surface as precipitation (rain, snow, hail, graupel, etc.). The distillation process in the conversion of some products into liquor with high alcoholic content – known better as "moonshine" and the equipment known as the "still" is the same principle but frowned on by some.

Gravity is the ever-present gravitational acceleration pulling every untethered or attached mass towards the center of the planet – such as precipitation in the forms of rain and snow. That precipitation provides the potable water vitally needed for agricultural growth, human health, and less vital as coolants in radiator systems cooling machinery systems for transportation vehicles and for residential and business facilities. The water that is converted into steam (the gaseous form of water vapor) for driving turbines to generate electricity must not have high salt-like impurities as that would corrode the boilers used in the water-to-steam conversion process. In the "big picture" most everything would have a short life if the corrosive salts had not been removed from the water taken into systems for use in human consumption, in transportation machinery, and in industrial processes.

The freshwater available to the human population on the planet is not distributed very usefully and is far less accessible than is the non-potable saline water in the oceans. An estimated 70% of the world's fresh water is in abundance (26.5 million cubic kilometers of potable water) on the Antarctic continent – but most is in frozen form on the Antarctic Ice Sheet. A smaller portion is on Antarctica's ice shelves and icebergs (which eventually are going to melt into the Southern Ocean and become non-potable) with a portion in glacial form above the Antarctic continent and in frozen lakes below the ice sheet. As this book was being written, there appeared an article in the news media that the world's largest known iceberg A23a – five times the size of New York City and as thick as the Empire State Building in New York had "calved" away from the ice shelf in West Antarctica. Its path is northerly in the Southern Ocean and with a possibility that winds and ocean currents may cause the gigantic iceberg to enter into southern Atlantic Ocean waters. The article shown next provides more information to the public on iceberg (from Germanic "eisberg" meaning mountain of ice) A23a (as it has been labeled for tracking purposes):

World's largest iceberg drifts toward South Atlantic

BY PAUL GODFREY
UPI.com

A gargantuan Antarctic iceberg five times the size of New York City and as thick as the Empire State Building is drifting away from the continental ice shelf toward the South Atlantic Ocean.

Designated A23a, the iceberg had remained in situ – grounded on the floor of the Weddell Sea since breaking from the Filchner Ice Shelf in 1986 - until earlier this year when it began slowly shifting northward.

It is believed A23a finally was able to break free from the seabed after decades of shrinkage.

Driven by wind and ocean currents, its passage recently has picked up pace, and satellite imagery now shows it is about to pass the northern tip of the Antarctic Peninsula.

Posting a map of the iceberg's movements over the past seven months on X, British Antarctic Survey mapping specialist Laura Gerrish said it was the world's largest iceberg.

"This giant berg calved in 1986 and has been grounded in the Weddell Sea since 1991. It started moving again last year, and now that A68 [another giant Antarctic iceberg] has broken up, it reclaims its title as the world's biggest iceberg right now at 1,544 square miles."

In 2020, there were concerns that A68 might run aground on the sub-Antarctic islands of South Georgia and kill marine life on the sea floor and subsequently affect the food supply for penguins and seabirds. Those fears were averted when the monster berg broke apart, which experts said was the likely fate of A23a.

Russia, which had built a research station on the Filchner Ice Shelf, was forced to mount a rescue mission in 1987 after the base found itself adrift on A23a. The base was quickly replaced with a new facility on nearby Cape Norvegia, partly using equipment and prefabricated structures salvaged from the old base.

(Source: United Press International)

Letting the human mind wander, it would be an amazing feat if that monster size iceberg – containing potable water (suitable for human consumption or agricultural irrigation) when thawed – could be tugged by a few sea-going tugs up to the coast of South America nearest the Amazon basin. The Amazon basin is currently suffering extreme drought conditions, and environmental scientists estimate that the ancient forest stores 123 billion metric tons of carbon; that is more than three times as much carbon as humans emitted in 2022. If scientists could devise an efficient way to transfer that much water in the A23a iceberg to irrigate the Amazon basin, the population of South America could benefit tremendously from that irrigation of the Amazon drought region. If nothing can be done with A23a, it will only melt into the Southern Ocean as it approaches warmer waters to the north and be wasted.

Since the A23a ice is already in a floating state, all melt-off will dissipate into the Southern Ocean and not raise the surface level of the Ocean – floating ice (including that on the opposite polar region in the Arctic Ocean) does not increase ocean surface levels when it melts.

A human-developed scientific process known as "reverse osmosis", basically to filter the unwanted salt from ocean water, has been applied in several large water purification applications throughout the world. However, that process has been found to be of limited use for only local human consumption – and not practical in the large scale for large agricultural needs (such as the Central Valley of California which suffers drought conditions for periods of forty years or more). The irony of that is that less than 80 miles away from the central valley of California lies an almost infinite amount of water – the Pacific Ocean. Granted, there is a mountain range – the Coastal Range – over which water would have to be pumped to make the potable water accessible for distribution onto the agricultural fields and groves of nut and fruit trees and cattle ranches, but solar energy collection (abundant solar energy is being deposited on the planet's surface in that region) could ease the energy needs for that pumping and distribution process. A few reverse osmosis facilities have been built on the California coast, but the process is not of a large enough scale to totally satisfy the high population of the state and/or the Central Valley agricultural needs.

Presently, the water for southern California comes from a few sources – a portion from the Colorado River east of the central valley as well as from lakes and reservoirs naturally formed or Man-constructed in southern California. A larger portion – serving the largest city in California, Los Angeles, with a population of 18.64 million per the 2020 Census - comes via the Los Angeles Aqueduct system built by civil engineer William Mulholland (1855 - 1935). The use of aqueducts to supply a city with water is documented in history, and perhaps the most famous is the aqueduct system in ancient Rome.

The Los Angeles Aqueduct system originates in the Owens Valley (above Yosemite National Park) in the northern part of the state and flow (north to south – i.e., "downhill" by gravity) in the Aqueduct relies on the melt-off of the snowpack in the Northern California mountain region. Where there are some long hills preventing water in the Aqueduct to flow over by gravity flow alone, there are pumping stations to lift the water over the hill and back into the downhill portion of the Aqueduct. Again, every region of the United States must develop a unique water system supply for its regional population. It seems that every small town in the United States has a water tank atop a steel tower (much like a gigantic golf ball on a tee – or as likened by the Author's children on driving tours to be "a giant spider with a cap") boasting the name of the town on its side. It is the responsibility of officials for any region where there is a large population to assure the public safety and health from whatever type catastrophe has the moderate to high possibility of happening.

While researching the topic area of desalination of ocean water, it was a pleasant surprise to become better informed on how much and what types of desalination are already in operation around the planet. An interesting illustration is shown next:

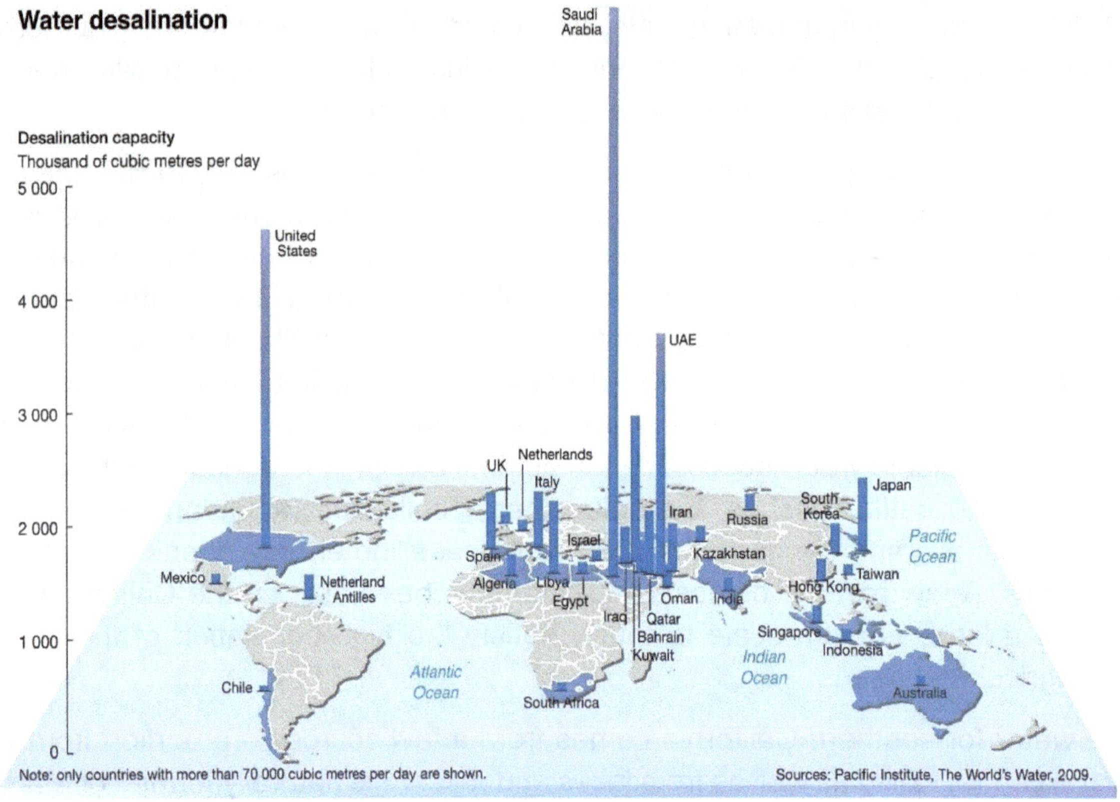

(Source: Pacific Institute, 2009)

The primary users of desalination are the countries in the Middle East countries (Saudi Arabia, United Arab Emirates, Bahrain, Dubai, etc.) using water from the Arabian Sea. The largest desalination facility using the reverse osmosis process of desalination is in Dubai: while the largest desalination facility using the distillation method is in Kuwait.

(Credit to unknown photographer in Dubai)

Desalination facilities in the United States are becoming more popular. Several exist on the west coast and most in California:

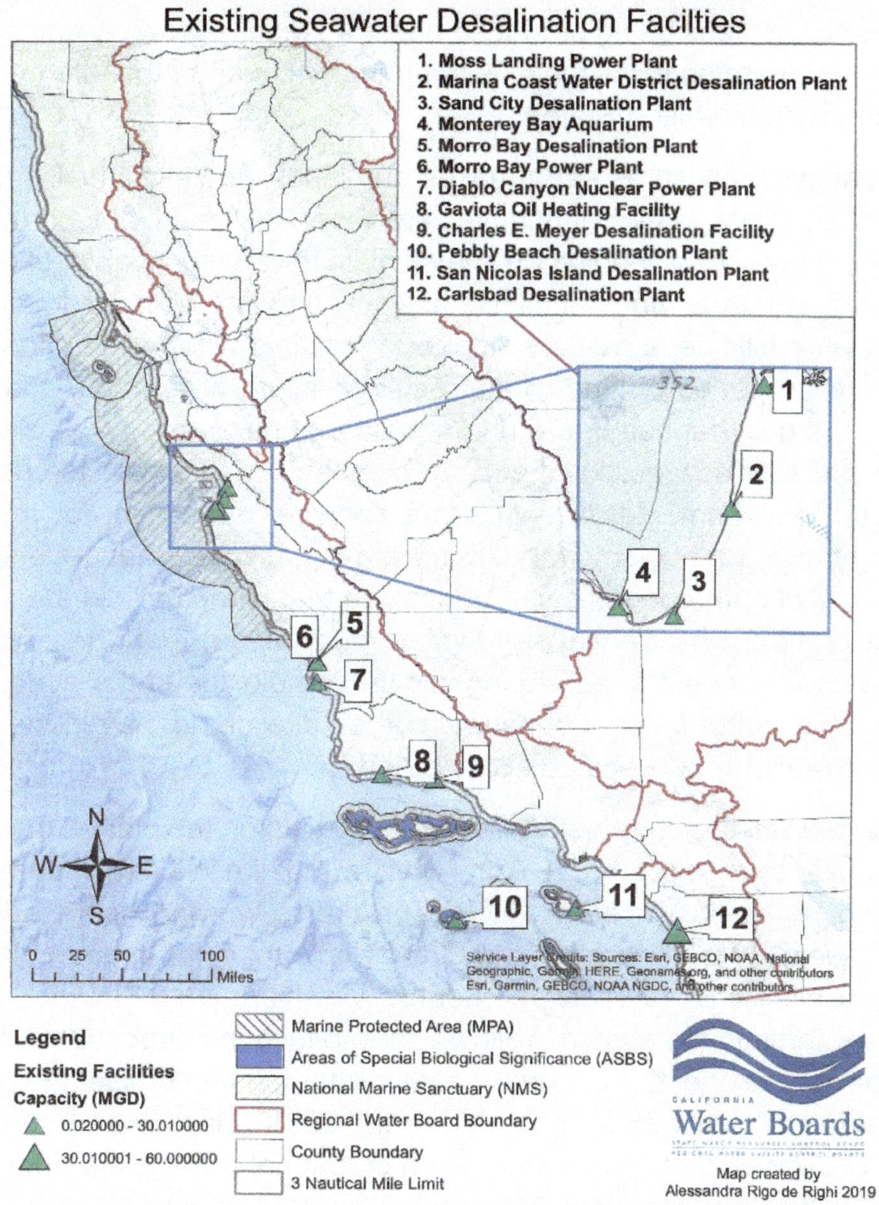

(Source: Water Boards of California)

Almost all known desalination facilities (in the USA and most of the planet) are used for water supply to the local human population and little for the agricultural supply. However, with the world population growing in number exponentially coupled with there being more drought areas, the need for larger volumes of desalinated water than currently developed (and using the slow reverse osmosis process) must increase use of the distillation process of desalination technology powered perhaps by solar energy. Two of the 24 weighted trade studies in the final part of this book assess and prioritize the desalination technologies.

The climate is changing in southern California where it once was boasted: “It never rains in California”. In recent years that boast is not heard very often as it has alternated between too much water, causing flooding in areas where no flood control channels were constructed and in turn causing mud slides that devastate residences in the hills for which the entertainment industry has drawn residents from other arid states, and too little water. The central California region which grows the agricultural products for much of the “48 lower states” is at the same time experiencing drought conditions.

Environmental scientists point the finger of blame at the Man-made "greenhouse gasses effect" but the amount of the greenhouse gasses being produced are in a steadily growing manner and fail to confirm that the "scourge of Man" (burning fossil fuels) is the primary cause of the alternating "too much or too little' water in the southwestern United States.

Short term geophysical events such as earthquakes and landslides are usually experienced with little or no warning and are over in a matter of minutes. However, the residual long-term after-effects may dwell for a generation. Longer term flooding on a level plain is very survivable, but flooding on a sloped land is highly dangerous to human and animal life as well as to any Man-made structure in the path of a flood rampaging down a hillside. It matters little to the victims whether the flooding was caused by too much rainfall in a short period of time or the result of failure of a Man-made structure such as a dam, life is much like the bell that has been rung and cannot be un-rung – death cannot be reversed either, no matter the cause or how quickly it occurs. The news media has little respect for ethical and complete reporting, but environmental scientists are showing disrespect for the honorable science profession by pointing the finger of blame for global warming, climate change, and the ocean levels to rise on the small amount of anthropogenic carbon dioxide that is factually in the atmosphere. There is over 100 times more of it from natural causes (water vapor alone) than there is from anthropogenic causes. How can the lower amount of anthropogenic carbon dioxide in the atmosphere be the cause of global warming and the higher amount of natural carbon dioxide not be? In the laws of chemistry, is carbon dioxide CO_2 no matter how that CO_2 was created?

Returning to how vast the volume of free-standing waters exists on the planet, the Author was given an "all-expenses-paid" cruise courtesy of the US Army who invited him to a beach party in the quaint harbor town of Inchon, South Korea, in the early-1950s. The cruise was on board a "luxurious" troopship (USMS Gen. William Weigel) for 15 days on the Great Circle (circular in geometry but not so great in ride quality) route in the Pacific Ocean between Seattle, Washington, to Sasebo, Japan, and then on a smaller landing craft up through the eastern Yellow Sea into Inchon Harbor. On a "buy one, get another one free" cruise (again courtesy of the US Army) a year after the first cruise, the fortunate survivors of the Korean War boarded a "luxury liner" troopship (e.g., USMS Gen. John Pope) and were returned across the center of the Pacific Ocean (passing south of, but not making port in, Hawaii) back through Puget Sound and into Seattle from whence the round-trip started over a year earlier. The return leg across the Pacific Ocean for the Author on that "Great Circle" route took 16 days.

It was those two cruises (totaling a full month "at sea") that gave the Author a first-hand sight of just how enormous the Pacific Ocean was/is. Knowing that the water is a few miles deep and exists from horizon to horizon and beyond is mind-boggling. The tabular statistics showing the immense quantities of water in and the surface areas of all the oceans and seas of the planet were shown at the start of this chapter.

Later during the Author's professional career, the opportunity arose to make 18 round-trip flights in commercial airliners over the Atlantic Ocean between the United States and western Europe – mostly to England, Germany and Switzerland. During their retirement years, the Author and his wife also took three cruises – aboard true luxury cruise ships. The first was from Florida through the Panama Canal and to California; the second from the French Cote d'Azur through the full length of the Mediterranean Sea, through the Tyrolean Sea and Aegean Sea, and through the Adriatic Sea back to Venice, Italy (which was shocking to see how much of Venice is already largely underwater). The third and last cruise was from Seattle up through the Inland Passage (and through Glacier Bay) of the Alaskan Coast to Skagway, Alaska and return to Seattle in the US Pacific northwest. Not only were those ocean and sea voyages very impressive as to the amount of water covering the planet, but it was the educational and visual experiences ashore in the many different countries and seeing how the local human residents adapted to their environment that differs in each country - especially depending on their latitude. (The Author also won a paper airplane contest aboard the ship in the Aegean Sea, but that experience does not lend much credence to the Author's capability for understanding the rules of aerodynamics.)

The Author and his family have memories and a wealth of interesting and learning experiences of having resided in or traveled by land-vehicle or train to/through every one of the fifty United States (the contiguous 48 and Alaska – fishing for King salmon and halibut - and snorkeling in warm Hawaiian waters). Those travels took the family on visits to all national and many state parks, to historical and natural geology museums and Presidential libraries, to military memorial cemeteries, or to visit with family or close friends. Included in those "bragging rights" is the fact that family homes were bought, or apartments leased in eight states (New York, North Carolina, Georgia, Texas, Ohio, California, Arizona, and Idaho in which the Author's family resided for more than two years (the longest in California for thirty-eight years). Personal observations of the different weather events and climate changes in different regions of the world and in the different states (all 50) of the USA have been found to be highly useful to the Author when interpreting the effects of those events and changes.

To see a photograph of a tornado is much less impressive than being physically in the fringes of one. Living in the different environments (weather, climate, and seasonal changes) in the different states and different countries in the Northern Hemisphere (no travel was made in the Southern hemisphere – although there are several family members in Australia, and New Zealand) as well as learning what the local public throughout the planet regards as being most important to their safety, health, and comfort equals the research done through books and papers and news articles. Those many experiences gave a wealth of background observations made as to the prioritized importance of the many factors that the trifecta environment problem (surface temperature extremes, climatic extremes, and ocean life in different temperature categories) has on the human population of many different regions of the planet (albeit all in the Northern hemisphere).

Understanding the factors in a multiple-factor, complex, database of information is made easier if/when there is close personal connection with most of the factors in that database. A major observation this Author has come to realize (from personal observation and with intensive research in the various databases available for scientific analysis) that the Earth's surface temperature is changing very much but more on a regional basis – severely in some regions and no significant change in fortunate others. Where the regions of elevated surface temperatures above the "normal average temperature - referred to as being temperature anomalies - for that region are located appears to have at least one boundary with a large mass of water.

The chart shown next shows how the different regions of the planet are experiencing different surface temperature anomalies (a temperature "anomaly" is the difference from "normal" (average annual) temperature for that region):

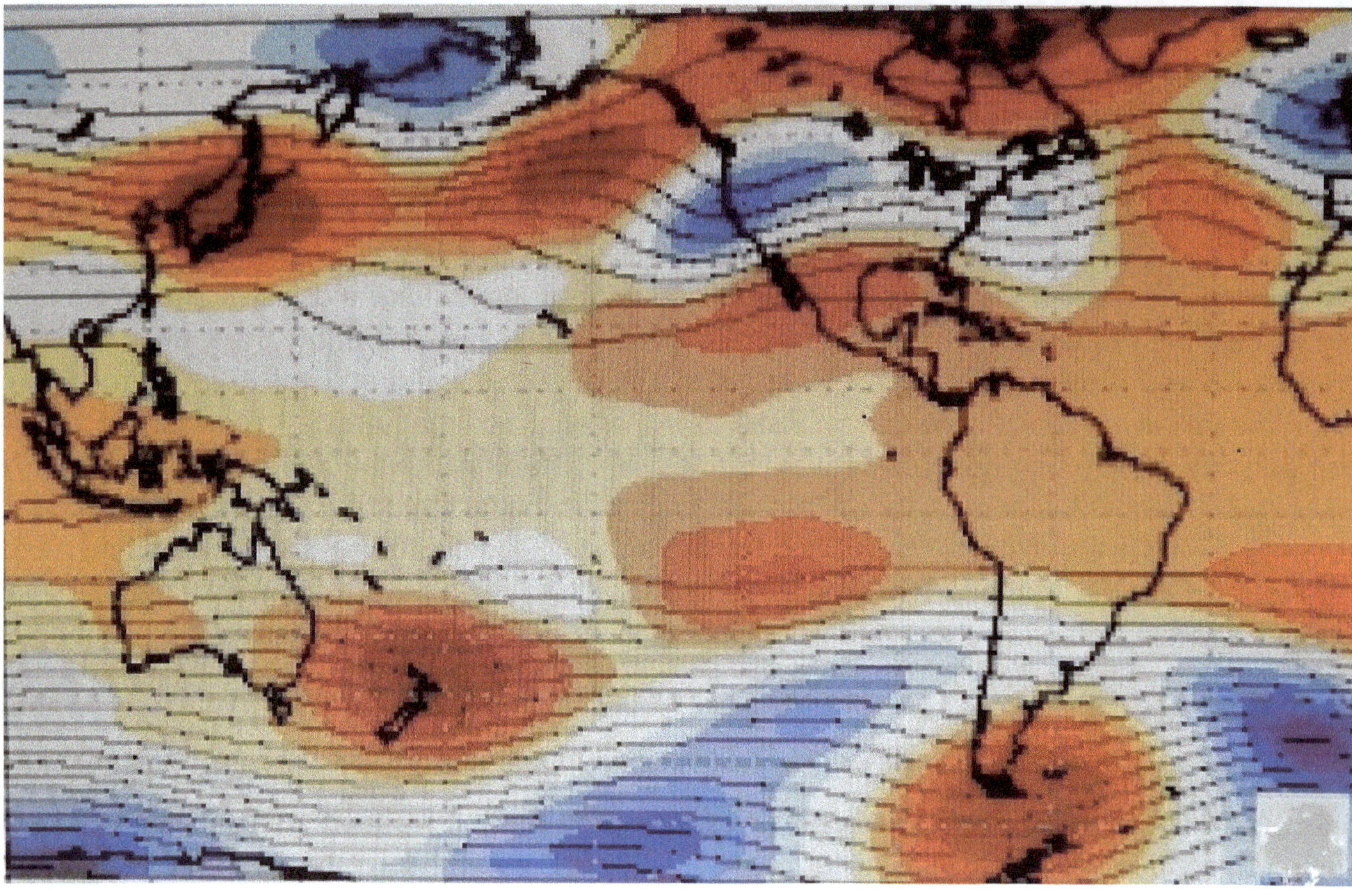

(Source: NASA)

The dark red and light red are above "normal", the white is about "normal", and the grey and shades of blue are below "normal" average annual temperature for that region.

If the environmental scientists can provide a scientifically valid explanation for public understanding as to how the "greenhouse gas effect" – which purportedly changes <u>global</u> average temperature - develops "pockets of anomalies" such as those shown above, it would be most helpful for the public to understand why there is a "crisis".

These forms of temperature observations on Earth's surface (land and water) reflects no scientific reason to believe that Man burning fossil fuels and creating "greenhouse gas effects" in industrial regions but not in non-industrial regions bears any scientific correlation with the measured temperature variations as measured by NASA satellites shown above. In the case of the temperature increasing at the Great Barrier Reef (on the east coast off Australia in the Coral Sea), it is seen that the reef is on the edge of the "temperature dome" centered almost directly over New Zealand. New Zealand's human population cannot be accused of generating that much "greenhouse gas" to warrant industrial "greenhouse gasses" generated in the upper Midwest of the USA being blamed by NOAA environmental scientists as the dominant cause for the bleaching of anemones on the Australian reefs.

A more plausible explanation, as shown earlier near the beginning of this book, is that the ocean currents are very warm in the Southern Pacific and affect the temperature of the coral reef which is not very deep below the Equatorial heating of the ocean water:

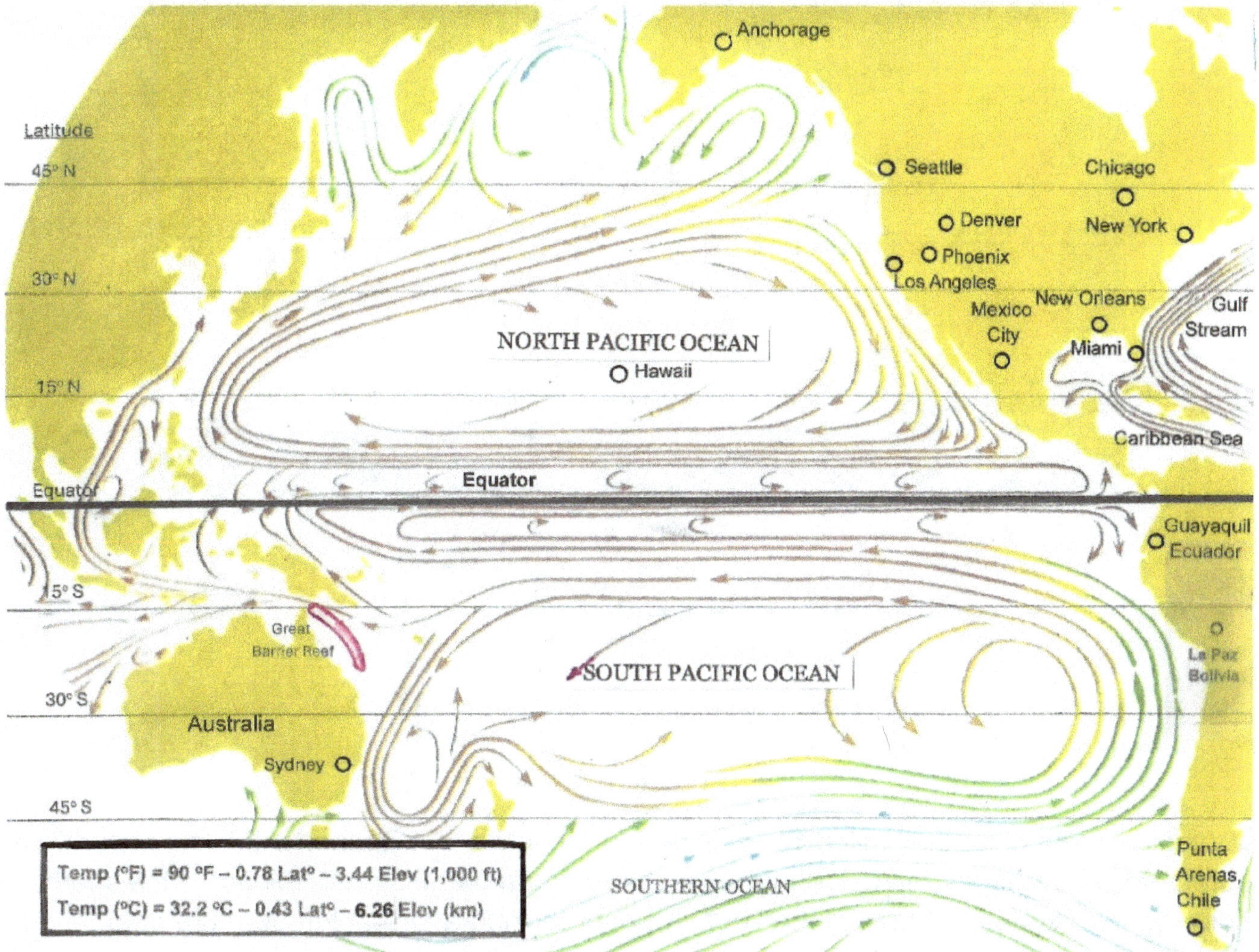

(Source: Rand McNally Atlas of the World, with annotation by Author)

The mathematical expression for temperature in both the Fahrenheit and Celsius scales may be helpful in estimating the average annual temperature of cities at different latitude due to latitude and elevation. Any effect from warm or cold ocean currents on that temperature estimate are guided by maps of the ocean currents being warmer or colder. The Great Barrier Reef bleaching is due to the very warm Equatorial current which is getting warmer – but not from the "greenhouse gas effect".

Readers are reminded that the illustration on the front cover of this book represents the "average" temperature (collected by satellites not the land-based "average global temperature database from questionable averaging of statistics from a non-uniformly distributed system of meteorological stations located in predominantly the northern hemisphere) against which the temperature anomalies shown above are based. Because of its extreme importance to understanding why this book continues to urge addressing the trifecta problem (global warming, climate change, and ocean levels rising) in a "big picture" perspective, the cover illustration is repeated here for Readers' convenience – and with a color scale shown below the illustration such that numerical temperature variations can be identified for their level of importance to human safety, health, and comfort regionally; i.e., as latitude varies:

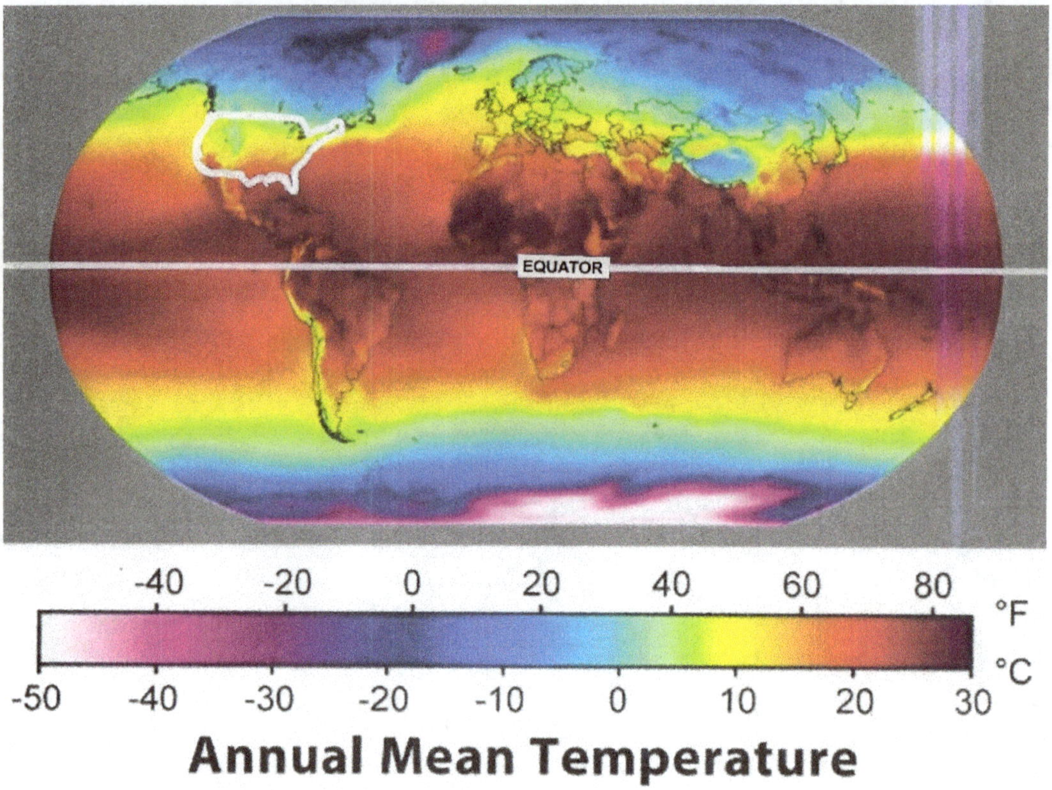

(Source: NASA)

Meteorological records for the United States region are consistent with the NASA "big picture" satellite measurements showing the red color (high average temperatures) is predominant in the southeastern and southwestern states comprising the USA. A closer-looking average temperature database for the "lower 48" states of the USA is shown as follows:

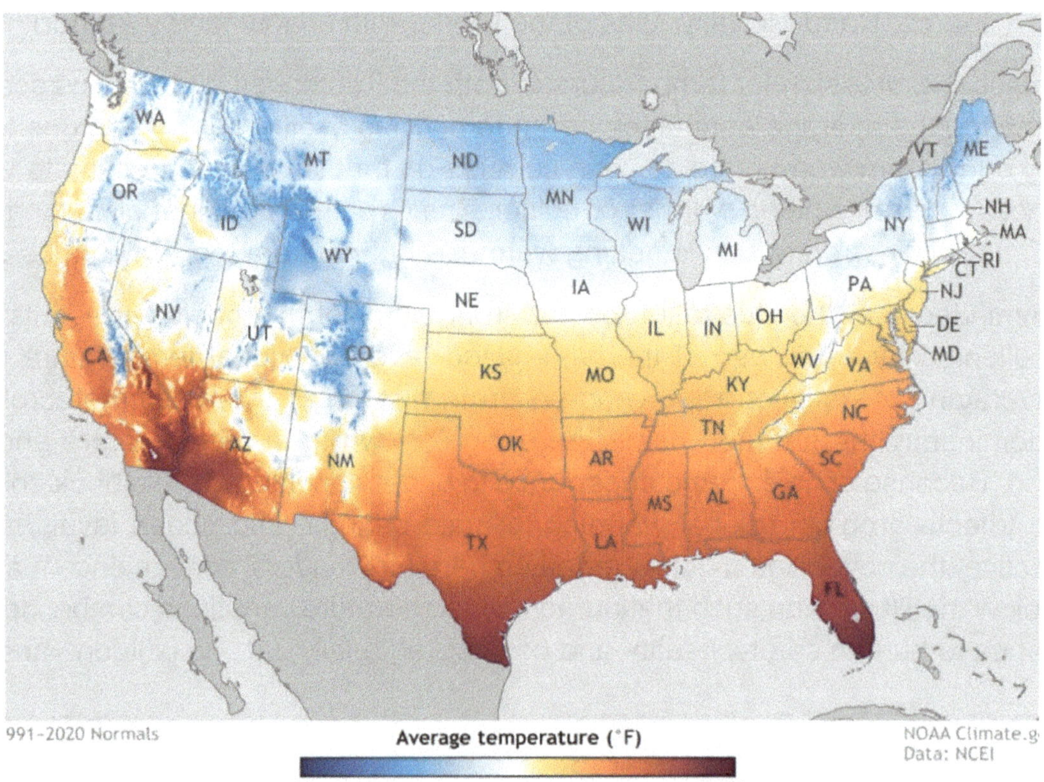

(Source: NOAA, National Weather Service)

The warming effect of the "Gulf Stream" (which originates in the Caribbean Sea) in the Atlantic Ocean on the eastern and southeastern states' coasts, the Gulf of Mexico on the south-central states, and the arid deserts of the southwestern states causes the surface temperatures to be the highest in those regions in deep red compared to the cooler regions of the New England region and the upper central states in light blue. The temperature difference from deep red to light blue on the color scale is seen to be on the order of 50 to 60 degrees Fahrenheit. The temperature in Miami, Florida may be 90 degrees Fahrenheit while at the same time the temperature in Milwaukee, Wisconsin, is at 30 degrees Fahrenheit). The mountain states in the west-central states (because of their high elevation) are the coolest of the states (except for the state of Alaska which is very much northwest of the "lower 48" states and at a much higher latitude between Ketchikan (55.3 degrees N) and Point Barrow (71.4 degrees N).

Within a region the size of the United States may also be characterized by identifying the extremes of its most southern latitude and its most northern latitude. For that purpose, the latitude of the southernmost point of latitude of the contiguous ("lower 48") United States is that of Key West, Florida at 24.5 degrees North (and longitude -81.8 degrees West). The latitude of the northernmost point of latitude (with a significant population density) of the contiguous United States is that of Bellingham, Washington at 48.8 degrees North (and longitude -122.5 degrees West).

Larger regions that have one or more boundaries with a major ocean, sea, gulf, or bay may also have an inland lake or series of lakes that total to be a highly significant influence on surrounding regions. The United States is one such region that is blessed to have three of its boundaries with ocean and gulf interfaces but has the series of lakes known as the Great Lakes. These five large lakes are the result of the last Little Ice Age when glacial ice a mile thick covered the northern half of the "lower 48" states. The glacial ice carved huge depressions in the surface with their immense power when they "flow" and produce kinetic energy of immense proportions. (The familiar scientific formula in Physics 101 for kinetic energy (K.E.) is: K.E. = ½ mass x Velocity^2.

Although the land speed V of the one-mile thick and hundreds of miles wide glacial ice is very slow, glacial ice mass m is so huge that the K.E. is equally huge.)

As the surface temperature of the region increased (due to astrophysical reasons) the glacial ice thawed and left the massive amounts of water in the depressions carved and the five Great Lakes were formed about 4,000 years before the present. The illustration shown next depicts the outlines and locations of the Great Lakes region of the USA:

(Source: U.S. National Geodetic Survey)

The statistics of the five Great Lakes and the amount of surface area in the United States they cover is as impressive when shown tabulated as the earlier tabulation of the major oceans and seas were shown earlier:

	Lake Erie	Lake Huron	Lake Michigan	Lake Ontario	Lake Superior
Surface area	25,700 km^2 (9,910 sq mi)	60,000 km^2 (23,000 sq mi)	58,000 km^2 (22,300 sq mi)	19,000 km^2 (7,340 sq mi)	82,000 km^2 (31,700 sq mi)
Water volume	480 km^3 (116 cu mi)	3,500 km^3 (850 cu mi)	4,900 km^3 (1,180 cu mi)	1,640 km^3 (393 cu mi)	12,000 km^3 (2,900 cu mi)
Elevation	174 m (571 ft)	176 m (577 ft)	176 m (577 ft)	75 m (246 ft)	182.9 m (600.0 ft)
Average depth	19 m (62 ft)	59 m (195 ft)	85 m (279 ft)	86 m (283 ft)	147 m (483 ft)
Maximum depth	64 m (210 ft)	228 m (748 ft)	282 m (925 ft)	245 m (804 ft)	406 m (1,333 ft)

Note: The total water surface area of the five Great Lakes is 244,700 sq kilometers (94,250 sq miles)

(Source: National Geological Survey database; tabulation by Author)

The lakes were created from the melt-off of the glaciers from the last "Little Ice Age" and were formed in shape by the advancing and retreating glaciers grinding depressions in the surface of Earth's surface in which the melting glaciers (fresh water) remained in liquid state. Pollution of the Great Lakes by human and industrial processes and inconsiderate practices is discussed in later chapters as being more of an issue than is the "greenhouse gas" crisis contrived by environmental protection scientists. <u>Far more attention and concern should be given to Man-made pollution than to Man-made global warming.</u>

When the states of Alaska and Hawaii) were entered into the United States as the 49th and 50th states – Alaska in January 1959 and Hawaii in August 1959 – the total United States land and water surface area was increased and became as shown next:

that generate the largest amount of "greenhouse gasses" in the United States - are not the regions that are exhibiting the highest temperature anomalies in the United States. If one also considers that the vast majority of the humans in the United States and their unquenchable consumption of electrical energy consumption are in the most densely populated regions of the USA – i.e., in the northeastern (New England) states and in California – neither region having the extensive amount of "Industrial Revolution" as the Great Lakes and southeastern regions, how do environmental scientists rationalize that it is the carbon emissions in the atmosphere caused by the Industrial Revolution processes and the human demands for energy that are the dominant cause for global warming?

Two other "big picture" questions that arise, if environmental scientists are blaming industry and humanity for causing climate changes that are feeding back to cause the frequency and intensity of both tornadoes and hurricanes to be increasing tremendously;

1) Why is it that tornadoes are formed in the southeastern Gulf states and proceed (on their seasonal "tornado alley" path from that low industrial activity and not extensive population activity and energy demands region northeasterly into the increased industrial activities and very high density populated regions of the USA?

2) Why is it that tropical depressions originate in the western North African desert regions, where there is almost zero industrial activity and no population concentrations with high demands for electrical energy, become of higher intensity as they pass over waters where again there is no industrial activity and population of humans, and then take Interstate A (west of the Florida peninsula) or Interstate B (east of the Florida peninsula) when turning northeasterly to batter the southeastern region of the United States where there again is almost no industrial activity and very low population density with low energy demands?

Neither the Tornado Reporting Center or the National Hurricane Tracking Center have a database being described openly and clearly to the public to explain those illogical circumstances. Instead, there is a continuing stream of news media articles that point the finger of blame on the public for being responsible for the tornadoes and/or hurricanes that reportedly are becoming more frequent and of higher severity due to climate change. As a point of information, hurricanes, typhoons, and cyclones have different names (depending on the latitudinal region where they are spawned) but have the same general adverse characteristics of causing flooding, property damage, and loss of human lives. For Readers' information, the following diagram indicates the regions where hurricanes, typhoons, and cyclones are formed and the typical variation in paths taken after formation:

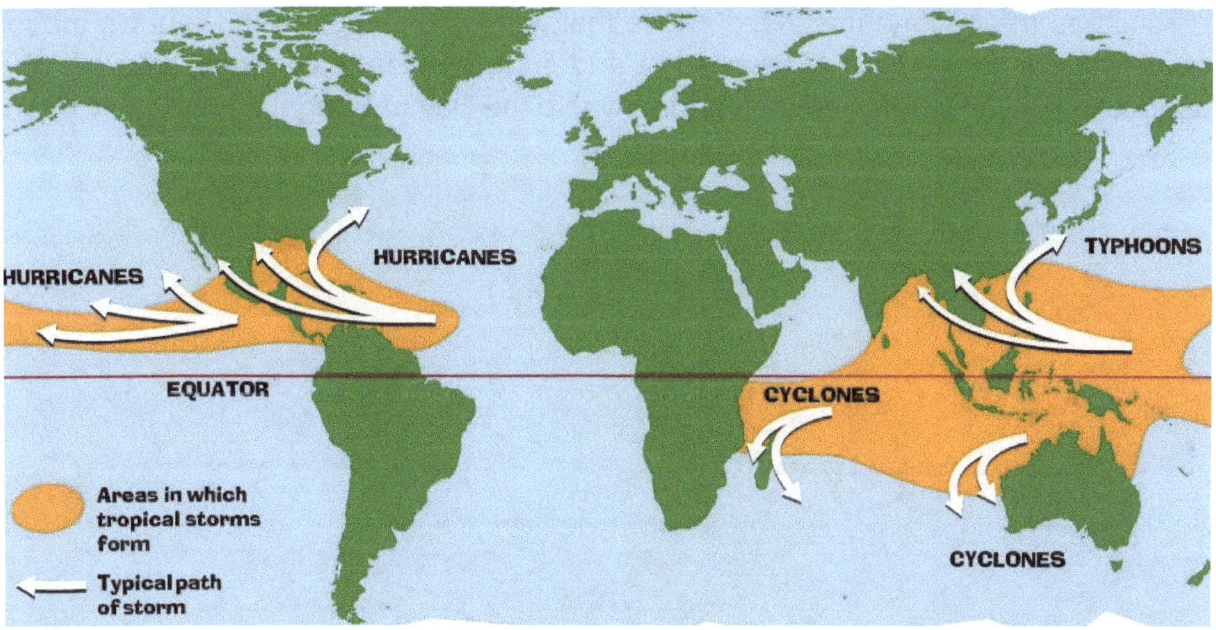

(Source: NOAA)

There has been a recent update to the above chart to add tropical storms that form in the Indian Ocean in the region of East Africa. Those tropical storms that form in that region have an unusual – and unexpected flow as cyclones moving northeasterly and entering the Bay of Bengal. If entering the Bay of Bengal, those cyclones turn northerly and impact the country of Bangladesh which has the highest density of human population on the planet. The similarity of the Bangladesh area (at the delta region of the Ganges River into the Bay of Bengal, and the situation of the city of New Orleans on the delta of the Mississippi River in Louisiana where the delta is formed on the Gulf of Mexico is uncanny in that similarity.

A view of the Category 5 unnamed cyclone that made landfall on Bangladesh on April 29, 1991, was track, size, and intensity-monitored by NASA satellite imagery and provided early warnings to Bangladesh:

(Source: NASA)

A chart of the currents in the Indian Ocean where cyclones (hurricanes but with a different name) are spawned – such as the cyclone described above that impacted Bangladesh – shows the location of those spawning grounds (north of Madagascar) and suggests the typical path that cyclones in the Indian Ocean take (northeasterly and then northerly into the Bay of Bengal).

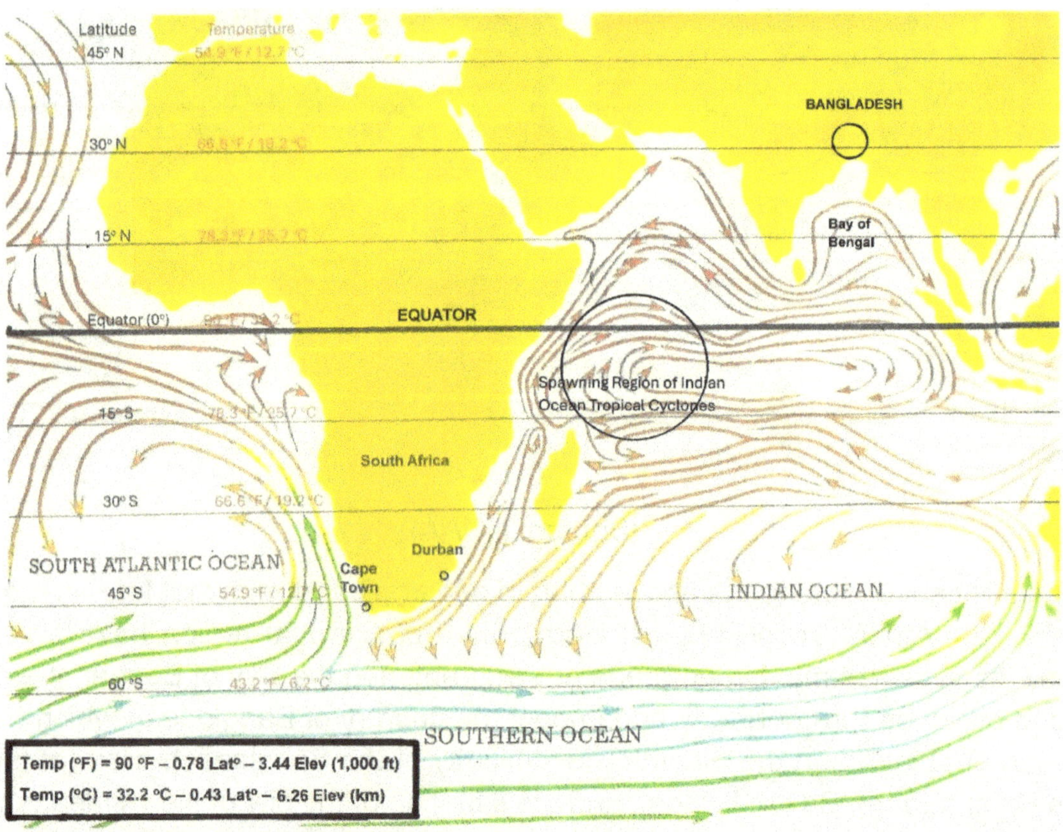

(Source: Rand McNally Atlas of the World, with annotations by the Author)

This event was the largest natural catastrophe in the world in the twentieth century. There were over 135,000 fatalities known to have occurred and over 10,000 people whose remains were never found. Over 10 million people were left homeless, and over 1 million cattle (chiefly Brahmans) were destroyed by drowning. The cyclone destroyed most of the agricultural crops in the region and the many millions of survivors faced a starvation crisis. To understand why the losses were so extensive, two maps of Bangladesh reveal some of the reasons (shown next:)

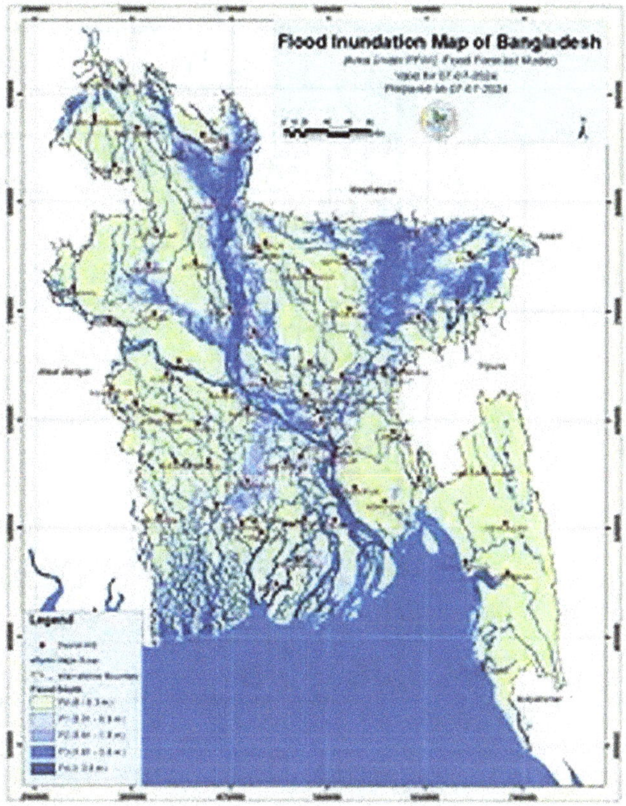

(Source: WorldAtlas.com)

The left map shows the topography (low elevation, hills, mountains) and the location of the major cities comprising Bangladesh. The total population of Bangladesh is 172, 954,319 in a 130,170 square kilometers (50,218 square miles) land area heavily laced with waterways (as seen in the left illustration above). The large Ganges River from India and Pakistan is merged by the Meghna River from Burma and there are literally dozens of smaller rivers that carry the drainage water of the mountains of Nepal to the North into the Bay of Bengal. That topography and high population density (1,329 persons per square kilometer – 3,444 per square mile) the perfect ingredients for cyclone disasters such as the one that struck Bangladesh on April 29, 1991.

Compounding the high population density on a land area that is extremely prone to flooding from heavy rainfall is the low elevation of its major cities. A tabulation of the largest ten cities of Bangladesh having high population density but low elevation is shown as follows:

Low elevations of Cities in Bangladesh

City	Elevation	Population
	meters (ft)	(As of 2023)
Dhaka (Capital)	4 (13)	23,935,652
Chittagong	11 (36)	5,379,660
Khulna	6 (20)	2,613,385
Rajshahi	21 (69)	2,909,622
Sylhet	35 (115)	999,000
Barisal	3 (30)	533,000
Comilla	13 (43)	1,364,172
Narayanganj	5 (16)	967,951
Mymensingh	14 (46)	576,927
Jessore	9 (30)	298,000

Note: Dhaka has the highest density of World cities at 23,234 people per sqare kilometer. Bangladesh has a total population (2024) of 172,954,319.

(Source: Tabulation by Author from United Nations Population Division database)

The capital city of Bangladesh is Dhaka and is seen to have a population of almost 24 million people residing on land at an elevation of 4 meters or 13 feet. Dhaka has 4 million people greater than are in New York City which is the largest city in the United States with a population of 19,034,000 in 2024. Of all cities that will be hit the hardest in the world by ocean surface level rising, the first largest world city to be impacted the most will be Dhaka if the ocean surface level exceeds a rise of 9-10 feet and the region were to be impacted by a hurricane and surge of 5 feet in water height.

An excellent series of articles by BBC covering the devastation of Hurricane Helene (in October 2024) that took an unusual track from the Gulf of Mexico north through the Pensacola region of Florida as a Category 4 on the Saffir-Simpson scale. The Saffir-Simpson scale is described as follows:

Saffir-Simpson hurricane scale explained

Category 1
Winds 74-95 mph (119-153 km/h)
Some damage and power cuts

Category 2
Winds 96-110 mph (154-177 km/h)
Extensive damage

Category 3
Winds 111-129 mph (178-208 km/h)
Well-built homes suffer major damage

Category 4
Winds 130-156 mph (209-251 km/h)
Severe damage to well-built homes

Category 5
Winds 157+ mph (252+ km/h)
Many buildings destroyed, major roads cut off

BBC

(Source: BBC News Service and NOAA)

From the perspective of human safety, the 2-3 days forewarning given by the National Hurricane Tracking Center of a Category X level (based on its wind velocity) is of immense value in saving human lives. Homeowners who choose to purchase or build a home in a "hurricane-alley" location should know to what extent of hurricane level has their home been built to withstand. Every community and depending on where the community is located – near or in a "hurricane alley", or "tornado alley", or known "flood zone" should have a minimum construction safety code to guide existing or potential residents planning to buy/build their home in that community capable of withstanding the natural wind forces and flooding potential known for that location. It may sound cruel or bureaucratic, but residents living in a trailer home not anchored securely should not be allowed to take the risk of knowingly placing themselves "in harm's way".

An accompanying graphic indicated the major ingredients of hurricanes, cyclones and typhoons), all in the category of tropical storms that have intensified to different levels of the Saffir-Simpson scale, to help the public understand why hurricanes and typhoons form, and that graphic by BBC/NOAA is also repeated next:

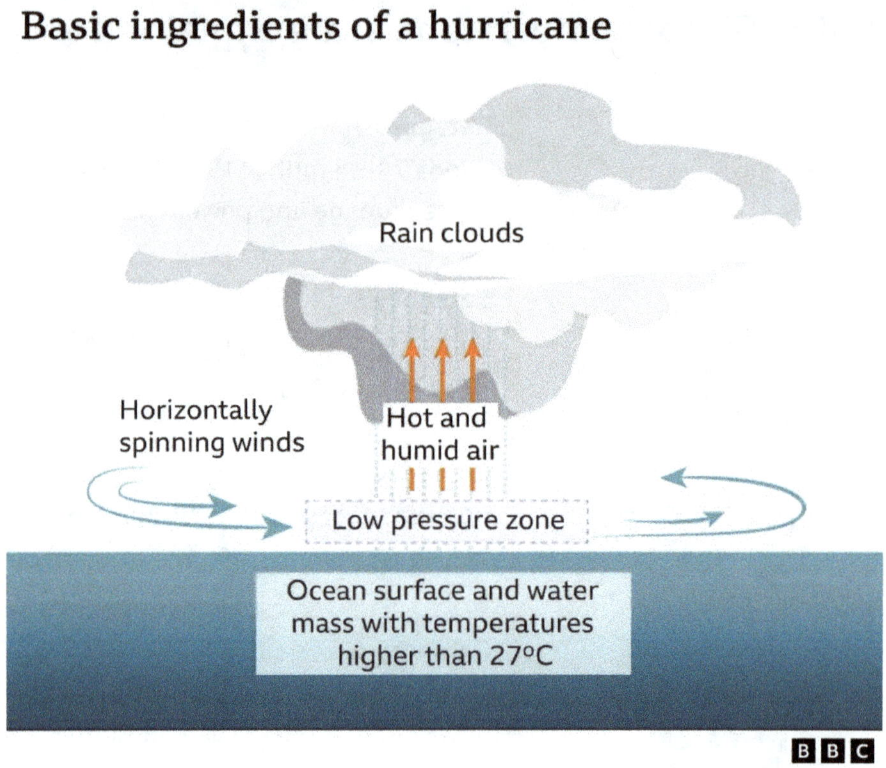

(Source: NOAA)

The temperature of 27 °Centigrade or Celsius is the equivalent of 80.6 °Fahrenheit. Those high temperatures are associated with the ocean waters in the Equatorial zone.

The ocean currents have a generally typical circulatory current within land coastal boundaries and move the heated waters as if within a large basin. The more regular prevailing wind currents in the atmosphere move unrestricted above the surface, regardless of being water or land, and in more regularly defined patterns:

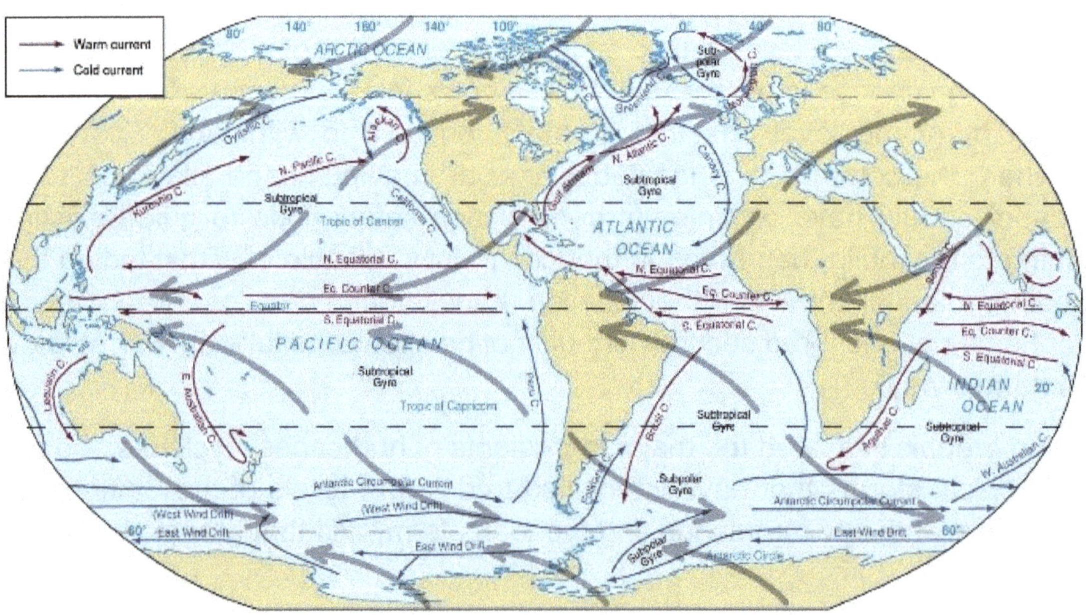

(Source: NOAA)

There is far more (by a ratio of 3% water vapor to 0.04% carbon dioxide = 75:1) heat-trapping content in the atmosphere – causing water vapor to be a much more causative variable on "average global surface temperature" than Man-made greenhouse gasses as Dr. S. Fred Singer revealed in his atmospheric chemistry series "A Closer Look at the Numbers". In that analysis, Dr. Singer compares the "settled science" view of "greenhouse gasses" (which excludes consideration of water vapor) with the ethical scientists' view of all heat-trapping components (by any name) and which includes water vapor:

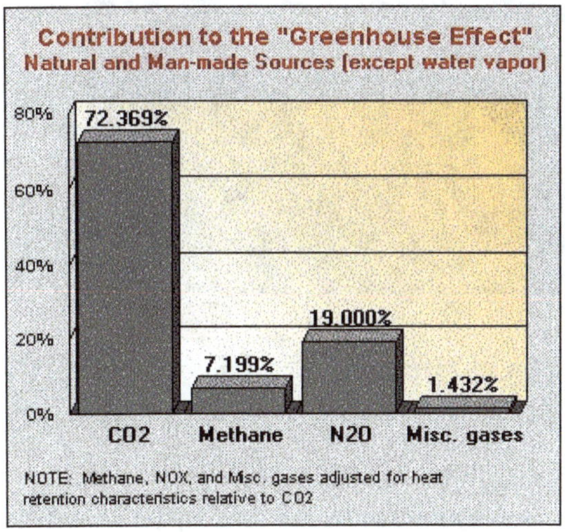

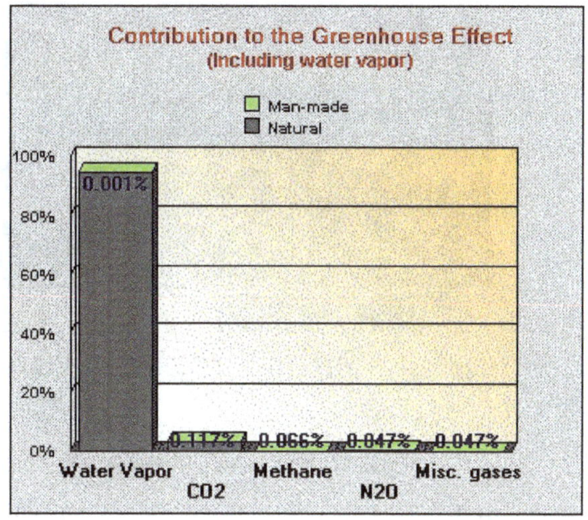

(Source: Dr. S. Fred Singer, University of Virginia, "A Closer Look at the Numbers")

This fact vs. fallacy will be shown more than once throughout this book because of its high significance to being factually accurate and revealing the fallacy of the "settled science" faction who disregard water vapor (as a part of the process of "when everything else is eliminated") despite water vapor being a more dominant effect (by a factor of 75:1) on global warming and climate change. The "settled science" set of Man-made "greenhouse gasses" without water vapor is a deliberate fallacy – not a hoax or lie – it is a scientific fallacy created for scientific convenience in modeling the atmosphere.

The combination of the major ingredients (independent and dependent variables) that govern how tropical depressions form into tropical storms (hurricanes, typhoons, and cyclones form), combined with the prevailing local (and not "averaged") ocean current temperatures and ocean currents irregular curvilinear flow paths, and winds above those ocean surfaces all mixing/combining to cause them to "grow" into hurricanes is a complex set of irregularly varying variables which are yet to be fully explained in a scientific manner. The inability to do so is a frustration to those attempting to "model" how it all occurs and to identify that it is caused by Man burning fossil fuels. That fallacy has been repeated so many times it has become factual in many gullible minds. The many examples of "propaganda" will not be repeated as the more they are presented to the public the more they are believed to be factual. The term "inconvenient" applies only to those who wish to misinterpret the facts for a non-scientific purpose.

Perhaps the most significant "eye-opener" diagram found during the research on truth in climate change for this book dealing with the natural events affecting the weather and climate is that shown next historically recorded for the different levels of severity of tropical storms that have developed into severity sufficient to be categorized on the Saffir-Simpson hurricane scale developed by the National Oceanic and Atmospheric Administration (NOAA):

Historically, including those before and since the Industrial Revolution, tropical storms of all variation in intensity have taken the following tracks:

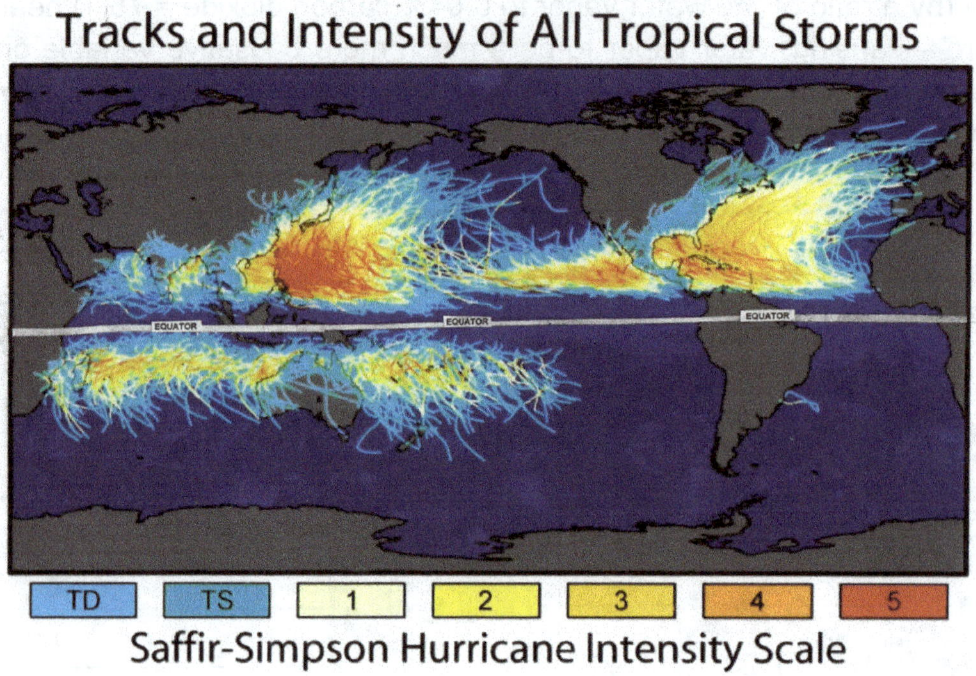

(Source: The National Hurricane Tracking Center, NOAA)

If the causes of those storm tracks were factually due to Man burning fossil fuels and introducing "greenhouse gasses into the atmosphere, it should show that the origins (the spawning conditions) should be over the Industrial regions most responsible for generating the "greenhouse gasses". The historical composite above shows that not to be the case. It is the combination of astrophysical, geophysical, and natural causes that generate larger tropical storms, and primarily on a seasonal basis not all year-long.

There is the possibility that the "greenhouse gas effect" is in small part – but definitely not dominant - due to Man-made emissions causing the waters north of the Equatorial zone such as those above the tropic of Cancer (north of the Equator by 23.5 degrees in latitude) where more than 90% of the heat-trapping "greenhouse gas" generation is produced. Neither natural nor Man-made "greenhouse gasses are being circulated southerly from the heavy industrial areas into the Equatorial zone by ocean currents or by winds. The Gulf Stream which is flowing clockwise in the Atlantic Ocean is northerly along the US East coast. The prevailing winds are also flowing northerly above the US East coast. Before ships were propelled by mechanical power, the sailing ships conducted trade by sailing with the wind current in the Atlantic Ocean.

That became known as "the trade route – from the British Isles southerly along the coasts of France, Portugal, Spain, and north Africa to the Palo Verde Islands where the ships changed course to westerly both with the winds and ocean current. They generally stopped at Barbados and other ports in the Caribbean region to offload manufactured goods from Europe and take on cargoes of sugar, molasses, and rum. From the Caribbean, the sailing ships sailed (again with the wind and ocean currents) to "the Carolines" where they offloaded more European goods and replaced them with cargoes of cotton and tobacco. From the 'Carolines", the sailing ships continued on the north-moving Gul Stream to ports in the New England area to pick up any passengers returning to the British Isles and took the north Atlantic wind and ocean current back to the British ports (primarily London). That clockwise circular route of the sailing ships is written in the Maritime Museum as being the "Trade Triangle". Any carbon dioxide concentration in the atmosphere – natural or Man-made – would be airborne above the same "Trade Triangle" route.

The previous diagram by the National Oceanic and Atmospheric Administration and the satellite images from NASA in the image above clearly indicate that the Equatorial zone of the planet contains the spawning grounds for the major tropical storms. Once spawned, the paths of the storms are influenced by the Coriolis Effect, acting on the heavier mass of the moisture-laden atmosphere, to travel in the direction of the ocean currents and prevailing winds. The author has modified the image with a white line (to contrast with the dark blue of the ocean waters background) indicating the Equator. It is readily seen that the major recorded storm tracks (from the 1700s to the 2000s) are typically clockwise north of the Equator and counterclockwise south of the Equator.

The news media, fed with the fallacious information from the environmental "settled science" faction, repeatedly refers to the Man-made "greenhouse gas effect such as in the BBC article (October 2024): *Assessing the precise influence of climate change on individual tropical cyclones is challenging. The storms are relatively localized and short-lived, and can vary significantly in any case. But rising temperatures do affect these storms in several measurable ways:*

Firstly, warmer ocean waters mean storms can pick up more energy, leading to higher wind speeds.

Record high sea surface temperatures were a key reason why US scientists forecast an above-normal Atlantic hurricane season for 2024.

<u>The high temperatures are mainly due to long-term greenhouse gas emissions.</u>

Secondly, a warmer atmosphere can hold more moisture, leading to more intense rainfall.

Climate change made the extreme rainfall from Hurricane Harvey in 2017 around three times more likely, according to one estimate.

The BBC article does show an increase in the Saffir-Simpson ratings varying from 1920 to 2013 as in the next diagram:

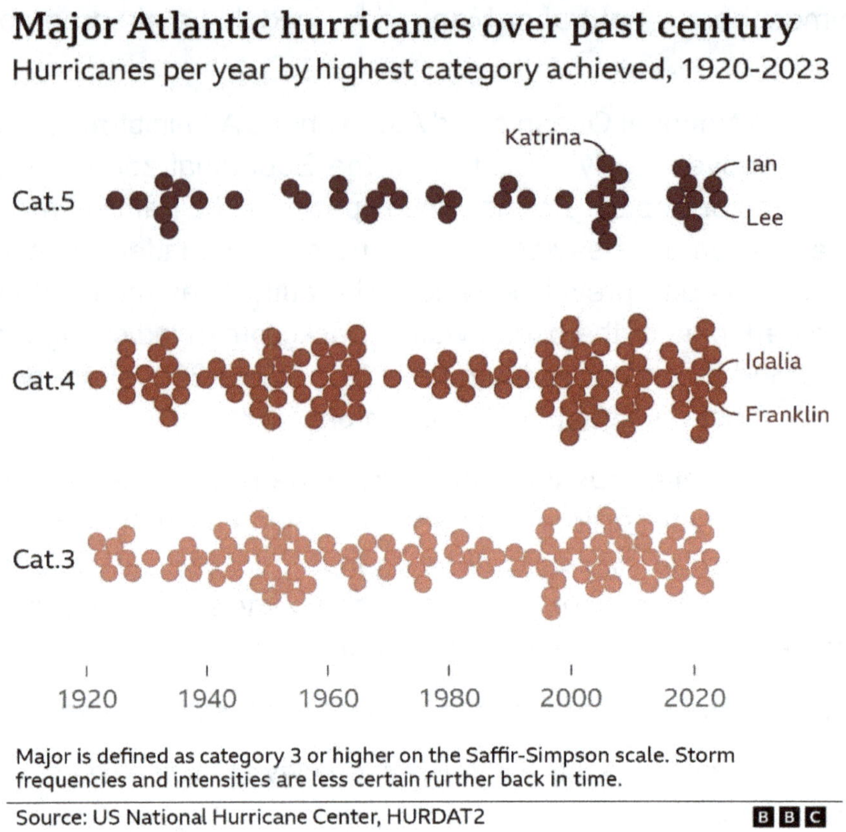

If the "settled science" has "eliminated everything else" except "greenhouse gas effect", the conclusion could be (and with "blinders-on" has been) reached – fallaciously – that it is "greenhouse gas-caused climate change" increasing the frequency and intensity of major Atlantic hurricanes. The spawning grounds of almost all Atlantic tropical depressions which grow into tropical storms and later into hurricanes are off the west coast (near the Palo Verde Islands) of the immense North African desert region where there is absolutely no industrial or human activity that creates an iota of Man-made "greenhouse gasses". The frequent formation of tropical depressions in this region has a well-known basis (from the atmospheric science associated with low-pressure cells and their wind patterns forming the commonly seen small "dust devils" in the southwestern US desert region):

(Photograph from Author's personal photography file)

With the assistance of improved imagery of satellites, photographs (taken in 2022) of the spawning grounds in the atmosphere west of North Africa reveal that the high winds and wind current not unexpectedly lifts enormous amounts of dust from the vast Sahara Desert into the atmosphere. Readers can envision the image below (with the dust appearing as a blurry region in the left half of the image) as being a very large magnification of the "dust devil" patterns of winds in a spiraling motion upward and carrying the dust particles not firmly attached to the ground surface to a higher elevation.

(Source: NASA)

There is not much kinetic energy expended in accelerating the small mass of one dust particle into motion in the plane of the Earth's surface, but to elevate the thousands of tons of dust particles from sea level to an elevation in the range of 1,000 to 5,000 feet is an indicator of the enormous amount of heat energy that is being drawn from the eastern North Atlantic Ocean in the region of the Cape Verdes island group. When the mass of the water vapor molecules (seen in the cloud formations) is added to the phenomenon, and the resulting total mass of dust and water vapor molecules and the mass of all of the gases contained in the atmosphere that is set into a westward motion to possibly become a tropical storm or hurricane, the power of Nature becomes apparent. The addition of any small amount of "greenhouse gasses" into the tropical storm/hurricane phenomenon is virtually of insignificance.

The ocean currents and the prevailing winds in the atmosphere turn due westerly at the Palo Verde islands and cross the southern Atlantic at about 10-12 degrees north latitude. After leaving the hot North African region, the hot winds increase the temperature of the water and continue doing so where they meet the hot waters from the Equatorial current. It is at that juncture that the warm water increases the intensity of tropical storms and they typically begin the climb on the Saffir-Simpson force scale as they approach Barbados. A graphic of the ocean currents in the Atlantic Ocean between the African continent and the North American Continent should be helpful to the public The spawning grounds of the tropical depressions and storm is indicated by a large hexagonal shaped figure and labeled as such:

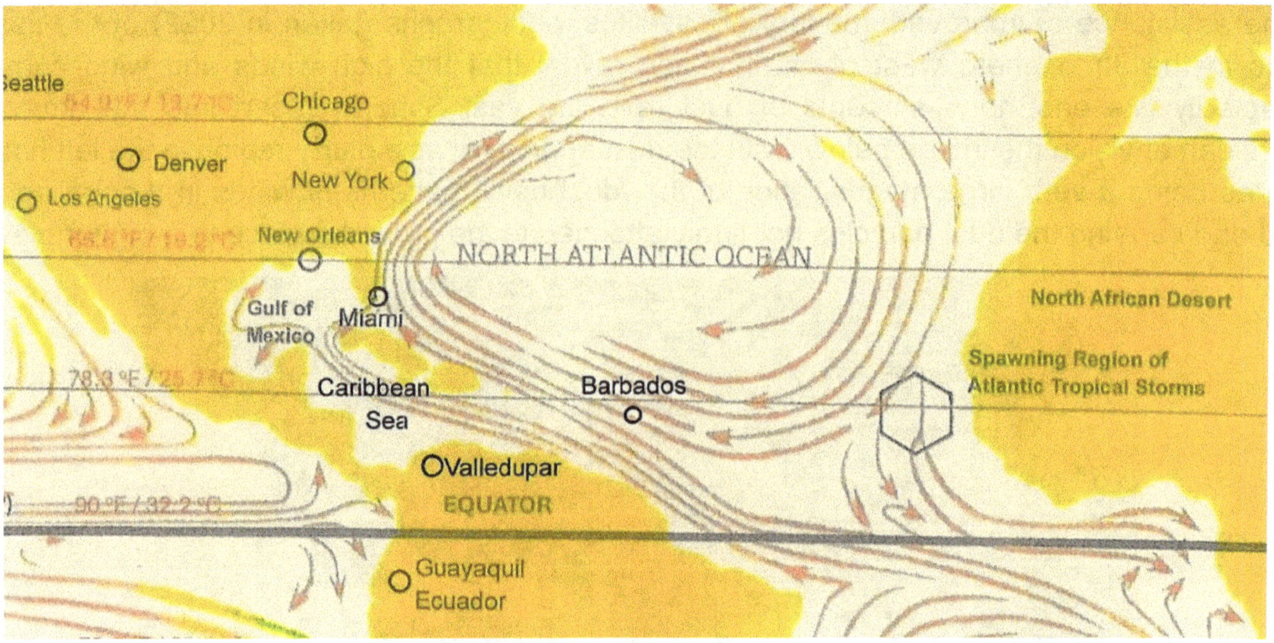

(Source: Rand McNally Atlas of the World, with Author annotations)

One of the largest hurricanes that devastated Barbados (which has a long list of hurricanes in its history since the island's discovery in the early 1600s) was that of the Great Hurricane of October 10, 1780 (before the Industrial Revolution creation of Man-made "greenhouse gasses"). There was no Saffir-Simpson scale to be assigned in 1780 and estimates from the extent of damages – every structure of wooden construction was blown down according to the parish church records on file in the Barbados Archives – the rating would have been most likely a Category 4 or 5. There was no Industrial Revolution in 1780; therefore the question becomes: "What caused that magnitude of hurricane without any Man-made "greenhouse gas emissions" causing or even contributing to that monster. The amount of carbon dioxide in the atmosphere from natural causes would have been below 200 ppmv (0.02%) as there is no ecologic growth creating or absorbing carbon dioxide above the Atlantic Ocean between the Palos Verde Islands and Barbados.

As another rebuttal of that "settled science" claim it has been shown in the previous few diagrams that the major hurricanes are at their highest intensity in the Equatorial Zone but lessen intensity as they progress into the regions above the tropic of Cancer. Since it is above the tropic of Cancer where the winds flowing the "greenhouse gasses" from the industrially generated emissions (90% of which are generated well above the tropic of Cancer) away from the Equatorial zone, how can the news media and/or "settled science" maintain the fallacious claim that Man-made greenhouse gasses are increasing the frequency and severity of hurricanes in the Atlantic?

It should be noted also that by far the highest concentration of the Category 5 cyclones have occurred in the west and southwest Pacific and not in the Atlantic.

The effect hurricanes and cyclones have when and where they make landfall is dependent on several factors – the elevation of the land in the region of the landfall, the extent of the tidal ocean level at the time of the landfall, the time of day, and of course the human and animal population "in harm's way". There cannot be enough plaudits and expressions of gratitude given to the organizations providing advance warning given to those in the projected path of major storms – such as the National Weather Service, the National Hurricane Tracking Center, and the NASA Earth Resources Satellites operators – for providing advance warnings and if local safety officials advise evacuations for advising the best evacuation routes to be taken and enough time in advance for safe evacuation. The same plaudits go to those rescue personnel who assist those families who remain to protect their property and animals as best they can during the storm and survive any flooding after the storm passes.

The USA has a Federal Emergency Management Agency (FEMA) that provides shelter and food and humanitarian services to those who lose their homes and "everything else but the shirt on their back" (such as from massive wildfires and floods), but the typical destruction of the infrastructure in the regions hit hardest by wildfires and floods deny access of FEMA – and all other relief organizations - from getting adequate food, water, and health care services to those residents affected the most severely. It takes days and often weeks to restore power losses to the regions affected and that is critical to the totals of fatalities and losses of homes how long the time period without power lasts.

On the preceding chart it is also seen that the largest concentration of red color (Saffir-Simpson Hurricane Scale 5) indicating that the cyclones (as they are named in the Eastern hemisphere) of highest intensity are most frequent in the Bay of Bengal (near India and Burma) off the Indian Ocean and which historically have devastated the country of Bangladesh on repeat occasions.

The statistics are such that the Bangladesh region has the highest number of cyclones than any other region, but the second most hazardous region prone to hurricane damage is the Louisiana Gulf Coast. The Gulf Coast has the highest number of hurricanes of all scales (including some up to scale 5) to make landfall.

Geographically, the northern section of the South American Continent serves as a deflector of the Equatorial current between Africa and North America after passing the island of Barbados, and the chain of islands (Haiti, Dominican Republic, and Cuba) split that deflected hot current in the southern part of the North Atlantic clockwise current with a segment flowing into the Caribbean Sea and Gulf of Mexico. (The Author's ancestral family were/are resident on Barbados, and on family visits, the Author is given the option to body surf in the Atlantic Ocean warm waves rolling onto the Eastern Coast beaches of Barbados or to swim on the placid West Coast of Barbados said to be in the even warmer Caribbean Sea. Some decisions in life are very hard to make.)

The tropical depressions that form (in the region marked with a hexagonal figure on the previous diagram) and develop into tropical storms and hurricanes typically flow toward Barbados. Those that pass north of Barbados generally flow with what is known as the Gulf Stream (with water temperature of about 80 °F (26.7 °C) that approaches the Miami part of Florida before turning northerly or northeasterly. The lower segment that passes south of Barbados mostly flows into the Caribbean Sea and then into the Gulf of Mexico. The Gulf of Mexico acts as a basin and can generate tropical storms within the Gulf that may further develop into named hurricanes with Force Level 1 to 5 and those are generally the ones that devastate the coast cities of Mississippi, Louisiana, Alabama, and Florida.

There are many in the "settled science" faction of environmental science that attribute the frequency and severity of the Atlantic hurricanes to being contributory to the frequency and severity. There does not appear to be any principle of atmospheric, thermodynamic, or hydrodynamic physics that supports the influence of the "greenhouse gasses" to change either the frequency or severity of ocean hurricanes, cyclones, or tornadoes. Hurricanes, cyclones, and tornados are seasonal which confirms they are primarily influenced more by astrophysical and meteorological factors and very little, if any, by Man-made atmospheric emissions.

A composite of Hurricane Scale Level 3 named and unnamed hurricanes that have made landfall in the New Orleans delta (of the Mississippi River) region from 1852 to 2015 (163 years) is shown as follows:

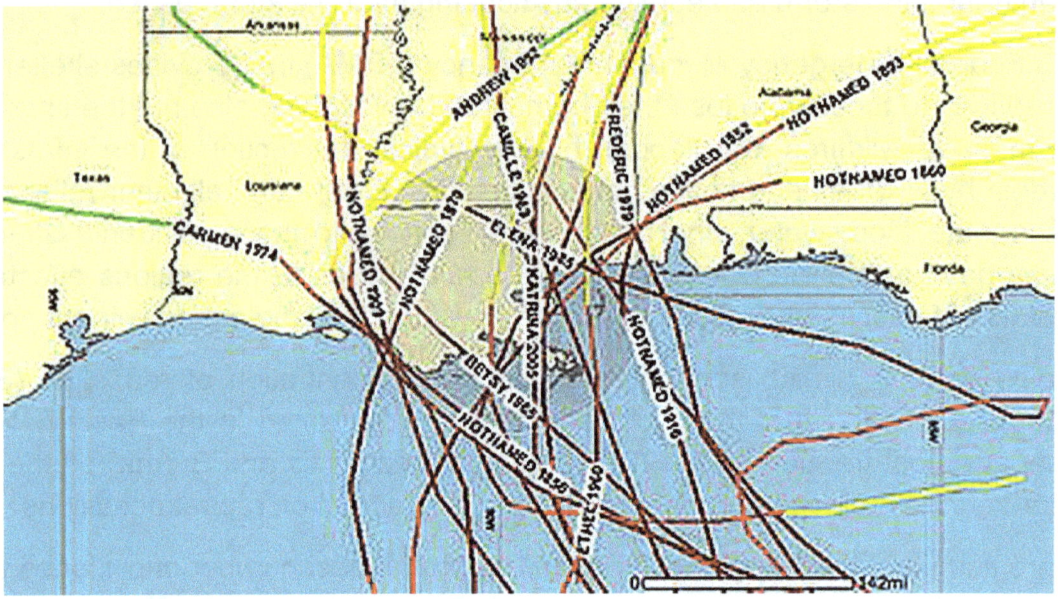

(Source: National Hurricane Tracking Center of NOAA)

The first major hurricane to make landfall on the United States and officially recorded was in 1852. Before hurricanes make their landfall on the United States, they seem to randomly enter the Gulf of Mexico and then turn northerly, or (for unknown natural reasons) to turn northeasterly before making landfall. The state of Florida has some influence on that variation in track, but it is unknown scientifically and factually why and how. Shown next is a composite of the Category 4 and Category 5 hurricanes that have crossed southern Florida in the period 1910 to 2018.

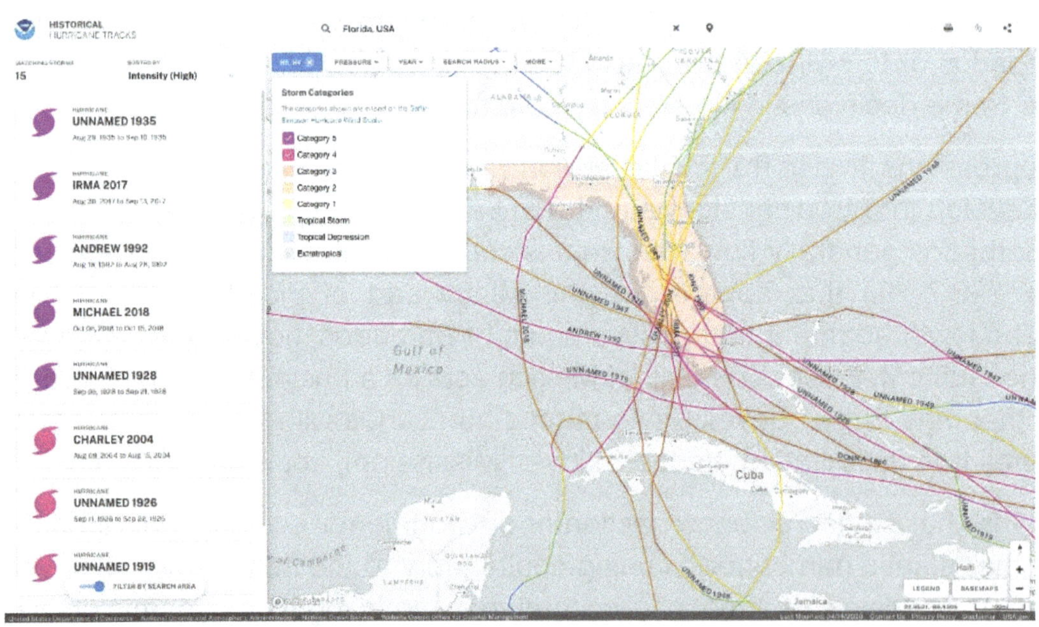

(Source: National Hurricane Tracking Center of NOAA)

One hurricane is seen not to have wanted to leave the Cuba region and made a farewell circular loop before heading toward its Florida target. The favorite city for hurricanes to "visit" in Florida is Naples, on the southwest coast (in the Gulf of Mexico) of the Florida peninsula.

The Atlantic hurricane season is generally in the six months from June through November of each year but there can be hurricanes occurring before June or after November. The bar chart shown next indicates the number of Atlantic hurricanes that occurred from 1923 to 2023:

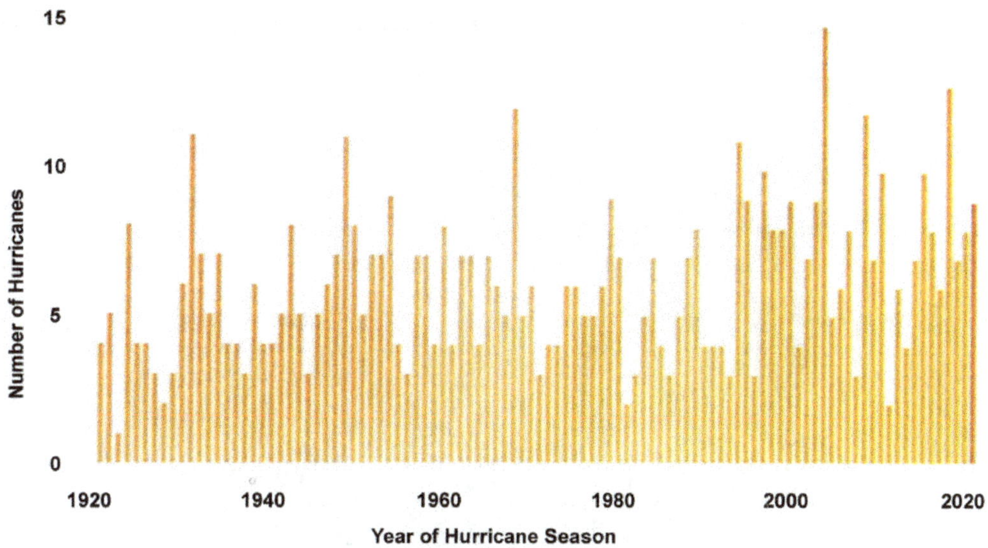

(Source: Colorado State University, Department of Atmospheric Science)

The trends being developed by the environmental science faction to reflect that Man-made "greenhouse gasses" are causing the intensity and frequency of the hurricanes is somewhat a non-scientific achievement to the human population who suffer the loss of lives and property whenever and wherever a major catastrophic event occurs.

The most devastating hurricane that has recently struck the United States was Hurricane Katrina which made landfall as Category 4 on the Louisiana Gulf of Mexico coastline on August 29, 2005. Shown next was a satellite image of Hurricane Katrina as the eye was over the New Orleans region:

(Source: NASA)

To be noted in the photo is that the fringe bands of cloud formations indicate the highest areas of rainfall, and it is these fringe clouds that have the highest winds and contain the most water that causes the wind damage and flooding across a much larger area of the land surrounding the eye. Hurricane Katrina caused 1,577 fatalities, mostly due to severe flooding and caused about $125 billion in damage.

A map of the state of Louisiana shows the number of Parishes and cities that are in the Louisiana Gulf Coast region prone to hurricane damage:

(Source: Atlas of the World, Rand-McNally)

Although Hurricane Katrina's eye came ashore in lower Plaquemines Parish of Louisiana, the resulting storm surge resulted in the failure of multiple levees of the Mississippi River in the New Orleans delta area, flooding approximately 80% of the city. Some areas of New Orleans were inundated by more than 15 feet (4.6 meters) of water. A photo of a flooded area of New Orleans (near the Superdome) is shown as follows:

(Credit to Unknown Photographer)

The failures of the levees in Orleans Parish were considered the worst engineering disaster in the history of the United States. The population of Orleans Parish in Thousands of people were stranded inside their homes or on rooftops and required rescue from boats and helicopters. Many buildings and homes were damaged, with 134,000 housing units – approximately 70% of residences in New Orleans – were impacted to some degree. The Superdome, which was sheltering many people who had not evacuated, sustained significant damage. Two sections of the Superdome's roof were compromised, and the dome's waterproof membrane had essentially been peeled off. The famous French Quarter and Garden District escaped flooding because those area are above sea level. A total of 573 deaths occurred in New Orleans alone.

Several other parishes were severely impacted by the storm. Many areas outside of New Orleans also suffered wind damage, especially St. Tammany and Washington parishes. According to the United States Department of Housing and Urban Development, in St. Bernard Parish, 81% (20,229) of the housing units were damaged. In St. Tammany Parish, 70% (48,792) were damaged, and in Plaquemines Parish 80% (7,212) were damaged. Nearly 900,000 Louisianians were left without electricity. Throughout the state, 1,577 fatalities were reported. Overall, it is estimated that Hurricane Katrina caused $125 billion in damage throughout the Southern United States, with much of the damage occurring in Louisiana and neighboring Mississippi.

The effect of natural catastrophes such as Hurricane Katrina on Orleans Parish in 2005 on the human population is not fully depicted by only the number of fatalities and cost of property damage. At the time Katrina impacted Orleans Parish in 2005, its population was 454,863. Most of those people had been given forewarning by the National Weather Service of the approaching monster (Category on the Saffir Simpson Scale) and evacuated low-lying areas further inland. Having lost their residences many decided not to return to Orleans Parish. In the aftermath of the Hurricane Katrina disaster, the Orleans Parish population in 2007 had declined to about 320,000. The population in the Parish stabilized in headcount and in the 2020 Census the population had re-grown to 383,997. The poverty in the Orleans Parish since the 2020 Census has worsened and caused another decrease in the population of Orleans Parish to presently (2024) be 358,355. Louisiana was once a very popular visitor location – especially at Mardis Gras time – but the visitor rate has also dropped considerable further in the past two decades affecting the entire southern Louisiana region economy. The trend of increasing frequency and increasing overall severity rating of hurricanes in the Gulf of Mexico region is having a profound impact of the economy of the Gulf Coast states.

The environmental scientists in the "settled science" fraternity are claiming that climate change is causing those changes in trends, but climate is the statistically dependent variable in the equation of what is happening with the independent variables in the Gulf region changing and it is mathematically a misnomer to assign blame for weather catastrophes to be due to climate change. The "tail does not wag the dog". The public has been insidiously misled into believing that Man burning fossil fuels is the dominant cause for climate change. The fallacious tangential inference fostered by most environmental scientists (including those in the United Nations Intergovernmental Panel on Climate Change, IPCC) has set a "one-size-fits-all" goal that Man-made emissions must be reduced to net zero by 2050 to bring climate change to a halt.

The analysis done for this book has shown that despite Man-made emissions being reduced (in the USA and Europe, but not in China and India) over the past two decades, the surface temperatures continue to increase, more in some regions than merely on an "average global temperature" basis. A very revealing illustration of that fact is shown in the following illustration (shown earlier but repeated here for emphasis of its importance):

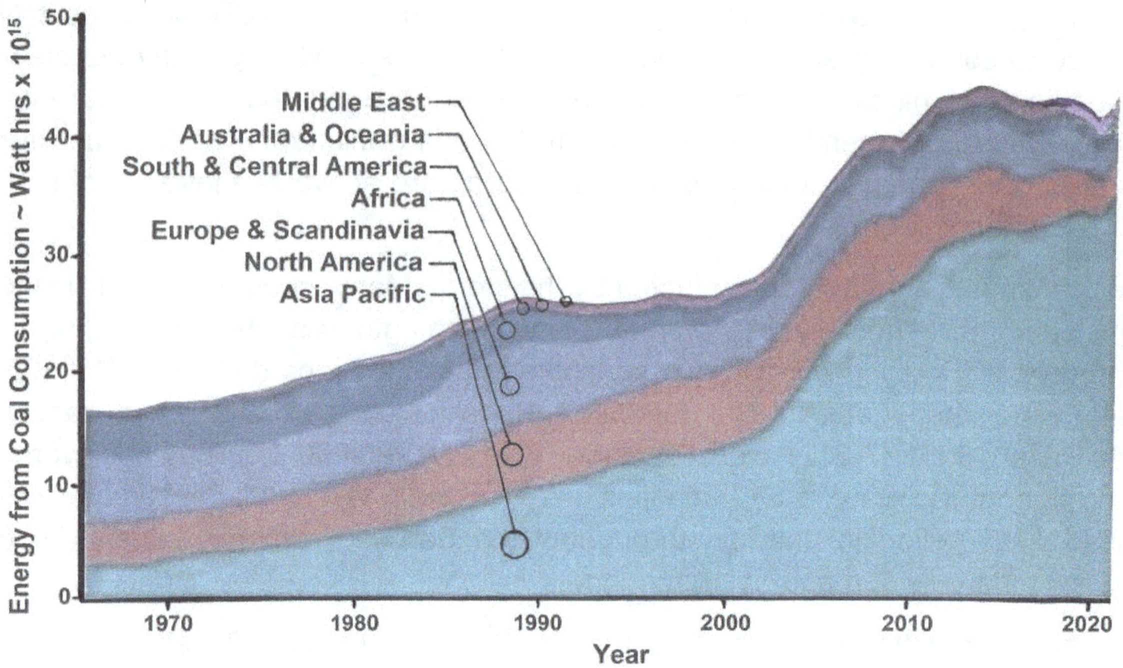

(Source: British Petroleum Corporation, LLC)

Articles in the news media have repeatedly stated that the frequency and the intensity of hurricanes are both in an increasing trend and those increasing trends are due to climate change making the Atlantic Ocean waters "soaring" in temperature. Those claims are not supported by any scientifically valid database found during the research and analysis of the hurricane database in the public domain. With the tangential claim that climate change is due to Man burning fossil fuels being the dominant cause for global temperatures rising, the Man-made emissions labeled as "greenhouse gas emissions" do not have the "spikey" frequency history shown in the above chart for the frequency of the Atlantic hurricanes and therefore Readers should not panic over the fact that even with the reductions in carbon dioxide in the atmosphere that are Man-made and being scientifically measured in a low concentration of 140 parts per million (0.014%) – with 295 parts per million of carbon dioxide (0.0295% being in the atmosphere before the Industrial Revolution – Man should not to be blamed as being the dominant cause for the naturally and seasonally occurring hurricanes that strike the United States southeastern coastal states.

Among the greatest leaps in the ability to monitor the track of hurricanes is attributed to NASA with the development of satellites, originally for communication, navigation, and military purposes, that are now also in the inventory of the National Oceanic and Atmospheric Administration for meteorological tracking of weather-related phenomena. With highly advanced optical imagery coupled with the satellite "big picture" perspective, the "eye" of the hurricane as well as the fringe waves, which typically cause the most flooding damage when passing over a land mass, are measured and tracked. The National Hurricane Center is a vital public safety National organization for providing days of warning to evacuate before an approaching "monster" attacks. An example of that photographic definition of a hurricane, that of Hurricane Dorian (Category 5 with 185 mph winds) which battered Abaco Island in the Bahamas off the east coast of Florida on September 1, 2019, is shown as follows:

(Source: The National Hurricane Center, NOAA)

Most importantly, the path of the tropical depression which grew into a hurricane being tracked quite accurately on a daily basis for location and strength of the "eye" is shown in the next illustration:

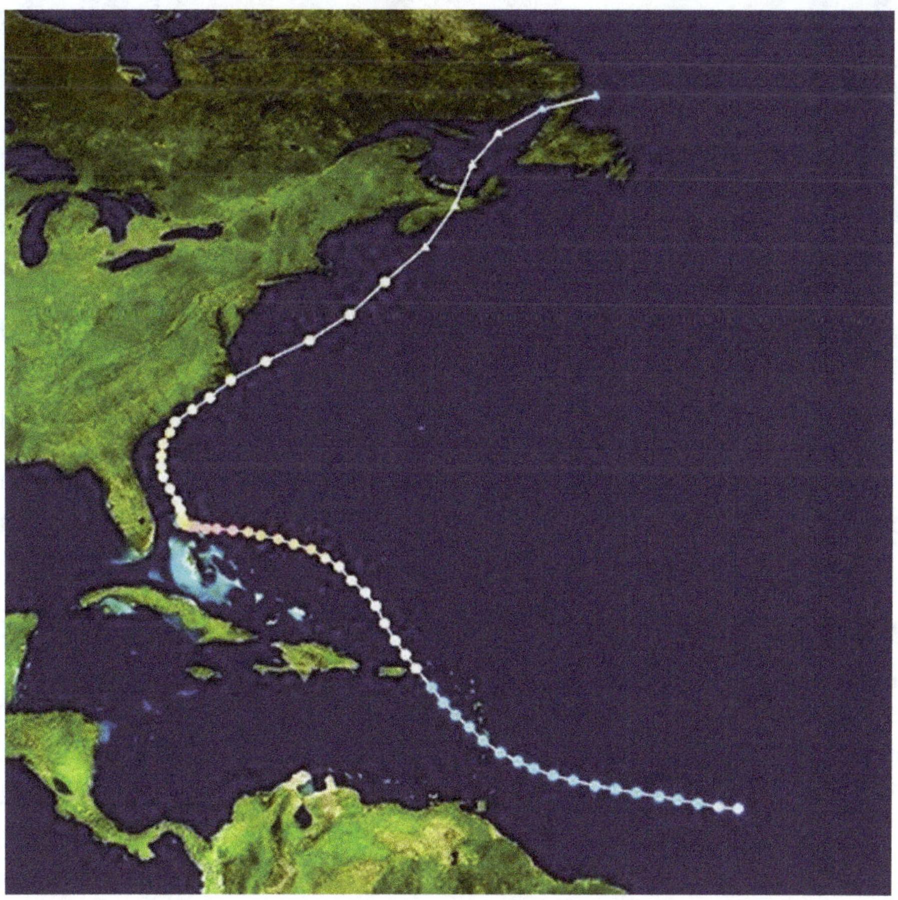

(Source: The National Hurricane Center, NOAA)

With this tracking information for Hurricane Dorian, human populations on Caribbean islands and in the southeastern United States were given as much advance warning as possible and far enough in advance that loss of life and property damage was drastically minimized. Even so, Hurricane Dorian took 84 lives, 245 still missing (and presumed dead) and caused over $5.1 billion in damages – most on Abaco Island in the Bahamas off the Florida coast on September 1, 2019 from Hurricane Dorian (which existed from August 24, 2019, to September 10, 2019). The colored circles reflect the intensity of the storm from its formation as a tropical depression in the warm south Atlantic Ocean waters to maximum Category 5 (in the hotter Caribbean Sea waters) and dissipation in the colder North Atlantic waters.

Still more impressive tracking with satellite imagery is that of an electro-optical thermal image showing the energy concentration of Typhoon Mawar which struck the island of Guam and caused damage to military facilities at Andersen AFB on Guam on May 24, 2023: (the terminology "typhoon" and "cyclone" are used interchangeably)

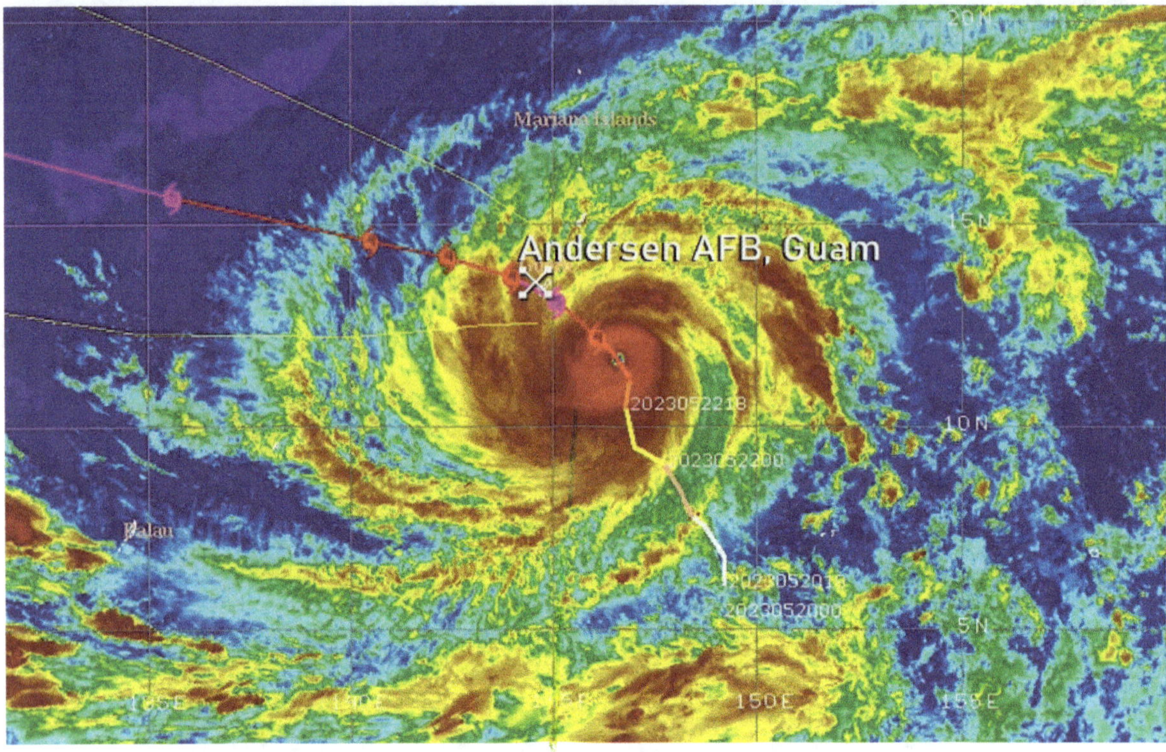

(Source: USAF Military Satellite Imagery, Typhoon Mawar at Guam)

Fortunately, thanks to advanced warning of the intensity (winds up to 175 mph) and accurate tracking of the super typhoon (Mawar), there were no deaths reported on Guam and less than $125 million in damage.

As this book was nearing completion, Hurricane Helene struck the United States Gulf Coast as a Category 4 (on September 25, 2024) and took a surprisingly inland route northerly through Florida, Alabama, Georgia, South Carolina, western North Carolina, Tennessee, Virginia, West Virginia, Kentucky, Ohio, Indiana, and Illinois. The path that Hurricane took - far inland and away from the Atlantic coast – left residents somewhat in disbelief that a hurricane would strike that far inland. Hurricane Helene also was unique in that it spawned near the Yucatan Peninsula and moved almost due northerly with only a few small serpentine-like deviations before deteriorating in the upper midwest. It also was a moderately slow moving hurricane which usually creates worse flooding than the faster-moving hurricanes. A map of the track Helene followed is shown next and how its intensity varied (the color dots representing the Saffir-Simpson scale):

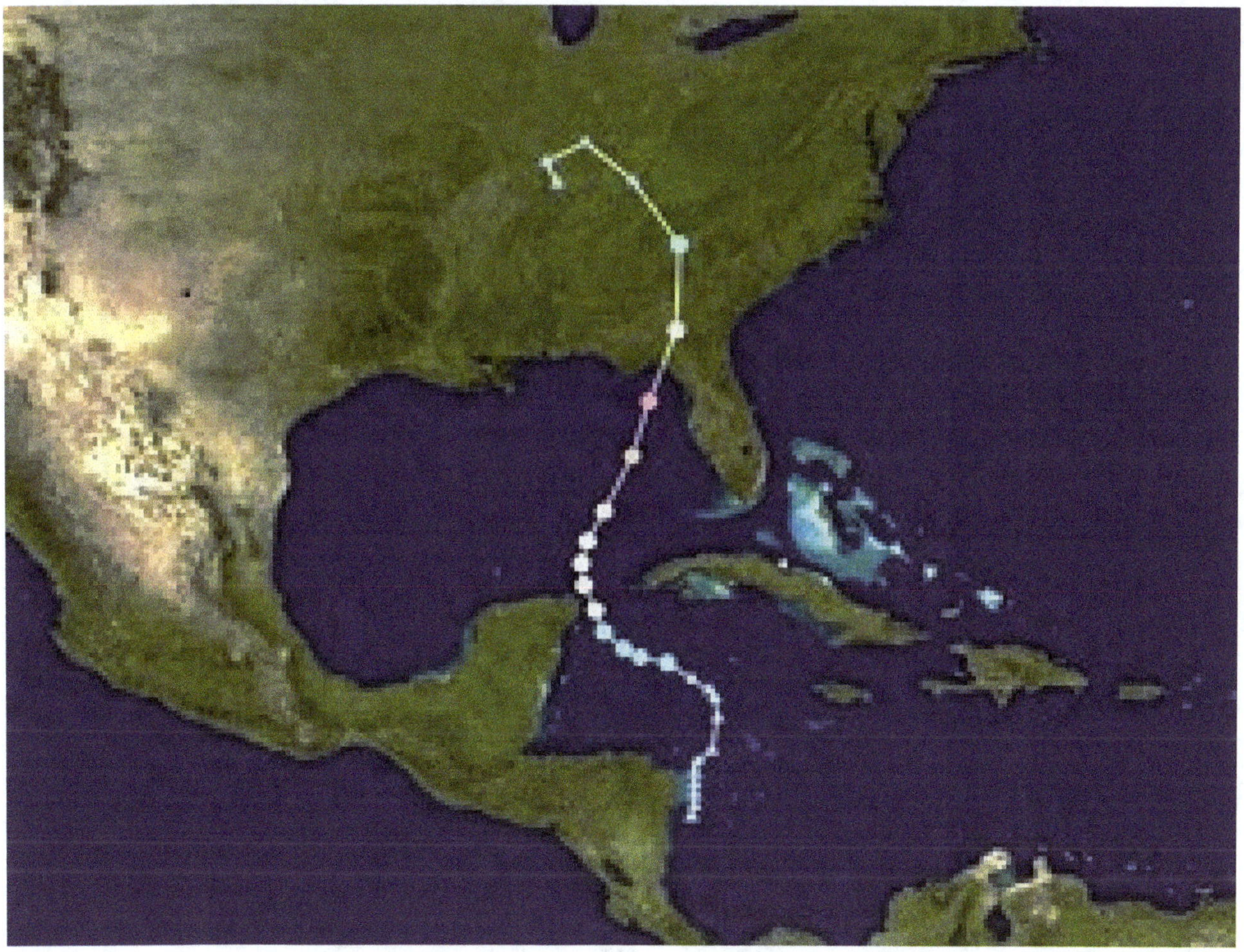

(Source: The Hurricane Tracking Center, NOAA)

The color-coding of the Saffir-Simpson scale is provided for reference:

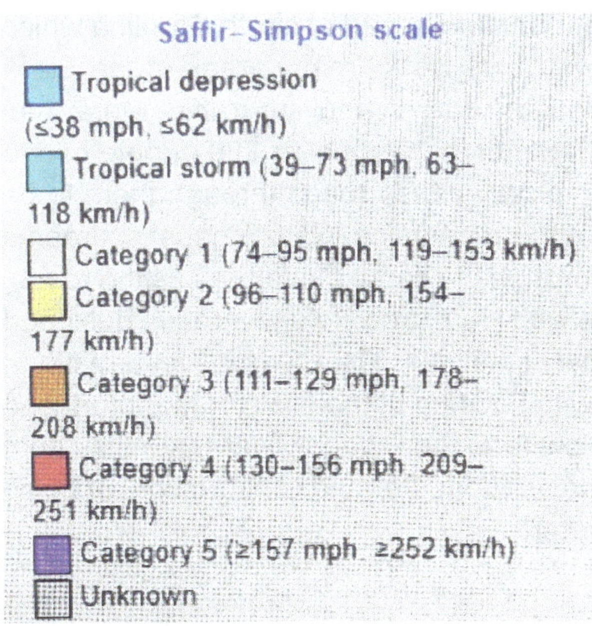

Storm type:

● **Tropical cyclone**

■ **Subtropical cyclone**

▲ **Extratropical cyclone**, remnant low, tropical disturbance, or monsoon depression

The impact (fatalities, missing, and direct property damage) by state is as shown as follows:

Impact of Hurricane Helene by State
(September 24-29, 2024)

State	Deaths (Missing)	Damage (US $)
Florida	26	>$21.1 billion
Georgia	33	>$6.88 billion
South Carolina	51	Unknown
North Carolina	102 (26)	$53.6 billion
Tennessee	18	Unknown
Virginia	2	Unknown
Indiana	1	Unknown
Total	233 (26)	>$87.9 billion

(Source: Wikipedia)

Hurricane Helene was the eighth named tropical storm, fifth hurricane, and second major hurricane of the 2024 Atlantic hurricane season. The hardest hit state was North Carolina (the Author's home state) and the damages and fatalities were primarily in the western mountainous part of North Carolina which received up to 19 inches of rain within a 24 hour period. Power was lost in many areas of the Black Mountains, and the magnificent Blue Ridge Parkway was closed for freely-moving traffic. More than 400 roads were closed (many that will not reopen for at least 18 months), and over 200 people had to be rescued from the floodwaters. The mountainous regions typically cause roads through them to be built following the low and very curvy lying land at the base of the mountains. Nature causes streams and rivers to also follow at that same low-lying elevation and through the same curvatures as the roads. When there is a once-in-a-hundred-years flood – as there was with Hurricane Helene – roads designed next to roadside streams are built for normally placid water flow conditions. The roads in western North Carolina have not been built for the flow of water associated with flooding of the scale of that produced by Hurricane Helene. The section of Interstate I-40 shown below (and the Pigeon River four miles from the North Carolina border with Tennessee) is an example of that inadequate road construction and also shows the erosive power of the fast moving very turbulent water:

(Credit to unknown photographer for North Carolina Dept. of Transportation)

Because of the high number of road closures into the area most impacted, the only access was by helicopter for food, water, and emergency medical supplies. Many of the homes had emergency generators but the fuel for those quickly was consumed and a large number of solar-powered generators were flown in to partially assist the emergency facilities in treatment of those injured or with pre-existing medical conditions.

The track that Hurricane Helene took in late September 2024 was most untypical of other hurricanes and no factual data or valid scientific explanations to either support or counter the claim that its cause, severity, and/or track could be related to Man burning fossil fuels has been uncovered. More than probably, it is not related since its origin was in a low population area and the earliest 3/4s of its track was through non-industrial “greenhouse gas emitting” regions. It is more likely that the generation of hurricanes (and tornadoes) in the southeastern United States are well known to be seasonal and the origin of Hurricane Helene was due to some unusual astrophysical anomaly in the causes of that seasonal effect.

A more detailed account of the causes of the deaths and the damages in each of the more severely impacted states from Hurricane Helene and from the five tornados that were spawned by the hurricane over land is beyond the scope of this book.

While continuing massive disruption of power caused by Hurricane Helene and with the road system into and from western North Carolina still impassable, another tropical storm that became a hurricane formed on October 4, 2024. It formed in an unusual location in the Gulf of Mexico but took a path of travel that was unlike that of Hurricane Helene less than two weeks earlier. Given the name Milton, the Hurricane Tracking Center of NOAA projected the path of what was to become Hurricane Milton as shown:

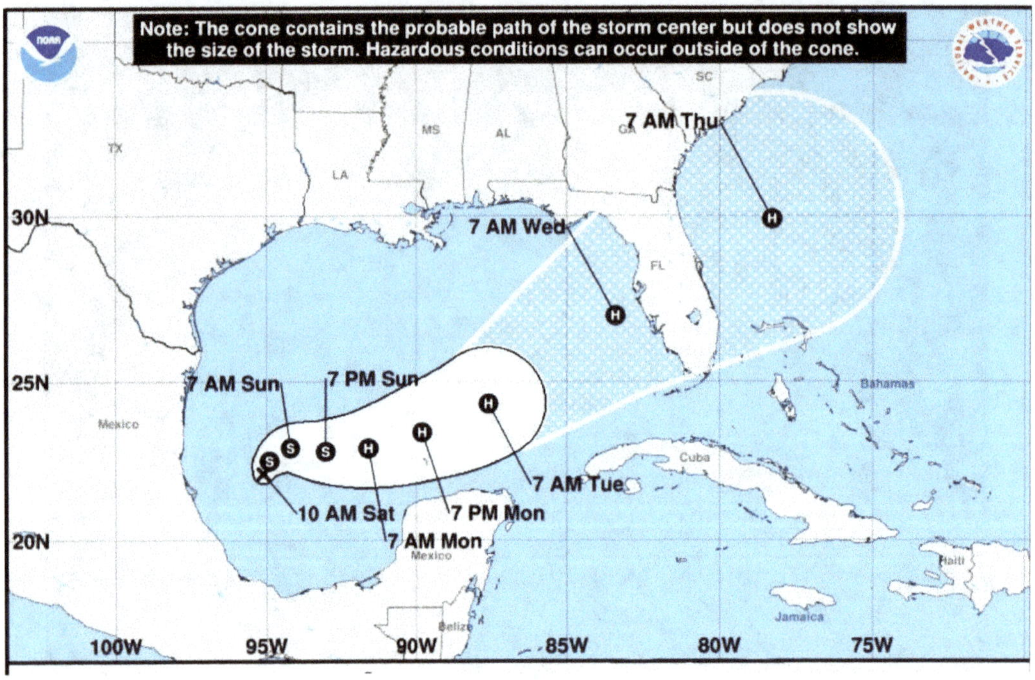

(Source: National Hurricane Tracking Center, NOAA)

The path that Hurricane Milton actually took during its seven day life span from October 5th to October 12th, 2024, is shown as follows:

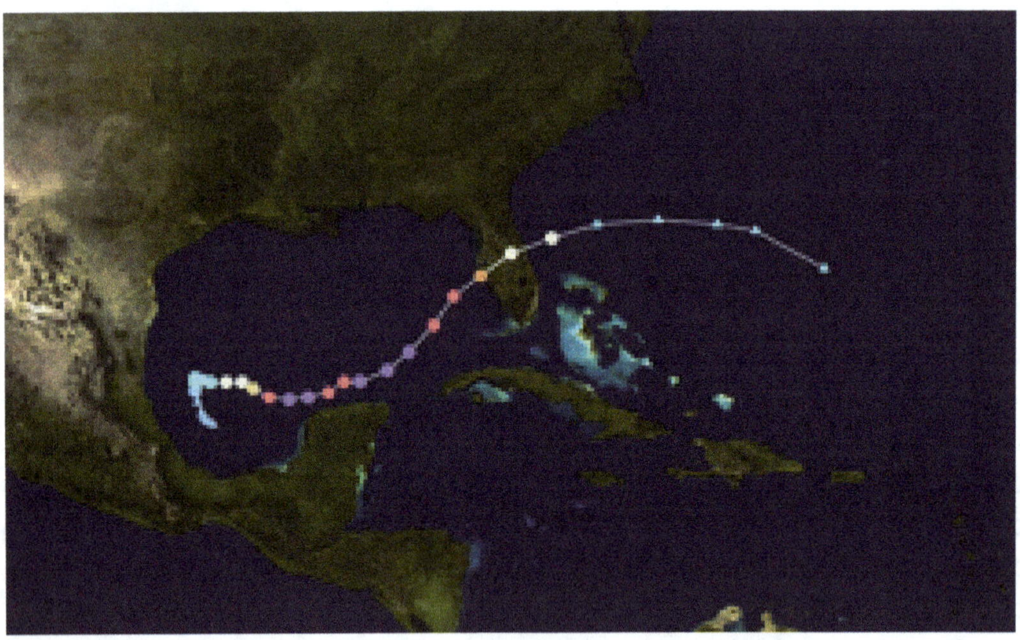

(Source: Hurricane Tracking Center, NOAA)

Hurricane Milton became a hurricane (Category 1) early on October 7th, but underwent an "explosive intensification" (NOAA description) and later that same day became a Category 5 with 180 miles per hour winds. Milton became the thirteenth named storm, ninth hurricane, fourth major hurricane, and second Category 5 hurricane of the 2024 Atlantic Hurricane Season. As can be seen, Milton weakened considerably to Category 3 as it made landfall on October 9th near Siesta Key, Florida, and continued weakening as it crossed the state of Florida, being rated as an extratropical storm on October 10th. Unlike Hurricane Helene, fewer deaths were incurred (35) and fewer persons were/are listed as missing (6) because of its low dwell time over populated land. Unofficial estimates of the damage – mostly on the coastal regions to boats and small beach-front residences – at about $85-90 billion. The damages are coincidentally almost the same as that of Hurricane Helene.

Waterspouts (a close counterpart of tornadoes on land) also have their origin in the warmer waters of offshore ocean currents as shown in the following diagram:

(Credit to unknown photographer)

It must be noted that there is a heavy low cloud cover from which the vortex flow extended down to the water surface and literally sucks the water upward as if in a big straw. The area surrounding the waterspouts is also seen to have much moisture content either as raindrops or a very high concentration of water vapor. Waterspouts, however, do negligible damage to property (only of occasional maritime vessels) or cause losses of human lives as do their land "cousins" tornadoes.

The most common cause for both tornadoes on land and waterspouts on water lies in the "shear" in the wind currents in the upper atmosphere, and by the pressure system between the surface and the upper atmosphere. The public sees only the letters H (denoting a high barometric pressure in a small region) and L (denoting a low barometric pressure) on weather maps in the newspaper - but that is because newspapers are two-dimensional and cannot indicate the more complete three-dimensionality of the phenomenon. A more scientifically described H and L system appears as follows:

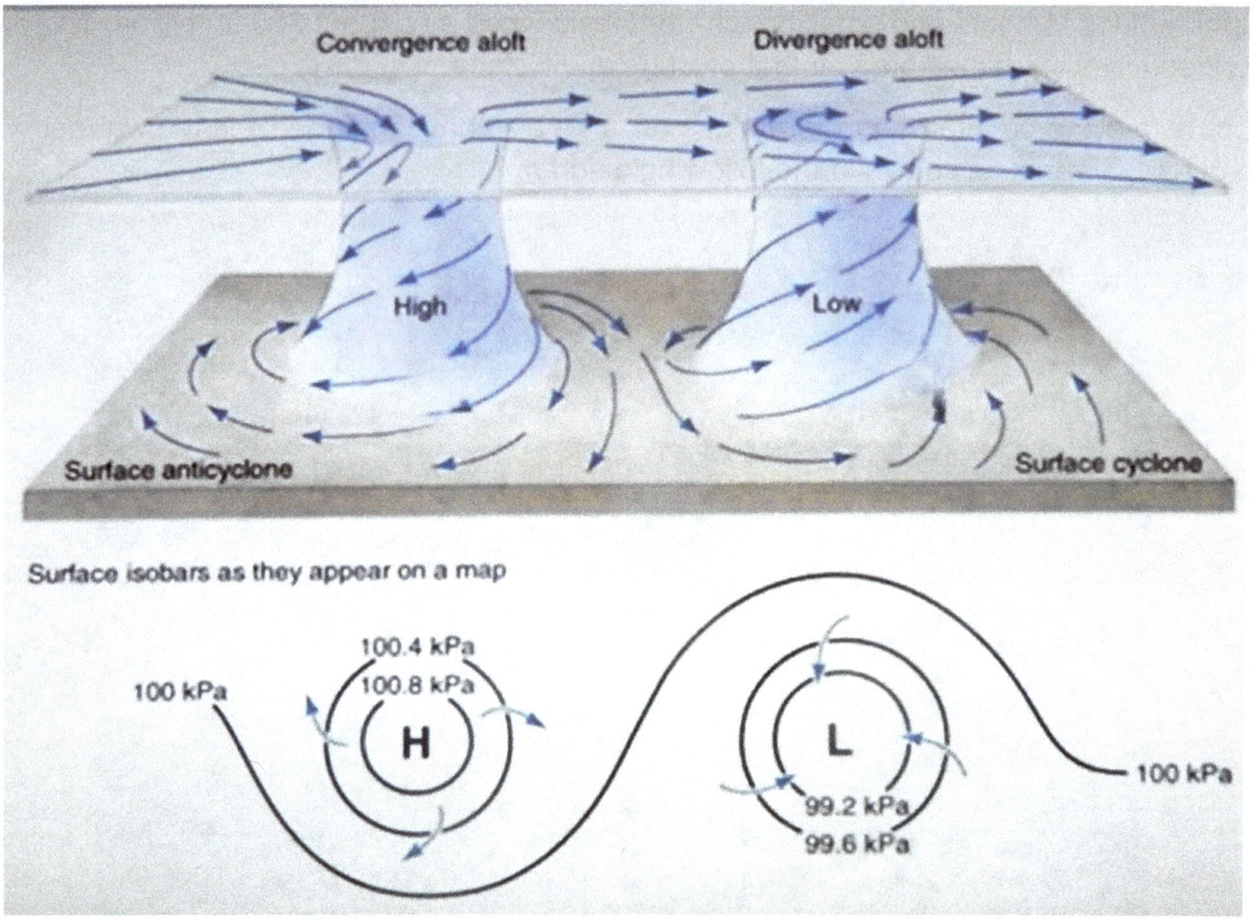

(Source: World Meteorological Organization)

On weather maps are also shown hot and cold "fronts" which in essence are lines of constant barometric pressure – either high barometric pressure or low barometric pressure. The curved line between the H and L systems shown above is a line (or "front") of that constant barometric pressure. In general, the weather associated with an H or along a warm front is an indicator of being fair (sunny with little chance of precipitation). Conversely, the weather associated with an L or along a cold front is an indicator of their being a high probability of adverse weather – cool or cold temperature, high winds, and high probability of precipitation. It is the rotational winds around the L which spawns hurricanes, cyclones, tornadoes, or just tropical depressions that will either increase or decrease in wind velocity and content of water.

It is rather well understood by meteorologists how tornadoes form and is due to the factor of wind shear. When a mass of wind-driven air moves past a different level of winds and direction of wind in an adjacent cell, there is a "shear" causing the mass of air at that juncture to transform from linear motion into rotational motion. The damages and fatalities incurred in the United States due to tornadoes is discussed in more detail in the next chapter, but the point to be made here is that the fallacious assumption apparently made by some environmental scientists is that the "Greenhouse gas emissions" (carbon dioxide, methane, etc.) are introduced into the global atmosphere in a homogenous manner such that emissions in the Northern Hemisphere will have global warming effects the same in the Southern Hemisphere as they do in the Northern Hemisphere.

Not to belabor the point, but environmental scientists using an "average global temperature" and the setting of a "one-size-fits-all" goal of reducing Man-made emissions to limit the rise in global surface temperature to 1.2 degrees Celsius by 2030 and achieve a net-zero in Man-made emissions by 2050 is unprofessionally fallacious. The wind current diagram above indicates just how fallacious the assumption of homogeneity of the atmosphere by environmental scientists is in the meteorological science profession. The medical profession takes severe sanctions against its members who commit malpractice (in which a fallacious act is intentionally perpetrated) against a huma, but fallacious acts against the environment are not subject to such human-protection sanctions.

Of the highest of significance to the weather, weather events, and climate of the United States Atlantic Coast states is the Gulf Stream shown in the next diagram:

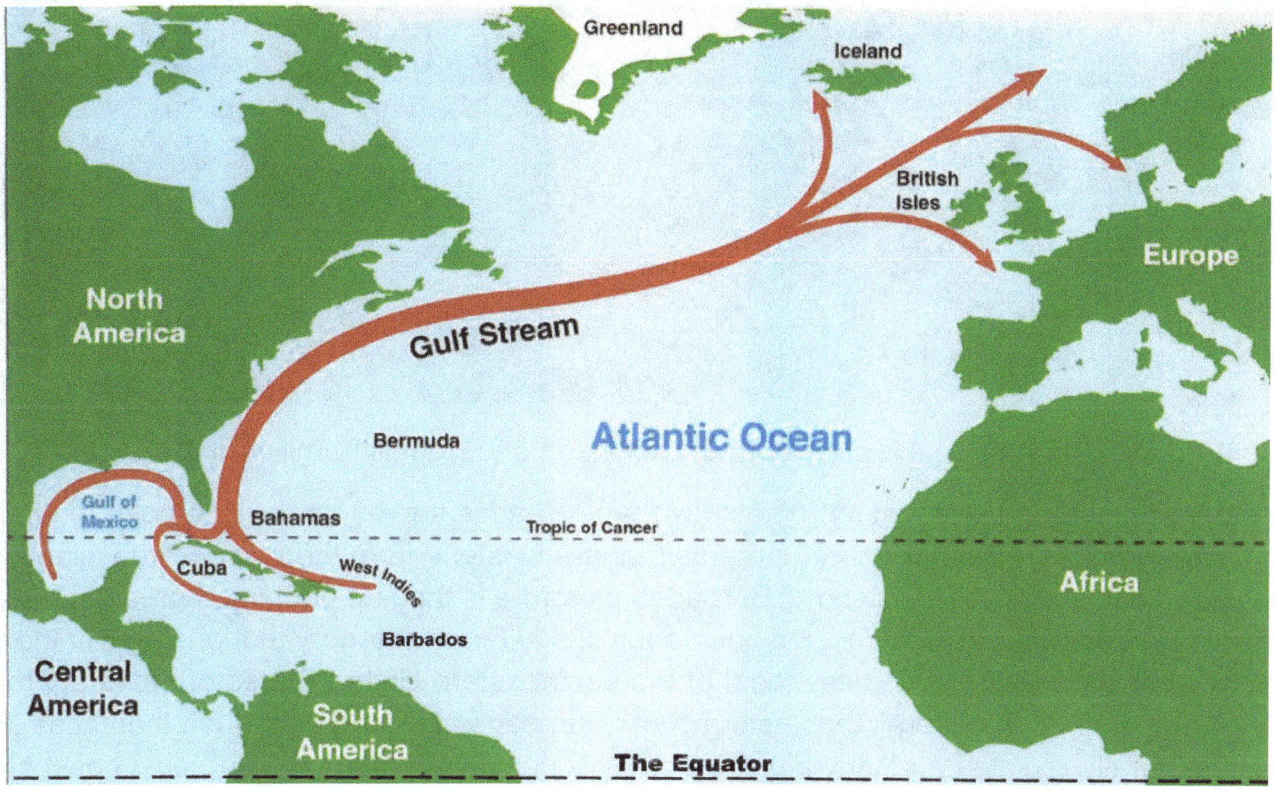

Source: National Oceanic and Atmospheric Administration)

The general ocean current of the northern Atlantic Ocean is a clockwise current (being north of the Equator). It is off the West coast of Africa that this current develops the pressure and temperature flowing over the vast desert areas of west central Africa and is the spawning grounds of tropical depressions that are formed by natural causes – without any influence of Man burning fossil fuels for industrial, residential, or transportation fuel combustion engines. When the ocean current and prevailing wind currents (labeled as the "westerlies" by old time mariners) make a turn at about the Cape Verde islands group, they proceed to about the island of Barbados (at the bottom of the Caribbean chain of mostly volcanic origin islands) in the West Indies.

Geographically, the west coast of Barbados is in Caribbean Sea waters and the east coast of Barbados is in Atlantic Ocean waters. The island is somewhat pear-shaped and is 14 miles from north to south tips and 21 miles from west to east beaches. (The Author's heritage is from this small island and on which the weather was found to be paradisical 99% of the time.) Unlike the other West Indian islands to its north being primarily of volcanic origin, Barbados is the exposed 165 square miles of land on top of a coral reef millions of years old that was thrust upward from the ocean floor.

(Source: "Genealogical History of the Hutchinson and Edghill Families of Barbados")

The geological significance of that set of facts is that Barbados serves as the first point of land mass for ships sailing on the ocean and wind currents that are westerly from the Cape Verde islands group off the West coast of Africa. The island of Barbados therefore is the first populated area to be exposed seasonally to the tropical depressions and hurricanes of any Force category that originate in the vicinity of Cape Verde and forewarns the east coast of the southeastern United States of those approaching hurricanes and storms. In general, those storms and hurricanes follow a path such that those north of Barbados generally proceed northerly along the heavily populated southeast and east coast states of the United States. Storm and hurricane paths passing south of Barbados are of highest probability to proceed into the Gulf of Mexico before making the turn to the north and northeast. On that more southern and westerly path, the storms accumulate more heat energy from the very warm waters of the Caribbean Sea and the Gulf of Mexico than their more northern "siblings" and when making their typical turn toward the northeasterly path direction the typically more damaging and injurious hurricanes pass over the Gulf coast states (southeast Texas, Louisiana, Alabama, and Florida Panhandle regions).

Hurricanes are given a name for record-keeping purposes and continuous tracking of each storm/hurricane is done for human safety warnings. (The naming originally was a female first name because their "personality" and direction of movement is as unpredictable as those of a misbehaving teenage female, but more recently are also given male first names to remove any sexist connotation.) Regardless of name, the advance warning of an approaching natural potential disaster is vital to minimize the adverse effects of the approaching monster.

(Author's Note: After retirement from a 40+ years career in aeronautical engineering – which entailed the analysis of large, complex databases of scientific and technical/technology information, the Author applied that analytical experience to generating a genealogical history of his family. That genealogical database revealed that his genealogy (starting with his parents) was from the past 14 generations of personages born on the island of Barbados from 1905 back to 1635. The genealogical database on the island is comprised of the Church Parish Registers (Barbados is composed of eleven parishes named after saints in the Church of England) for births, baptisms, marriages, and deaths.)

In addition to the vital statistics of the parishioners – later used as the genealogical history of the island - the Parish cleric typically wrote accounts of the devastation done to the residences and government and public buildings on the small island when hurricanes made landfall on the island "in harm's way". On October 10, 1780, an extremely powerful (there was no severity rating system at that time) ravaged the small island of Barbados. The parish cleric (John Orderson) of St. Michael parish recorded in St. Michael Parish Register RL1/5 an unimaginable description of the "1780 Great Hurricane".

The devastation done to the wooden construction buildings on the island and the 4,500 deaths on Barbados alone due to that hurricane (now estimated by scientists at the Hurricane Tracking Center of the National Oceanic and Atmospheric Administration) to have been larger than a Force 5), prompted John Orderson to scribe in the Parish Register: *"To give a description of the terrible storm is beyond the abilities of man; but in order that posterity may be put in mind of the dismal scene which was before us, it is here faintly recorded. Adequate description of the accumulated horrors of the night is impossible, but the roaring of the tempest, the noise occasioned by the fall of groans of the wounded, who were hurt by, and buried under the ruins, the insipient flashes of lightning, and unabating torrent of rain, formed together such a scene of terror as has seldom been equaled in any part of the world."* He added (in Latin): *"Crudelis ubique luctus, ubique pauor, eplurima mortis imago"* meaning *"Suddenly everywhere there was mourning, everywhere great terror, and many images of death appeared."*

The Great 1780 Atlantic Hurricane, which formed on October 9, 1790, and dissipated on October 20, 1780, continued on a path over the other islands of the Lesser Antilles and over the larger islands in the Caribbean Sea. The total death toll of that singular hurricane was estimated at a minimum of 22,000 with many humans missing (presumably their bodies taken out into the ocean) to a total of possibly 28,500, making it the deadliest hurricane in the recorded history of the Atlantic hurricane basin. Man was not burning any significant amount of coal in 1780 to have affected that hurricane.

For Reader interest, a tabulation below reflects the Top 10 Deadliest Atlantic Hurricanes recorded starting in 1775 and to the second deadliest Hurricane "Mitch" (Force 5) in the 1998 Atlantic hurricane season (with 11,374+ deaths):

Deadliest Atlantic hurricanes			
Rank	Hurricane	Season	Fatalities
1	? "Great Hurricane"	1780	22,000–27,501
2	5 Mitch	1998	11,374+
3	2 Fifi	1974	8,210–10,000
4	4 "Galveston"	1900	8,000–12,000
5	4 Flora	1963	7,193
6	? "Pointe-à-Pitre"	1776	6,000+
7	5 "Okeechobee"	1928	4,112+
8	? "Newfoundland"	1775	4,000–4,163
9	3 "Monterrey"	1909	4,000
10	4 "San Ciriaco"	1899	3,855

(Source: US National Oceanic and Atmospheric Administration)

It must be said – from personal experience – that from every problem, there is often some good that results. In the engineering profession it is referred to as "Lessons Learned". After the population of Barbados recovered from their massive losses in life and property caused by the "Great Hurricane of 1780, Barbados rebuilt the buildings all over the small island, but they were rebuilt from coral blocks cut from the coral rock quarries on the eastern half of the island and not from the soft wood structure that could not withstand the hurricanes that have impinged on the island almost every Atlantic hurricane season. Except for the Parish church records (Barbados is comprised of 12 Parishes named after saints in the Christian Bible, and each Parish has a church serving that parish), detailed statistics are unknown about the Great Hurricane of 1780.

The second deadliest hurricane shown on the list above was Hurricane Mitch in 1998 and that is one of the best documented hurricanes from when it was created as a tropical depression. Hurricane Mitch was singled out by this Author, whose background includes several aircraft accident investigations for accidents involving multiple fatalities, to draw a parallel if any between the root cause for high number of fatalities accidents and with the objective to identify what can be done to prevent such massive fatalities from recurring. It was felt that understanding hurricanes better might help to understand the largest and most frequent natural event that kills so many humans every year and gain a better appreciation for the hazards that Man encounters living on the planet.

The US National Hurricane Tracking Center of NOAA is one of the best public service groups in the United States and it is from their database of information that enabled a rather fascinating history of the path that Hurricane Mitch took in October and November 1998 in the Central American region before passing over southern Florida and into the “hurricane graveyard” of the north Atlantic Ocean. The track of Hurricane Mitch is shown as follows:

(Source: Atlas of the World (Rand McNally) and National Hurricane Center database)

Most hurricanes originate as being an unnamed Tropical Depression (TD), and on October 22, 1998, the NHTC detected a tropical depression in the southwestern Caribbean Sea near the cities of Cartagena and Valledupar in Colombia. The TD began moving westerly and turned northwesterly in the same manner most Atlantic hurricanes do, but it gained hurricane Category 5 so quickly that even hurricane trackers were astonished the outer fringes of the eye were extensive and deluged the Panama and eastern Cost Rica with over 24 inches of rain in a two-days slow-moving but very water-laden precipitation. To avoid a lengthy narrative, a table was constructed showing the countries that were impacted, the number of fatalities that were incurred (mostly due to flooding and mudslides) in densely populated regions and losses of many thousands of homes.

Regional Fatalities and Damage from Hurricane Mitch 1998

	Region - Date(s)	Fatalities	Damage Costs (Structures & Infrastructures)	Rainfall Amount ~ inches	
1	Panama - October 22-23	3	$50,000	24+	High Winds; Flooding
2	Costa Rica - October 24-25	7	$92 Million	24+	Flooding; Mudslides
3	Jamaica - October 25	3	Unknown	15+	High Winds; Flooding
4	Nicaragua - October 25-26 (Cat 5)	3,800	$1 Billion	50+	High Winds; Flooding; Mudslides
5	Belize - October 26-27 (Cat 5)	11	$50,000	15+	Flooding
6	Honduras - October 27-29 (Cat 4)	7,000	$3.8 Billion	36.5	High Winds; Flooding; Mudslides
7	Offshore - October 28	31 (Crew)	$1 Million	Unk (50 ft waves)	Sank Windjammer "Fantome"
8	El Salvador - October 30 (Cat 3)	240	$400 Million	24+	Flooding; Crop & Livestock
9	Guatemala - October 30-31 (Cat 2)	268	$748 Million	24+	Flooding; Mudslides
10	Mexico - November 1- 2	9	$1 Million	24+	Flooding; Landslides
11	United States - November 5	2	$40 Million	10	5 Tornadoes F1 (4) - F2 (1)
	Total	**11,374**	**$6.8 Billion**		

Note: Tropical Depression formed on October 22, 1998, in SW Caribbean Sea near Cartagena

(Source: Tabulation by Author of multiple databases of countries involved)

It was noted that a common hazardous factor was the slow-moving hurricane and tropical depression inundating each of the ten countries in Central America and in Mexico with water depths varying from one foot to five feet. To ascertain if there was any correlation between the number of fatalities and extent of damages with the depth of precipitation, the Author constructed the diagram shown next (the fatalities shown in red, and the damage costs shown in blue as they increased with increased water depth.

Fatalities and Damages Proportional to Rainfall Amount

(Based on Hurricane Mitch in October 1998 in Central America Region)

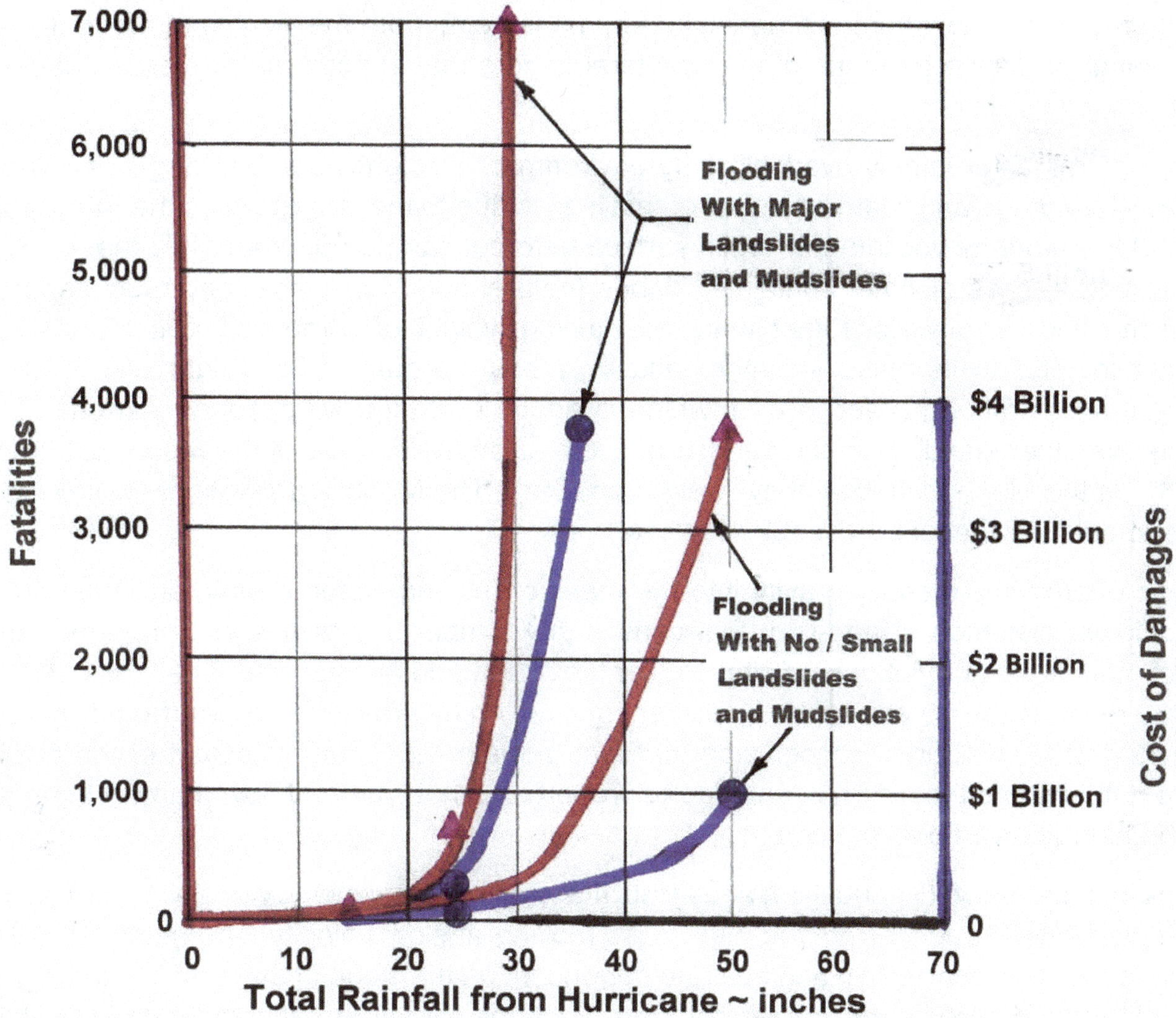

(Source: Author generated from previous tabular database)

The primarily conclusion drawn, not unexpectedly, was that the most useful "Lesson Learned" was that the soil composition on slopes of hills and low mountains can handle up to about fifteen inches of rainwater before causing massive flooding but at about twenty inches of rainwater in less than two days will create landslides and mudslides in hilly regions which are the main cause for high numbers of fatalities and high costs of property damages. That obviously may be different for flatter terrains where flooding becomes more of the hazard to which humans, animals, and properties are exposed. There may be some correlation that the red and blue lines that are at the left are deaths and damages due to flooding and with major landslides and mudslides, whereas the red and blue lines at the right are lower deaths and damages due to large depths of floodwaters but with no or only minor landslides and mudslides accompanying the flooding.

With reference to the deaths and damages in the United States caused by Hurricane Mitch, the database showed that there were 5 tornadoes spawned from Hurricane Mitch – four at Force 1 category and one at Force 2 category. As noted by the National Hurricane Tracking Center, Hurricane Mitch indicates that a slow-moving tropical storm can cause more fatalities and damages than a fast-moving higher Force-rated hurricane because of the lesser depth of water being able to be absorbed more gradually and not causing landslides and/or mudslides.

This book has thus far emphasized the many causes for natural disasters, but the other “side of the coin” is the more important humanitarian side – humans being both the victims of those disasters and perhaps being the cause for some of the collateral factors that exacerbate the losses caused by the disasters.

A primary natural factor largely overlooked by environmental scientists is that of gravity. The flooding causing much damage and a large number of fatalities is due to the rate at which the rain saturates the ground and therefore moves laterally on its surface before it can be absorbed. It is the acceleration of gravity causing a mass of water containing debris (natural and Man-made) and most importantly the mud with the flow. In physics, rather than refer to the weight of something it is more scientifically appropriate to refer to its mass – mathematically, mass is equal to its weight (W) divided by the gravitational acceleration (g) acting on it which is about 32.2 feet per second^2. Astronauts have an equal mass whether standing on the surface of Earth or on the surface of the Moon, but they “weigh” about 1/6th on the Moon than what they “weigh” on Earth. The Moon’s gravitational acceleration (5.34 ft/sec^2) is about 1/6th that of Earth’s (32.17 ft/sec^2).

The kinetic energy imparted by gravity into the mass of moving water, debris, and mud and moving towards a lower elevation is the factor that causes the number of fatalities and property damages to soar – not merely the amount of precipitation that has fallen and at what rate. The mass of debris/mud adds to the mass of the water and the kinetic energy, driven by gravity, is known (from Physics 101) to be KE = ½ mass x Velocity^2. Recognizing that fact, a major factor in the extent of the fatalities and injuries to humans and the extent of damage to property is what type and how many of them are in the path of the high energy flood movement.

The debris in a “mudflow” or “landslide” adds to its mass as it progresses. Even a slow-moving mudslide can have high kinetic energy from the high mass moving at slow velocity. The mudslide will move in the direction over the local terrain and at a rate dependent on the slope of the terrain in that topography. A mudslide – or avalanche – moving rapidly down a hillside is of no great concern to the public (but is to environmental protectionists) if there is no sizeable population or large number of properties in its path.

A tragic example of those combinations of different factors – geophysical, natural, and human – is seen in the mudslides that occurred in the regions of eastern Spain on the northern Mediterranean Sea in the nine days of late 2024 (October 29 to November 8). A photograph of the city of Valencia (population about 808,000) and inland area above Valencia is shown as follows:

(Credit to unknown photographer

This is the hillside above the city of Valencia (seen in the background of the photograph with multi-story buildings), at latitude 29.47° N in the central east coast of Spain and adjacent to the generally placid Mediterranean Sea. There are only warm currents in the northern Mediterranean Sea in the coastal regions of Spain and southern France. There are occasional low-pressure systems throughout the lengthy Mediterranean Sea but they are not seasonal in nature. These depressions are termed as "Medicanes" and typically are not as strong as Atlantic Ocean hurricanes or Indian Ocean cyclones. The photo shows the origin of the mudslides in the region of Valencia – which caused 226 deaths, 14 still missing, and property damage over US $3.8 billion (about 3.5 billion euros).

Rather than attempt to describe the details of the Valencia mudslide, the Author has included an article by the Reuters News Service that is very credible and presents a large number of photographs which describe the physical destruction, and the harm caused to residents in the city regions. The article is included in its near entirety with over 40 emotionally "wrenching" photographs in Appendix 2 of this book. The mudslide flowed rapidly and with so much kinetic energy from the debris and mud in the flowing mass that it caught residents in underground parking facilities and literally buried them alive. Two photographs are extracted to attach a visual example of the problem of debris and mud being mixed in with the excessive flood waters forming a "battering ram" effect more destructive than just the hydrodynamic pressure of the flow of water without any debris:

(Credit for Photos Above to Reuters News Service Photographer)

The debris in the high kinetic energy mudflow not only added to the magnitude of the clean-up of the mud and debris but denied access to the impacted area by rescue/recovery and clean-up crews to the scenes of the mess within the city of Valencia. There is not much preventative action that can be taken to avoid such problems in cities like Valencia, but municipal leaders in cities everywhere should ensure that, when such unavoidable tragedies do and will strike, the resources needed to recover from the tragedies are available on an emergency basis. Debate over what caused the mudslides is of secondary importance – human health and safety are of far higher priority than where to point the finger of blame.

There is also a very humanitarian aspect of tragedies that are caused by rogue natural events in a region. Neighboring regions should assist in the rescue and recovery efforts as a humanitarian act and not bill the impacted region for those already insurmountable loss of lives and properties. Homeless families need and deserve a place to take shelter from merciless natural elements that ensue tragedies such as that in Valencia. Reduction of "greenhouse gasses" in the atmosphere may be of primary importance to environmental scientists to protect the environment, but it is a far higher priority to protect the human population of the planet.

For the purposes of this book, the occurrence of hurricanes being seasonal and/or regional is yet another confirmation that environmental scientists should no longer use the "average global temperature" as the entire basis for setting goals to control Man's emissions of "greenhouse gasses". In a later chapter of this book, the "Lessons Learned" from Hurricane Mitch and other Atlantic hurricanes will be included as the basis for one conclusion and recommendation made to improve the safety of the human population of the United States region. It is strongly recommended that the building codes of future structures (residential and commercial) be changed to withstand those excessive forces of nature unleashed in hurricanes and tornadoes that will most likely get worse than before they get better.

A composite of several NASA satellite photographs tends to verify the wind currents near the Equator with the cloud formations:

(Source: NASA)

From the observation that the clouds are not homogenously distributed in the atmosphere, there should also be a question raised about Man-made "greenhouse gasses being homogenously mixed into the atmosphere at all latitudes. There can be little disagreement by any branch of science that near the surface of the planet – where the humans inhabit the land area and industrial processes and the transportation in land-based vehicles travel - that whatever emissions are created by any process (natural or anthropogenic) in the Northern hemisphere are not transferred into the Southern hemisphere and vice versa. There are very few coal burning facilities south of the Equator as there is almost negligible heavy industry and much less movement of vehicles on roads in the infrastructure below the Equator. This factor alone would cast great doubt on the environmental scientists claim that carbon dioxide emissions that are Man-made in the industries in Europe and the United States causing the coral reefs near Australia to be bleaching white from stress or causing such environmental chaos in Bangladesh. If stating factual data is a form of "naysaying" then there is little hope that environmental scientists will ever admit to the public and to Government environmental regulators the fallacy that Man-made "greenhouse gas emissions" are not the dominant or even highly significant contributors to the planet's "average global temperature" not being a credible scientific indicator of climate change.

The Equator is a most important dividing line in earth and climate science. Accordingly, the Author will repeat again for emphasis that the planet cannot be analyzed for global warming, climate change, and/or ocean levels rising with the "average global temperature" reference from which temperature and climate anomalies or present ocean levels are rising but must be analyzed in logically chosen regions (of maximum size being the continents, and of minimum size being but preferably within latitudinal zones. Logically, there can be no region on one side of the Equator being affected by natural-caused effects and/or anthropogenic lack of consideration of the environment of a region on the other side of the Equator. By the same logic, the consideration of gravity, the principles of the Coriolis effect, and the principles of water vapor in the atmosphere cannot be different between regions on the same side of the Equator.

Historical records and documents show that there are about 1,400 reported tornadoes annually. Every continent except Antarctica has experienced tornadoes. With about 300 tornadoes reported in Europe annually, the United Kingdom having about 33 reported tornadoes a year has the most of any European country. Tornadoes can occur at any latitude on the rotating spherical mass called Earth., and a photograph shown next was taken of a tornado in Australia:

(Credit to unknown photographer in Australia)

A rogue tornado in Bangladesh in 1989 left 1,300 people fatally injured by the tornado and another 80,000 displaced from their homes – built without construction safety standards - which were never generated and still do not exist in too many cities in lesser developed regions. However, it is the United States where tornadoes occur with disturbing regularity and high severity and predominantly in the eastern half of the United States annually that is the focus of this book.

Various scientific interpretations exist for the reason why the eastern half of the United States has the historical reputation for having the most frequent and most severe tornadoes (with the greatest property and human lives losses). The illustration shown next for the tornado history in the United States clearly indicates where the highest annual frequency of tornadoes has been reported:

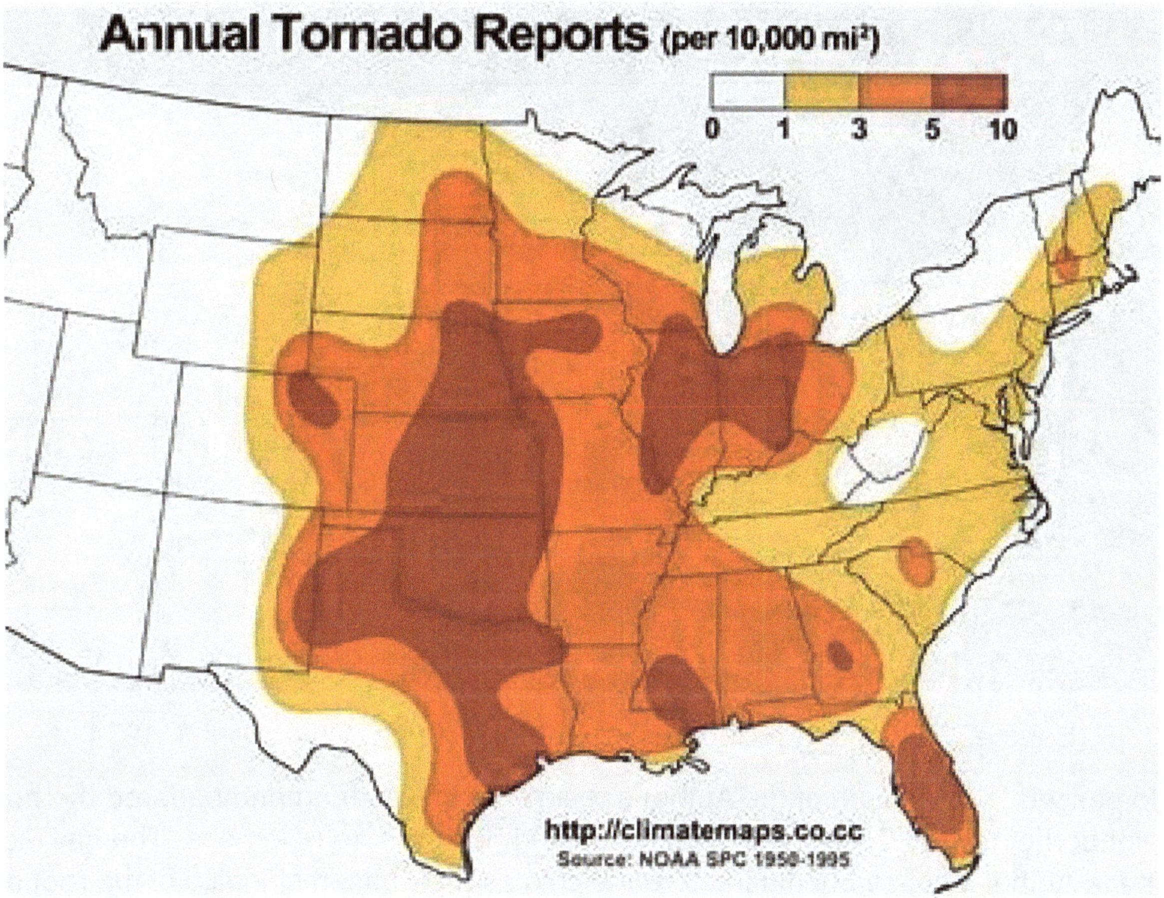

(Source: NOAA)

The rotation of tornadoes and hurricanes are opposite in the Northern and Southern hemispheres, but they have the common ability to cause weakly constructed structures to collapse or be torn apart and the flimsy roofing and wall covering debris (typically wood planks) to be lifted into the atmosphere and deposited up to hundreds of miles elsewhere in the path of the tornado. Generally, but certainly not exclusively, the most severe (i.e., highest energy content) tornadoes are again seen to occur in the states of Texas and Oklahoma. Also, generally most tornado damage is done by the high wind energy and the very low atmospheric pressure contained within the highly vortex shaped core of the tornado. There is not normally any extreme amount of precipitation associated with tornadoes except for the possibility of large hail stones on the hurricane's outer fringes which fall more vertically rather than being driven horizontally. (This Author and his family resided in Kettering, Ohio, for twenty years and were property damage victims from the hail in the periphery of the disastrous Xenia Tornado on April 3, 1974. The town of Xenia was/is located 15 miles directly east of Kettering in the extreme southwestern corner of Ohio. A photograph (from an Ohio friend's memorabilia) of the Force 5-6 tornado as it was entering Xenia (passing through the Arrowhead subdivision of Xenia) is shown as follows:

(Source: Co-worker (J. Shoates) of Author – Xenia, Ohio, April 3, 1974)

The tornado passed ½ mile south of the Author's residence in Kettering and battered the homes in the Author's neighborhood (Oak Creek) with high winds and hail the size of lemons. The hail did not break any glass in the Author's house but did fracture the cold asphalt-based shingles of the roof and caused the entire roof (and of all neighbor's) of the house to need replacement. In addition, it was popular to have corrugated aluminum panels over the outdoor patios of the houses and the Author's patio cover resembled having been struck by a thousand ball peen hammers damaging the patio cover so that it too needed replacement. But that is minor annoyance compared to the 33 persons who lost their lives in Xenia and the losses of over 300 family homes in the Xenia area – the homes were rebuilt but the deaths could not be reversed. By good fortune, the tornado struck in the late afternoon and after the school children had vacated the school and grounds. The school was demolished 1 ½ hours later by the merciless tornado. The latitude of both Kettering, Ohio, and Xenia, Ohio, is approximately 39.7 degrees North.

Tornadoes have the unpredictable variables of width of the swath as well as the length of "dwell" time the swath will exist on the ground. The Xenia tornado is an example of those that have very wide-swaths and "bull-doze" through neighborhoods such as that of the Arrowhead sub-division of Xenia. A photograph, with annotations by the Author to illustrate the swath width and the direction of the Xenia tornado, is shown as follows:

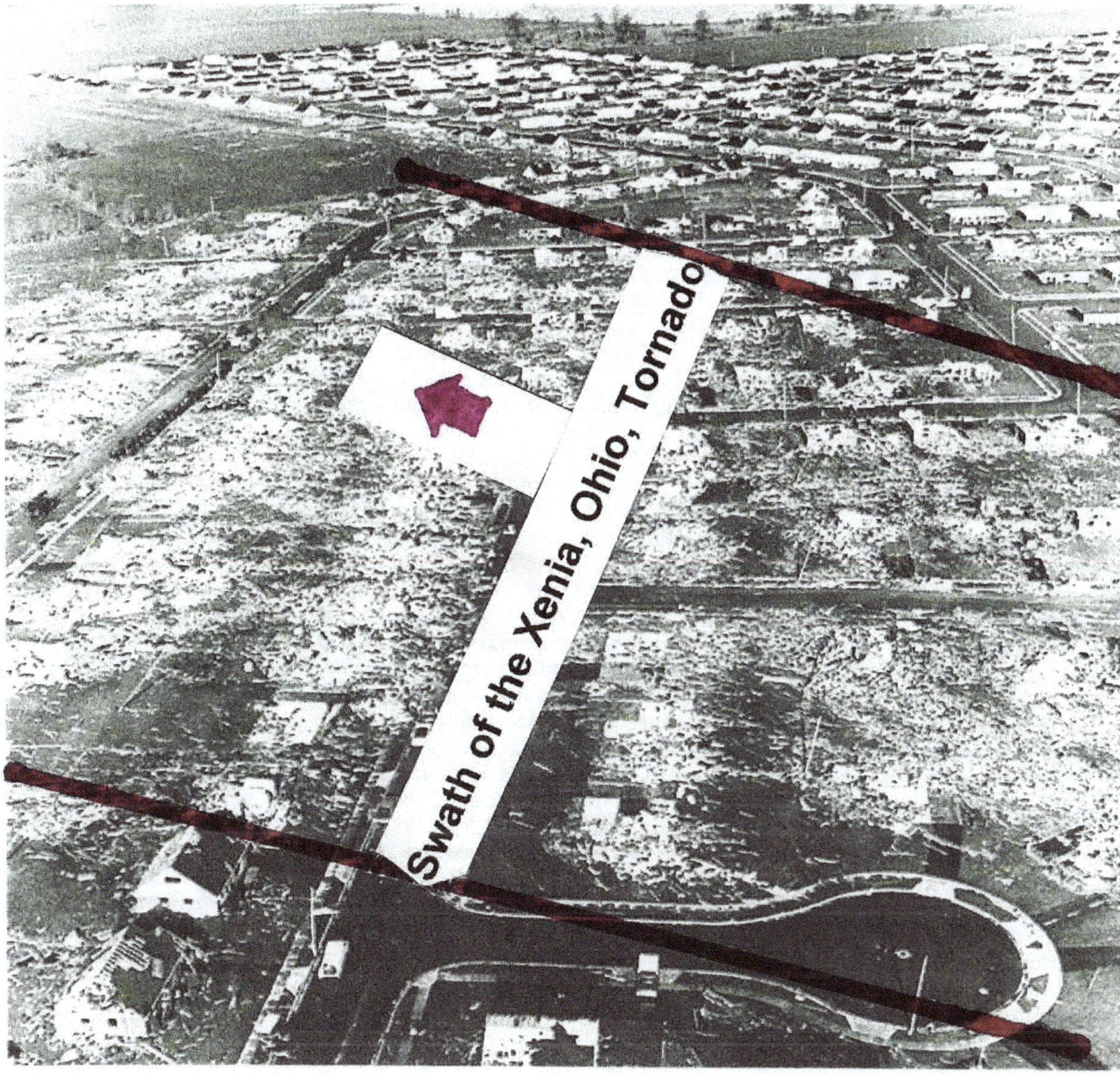

(Source: Author's Personal Photographs file with annotations)

The "misbehaviors" of water on the planet were accounted in the news media, most being largely factual as to the fatalities caused and the vast amount of damages (and high cost thereof to repair and replace). But many – and increasingly more fallacious - report the misbehaviors and weather-related catastrophes to be due to the Man-made "greenhouse gasses" in the atmosphere as the <u>dominant</u> factor causing a heat energy imbalance that instigated the weather to misbehave.

An example of a typical article – one of the better-written ones - in a reputable newspaper is shown next in its complete form that illustrates that the public is assumed to have accepted the continuously repeated mantra of the environmental science profession that it is "settled science" that humans creating "greenhouse gas emissions are the dominant cause of climate change". Any fallacious claim (to any degree) that is repeated often enough and widely enough for a lengthy period of time eventually becomes factual in the public interpretation.

Record heat unleashes deadly floods from New York to Libya

BY BRIAN K. SULLIVAN
Bloomberg News

Extreme heat is usually associated with drought and wildfires. But across five continents this year, it's also unleashed deadly flooding.

Cities around the world have seen record rainfall 139 times in 2023. A rare hurricane-like storm inundated Libya last month, killing thousands. More than 100 people died across Asia during an intense monsoon season in July. After fatal floods pummeled the U.S. Northeast over the summer, torrential rains paralyzed New York City last week.

Soaring temperatures were at the root of all these weather calamities. June, July and August were the hottest ever for that period. Higher air temperatures mean the atmosphere can hold more moisture, leading to heavier rainfall, while record-hot oceans provide fuel for storms. And as the climate warms, flooding is poised to get worse.

"We are now seeing record-smashing ocean temperatures, record-smashing global temperatures and record-smashing floods," said Jennifer Francis, a climate scientist at the Woodwell Climate Research Center. "It's all connected."

In the U.S., two major flood events this year – in California from January to March and the Northeast in July – killed 32 people and caused $6.7 billion in damages, according to the U.S. National Centers for Environmental Information. The damages are the highest since 2019, when accounting only for flood incidents totaling more than $1 billion.

Global damages are difficult to tally because not all countries keep accurate records, but flood events have left more than 17,000 people dead or missing in 2023. On average, there were 216 floods or landslides worldwide per year from 2019 to 2022, compared with an average of 160 over the previous four years, according to the International Disaster Database.

In large part, the ever-worsening downpours are caused by what scientists call a "positive feedback loop."

The Earth's temperature rises in response to greenhouse gases like carbon dioxide and methane. Hotter air holds more water vapor, which then absorbs heat and prevents it from escaping the atmosphere – resulting in even more moisture in the air and heavier rainfall.

As the atmosphere gets warmer, so have the world's oceans. Higher water temperatures heat up the air above the sea surface, generating energy for hurricanes and tropical storms to form. The Atlantic saw 10 storms develop from Aug. 20 to Sept. 16, the most on record for that period, according to Phil Klotzbach, a Colorado State University hurricane researcher.

The storm that devastated Libya last month was a "medicane," a name that's an amalgamation of Mediterranean and hurricane. These weather systems are smaller than large Atlantic storms but can strike with Category 1 force on the five-step Saffir-Simpson scale, bringing winds greater than 74 mph and flooding rains.

(Source: Durham, NC "News & Observer" October 9, 2023)

Three factors seldom, if ever, mentioned in news media articles are 1) that it is high heat energy that generates water evaporated from the water surface of the planet into the atmosphere; 2) the ever-present gravitational acceleration (gravity is not a force but an acceleration factor) which causes the massive amount of water in the atmosphere to fall back (i.e., be accelerated down toward the center of the Earth) onto the Earth's surface; and 3) the human-built infrastructure that was designed for "normal" environmental conditions but is not sufficiently strong to withstand anomalistic weather and climatic extreme events (seasonal or random).

Environmental scientists seem to also be "settled" on there being a substantial "feedback" cycle resulting from "climate change", as indicated in the chart shown as follows:

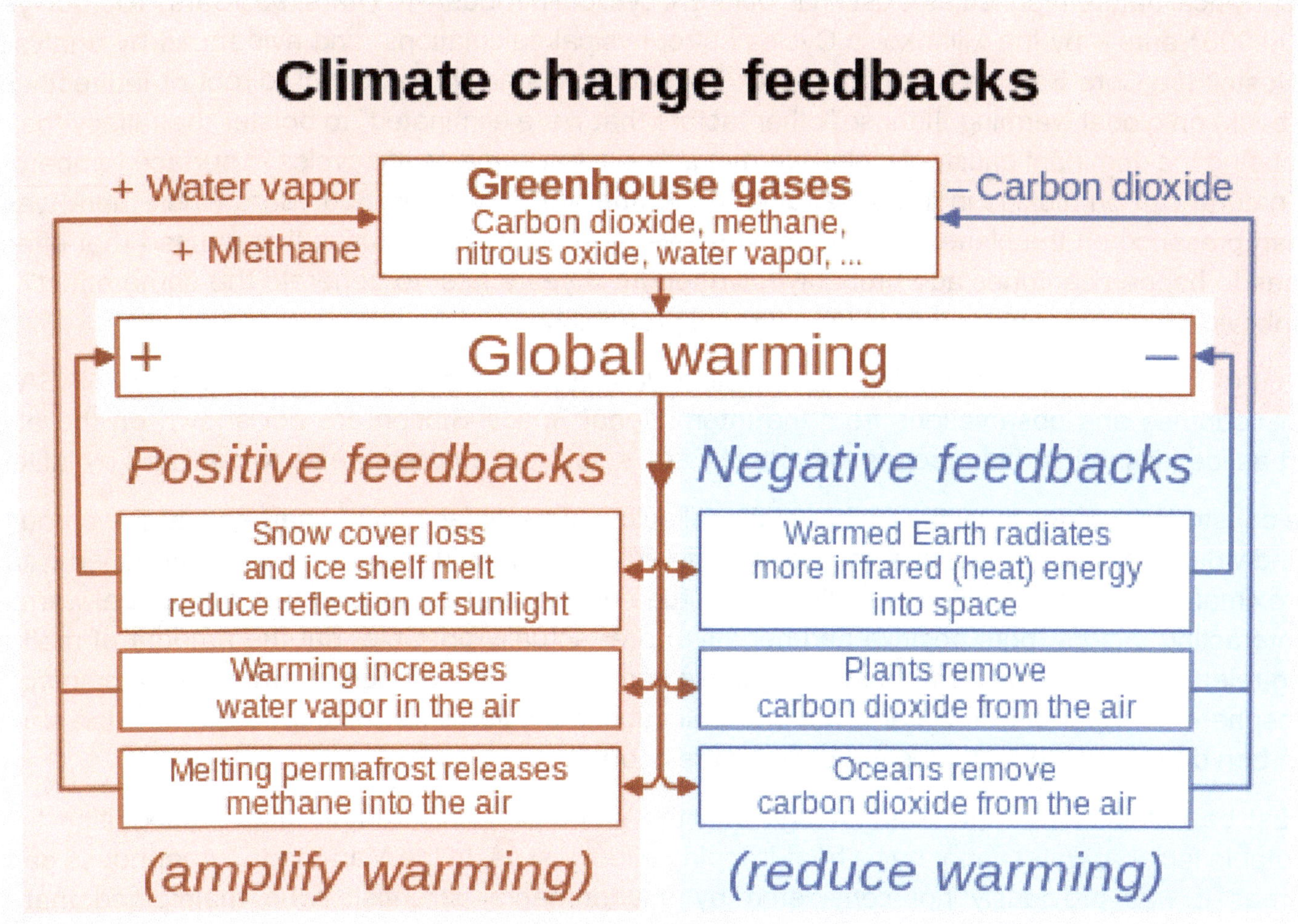

(Source: NOAA)

It should be noted that the chart indicates that the feedback described (both positive and negative) are on "Greenhouse gases" and "Global warming", but the title of the chart is "Climate change feedback". It appears that the conceiver of the chart has mistakenly assumed that "global warming" and "climate change" are one and the same. (In an earlier chapter, the variables of surface temperature and precipitation are defined individually as "weather" in the short term (days) and as a combination over a long term (minimum 25 years) produce a database for calculating "climate". If the chart is mislabeled (should not the title more correctly be "Global warming interactions"?) that would explain part of the mistitling.

But how can the "after the fact" climate average of more than 25 years of weather factors effect (as "feedback" or any form of "effect") those weather factors that occurred as much as 25 years earlier? That doubt of scientific inconsistency in the environmental scientists' hypothesis causes the probability of scientific validity of the hypothesis being low to nil. There can be an "interaction" between the weather and the atmosphere and vice versa but they are contemporary actions, not "feedback" from a future state (of the climate) to a current state (of the weather). The "tail does not wag the dog". Using the term "feedback" in electronics may seem appropriate because of the speed of electrons – contrarily, the time period between "weather" and the "climate" resulting from the weather is measured in years of time. The weather today does not define the climate tomorrow or next month.

It should also be noted on the chart that the chart originator does not show any effect of astrophysical nature on global warming (as discussed in the earlier chapter of this book). The other databases of astrophysical nature (verified as existing in different cyclical periods - 40 years, 90 years, 100,000 years to 400,000 years – by the Milankovic Cycles astrophysical calculations, and evidenced by analysis of the Vostok Ice Core Samples) are not shown in the chart as having an effect (direct or indirectly as a feedback) on global warming. If those "other factors that were eliminated" to bolster the fallacy that it is Man being the dominant cause of global warming, then what explains the cycles in surface temperature and natural carbon dioxide in the air lagging the temperature by about 800 years when there was no human presence on the planet? Would not whatever combination of causes that caused that effect of causes to happen centuries ago probably have recurred years later to generate the same effect? The Milankovic calculations show that to be a distinct probability.

The current database of information from satellites launched into orbit around the Earth by NASA and other countries and observations from the International Space Station are not shown on the above chart as identifying the other factors that affect Earth's surface temperature regionally and by latitude.

As a constructive criticism by the Author of the fallaciously titled 'feedback" chart above, the amount of the individual positive interactions and negative interactions should each at least be identified with an approximate percentage of their individual magnitude on the total interactive loop with global warming. All interacting factors (both positive and negative) are factually possible, but the amount of methane being released by melting permafrost is negligible compared to the amount of water vapor being in the atmosphere. The increase in water vapor in the air caused by global warming far outweighs the amount of carbon dioxide in the air that is removed by the oceans.

It must be noted more importantly though that the interaction chart above gives recognition to the irrefutable fact that water vapor has a heat-trapping effect just as do the Man-made "greenhouse gases" whereas it was previously not considered by environmental scientists who maintained that "by eliminating all other contributors (which included water vapor), settled science concurs that Man burning fossil fuels is the dominant cause for global warming". The charts shown earlier in this chapter - generated by Dr. S. Fred Singer (Professor Emeritus of Climate Science at the University of Virginia) - dispels that original hypothesis that was first stated in the mid-1950s yet fallaciously persists to the present time. The environmental scientists, the "green" activists, the news media, and worst of all the governmental organizations that are spending taxpayers' monies to "combat climate change" have "circled the wagons" and have totally misled and confused the public thinking. It is hoped that this book will cause common-sense to prevail once again in the scientific records of the Earth's turbulent history.

Nothing humans can do will change the effect of gravitational acceleration when it moves a flood of water downhill, other than perhaps to create diversion channels that change the rampart water to flow into reservoirs or catch basins for later use to offset any drought episodes that may and probably will occur. But there is much the human population of a region can do to make their infrastructure able to withstand earthquakes, floods, tornadoes, and hurricanes. It is disturbing when reading that weather-related catastrophes have taken thousands of human lives and caused billions of dollars in property damage, but the news media seldom reports all the factors that contributed to the causes of the catastrophe. In the zeal to tag humans and their dreaded greenhouse gas emissions with the responsibility for the catastrophes reported above, some of the more important contributing factors – far more consequential than the small amount of greenhouse gasses created in Libya for example was the fact that the pressure of the flood waters had caused a poorly constructed dam to burst when excess water behind the dam exceeded the design structural strength of the dam (to withstand the pressure of the water behind the dam). The collapsed dam in Libya sent millions of gallons of water racing downhill towards the coast of the Mediterranean Sea and washed thousands of victims residing below the dam into the Mediterranean waters.

The news media however contends that all the flooding in Libya was triggered by Man burning fossil fuels and causing "greenhouse gasses" to trap all the heat of the Libyan desert and lift that much water into the atmosphere above Libya. Repeating a photo from satellite optical records of the planet shows that the atmosphere above the major deserts of the planet – including Libya in North Africa – is "normally" devoid of cloud cover and significant water vapor:

(Source: NASA)

It is more probable that winds at altitude had blown the water-saturated clouds seen just south of the North African region over onto the location above Libya that experienced the natural catastrophe. The news article inferred that the cause of the natural and Man-made error in dam design was "The Earth's temperature rises in response to greenhouse gases like carbon dioxide and methane." And subliminally causes the public to accept the scientific hypothesis that "Man is to blame for having burned a massive amount of fossil fuels in that region".

Before continuing with other pertinent topics, a very noticeable coloration in the above photographs indicates the extensive areas of the planet that are desert. A table below indicates where the largest deserts of the world are located and their vast size:

Largest Deserts of the World

(Arranged by Planet Quadrant and Surface Area)

Desert Name	On Continent	In Quadrant on Planet	Desert Area ~ Sq. Miles	Desert Area ~ Sq. Kilometers	% of Quadrant Surface Area
Sahara	Africa (Northern)	Northeast	3,552,140	9,207,502	7.21%
Arabian	Asia (West)	Northeast	1,000,000	2,592,100	2.03%
Gobi	Asia (East)	Northeast	500,002	1,296,055	1.02%
Rub' Al Khali	Asia (Middle East)	Northeast	250,000	648,025	0.51%
Syrian	Asia (Middle East)	Northeast	193,051	500,407	0.39%
Karakum	Asia (Central)	Northeast	135,136	350,286	0.27%
Kyzylkum	Asia (Central)	Northeast	115,831	300,245	0.24%
Taklamakan	Asia (China)	Northeast	104,248	270,221	0.21%
Thar	Asia (Indian Subcontinent)	Northeast	77,220	200,162	0.16%
		Total	5,927,628	15,365,003	
		Quadrant Surface Area	49,240,436	127,636,134	**Total 12.04% (Sand)**
Great Basin	North America	Northwest	189,845	492,097	0.39%
Sonoran	North America	Northwest	119,692	310,253	0.24%
Mojave	North America	Northwest	47,877	124,102	0.10%
		Total	357,414	926,452	
		Quadrant Surface Area	49,240,436	127,636,134	**Total 0.73% (Sand)**
Kalahari	Africa (Southern)	Southeast	347,492	900,734	0.71%
Great Victoria	Australia	Southeast	134,650	349,026	0.27%
Great Sandy	Australia	Southeast	110,036	285,224	0.22%
Namib	Africa (Southern)	Southeast	61,776	160,130	0.13%
		Total	653,954	1,695,114	
		Quadrant Surface Area	49,240,436	127,636,134	**Total 1.33% (Sand)**
Patagonian	South America	Southwest	259,847	673,549	0.53%
Atacama	South America	Southwest	54,054	140,113	0.11%
		Total	313,901	813,662	
		Quadrant Surface Area	49,240,436	127,636,134	**Total 0.64% (Sand)**
Antarctica	Antarctica	South Polar	5,500,024	14,256,612	**Total 100% (Ice)**

Notes:
Planet is assumed to be spherical with average diameter of 7,918 miles (12,742 km). Area is 196,961,745 sq mi (510,065,665 sq km)
Quadrants are defined relative to Equator (0 degs latitude) and Prime (Greenwich) Meridian (0 degs Longitude).
The total area of all major deserts on the planet is 12,752,921 sq miles (33,056,843 sq kilometers).
6.47% of the planet's surface area (196,961,745 sq miles) is covered by desert sand (or desert ice).

(Source: Author tabulation from databases in Encyclopedia Brittanica)

Scientifically, a desert region is classified as being a desert due to lack of annual precipitation. By that definition, the entire continent of Antarctica is classified as being a desert due to almost zero precipitation in that polar region where the average temperature at that southern latitude is about -76 degrees Fahrenheit in the high elevations and 14 degrees Fahrenheit in the coastal regions (Southern Ocean). The coldest temperature ever recorded on Antarctica was -128.6 degrees Fahrenheit (-89.2 degrees Celsius) at the Vostok Ice Research Station at latitude -75 degrees South in July 1983.

Technically, because of its dry climate (having an average temperature year-round of below -20 degrees Centigrade (- 4 degrees Fahrenheit) and therefore no water vapor in its local atmosphere), the Antarctic Continent is classified as a desert. Much like the numbers in the table of oceans being huge, the total of the desert areas (to be shown in tabular numerical format) are slightly smaller than are the oceans, but with the surface of the planet being over 80% water, and the deserts (sands plus Antarctica's ice) covering another 6.4% of the planet's surface, the world population of 8 billion humans have less than 13.6% (about 1/7th) of the planet's surface area on which to reside.

A perhaps better visual indication of the percentage of area of land on the planet's surface than that of the large table of numbers listed earlier is reflected in the graphic diagram shown as follows:

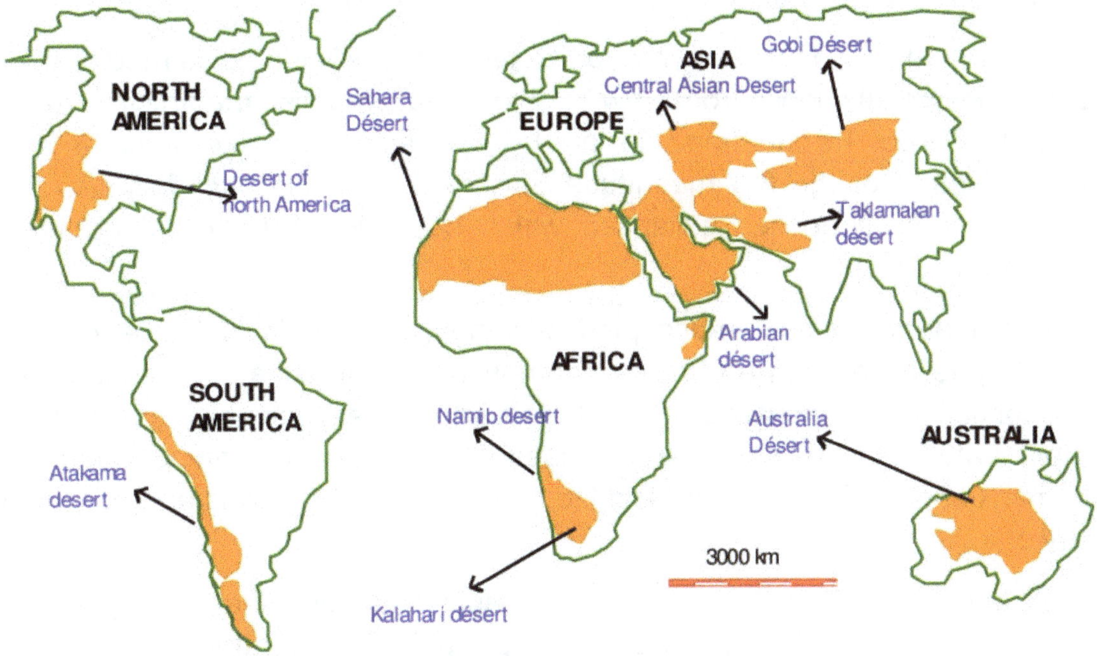

(Source: National Geographic Society)

While there are relatively few population centers in the deserts, there are populated areas on the edges of the deserts especially those on the shores of large bodies of water. A scene (in the Arabian Desert) from a National Geographic Magazine shows the means of travel across desert sands for trade purposes and for supplying of food from one region to another:

(Source: National Geographic Society)

There is a distinct lack of water in the vicinities of deserts and there seldom are any clouds bringing rain to the deserts. For reference, Antarctica is also considered as a desert as there is no precipitation in Antarctica as the temperature is on average well below freezing.

On page 8A in the Durham, NC News Observer on October 9, 2023, the news media article describing "Record heat unleashes deadly floods from New York to Libya" was also the article "Quake devastates W. Afghanistan; 2,000 killed, Taliban says". In the Afghanistan earthquake article is stated: "Afghan officials said the epicenter of the 6.3 magnitude earthquake was in two districts, Zinda Jan and Ghurianan, where mud brick houses collapsed within seconds of the initial earthquake, leaving residents with no time to escape."

Poor construction standards of human-occupied buildings and proximity to the origin of the force creating the catastrophe have been and are a contributing factor in natural-caused catastrophes of hurricanes, monsoons, flooding, landslides, tornadoes, earthquakes, and volcanic eruptions throughout the planet's history. Later in this book will be described the choice of residency (type construction and location) that humans choose for economic and lifestyle enjoyment reasons and not for best safety or health considerations.

Few articles ever attribute any of the climate changes being due to changing astrophysical and geophysical changes over which Man has no control to abate – most attribute climate change to be caused by Man-made "greenhouse gas (GHG) emissions". In the article above, water vapor in the atmosphere is mentioned in passing, but no mention is made that it is the water vapor that is retaining most of the heat energy in the atmosphere in a manner about 90 times more effectively than does the relatively small concentration (total 0.0435% - of which 0.0295% is of natural origin and 0.0140% is Man-made) of greenhouse gasses (such as carbon dioxide and methane) in the atmosphere. The "settled science" hypothesis that Man burning fossil fuels is the dominant cause for global warming is clearly fallacious – but it must be recognized that the environmental protection scientists generating that fallacious claim did so without the benefit of many of the updated databases researched by the Author and which came into existence after the fallacious "settled science" conclusion was derived.

Returning to the region of the United States, it is generally known that the geophysical effect of the cool Pacific Ocean current flowing clockwise from Aleutian waters southerly down the West Coast states (Washington, Oregon, and California) generates lower anomalies in temperature in those three western states than in the mountain and central states. The West Coast Pacific Ocean waters flowing clockwise cause the water on "cool" (anything "cool" in California is not all that bad) southern California beaches to range from a cool temperature of 55 degrees Fahrenheit to a hot temperature of 65 degrees Fahrenheit – slightly warmer when an "El Nino" much warmer than the remainder of the Pacific Ocean current flows southerly along the Washington and Oregon coasts and the entire California coast. (It is for that reason primarily why all surfers on the West Coast wear thermally insulated "wet suits" to prevent getting hypothermia during long "hanging-ten" sessions. For unknown reasons (probably for warmth rather than for aesthetics), the wet suits are most often produced in dark color giving the surfers an appearance of resembling an elephant seal or walrus which are high on the menu for great white sharks – but that aquatic hazard to humans is not directly related to climate change.)

The East Coast Atlantic Ocean current is also clockwise but brings the very warm "Gulf Stream" current from the southern Caribbean region and from the Gulf of Mexico currents flowing northerly in parallel with the East coast of the United States. The water temperature of the Gulf Stream ranges from an almost never low of 75 degrees Fahrenheit to sometimes over 85 degrees Fahrenheit and almost all year round. It is little wonder why there is so much travel by human "snowbirds" during the colder months in the northern US and southeastern Canadian provinces to the Florida beach cities and towns. Wind surfers in Florida have no need for wet suits as do the "cool" dudes in California.

The upper central states near the Great Lakes experience more and different problems than do the southeastern states. The winters in the upper central states are normally very cool – due to the "Lake Effect". However, those states just east of the Great Lakes experience blizzards and very heavy snowfalls by the warmer and more moist winds flowing to the east and encountering the warmer Great Lakes waters where they pick up additional moisture (water vapor in the air) and that high water vapor content condensing out as heavy snowfall when the moisture laden winds encounter the very cool surface temperatures of western Pennsylvania and "upstate" New York. Buffalo, New York, is notorious for having multiple blizzards of greater than 36 inches of snow during the Winter and with drifts of blown snow of 6 feet or greater. The high winds cause the snow to drift and bury cars on the highways – presenting life-threatening hazards to motorists stranded for sometimes 12-15 hours before being rescued.

Shown in the following photograph is what that type of scene is like all too often in the northern states of the USA (those having a "rush hour") during a winter season:

(Source: Automobile Association of America)

Resemblance of the above scene to that of a herd of lemmings headed for a catastrophic result is coincidental.

The northeastern states (from Maine to Delaware enjoy the "four seasons" but are clearly not paradise in the hot humid summer season. The color of the leaves on the trees change to beautiful "Autumn leaf colors" but snow is in the forecast soon after. The temperatures are often in the 80-85 degrees range during the summer but the high humidity 75-90%) associated with the high temperature can be miserable (as attested by the Author who grew up in New York City and with much time in the "upstate" region.) A scene such as above will never be seen on the Florida Turnpike – unless of course it is far in the future (10,000 years or more) when the Antarctica ice core samples indicate there will likely be another "Little Ice Age". Could it be possible that future Man may need to burn everything that can burn to survive the ice age low temperatures? Once again, the public and the environmental scientists are reminded that it is Man who must survive the extremes of the weather and the climate not the average of those extremes. The Author is not a botanist but is well aware that orchids do not bloom in outside gardens in the colder climates of the United States, but tropical flowers do bloom in "greenhouses" almost anywhere. Many "greenhouse" operators inject carbon dioxide inside the greenhouse to facilitate the chlorophyl process.

Scientists differ from the average member of the public in how to resolve a large problem being encountered such as that of the causes and effects of changes in the natural environment and by/on the activities of humans populating that environment. The average human has not been so fortunate (by choice or because of lack of funds) to possess the level of education needed to do a scientific analysis of the very complex problem that involves many factors. Perhaps this book will serve two purposes: 1) to provide the public with a simplistic "big picture" set of information, and 2) to provide the scientific profession with a different perspective from using a "one-size-fits-all" approach and "average global temperature" variable to "combat climate change". A subset of 2) above is to inform teachers of science to educate the young people in the public-school systems with the more scientifically correct causes and effects of climate change than the fallacious present set of information being taught in schools as "settled science" based on the fallacious hypothesis that "Man burning fossil fuels is the dominant cause for the climate to be changing."

Credit for the concept of analyzing the climate on the planet by regions rather than by "global averaging" should be given to the German-Russian climatologist and botanist Wladimir Köppen (1846 – 1940) and joined later by Rudolf Geiger (1894 – 1981). Without the benefit of any "big picture" satellite imagery of the climate being regional in nature, Köppen and Geiger developed a climate classification scheme for the planet but with attention to how the climate affected botanical "victims" not the human population. Their climate zone classification scheme also did not consider any effect of human-generated "greenhouse gasses" on the environment or on vegetation. Their system was much like a 3-column tabular arrangement where the first column designated a climate zone (with five climate zones generated during the "Age of Discovery" 500 years ago). The second column designated the type of precipitation that affects botanical vegetation (with four levels of "wetness"). The third column designated the general warmth of the seasons (with nine types of seasonal temperature variation).

The tabulation shown next illustrates the three-letter "code" system that Köppen and Geiger used in analyzing the five different climate groups of the planet in regions that had a unique three-letter code and color assigned to each region.

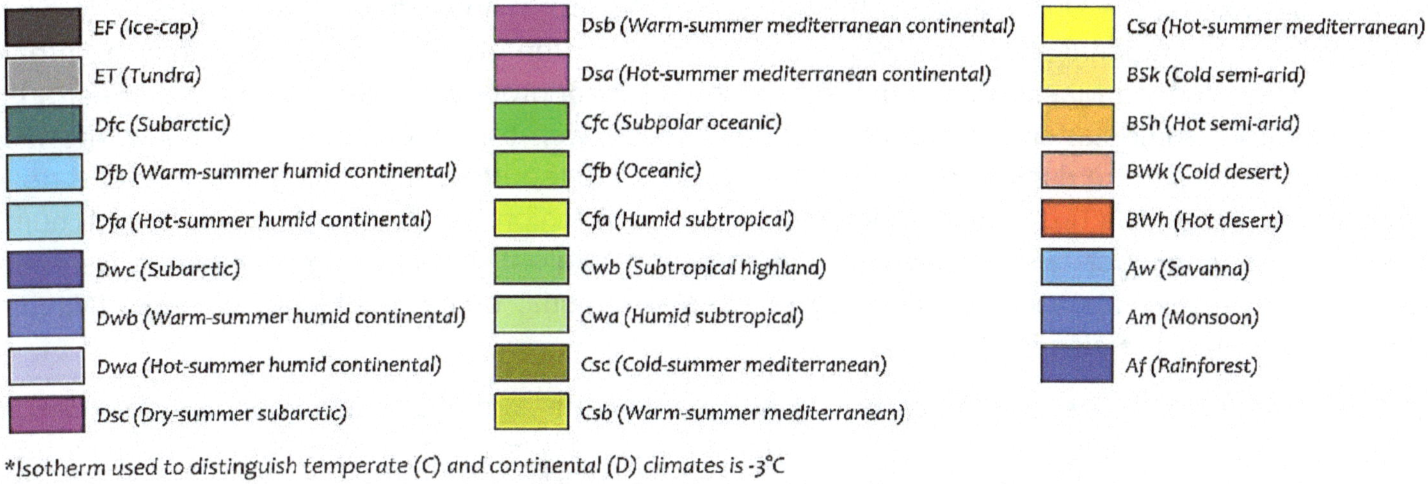

(Source: Britannica database of Köppen - Geiger Climate Classification)

The associated map with the regions color-coded per the Köppen Geiger Climate Classification method is shown next:

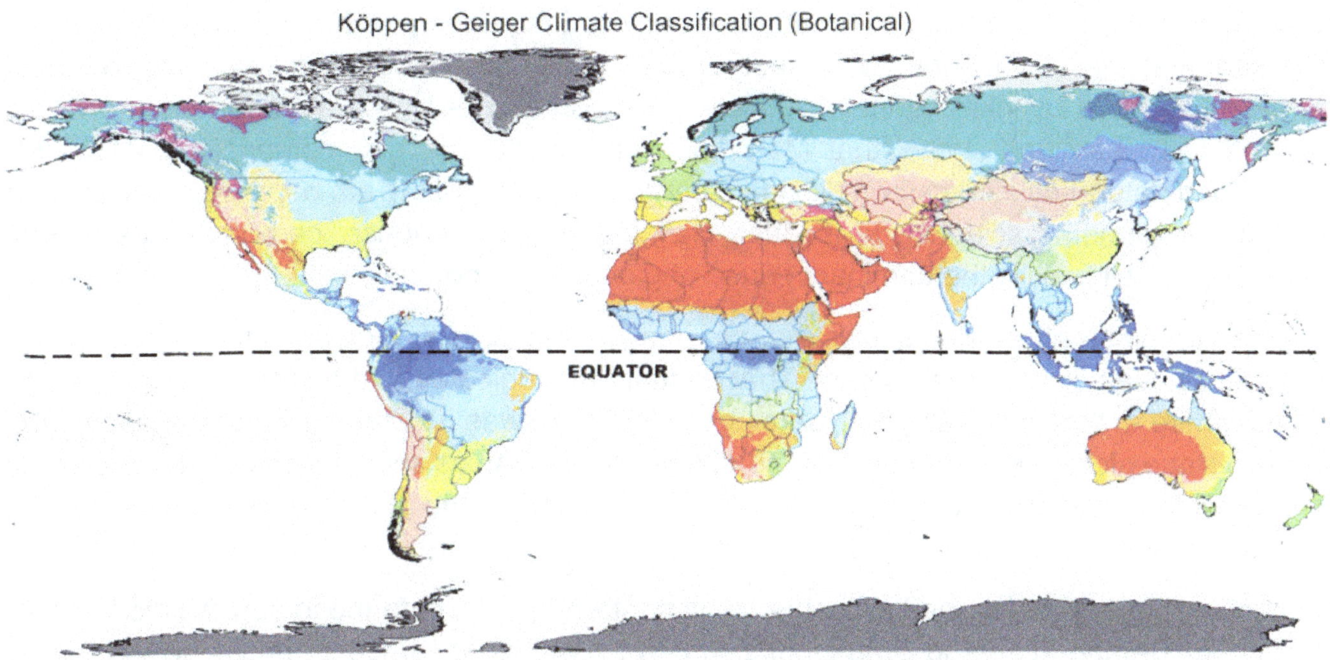

(Source: Encyclopedia Britannica)

Readers are reminded that this climatic- regions chart was generated for botanical purposes and not for any implications of the effect of climate change on the human population of the planet. However, it reveals that reputable scientists recognize the value of classifying the regions of the planet for scientific needs into regions rather than by the "one-size-fits-all" approach by environmental scientists using the "average global temperature" method.

Another observation made from the Köppen - Geiger colored regions map points out the large proportion of the map being in the red. The red area designates the arid – i.e., the lack of water for growing vegetation - conditions of the desert. A later chapter describing why humans populate favorable climatic conditions in many different regions of the planet shows the Köppen – Geiger climate maps for six different regions of the planet in greater detail than that shown above.

It would be wise for the United Nations Intergovernmental Panel on Climate Change (IPCC) to be the "Maestro" for the regional assessments to be done by all of the member countries within the present IPCC purview. A good Maestro does not "direct" the different groups in the symphonic orchestra how to play their instruments, only to "harmonize" their music with the other groups. Similarly, the IPCC would serve to minimize duplication of efforts and waste of resources in pursuing activities that do not apply to each country's unique set of circumstances. Each region has its natural environment, human needs and demands (for safety, health, and comfort), and different priorities, regional alternative energy resources available, absence or presence of coastal regions, being located in different latitudinal climatic regions of the planet, etc.).

The IPCC should not be setting "one-size-fits-all" goals for the control of any human activities needing to be met by an arbitrarily chosen date as a "tipping point". The goals for regions can be set by a group of countries in a common region with IPCC acting as the mediator but not as the director. The strategy of "combatting climate change" must be revised to being one of "détente" and adaptation between the human population residing in a region to be in harmony with the environment of that same region.

Humans being in harmony with their environment includes not polluting the air, water, or land of the environment. Environmental scientists have resorted to identifying Man-made carbon dioxide as the dominant cause for global warming to fallaciously capture the minds of the public, but the analysis in this book disproves that claim. Their own measurements of the "greenhouse gasses" in the atmosphere and the measurements by international scientists of carbon dioxide in the atmosphere shown by the Antarctic Ice Core samples shows that Man-made carbon dioxide gas comprises 0.013% of the atmosphere and natural-caused carbon dioxide comprises 0.029% of the atmosphere. That contradicts present day environmental scientists fallaciously claiming that "greenhouse" carbon dioxide generated by Man burning fossil fuels is the dominant cause of global warming.

The ice core sample analysis also shows that the peak in total carbon dioxide in the atmosphere lags the temperature peak by about 800 years. That factual data indicates that it is not the carbon dioxide in the atmosphere at that time (120,000 years ago when there was no Man on Earth burning coal) that drives the average global temperature. Instead, the ice core samples show that it is the average global temperature which influences how much carbon dioxide will be in the atmosphere – a contradiction of the claim by present day environmental scientists.

Environmental scientists may – and probably will very strongly – dispute those contradictory facts that the environment (temperature and atmospheric content) at 40° North latitude (amidst the highest World population density where the global warming is of highest concern to present day environmental scientists) is not the same environment as at 78° South latitude (at the Vostok Antarctica Ice Research Station. The effect of latitude on the temperature on cities at any latitude is factually a strong cause and effect factor as shown in an earlier chapter (describing geophysical factors affecting global warming and climate change.) When the average annual temperature for 40 major cities (as recorded by the World Meteorological Organization) is properly analyzed, it was shown that the latitude and elevation of the cities varies very strongly with latitude. A repeat is made of the graphical illustration for the effect of latitude on eight cities in the USA which differ by only 22° in latitude (Miami, Florida, at about 26° N, and Seattle, Washington, at nearly 48° N):

Average Temperatures In United States Cities (8)
(US National Weather Service, NOAA)

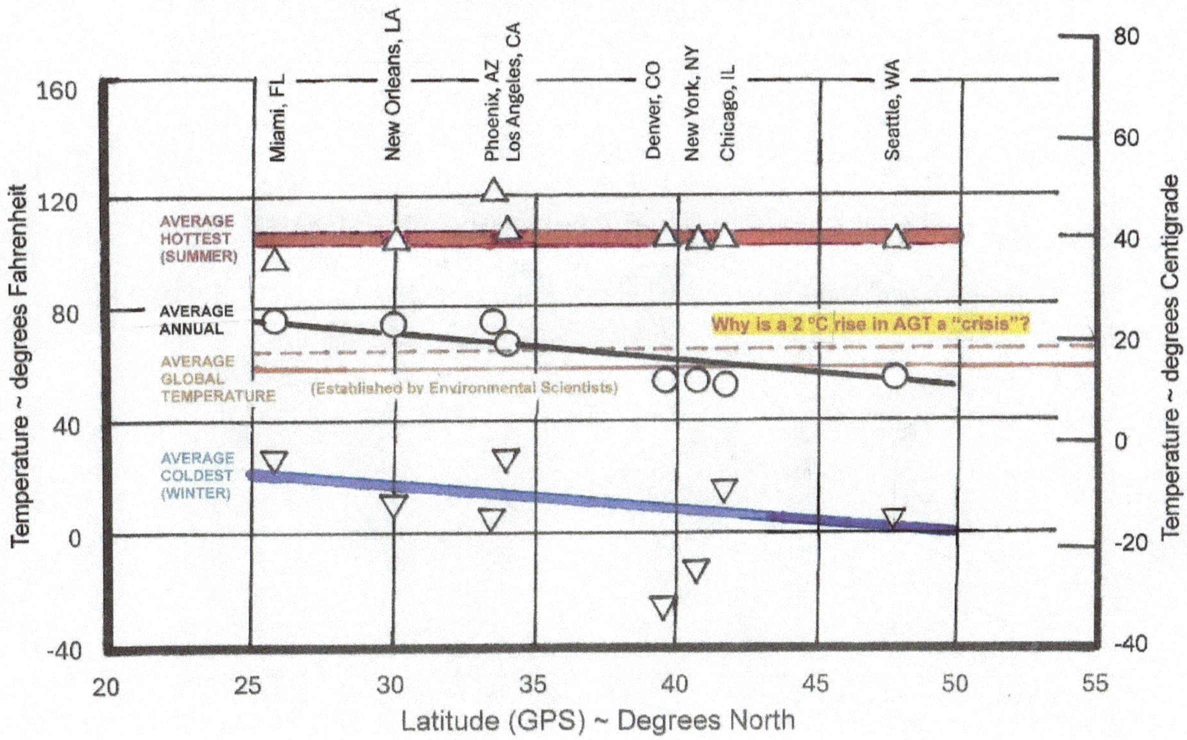

(Source: National Weather Service, NOAA)

As seen above (the data with a solid black line), the human residents in Miami near the lowest latitude of the USA are exposed to an average annual temperature of 76 °F (24.4 °C) but the residents of Seattle near the highest latitude of the USA are exposed to an average annual temperature of 55 °F (12.8 °C). That is a 21 °F (11.6 °C) difference in average annual temperature primarily due to the latitude difference of 22°. The first of three questions that environmental scientist have yet to answer clearly to the public is: **1a)** Why does a 2 °C rise in average global temperature (the dashed red horizontal thin line at the center of the illustration and goal of the UN IPCC to be not exceeded from any cause including Man burning fossil fuels) constitute a near-term global "crisis" when all cities in the USA at or above 38° latitude are already more than 2 °C (3.6 °F) cooler than the environmental scientists' "crisis" threshold average global temperature? **1b)** Why has there not been a "crisis" declared for all US cities below 38° latitude and having an average annual temperature already a few degrees above the environmental scientists' "crisis" threshold temperature of 18 °C (64.4 °F)?

The average hottest temperatures (bold red line) for the eight cities are shown in the illustration and surprisingly (to this Author) are seen to be a flat-line variation at about 105 °F (40.6 °C) with no significant effect of latitude. The city of Phoenix, Arizona, is seen to be the furthest difference (hotter) from the solid bold red line, but that is most probably due to Phoenix being in the very hot and arid Southwest Desert region of the USA. A photograph of the Phoenix area as seen from space shows the absence of green color and the predominant tan color of desert sand:

(Source: NASA) Phoenix, Arizona

The Author was residing in a southern suburb (Chandler) of Phoenix in the summer of 2023 when the temperature was at or above 100 °F (37.8 °C) for 23 consecutive days, above 105 °F (40.6 °C for 7 consecutive days, and reached 115 °F (46.1 °C) as the peak maximum temperature in the Phoenix area. Humans do survive extreme temperatures such as those by adapting to them, not "combatting" them. The second question the environmental scientists have yet to answer clearly to the public is:

2) With the average hottest temperature (which occurs in the summer season) of the entire USA being at 105 °F (40.6 °C) and which is 40 °F (22.2 °C) hotter than the environmental scientists' "crisis" temperature threshold of 18 °C (64.4 °F), why has the entire population of the USA not already succumbed to the hyperthermia "crisis?

The use of "average global temperature" and "average annual temperature" for a city masks the seasonal effects on temperature that occurs on all parts of the planet's surface area. The prior graphical illustration shows the two extremes of heat and cold that each city undergoes and within which human residents must adapt to those factual extremes - not their average - to survive. Averaging of any of the meteorological variables (temperature, precipitation, wind speeds, humidity, % cloud cover, H and L cells, etc.) may be convenient for scientists to attempt to be modeled for predictive purposes but that averaging process is not what humans and all living forms of life experience daily in the real world. Fatalities of humans can and do occur both from hyperthermia (excessive heat) and hypothermia (excessive cold) but almost zero fatalities due to heat at the fallacious "average global temperature".

The average coldest temperatures (which occur during the winter season) for the cities considered in the graphical illustration above are shown (in the solid blue line category). The difference in temperature between the hottest and coldest extremes are 85 °F (47.2 °C) at latitude 25° N and 105 °F (58.3 °C) at latitude 50° N. The third question for environmental scientists yet to answer clearly to the public is:

3) Why is a 2 °C (3.6 °F) rise in average global temperature considered a "global crisis" despite there being more deaths of humans from hypothermia than there are from hyperthermia in the World in regions at very different latitudes?

To change the tone of this book from being a report on the many problems that are caused by too much water, or not enough water, or water that is abundant in quantity but unsafe for human or agricultural use because of its high saline or other pollutant content, or water rampaging in a flash flood washing away vehicles that are occupied, there is an aspect of water vapor and rain precipitation in the atmosphere that always brings a pleasurable sight to human eyes. Seeing a rainbow in the sky is one of those pleasures in life that cannot be bought – as shown by a photograph taken in Prescott Valley, Arizona, in September 2023, as a weather front was approaching:

(Credit to Diane Stilson, Author's daughter; Prescott Valley, Arizona)

The occasion of seeing a rainbow is made more enjoyable by seeing a double rainbow as shown in the photograph. There are many stories told about rainbows – leprechauns guarding a pot of gold at the end of the rainbow, etc., but the scientific reason for a rainbow to be formed is much like the theme of this book – there are facts and fallacies when seeing a phenomenon of Nature. The fact is that, unlike some environmental scientists who choose to disregard water vapor in the atmosphere, there occasionally are small droplets of water in the atmosphere in the air ahead of heavy rainfall. When the rays from the Sun, which contains light having a spectrum of colors having a unique wavelength for each color, encounter a droplet of water, the solar ray is refracted into its spectrum of colors according to each color's wavelength. The phenomenon can be duplicated in a laboratory with a prism of glass, but in the atmosphere the prism is replaced by the billions of water droplets. The simple arrangement of the Sun, the rain droplets, the angles, and the position of the human's eye are all factors – and the factors must be in the correct positions for the rainbow to be seen. The arrangement of those factors appears as follows:

(Source: Sciencetalk.com)

The illustration shown next shows the process with one droplet of water, but the sun's rays are so widespread the rainbow appears to arc across a large portion of the sky.

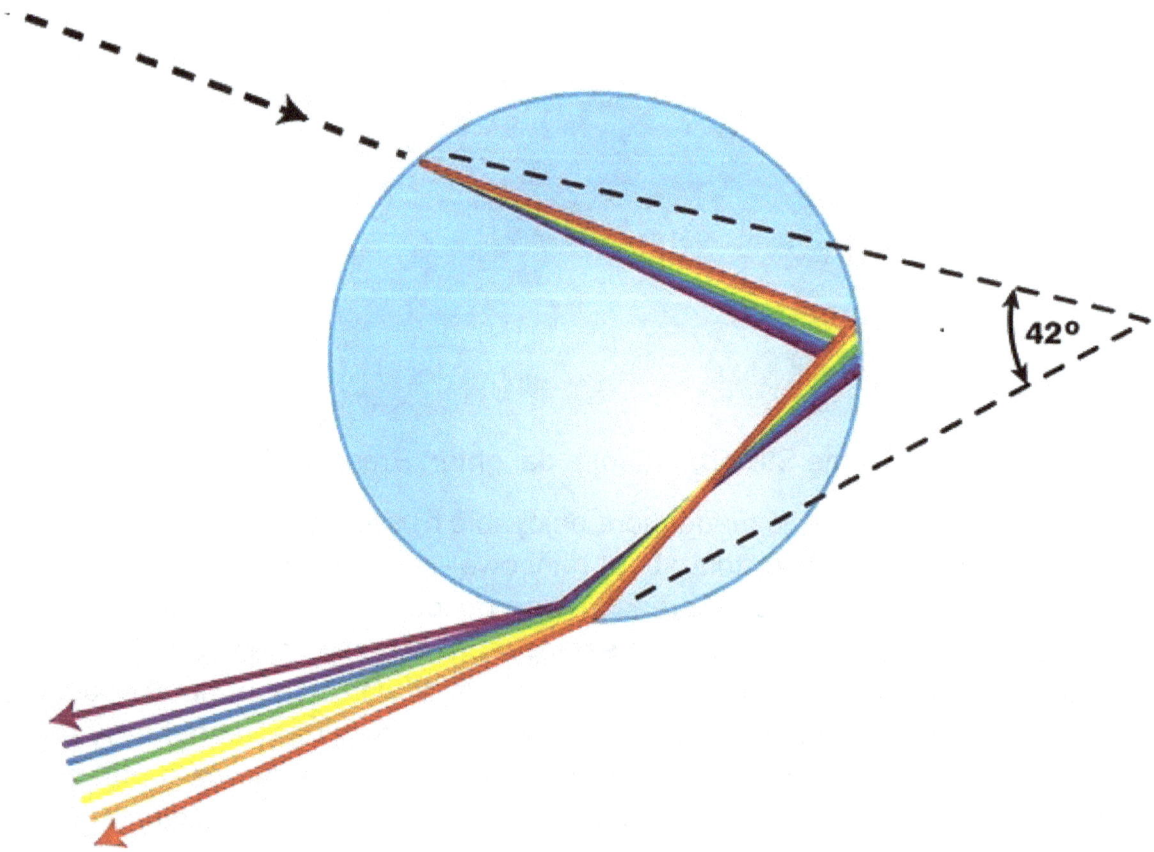

(Source: Sciencetalk.com)

The colors in the spectrum of a rainbow are always in the same variation – which the Author was taught in eighth grade by his science teacher, Mr. Roy G. Biv. (Readers are asked to pardon the Author's attempt at humor but the Author's science teacher's name in eighth grade actually was Mrs. Margaret Thompson and who made it easy for her students to remember the color spectrum using the letters in the name of the mythical character (perhaps a leprechaun's name?) Roy G. Biv – red, orange, yellow, green, blue, indigo, and violet. There are some shades of color before the red, and after the violet but they are in wavelengths shorter than red and longer than violet that are not discernible by the human unaided eye, and of course there is black (no color) and white (all colors). If any Reader doubts that the color spectrum is always in that same sequential order (in the same order of alphabetic characters as Roy G. Biv spells "his name") the Reader is invited to compare the color sequence in the illustration above with the color sequence shown in the actual double rainbow presented in the previous photograph taken in Prescott Valley, Arizona.

It is also important to note that there is no space between each different color as the wavelengths are a continuous variation and not in steps of increasing wavelength to give the separate colors in the diagram above. The rainbow photograph shows the rainbow colors variation without there being any distinct color boundary line between them. From a climate perspective, while the rainbow is a beautiful sight, the more important aspect of sunlight impinging on a water molecule in the atmosphere is that a small amount of solar radiation is also trapped in the molecule as heat energy. The effect of water vapor in the atmosphere is far more important as a heat-trapping phenomenon than is that which is attributed to very small concentrations of "greenhouse gasses" in the atmosphere.

It was mentioned earlier that when gravity accelerates a mass of water downhill, that mass of water possesses kinetic energy (which Sir Isaac Newton developed a set of mathematical laws for "how stuff moves" hundreds of years ago). It is kinetic energy when it strikes a solid object – as flood waters often do - causing physical damage to that solid object and can cause injuries to a human or other properties also that are in its path of motion. The combination of gravity and water in nature can be fearsome at times - but for most of the time as historical records document, gravity and water can do wondrous things when harnessed. For many centuries, the flow of water in a river – the flow being always from a higher elevation to a lower elevation – has been harnessed with water mills of many varieties and purposes. An example of the two most common water mills is shown as follows:

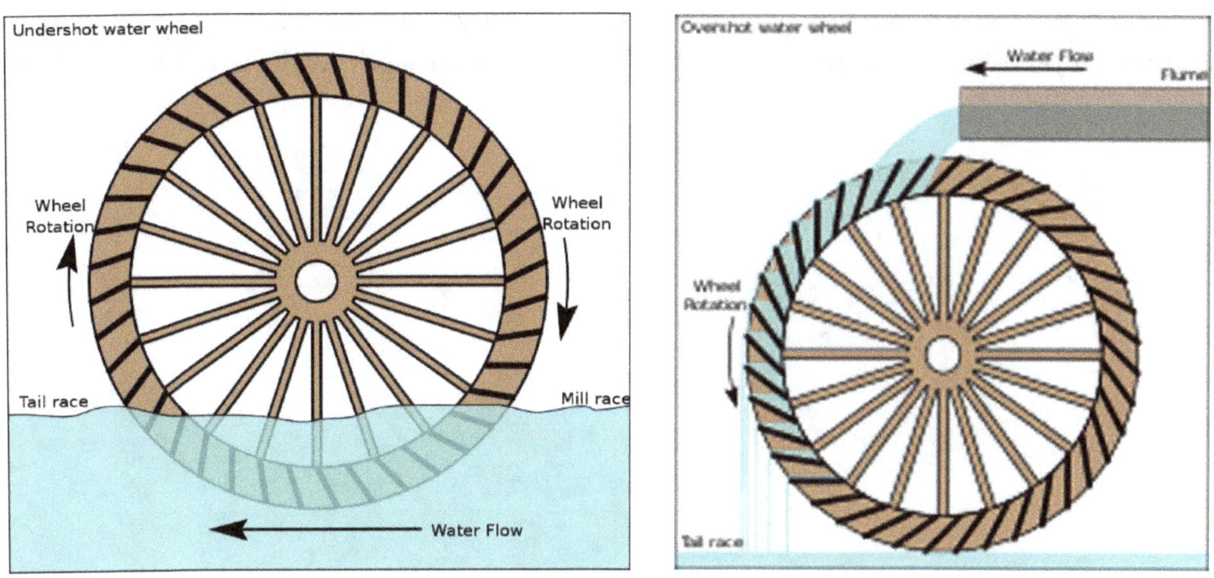

(Credit to Unknown Graphic Illustrator)

The working water mill wheel is mounted on the side of a facility housing the mechanical system (the kinetic water energy being generated by gravity to be converted into mechanical energy to power the grinding system) as shown in the next two illustrations:

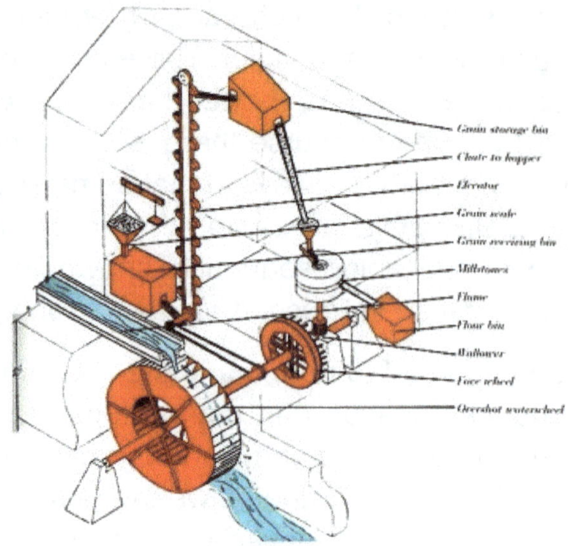

(Credit to unknown photographer and unknown graphic illustrator)

Water powered mills were used for grinding grains into flour, sawing tree logs into planks of lumber, drawing hot metal into wire, separating seeds from cotton, and other useful purposes that contributed to the development of the country where used. As electric power became available to drive the mechanisms, the water-powered mills were obsoleted and water-powered mills entered the annals of history. The current generations have little recollection of them, but they exemplified that Man can use the elements of nature (water and gravity) without damaging his environment.

This technology led to the much larger practice of using flowing rivers and the creations of dams near the bottom of which the water pressure was converted into mechanical and hydrodynamic energy for driving turbine machinery used to generate electric power on a huge scale. An unverified estimate is that there are over 800,000 dams of significant size in use on the planet.

The technology of the large-scale use of hydrodynamic energy is described more fully in a later chapter, but a summary of how hydrodynamic power has grown in use world-wide is evident in the following table.

Hydrodynamic Power Useage by Top 22 Countries

(As of 2022)

	Country	Power Generated		Capacity (GW)	Capacity Factor (%)	Number of Dams Over 3,000 MW (see Note)
		Hydro Power (TWh)	Per Cent of Total Country Power			
	World	4183.41	14.2%	1267.90	38%	38
1	China	1226.00	13.0%	370.60	38%	17
2	Brazil	428.65	60.4%	109.90	50%	8
3	Canada	364.20	57.5%	83.31	50%	2
4	United States	233.96	5.5%	86.66	31%	2
5	Russia	200.87	17.1%	50.57	45%	4
6	India	149.17	7.6%	47.33	36%	0
7	Norway	135.96	88.5%	34.40	45%	0
8	Vietnam	80.90	29.3%	22.64	41%	0
9	Japan	74.50	7.4%	28.22	30%	0
10	Sweden	66.07	39.7%	16.40	46%	0
11	Venezuela	65.68	77.6%	16.81	45%	1
12	Turkey	63.72	19.9%	31.78	23%	0
13	Colombia	54.24	62.5%	13.21	47%	0
14	France	53.19	10.4%	24.14	25%	0
15	Paraguay	43.87	99.7%	8.81	57%	2
16	Austria	39.79	59.4%	14.71	31%	0
17	Switzerland	39.00	54.8%	15.28	29%	0
18	Italy	37.94	14.5%	18.85	23%	0
19	Pakistan	37.90	23.5%	10.64	41%	1
20	Laos	33.40	72.7%	9.65	40%	0
21	Germany	19.47	3.9%	5.74	39%	0
22	Tajikistan	18.66	89.4%	5.76	37%	1
	58 Other	716.27	1.8% - 100	0.81 - 8.44	15 - 85%	0

International Energy Agency (IEA)

Note: 6 large dams (over 3,500 MW) are under construction: 2 in Myanmar (7,110 MW and 3,600 MW); 2 in Pakistan (4,500 MW and 4,320 MW); 1 in Ethiopia (5,150 MW); 1 in Tajikistan (3,600 MW)

(Source: International Energy Agency (IEA)

Very few accounts exist in the current news media describing how Man has used water and gravity ingenuously in past decades, but the Author's favorite account of a amazingly "green" project that both of his grandfathers worked on constructing in the first decades of the 1900s is hard to describe its immeasurable importance to the World in a few sentences. That "wonder of the world" project was/is the Panama Canal.

The Panama Canal is high on the list of the "greenest" ventures successfully accomplished by humans in modern years, but it was built more for ocean transportation improvement not for improving the environment. A magnificent accounting of the unusual history of the construction of the Panama Canal is available for Readers of this book in "The Path Between the Seas" by the renowned author David McCullogh. (Author's Note: The Author of this book on global warming is a contributor to the historical documents on file at the Barbados Museum and Historical Society on the island of Barbados in the

Caribbean Sea and Atlantic Ocean. The Author donated a complete US Postage stamp collection of Panama and Canal Zone postage stamps, a large collection of post cards of scenes while the Panama was being constructed (by the French initially and then completed by the United States), and his large collection of postage stamps for the 400 years old island of Barbados (his family's ancestral home). In addition, the Author has written articles about the Barbadians who were part of the Panama Canal creation team. The article "Commemorating the Barbadians who Excavated the Panama Canal" appeared in the December 2008 *Journal of the Barbados Museum and Historical Society*. Subsequently, in collaboration with historians on the Barbados Museum Staff, the Author was instrumental in causing the Panama Canal to be designated as a UNESCO World Heritage Site.)

The importance of water and gravity on the planet cannot be better captured than describing the example of the Panama Canal. The Panama Canal story – for this book – starts with the Panama Railroad. Before the Panama Canal was constructed, the Panama Railroad was constructed across the Panama Isthmus. Passengers and cargo in the Atlantic Ocean waters wishing to reach a destination on the West Coast of the USA (a popular route inasmuch as there was no Transcontinental Railroad yet between the coasts of the USA) would debark at the port of Colon in Limon Bay, load onto the Panama Railroad to travel across the isthmus past the towns of Gamboa and Miraflores to Panama City. There they off-loaded from the railroad onto a passenger sailing ship in Panama Bay for completion of their trip Northerly in the Pacific Ocean to the West Coast USA – principally San Francisco (where they bought mining equipment in preparation for search of California gold during the Gold Rush years). The eventual Panama Canal shipping route was/is still very much aligned with the original railroad track route. (One point of trivia though that must be stated is that in traveling from the Atlantic Ocean (at Gatun) to the Pacific Ocean (at Panama City), a ship passenger is unexpectedly confused when subjected to traveling from northwest to southeast due to the reflex curvature of the isthmus at that location).

While the original attempt – which was doomed to failure - to construct a canal across the Panama Isthmus goes back to the 1860s and with the International Canal Commission issuing a construction contract to the French entrepreneur Ferdinand de Lesseps (credited with creation of the Suez Canal), this book fast forwards into the years in the first decade of the 20th century (1900) as the United States Canal Zone project under President Theodore Roosevelt. President Roosevelt saw that a canal at Panama was a boon not only to commercial shipping trade and passenger travel efficiency but was a necessity for military purposes as well. The US Navy has maintained large fleets of warships in both the Atlantic Ocean and the Pacific Ocean since before World War I (which started in 1917), and President Roosevelt saw the canal as a strategic war asset.

The route the Panama Canal was selected to take across the Panama isthmus connecting Central America and the North and South American continents, for good reason follows the path of the trans-isthmus Panama Railroad.

(Source: Panama Canal Museum, Miraflores, Rep. of Panama)

Virtually all of the water used in the functioning of the Panama Canal and its system of locks is provided by rainfall in the mountainous region of North Panama – none of that water is pumped anywhere in the system by electric-driven pumps. The flow is gravity accelerated only. The collected rainwater starts in the upper right corner of the above map in the higher elevations of Panama and flows, as the Chagres River, southwesterly through the town of Gamboa and into the immense Gatun Lake. The Chagres River - before the Gatun Dam and Gatun Lake were created - flowed northwesterly toward the Atlantic Ocean end of the Panama Canal at the city of Gatun. The Gatun Dam was created across the Chagres River exit into the Atlantic Ocean and eventually the water level of Gatun Lake would be 85 feet above the sea level of the Atlantic Ocean.

Before showing the scenes of the Panama Canal, a profile is necessary to be shown explaining how ships enter the system of locks at one end of the canal into and up a "water ladder", travel across the isthmus at the upper level of the lakes (Gatun Lake and Miraflores Lake) and then descend down the "water ladder" to exit into the ocean at the other end 50 miles away. That "Cross Section diagram" is shown next:

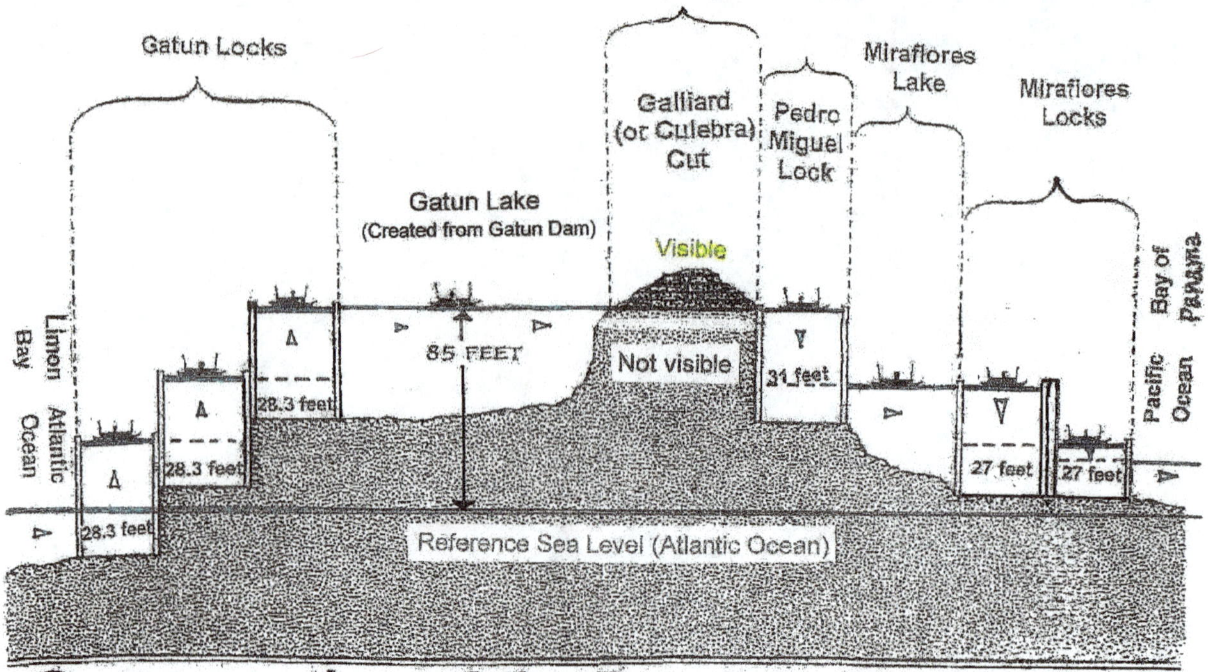

Cross section of the Panama Canal

(Source: The Panama Canal Museum at Miraflores, Rep. of Panama; with Author modifications)

The green arrows below the small ship icons reflect the direction of motion the ship takes during the several different phases of transit through the Panama Canal. The diagram shows only a small portion of the terrain that had to be excavated to create a watery path on which each ship takes to cross the isthmus at Panama. It is the manner in which the designers and construction/excavation team endured and captured the forces of nature and physics to make the Panama Canal a reality not merely a wishful dream that is the wonder of its existence – for the benefit of humanity. The Panama Canal exists in upgraded (with a different technology for the lock system) form at present and is projected to be an enduring transportation feature for many decades and perhaps centuries to come.

Among the earliest activities at Panama was the construction of the Gatun Dam across the Chagres River at Limon Bay in the Atlantic Ocean – while the six locks were each being constructed as a pair of chambers, 12 chambers in all, to permit ships passing in either direction (Atlantic to Pacific or Pacific to Atlantic). The initial purpose of the Gatun Dam was to create Gatun Lake, and then subsequently to generate all of the electrical power for the functional operation of the canal's locks and lighting system.

As a unique approach to obtaining the construction materials for the construction of the dam and the several lock chamber walls and center island, some of the "spoil" from the excavation blasting of the Culebra Cut through the mountain peaks of the Continental Divide was used (after being crushed and sieved to the needed size). The clay-like "spoil" mixed with rocks was used to fill the cavity between the Gatun Lock face and interior wall, while the enormous boulders that rolled down into the cut were hauled away on railroad flat cars to be used in construction of the breakwaters at both ends of the canal (in Limon Bay and in Panama Bay).

The Miraflores Lake was an existing natural lake and was used in conjunction with the Pedro Miguel single lock as the intermediate path leading to the pair of Miraflores Locks at the Panama Bay end of the Panama Canal.

A photograph of the completed Gatun Dam, Gatun Lake behind it, and the Gatun Electricity Generation Powerhouse below it is shown next.

Gatun Lake, Gatun Dam, and the Gatun Hydroelectric Power Station
(Photo from Postcard Collection of Author on File at Barbados Museum and Historical Society)

The Gatun Lake was/is formed and is maintained full of the heavy rainfall in the Panama Mountains, with the creation of the upper surface of the lake 85 feet above the seal level of the Atlantic Ocean at Limon Bay. The Gatun Dam is a magnificent example of the use of renewable energy to operate a facility as important as the Panama Canal. The electricity generated at the Gatun Hydroelectric Power Station makes the Panama Canal system of locks for transporting ships across the Panama Isthmus between the Atlantic Ocean and Pacific Ocean totally a self-powered operation. By creating the Gatun Lake 85 feet above sea level, the three Gatun Locks (each 28.5 feet high) of the Panama Canal form a 3-step "water ladder up" to lift and 3-step "water ladder down" to descend massive ships using gravity alone to fill a lock chamber from the chamber or lake above it or to empty into a chamber next to or below it.

Gravity flow through the successive three chambers constitutes the "water ladder" but the gates at each chamber end are electrically opened and closed to permit the ship to pass from one chamber to the next. There are no electric-driven pumps pumping water up or down in operation of the "water ladder" system of locks. In the center wall of each pair of lock chambers, a large culvert having multiple gates (manually controlled in the gates control house and electrically operated by power from the Gatun Dam hydroelectric power station) can be opened to allow gravity to flow the water either from one chamber laterally to its twin chamber or from a higher chamber to a lower chamber. The center culvert is about 20 feet in diameter and is shown next when the lock was under original construction: (The culvert is not visibly seen by humans as the water level never gets below the keel level of ships).

(Source: The Panama Canal Museum, Miraflores, Rep. of Panama)

An ingenious design for the filling and emptying the chambers in each lock – i.e., the "plumbing system" – explains how ships weighing tens of thousands of tons when fully loaded with vital cargo – are raised up a "water ladder" at one end of the canal, cruises across man-made and a natural lake, and is then lowered again on a "water ladder" to the sea level of the ocean on the other side of the isthmus. The water in the system is entirely from the heavy rainfall prevalent in the climate at the isthmus of Panama – at latitude 9.38° North.

Cross-section of Lock Chamber and Walls, Panama Canal

A section across the width of the locks, showing the culverts for filling and draining the chambers. One side is shown; the other is the same.

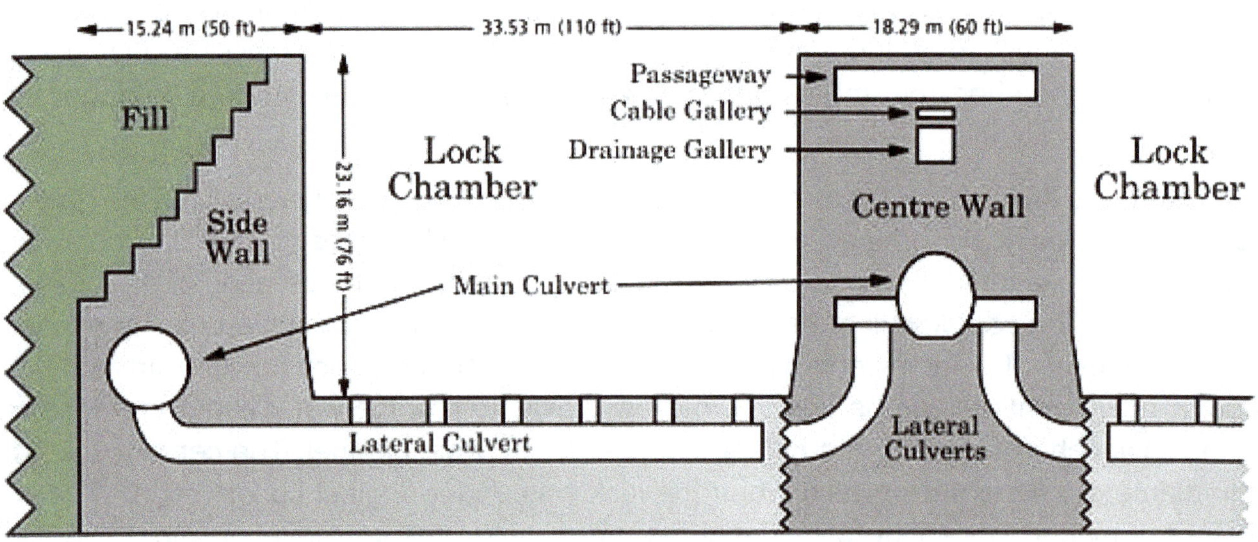

(Source: The Panama Canal Historical Museum at Miraflores)

The opening and closing of each of the massive lock gates at the ends of each chamber (12 pairs of gates in all) are electric powered using the power generated at the Gatun Hydroelectric Power Station. A photograph of a ship descending the Gatun locks – with the gates in the two chambers in an opening mode to allow the ship to pass from one chamber to the next – is shown next. The gates can be opened only when the water level is the same level to prevent a rush of water into the chamber with the lower surface. The gates were designed to be flush with the inner walls of each chamber so that, when fully open, the gates do not present an obstruction for the side of the passing ship to snag the gate and damage it. Regrettably, the page constraints of this book do not permit a fully detailed description of the ingenuity of the Panama Canal system and down to the individual gate controls in the gate control house (with a single person moving all those large ships, and chamber gates, and multi-millions of gallons of water in a perfectly orchestrated manner).

A scene from a high deck on a ship transiting through the Gatun Locks is shown next:

(Source: The Panama Canal Museum – Gatun Locks in Operation)

The electric power to operate all the gates in the12 chambers of the lock system is generated by the singular hydrodynamic electric generation facility at the base of the Gatun Dam (shown in an earlier photograph and not seen in the photograph above - only a short distance across the original path of the Chagres River to the left of this photograph). That generation facility also provides the electric power for the controls operating the gates, for the extensive lighting system, and to power the electric "mules" that pull the larger ships through each chamber to give the Panama Canal a 24 hours per day full operating capability. Small ships with less draught are self-powered through the series of six chambers in each transit of the Panama Canal – three chambers up and three chambers down.

Since all water to operate the six chambers for each full transit of the Panama Canal by a ship of any size is drawn from the fresh water Gatun Lake or Miraflores Lake, each full transit of the ship flows a total of 53 million gallons (200 million liters) of fresh water (from rainfall alone) into the saline oceans (Limon Bay of the Atlantic Ocean and Panama Bay of the Pacific Ocean). The barnacles that grow on the keels of ships while in saline ocean waters die off when entering the first freshwater chambers and drop to the bottom of the first two chambers – requiring periodic maintenance (cleaning) of the barnacles' debris from those chamber floors. With an average of close to 40 ships per day transiting the Panama Canal (in both directions), if each ship causes 53 million gallons of freshwater to be transferred out into the oceans at its ends, 40 ships per day would cause the loss of about 2.12 billion gallons of water per day. In recent months the rainfall in the upper Panamanian mountains has decreased somewhat (natural causes) and as this book is being written, a limit has been placed on the transit of ships per day to be an average of 32 transits to preserve the level of Gatun Lake as being 85 feet above the Atlantic Ocean surface level. If the ocean levels rise 20 feet, this would obviate the need to use the lowest of the three Gatun Locks and the two Miraflores Locks.

After President Jimmy Carter granted ownership of the United States Canal Zone (and the Panama Canal in its entirety) to the Panamanian Government in 2000, the design of "super-size locks" to handle those ships designated as Super locks size (i.e., too large for the original lock chambers dimensions) were conducted at both ends of the Panama Canal. Those new Super locks featured three water reservoirs located to the side of the original locks and conserved on the amount of water that was lost by each transit of a ship through the original locks. This book will not describe any further details of the new Super locks as they use the same principle of no electrically-driven pumps – only gravity – to transfer water into the "water ladders" at both ends of the Panama Canal.

The most labor-intensive portion of the Panama Canal construction was the excavation through the hill (part of the Continental Divide) known as the Culebra Cut (later renamed to be the Gaillard Cut in honor of Major David Gaillard in the US Army Corps of Engineers who orchestrated the cut). The "cut" was excavated several times in the same place because the heavy rainfall at Panama would cause mudslides down into the cut and bury men and/or their equipment without ample warning. On many days, the excavation crew would complete a section of the excavation cut by the end of the workday, but when they returned the next morning for another day at the "big dig" they found all the previous day's excavation was filled in by an overnight mudslide. Gravity can be a wonderful assistant or a formidable challenge, changing from one event to the next.

Shown in the following photograph is the ill-fated US Navy battleship *Arizona* on a passage (circa 1936) past the Culebra/Gaillard Cut sailing towards its ignominious fate at Pearl Harbor, Hawaii, on December 7, 1941. Culebra is the Spanish word for "snake" as the shape of the canal "snakes" through that cut.

(Source: The Panama Canal postcard collection of the Author) Gaillard Cut

The canopy over the bow section of the Arizona was to allow the crew to witness the historic transit through the Panama Canal without being burned severely by the Sun's rays and the extremely high humidity in the hours.

The massive amount of coal used to power the steam powered excavation equipment and the steam engine trains to haul away the debris, to construct the Gatun Dam, the Gatun Hydroelectric Power Station, the five pairs of major lock chambers, and to power the "mules" that originally towed the ships through the locks will be described later. The many millions of tons of explosives used in the construction of the Panama Canal would probably be prohibited by today's environmental protection rules. More important, if the burning of several hundreds of thousands of tons of coal powering steam shovels and locomotive engines had not been allowed to be used, the Panama Canal would not exist.

It is the nature of technology development that the capabilities and limits – the "pros" and "cons" – of a technology existing at a period in time can be and typically are improved as upgrades are designed. In addition, the needs and demands for the original technology may be expanded as the world population and international commercial trade dictates. The original design of the Panama Canal in the early 1900s was to improve the strategic capability of the US Navy to move a fleet of warships from one ocean to another. The commercial transportation ships were designed accordingly to fit within the design limits of the Panama Canal

This is perhaps the best example of expending the huge amount of fossil fuels to avoid the many millions of tons of coal to power steamships the shorter route between the two major oceans and avoid the huge amount of coal burned for those steamships by not having to sail over 2,000 miles on a ship route down and around the southern tip of South America (sailing in the hazardous Cape Horn region

– "round the Horn" in old-time maritime historical annals). There was no concern of the creation of "greenhouse gas emissions" in the early 1900s and affecting global temperature – only the concern of building the Panama Canal to allow passage of commercial freighters and warships to transit the Panama Isthmus in the minimum time and expenditure of coal fuel.

The details of the technological achievement of the Panama Canal as a "green" venture are able to be described by this Author because both Author's grandfathers were part of the construction team that created the Panama Canal. Their first-hand accounts of the ingenious planning and design to following generations is a resource that is no longer available. (Author's Note: The Author of this book on global warming is a major contributor to the holdings on file at the Barbados Museum and Historical Society on the island of Barbados in the Caribbean Sea and Atlantic Ocean. The Author has donated a complete stamp collection of Panama Canal and Canal Zone postage stamps, a large collection of post cards of scenes while the Panama was being constructed (by the French initially and then completed by the United States), and his large collection of postage stamps for the 400 years old island of Barbados. In addition, the Author has written articles about the Barbadians who were part of the Panama Canal creation team. Those articles appear in the *Journal of the Barbados Museum and Historical Society*. In collaboration with historians on the Barbados Museum Staff, the Author was instrumental in causing Barbados to be designated as a UNESCO World Heritage Site and to perpetuate the family stories told by those who know the story of the Panama Canal.)

The Panama Canal could well serve as a model for present day scientists and engineers and physicists on how to use the climate of a region, the material resources needed for that project being all naturally available on site, the acceleration of gravity endlessly both assisting and hindering the process of constructing the elements of the Panama Canal system, and using the persistence and ingenuity innate to some humans to create a "green" project before "green" was the challenge on every member of the public's mind. The irony however, is that without humans burning fossil fuels to power the machinery to excavate the canal path is of minor concern when compared to the huge amount of fossil fuels that were avoided being burned by thousands of ships not having to travel an extra 8,000 miles (and many days) at sea (via the southern tip of South America) to propel themselves from the East coast Atlantic Ocean ports of the United States and of the West coast of European countries to West coast Pacific Ocean ports and to ports everywhere around the Western coastal regions of the Pacific Ocean (e.g., to Japan, China, and the regions of Southeastern Asia). But a picture is worth a thousand words and shown below is a map illustrating the savings for ships transiting between the port of New York in the Atlantic Ocean and San Francisco in the Pacific Ocean:

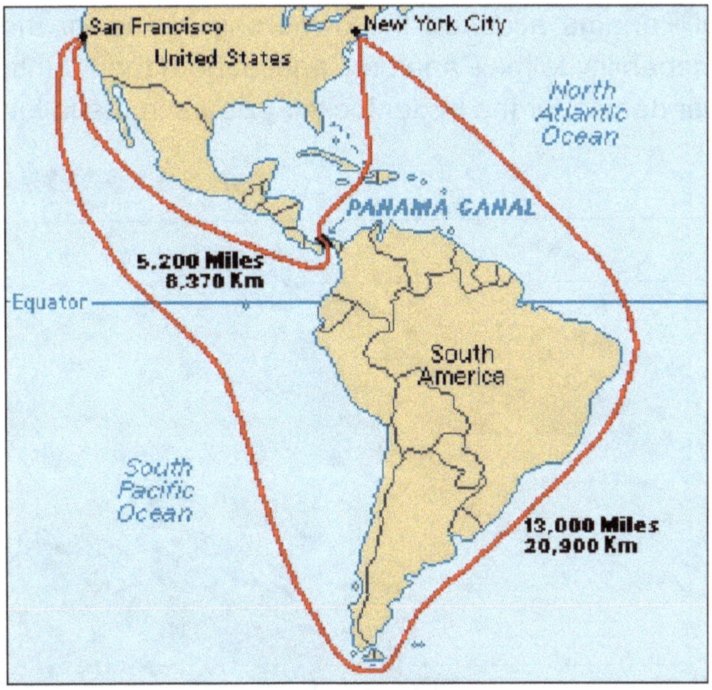

(Source: The Panama Canal Museum at Miraflores, Rep. of Panama)

To be factually correct, the Panama Canal was seen needed by President Theodore Roosevelt, a former Secretary of the Navy, to provide the shortest distance and shortest time to move US Naval vessels from one ocean to the other for wartime needs. Perhaps the most historically important event of a single ship passing through the Panama Canal was that of the USS Arizona on her voyage from the Atlantic Ocean (where she was part of the Atlantic Fleet) to her new addition to the US Naval Pacific Fleet headquartered at Pearl Harbor, Hawaii (before Hawaii was a State of the USA). That photograph of the battleship USS Arizona at the Panama Canal was shown a few pages earlier.

The transportation technology found that the efficiency of shipping cargo was increased by increasing the amount of cargo per ship – dictating that cargo ships should be constructed with larger dimensions. That technological change has resulted in behemoth ships being designed to carry the larger cargo loads (regardless of the type of cargo) and those larger ships were titled as being Panamax size.. Shown next is the extent to which container ships have been resized:

(Credit to Unknown Photographer)

Soon after the Republic of Panama acquired the ownership rights for the Panama Canal, planning began to add a larger lock capability to the Panama Canal at both the Atlantic Ocean and Pacific Ocean ends. The new lock chamber design for the larger locks appears in the following conceptual illustration:

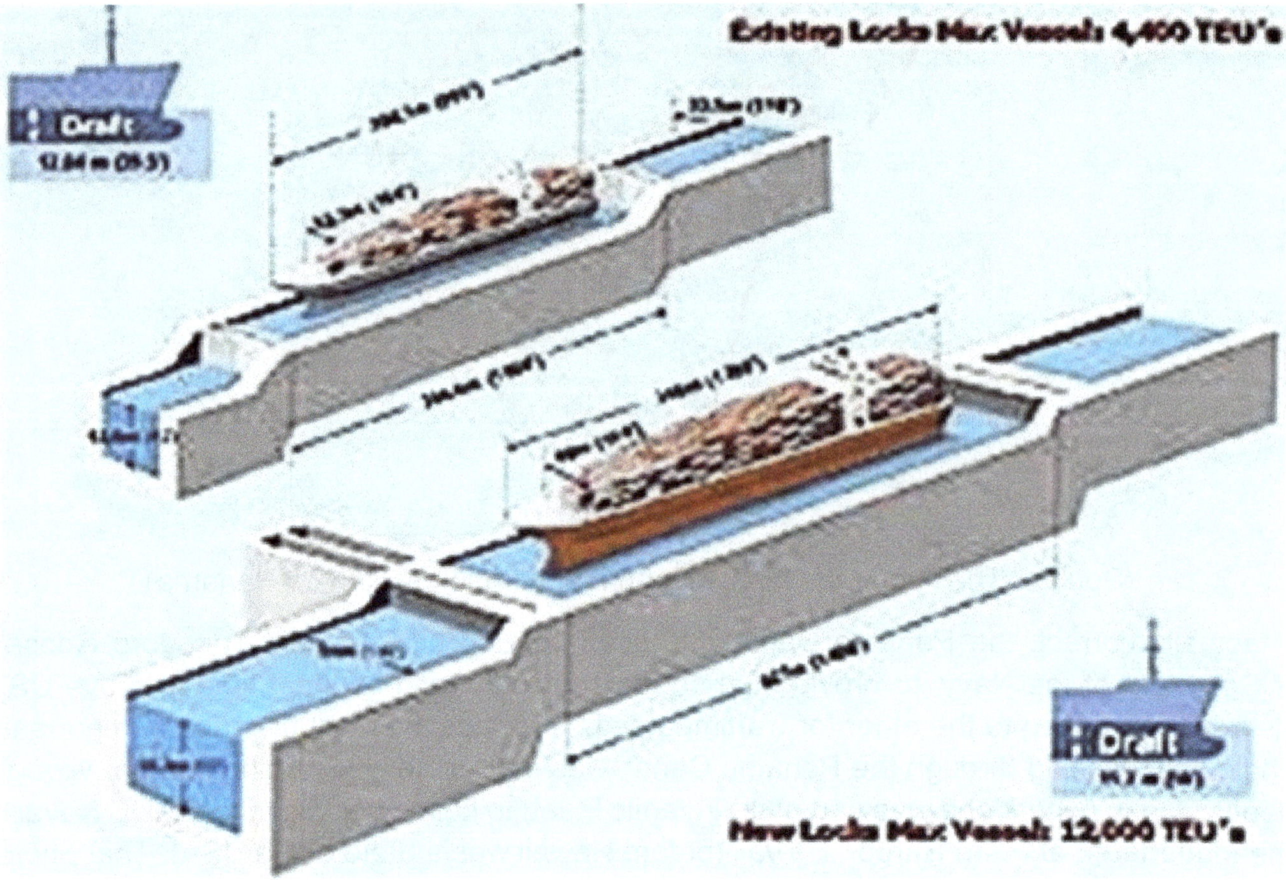

(Source: Panama Canal Historical Museum at Miraflores)

As indicated by the pair of narrow slotted chambers near the lower left corner of the illustration. Whereas the original gates of the chambers in the Panama Canal were of a rotating design in two segments and being embedded in the chamber wall, the new Super locks for post-Panamax ships would be single gates that slide laterally into those narrow chambers' slots shown.

In the late 1990s, drought conditions were experienced in Panama (as well as in other regions) and the decrease in rainfall spurred the need for a lock system that would perform the same "water ladder" concept of operation, but which would cause less water to be wasted when emptying the last lock chamber for the ship exiting into the ocean after completing an entire transit from one ocean to the other. That change in design resulted in a triple water basin concept in lateral location to each of the new Super locks and was conceived to reduce the amount of freshwater expelled into the ocean at the exit end of each ship transit.

The concept of water being kept in lateral basins for the emptying and filling of the new Super lock chambers is indicated in the following illustration:

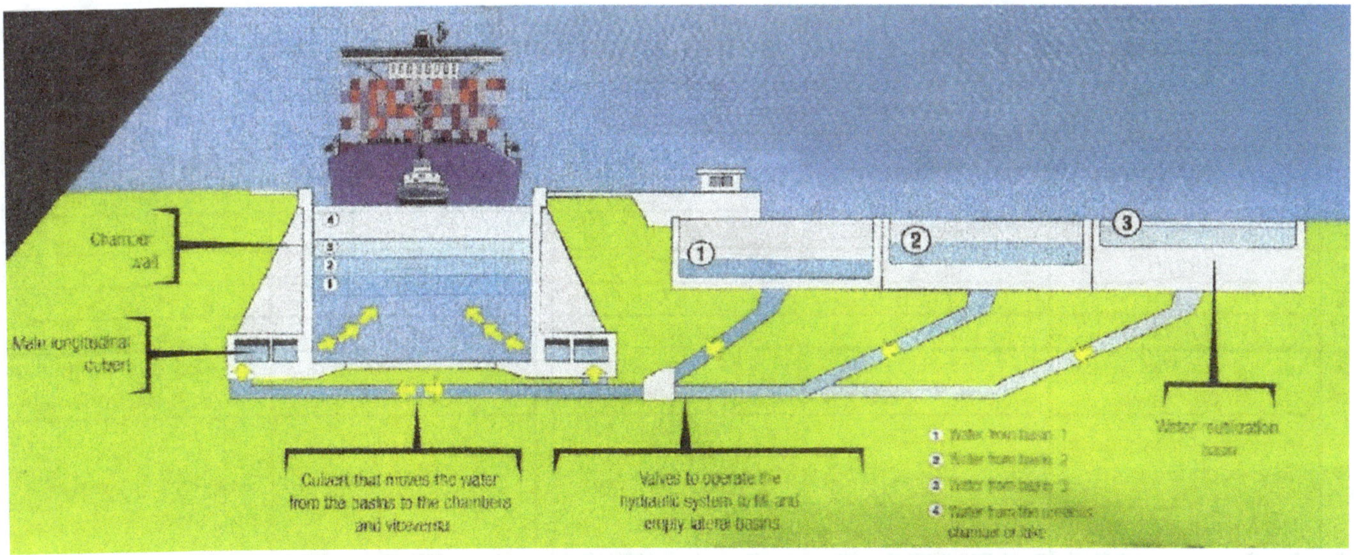

(Source: Panama Canal Historical Museum at Miraflores)

The composite of those three major elements (larger chamber dimensions, sliding gates, and lateral water basins) of re-design in the NeoPanamax Lock System is shown in an artist's rendition as follows:

(Source: Panama Canal Historical Museum at Miraflores)

The plan was to install the three Super lock arrangement at the two ends of the original Panama Canal – the Agua Clara set in parallel with the three locks at Gatun, and another three Super lock arrangement at Cocoli (a nearby town) in parallel with the single Pedro Miguel Lock and the pair of Miraflores locks. Some interesting numerical statistics are provided next to satisfy readers interested in the enormity of the new Post Panamax Locks:

Comparison of Original and NeoPanamax Locks Dimensions

(and Allowable Dimensions for Ships Transiting Locks)

		Original Locks		NeoPanamax	Super Locks (3 lateral basins)
	Dimensions	(Imperial)	(Metric)	(Imperial)	(Metric)
Chambers	Length	1,050 ft	320.04 m	1,400 ft	426.72 m
Chambers	Width	110 ft	33.53 m	180 ft	54.86 m
Chambers	Height	41.2 ft	12.56 m	60 ft	18.29 m
Chambers	Volume of chamber	4,758,600 cu ft 35,596,707 US gal	134,805 cu m 134,805,000 liters	15,120,000 cu ft 113,105,160 US gal	428,329 cu m 428,329,000 liters
Chambers	Volume of freshwater lost each full transit (when Pacific Ocean is at median of 20 ft tide)	26,700,000 US gal	101,060,000 liters	10,860,000 US gal	40,423,800 liters
	(3.28 ft / m)	(7.4805 US gal / cu ft)	(1,000 liters / cu m)	(35.3 cu ft / cu m)	(3.785 liters / US gal)
Ship	Maximum allowable ship length	965 ft (self-propelled or pulled by "mule")	294.13 m (self-propelled or pulled by "mule")	1,200 ft 9 in (being pulled or pushed by tug boat)	366 m (being pulled or pushed by tug boat)
Ship	Maximum allowable ship width (beam)	106 ft	32.31 m	160 ft 9 in	49 m
Drafts	Maximum draft of ship (in freshwater)	39.5 ft	12.04 m	49.9 ft	15.2 m
Drafts	Minimum allowable depth below keel	5.0 ft	1.52 m	5.0 ft	1.52 m

(Source: Tabulation compiled by Author from multip0le databases of information)

In technology development, it is often found that one technology will interact with another technology to be 1) synergistic (the net capability of the combination of two technologies being greater than merely the sum of the two capabilities); 2) independent (to have no effect of one on the other), or conflicting (to have a net capability of the combination that is less than the sum of the two individual capabilities).

Another principle of system engineering (unlike that of “pure” science) is that the technology must function safely and efficiently within the external environment of the location where the technology is installed. A third principle is that the subsystems comprising a technology system must also function in harmony – i.e., one subsystem may perform safely and efficiently but have emissions (acoustic, thermal, vibration, or radiative) that degrades the efficiency and/or safety of a neighboring system.

The NeoPanamax Canal technology has outstanding synergism with the transportation technology of other forms of energy to satisfy the energy needs of the world population and the industrial practices of entire regions of the planet. A very compelling visual indicator of the Panama Canal transportation technology being synergistic with the technology of natural gas energy is illustrated in the following photograph:

(Credit to Unknown Photographer)

The capability for the entire transportation and supply system to provide life-sustaining services and supplies to the world population in all regions of the world – and do so without endangering the health or life or functioning of all forms of life on the planet's surface is proportional to the size of the population being serviced and supplied. The following illustration from the Department of Energy illustrates how the elements in the "big picture" relate to each other synergistically, independently, or conflicting.

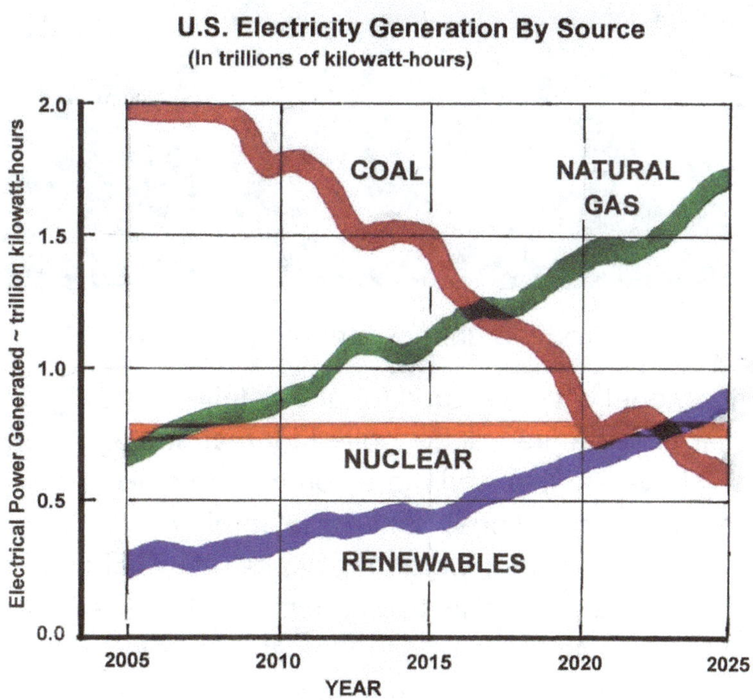

(Source: US Department of Energy)

The reduction of the use of one form of energy – with any goal after achieving the reduction – requires the replacement of that form of heat energy with an alternate form of energy. In addition, for technologies that require shipping of fluid from one point to another, the problems of spillages and leakages when using long pipelines and polluting the environment are lessened (to near elimination) if the fluid can be shipped in very large spherical tanks. The shipment of liquid natural gas (LNG) by large LNG tankers – similar to the one shown in the previous photograph passing through the Panama Canal is a good example of changing the form of transfer of the energy material (i.e., changing the form of its state – solid, liquid, or gas). Liquifying natural gas reduces its volume considerably and produces a more economical transportation option - by tanker ship in large spherical tanks vs. lengthy large diameter cylindrical pipeline. Pipelines traversing areas where native American burial grounds exist has doomed several proposed pipeline projects because of potential environmental damage and/or disrespect for the departed.

Those tangents of technology development are addressed in a sequel to this book and published under the title of "Dealing with the Real Climate Change" by the same Author as this book.

All regions on the planet that have a significant growth of trees are vulnerable to there being wildfires that destroy the natural resources that touch many aspects of human life. Some wildfires are started by Man (intentionally and unintentionally), but the vast majority are started by lightning strikes. Lightning strikes have their origin in the clouds and their termination on the surface of the planet, as shown in the next pair of photographs:

(Credit to unknown photographers)

There has been no scientific model yet developed for accurately predicting where and when a lightning strike will occur and whether or not a fire will be ignited by that strike. It is most common for lightning to strike from a cloud mass above an area and more common to strike a taller feature within that area. Anyone who has visited the Sequoia National Park in California cannot help but feel saddened when viewing the giant sequoia trees with their top portion decapitated by lightning strikes. The best estimate for predicting a lightning strike is the probability that lightning strikes are far more common in the hotter months than in the colder months, and the location of the strikes being in the southeastern quadrant of the contiguous US (Conus). A distribution of lightning strikes in the United States is shown on the following databases (the frequency distribution in the USA shown first):

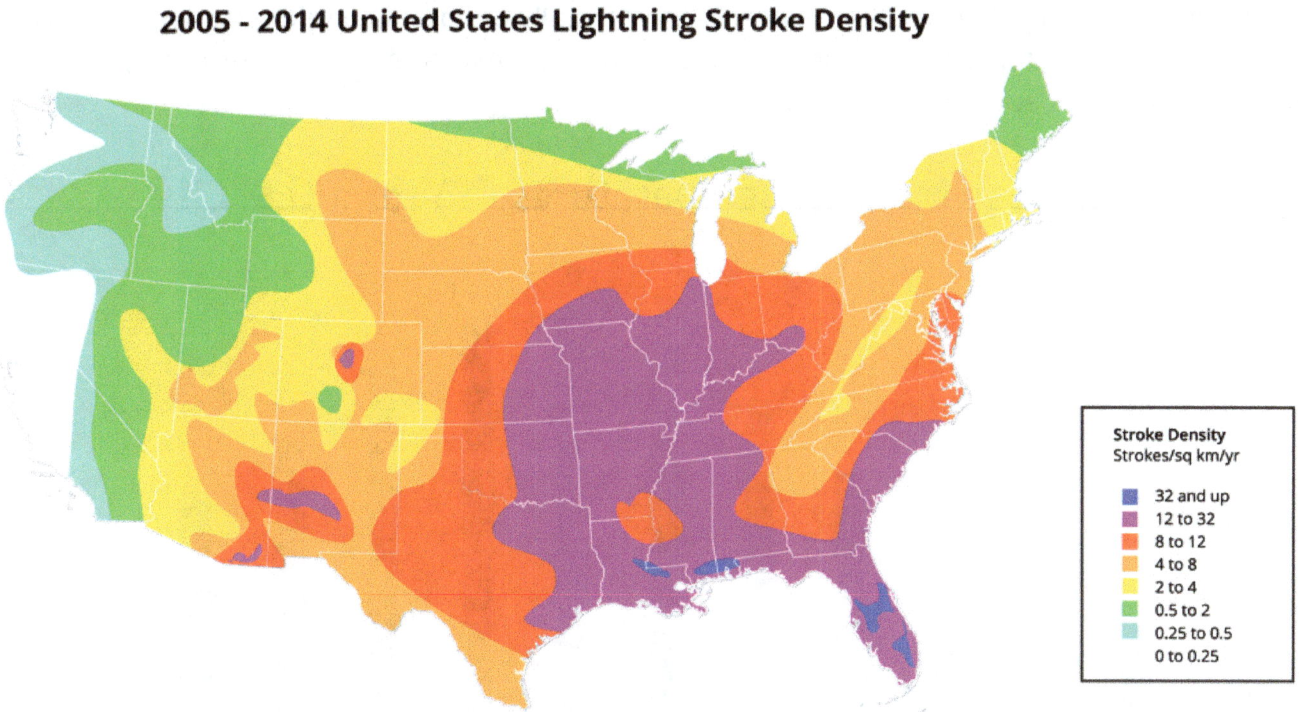

(Source: NOAA)

The time of year – i.e., the seasonal effect - drives the highest probability of lightning strikes. The National Weather Service meteorology office has declared 2023 to be the hottest year on record for Conus. In 2023, the year in which the NWS recorded the highest average monthly temperatures for Conus, the US National Lightning Detection Network recorded an astonishing number – in the millions - of lightning strikes:

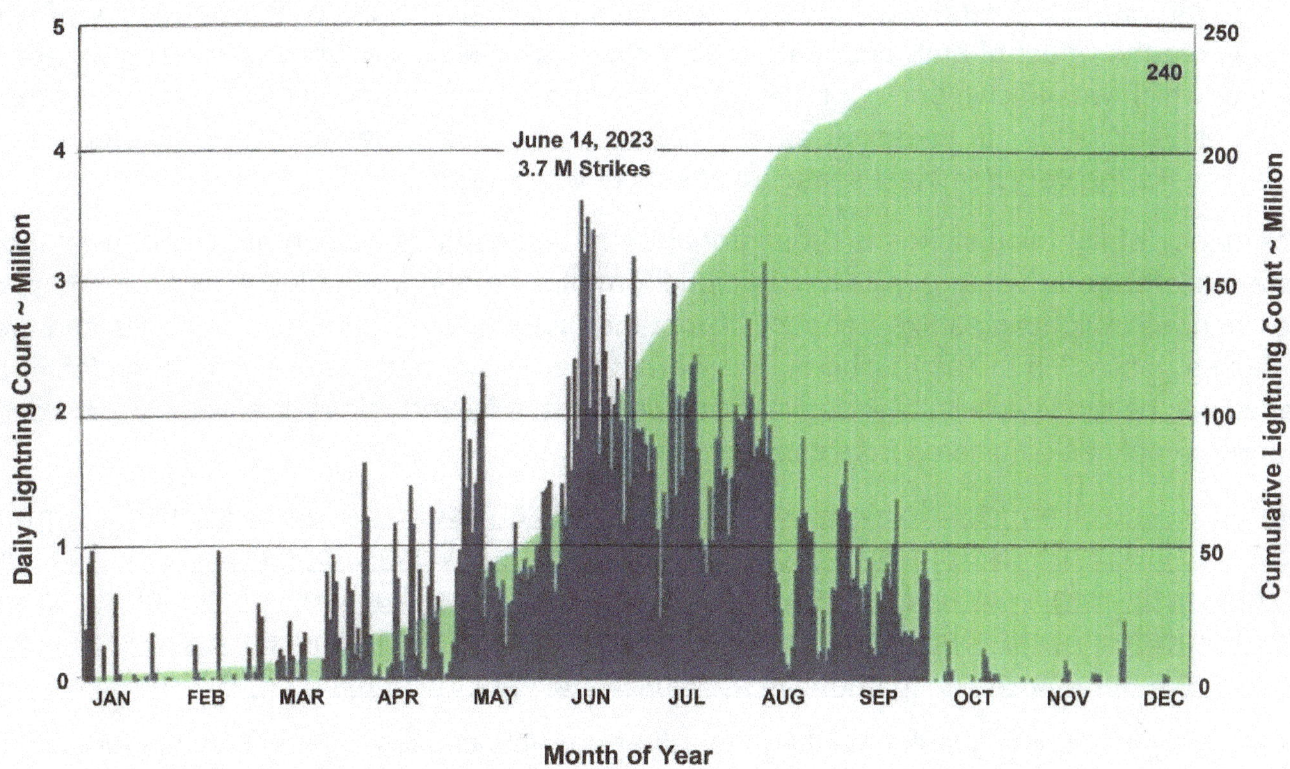

(Source: US National Lightning Detection Network)

The distribution of the lightning strikes is not mathematically uniform in distribution with time of year but the preponderance of strikes very closely correlates with the average temperature per year in US cities (including those in the southeastern US states:

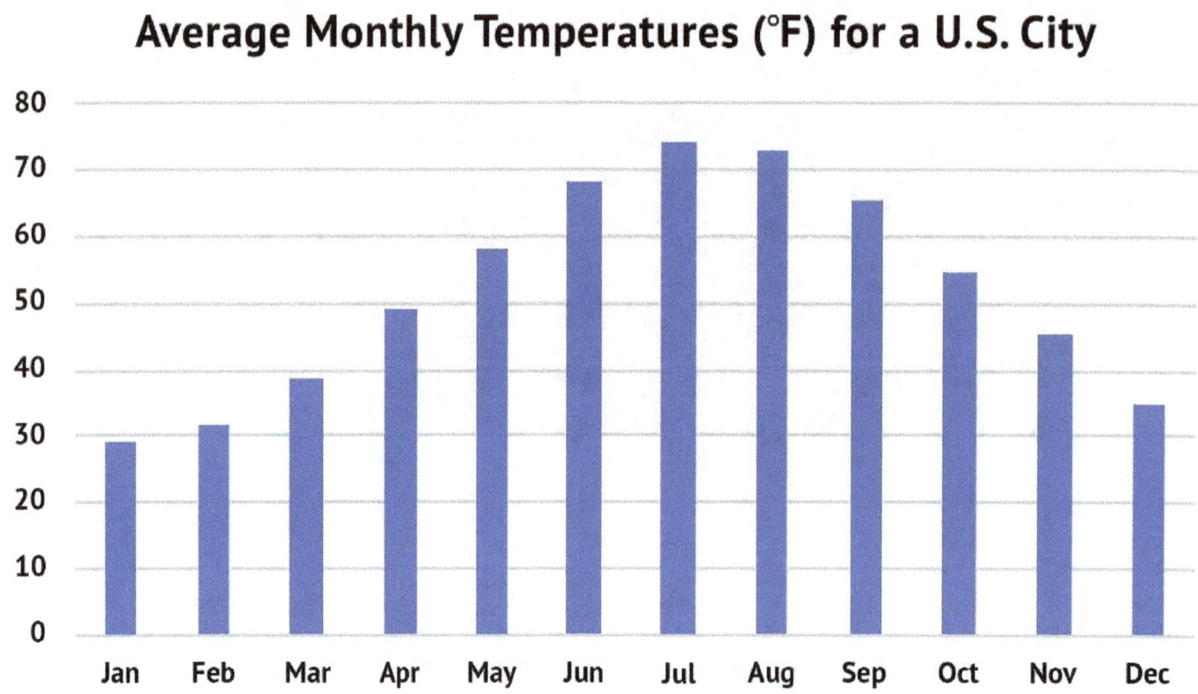

(Source: US Meteorological Service, NOAA)

It should be noted that the overall shape of the average maximum temperature variation in the CONUS temperature database is much the same overall shape as the frequency of lightning strikes. There can be little doubt that the seasonal temperature variation is a larger driver of the frequency of lightning strikes than is the effect of anthropogenic greenhouse gas emissions which are being measured by NOAA and which was shown earlier (in the Keeling Curve) to be a smoothly increasing level (at about 0.008% per year) after the seasonal effects are "adjusted" from the Keeling measurements. That adjusting masks the very dominant seasonal effects of temperature.

The frequency of lightning strikes is proportional to the atmospheric temperature and existence of a high presence of water vapor (typically with cloud formations near ground level) as the primary cause for the changes in lightning strike frequency – not the factor of Man burning fossil fuels and creating "greenhouse gasses" in the atmosphere. The water vapor in the atmosphere is a strong factor that is too variable to be modeled with high confidence by scientists. Nevertheless, the water vapor does exist and cannot simply be eliminated from consideration.

Temperature – resulting from astrophysical and geophysical factors (reflected in the diagram on the front cover of this book) - is a primary factor (with the presence of water vapor as a secondary causative factor) driving the frequency and intensity of lightning strikes as well as most of the natural hazardous events (hurricanes, tornadoes, flooding, landslides/mudslides, etc.). The environmental scientists' hypothesis that anthropogenic greenhouse gas emissions from the burning of fossil fuels is fallacious.

In researching the lightning strike database for other possible causes, a recent issue (June 15, 2024) of the Science News magazine reflects the findings of a scientist 50 years ago in 1974 having correlated the lightning strike phenomenon with solar activity ("sunspots"). In the June 22, 1974 issue of Science

News, the link between solar activity and lightning strikes was as follows: *"Using data from 1930 to 1973, a researcher reports (May 24, 1974 issue of Nature) that in spite of year-to-year variations in the incidence of lightning, there is an underlying cyclic variation* (in sunspot activity) *with a period of about 11 years which is in phase with the solar cycle."*

The entire topic area of solar activity and the electromagnetic field surrounding the planet are beyond the scope of this book to be analyzed. The planet is known to have an electromagnetic field existing in its exosphere as indicated next:

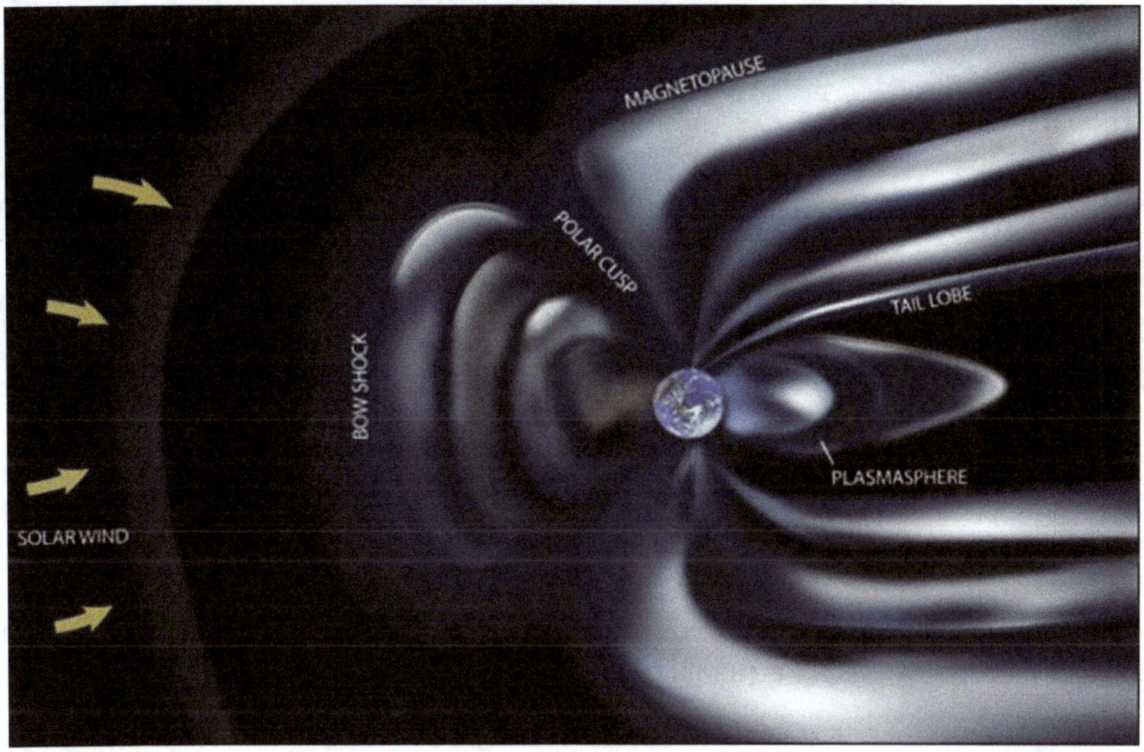

(Source: National Science Foundation)

The effect that the electro-magnetosphere has on the ozone layer and on the climate of the planet's surface is beyond the scope of this book to describe (the Author knows far better how air molecules behave than how electrified electrons behave), but again the physics of electromagnetism is left to those whose careers have specialized in the mushrooming fields of science based on electromagnetism. That admission is necessary to serve as a "Lesson Learned" from several complex accident investigations in which the Author has participated. A primary "lesson learned" is that when confronted with an unknown factor in solving a problem, it is best to consult with others having expertise in that aspect of STEM – science, technology, engineering, and mathematics. By no means is it acceptable for eliminating a factor from further consideration simply because that factor is not fully understood – and what it factually affects elsewhere in the system.

The public is given a glimpse of the artistry of Nature – which results from the electromagnetic field surrounding the planet - when in the northern latitudes of the planet and are treated to a display of aurora borealis known also as the "Northern Lights" and as "the Aurora". In those northern latitudes (Scandinavia and Iceland and Greenland), aurora tourism is one of the largest tourist attractions and source of much welcomed visual entertainment. The awesome sights of volcanic and geothermal activities is commonplace on Iceland. A photograph of the aurora borealis is shown next illustrating the beauty – and the geologically chaotic environment – Nature causes on the planet.

(Credit to Unknown Photographer in Iceland)

Returning to what is known about the effects lightning strikes have on the environment (and indirectly on the safety of the human population), in areas of flat or nearly flat terrain, lighting tends to strike a high protuberance on the flat terrain – either of natural origin (tall trees being a primary "lightning rod") or Man-made construction (such as a metallic tower or a tall building). It was shown earlier that wind turbines on top of high poles are a frequent lightning attach point.

Shown next is a lightning strike that typifies the high voltage level to cause a large lightning bolt to flash in a matter of a second or two during its short-lived existence:

(Source: The National Interagency Fire Center and of the National Weather Service, NOAA)

While the above photographs do not reveal any fire having resulted from the lightning strikes, when a lightning-started wildfire does get ignited it rapidly grows depending on many variables of the local weather and environmental surroundings.

The numerical statistics indicate the large number of wildfires that are started by lightning strikes in forested areas of the United States as shown as follows:

Lightning-Started Wildfires Reported by Geographic Area - United States

YEAR	Alaska (partial)	Northwest	Northern California	Southern California	Northern Rockies	Great Basin	Western Great Basin	Southwest	Rocky Mountains	Eastern Area	TOTAL (US)
2001	29	159	2,238	832	1,041	2,405	1,697	2,298	2,114	889	13,702
2002	165	1,797	301	179	1,130	1,602	556	2,469	2,039	372	10,610
2003	72	1,605	966	428	1,921	2,004	569	2,702	1,918	102	12,287
2004	270	2,042	670	323	1,090	1,760	781	2,062	1,340	88	10,426
2005	311	901	186	272	748	1,345	536	1,935	1,398	175	7,807
2006	54	2,170	948	409	1,970	2,259	943	3,220	2,479	256	14,708
2007	201	1,486	574	291	1,363	1,434	463	1,869	1,672	330	9,683
2008	75	1,624	1,400	174	679	835	224	1,027	941	171	7,150
2009	199	1,843	890	179	1,212	1,086	478	1,546	1,090	62	8,585
2010	330	1,110	441	216	633	1,036	273	947	941	169	6,096
2011	138	808	301	258	944	1,168	432	1,678	1,328	161	7,216
2012	141	856	180	266	1,328	1,259	577	1,224	1,992	205	8,028
2013	212	2,282	596	274	1,305	1,441	542	1,238	1,166	54	9,110
2014	55	2,417	477	269	1,019	937	284	840	689	34	7,011
2015	417	1,705	785	397	1,571	1,111	N/A (~300)	986	650	86	7,708
2016	229	437	97	96	916	812	N/A (~300)	1,413	1,247	107	5,354
2017	155	1,254	728	188	1,254	1,139	N/A (~300)	922	1,014	59	6,713
2018	138	977	174	131	604	1,145	N/A (~300)	1,216	1,086	69	5,402
2019	371	1,430	420	76	686	921	N/A (~250)	786	736	38	5,093
2020	168	746	430	124	625	804	N/A (~250)	746	789	171	4,603
2021	122	1,000	341	241	1,071	1,087	N/A (~250)	747	1,170	130	5,709
2022	282	1,225	185	115	1,061	949	N/A (~250)	754	808	107	5,485
TOTAL	4,134	29,874	12,788	4,896	24,546	28,539	8,355 (14)	32,625	28,607	3,835	178,486
AVG	188	1,358	581	223	1,116	1,297	597	1,483	1,300	174	8,317 / YR

Legend:

>AVG	>210	>1,400	>650	>250	>1,200	>1,400	>650	>1,600	>1,400	>200	>8,400
AVG	160-210	1,300-1,400	550-650	200-250	1,000-1,200	1,100-1,400	550-650	1,400-1,600	1,200-1,400	150-250	8,200-8,400
<AVG	<160	<1,300	<550	<200	<1,000	<1,100	<550	<1,400	<1,200	<150	<8,200

Note: The statistics for Alaska are only those reported near inhabited areas. Wildfires in uninhabited areas are occurring but are not reported.

(Source: The National Interagency Fire Center of the US Bureau of Land Management)

The color coding red and green of the table (described in the footnote of the table) reveals an interesting statistic that rebuts news media statements (fed to the news media by over-exaggerated observations of some environmental protection scientists and activists) that Man-generated "greenhouse gasses" have increased the frequency and severity of wildfires with time. By comparing the trend of above (red) and below (green) relative to average (in white) it is seen that there is a trend just the opposite to what the public is being told. There are clearly more wildfires numerically above average in the decade of 2000 to 2010 and fewer wildfires numerically below average in the decade of 2011 to 2022.

A graphical presentation of the wildfire database is shown as follows:

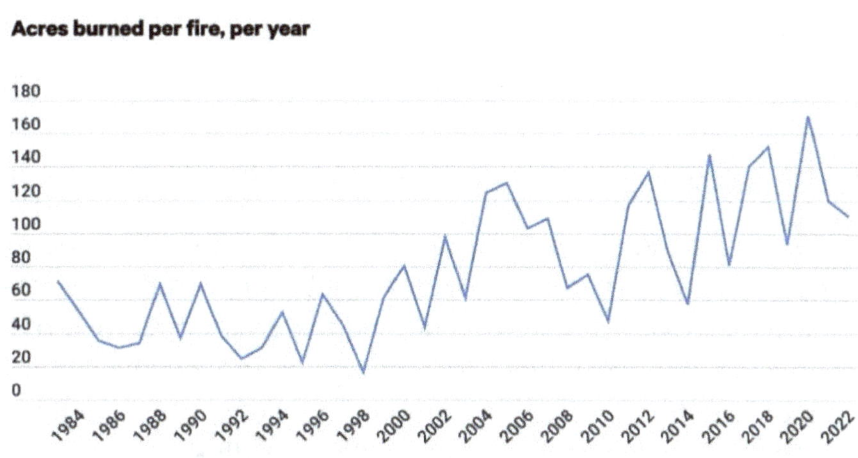

(Source: National Interagency Fire Center (NIFC))

When viewed in graphical form, the trend in wildfires is seen to have been somewhat flat line from 1984 to 1988, but there is a distinct increase in the average trend of acreage burned per fire from 1992 to 2022 (as the tabulated data also showed by the color-coded data). It would be considered statistically absurd to correlate the small increase (0.008% per year) of anthropogenic carbon dioxide in the atmosphere during the same 30-years span with the many times larger percentage increase (3.33%) in acres burned per fire per year.

Each wildfire consumes a widely varying number of acres (square miles or square kilometers), but when their frequency is as high as shown in the table presented above, the totals of areas burned are incredibly high:

Wildfires Area Burned Cumulative USA 2001 to 2022

Year	Number of Wildfires	Cumulative Number of Wildfires	Number of Acres Burned ~ millions	Square Miles Burned	Cumulative Square Miles Burned	Square Miles Burned per Wildfire
2001	13,702	13,702	7.4	11,840	11,840	864
2002	10,610	24,312	3.6	5,760	17,600	543
2003	12,287	36,599	7.2	11,520	29,120	938
2004	10,426	47,025	4.0	6,400	35,520	614
2005	7,807	54,832	8.1	12,960	48,480	1,660
2006	14,708	69,540	8.7	13,920	62,400	946
2007	9,683	79,223	9.8	15,680	78,080	1,619
2008	7,150	86,373	9.3	14,880	92,960	2,081
2009	8,585	94,958	5.3	8,480	101,440	988
2010	6,096	101,054	5.9	9,440	110,880	1,549
2011	7,216	108,270	3.4	5,440	116,320	754
2012	8,028	116,298	8.7	13,920	130,240	1,774
2013	9,110	125,408	9.3	14,880	145,120	1,633
2014	7,011	132,419	4.3	6,880	152,000	981
2015	7,708	140,127	3.6	5,760	157,760	747
2016	5,354	145,481	10.1	16,160	173,920	3,018
2017	6,713	152,194	5.5	8,800	182,720	1,311
2018	5,402	157,596	10.0	16,000	198,720	2,962
2019	5,093	162,689	8.7	13,920	212,640	2,733
2020	4,603	167,292	4.6	7,360	220,000	1,599
2021	5,709	173,001	10.1	16,160	236,160	2,831
2022	5,485	178,486	7.1	11,360	247,520	2,071
	Avg 8,317		Avg 7.0	Avg 11,251		Avg. 1,528

1 million acres = 1,600 sq miles

(Source: The National Interagency Fire Center of the US Bureau of Land Management)

There is not a conclusive trend in the above table showing a decrease or an increase in acreage burned by wildfires but the acreage per wildfire does show a slight trend toward an increase in acreage burned per wildfire. That slight trend is somewhat surprising and disturbing because of the large improvement in observing wildfires in their infancy (soon after being ignited by the lightning strike(s) and with continuing improvement of forestry management (more intentional burns to thin out fuel on the forests floors) should show a trend toward less acreage burned per wildfire.

A photograph that adds realism to the effect wildfire has on the environment is shown next:

(Source: The National Interagency Fire Center and of the National Weather Service, NOAA)

From a geographical viewpoint, the next illustration (bar chart) indicates the states that have the most damage resulting from wildfires:

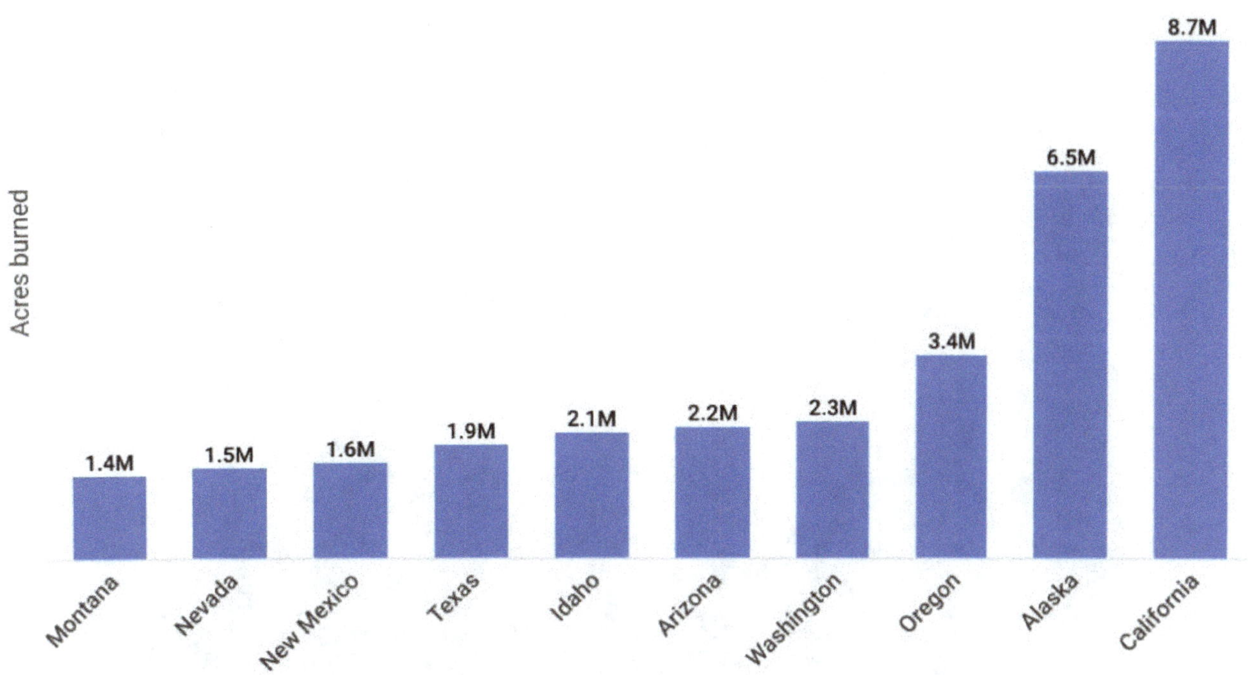

(Source: National Interagency Fire Center (NIFC))

From an impact on the amount of property damage and number of human fatalities perspective, but not to underplay the effect wildfires have on the environment, this entire chapter could be filled with praise for those heroic individuals who volunteer to be on firefighting, rescue, and recovery teams. There can be no labeling of their inefficiency or lack of bravery to be on the front lines with attempts to control the wildfires and assisting humans with recovery of their lives during the aftermath.

The magnitude of losses shown on the table as "Cumulative Number of Wildfires" (total of 178,486 during the period 2001 to 2022) and "Cumulative Square Miles Burned" (total 247,520 during the period 2001 to 2022) do not adequately portray the type of damages done by wildfires differently from those caused by tornadoes and hurricanes. The worst example is that of the wildfire which ironically devastated the town named Paradise, in northern California, in 2018. On the morning of November 8, 2018, a strong downslope windstorm in Butte County in Northern California blew down a high voltage (115 kilovolts) electric transmission line which ignited a fire in the brush below the transmission tower. This account will not highlight the reportedly poor maintenance of the suspension system (e.g., C-hooks) holding the electrical transmission lines in place on the tower; suffice it to say that the massive damage and cost of the fire resulted in the power company filing for bankruptcy. But it does point to the need for power distribution companies to assure that power transmission lines are safe for the extreme environmental conditions possible everywhere along the high voltage distribution route. There also is a need to study and enforce the transmission technologies as more demand for electricity increases due to population increases, especially the need to route high voltage lines at ground level (and safely insulated) through heavily forested regions.

The effect of wildfires on forestry may be the highest concern by environmental scientists; however, the larger concern by far is the effect wildfires has on human (and animal) life in the affected area.

The already high winds blew the fire (titled as the "Camp Fire" as it was ignited near Camp Creek Road – not by embers from a human-made camp fire) through the dense forestry downwind and toward several small communities. Wildfires create their own local weather and especially add wind velocity from the heat expanding the surrounding air and causing the unusual – but common in wildfires – presence of a fire tornado. A photo of a fire tornado is shown next:

(Source: Photograph by photographer of Butte County Fire Department)

It can be seen by the bending of the tree foliage that the velocity of the winds was high, and in the direction ahead of the advancing fire toward doomed residential construction. Some early warnings were given to local residents to evacuate immediately but many of those warnings were either not received or foolishly went unheeded. Compounding the problem was that communities situated in forested areas have but one two-lane route servicing the area – and those evacuating on that single one lane escape route are sometimes jammed by a failure of one vehicle causing a traffic jam. A photograph of the vehicles burned on the escape route of the Camp Fire is shown next:

(Source: Photograph by Photographer of Butte County Fire Department)

The burned vehicles shown (many had their tires burned by the heat as they were jammed in escaping traffic) have been moved from the traveled lane to the shoulder by tow trucks. Occupants of these vehicles tried to flee by foot, and some were rescued in fire engines; others were fatally destroyed in their vehicles by the rapidly moving flames.

Shown is a scene of firefighting equipment unable to get into the towns as the fire had overrun the single lanes into and out from those communities:

(Source: Butte County Fire Department)

Photos of several communities all show the total devastation of the entire community as the fire raced through the communities as in a row of falling dominos.

(Source: Butte County Fire Department)

(Source: Butte County Fire Department)

Damage of a residence by wildfire is more extensive than that caused by a tornado since in a tornado the damage (loss of rooftop or collapse of the structure) is done in a matter of seconds whereas in a wildfire the fire consumes everything for minutes and hours until there is no fuel (anything consumable by flame or heat) left to be burned. Nothing is salvageable or recyclable when burned into ash.

(Source: Butte County Fire Department)

The loss of 85 lives, loss of 18,661 structures (predominantly residential), the loss of 154,000 acres of forest to the Camp Fire were/are incredibly steep; however, the loss to the animal denizens within the forest as well as the loss of the carbon-dioxide absorbing leafage on the trees and plants in those 154,000 acres burned is unmeasurable. As seen in some of the foregoing photographs the smoke pollution into the atmosphere from wildfires is an unseen hazard – worse than the emission of carbon dioxide - that lasts for weeks and months after the wildfire has been contained or extinguished.

Stating (from personal experience of having resided in California for 38 years) a lesser- known fact to the public that living very close to a wildfire is less hazardous to the respiratory system of humans than it is to those living 3-8 miles downwind of the burned area. The reasoning is based on the larger particles of soot and ash that fall back to the ground close to the fire. The smaller particles of soot and ash – the type that can do serious harm to the alveoli (air sacs) in human lungs – travel for a downwind distance of 3-8 miles from the fire zone depending on the winds aloft.

When humans breathe in, air containing oxygen enters their windpipe, passes through the bronchi and then reaches the air sacs. These air sacs, called alveoli, look a bit like tiny grapes at the end of the bronchial branches. Healthy lungs have about 300 million air sacs in them. Each air sac is surrounded by a network of fine blood vessels (capillaries). The oxygen in inhaled air passes across the thin lining of the air sacs and into the blood vessels. This is known as diffusion. The oxygen in the blood is then carried around the body in the bloodstream, reaching every cell. When oxygen passes into the bloodstream from the air sacs, carbon dioxide leaves the blood and passes into the air sacs. Carbon dioxide (CO_2) is a waste product of reactions in the cells. Humans get rid of it when they exhale. Because these two gases pass between the bloodstream and air sacs in opposite directions, this is referred to as "gas exchange" in the lungs.

A consideration also about wildfires (as well as the other forms of natural catastrophic events – seasonal and anomalistic extreme weather conditions, tornadoes, hurricanes, floods, volcanic eruptions, landslides, etc.) is the survivors of most natural catastrophes losing their places of residence and often a loss of their food supply. This adds to the already disturbing fact of homelessness in urban areas, which is exacerbated if the weather outside is in an extreme state of cold, hot, or precipitating high amounts of rain or snow. In developed countries such as the USA, there have been established various organizations and governmental programs providing emergency services for victims in communities (such as that of Paradise, California), but many undeveloped countries do not assist their residents with loss of residences or loss of food supply.

Where there is a lack of firefighting services to quickly extinguish fires from spreading in the entire region when driven by high wind conditions, the loss of lives and cost of fire damage grow in proportion to the number of facilities involved. The fires that destroyed the beautiful town of Lahaina on the Island of Maui in the state of Hawaii on August 8th through the 11th, 2023, killed at least 98 persons and left 31 persons missing (presumed either destroyed in the fire or from jumping into the nearby ocean and drowning or being taken by sharks). Damage to property in the 2,170 acres burn zone was almost entire. This fire at Lahaina is especially tragic to the Author who, with his wife, enjoyed visiting the small but hugely interesting Whaling Museum section of the Lahaina Heritage Museum in the Lahaina Historic District on Front Street in Lahaina on two occasions. In front of the Lahaina Heritage Museum was a large old, iron anchor from a sailing ship which was seen on a television news report of the Lahaina Fire and being the only recognizable feature surviving the firestorm in Lahaina.

Shown next is the town of Lahaina with the fire shortly after its outbreak. The fringe winds (30-45 mph with gusts to 60 mph) caused by Hurricane Dora (Category 4) over six hundred miles south of Hawaii drove the fire storm quickly across the entire town seen blowing from the initiation area (where a downed power line – blown down by the high wind gusts - started the fire) virtually horizontally very close to the ground.

(Source: The Federal Emergency Management Agency (FEMA))

Firefighting equipment was unable to prevent the fire from spreading rapidly and trapping humans because of the high winds and extensive area involved. The homes had no built-in water sprinkler systems and were built of almost all wooden construction.

In other more common circumstances, that of fire in a tall commercial structure, putting water on the fire can be a formidable challenge:

(Credit to Unknown Photographer)

The humans least hazarded by wildfire are those living downwind from the main fire and being exposed to the particulate matter in the wildfire smoke – a form of natural pollution of the atmosphere. People living within about 3 miles from the wildfire when active are exposed to large soot particles which are less hazardous to human respiratory systems because they fall to the ground due to gravity and are not inhaled. People beyond that distance are exposed to the fine particulates in the atmosphere that are unhealthy. Two exposures to smoke from wildfire affecting mass populated areas were the wildfires in Northern California that lingered in the Los Angeles area for a few days (starting June 20, 2020) and caused uncounted health injuries, some undoubtedly fatal to persons in already abnormal respiratory ill health. The Los Angeles skyline resembled that in the early twentieth century before stringent EPA smog-reduction regulations were enforced on motor vehicles.

(Source: Los Angeles Times, Unknown Photographer)

The Author and his family resided near Los Angeles for 34 years and are enormously grateful to the EPA for their smog-reduction activities in southern California. A later section is discussed dealing with the human transportation system (with the transition to all-electric vehicles) expands on the pros and cons of what is being done in the human transportation systems globally.

Another example of wildfire smoke and hazardous particulate matter in the atmosphere downwind was that in New York City in September 2023 caused by wildfires in southeastern Canada. The iconic Empire State Building appeared as shown in the next photograph:

(Source: New York Journal American, Unknown Photographer)

Numerous business offices in the downtown New York City area directed their employees to stay at home in outlying residential areas (in Brooklyn, Queens, and on Long Island). Again, there are no medical statistics for the number of fatalities among the 20 million inhabitants in the Greater New York area. The Canadian wildfires were naturally caused as were the winds that carried the smoke toward the USA, and therefore the fatalities incurred cannot be directly attributed to Man burning fossil fuels changing the climate as having been the cause for them.

A not conducive to health status report caused by air pollution was recently posted (April 25, 2024) by CBS News:

"Almost four out of every 10 people in the United States live in a place where air pollution is considered bad enough to put their health at risk, the American Lung Association warned in its latest "State of the Air" report released on Wednesday. That proportion of people — about 39% of the population — had risen sharply since earlier rounds of pollutant data were analyzed for the annual report last year, and the trends were especially pronounced in certain parts of the country.

This year's air quality report was based on pollution data collected in 2020, 2021 and 2022 by the Environmental Protection Agency.

The American Lung Association, a charity organization focused on improving lung health and addressing lung disease, has released "State of the Air" reports annually since 2000. The new report's findings show that roughly 131 million Americans were breathing unhealthy air during the three-year monitoring period. That number had jumped by almost 10% since the lung association issued its "State of the Air" report in 2023, when data showed 11.7 million fewer people had been regularly exposed to toxic and potentially deadly pollutants. Echoing longstanding concerns from experts and advocates across the board, the newest findings highlight a disproportionate environmental threat to people of color. Since the Clean Air Act was signed into law in 1970, emissions of the six most common air pollutants have fallen by 78%, according to the EPA. But experts warn that climate change has made air quality harder to manage despite policies designed to protect it, especially as growing sections of the country grapple with soaring temperatures, longstanding drought and unprecedented wildfires."

The report added: *"High ozone days and spikes in particle pollution related to extreme heat, drought and wildfires are putting millions of people at risk and adding challenges to the work that states and cities are doing across the nation to clean up air pollution,"* the report's authors wrote. Two groups of pollutants were examined: fine particles, which are emitted directly into the air from combustion and some chemical reactions, and ozone, which is usually produced when other air pollutants react with sunlight in the atmosphere. For fine particles, the pollution trends were broken down into short-term and long-term patterns. The American Lung Association ranked U.S. cities with the worst air quality based on daily particle pollution, annual particle pollution and ozone between 2020 and 2022. The report assigned many cities a failing grade when repeatedly poor readings on the Air Quality Index pushed an area outside the bounds of accepted general health standards."

Except for those wildfires that are caused by Man (by unintentional carelessness with camping fires not being properly extinguished, or intentionally by arson), the Author found no database that would challenge the annual "State of the Air" report. However, the "State of the Air" report does not associate Man creating "greenhouse gasses" to be a leading cause for the wildfires or for the ozone concentration in the near atmosphere "which is usually produced when other air pollutants react with sunlight in the atmosphere". It is beyond the scope of this book to delve into the ozone increasing or decreasing phenomenon – that is left to atmospheric physicists and chemists to further investigate and to control if possible.

Water as a resource has had a pronounced effect on where a human chooses to establish his/her residency (although most humans remain within 100 miles of the place where they were born for family purposes). The larger port cities of the world were developed by sailing ship trade routes for cargo and/or passengers, and countries that have a large body of water (coastal or inland) suitable for recreational as well as for aquatic food sources also tend to have a higher density of their human population near those oceans, lakes, rivers, etc. Humans are humans and many prefer to prioritize their lifestyles and recreational needs over health, safety, and comfort, regardless of how large bodies of naturally placed water sometimes "misbehave". Homeowners pay extra insurance premium fees to reside in known flood zones taking the gamble that no flooding will occur.

There has been found no database of factually valid (not over-exaggerated, not "cherry-picked", or based on creating fear) information that supports the environmental scientists claims that climate change is having a feed-back effect causing the onset, frequency and/or severity of natural catastrophes (tornadoes, hurricanes, wildfires, volcanic eruptions, earthquakes, floods, droughts) to be increasing – either globally or regionally. Similarly, the meager database of measurements of the ocean water temperatures rising (a few millimeters change which appears to be equally seasonal as it is to water temper volume increase due to all the combined effects causing atmospheric warming) and discounting the cyclic nature of El Nino and La Nina ocean phenomena), or the meager database on ocean levels rising more than should be expected. The probability of the glacial ice sheets on Greenland and in Antarctica suddenly melting to raise ocean levels to a crisis level is very low – but some planning should/must be started to relocate human residences, commercial, and transportation on the infrastructure to higher elevations further inland in some regions of the USA and world-wide should it eventually become necessary.

The inner core regions of the Earth are made of molten iron and nickel, and the Earth is known to be a rotating sphere. Anything having a ferrous (iron) content when rotated in an electric field is known to generate electrical current flow in the mass of iron. If the electric field shown in the aurora borealis is applied to the planet having a large core of iron (ferrite), volcanologists are theorizing there may be a resulting magnetic force causing the molten ferrite to be attracted toward the electromagnetic field – and resulting in a volcanic eruption. Once again, the Author will leave that possible electromagnetic phenomenon to be researched for its possibility of causing ("aiding and abetting") the eruptions that recently occurred in Iceland (as shown next):

(Credit to Unknown Photographer in Iceland)

That flow of magma in Iceland was the fastest ever seen in modern volcanology. The flow of magma into a 15-kilometer-long crack ahead of the recent volcanic eruptions in Iceland occurred at the highest rate observed anywhere in the world for this kind of natural event (bearing no influence on or of Man-made carbon dioxide emissions into the atmosphere from the burning of fossil fuels). This volcanic activity began with magma accumulating several kilometers beneath the Svar sengi region in Iceland and not far from the site of a geothermal power plant that supplies hot water to the capital city of Reykjavik. More discussion of geothermal power appears in a later chapter on the current and potential use of geothermal power as an alternate and clean energy source which is abundant in many regions of the world.

During the research on this topic area, much has been found on the severe problem of pollution (air, water, and land) that is clearly Man-made, and which poses near-term sanitation, health, and safety problems to humans that are more ominous and urgent than whatever the valid scientific concerns about global warming and climate change portend.

To conclude this chapter, it is a general conclusion based on all of the factual scientific information in the preceding chapters (updated to 2024) on factors (astrophysical, geophysical, natural, and anthropogenic) that most significantly affect the human population and the atmospheric environment that the opening hypothesis of environmental scientists: "When all else is eliminated, Man burning fossil fuels is the dominant cause of global warming and climate change" is unprofessionally fallacious. It remains that water vapor in the air is scientifically factual to be a far greater partial cause of the surface temperature of the planet to be increasing. The foregoing chapters have shown that the "science is settled" claimants do not acknowledge that fact and are wastefully expending much effort and an incredibly size number of financial resources – taxpayers' monies – to "combat" climate change by eliminating the use of fossil fuels. Furthermore, the miniscule effect of burning fossil fuels on disrupting or displacing the astrophysical, geophysical, natural causes of "global warming" is not of "crisis" level of urgency.

It is hastened to be added that this book does not advocate that the use of fossil fuels to satisfy Man's need for heat energy should continue as it has been for the past century. As with virtually all technologies developed by Man – including the technology of combusting fossil fuels as an energy source – if the technology is generating any form of pollutant to the planet's essential resources (air, water, or land) the technology must be improved or abandoned. The technology of combusting coal is different from the technology of combusting oil, and natural gas in that the combustion of coal (which contains variable amounts of impurities when in solid state) generates both heat and an ash residue which contains toxic and carcinogenic components in its ash residue. Disposal of post-combustion coal ash is known to be polluting the planet's water and land environment, and it is for that reason that the cessation of the coal burning technology be rapidly set in place by United States government anti-pollution regulations. Other regions not under United States jurisdiction should make the same type of assessment that has been made for the United States technology developments to eliminate pollution of their regionally environment.

The next chapter adds a more complete set of interim conclusions from the foregoing "big picture" analysis of what is – and what is not – causing planet surface temperatures to be increasing, the climate (regionally) to be changing, and the oceans' surface level to be rising.

Chapter 5
Other factors and the LEO Temperature Model

The illustration in the Preface of this book gave readers a preview of the many factors considered in this book that may be causing or influencing the planet surface temperatures rising, the climate changing, and/or the ocean levels to be rising. Those factors are graphically repeated as a "big picture" below for Readers' convenience (and used as a "score card" for the material presented in the preceding four chapters).

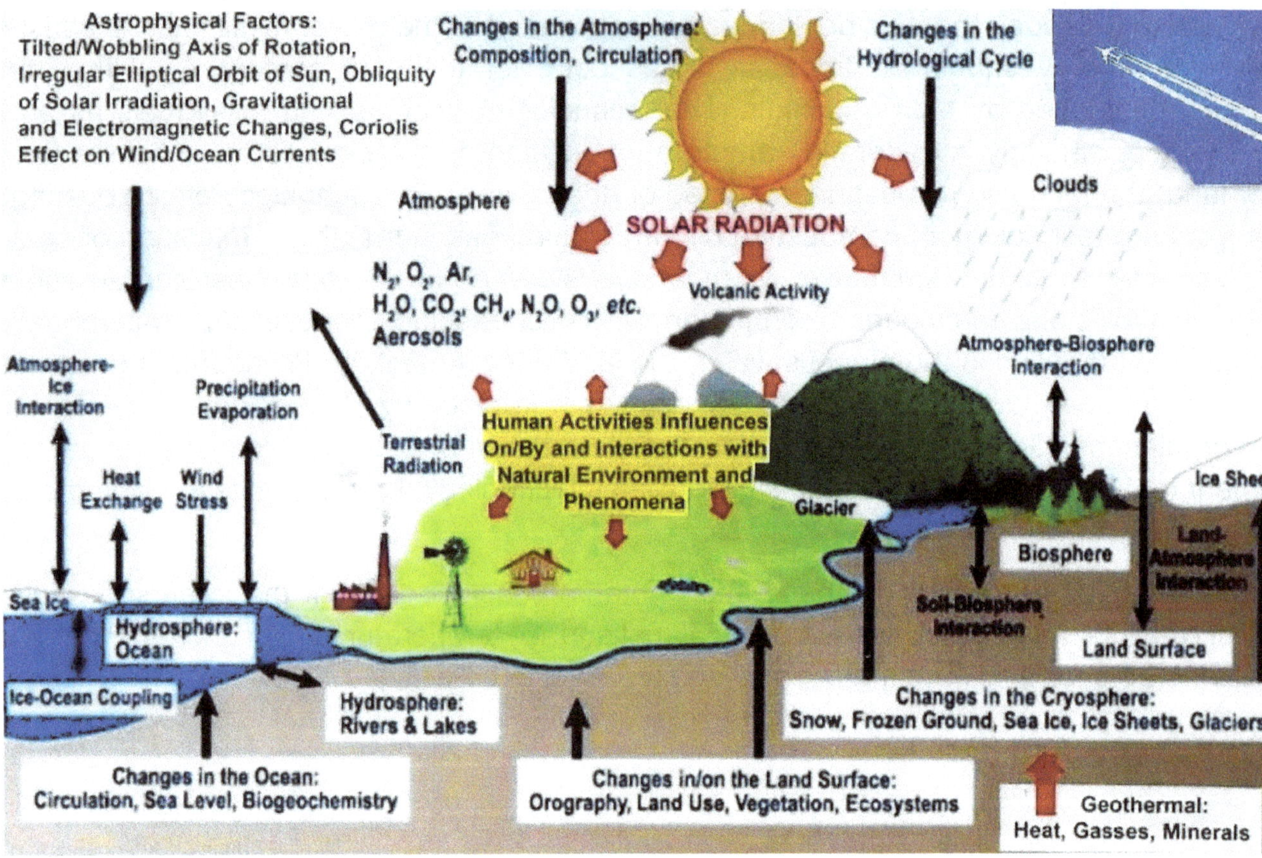

Source: Diagram created mostly by Author)

The environmental science zealots hypothesized and fallaciously concluded that it is the anthropogenic (Man-made) gasses that pervade the planet's atmosphere and is the dominant cause for surface temperatures to be rising. By "eliminating everything else" (quoting an environmental scientist) the "science is settled" and the following chart is "simplified" to reflect the primary "picture" of the problem declared by many environmental scientists to be a "crisis":

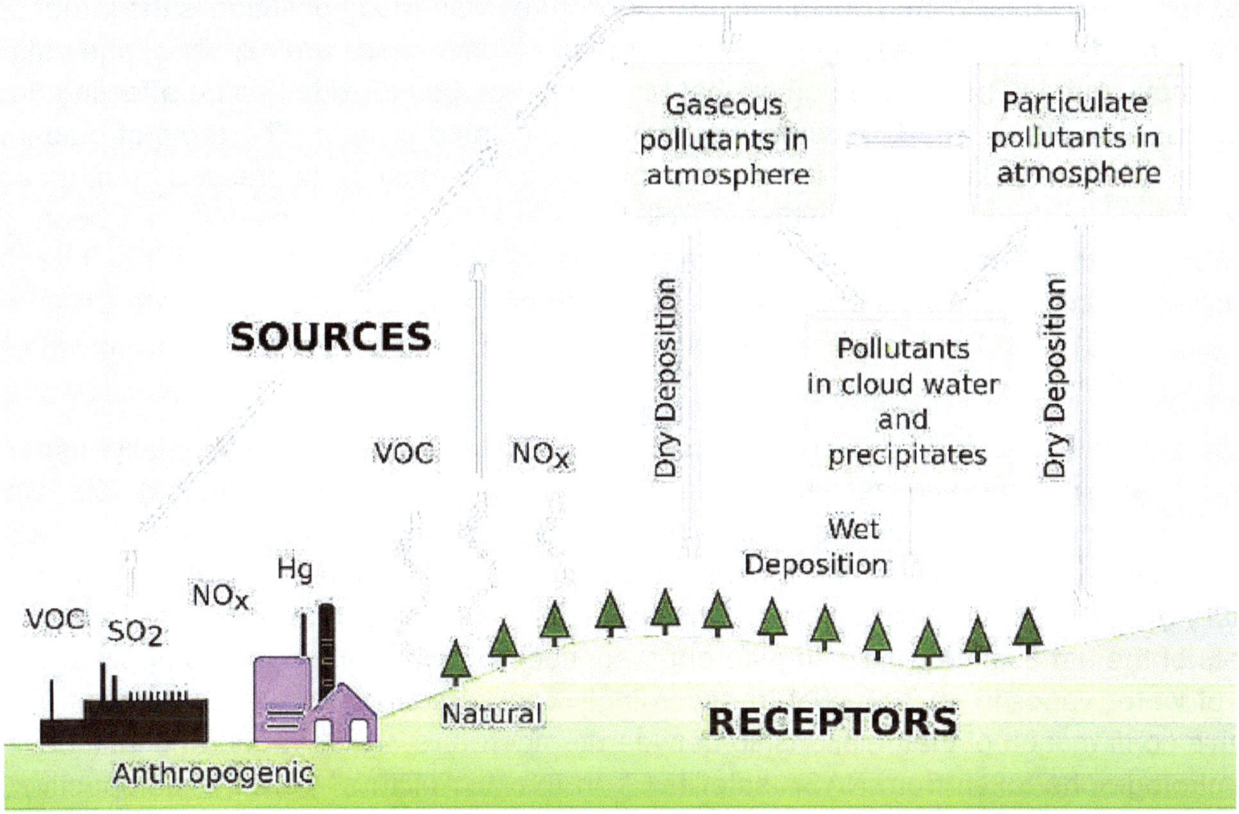

(Source: National Oceanic and Atmospheric Administration)

It should be noted, however, the illustration does highlight that pollutants from industrial process are polluting the atmosphere. The terminology of carbon dioxide (CO_2) is not directly reflected in the illustration but highlights the VOC and other gaseous pollutants from industry. What then is VOC?

Volatile organic compounds (VOCs) are organic compounds that have a high vapor pressure at room temperature. High vapor pressure correlates with a low boiling point, which relates to the number of the sample's molecules in the surrounding air, a trait known as volatility. VOCs are responsible for the odor of scents and perfumes as well as being pollutants. VOCs play an important role in communication between animals and plants, e.g., attractants for pollinators, protection from predation, and even inter-plant interactions. Some VOCs are dangerous to human health or cause harm to the environment. Anthropogenic VOCs are regulated by law, especially indoors, where concentrations are the highest. Most VOCs are not acutely toxic, but may have long-term chronic health effects. Some VOCs have been used in pharmacy, while others are target of administrative controls because of their recreational use.

The lower illustration also points out that it is the particulate matter content – pollutants - that interacts/mixes with the gaseous emissions from anthropogenic activities that causes concern – to environmental scientists and should be and is of high concern to the public (including this Author and his family).

The review and analyses of the available databases as shown in the previous chapters indicate that Man burning fossil fuels is not the dominant or even a comparatively significant factor affecting the surface temperature or climate of the planet. That realization surfaces the logical “big picture” system engineering question “What then is/are the cause(s) for the surface temperature and the climate on the planet – regionally not globally – to be changing?”

This book repeatedly has pointed out that in the "big picture" Man-made pollution is the larger challenge needing urgent attention in the set of realistic hazards to which humans are exposed, and not the much smaller (almost, but not, negligible) portion that Man-made carbon dioxide has on affecting (increasing or decreasing) the planet's surface temperature and associated climate. The topic of global pollution has an entirely different database of facts and information and must be addressed by other scientists, human behaviorists, and system engineers. Pollution therefore has purposefully not been given the level of attention that global warming and climate change has in the analysis that has been done by the Author in his trilogy of books published in 2025: "Facts and Fallacies of Climate Change on Our Planet", (this book) "The Big Picture of Global Warming on Planet Earth", and "Dealing with the Real Climate Change".

It is with the fullest expectation, drawn from the various databases (especially astrophysical, geophysical, and natural, and as reflected in the ice core samples in Antarctica), the climate will continue to get warmer in the near-term decades and as it does there probably will be an associated increase in precipitation. Readers are reminded that climate is determined not only by temperature but additionally by the amount of precipitation combined with temperature. The warming of the atmosphere in pockets of the atmosphere and with different magnitude of temperature in each pocket causes the amount of water vapor in each pocket to accordingly vary regionally as well as latitudinally. A most informative confirmation of that relationship between temperature, water vapor, and latitude is seen in the two photographs taken from NASA satellites – in the "big picture" from the perspective of outer space:

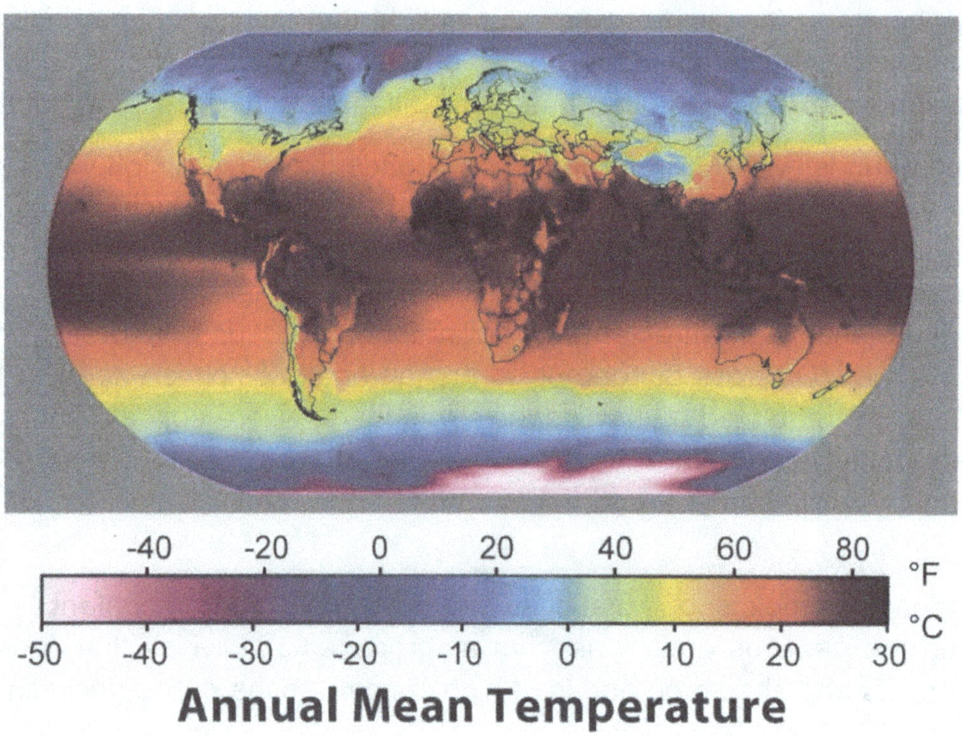

(Source: NASA)

The above photograph may be recognized as the image appearing on the front cover of the book "Facts and Fallacies of Climate Change on our Planet" but without the temperature color code scale on the cover. The amount and distribution of water vapor seen also in the "big picture" of the planet is measured to be as follows:

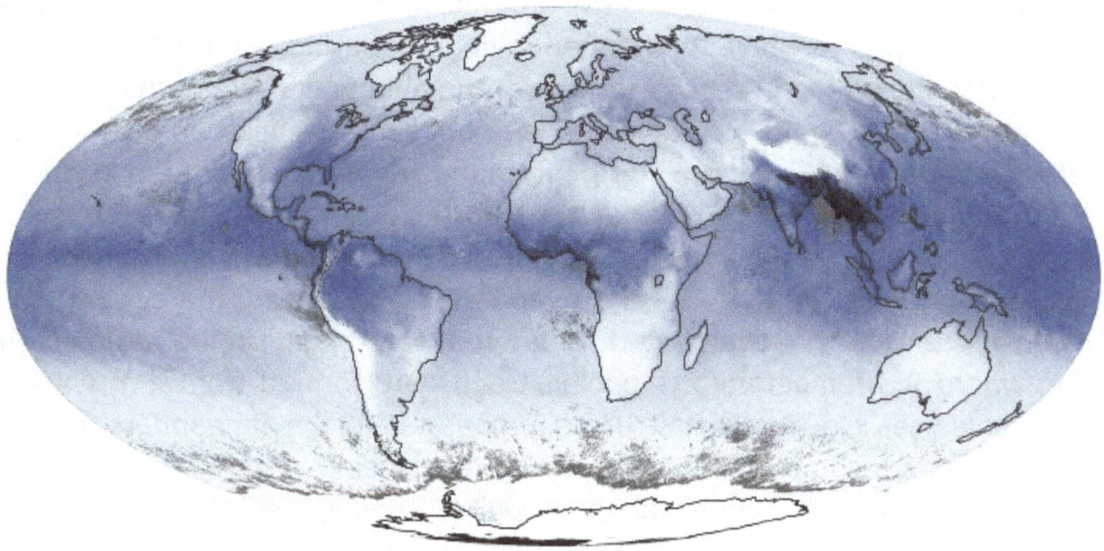

(Source: NASA)

The scale for the amount of water vapor in the atmosphere is not shown below the image above; however, the relationship between temperature and water vapor has been determined scientifically to be as shown in the following graph:

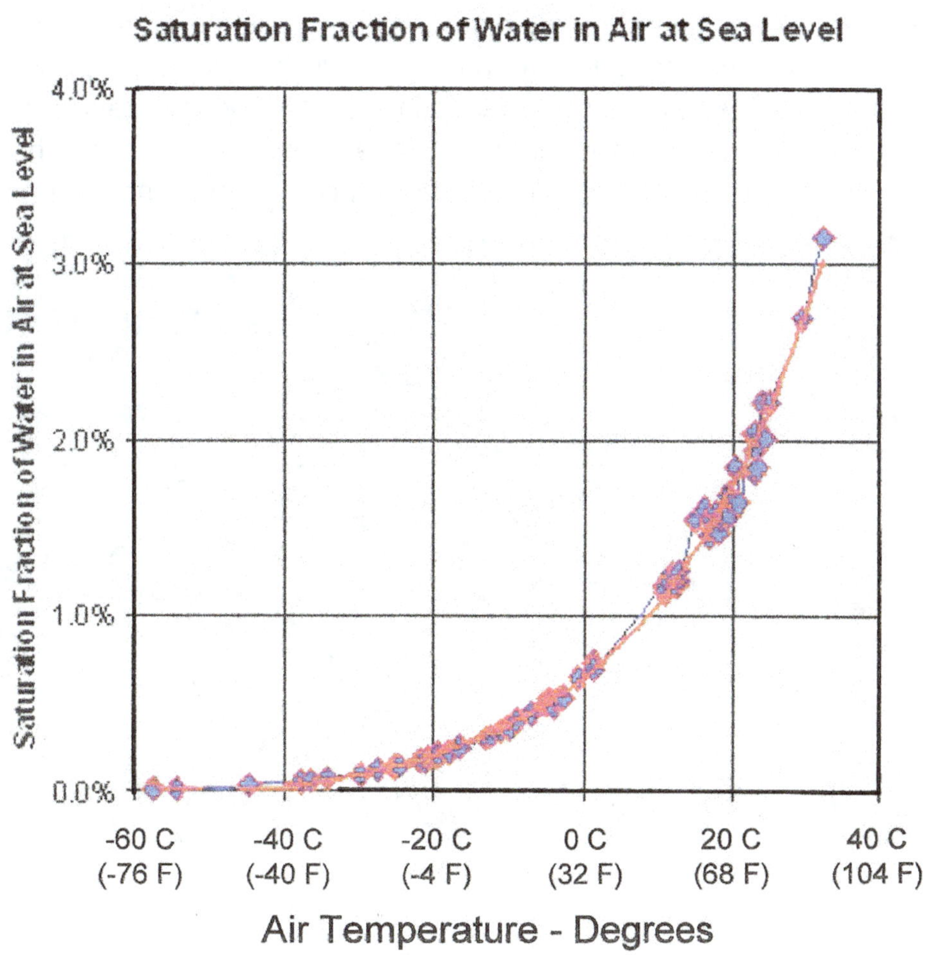

(Source: Smithsonian Institute for Science

A major portion of this book has been presented showing the natural hazards that result from the large amount of water vapor in the atmosphere. That fact is most important in life as without water, there would be virtually no life on the planet. Earth is the only known planet to presently have an abundance of water on its surface as well as above and beneath its outer surface. In the "big picture", Earth has developed an enormous distillation process to provide potable water for all the forms of life on the planet. The oceans cover over 2/3rds of Earth's surface but contain a high percent of sodium chloride – commonly referred to as "salt". The heat energy from the Sun radiating on the surface of all bodies of water on Earth's surface raises the temperature of the water and cause the surface water to evaporate – in the form of water vapor in the atmosphere. By the wondrous property of the evaporative process, only the water molecules – H_2O – are evaporated and leaving the sodium chloride and all other forms of salt in the water not evaporated. When the water vapor in the atmosphere rises to an altitude/elevation where the temperature is cooler, clouds are formed from that water vapor which contains only H_2O molecules in gaseous state. If the temperature of the atmosphere is too low to sustain the amount of water vapor (according to the preceding relationship in the graph with red data points) the excess water molecules fall (by the accelerating pull of gravity) toward the center of the Earth. In that distillation process, potable water is provided naturally to all forms of life on the planet (animate or inanimate, organic or inorganic) that require water for sustenance of that life form.

Accompanying that ever-continuing process – and bearing in mind that it is the combination of temperature and precipitation that defines climate – are the winds that are also generated by the difference in atmospheric pressure occurring in "pockets" over different regions and latitudes of the planet. When the conditions of those elements of nature are combined there can be the occurrence of rainstorms. When the winds become extremely high, they can form into storms that typically have high wind velocities and an excessive amount of rainfall – to wit hurricanes, typhoons, monsoons, and cyclones. The following three illustrations indicate the historical "documentation" of the regions where tropical storms of varying intensity and severity level have occurred on Earth. The first is the satellite image showing the temperature distribution on the different continents (and their population):

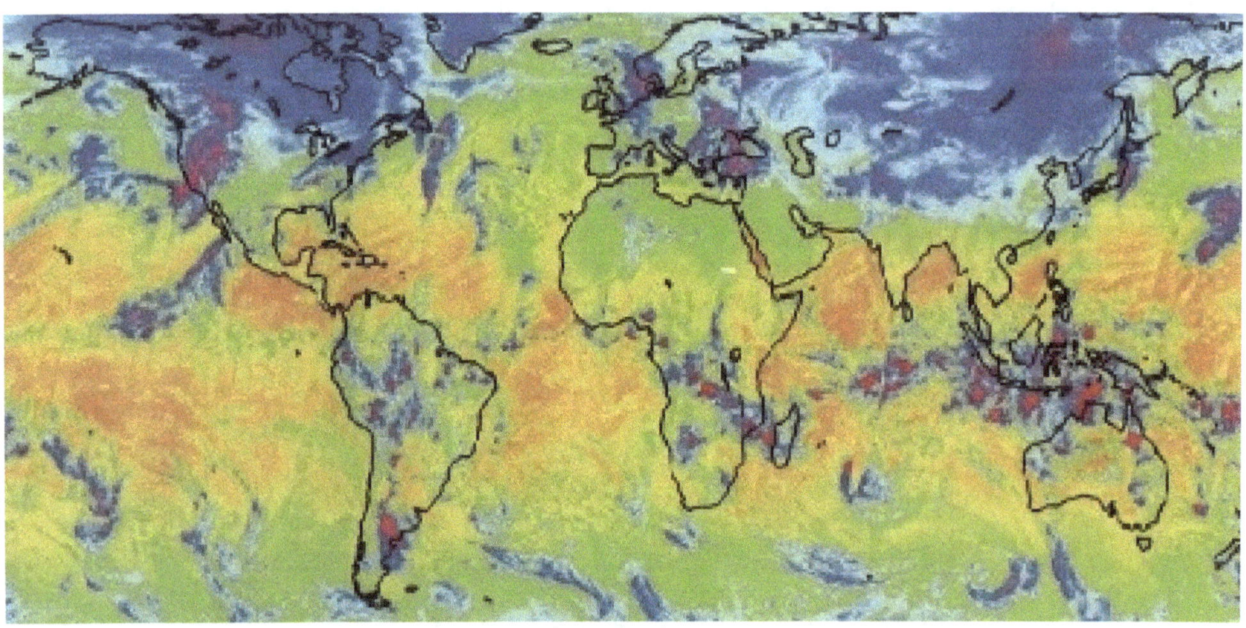

(Source NASA)

Next is shown the "big picture" of where tropical storms of varying severity have occurred: To be noted is the somewhat mirror image of the storms relative to the Equator which is the imaginary horizontal line in the center of the diagram – at latitude 0°.

The storms may appear to be stationary in location in the above illustration; however, each storm has taken a semi-predictable path of travel generally following the flow of warm currents in the ocean beneath them:

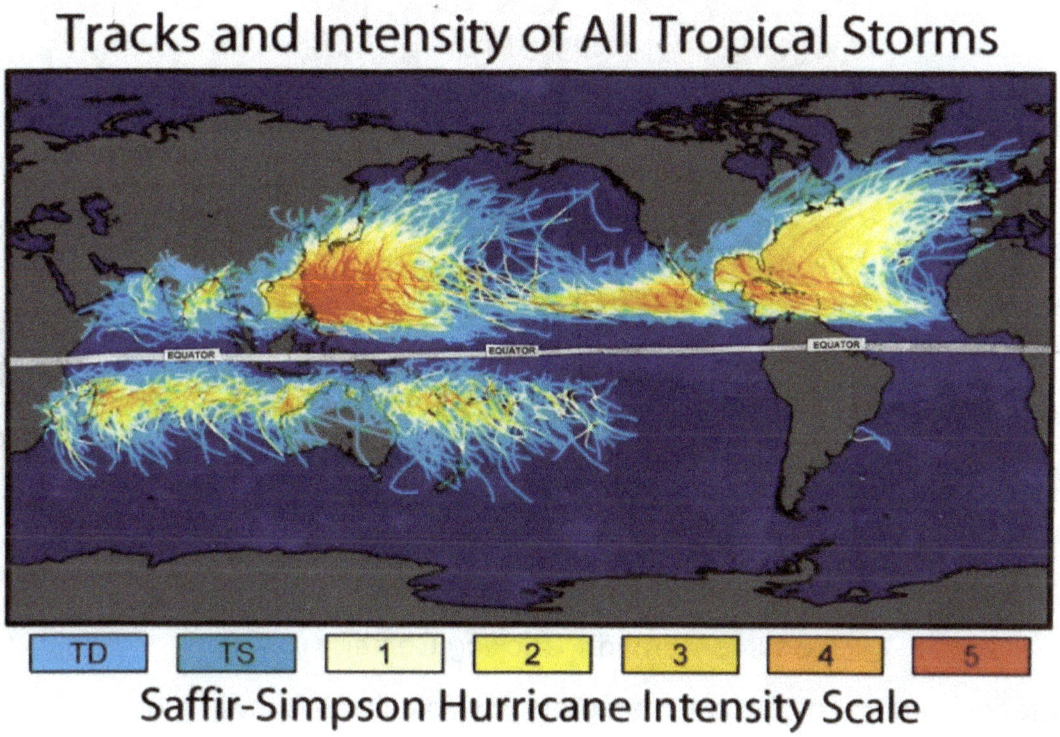

(Source: National Hurricane Tracking Center, NOAA)

A "cousin" of the hurricane is the tornado, but which has a tighter formation resembling a vertical vortex. Tornadoes can form anywhere in the world, but the central and southeastern United States is historically the tornado center of the world – largely due to the inland flow of the warm, moist northerly flow of the atmosphere (with high wind vorticity content) flowing over generally flat terrain: The colored region shown has been named as "tornado alley".

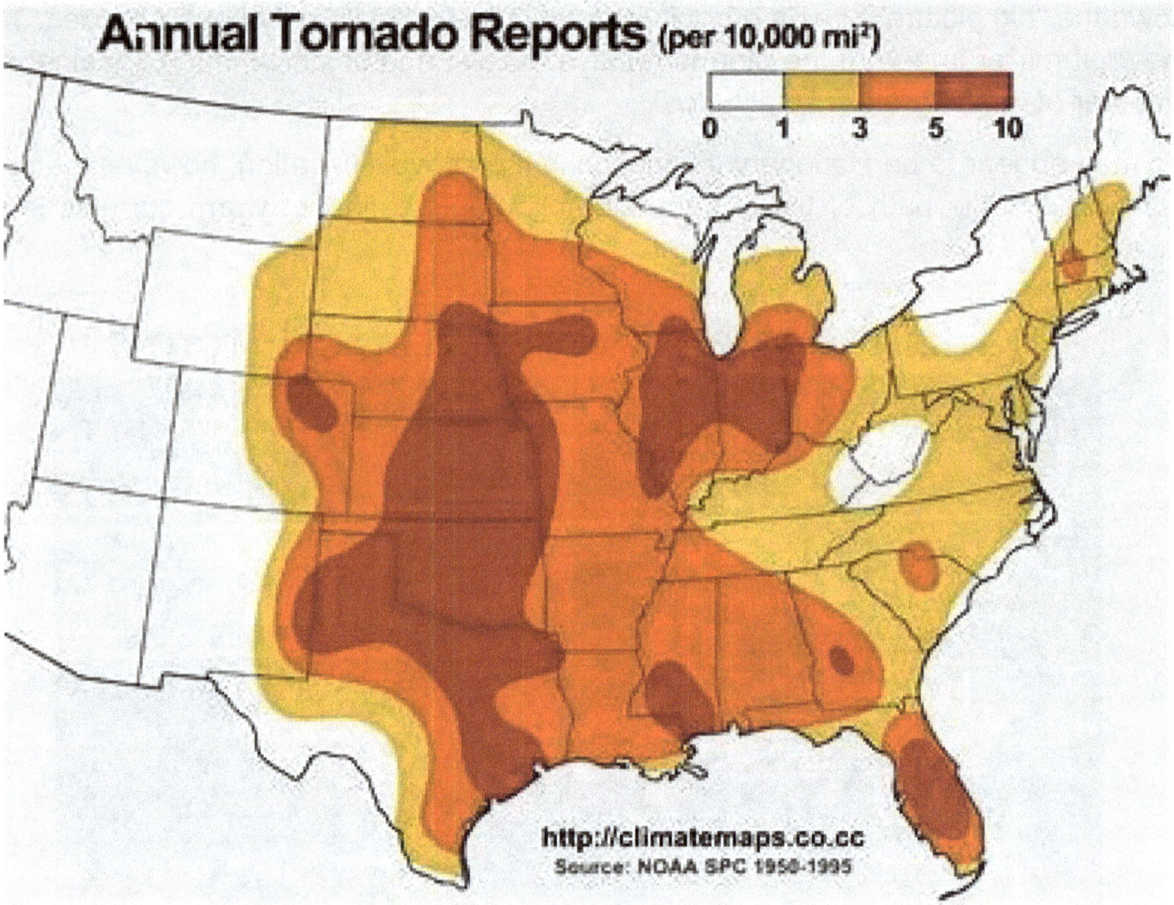

(Source: Tornado Tracking Center, NOAA)

Despite the preponderance of factual meteorological statistics showing otherwise, environmental scientists adhere to their fallacious hypothesis – which has been embedded in the public as being factually and scientifically proven (often referring to the common statement that "the science is settled") – Man burning fossil fuels is the dominant cause for global warming and climate change.

Major ocean levels will probably rise in the future years of the planet, but the rate and to what higher level and how much of the human population will be endangered by both the warmer surface temperatures, the increase in precipitation, and by the level to which the water will rise, remain uncertain. The magnitude and rate of endangerment to humans and other life forms living in low-lying, flood-prone regions remains unknown. The prevailing ocean currents and wind currents may remain largely unchanged in most regions but likely will become more severe in magnitude in some regions. It remains to be assessed factually why the ocean surface levels are changing, in what regions, and at what rate. Once that in-depth assessment (by environmental scientists and human factors specialists) has been completed and without bias, it will be shown what regions (with significant human population) will expose the humans to the most serious aspects of climate change and need to be mitigated.

A search for scientific information on water temperature produced a graphical presentation of ocean temperature measurements collected by a research organization associated with the National Oceanic and Atmospheric Administration (as shown in the following diagram):

CLIMATE

Grim results of the ocean's hot streak

Off-the-charts temperatures impact corals, sea ice and more

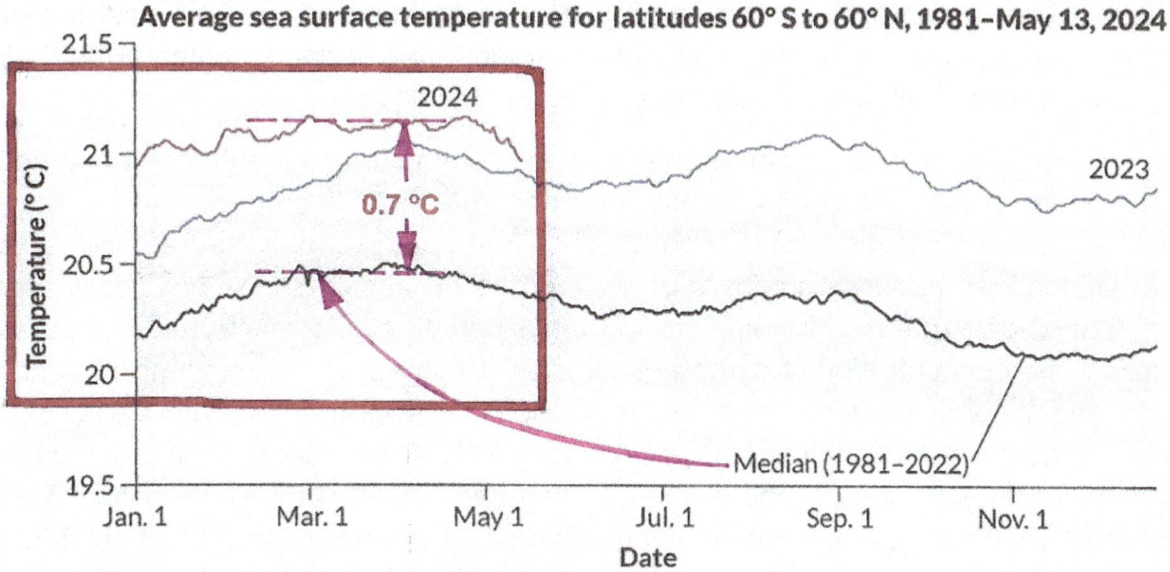

Soaring sea temps Since March 2023, the average sea surface temperature of the extra-polar ocean, at latitudes from 60° S to 60° N, has broken the daily record every single day, NOAA data show. And recent daily temperatures have shattered records that were set last spring.

SOURCE: CLIMATE REANALYZER/UNIV. OF MAINE; NOAA

(Source: Climate Reanalyzer / University of Maine; NOAA)

The chart (with Author's emphasis markings in red) is an example of how the public has been misled by scientific information that does not clarify what are the causes of the ocean levels rising. Instead, the research organization reports that the ocean waters are becoming warmer at a rate that is considered to have *"Grim results of the ocean's hot streak"*. The note at the bottom of the chart is seen to state: *"Soaring sea temps"* followed in very small font *"Since March 2023, the average sea surface temperature of the extra-polar ocean, at latitudes from 60° S to 60° N, has broken the daily record every single day, NOAA data show. And recent daily temperatures have shattered records that were set last spring."* Those descriptors ("grim" and "off-the-charts") and superlatives ("soaring", "broken the record", and "shattered records") must have been written by an eager, but over-emotional, new scientist in oceanography, not by an ethical professional scientist. Why a "soaring" 0.7 °C rise in the ocean water average surface temperature in a span of 43 years may be considered as "grim" to coral reefs health is not questioned but the inference that it is an ominous trend on the planet's climate is a gross over-exaggeration. To other more ethical science and engineering professionals, the "grim" revelation reveals that environmental scientists have resorted to "crisis" tactics to bolster their reputation among their peer group and among the national science organizations. That type of tactic is further magnified by scientists in the Environmental Protection Agency and 'snowballs" by sensational news media reporting to the public. Governmental regulators created legislation (under the Federal regulatory veil of an "Inflation Reduction Act" to "combat the climate crisis" – and the snowball has become of avalanche proportion requiring United Nations action to "save the World". This book is an attempt (as a public service) to restore common-sense and reality to the out-of-control process and avoid further wasteful use of taxpayers' monies from which they derive minimal benefit.

Readers should again be made aware that any data collected with accurate measurement instruments can be overstated in significance by selecting the increments of the vertical scale assigned to the dependent variable ("average sea surface temperature" in the "grim" illustration above) to be in magnified increments. The smaller the increment, the greater the variation measured will appear emphasized – much like the use of a laboratory electro-microscope to enhance biological phenomena at the cell level. Scientifically, any scale is appropriate for detecting trends in a scientific variable, but the interpretation of the magnitude of the absolute change in the variable scaled up to that existing in their surrounding environment is of primary interest to the public.

That has been displayed by environmental scientists expressing the concentration of carbon dioxide in Earth's atmosphere in parts per million (measured accurately as 415 ppmv) and a smaller trace of methane in parts per billion (measured accurately as 300 ppbv). Those may appear to the average member of the public to be large numbers of carbon dioxide and methane physically present in the atmosphere, but when expressed in percentage form – percentage is much better understood by the general public – the concentration of carbon dioxide is 0.0415% and the concentration of methane is 0.00003%. To attach some "relative importance" to those small scientific percentages, a common penny coin is 1% of a dollar. What relative significance is 0.0415% (the concentration of carbon dioxide equivalent to 415 ppmv) of a penny whose value is 1% of a US dollar? More dramatically, what relative significance is 0.00003% (the percentage concentration of methane equivalent to 300 ppbv) of a penny? On what basis do environmental scientists consider the almost insignificant levels of Man-made carbon dioxide and methane in the atmosphere to be trending to a "crisis" level?

When grid lines are added to the preceding chart, it reveals that *"the average sea surface temperature set last spring"* (April 1, 2023) was 21.0 °C (69.8 °F) and on May 1, 2024 (13 months later) was 21.15 °C (70.07 °F). If a 0.15 °C (0.27 °F) temperature rise in average sea water temperature *"shattered records that were set last spring"*, what must the environmental scientists' superlative be to describe that the cooler water temperature in the Pacific Ocean (Aleutian current N-S) on the California Pacific coast averages 59 °F (15.0 °C)? *"Freezing"?*

Similarly, the water temperature of the Atlantic Ocean (Gulf Stream current S-N) on the East Coast of the USA averages 80 °F (27.0 °C). Is the scientific terminology that compared to the "average sea surface temperature, the waters on the US East Coast are "boiling"? It may be scientifically correct to "average out" the difference of 12.0 °C between the coasts and consider both coasts to be at an "average" temperature 6.0 °C cooler than what the temperature is factually at in the Atlantic Ocean and 6.0 °C warmer than what the temperature is factually at in the Pacific Ocean? That "averaging" practice is tantamount to a deceptive practice known in "street language" as "cooking the books".

The ocean waters are becoming warmer (as they factually are in certain regions but to different magnitude from "normal"). Warming a volume of water will cause the water to expand. Does that deserve the title "Grim results of the ocean's hot streak" and a second subtitle "Average sea surface temperature for latitudes 60° S to 60° N 1981-May 13, 2024"? It is not apparent that the data for the weather or climate or ocean levels addresses the effect rising surface temperatures of the ocean water temperature has on the global weather and/or global climate. Instead, as eventually noticed (but almost over-looked due to the small print of the line below the large *"Grim results …"* line).

It is also questioned how are *"off-the-charts"* temperatures – are they the high end of the ocean buoy measurements that are averaged down to "average sea surface" temperatures or are they the low end such as shown for the East Coast and the West Coast? If those off-the charts temperatures are the ones on the high end (above average), shouldn't the NOAA scientists be concerned more about the aquatic animals and food fish life on the US East Coast in the 27.0 °C waters – being about 60 °C above the "grim results" chart – instead of the "soaring" average sea surface temperatures on the Great Barrier Reef coral in Australia)?

It was noted in the citation of the Source that the "*Grim results*" chart was created by the Climate Reanalyzer organization at the University of Maine (sponsored by NOAA near the New England Coast), and that a concern was also for the *"sea ice and more"*. Here again, the *"grim results"* concern on the sea ice appears not to be for the effect the melting of sea ice has on the ocean surface levels rising. When sea ice in the Arctic region (well above 60° N latitude) melts, it does not add to Arctic Ocean surface levels rising and thus it is surmised the melting of sea ice at the University of Maine Climate Reanalyzer organization is of more concern (as was/is to former Vice President Al Gore) about the indirect threat global warming has on the animal population in the Arctic. Polar bears are being threatened by a diminishment of the sea ice on which unwary seals crawl up to snooze. The less sea ice, the less available does seal meat become, in the polar bear food supply. The polar bears are smart animals and are adapting by traveling larger distances on the land portion of the arctic region and raiding the trash and garbage dumps of humans residing in the Arctic region.

Shortly after the "*Grim results of the ocean's hot steak*" article was published, there was a companion article in the Health and Science Section of The Week magazine (May 3, 2024) entitled "*Hot oceans are killing corals worldwide*". A marine biologist (associated with NOAA) stated: *"Record high ocean temperatures have triggered a mass bleaching event in coral reefs around the world. Over the past 14 months, bleaching has been documented in every major ocean basin, from the Florida coast to Australia's Great Barrier Reef. In some places, the water is so hot that the corals haven't had time to bleach; they've simply died, as if cooked."*

Environmental scientists apparently are so focused on the coral reefs being endangered by temperature rise caused by Man burning fossil fuels they ignore all other possible factors such as there being geothermal hazards that are transferring heat into ocean waters. For those scientists at the Climate Reanalyzer program (University of Maine, for NOAA), it might be of value to remind environmental scientists that there is a well-known "Ring of Fire" in the Pacific tectonic plate:

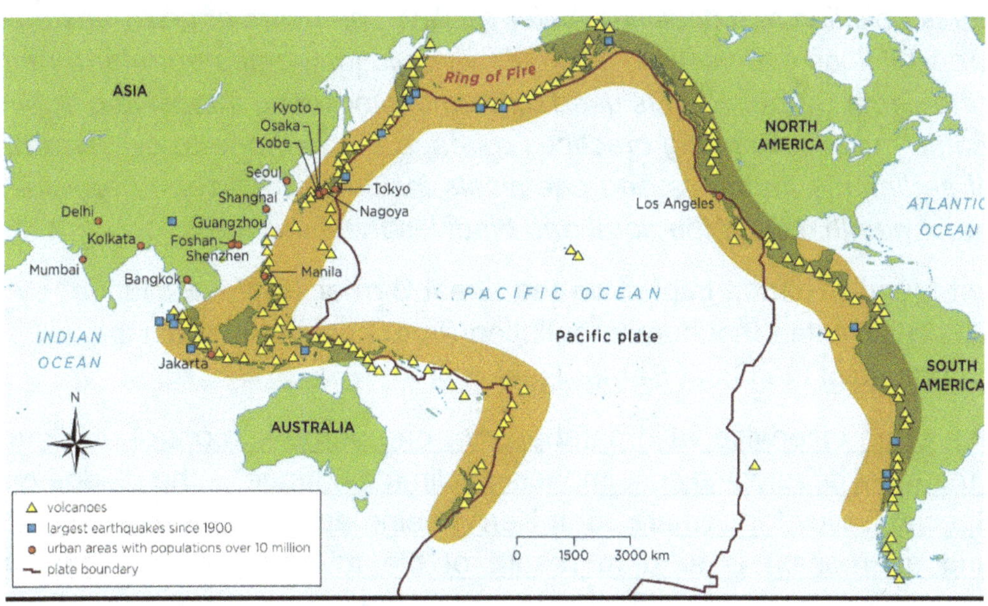

(Source: Encyclopedia Britannica) The Pacific "Ring of Fire"

Is there no possibility that the anemones on the Great Barrier Reef coral (off the northeastern coast of Australia) are bleaching from "soaring" ocean temperatures because of their proximity to several volcanoes known to still be considered active in the Ring of Fire which closely passes the entire northern and eastern coastline of Australia? An earlier chapter showed the Equatorial current of the southwest Pacific to be distributing both the solar heat-related Equatorial current and the geothermally warmed waters over the Great Barrier Reef. Should not all effects be considered rather than pointing to Man burning fossil fuels as being the primary cause of the coral reef heat stress?

If that average water surface temperature rise of 1.2 °C in 43 years is of such "grim" level of concern (to corals but not to humans?), is that why NOAA and/or EPA are using fear-mongering tactics in coaxing the public to believe and support their hypothesis that reducing Man-made carbon dioxide emissions will bring climate change under control? The Intergovernmental Panel on Climate Change (IPCC) has set a goal (in the Paris Climate Accord) that Man-made emissions of "greenhouse gasses" must be reduced to "net zero" to achieve not more than 1.2 °C "average global temperature" rise by 2050?

It is also not clear if/how the temperature on the surface of the oceans changing by such a small amount (a temperature gradient of 0.046 °C per year) factually affects ocean floor life-forms – i.e., does that same temperature gradient exist at the depth of the coral reefs as it does on the surface? A parallel question needing valid scientific analysis: "Is the water surface temperature rise caused by global warming in the atmosphere or is there a geothermal heat source located beneath the coral reef that has rapidly become warmer – or perhaps both?" (Author's Note: As a younger high school student in Brooklyn, New York, a high school science teacher quipped after a laboratory experiment: "Don't you wonder if/how that affects the tea in China?").

So as not to "put all the eggs in one basket" over one misleading database, the Author researched other valid databases for the ocean waters that might exist that are less "simplified" (by oceanographers averaging a set of water surface temperature measurements recorded with a set of ocean buoys distributed in some unknown pattern and at an unknown set of latitudinal locations). The Author discovered a finding (in the "Final Report – Great Barrier Reef Water Science Taskforce" under the auspices of the Queensland Government, May 2016) the finding: *"Another key threat faced by the Great Barrier Reef is pollution and declining water quality. The rivers of north-eastern Australia pollute the Reef during tropical flood events. Over 90% of this pollution comes from farm runoff.[73] 80% of the land adjacent to the Great Barrier Reef is used for farming including intensive cropping of sugar cane, and major beef cattle grazing. Farming practices damage the reef due to overgrazing, increased run-off of agricultural sediments, nutrients and chemicals including fertilizers, herbicides and pesticides representing a major health risk for the coral and biodiversity of the reefs."*

Another document found ("Human Impact on the Great Barrier Reef – Pollution" by the University of Michigan, circa 2013) indicated that human pollution (from household trash including items of plastic) was also coating portions of the Great Barrier Reef

The misapplication of an otherwise valid database for one purpose (control of climate change "if the world is to avoid the most deadly and cost impacts of climate change" – the pledge made by President Joe Biden and Special Envoy for Climate John Kerry) being applied to a totally unrelated purpose (to protect corals and anemones) is a "grim result" of the misinformation being transmitted to the Government by some less-than-professional elements of the science profession. Pollution and climate change have no known causative interaction except for excessive rainfall polluting underground aquafers and animal e coli bacteria being introduced into water reservoirs intended for human consumption.

Environmental scientists in the National Oceanic and Atmospheric Administration have earned an exemplary reputation for their research in aquatic biology and marine biology. NOAA (and the Scripps Institute for Oceanography). Their scientific focus has long been on the preservation of the oceans and ocean life-forms (vertebrates and invertebrates, kelps and other aquatic plants). It should come as no surprise that NOAA is and will defend that interest in oceanography and marine biology as a higher priority than how changes in the climate (on land areas) and pollution are affecting the human population. More ethical scientists in other branches of science have the opposite perspective. Effects of the changes in climate and the increasing Man-created pollution on the world population has a combined effect that need to be addressed in the "big picture", not as isolated problems.

A NASA satellite image also was found that more clearly shows that the temperature anomalies (i.e., the difference from "normal" for that location – shown in an earlier chapter) are not distributed in an "average" manner and have a wider distribution of temperatures measured in degrees (seen as different colors on the illegible scale at the bottom of the image). The non-uniformity of ocean temperature is much closer to factuality than is the "average sea surface temperature" calculated to an accuracy of tenths of one degree (precisely incorrect) as shown in the previous NOAA chart.

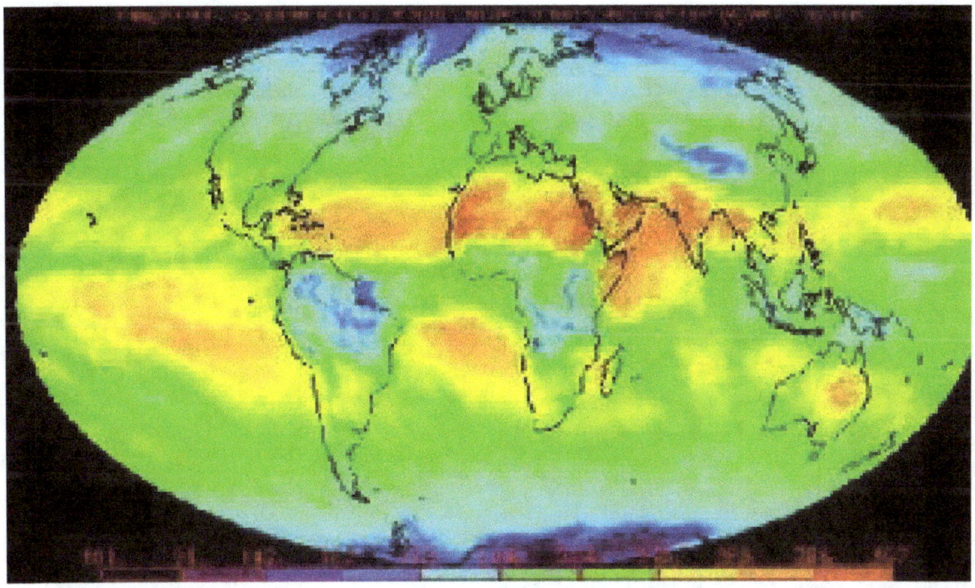

(Source: NASA)

The value of the satellite thermal imagery is that it is not restricted to only water temperature anomalies but also land surface temperature anomalies at the same time. The land surface temperature anomalies over the North African, Middle-Eastern, and Indian land areas are moderately high. The land areas of the North American, South American, European, and South African regions are mostly clear of the yellow or orange-tinged colors reflecting that in those regions the temperatures were near normal.

Relative to the concern about Man burning fossil fuels and destroying the coral reefs, a far more reliable and factually more accurate database for ocean water temperature in the regions where the water surface temperature has been directly measured by thermal sensing instruments on NASA satellites and the International Space Station reveal a very different story about the ocean waters' surface temperatures than what the outdated "average water temperature" calculations show from a set of irregularly spaced ocean buoys.

It is to be noted in the satellite image that off-shore from the eastern coast of Australia where the Great Barrier Reef is located is an area of thermal anomaly. Do environmental scientists (in NOAA especially) factually believe (i.e., "settled science") that all of the "greenhouse gasses" emitted from the Industrial nations in the United States and Europe that burn fossil fuels to satisfy their energy needs are concentrating in eastern Australia (where there is no significant industry or human population) and thereby causing the coral reefs to commit anemoneicide?

Conversely, the satellite imagery is admittedly for the temperature anomaly distribution in a shorter period of time (on the order of perhaps a week or two) with the satellite imagery whereas the NOAA data is the average sea water temperature (not the surface temperature anomaly) for a span of years from 1981 to 2024. Since climate should be calculated with a time period of about 25 years, satellite imagery used for climate assessments needs to span at least 25 years of coverage to deemphasize seasonal or extreme sporadic variations of the variables causing weather extremes in one year but not in the other several years prior to or after the anomalistic year.

Researching the databases of global warming, climate change, and ocean levels rising is well characterized by the quip: "I turned over a stone to see what was under it, and I found more stones to be overturned." Another database found of the ocean water measurements was that of ocean surface level height as the following diagram shows:

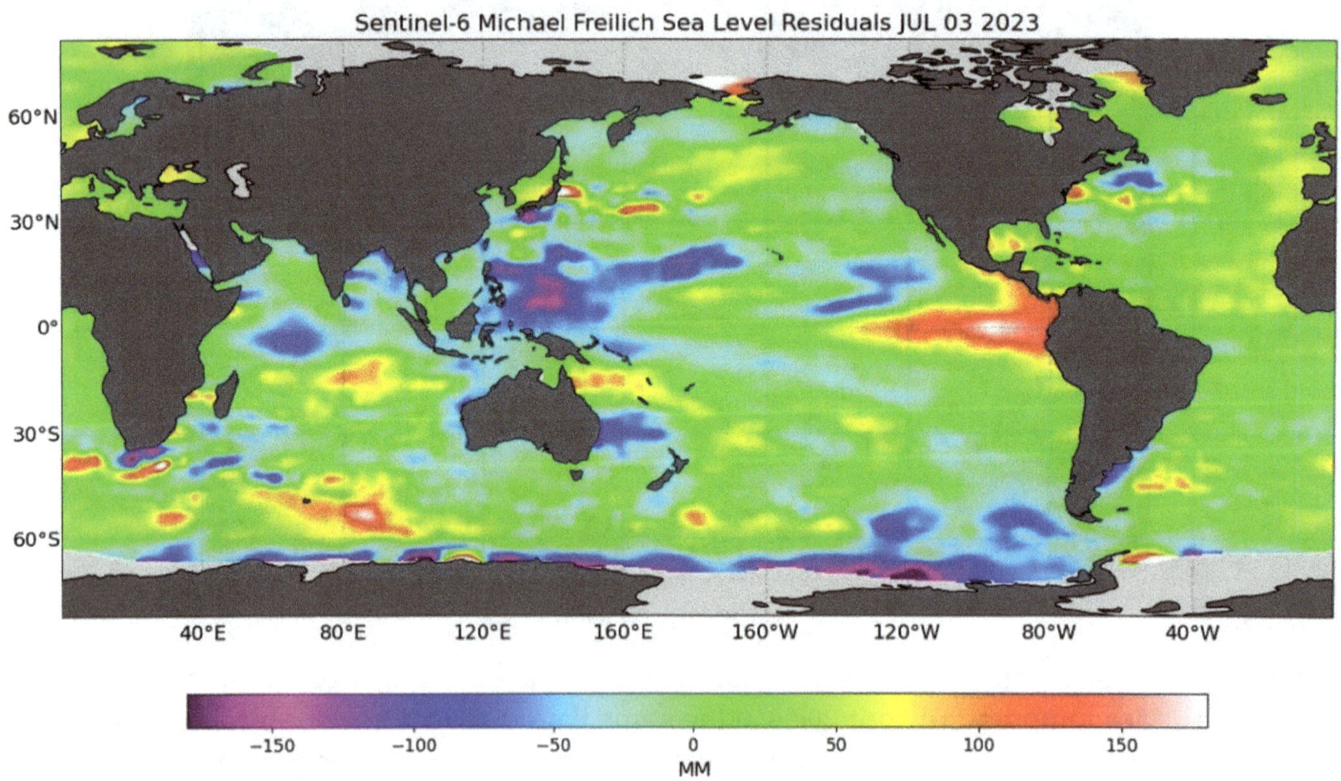

(Source: NASA satellite imagery)

This image shows the ocean water levels (measured in the millimeter color-coded scale seen at the bottom of the image) being variable in several small localized regions in all of the oceans. The discussion in the source document associated with the above chart indicated that the satellite image shows a "La Nina" effect on water level increase off the northwestern coast of South America and western coast of Central America and the lower Baja Peninsula of Mexico. There also can be seen a region in the waters of the Coral Sea above the Great Barrier Reef off the east coast of Australia displaying a height difference of about 125 millimeters (4.92 inches) compared (red color to green color) to nearby surface levels. The "average sea surface level" and "average seawater temperature" in the "grim results" database shown earlier totally ignores whatever is the factual cause of local water height levels of the type shown in red color in the above satellite image.

It should be noted that there are also some dark blue areas in the ocean which are below the green near average level. Estimates differ as to how much of the ocean floor remains unexplored but the estimates range between 75% to 85%. Considering that the oceans cover over 70% of the planet's surface area, If only 75% to 85% of the ocean floors (comprising 70% of the planet's surface) have been explored, that equates to half (about 50%) of the planet's surface remains to be explored. There has been an understanding of the general nature of the topography of the ocean floors, mostly by the naval military organizations, but there remains much unknown as to the geothermal activity from below the ocean floor and warming the ocean waters from below not just at the surface.

It should also be noted that natural ocean tides are a result of the Moon (which orbits the Earth approximately every 30 days) and Earth's gravity causing tides from negligible increase to over 20 feet tides in different regions. Schematically shown, the gravities of Moon and Earth cause the ocean water levels to bulge twice a day and to be at their maximum at the nearest and furthest points to the Moon.: The bulge is seen to be dependent on latitude.

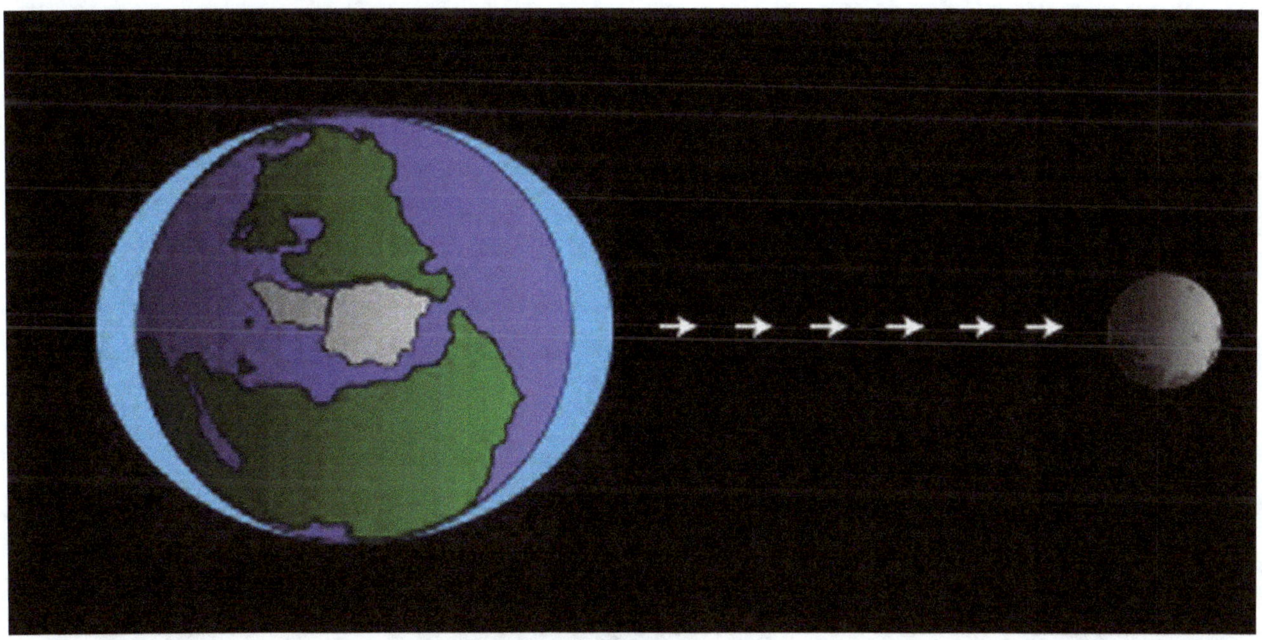

(Source: Sciencetalk.com)

In reality, the surface of the Earth has seven major continents which "get in the way" and the bulge is of course not evident on the land portions of the surface of the rotating planet.

This schematic shows a "static" Earth and Moon when in fact the Earth is in a rotation about a tilted wobbling axis that changes in a cyclic manner defined by the earlier-described Milankovic calculations in the early 1900s. Each harbor has the schedule for the high and low tides to advise ships entering and leaving how much depth is below the ship's keel to the harbor bottom. A port with an extreme tidal action is that of Inchon, Korea, and an example of the high tides and low tides are as shown next (for two days August 12 and 13, 2024). The scale at the left is in meters (and there are 3.28 feet in a meter):

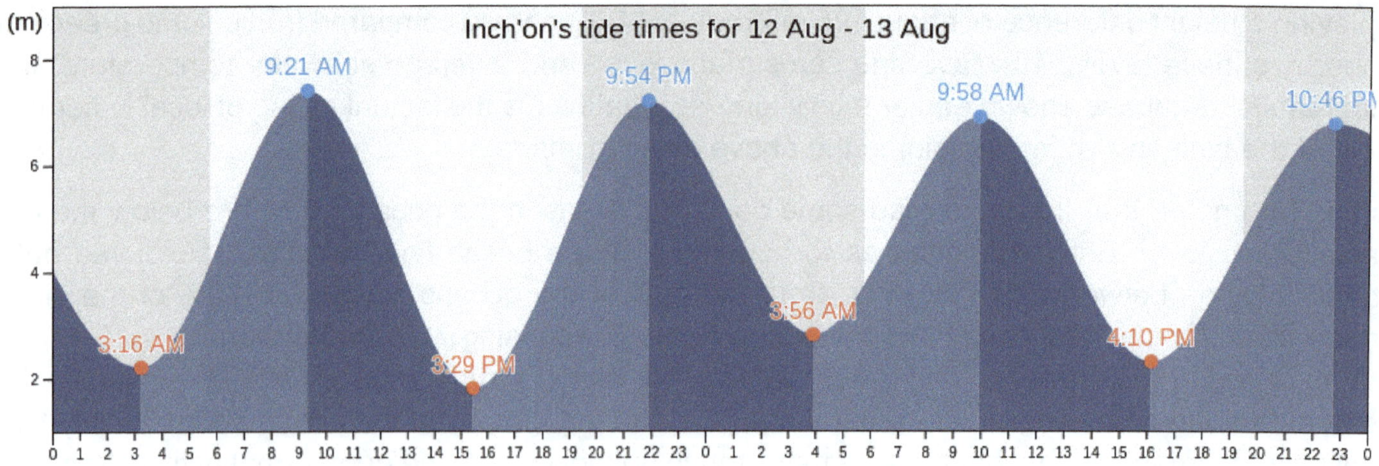

(Source: World Geological Association)

With high tide being about 7.5 meters (24.6 feet) and low tide being about 2.0 meters (6.6 feet), the difference is 5.5 meters (18 feet) between the peak of high tide and the valley of low tide. Numbers do not provide much significance, but the Author was a "guest" passing through Inchon Harbor on a US Army LST - seen in the center of the photograph below - during the Korean War and took a photo from a US Army artillery-spotting aircraft of the Inchon Harbor (just below 38 degrees North latitude) at near low tide:

(Source: Author's Personal Photographs file) Inchon Harbor, 1953)

Readers should note the extensive area of mud flats on which boats and ships are trapped when the tide recedes in two cycles each day. Inchon is not a large industrial port, and the tidal effect is not quite as disruptive in most large ports (New York, Boston, Miami, New Orleans, Houston, Los Angeles, San Francisco, and Seattle). Once again, tidal effects must be considered by the region where its ports and its population is hazarded by the local tidal actions – over which Man has no control.

If the NOAA "average ocean surface level rise" measured at 150 millimeters rise (0.0394 inches per millimeter equals a rise of o.591 inches or 0.049 feet) caused by ""average ocean water temperatures soaring" is of such ominous proportion in environmental scientists' perspective, it is wondered what descriptor is used for the 5.5 meters (18 feet) tidal action that is at Inchon Harbor every day twice a day? Other harbors are generally of lesser magnitude than that of Inchon Harbor because of their latitude being generally above 38 degrees North latitude). At the Panama Canal, the Pacific Ocean (Panama City end) tidal action is about 20 feet whereas the Atlantic Ocean (Gatun end) tidal action is only 3 feet.

There have been studies to use the natural tidal action in the oceans at harbors and ports to be used as an alternative energy source and the feasibility of tidal energy is discussed later in this book.

The use of an "average sea surface temperature" to establish the goal of limiting the "average global temperature" is very troublesome to this Author who has had the benefit of travelling long distances on the oceans and seas in the Northern hemisphere. It is troublesome because it is not realistic. The longest (16 days) and furthest of those voyages were on the Great Circle route from Seattle via the Aleutian Island chain and to the lovely port of Inchon, Korea (on board the USMS Gen. Wiliam Weigel troopship) to participate in Army activities in South Korea in the early 1950s. That was not the most enjoyable voyage as the waters along the Aleutian Islands are very cold and cause ground swells that sickened most of the young men from the central US who had never seen an ocean-going ship before then. A lengthy book could be written about that experience. However, the return from Inchon to Seattle (aboard another troopship, the USMS Gen. John Pope) was via a more southern balmy route and passing – but not making port in – the Hawaiian Islands. The water temperature was warm and the weather perfect during that 15 days voyage taking advantage of sailing with the North Pacific Ocean current rather than opposing it. Several other voyages were made in the North Atlantic between Barbados in the Caribbean Sea and Port of New York in the Atlantic Ocean, and voyages in the Caribbean Sea through the Panama Canal, in the Mediterranean, Aegean, and Adriatic Seas on ships much nicer than troop ships.

The Preface to this book indicated that the ocean currents vary in temperature, direction and current rate, and most especially within the constraints of the continental land masses. Most human populations have developed communities and cities and business centers around the coastal regions of the continents – a carry-over from the age of travel and trade by sailing ships. For this book, forty world cities, most of which were on the coasts of large oceans with both cold and warm currents in both the Northern and Southern hemispheres, were analyzed for their annual average temperatures. The cities were chosen to give a representative effect of warm ocean current, cold ocean current, at different latitudes between 64.13° N (Reykjavik, Iceland) and 53.17° S (Punta Arenas, Chile), and at average elevations from -7 feet (New Orleans, Louisiana, USA) and to 12,970 feet (La Paz, Bolivia).

A tabulation of the average annual temperatures for twenty locations at different latitudes and average elevations in the Northern hemisphere was made and is shown as follows:

World and US Cities Showing Latitude Effect on Temperature

(Northern Hemisphere)

	City	Elevation	Latitude	Avg. Ann. Temp.
1	Rejkjavik, Iceland	26 ft (8 m)	64° 08'	42.5 °F (5.8 °C)
2	Oslo, Norway	610 ft (186 m)	59° 55'	42.3 °F (5.7 °C)
3	Stockholm, Sweden	85 ft (26 m)	59° 20'	43.9 °F (6.6 °C)
4	Copenhagen, Denmark	30 ft (9 m)	55° 40'	48.4 °F (9.1 °C)
5	London, United Kingdom	138 ft (42 m)	51° 30'	52.3 °F (11.3 °C)
6	Paris, France	114 ft (35 m)	48° 51'	54.1 °F (12.3 °C)
7	Seattle, WA USA	131 ft (40 m)	47° 48'	55 °F (12.8 °C)
8	Chicago, IL USA	579 ft (177 m)	41° 53'	51 °F (10.6 °C)
9	New York, NY, USA	62 ft (19 m)	40° 43'	53 °F (11.7 °C)
10	Denver, CO, USA	5,525 ft (1,684 m)	39° 44'	54 °F (12.2 °C)
	"Average Global Temperature" (pre-IR)	0 ft (0 m)	Everywhere	59 °F (15 °C)
	Durham, NC, USA	404 ft (123 m)	35° 58'	63.9 °F (17.7 °C)
11	Los Angeles, CA, USA	791 ft (241 m)	34° 03'	67 °F (19.4 °C)
12	Phoenix, AZ, USA	1,086 ft (331 m)	33° 27'	76 °F (24.4 °C)
13	New Orleans, LA, USA	-7 ft (-2 m)	29° 58'	75 °F (23.9°C)
14	Miami, FL, USA	6 ft (2 m)	25° 47'	76 °F (24.4 °C)
15	Guadalajara, Mexico	5,062 ft (1,543 m)	20° 40'	69.6 °F (20.9 °C)
16	Mexico City, Mexico	7,347 ft (2,240 m)	19° 26'	63.5 °F (17.5 °C)
17	Managua, Nicaragua	636 ft (194 m)	12° 8'	81.1 °F (27.3 °C)
18	Valledupar, Colombia	538 ft (164 m)	10° 28'	85.6 °F (29.8 °C)
19	Panama City, Panama	20 ft (6 m)	8° 59'	82.6 °F (28.1 °C)
20	The Equator	0 ft (0 m)	0° 0'	85.0° F (29.6° C)

Note: The "average global temperature" calculated by WMO in 1990 became 16 °C (60.8 °F)

(Source: World Meteorological Organization database, Author compilation)

It should be noted that there are two other geological variables besides latitude that are suspected to be influential on the surface temperature and therefore climate of each of the respective cities. More will be presented on those effects in later sections.

A most pleasant surprise – and an important one – was seen when a graph was made of the average annual temperature from the World Meteorological Organization database for each of twenty selected cities (eight in the "lower 48" United States) in the Northern hemisphere when the annual average temperature was charted as a function of their latitude.

Effect of Latitude on Average Annual Temperature (20 World and US Cities)

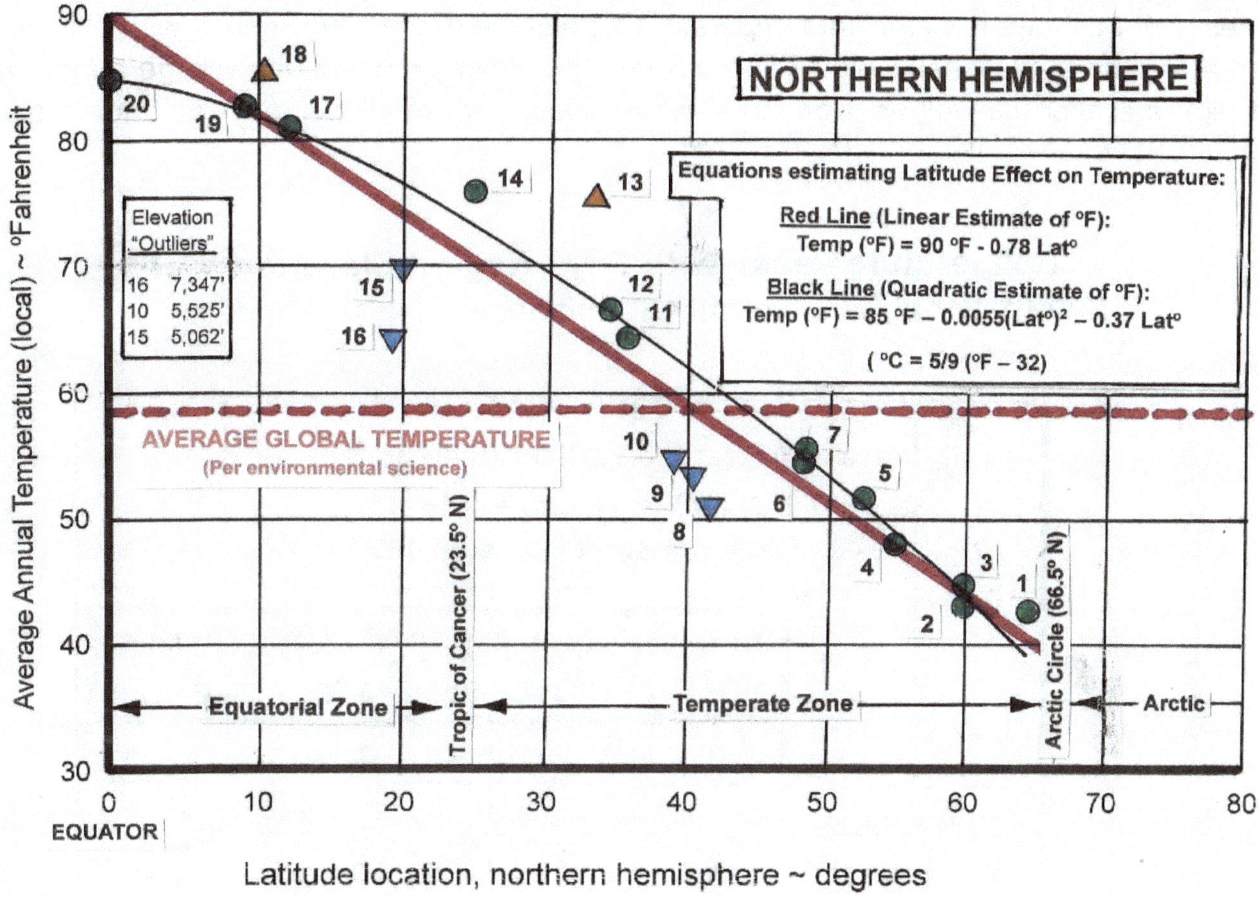

(Source: Author's Analysis)

A curvilinear (solid black line) was drawn through the circular data points in a "least square" manner of data analysis and can be seen to have a quite good approximation of those data points. A linear (solid red line) was then drawn through that first black line and indicated a good approximation of the circular data points. A simple point-slope linear expression for the red line representing the latitude effect was mathematically generated and that effect is shown by the small equation in the inset box near the upper right corner of the figure. A corresponding mathematical expression for the curvilinear solid black line was approximated by a 3-point (cities numbered 20, 12, and the average of 2 and 3) quadratic function and is also shown in the inset. Environmental scientists are well-advised to use that form of modeling the effects of latitude on cities dispersed on the globe instead of the one-size-fits-all "average global temperature" in their myopic activities.

Seven of the data points (shown in triangular symbols) are considered as "outlayers". Two cities (13 – New Orleans, and 18 – Valledupar) that were hotter than both the solid linear red line and the solid black line were/are seen to be significantly warmer than the red and black lines for their latitudes. These will be shown in the next chart to be strongly affected by an unusually warm ocean current. Three of the five downward pointing triangles (16 – Mexico City at elevation 7,347 ft. (2,240 m.); 1o – Denver at 5,525 ft. (1,684 m.); and 15 – Guadalajara at 5,062 ft. (1,543 m.) were/are seen in the listing to be of very high average elevations. Based on curiosity if elevation has a significant effect on surface temperature of cities, a cross-discipline reference was made to aeronautical engineering (this Author's profession).

There is a standardized (adopted by the International Civil Aircraft Organization, ICAO) relationship in aircraft design that shows the ambient temperature for altitude decreasing linearly with altitude. It was reasoned that such a relationship would approximate the elevation effect on average temperature for cities at high elevation as well as it does for air vehicles. That relationship is graphically shown as follows:

Temperature Lapse with City / Region Elevation

(International Civil Aircraft Organization, ICAO) Standard)

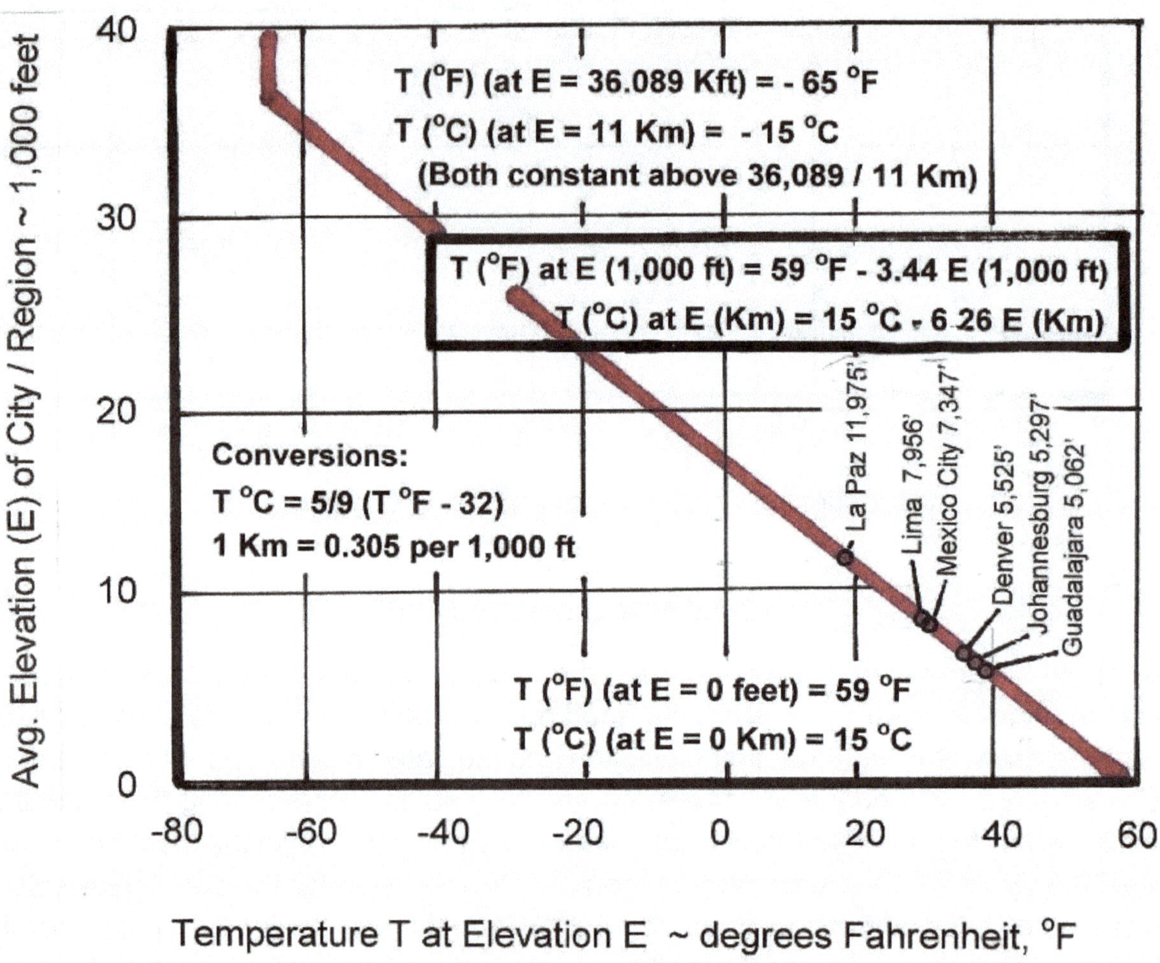

(Source: International Civil Aircraft Organization, ICAO standard)

A mathematical point-slope method of approximation (applicable for cities at elevation below 36,089 feet (11 kilometers) is shown for the temperature at that elevation (in °F and °C).

To examine whether those two warmer-than-predicted cities were warmer because of a nearby warm ocean current (and not by fallacious "greenhouse gas effects"), the next chart was generated (using the ocean current database contained in the Rand McNally "Atlas of the World":

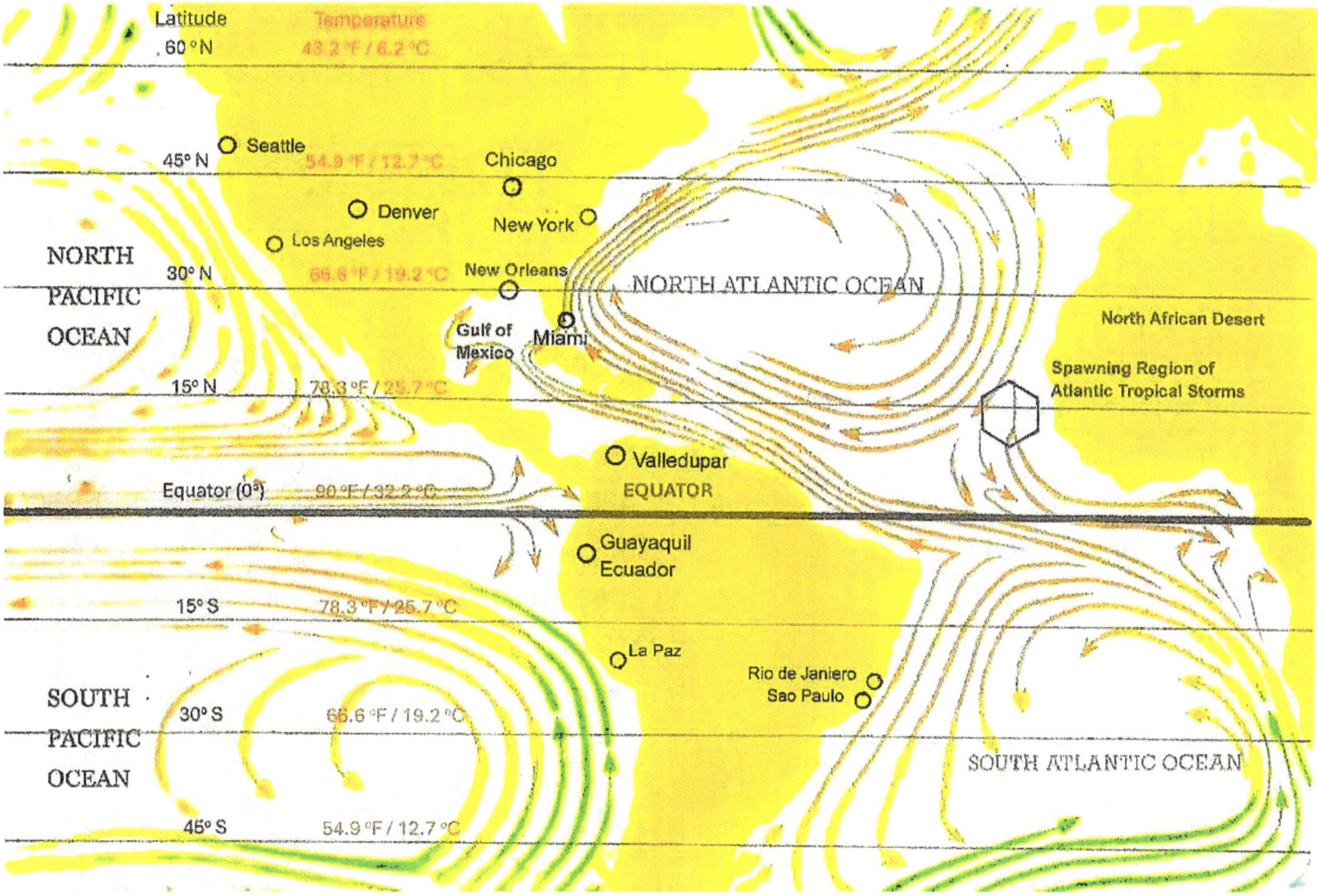

(Source: Ocean Currents from Rand McNally Atlas of the World, Annotations by Author)

The two hotter-than-predicted cities: (13) New Orleans, Louisiana, USA and (18) Valledupar, Colombia, are seen to be warmed by the hot Equatorial current of the Caribbean Sea and the Gulf of Mexico. The two colder than predicted cities: (8) Chicago Illinois USA – with the prolonged cold temperatures from Lake Michigan, and (9) New York, USA – with the prolonged winter cold temperatures from Lake Erie - are explained for their deviation from predicted.

A reasonable question should be whether that temperature analysis done for Northern hemisphere cities applies to cities in the Southern hemisphere? The same sequence of tables and graphics charts for 20 other cities in the Southern hemisphere are presented for completeness and to verify that the combined effects of latitude, elevation, and ocean current proximity and temperature are the same in the Southern hemisphere.

World Cities Showing Latitude Effect on Temperature

(Southern Hemisphere)

	City	Elevation	Latitude	Avg. Ann. Temp.
1	Equator	0 ft (0 m)	0° 0'	85.0 °F (29.6 °C)
2	Belem, Brazil	30 ft (10 m)	1° 27'	87.4 °F (30.8 °C)
3	Guayaquil, Ecuador	16 ft (5m)	2° 11'	78.1 °F (25.6 °C)
4	Papua, New Guinea	2,188 ft (667 m)	5° 30'	81.5 °F (27.5 °C)
5	Jakarta, Indonesia	26 ft (8 m)	6° 10'	82.5 °F (28.0 °C)
6	Lagos, Nigeria	39 ft (12 m)	6° 27'	84.8 °F (29.4 °C)
7	Lima, Peru	7,956 ft (2,426 m)	12 ° 04'	63.5 °F (20.5 °C)
8	Darwin, Australia	26 ft (8 m)	12° 26'	81.3 °F (27.4 °C)
9	La Paz, Bolivia	11,975 ft (3,650 m)	16° 30'	70.0 °F (20.9 °C)
10	Rio de Janiero, Brazil	1,047 ft (319 m)	22° 54'	74.5 °F (23.6 °C)
11	Sao Paulo, Brazil	2,621 ft (799 m)	23° 33'	65.0 °F (18.0 °C)
12	Johannesburg, S. Africa	5,297 ft (1,615 m)	26° 12'	60.6 °F (15.9 °C)
13	Durban, S. Africa	272 ft (83 m)	29° 53'	77.2 °F (25.1 °C)
14	Santiago, Chile	1,887 ft (575 m)	33° 26'	61.7 °F (16.5 °C)
15	Sydney, Australia	190 ft (53 m)	33° 52'	64.5 °F (18.0 °C)
16	Cape Town, S. Africa	174 ft (53 m)	33° 55'	61.0 °F (15.9 °C)
17	Buenos Aires, Argentina	82 ft (25 m)	34° 36'	60.0 °F (16.0 °C)
18	Auckland, New Zealand	105 ft (32 m)	36° 51'	64.7 °F (18.2 °C)
19	Melbourne, Australia	607 ft (185 m)	37° 49'	59.5 °F (15.3 °C)
	"Average Global Temperature"(pre-IR)	0 ft (0 m)	Everywhere	59 °F (15 °C)
20	Punta Arenas, Chile	112 ft (34 m)	53° 10'	44.5 °F (6.9 °C)

Note: The "average global temperature" calculated by WMO in 1990 became 16 °C (60.8 °F)

(Source: World Meteorological Organization and World Geological Services database)

Data presented in tabular form is difficult to envision its meaning in a "big picture" perspective. Colorizing is sometimes helpful. A point to be made in the above table is to compare the actual average temperatures of a number of world cities – shown in red - with the average annual global temperatures (15 °C / 59 °F) – shown in yellow - used by environmental scientists when informing the public that the globe is warming and the climate is changing at a "crisis" rate. It is seen that 18 of the 20 cities listed above are already at a higher than the environmental scientists' average annual temperature by a range of 0.5 °F (0.3 °C) to 28.4 °F (14.3 °C) and none are reported to be having an average annual temperature or climate change "crisis".

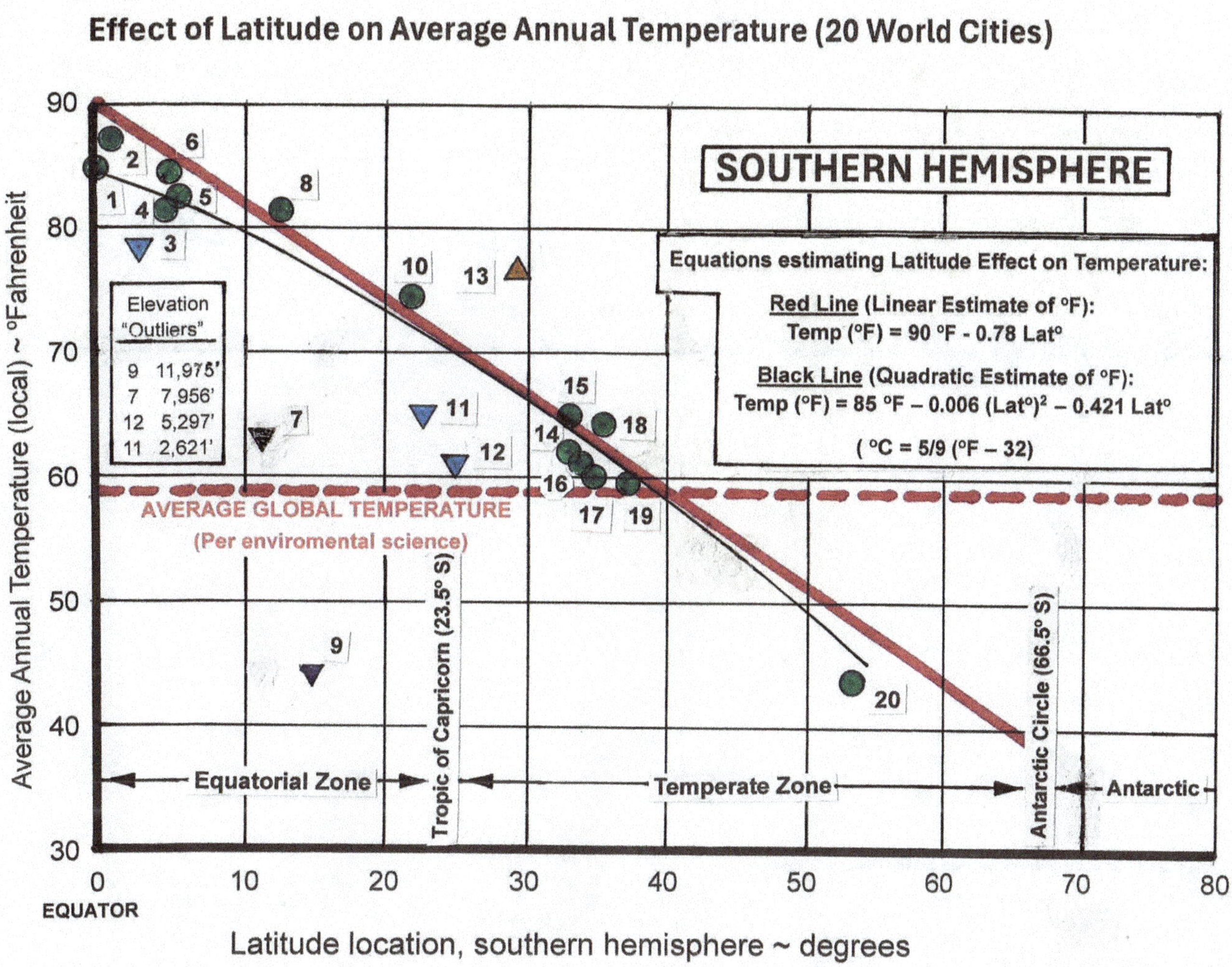

It is seen in the Southern hemisphere that there are four cities much cooler than is predicted by both the linear red line and curvilinear black line. The cause for this is the elevation of the four cities being at high average elevations. In order of largest deviation,

City 9 (La Paz, Bolivia) is at an elevation of 11,975 feet (3,650 meters)

City 7 (Lima, Peru) is at an elevation of 7,956 feet (2,426 meters)

City 12 (Johannesburg, South Africa) is at an elevation of 5,297 feet (1,615 meters)

City 11 (Sao Paolo, Brazil) is at an elevation of 2,621 feet (799 meters)

The city 13 (Durban, South Africa) is only at an elevation of 272 feet (83 meters) and much hotter than predicted by about 10 °F (5.5 °C) while the city 16 (Cape Town, South Africa) - at less than 4° of latitude difference - is cooler than predicted despite being at a lower elevation 174 feet (53 meters). For a plausible explanation, the ocean currents in the South African region are shown as follows:

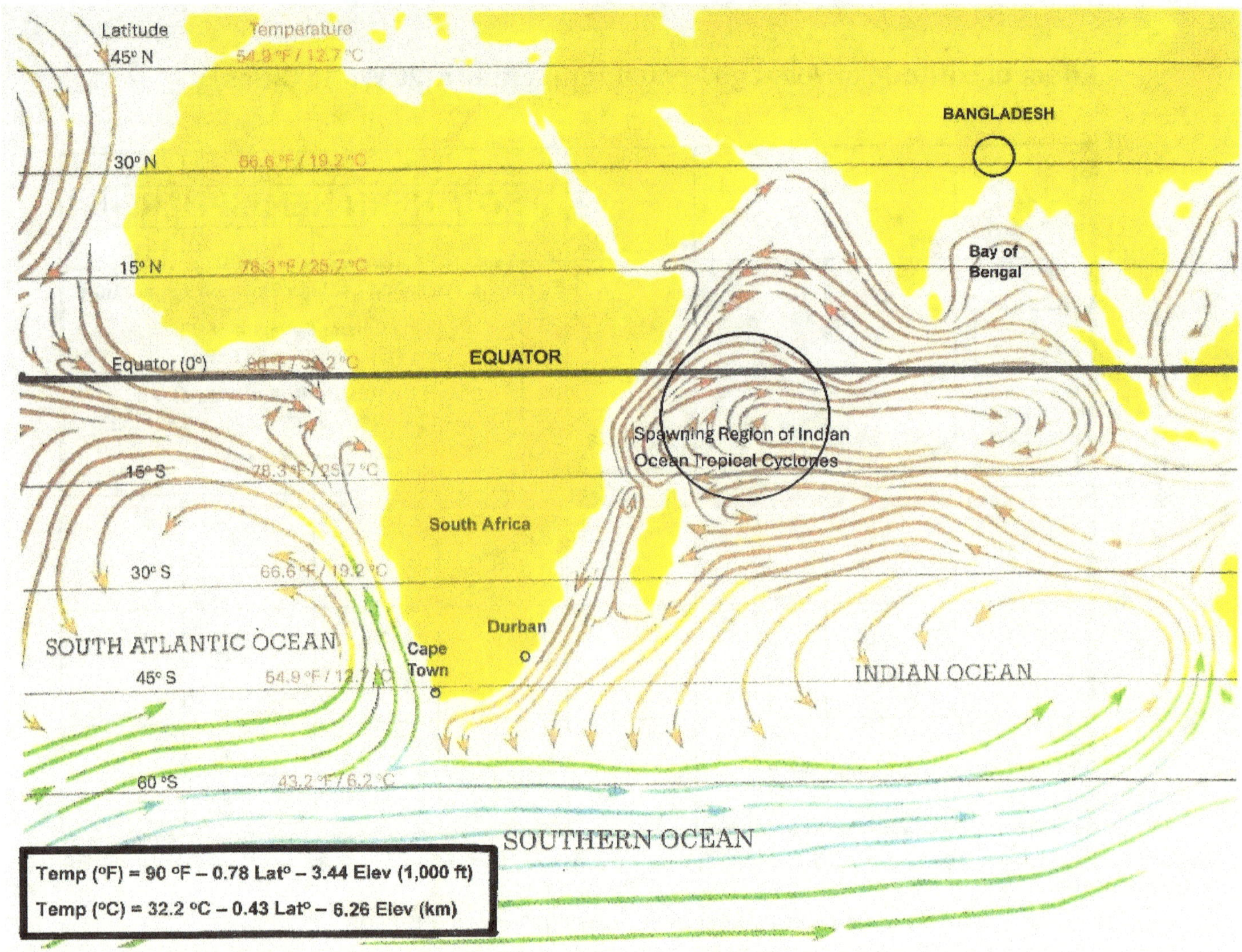

Here the 15 °F temperature difference between Cape Town and Durban is most likely to be due to the ocean current effect being far greater than the latitude effect. Cape Town is subjected to the very cold Southern Ocean current flowing northerly from the Antarctic region while Durban is subjected to the much warmer Madagascar current in the warm Indian Ocean flowing southerly toward the Southern Ocean.

In the South Pacific, the ocean current affecting the Great Barrier Reef (reportedly bleaching white because it is being subjected to hot water temperature from the Man-made "greenhouse gas effect") is seen to most probably be suffering heat stress from the Equatorial current (in the Indian Ocean) effect far more likely than from the Man-made "greenhouse gas effect":

City 20 (Punta Arenas, Chile) is seen on the latitude effect chart to be below both the red line and the black line reflecting being cooler than predicted, and (as shown on the next chart) are subjected to the cold Southern Ocean effect in Antarctica regional waters.

Added to the ocean current map in the lower left corner inset are the temperature relationships for estimating the average annual temperature on any city accounting for its latitudinal location and its average elevation – and expressed in either Imperial units (feet and degrees Fahrenheit) or metric units (kilometers or degrees Celsius). A small adjustment (warmer or colder) for a nearby large body of water (ocean, Gulf, or Great Lake) in that predicted temperature (already adjusted for latitude or elevation) can be estimated from the ocean current chart shown above. This method is considered to be far more valid than using the one-size-fits-all "average global temperature" which does not consider latitude, elevation, or proximity of the temperature of an ocean current or heat sink such as a Great Lake.

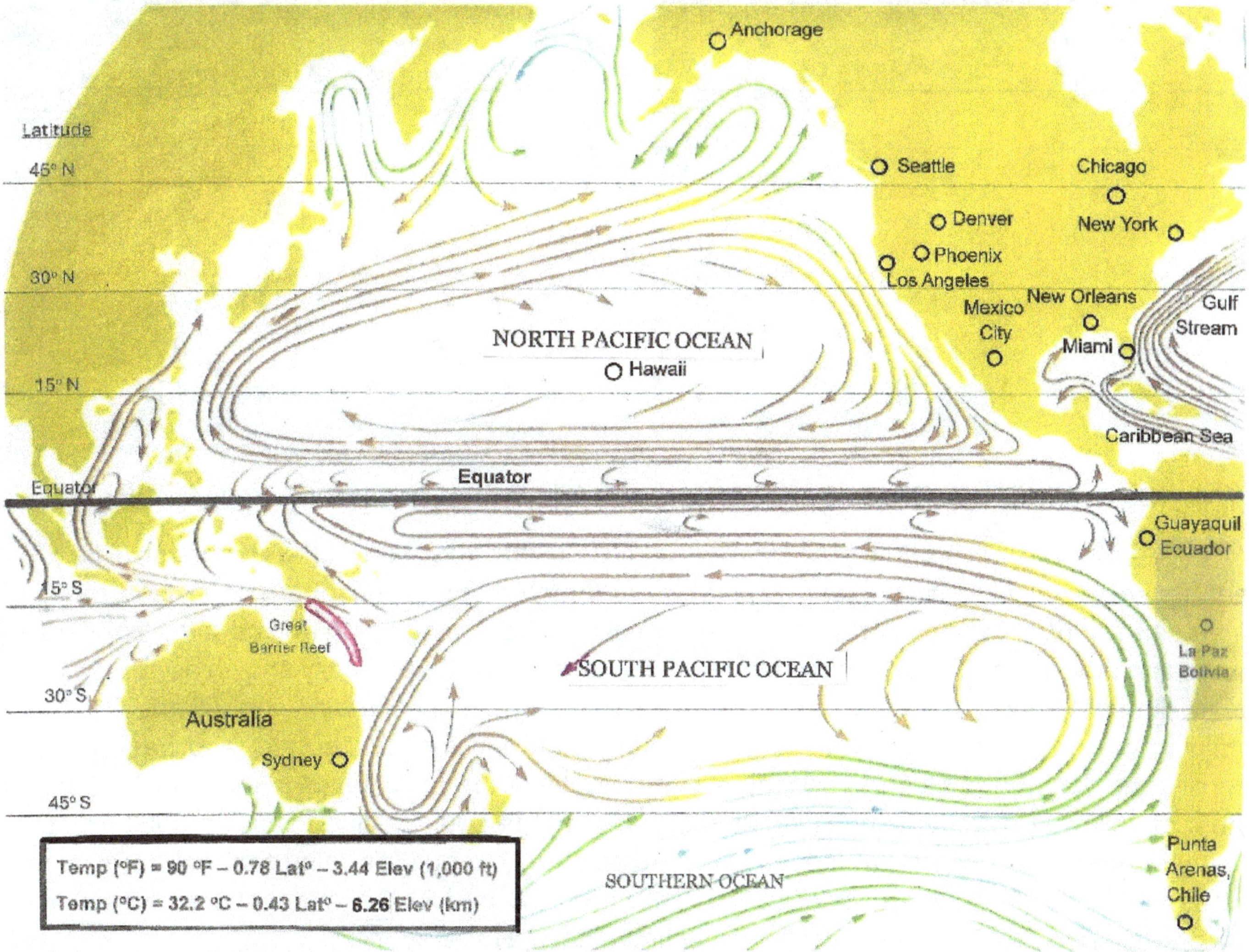

The preceding series of charts and tabulations of valid meteorological data for atmospheric and ocean water temperatures realistically indicate that <u>the combination of latitude effect, elevation effect, and ocean current temperature effect are far more influential on the annual average temperature in the cities of the world at any latitude than might be caused by the hypothetical "greenhouse gas effect" resulting from Man burning fossil fuels</u>.

During the research and analysis of the many factors that had a possible effect on global warming, climate change, and ocean surface levels rising, the recurring question coming into mind was how influential – either comparatively to all the other known factors or as an isolated factor - is to what extent is carbon dioxide (the dominant "greenhouse gas" in the atmosphere affecting the temperatures of the planet's surface? A surprisingly high number of diagrams placed by environmental scientists into the public domain "explaining" how carbon dioxide causes the global warming effect described the effect graphically in a well-intentioned manner but also in an intentionally misleading manner. Examples of the type of "mis-explanation" – which the public has come to accept as being correct since it is being presented by the National Oceanic and Atmospheric Administration Climate Program Office – are provided here without any alteration by the Author:

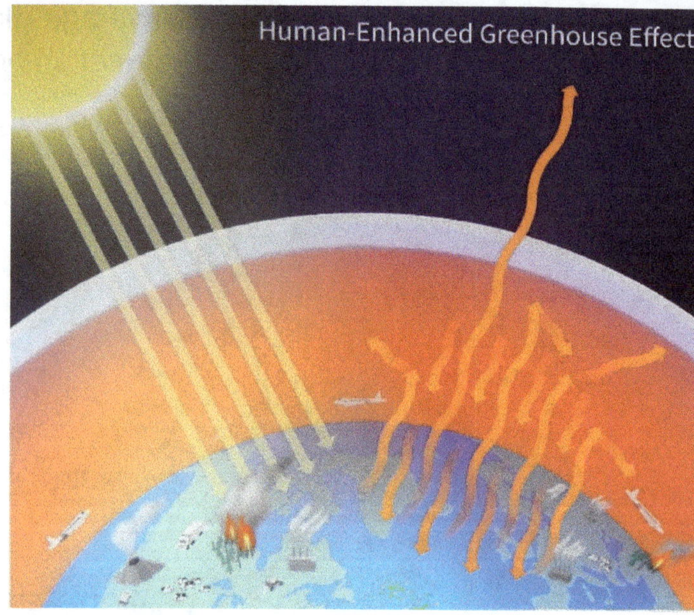

(Source: National Oceanic and Atmospheric Administration Climate Program Office)

The NOAA diagram on the left schematically indicates that the "Natural Greenhouse Effect" – the "greenhouse effect" indicating heat-trapping and warming of the atmosphere - causes most of the Sun's insolation (incoming solar radiation) wavelengths to be impinging on and causing Earth's surface to warm and emit (reflect) back into the atmosphere where some (graphically shown as squiggly orange lines) of the heat energy is trapped in the atmosphere. The majority of the squiggly orange heat energy containing wavelengths are shown to pass unaltered through the atmosphere and pass into space. Accepting the diagram to be only conceptual in nature, the impression given is that there are the same number of yellow straight rays incoming as there are orange squiggly lines passing unaltered through the atmosphere into space. That leaves the impression in the public's mind that natural greenhouse effect is very little. Since there ostensibly is little heat trapped in the atmosphere, the color of the atmosphere has a purple tint but not red (reflecting higher temperature).

Before clarifying the facts for the Natural Greenhouse Effect diagram on the left, the diagram shown on the right is entitled "Human Enhanced Greenhouse Effect" and with the intent of NOAA to graphically illustrate that human enhanced greenhouse effect traps more heat in the atmosphere and as a result causes the global warming of the Earth's surface below the atmosphere. The number of squiggly orange lines are shown to be in a chaotic arrangement in the atmosphere with only one of the squiggly orange lines escaping into space. A very distinctly different tint – more red and therefore much warmer - of the atmosphere is evident. The public has accepted that the "human enhanced greenhouse gas effect" is obviously greater than is the "natural greenhouse effect" on global warming on the planet.

There are a few basic flaws however in the simplistic diagrams that are presented in the NOAA diagrams which give the appearance of being the results of a professional analysis considering the several major factors that affect global warming and climate change. The largest of the flaws is that the environmental scientists do not include the scientific fact that gases having a heat-trapping "greenhouse" effect on the temperature of the atmosphere are not constrained to be all generated from anthropogenic activities. To prove that contention, an analysis done by several international scientists of the ice core samples collected by drilling into the approximately 3,500 meters (almost 5.6 miles) deep ice sheet formed on the top of the Antarctic continent have revealed the following facts shown graphically:

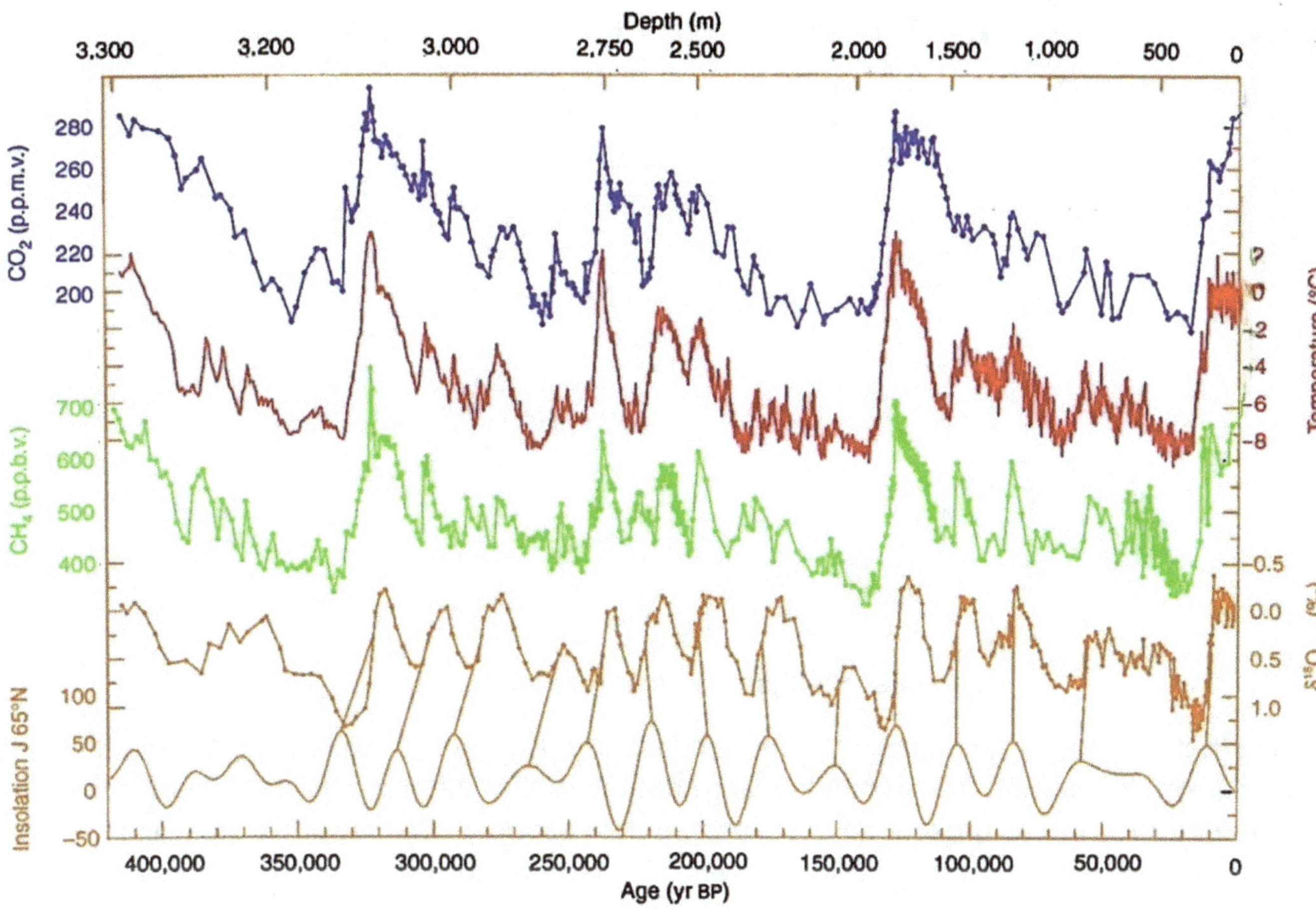

(Source: Smithsonian Institute for Science)

The ice core samples indicate that temperature (in red) and the two largest "greenhouse gases" – carbon dioxide CO_2 (in blue) and methane CH_4 (in green) are somewhat "spikey" – low and high – but the spikes are contemporary. The most important factor of interpretation of the Vostok ice core data are that they were formed by factors other than of anthropogenic origin – Man was not present on the planet many thousands of years ago. That conflicts with the environmental scientists' hypothesis that it is Man burning fossil fuels being the dominant cause for global warming and climate change.

The Vostok ice core graphics are also not to be interpreted that the peak highs and peak lows of the two most important variables – temperature and carbon dioxide - are occurring at the same point in time. A forensic analysis of the Vostok Ice Core database conducted by the International Ice Core Research team (Dr. Nicolas Caillon et al) found that the peak of the carbon dioxide trace lagged the peak of the temperature trace by about 800 years:

CO_2 lags temperature rises in ice core data

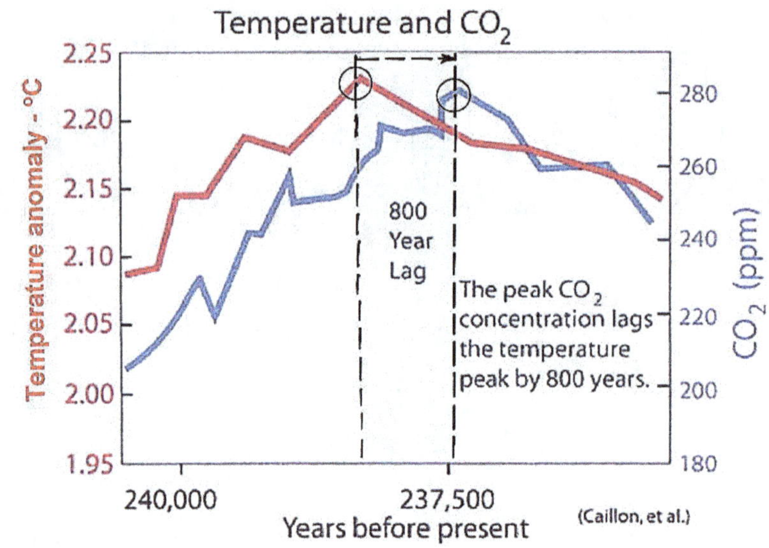

(Source: Smithsonian Institute for Science)

This finding that carbon dioxide lags temperature by about 800 years is of equal importance to the most major conclusion drawn from the Vostok Ice Core database that natural carbon dioxide existed in the atmosphere before Man was present on Earth and peaked at about 285-290 parts per million by volume (ppmv). There can be but one interpretation that temperature (or some other astrophysical, geophysical, or natural factor or combination thereof) is the driver of the amount of natural carbon dioxide in the atmosphere.

Referring to a recent set of carbon dioxide measurements collected by the Mauna Loa Observatory (Atmospheric Laboratory) operated by NOAA, copy shown in the following graphical illustration, there is also an important conclusion derived regarding the contribution anthropogenic carbon dioxide makes to global warming and climate change:

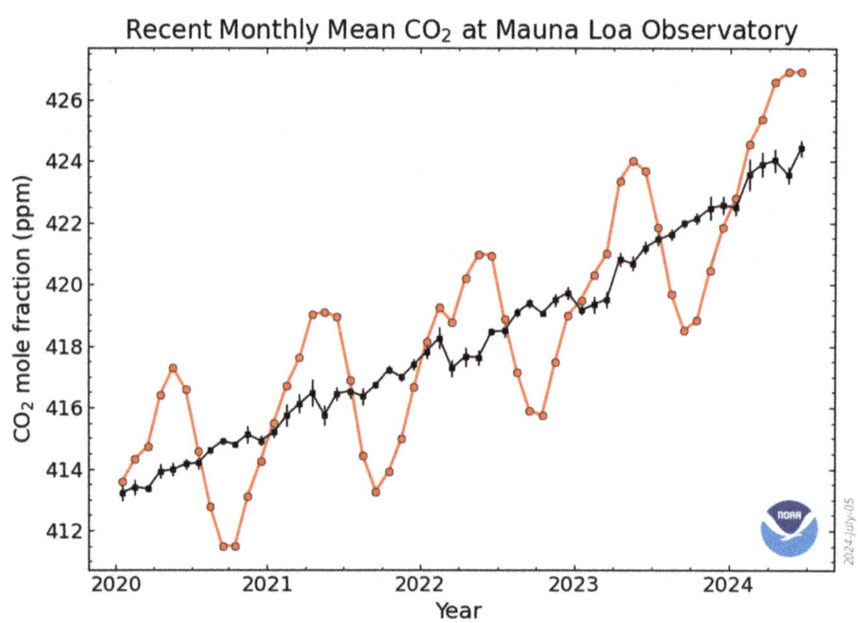

(Source: NOAA)

There are factually two factors embedded in the above database, one being that the red line data – the "raw" (non-adjusted) measurements of total carbon dioxide (which also includes water vapor) – is adjusted to remove the water vapor contribution and leaving only the total carbon dioxide (black line) variation in the atmosphere during the time span 2020 to 2025. The black line total carbon dioxide contains both natural carbon dioxide and anthropogenic carbon dioxide. Reflecting on the Vostok Ice Core database, there are four peaks in the natural carbon dioxide of magnitude 285 ppmv before Man was present on Earth (i.e., pre-Industrial Revolution by about 130,000 years. The fifth peak of the Vostok Ice Core database at the Age 0 (i.e., "Now") is not representative of the natural CO_2 because it does contain some anthropogenic CO_2 but also contains some unreliable ice core sample data that was contaminated by the new drilling process in its still experimental state of development. The trace above for CO2 that is fallaciously claimed to be exclusively anthropogenic CO2 is therefore 285 ppmv less than the 413+ ppmv shown. In 2020, the anthropogenic CO2 concentration in the atmosphere was 128 ppmv (0.0127%) and 139 ppmv (0.0139%) in 2025.

Why environmental scientists maintain the claim that the anthropogenic "greenhouse gas emissions from Man burning fossil fuels" constitutes "a global warming crisis" is beyond scientific comprehension. Their reluctance to accept that water vapor in the atmosphere – averaging 1.4% at the average annual global temperature of 59 °F (15 °C) - is used as an estimate of the ratio of water vapor in the atmosphere compared to the environmental scientists' much overexaggerated concern over 139 ppmv (0.0139%) of anthropogenic carbon dioxide being at a "crisis" level. The 1.4% of water vapor over the 0.0139% of anthropogenic carbon dioxide is a ratio of 101 to 1. There can be no factual claim that anthropogenic carbon dioxide is the dominant cause for global warming when the concentration of anthropogenic in the atmosphere is 106 times lesser than the natural greenhouse gas of water vapor. When water vapor is scientifically correctly considered, there is much less warming than that shown escaping to the outer atmosphere than there is when due to "human-enhanced greenhouse effect. The two prior illustrations by the NOAA Climate Program Office are in effect reversed.

The Author also conducted an independent statistical analysis of the Vostok Ice Core database, and the analysis has revealed another observation that is of much greater significant importance to the subject of Earth's surface temperature and climate change. Shown next is the Author's analytical mark-up of the published ice core database for the past 420,000 years:

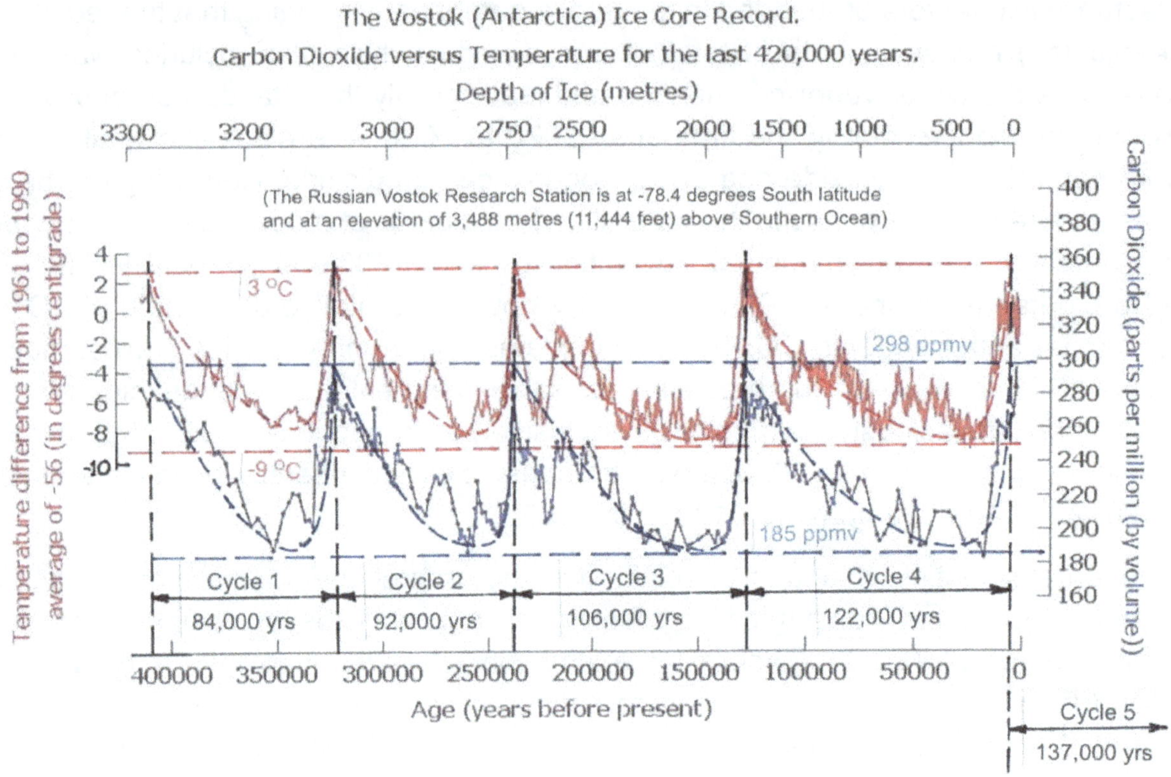

(Source: Smithsonian Institute for Science database with analysis mark-up by Author)

Readers are reminded that there was no presence of Man or industry generating “greenhouse gas emissions” on the until only the most recent 120,000 – 150,000 years such that the changes in the temperature and carbon dioxide levels shown in the database above are entirely those generated by all other probable causes: astrophysics, geophysics, and natural (including geothermal – which have been revealed by further improved instrumentation measurements to exist under the Antarctic ice sheet and in the form of many warm lakes of water.

Of highest importance in this book, the Author has shown that the ice core samples reflect the temperature (and carbon dioxide) to have undergone four major cycles during the past 420,000 years, with a temperature anomaly ranging from about 3 °C above (hotter) average temperature at that latitude (78.4° S) and 9 °C below (colder) than average temperature at that latitude. The period is not equal for each of the four cycles but are becoming longer in period (+ / - about 1,000 to 1,500 years) with the following trend:

Cycle 1 84,000 years
Cycle 2 92,000 years
Cycle 3 106,000 years
Cycle 4 122,000 years (ending at about the present time 21st century)

The last cold peak (valley) identified as “a little ice age” occurred approximately 30,000 years ago with a temperature at about 9 °C below the average for that latitude and the carbon dioxide in the atmosphere at about its low level of about 185 ppmv (0.0185%) – fairly consistent highs and lows in both temperature anomaly and carbon dioxide concentration in all four of the major cycles. During the last “Little Ice Age” about 11,000 years ago, there was glacial ice covering the Northern hemisphere from the Arctic region to below about latitude 37° N (which included New York City):

(Credit to Graphic Artist April S. Dalton – Affiliation Unknown)

Much of the Great Lakes were formed by the melt off of the ice sheet as it retreated.

The earlier forensic analysis of the irrefutable Antarctica ice core sample database by international scientists reflects a lag of about 800 years in the carbon dioxide concentration (at about 298 ppmv (0.0298%) peaks after the temperature peaks 3 °C hotter than average temperature for that latitude.

Transitioning from the surface of the planet and the atmosphere above that surface, some of what is known of the topography of the oceans' floors can be described for completeness of the analysis. The floor of the Atlantic Ocean is shown first and that of the Pacific Ocean is shown second.

(Source: World Geodetic System Committee)

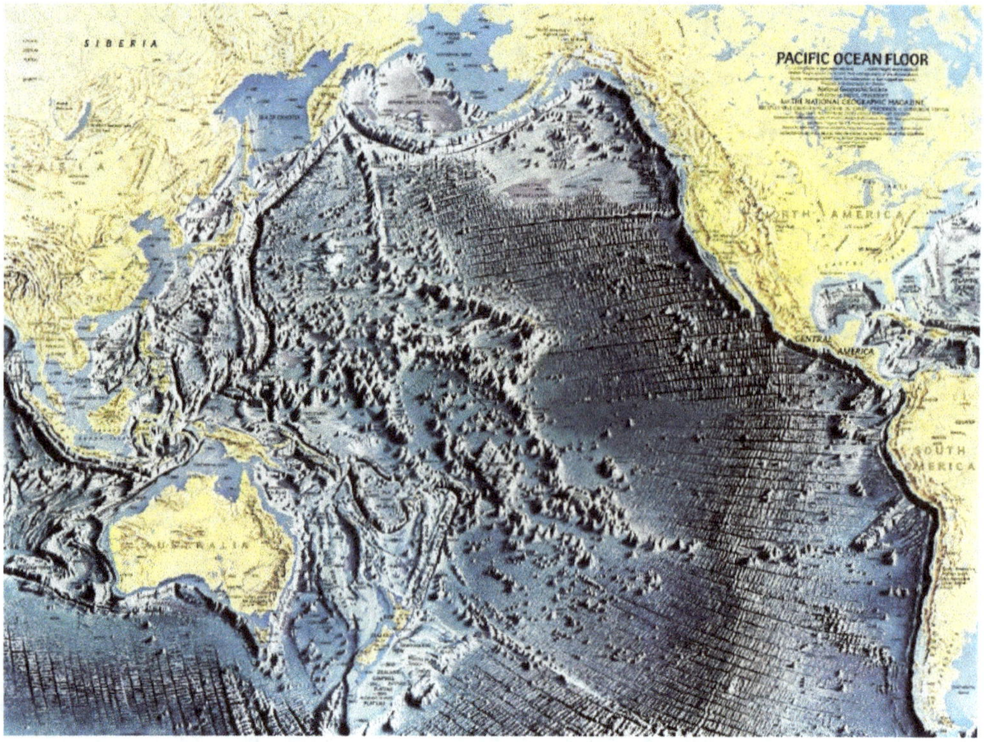

(Source: World Geodetic System Committee)

By observation, it is seen that the floors below the Atlantic Ocean and Pacific Ocean have a distinct difference – the Atlantic floor is divided into two portions (west and east) by the mid-Atlantic Ridge and which are somewhat "mirror image"; whereas, the Pacific floor has a western and an eastern topography totally opposite in topography from each other. The common fact is that both oceans have floors that are formed by tectonic plate origin (post Pangaea Age) and have a relative motion – the energy coming from geothermal energy below the outer crust of the planet to move the massive tectonic plates. Moving the hug masses of the tectonic plates requires a little bit of scientific "mumbo-jumbo" about energy. Kinetic Energy is mathematically defined as one half of the mass times velocity squared – the velocity of each tectonic plate when moving is very miniscule but the mass of each tectonic plate is too large to be comprehended. The end product is massive kinetic energy expended to cause the motion of the huge mass, and that is not obtained from the Sun.

No amount of decrease or increase in carbon dioxide emissions will alter that transfer of heat energy from within the planet to its surface. Tectonic plate movement is another one of those factors that Man must adapt to regardless of climate change (either warming or cooling). Man burning fossil fuels cannot be blamed for tectonic plate movement or the transfer of heat energy from below the planet's outer crust causing the planet's surface temperature to rise or fall. The interconnection however is the human population in the East Coast states must adapt to East Coast climate and the human population of the West Coast states must adapt to the West Coast climate. Once again, the environmental protection scientists of the planet set a "one-size-fits-all" goal of net zero emissions by 2050 which is totally meaningless to the human population.

To better understand the significance of the ocean floor topography becomes clearer when the "big picture" focusses closer to the topography offshore of the US East coast and US West coast. Using ocean floor topographical maps from the "Atlas of the World" created by the Rand-McNally cartographers, the map of the topography of the western Atlantic Ocean floor and the portion of the East coast located on the continental shelf is seen in the next illustration:

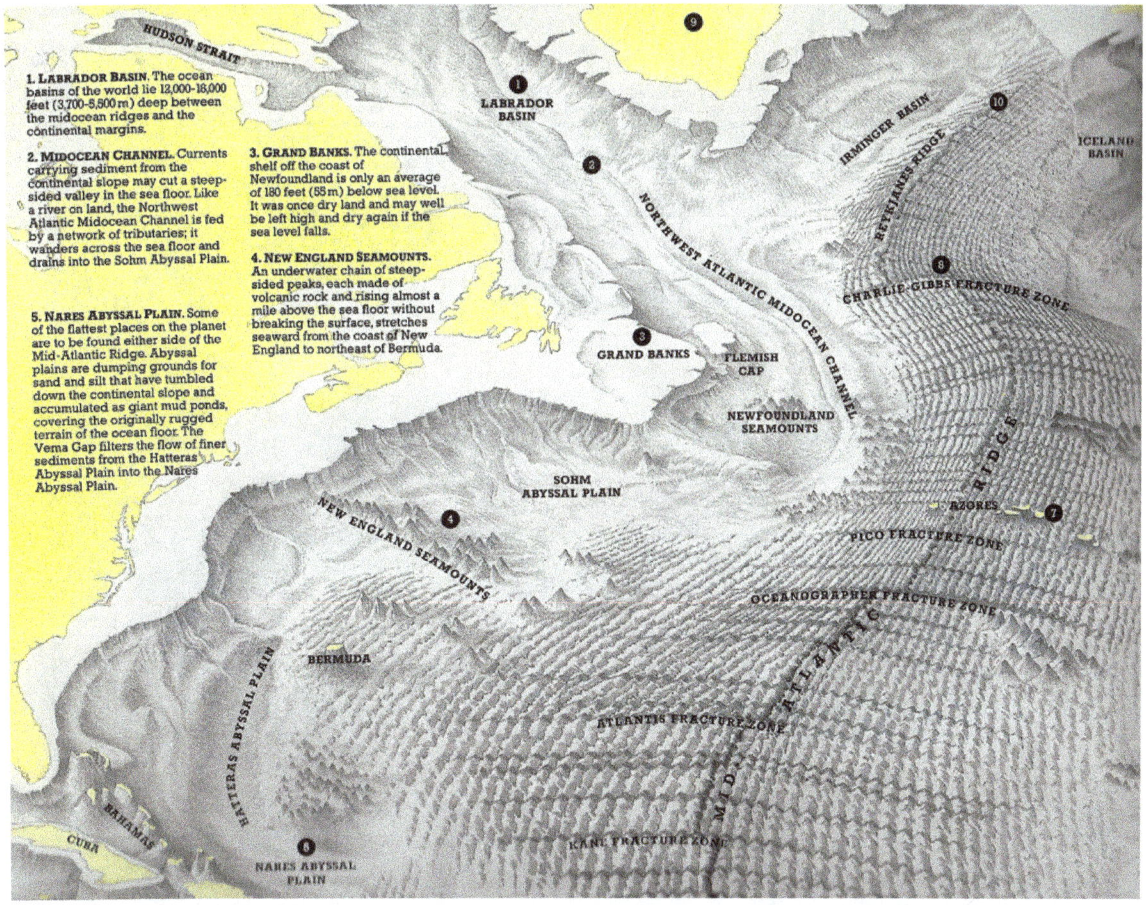

(Source: "The Atlas of the World", by Rand McNally for Readers' Digest)

The very warm waters of the Equatorial region north of the Equator have the characteristic clockwise current flow and the Gulf Stream current (described in earlier chapters) is channeled northeasterly in that trough seen to be formed between the west slope of the Mid-Atlantic Ridge and the edge of the continental shelf of the Eastern United States. Hurricanes in "Hurricane Alley" fake paths parallel to the very warm waters of the Gulf Stream current as their source of energy. More research is necessary to assess how much of the heat energy in the Gulf Stream current comes from solar heating of Equatorial zone waters and from the geothermal activity below the ocean floor that formed the Mid-Atlantic Ridge.

The climate of the United States West coast has a different set of temperature drivers than that of the East Coast. The topography map of the eastern Pacific region is shown next:

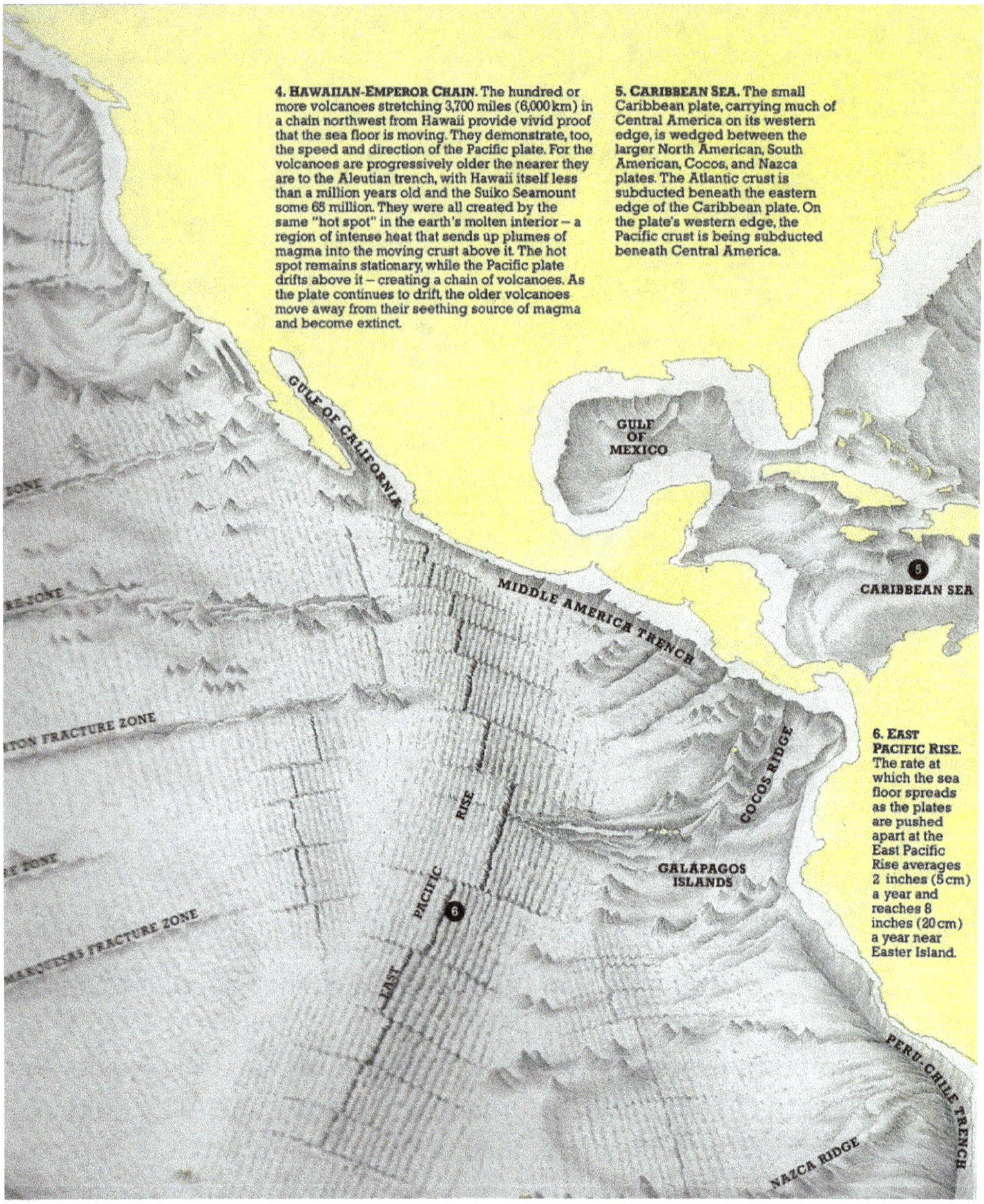

(Source: "The Atlas of the World", by Rand McNally for Readers' Digest)

The above illustration reflects the west coasts of North America, Central America, and South America to be congruent with the eastern sector of the Pacific "Ring of Fire" (and containing hundreds of volcanoes active and inactive). The San Andreas Fault in Earthquake history fame stretches the entire west coast of the United States where small earthquakes below 3.0 magnitude (typically 40-60 per day) are common in the Los Angeles area. If/when the "Big One" – above 9.0 magnitude - strikes, there will be losses of human life and property damage in the billions of dollars in the Los Angeles area alone. Earthquake-proofing of structures is required for building structures in Southern California and the network of freeways and interstate highways have also been earthquake-proofed but not to the extreme of withstanding earthquakes of above 9.0 magnitude.

The earlier photos show a major difference in the ocean floor topography in the Atlantic Ocean and the Pacific Ocean. The Mid Atlantic Ridge displays all the characteristics of a mountain range that is upheaval of land mass. The Pacific Ocean floor has many seamounts which are of volcanic origin, with an estimate of 34,000 seamounts (geothermal vents) under the West Pacific Ocean.

(Source: "The Atlas of the World", Rand McNally for Readers' Digest)

Another form of heat from below tectonic plates – the majority being near tectonic plate edges – are hydrothermal vents that were discovered in the late 1970s. There has been no estimate of the number of hydrothermal vents under all oceans waters but one estimate is "there are probably many hundreds of thousands of them". An excellent visual aid to understanding hydrothermal vents is shown as follows:

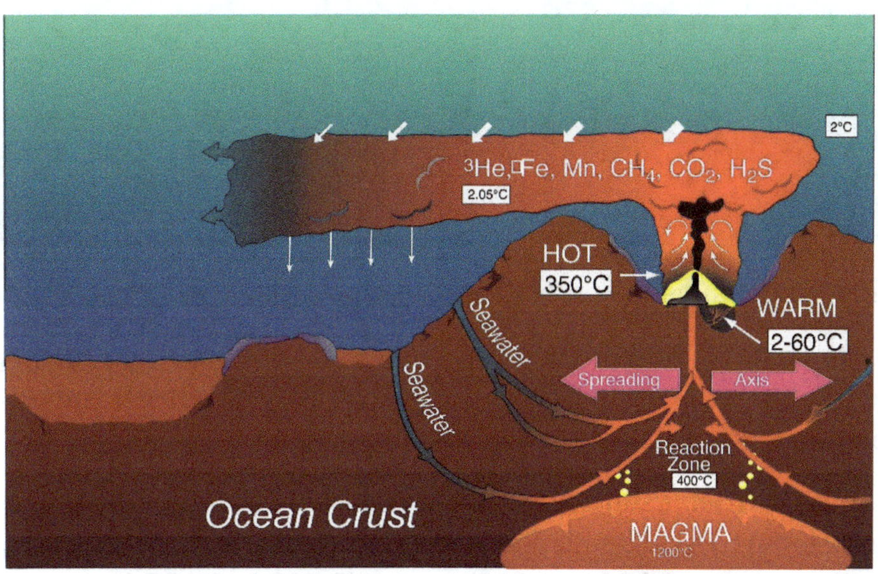

(Source: Smithsonian Institute for Science)

The heated water from these vents is in the temperature range of 350 $^{\circ}$C to 450 $^{\circ}$C (662 $^{\circ}$F to 842 $^{\circ}$F) and with the added property of being at high pressure to move the "smokey-colored hot liquid from the ocean floor to the ocean surface. A technology is envisioned that a funnel-like structure (heavy and made to withstand the high temperature and pressure) be placed over the vent. There could be a supply pipe of cool ocean water passing down into the funnel, the natural heat transferred in a coil to heat the supply water, and the heated high-pressure fluid passed through a return pipe to the surface. At the surface, the hot high-pressure mixture can be passed through a heat exchanger, the heat energy removed to produce electricity, and the now-cooler liquid returned to the ocean. Any leakage in the closed circuit will be into the ocean and not pollute any land area. The effect on the local ecology would be minimal if any.

Underwater photos reveal the "smoke", flowing at a higher pressure than the surrounding pressure at that depth, to be dark grey to near white, and their venting substructure to be very fragile tubes, some very colorful. Three examples of hydrothermal vents (which were photographed in three different oceans - Pacific, Atlantic and Indian) are shown as follows:

(Source: Scripps Institution of Oceanography, Unknown Photographers)

The hydrothermal vents do not create a hazard to the human population of the region. However, there are so many of them in all oceans and all have a constant flow of heated water at about 350 $^{\circ}$C (662 $^{\circ}$F), the total effect on heating ocean waters cannot be ignored.

There exists the potential for "harvesting" the heat energy from some of the larger hydrothermal vents (much like small sea mounts) to provide an alternative source of geothermal energy. The hydrothermal vents are at typically deep depths 2,500 to 3,000 meters (8,200 to 9,800 feet) on the ocean floor which poses a formidable accessibility challenge to harvest that heated water energy. One of the 24 weighted trade studies done in this book (included as worksheets in Appendix 1) is to investigate the feasibility of harnessing the reliable, clean, relatively low cost to capture and utilize the hydrothermal vent energy.

The "black smokers" hydrothermal vents have the dark color indicating having a large content of sulfides and carbon dioxide in the super-heated steam. The element carbon is black in color when in solid state or when in particulate form in a colloidal suspension in a liquid and may have some presence as seen in the "black smoker" photograph above. Environmental protection scientists are identifying the source of the increased carbon on the ocean floors as being from the carbon dioxide gas emissions generated by Man burning fossil fuels and introduced as "greenhouse gasses" into the atmosphere. However, oceanographic scientists apparently recognize that the increasing carbon on the ocean floors is a more likely result of the constantly flowing many hundreds of thousands of hydrothermal vents ("black smokers" and "white smokers") that have yet to be explored in the oceans' floors.

From the little known about hydrothermal vents (which were discovered in 1979 by scientists from the Scripps Institution of Oceanography), they are found mostly in hydrothermal fields at depths from 2,500 to 3,000 meters (8,200 to 9,800 feet). While their individual composition of minerals and gases varies considerably, they are all superheated to about 400 °C (752 °F) or more. If the efflux of a thermal vent is collected, the heat energy content could provide the valuable heat energy resource being sought. After the heat energy is extracted, the residual fluid and contents should be returned to the ocean so as not to create any noticeable effect on the ocean ecology.

A document researched for better understanding of the importance of the deep ocean (referred to as "deep sea") provided the following narrative (by an environmental scientist): *"Despite its remoteness, the deep sea is already under threat from human activity. Cold seeps provide oil, fueling our cars, and hydrothermal vents are threatened because they contain highly valuable mineral resources, such as lithium for our batteries. How animals have adapted to these ecosystems is important to understand the impact that human activities will have, and whether the animals in these habitats can recover from any damage to the habitats caused by human activities. The deep sea is challenging to reach and experimentation on its many and varied inhabitants is difficult. It is only through the dedication of deep-sea scientists and the use of submersibles, underwater robots, and specialized equipment that we are beginning to understand the basics of deep-sea biology. The deep sea is one of the last true wildernesses on Earth. Hydrothermal vents were discovered 8 years (in 1977) after humans first set foot on the moon. This is why we understand so little, and it emphasizes why we must protect the deep sea, from which come some of the most beautiful and bizarre organisms known to science."*

The alternate perspective to that aspect of the Man-made hazards to the ecology on the ocean floor is the perspective of what and how the deep ocean is affecting itself and other life forms on the planet. The heat energy in hydrothermal vents is perhaps the most useful potential factor – and causes this deep ocean phenomenon to be named "hydrothermal venting". The "smoke" shown in the three earlier photographs is not only categorized into being "white smokers" or "black smokers" in color but distinguishes their wide variation in chemical composition. The metals and other elements vary with each thermal vent but it is most common that the vented heated water contains much carbon dioxide. After visualizing several photographs and illustrations of the oceans' floors, anyone with an interest in preserving scientific integrity should question the environmental scientists' hypothesis that *"by eliminating everything else, settled science has concluded that Man burning fossil fuels is the dominant cause of global warming and climate change"*. If the elimination had been done with a logical and scientifically valid reason for doing so the hypothesis could be regarded as credible. Elimination of contradictory facts from consideration without valid reason causes any hypothesis to not be scientifically credible.

Oceanography scientists admit there is little known about the ocean floor because of its remoteness and forbidding pressure at large depths. A tragic example of the hazard of the pressure existing in the ocean depths was the implosion of the deep ocean submersible *Titan* that was on a tour of the wreckage of the behemoth "unsinkable" *Titanic* that sank on April 15, 1912. The *Titanic* struck an iceberg (shown in the next photograph) that opened seams in its underwater plates and "went down by the nose". Over 1,500 persons died from drowning or hypothermia in the cold waters of the North Atlantic.

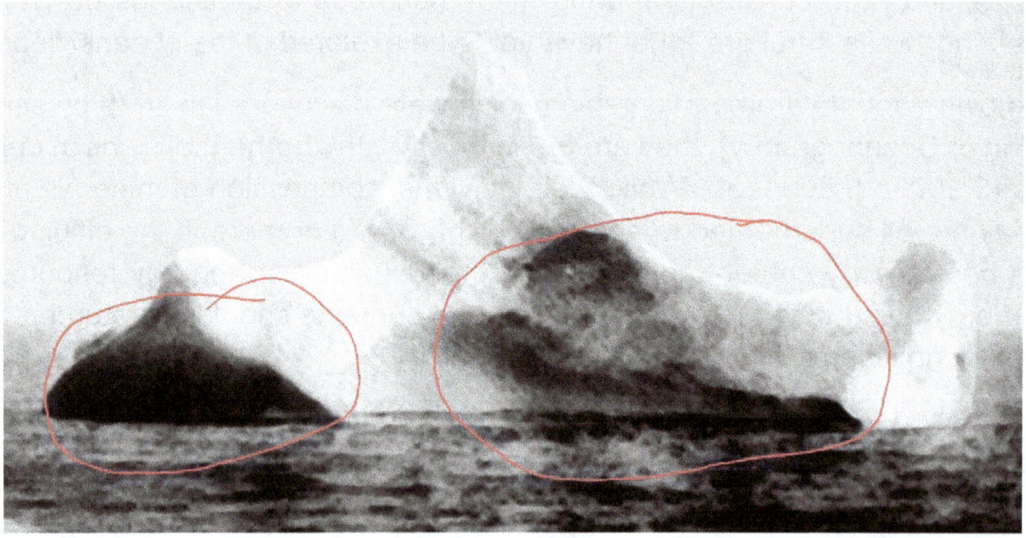

(Source: Federal Maritime Commission, Photographer unknown)

Paint from the *Titanic* is seen (encircled in red) where it collided with the face of the iceberg which had floated well below the common latitudes for icebergs in the North Atlantic. The accident investigation board concluded that the *Titanic* should not have been travelling at such a high speed (over 20 knots) in waters where icebergs had been reported to have drifted down from Arctic waters.

More than a century later, the submersible *Titan* was at a depth of 3,810 meters (12,500 feet; 2,080 fathoms) to offer tourists a view of the sunken *Titanic* when it imploded on June 18, 2023, and was moving at a very slow speed. Five persons in the Titan were killed by crushing when the outer structure of the *Titan* failed (cyclic fatigue resulting from several exposures to the extreme loading and unloading of the pressure cycles on its outer structural materials). The pressure at a depth of 12,500 feet (3,810 meters) below saline ocean water is estimated to be about 15,750 pounds per square inch. The five humans in the *Titan* subjected to that amount of pressure had no possibility of surviving the implosion accident.

Because of the enormous natural pressure at the depths of the ocean floors, it is understandable why more than half of the planet's surface area remains unexplored. That fact of inaccessibility does not condone any branch of science - claiming "the science is settled" of the effect of "greenhouse gasses" on its land surface or in its atmosphere, and thereby to ignore the effect of many unknowns in an important area of the environment such as the ocean floors. The human population should absolutely not be dumping its commercial waste and household garbage into open ocean waters, and environmental protectionists rightfully should complain that human garbage is polluting the environment. Conversely and ideally, the environment should not create so many hazards (tornadoes, hurricanes, lightning-strike ignited wildfires, flooding, droughts, volcanic eruptions, earthquakes, etc.) that endanger human life. “Two wrongs do not make a right.”

The ocean waters have become saline during past eons and unfit for humans to drink. However, nature has provided the evaporative process to distill the saline from the water before it ascends and eventually return as precipitation of fresh water. That distilled water when aerated is suitable to be used by Man and to irrigate everything that grows (including the agricultural products in Man’s food supply). In a sense, water is the vital lubricant in the machine that Man calls Earth. Later in this book will be described the potential for Man using solar energy to distill saline ocean water into potable water to abate drought areas near the coastal regions of the USA. A weighted trade study has been generated to help in prioritizing the effort and resources that could be applied for this purpose.

Another aspect of water in the ocean currents being constantly in a predictable rotational motion is the factor identified by the brilliant French scientist, mathematician, and mechanical engineer Gaspard-Gustave de Coriolis (1792 - 1843) in 1835. The Coriolis Effect is a result of the planet being in rotational motion about its tilted, wobbling axis of rotation and having gravitational acceleration interacting with that rotational motion to cause the liquid water to swirl on Earth’s surface. Each of the major oceans and in fact every large body of pooled water is seen to develop the swirl The direction of rotation, depending on latitude relative to the Equator, is visually seen in the ocean currents in the following diagram

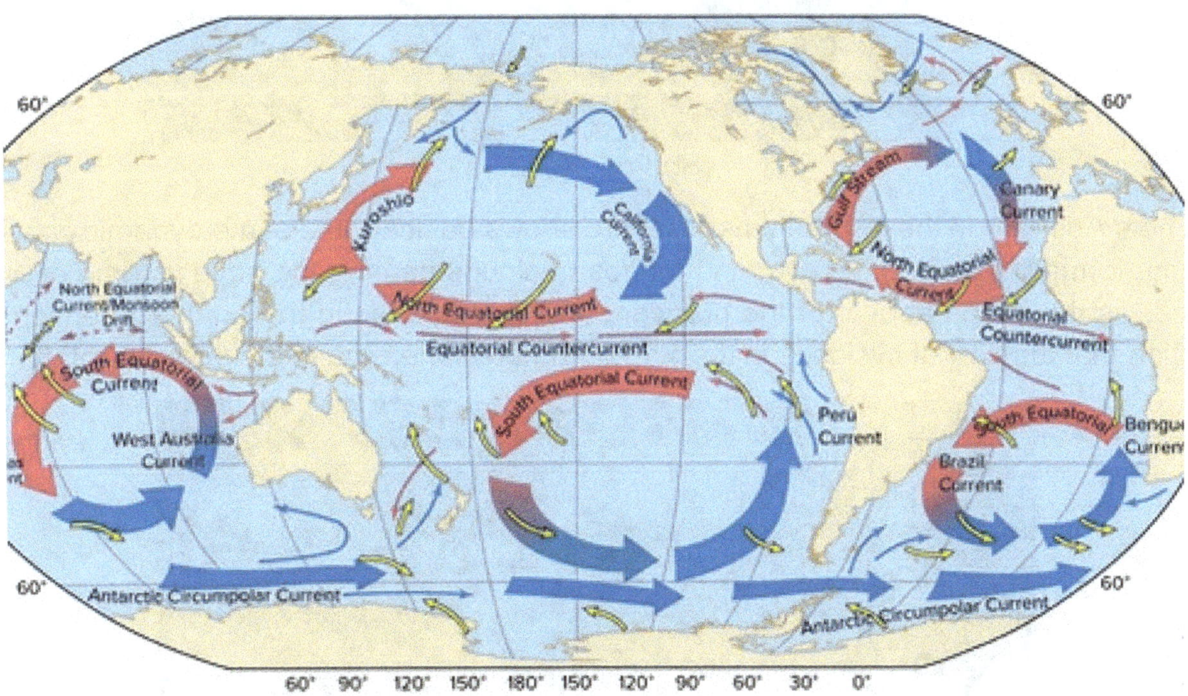

(Source: National Oceanic and Atmospheric Administration)

Readers interested in scientific factors can do a simple experiment at their place of residences to confirm the Coriolis Effect. The experiment consists of first filling a wash basin (having a drain at the bottom) with water (of any temperature between freezing and boiling). When the drain valve/stopper is opened, the water will drain downward initially and then develop a swirling motion in combination with the downward motion. Adding a small circle of tissue paper on the water surface will help to visualize those motions the water takes as it drains. For those in the latitudes of Earth north of the Equator, the water will swirl in a clockwise manner as seen from above due to the Coriolis Effect. For those in the latitudes south of the Equator, the water will swirl in a counterclockwise manner.

It is the same Coriolis Effect which causes hurricanes in the northern hemisphere to have counterclockwise motion whereas monsoons (a "cousin" of hurricanes) in the southern hemisphere to have clockwise motion. A satellite photo of Hurricane Ida in 2021 as it slowed over the Bahamas and caused massive damage illustrates that fact quite well: (Note that within the walls of the "eye" of a hurricane there is very little rotational velocity and water content, but the fringes of the hurricane have high winds and much water content. It is the fringe regions of hurricanes (with winds blowing the high cumulative mass of water drops at high speed) which typically produce the heaviest direct damage and indirect human injuries/fatalities from flooding. Conditions within the "eye" are comparatively benign, and the "eye" usually produces little or no significant damages directly to property or injuries to living lifeforms.

(Source: NASA)

A cyclone near Australia in the Southern hemisphere shows basically the same dynamic characteristics (eye, circular formation around the eye, and fringes of clouds bearing the huge volume of water to be precipitated) as the image above (of the hurricane in the Bahamas in the northern hemisphere) with the exception that the flow of cyclones in the Southern hemisphere have a clockwise rotation.

(Source: NASA)

When observing a “big picture” diagram of the prevailing (“normal”) wind patterns on Earth’s surface and lower atmosphere, and as affected by the warm currents of water flowing below them, the prevailing wind patterns on Earth are seen to have the same curvy patterns emanating about the Equator (the hottest region of the planet) as do the ocean currents:

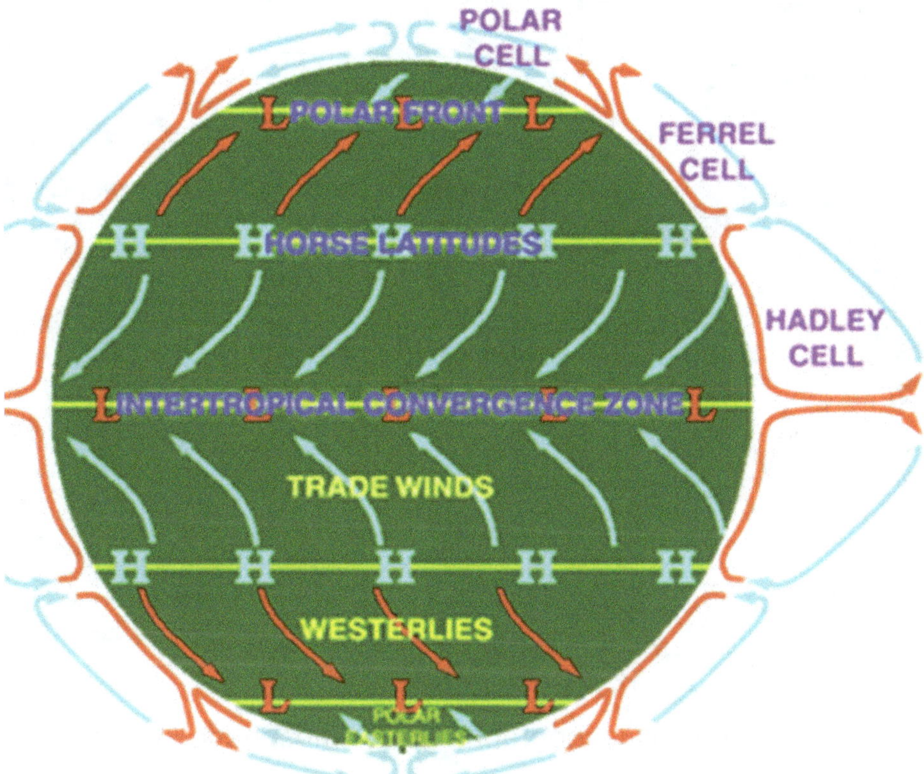

(Source: NOAA)

Three most important observations that should be drawn from the diagrams above are:

1) The ocean circular currents north of the Equator (therefore in the northern hemisphere) rotate counterclockwise whereas the ocean circular currents south of the Equator (in the southern hemisphere) rotate clockwise. (This phenomenon - direction of rotation being latitude dependent – is identified as being due to the Coriolis Effect.)

2) Each of the five main ocean current circulations have a large segment that contains water heated to its maximum temperature by circulating through the middle of the Equator. The front cover of this book illustrates that surface temperature (measured by NASA satellites circumventing the planet and measuring the surface temperatures accurately and globally without any “averaging” by scientists from data collected from randomly located meteorological stations) being at its highest degree – shown in dark red – at the Equator. That fact is due to the Equator being the part of the rotating spherical planet Earth at the closest distance to the Sun – much like a spherical mass of meat rotating on a rotisserie spit. The angle of the Sun’s heat energy waves are also at their most orthogonal (perpendicular angle) on the Equator and not deflected in a large lateral acute or obtuse angle (as they are in the polar regions).

3) From observations 1) and 2), an important continuing observation and understanding of the ocean currents should be that the amount of heat energy added to those ocean waters being heated by rising temperatures in the Equatorial region are being transferred by the currents to latitudes away (north and south) from the Equator.

The following diagram illustrates that transfer process of the Gulf Stream current conveying heated ocean water (to above 8o degrees Fahrenheit) from the Caribbean Sea and Gulf of Mexico into the frigid waters of the North Atlantic:

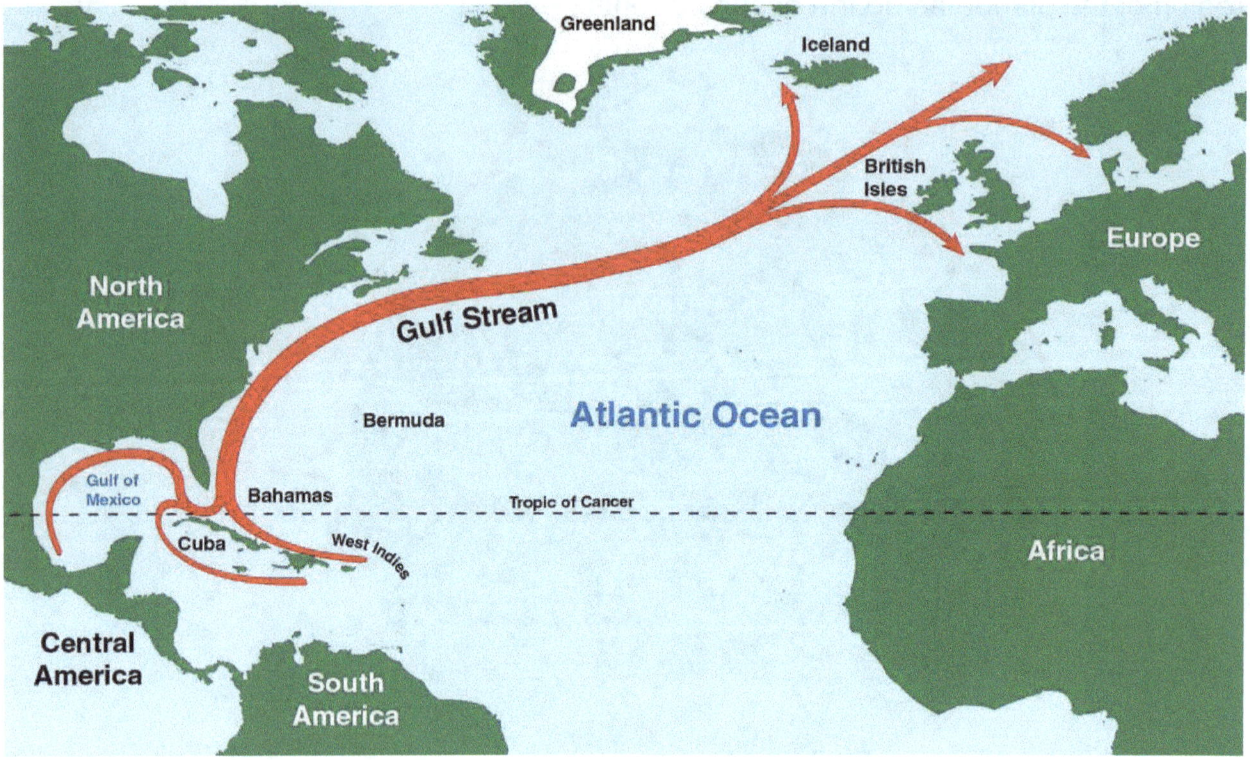

(Source: The National Oceanographic and Atmospheric Administration)

The legendary "London fogs", occurring in long before the onset of the "Industrial Revolution" but still occurring, emitting "greenhouse gas emissions" were and are visual indicators of the extremely high concentration (above 4-5%) of water vapor in the atmosphere above the Gulf Stream finger of current circling the British Isles (England, Wales, Scotland and Ireland) and merging with the much cooler waters of the North Sea and waters in Scandinavia.

Reverting to the topic discussed earlier as to whether it is Man creating "greenhouse gas" warming of the Great Barrier Reef off the eastern coast of Australia being a major cause for the coral to be bleaching white ("cooking", from the "soaring" (0.01° C per year) sea surface temperature), the Southern Pacific Equatorial ocean current (drawing heated water from the Equatorial region toward the east coast – and Great Barrier Reef region - of Australia) is a more direct cause for the warm waters causing the coral reef distress than is the indirect effect of "greenhouse gas" atmospheric heating from above the water surface.

When considering the movement of ocean surface waters, the shape of the ocean floor beneath the currents also must be considered. That is especially true for that portion of the planet which affects the weather and climate of the United States (those on the main continent). Much attention has been and is being given by environmental scientists to human activities affecting the weather and climate, but too little information has been provided to the public as to how heat from within the planet is affecting the terra firma and the waters interfacing with the atmosphere and the heat transfer (both directions) between the surface (land and water) and the atmosphere. The prevailing current in the northern Atlantic Ocean (above the Equator) is seen in the two preceding diagrams above.

It is the southeastern portion of that Atlantic Ocean current that has a large effect on the Gulf Coast and Atlantic coastal states, and in fact is the spawning grounds for the seasonal hurricanes that ravage those southeastern US states. The heat energy contained in the warm current is not just from the top few inches of the water surface but from the water movement well below the surface. It is a fascinating scene as to what path tropical storms and hurricanes take when moving from the west coast of Africa (at about the Cape Verde islands) and use the Caribbean islands chain much like steppingstones on the upper edge of the Caribbean Sea "cup" of water before making landfall on the US southeastern coastal regions.

(Source; National Geographic Society)

What is most interesting (at least it is to this Author) is that there is a distinction between the larger islands (named as the Greater Antilles) and the smaller islands (named as the Lesser Antilles) in the arc-like formation chain. The Lesser Antilles islands are mostly all the peaks of volcanoes (most inactive but some still active) with the exception of the most eastern island in the West Indies (formerly British West Indies prior to 1965) of Barbados. The Greater Antilles are not of volcanic origin. While ancient mariners only sailed in sailing ships among the islands, recent marine geologists sailing in motorized vessels equipped with high technology sensors have mapped the floors of the oceans and have revealed a fascinating underwater scene of the Lesser Antilles region off the east coast of Venezuela as shown in the following diagram:

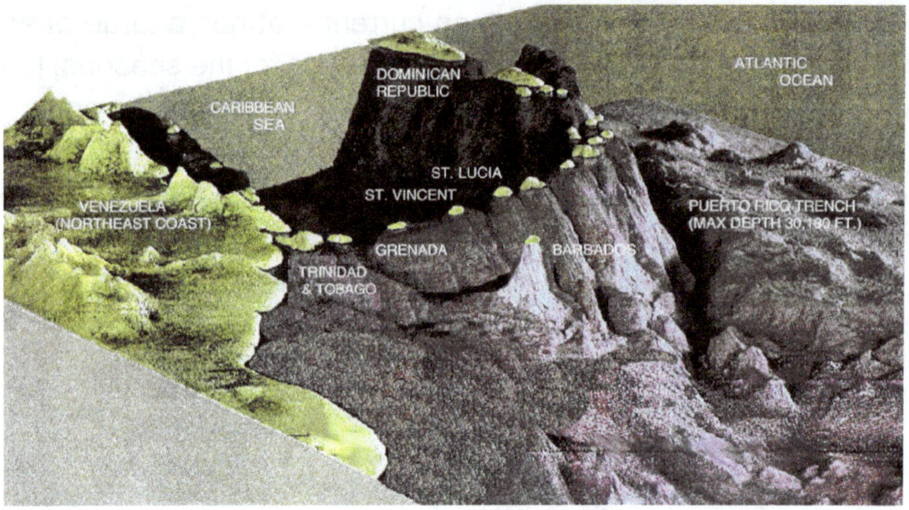

(Source: The Barbados Museum Historical Society)

As a frame of reference for scale, the dimension of pear-shaped Barbados is 14 miles East to West and 21 miles North to South with an area of approximately 165 square miles of land above sea level. The Lesser Antilles and Greater Antilles islands defines the edge of the Atlantic tectonic plate and the South American plate in tectonics science. Any warmth of the water in the lower portion of the Atlantic Ocean North Equatorial current is deflected upward along the slopes of the volcanic cone-like formations. That upswell of warm water probably adds heat energy to the frequent hurricanes that pass over the Eastern Caribbean region at causes the severity of hurricanes to increase accordingly. The amount of carbon dioxide in the atmosphere is not of comparative significance to affect the severity of hurricanes.

With a little imagination, the mountain peaks of the northeastern coast of Venezuela (South American continent) can be regarded as resembling lower right molar teeth in the human jaw, the arced chain of the Lesser Antilles imagined to be the front lower teeth in that imaginary jaw, and the Greater Antilles islands as lower left molars completing the imaginary lower jaw of humans. A most unusual formation is the Puerto Rico Trench seen in the next photograph and which has been measured to have a depth of 28,232 feet below current sea level – as the deepest depth of the Atlantic Ocean.

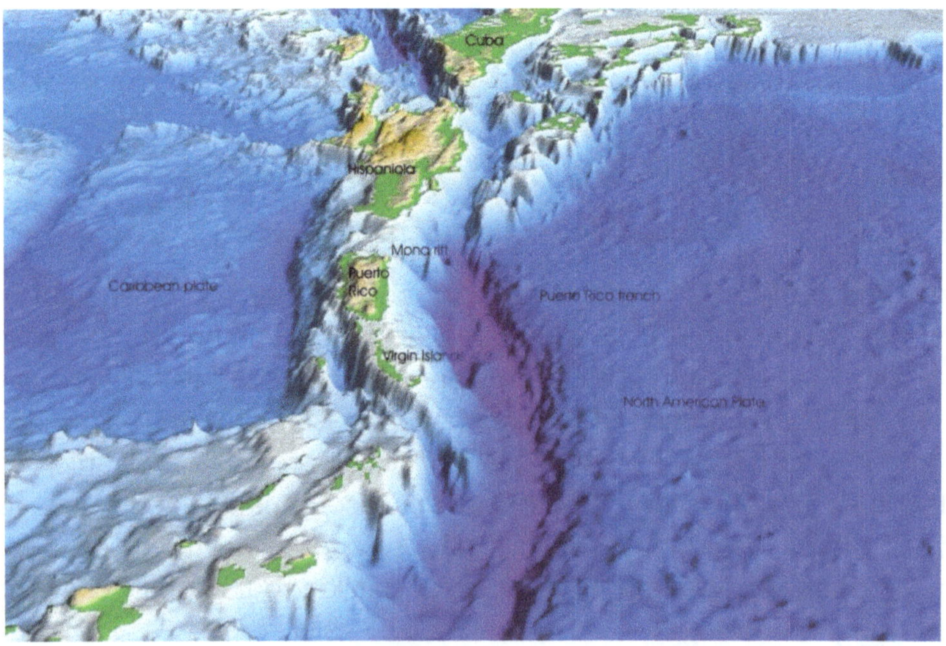

(Source: US Geological Survey)

There are several other trenches than the Puerto Rico Trench, with the deepest trench being the Mariana Trench in the Pacific Ocean at 36,197 feet (11,036 meters) in depth. The pressure at the deepest point in the Mariana Trench was estimated to be 1,086 bars (15,765 pounds per square inch. The tabulation of the 20 deepest trenches in all oceans is shown next, and with an estimate of the pressure in each trench based on its depth measured by the World Geological Survey.

Depth and Pressure in the Deepest Ocean Trenches

			Depth		Pressure	
	Trench	Ocean	feet	meters	bars	lb/in² (psi)
1	Mariana	Pacific	36,197	11,036	1,086	15,765
2	Tonga	Pacific	35,702	10,885	1,071	15,548
3	Philippine	Pacific	34,580	10,543	1,037	15,055
4	Kuril-Kamchatka	Pacific	34,449	10,503	1,033	14,997
5	Kermadec	Pacific	32,963	10,050	989	14,358
6	Izu-Ogasawara	Pacific	32,087	9,783	963	13,981
7	Japan	Pacific	29,527	9,002	886	12,865
8	Puerto Rico	Atlantic	28,232	8,607	847	12,299
9	Yap	Pacific	27,976	8,529	839	12,183
10	South Sandwich	Atlantic	27,651	8,430	830	12,052
11	Peru-Chile	Pacific	26,456	8,066	794	11,530
12	Diamantina Fracture Zone	Indian	26,401	8,049	792	11,501
13	Romanche	Atlantic	25,460	7,762	764	11,095
14	Cayman	Caribbean Sea	25,238	7,722	760	11,037
15	Aleutian	Pacific	25,194	7,681	756	10,979
16	Java	Indian	24,460	7,457	734	10,660
17	Middle America	Pacific	21,880	6,671	656	9,529
18	Puysegur	Pacific	20,700	6,311	621	9,021
19	Vityaz	Pacific	20,177	6,152	605	8,789
20	Eurasian Basin	Arctic	17,881	5,452	536	7,788

Note; Depth measurements are from NGS sonar database. Ocean water is incompressible; therefore, pressure at depths of trenches are ratioed to those in the Mariana Trench: $P_n = P_1/h_1 \times h_n = 0.0984\ h_n$ with h in meters

(Source: Tabulation by Author with World Geological Survey database)

Volcanoes frequently exist around the edges of tectonic plates, but there also are mountain peaks not of volcanic origin caused by the subduction of one plate moving under an adjacent plate. Most of the islands in the Caribbean region are of volcanic origin and are totally extinct, inactive, or active. Those that are of upheaval origin were submerged and have a coral crust forming a shell over limestone substructure. Most have a soil content that is favorable for growing different food and fruit products depending on the type of soil.

The island of Barbados has a somewhat unique geological origin for that region and its soil composition, high in limestone (calcium carbonate) formed by upheaval of coral reef formations, was found extremely favorable for the growing of sugar cane. The three main products derived from sugar cane are raw sugar (in crystallized form), molasses, and rum. The process is an absolutely environmentally "clean" one. The extraction of sugar syrup from the crushed cut lengths of sugar cane leaves a woody pulp known as *bagasse* (from the French *baggage* meaning "trash"). The power for the crushing millstones was early in Barbados history of the Dutch windmills design powered by wind power. The pulpy bagasse is spread to dry in the typically hot sun (Barbados is at 11 degrees North latitude), and when sufficiently dried, the bagasse is combusted to heat the extracted syrup to convert it into a thick syrup. It is from that thick raw brown syrup which (by centrifuges) is drawn the light brown colored sugar crystals (near the top), the syrup (in the middle) sent factories to be distilled and fermented into rum, and the thickest fluid molasses from the bottom. The ashes from the burned bagasse are returned to the cane fields to fertilize the next crop of sugar cane.

Coupling the desire for the three commodities of sugar, molasses, and rum in the British Isles (where the cool European climate does not support growing sugarcane) with the ocean currents and prevailing wind currents caused sailing ships from England to load cargoes of sugar, molasses, and rum at the port of Bridgetown, Barbados. The ships then proceeded on the ocean and wind (known as the "Trade Winds") currents to the southeastern ports (primarily Charleston, South Carolina) of the Colonial Americas. At Charleston (in the "Carolines"), additional loads of cotton and tobacco were taken on. The fully loaded ships completed the clockwise circuit of what was referred to as "the Trade Triangle" back to the point of origin – the ports of the British Isles. The "points" of the trade triangle were the British Isles, Barbados, and the port of Charleston, South Carolina.

Those prevailing winds still blow the same and became known to mariners as "the trade winds", but it is those same winds that follow the same path as the hurricanes that originate at the Cape Verde Islands region, track past Barbados, and follow the heat energy path of the warm current which splits into two sub-currents. One portion proceeds into the Gulf of Mexico, and the other portion (the Gulf Strean) turning northerly toward the path and paralleling the US East coast states. Each of the Caribbean islands has a similar story which is outside the scope of this book, but all depended on ocean travel for their development and are still a magnet for international tourists to bask in the warm Barbadian sun (while imbibing Barbados dark rum on a beach in the shade of a coconut palm).

For the Caribbean region shown graphically above, a tabulated set of statistics provides the demographic as well as the topographic and geologic nature of each of the inhabited land masses shown in the light green color above. It should be noted that when/if/how much the ocean levels rise will have considerable effect on the population of the islands and their future existence for residency, trade, and tourism.

Caribbean Islands Most Affected by Ocean Levels Rising
(Coastal beaches and population at low height above Sea Level)

Island	Population (Census Year)	Area of Exposed Land (sq miles)	Highest Point of Island(s)(feet); (Type geologic origin)
Trinidad & Tobago	1,367,558 (2021)	1,980	3,080 (Trinidad non-volcanic; Tobago inactive volcano)
Grenada	124,610 (2021)	135	2,755 (multiple inactive volcanoes)
St. Vincent & Grenadines	110,872 (2022) 10,900 (2012)	142 17	4,049 (La Soufriere active volcano)
Barbados	277,821 (2010)	169	1,120 (upheaval, not volcanic)
St. Lucia	180,000 (2018)	238	3,120 (semi-active volcanoes)
Martinique	349,925 (2024)	436	4,583 (Mount Pelee' active volcano)
Dominica	71,293 (2011)	290	4,747 (two inactive and 3 active volcanoes)
Guadaloupe	395,925 (2024)	629	4,813 (Grand Soufriere active volcano)
Montserrat	4,390 (2022)	102	3,440 (inactive volcano)
Antigua & Barbuda	97,120 (2019)	170	1,319 (Antigua inactive volcanic; Barbuda upheaval)
St. Kitts & Nevis	54,338 (2023)	101	3,793 (Both inactive volcanic)
Anguilla	15,753 (2021)	35	240 (Of upheaval limestone)
US Virgin Islands	108,612 (2000)	134	1,555 (Mix of inactive volcanic and upheaval)
Puerto Rico (US)	3,285,874 (2020)	3,424	4,390 (inactive volcano)
Total:	6,454,991	8,002	

(Source: Tabulation of Statistics by Author from multiple databases)

"Settled science" has concluded that it is those same Man-made "greenhouse gasses" in the atmosphere causing the melting of ice shelves on the Antarctic Continent in the south polar region. (Those fallacious conclusions were expressed in a quote attributed to an environmental scientist possessing a Ph.D. degree at a major Government Earth Resources Center in Texas). It was found that little or no consideration has been given (probably because that consideration was "eliminated" earlier) of the ocean waters warming in the north and south polar regions being probably more due to the ocean currents (shown in the preceding diagrams) directly transferring exceedingly warm waters from the Equatorial zone. Measurement of the water temperature of the Arctic Sea waters from satellites does not distinguish how much effect of the temperature is due to the "greenhouse gas" effect (natural and Man-made) in the atmosphere or due to the ocean currents transferring huge quantities of warm water from the Equatorial region into the North Atlantic region.

With accurate scientific measurements (formerly only land-based and since 196o satellite-based) of carbon dioxide in the atmosphere indicating a total (natural and Man-made) concentration of carbon dioxide in the atmosphere of less than 0.05%. and of that total, Man-made "greenhouse gasses" in the atmosphere are estimated to be in a concentration of only 0.02% (everywhere including the British Isles). Many scientists in the climate science profession have disassociated themselves from the environmental "settled science" cult and their claim that "greenhouse gasses" and excluding water vapor are the primary cause for climate change.

Two diagrams have been singled out from the database compiled by Dr. S(iegfried) Fred Singer (1924-2020), a Professor Emeritus of Environmental Science at the University of Virginia. He is the author of three books on the topic of climate change: "Global Climate Change" (1989), "The Greenhouse Debate Continued" (1992), and "Hot Talk, Cold Science" (1997). He also co-authored (with Craig Idso) "Climate Change Reconsidered" in 2009. His series of papers and lectures entitled "A Closer Look at the Numbers" provided this Author with invaluable scientifically factual data and information about the effect of water vapor in the atmosphere on the environment of the planet. These diagrams were shown previously in this book, but because of their major significance are repeated here for Readers' convenience: Environmental scientists have misled the public to believe that the only "greenhouse gases" in the atmosphere are those that are natural (except water vapor), human generated, and industrial generated origin.

The first diagram portrays the misleading information environmental scientists provided to the public, news media, and government officials to cause them to believe that it was only carbon dioxide (natural and anthropogenic) methane, nitrous oxide and a few other less important "greenhouse gases" in the atmosphere that were the predominant causes for global warming and climate change. Dr. Singer typically started his lectures with this chart:

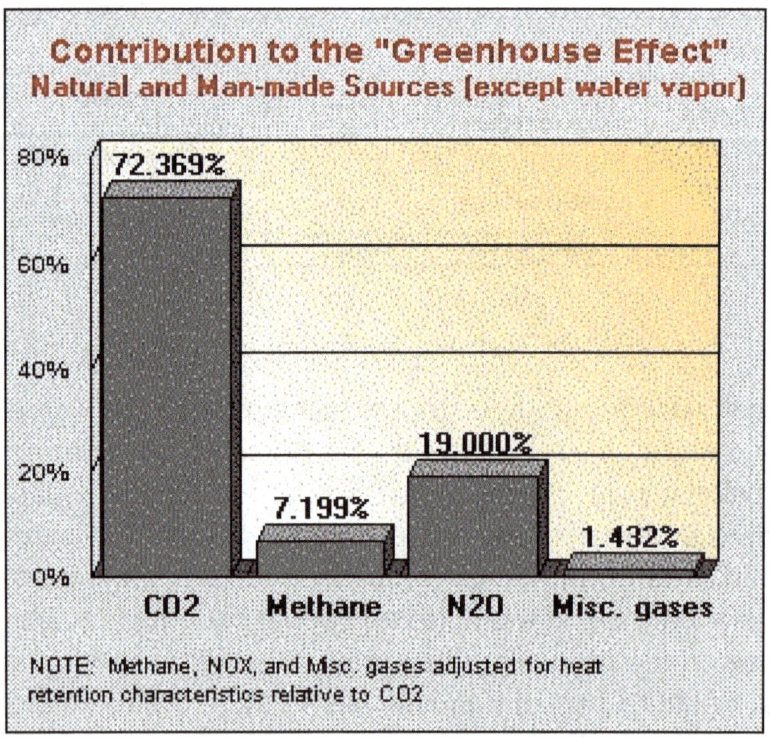

(Source: Dr. S. Fred Singer, Professor Emeritus University of Virginia)

His second chart – from his "Closer Look at the Numbers" analysis – quickly pointed out the critical omission of water vapor in the environmental scientists' fallacious hypothesis, and is shown as follows:

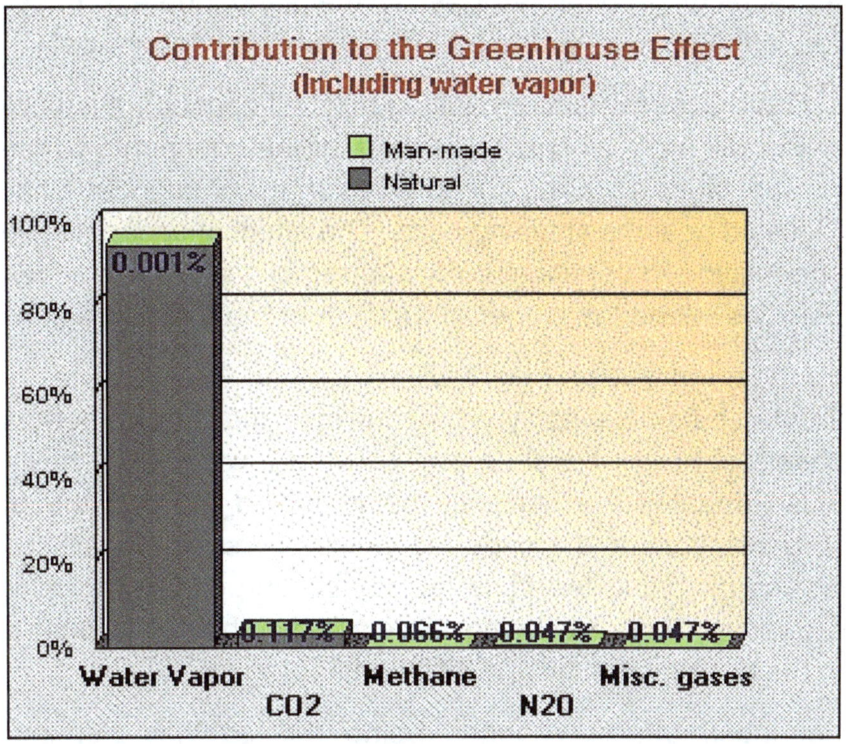

(Source: "A Closer Look at the Numbers", Dr. S. Fred Singer, U. of Virginia)

Not surprisingly, the "settled scientists" were joined by a large number of the environmental science profession – everyone and anyone supporting the continued sensational, but fallacious, hypothesis that "greenhouse gas" emissions are the dominant cause of global warming, climate change and ocean levels rising. The careers of those adhering to the "settled science" propaganda is sustained by the funding resources being supplied in the form of Government grants. Anyone and everyone who questioned that conclusion was and is labeled as being a "naysayer", as Dr. S. Fred Singer was (up until his passing in 2020). The news media has yet to publish any of Dr. Singer's findings (compiled in a series of environmental research papers as "Climate Change – A Closer Look at the Numbers") because of his anti-sensational research and factual analysis of global warming and climate change.

Most of the news media services relies on sensationalism of any topic and on a contrived "crisis" level of importance to frighten the public with fearmongering ("the sky is falling") type of rhetoric and to increase sales of the media containing the fear-creating reporting. How the public's taxes collected to sponsor Government research on any topic affecting the environment or public safety, health, or comfort are being expended typically sounds very convincingly if regarded as a "last resort". Not being told is that their taxes could – and should – be expended on more efficient activities to protect human safety, health, and comfort is the "normal" for the news media and Government environmental protection agencies. The United Nations formed an Intergovernmental Panel on Climate Change (IPCC) in the early 2000's (composed of environmental scientists and "experts" from all over the developed world) and which have established emissions reduction goals aimed at reducing the "climate change crisis". But that is not a tangent this Author will follow at this point in this book. (The fallacies of the "greenhouse gas" emissions crisis (which is an unprofessional fear-producing tactic to enlist public support) are treated in other books by holders of a Ph.D. degree in environmental science and atmospheric physics and it is those other peer criticisms and constructive criticism that cause this

Author to defer any challenges to the scientific fact that water vapor cannot summarily be eliminated from consideration in any analysis of global warming and climate change. Water vapor remains as the far most effective heat-trapping "greenhouse gas" in the atmosphere – causing the anthropogenic portion of carbon dioxide in the atmosphere to be insignificant in importance.

Returning to the reality of the oceans rising (resulting from all causes), this book places priority of the effect the ocean level rises will have on coastal regions of the continents where human development of the coastal regions and shipping ports for international commercial trade results in the most populated areas being located near the shipping ports. The effect of ocean rise therefore will have more impact on the human population in continental coastal regions and have much lesser impact on the population of towns and cities inland (on higher ground) from the coastal regions.

The change in timing of the surface level height will vary along the coastlines of the major continents and scientific and political debates inevitably will occur over the rate of increase of the ocean levels with time to different levels of height increase and to what level will be their ultimate distant future increase in height. Those debates are currently being fueled (pardon the pun) by environmental scientists and activist organizations and groups (as well as by sensationalistic news media reporting and Government politicians distorting the facts for obtuse purposes) for the effect the weather, the climate, and the ocean levels rising are being projected for their individual and interactive effects (present and future) on the ecology and the environment.

Focusing the cause of those trifecta changes being due to Man burning fossil fuels creating "greenhouse gas" emissions (principally carbon dioxide) results in massive effort and funding activities that are well-intentioned but aimed at a very small influence factor on weather, climate, and ocean dynamics that of man-made carbon dioxide at a concentration of about o.o15% (15o parts per million) of the atmosphere.

Two illustrations are presented next to assist Readers in visualizing the amounts of ice on Antarctica and in the Arctic:

The Arctic Region
(North Polar)

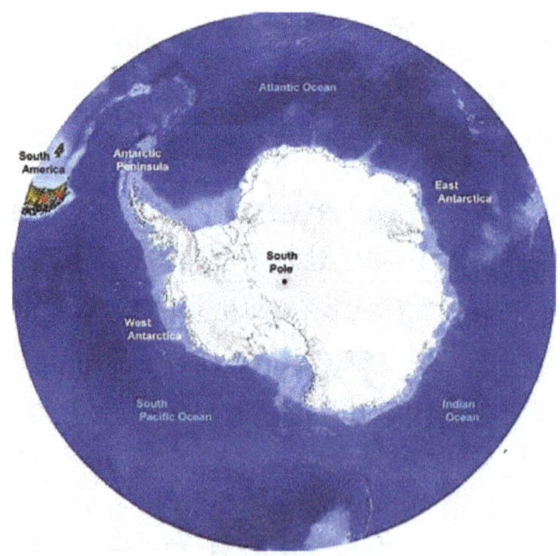

The Antarctic Region
(South Polar)

(Source: National Geographic Society)

The latitude of the geometric North Pole is 90° North and the latitude of the geometric South Pole is 90° South. There is a major difference in the topography and geology of the Arctic Region and the Antarctic Region. (As a popular item in the game of Trivia, the Greek name Arktos (or Anglicized Arctic) translates to "bear" and is derived from the Greek astronomical constellation Arktos which had the shape of a bear. The Greek name Antarktos (Anglicized to Antarctic) translates loosely to "without bear" as there are no polar bears in Antarctica. Polar bears have adapted to the availability of their primary food source - seals, abundant in the floating ice floes of the Arctic Ocean. At the South Pole, there is only the land mass of Antarctica covered in more than a mile thick of ice and there are no seals – the primary food source for polar bears.)

During the research of information about ocean levels rising for this book, three excellent reports were found that described the effect that global warming was having on the ocean levels and the many associated factors caused by and resulting from climate change. Among the three most interesting and informative database of scientific regarding the oceans' surface levels rising and due to the melting of glacial ice in the Arctic and Antarctic Regions with their highly commendable scientific explanations and graphical illustrations are: "Understanding Earth: The Icy Arctic" presented by NASA and the staff of the Earth Resources Department at the Goddard Space Flight Center et al; "Sea Ice" in the Arctic Report Card 2021 presented by the National Snow and Ice Data Center, Boulder, Colorado, and by the National Oceanic and Atmospheric Administration, et al; and "Greenland Ice Sheet" written by Sanat Pai Rankar and presented for the Geography and Travel research library of the Encyclopedia Brittanica.

From other perspectives by lesser disciplined scientists, environmental scientists (and many misinformed environmental protection activists – most notably former Vice President Al Gore, Swedish teenager Greta Thunberg, and many misinformed government regulators - have forewarned that because the "average global temperature" is on a warming trend, the ice in both the Arctic region and the Antarctic region will eventually be melted and merge into the surrounding ocean waters causing horrendous impact to human activities on the coastlines of the continents.

While the scientific principle is valid that when ice melts its volume will transform into the same mass but in a liquid state, the environmentalists have developed a well-intentioned but perhaps unnecessarily alarming caution to the public that ocean levels will rise to a hazardous level (more than 5 feet and to as much as 24 feet) in the time period from at the end of the twenty-first century to as long as in 20,000 years from the present. The Russian Vostok Ice Research Center in Antarctic has collected and analyzed ice core samples (described in an earlier chapter) suggesting that the current warming trend will soon reverse its warming trend and move towards another "Little Ice Age". There is little Man can do to avert that distant future eventuality, but there is much Man can and should do to prepare for that eventuality. In the near term, Man can and should concentrate on reducing or eliminating all practices that creates pollutants into the land, sea, and air. (It is hoped that those anti-pollution practices will not fall by the wayside as have the efforts – not considered in this book - to reduce felonious crime, homelessness, lack of sanitation, major accidents during transportation of humans, mass shootings in schools and places of religion or in commercial facilities, the ban of assault weapons in public hands, sensible abortion controls, etc.).

To assess the ominous prospect of the ocean levels rising and at what rate, the volume of ice existing in both the Arctic region and Antarctic region must be estimated in its entirety and then an estimate made of the rate – which is likely to be a non-linear rate with time (for several reasons) – must be generated. In conjunction with the ocean surface levels rising (by any cause and any rate), the effect which the rising water levels will likely have on the coastal regions of the several continents will widely vary just as it will on land masses of less than continental size and height above the present sea level. A critical factor as to the cause-and-effect relationship is the question of what segment of life forms will be affected, and by what scientific manner. It is human nature to protect the safety, health, and comfort of the human population prior to being equally concerned over protecting the many billion animals occupying different regions and protecting the agronomy of the regions. Those animals that form a part of the food supply (protein and dairy products) of the human population must be given priority of attention ahead of those animals that are carnivorous, herbivorous, or omnivorous. Attention must be given in parallel to that of the ecology of the planet again divided into the category of being a part of the food supply system for humans and domesticated animals.

The illustrative comparison of the two polar regions is intended to emphasize the differences – much like the positive and negative images on a frame of film – in topography and type of ice (floating or in ice sheet form) within the two polar regions. The ice (shown in near white color) in the Arctic Region is almost completely in floating form ("sea ice") with the glacial ice on Greenland (shown in green with ice shelves shown in dark green). The ice in the Antarctic Region is primarily in a thick sheet form ("glacial ice") above the terra firma of the Antarctic Continent with "ice shelves" (shown in pale blue). Ice that is floating will, when melted into liquid form, combine with the saline ocean water on which it was floating and will not cause a rise in the surface level of the oceans. Conversely, ice that exists on a terra firma (such as that on Greenland and on almost all of the Antarctic Continent) will, if/when melted, cause run-off into the surrounding ocean water and raise its surface level.

Numerically, (and adding the third dimension) the average ice thickness in Greenland is 1.67 kilometers (0.964 mile) and is on a land area of 2,167,851 square kilometers (836,330 square miles). If that volume of land ice on Greenland is completely melted, it is estimated (by scientists) that the surface levels of the world oceans would rise about 23.5 feet. The probability of that happening to that extent is very low, and at the present rate of ocean waters surface level rising (1.2 mm in 50 years) per NOAA, that complete transformation of ice to liquid state would not happen for approximately 300,000 years from now (2024).

For Antarctica, the average ice thickness on the Antarctic continent is 2 kilometers (1.2 miles), and is on a land area of 13,997,340 square kilometers (5,400,000 square miles). If that volume of land ice on Antarctica is completely melted, scientists estimate that the surface levels of the world would rise about 207.5 feet. The probability of that happening is extremely low (based on the Antarctica Ice Core samples history), and at the present ocean levels rising at the rate of 1.2 mm in 50 years, that complete transformation of Antarctica ice to liquid state would not happen for approximately 2,647,750 years from now. That too is not likely to occur, as the Antarctica Ice Core samples show that there is a cycle from the planet being at its peak hot temperature to its peak low temperature (at which no ice melts) in remarkably similar gradients patterns approximately every 102 years. In essence, the planet would have gone through 2,647 cycles of extreme high temperatures to extreme cold temperatures.

Returning to the surface and atmosphere of the planet, environmental scientists discount all other factors except Man burning fossil fuels to be the dominant factor in determining surface temperature. The cold Aleutian current in the higher latitudes of North America flowing clockwise in the Northern hemisphere mixes with the warmer waters of the Eastern Pacific Ocean. When that cool current flows past the West Coastal region (the states of Washington, Oregon, and California) of the United States, it has a very significant moderation of the temperatures and climate of the US West coast. (Author's Note: Anyone who has driven the Pacific Coast Highway through the Malibu Beach and Santa Monica Beach communities of Southern California will have seen hundreds of young people wearing scantily-clad swimming attire (to sunbathe – "catch some rays" - on the warm sands) and wet-suits (to surf – "hang ten" -in the cold water) thronging the beaches at 8o °F (26.7 °C) while the waters 1oo feet from the beach are flowing at 59 °F (15 °C).)

By contrast, people living in Phoenix, Arizona, must stay inside their air-conditioned homes when the temperature is over 100 °F (37.8 °C) and may have been for 16 straight days. The two years that the Author resided in a suburb of Phoenix. Pets being taken for a "walk" had to be carried to the green areas to avoid their paws being burned walking on the hot concrete pavement. The humidity is mercifully low (10-15% most of the year) in Arizona otherwise the heat index would be a major concern for people living in the desert southwest US. That Heat Index table, developed by the National Weather Service of NOAA is shown again here for Reader's convenience:

NWS Heat Index — Temperature (°F)

Relative Humidity (%)	80	82	84	86	88	90	92	94	96	98	100	102	104	106	108	110
40	80	81	83	85	88	91	94	97	101	105	109	114	119	124	130	136
45	80	82	84	87	89	93	96	100	104	109	114	119	124	130	137	
50	81	83	85	88	91	95	99	103	108	113	118	124	131	137		
55	81	84	86	89	93	97	101	106	112	117	124	130	137			
60	82	84	88	91	95	100	105	110	116	123	129	137				
65	82	85	89	93	98	103	108	114	121	128	136					
70	83	86	90	95	100	105	112	119	126	134						
75	84	88	92	97	103	109	116	124	132							
80	84	89	94	100	106	113	121	129								
85	85	90	96	102	110	117	126	135								
90	86	91	98	105	113	122	131									
95	86	93	100	108	117	127										
100	87	95	103	112	121	132										

Likelihood of Heat Disorders with Prolonged Exposure or Strenuous Activity

Caution | Extreme Caution | Danger | Extreme Danger

(Source: National Weather Service, NOAA)

The weather and climate of the coastal regions (western /Pacific and eastern/Atlantic) of the United States during the cold months of October through March are spared the harsh extreme cold and possible blizzards (high winds blowing snow) that prevail in the upper central portion of the USA. The extreme hot hazards shown in the Heat Index (high temperature and high humidity) above are replaced by the Chill Index (low temperature and high wind velocity) shown below:

National Weather Service Wind Chill Chart

Temperature (°F)

Wind (mph) / Calm	40	35	30	25	20	15	10	5	0	-5	-10	-15	-20	-25	-30	-35	-40	-45
5	36	31	25	19	13	7	1	-5	-11	-16	-22	-28	-34	-40	-46	-52	-57	-63
10	34	27	21	15	9	3	-4	-10	-16	-22	-28	-35	-41	-47	-53	-59	-66	-72
15	32	25	19	13	6	0	-7	-13	-19	-26	-32	-39	-45	-51	-58	-64	-71	-77
20	30	24	17	11	4	-2	-9	-15	-22	-29	-35	-42	-48	-55	-61	-68	-74	-81
25	29	23	16	9	3	-4	-11	-17	-24	-31	-37	-44	-51	-58	-64	-71	-78	-84
30	28	22	15	8	1	-5	-12	-19	-26	-33	-39	-46	-53	-60	-67	-73	-80	-87
35	28	21	14	7	0	-7	-14	-21	-27	-34	-41	-48	-55	-62	-69	-76	-82	-89
40	27	20	13	6	-1	-8	-15	-22	-29	-36	-43	-50	-57	-64	-71	-78	-84	-91
45	26	19	12	5	-2	-9	-16	-23	-30	-37	-44	-51	-58	-65	-72	-79	-86	-93
50	26	19	12	4	-3	-10	-17	-24	-31	-38	-45	-52	-60	-67	-74	-81	-88	-95
55	25	18	11	4	-3	-11	-18	-25	-32	-39	-46	-54	-61	-68	-75	-82	-89	-97
60	25	17	10	3	-4	-11	-19	-26	-33	-40	-48	-55	-62	-69	-76	-84	-91	-98

Frostbite Times

30 minutes | 10 minutes | 5 minutes

(Source: National Weather Service, NOAA)

Some environmental scientists claim that more deaths are caused by extreme hot temperatures than by extreme cold temperatures. In contrast, the Center for Disease Control (CDC's National Center for Health Statistics Compressed Mortality Database, which is based on death certificates), indicates the reverse—about twice as many people die of "excessive cold" conditions each year than of "excessive heat." According to a 2014 study by the CDC, approximately 1,300 deaths per year from 2006 to 2010 were coded as resulting from extreme cold exposure, and 670 deaths per year from extreme heat. However, the two sets of numbers are both likely to be underestimated. According to the 2016 study, The Impacts of Climate Change on Human Health in the United States, "It is generally accepted that direct attribution underestimates the number of people who die from temperature extremes." For example, during the 1995 Chicago heat wave, only 465 death certificates had heat as a contributing cause, while excess mortality figures showed that close to 700 people died as a result of the heat.

Regardless of which temperature extreme is the more dangerous, such are the factual extreme weather conditions that cause concern for humans and pet animals – they must adapt to the weather and climate extremes in the region in which they reside, not just to the scientifically convenient "average global temperature" or to the "average sea surface temperature". An interesting and more valuable contribution for the public is to compare the deaths of humans by the differences in being exposed to the natural weather-related hazards on the planet. The National Weather Service has compiled those statistics and that compilation is shown as follows:

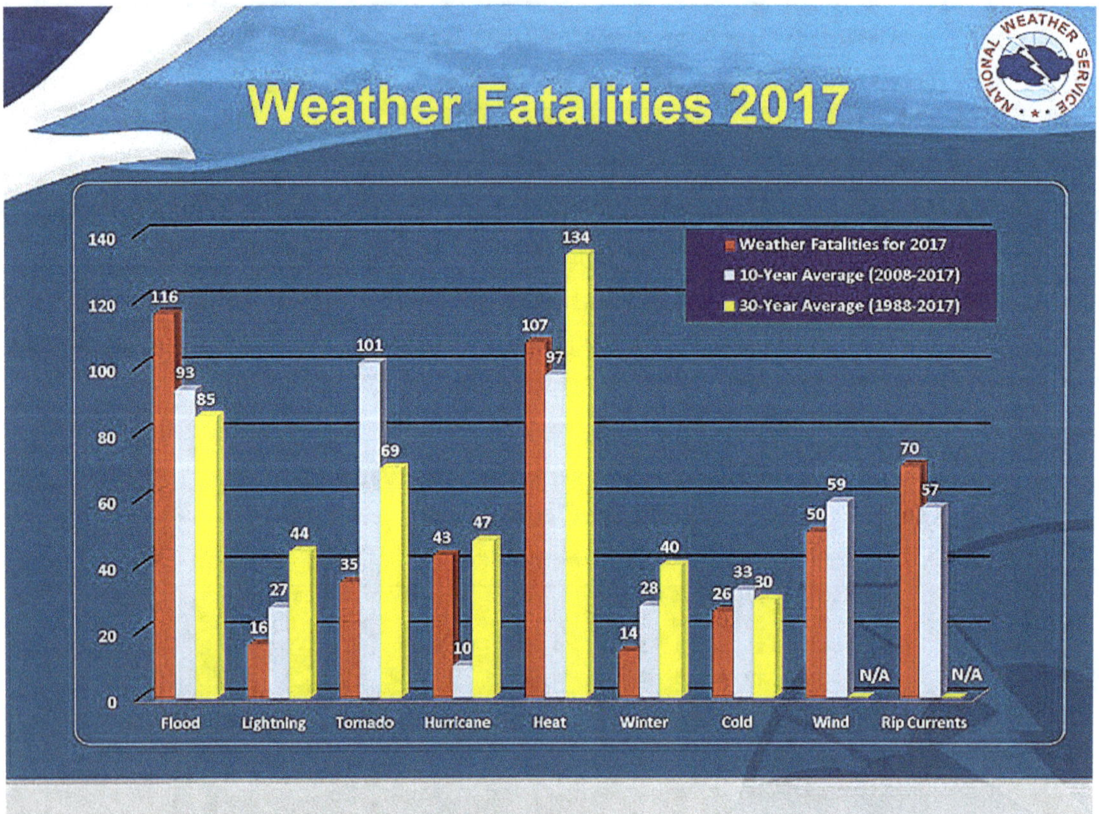

Source: National Weather Service, NOAA)

It should be apparent to Readers that the fatalities caused by uncontrollable natural causes of floods, lightning, lightning-caused wildfires, tornadoes, hurricanes, snow blizzards, ice storms, winds, rip currents, earthquakes, landslides, mudslides, avalanches, and non-Man caused daily and seasonal temperature and climate extremes are the most hazardous phenomena to humans on the planet.

When the heat and cold characteristics of the many hazards are separated from those for the measured natural carbon-dioxide and water vapor in the atmosphere, and the concern about "greenhouse gasses" that are Man-made (a natural to anthropogenic carbon dioxide ratio of about 90 to 1) is set aside, it becomes still more apparent that there are several more ways in which the environment as well as Man are introducing carbon dioxide (officially not a pollutant) and other carbon-containing polluting chemicals into the planet's surface and atmosphere.

The world population lives in latitudinal and geophysical regions where temperatures can vary greatly from one region to another. The latitudinal extremes of a region determine the temperatures to which the population of the region must adapt. For the United States, its latitudinal extremes are 21.3° North (Honolulu) to 47.6° North (Seattle). So as not to offend the Alaskan people, it is known that the latitude of the most northern city in the entire United States is Barrow (now Utqiagnik since 2016) , Alaska at latitude 71.3° North but the population of Utqiagnik is under 5,000. Alaskans are generally noted to be a more cold-tolerant people than those in the contiguous 48 states of the USA and their wardrobes in Utqiagnik reflect that necessity for safety as well as for comfort. The average annual temperature is not a good measure of a region but instead the seasonal variation of temperatures is a far better portrayal of the temperatures that people must adapt to – and choose their personal wardrobe to maintain their body temperature in a temperature range that the person can be considered as "comfortable" 24/7/365 and under all possible weather and climate conditions for that region.

There can be little disagreement that the oceans are the most important physical feature on the planet. The oceans are one of the elements of the Earth as a system that is in a complex set of motion in all three dimensions. Much attention is given to the surface conditions of the oceans and their temperatures, currents, and salinity, as well as with the interactions between the oceans' surface areas with the atmosphere above them. Not enough attention has been given to the escape of geothermal energy through perforations in the outer crust; assuming there is none below the surface of the oceans is a huge error in the analysis process of global warming and climate change, and especially in ocean levels rising because the ocean waters are being heated from below as well as from above.

A large percentage of humans (those with no current respiratory and cardiac health issues) could probably survive the extremely high temperature (12-14 degrees above the current "average global temperature" as calculated by environmental scientists and the hyperthermia) and the hazards of associated ocean surface level rises. A fraction of the human population could possibly survive the extremely low temperatures 1o degrees below the current "average global temperature" in the Equatorial regions (escaping hypothermia effects on respiratory and cardiac issues). Unfortunately, temperatures that low would create "Little Ice Age" conditions where the water on Earth (probably not in the Equatorial regions) would freeze and glacial conditions would destroy agricultural and animal protein food supply for the human population.

Readers of this book are advised that the facts and fallacies of global warming, climate change, and ocean levels rising are as complete in the first portion of this book as deemed appropriate. The foregoing facts distilled from the several different databases are the "Inconvenient Truths" about global warming, climate change, and ocean levels rising. Readers are requested to please "Do not shoot the messenger", at least not until it is shown (in the ensuing chapters) what are the mitigation and planning needs for the future generations' safety, health, transportation, and comfort needs. Instead of Man wringing his hands in despair and fear, those hands can/must be put to better uses.

The "big picture" of Man and his planet, the planet portion – derived from scientific databases that have been updated from the original decision-making time-period of the pre-1970s – has shown the planet's natural effect it has had on itself, and on all the other "tenants" on the planet. Statistics on human fatalities and property damages incurred by the extremes and catastrophes that are documented in history are added to those of nature to generate the reality of what is going on within the big system. Man may be thought of as being "just along for the ride".

It is hastened to be added that during the analysis of the several scientific databases, there was much evidence found of Man polluting his environment. It is that pollution which must be better controlled, if not outrightly eliminated. The evidence also showed that the time has come to stop the use of coal in the energy generation/transformation process because of the coal ash residue containing (from when it was originally mined) toxic and carcinogenic components in the ashes. In any and every technology development, the technology that was appropriate at its start may be overshadowed by the detriments that became known only after using that technology for a long period of time.

The environmental protection scientists' fallacious claim that Man burning fossil fuels is the cause of climate change is regrettable and brings to mind the expression: *"Oh what a tangled web we weave, when first we practice to deceive*". The familiar quote is from Sir Francis Scott's epic poem "Marmion: A Tale of Flodden Field", a historical romance in verse published in 1808. The meaning is accepted to be that when you lie or act dishonestly you are initiating problems and a domino structure of complications which eventually run out of control. There is no inference that Dr. Charles Keeling's dedicated scientific measurements of carbon dioxide in the atmosphere were intended to unethically deceive the public. It was the overexaggeration of Keeling's brilliant works by associates with lesser ethical reasons that the carbon dioxide being emitted from industrial practices since the mid-1850s had profoundly changed and continues to change the climate. Analysis of the Antarctic Ice Core Samples indicates the fallaciousness of the "settled science".

The climate coincidentally had been validly observed to be warming in many regions, but as this book has proven, the warming climate was due more to astrophysical, geophysical, and natural factors primarily and only imperceptibly by the introduction of Man-made "greenhouse gasses". Many photographs were shown of very ominous smoke from coal-burning smokestacks to delude the public. What the smoke exhibited however was particulate matter (and water vapor in the form of steam) being emitted and rightfully needing to be brought under control as pollutants harmful to human health. Carbon dioxide is a colorless, odorless, non-visible gas (to the human sensory organs) and is not judged (in a US Supreme Court decision since carbon dioxide is a naturally made gas vital to plant growth) as being a pollutant.

Further supporting that contention that it is not carbon dioxide that is altering climate is a recently presented satellite image database showing the tracks and intensity (rated on the Saffir-Simpson Hurricane Intensity Scale) of all tropical storms in the past forty years. There is no scientific by-product of any technology that can or will change those tracks and severity levels. But there is much that can be done to mitigate the effects of them and most other natural extremes and calamities on the human population. It is a non-critical priority to fully protect environmental and ecological from natural effects (the coral reefs will have to evolve to survive).

Tracks and Intensity of All Tropical Storms

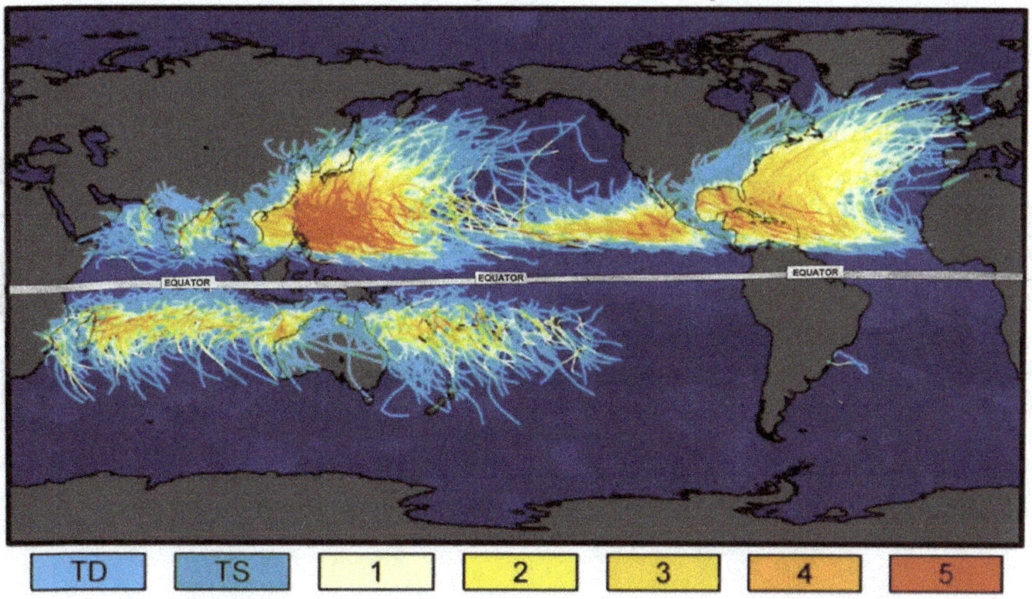

Saffir-Simpson Hurricane Intensity Scale

(Source: NASA)

The Author added the almost horizontal white line denoting the Equator. As a general observation, the Equator is acting somewhat like a double-sided mirror of the tracks of tropical storms. Although the origins for the tropical storms seems to differ in their region of spawning, and the intensity of the tropical storms is not a one-for-one mirror-image, there can be little doubt that once a tropical depression is initiated and increases in intensity as it passes over heated water currents in the oceans, the rotation of the storm is counter-clockwise in the Northern hemisphere and clockwise in the Southern hemisphere. Unless scientifically proven otherwise, a conclusion reached for this book is that the Coriolis Effect and the dynamics and inertial forces caused by the rotation of the planet about an axis normal to the plane containing the Equator are combined to create that difference in rotation direction of the storms rotating in opposite direction in the Northern and Southern hemispheres. The tracking paths are congruent with the ocean currents and prevailing winds.

The largest red blob in the diagram above suggests that the waters of the western Pacific Ocean are warmer than those in the eastern Pacific and of the Atlantic Ocean. A chart shown earlier in this chapter showing the floor of the western Pacific Ocean having many sea mounts and a proliferation of hydrothermal vents suggests the possibility that it is geothermal energy entering the western Pacific waters in addition to the heat energy from solar irradiation close to the Equator that might explain that higher intensity. It is most unlikely that the intensity of the hurricanes is due to Man-made "greenhouse gas emissions" in the atmosphere.

There was no database found, and no logical scientific explanation formed that supports any claim of the heating of the atmosphere (due to the "greenhouse gas effect" or any other natural effect) causing the direction of rotation to be a mirror image about the hot Equatorial region. There is also no yet proven scientific basis for the fallacious claim of "feedback from Man-influenced climate change" is causing the frequency or intensity of the storms to be changing. The seasonal effects are still very evident, although there is starting to be observations and events factually happening that the "season" (for hurricanes and tornadoes) is starting earlier as the ocean water temperatures rise. In the opposite trend is that the leaves falling from trees in the autumn are seemingly changing color and dropping later in the "Fall" season. The "Farmers" Almanac" also indicates the "Wooly-bear" caterpillars are getting fuzzier later in the year because of the summer temperatures trending higher.

It may be worth scientifically pursuing if/how the fact that almost all the "Industrial Revolution" factors and human population factors – both being far more present in the Northern hemisphere than they are in the Southern hemisphere – are contributing to some small degree to explain the difference in the "mirror image" pattern of the tropical storms shown in the satellite imagery. If that Man presence/activities fact is considered, then the fact that there is also a huge difference (by a factor of 9 : 1) in the volume of ice in Antarctica compared to the volume of ice in the Arctic serving to cool the ocean waters in the Southern hemisphere more than the cooling of the ocean waters in the Northern hemisphere. The presence of more clouds and water vapor (based on NASA satellite imagery) – aggravated by the larger presence of geothermal heat energy in the Northern hemisphere causing the ice in the Arctic to be melting faster than the ice in Antarctica in the Southern hemisphere which has less geothermal heating - also is a natural factor having a more dominant effect than any Man-made factor.

Hurricanes, tornadoes, monsoons, earthquakes, floods, wildfires (ignited by lightning), volcanic eruptions, tsunamis, etc. have historically been observed since before humans inhabited the planet, are happening (as this book illustrates), and they will continue to happen regardless of any human action to cause them or to prevent them. What is more important is that humans should pro-actively place themselves in circumstances to not be in "harm's way". The following photograph illustrates the lack of common-sense decisions being made by too many inconsiderate members of the public: (It is they who use the false logic that no matter what levels the ocean may rise to be, the stilts supporting the house can be extended.)

(Credit to unknown photographer – Outer Bank of North Carolina)

Where humans choose to reside is their own personal choice to be made as they are the ones who must make decisions that best suit the safety, health, comfort, transportation, employment, recreational, and economic status of/for their families. However, humans must not reside where there are high risks of the effects of natural catastrophes and requiring endangering the safety of emergency rescue, and recovery personnel in the event of such misfortune. The next photograph reflects such a choice having been consciously made (assuming that this is not the result of having been found guilty of a crime to society and being sentenced to banishment to an isolated island from which there is no escape):

(Credit to unknown photographer)

It is the type of Man-made activity which endangers others that must be reduced and eliminated – the pollution created by Man. An excellent quote from one of the greatest scientists of all time is that from Albert Einstein: "Two things are infinite; the universe and human stupidity; and I'm not sure about the universe."

(Author's Note: It is nearly impossible to describe the experience of interacting with humans and adapting to the different types of housing and population lifestyles and everyday weather conditions (seasonal and extremes) found during the past many decades of the Author's life after residing in the six climatically different regions of the United States: Northeast (23 years in New York City); Southeast (7 years in Atlanta; Georgia, and 2 years in Raleigh/Durham, North Carolina); Southwest (3 years in Fort Worth, Texas, and 2 years in Phoenix, Arizona); Upper Midwest (21 years in Dayton, Ohio); Coastal Southwest (38 years in Los Angeles, California); and heavily forested Northwest (1 year in Bonners Ferry, Idaho). It is said that "Experience is the best teacher" and that is very wise philosophy. That nomadic experience alone does not scientifically qualify the Author to expound on how global warming, climate change, or oceans levels rising affects the human population – but it has qualified the Author to understand (in a "been there, done that" mode) what is important to humans. (Another tongue-in-cheek expression frequently heard at high-level technical meetings – regarding the definition of the "experts" in attendance: "An expert is one who is more than 50 miles from home.")

The environment can be paradisical much of the time or terribly cruel some of the time to humans. Humans can be both constructive to, and very destructive to, the environment in which they live and survive.)

While conducting the analysis of the databases of information and exploring the many possible causes for global warming, climate change, and oceans' surface levels rising, an open-mind was kept to not affix responsibility or blame on what humans are doing to their environment, or what the environment is doing to the human tenants on the planet. Having arrived at several conclusions as to "who shot who and where, and with what motive", the scientific facts (the pieces in the jigsaw puzzle) became clear that the US public has been intentionally misled (not just by a few unethical scientists with a different agenda of his own, but by the news media, governmental regulators, politicians, activists, and by unqualified teachers of science to younger members of the population). It is that set of observations and interim conclusions that were factually determined during the review of the large amount of available scientific information from many different sources that will be shown next – before proceeding to present well-founded suggestions as to "what can be done and at what rate of accomplishment?" in the remainder of this book.

A few interesting charts (from the US Energy Information Administration) illustrating the status of where the US has been and is projected to be in the near future are shown next. The first "pie chart" gives a concise "picture" of the percentage of each of the type of fuel technologies that were being consumed in 2019 in the United States.

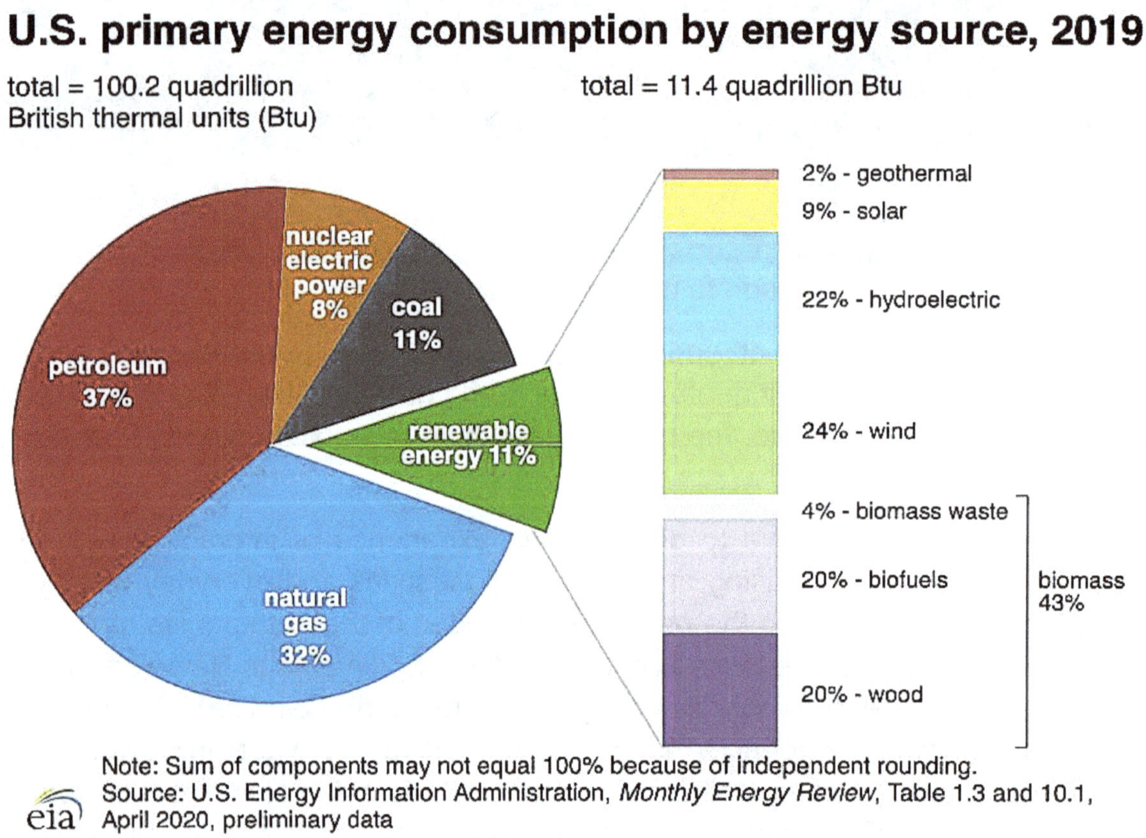

(Source: US Energy Information Administration)

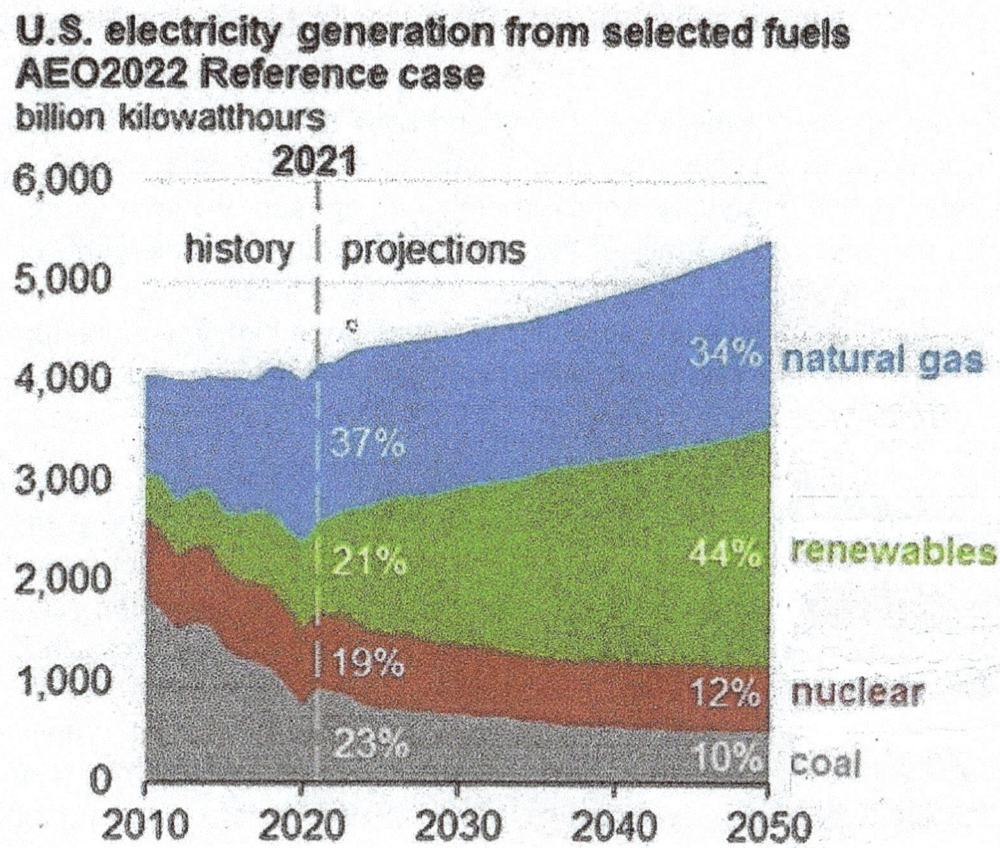

(Source: US Energy Information Administration)

The second chart may be a bit misleading when compared to the first chart (although both are from the same source the Energy Information Administration) in that the large use of petroleum (34% of the "pie") is not shown as a category in the second chart. That is either an administrative error or a signal that by 2050 all petroleum fueled forms of transportation will have been replaced by all electric vehicles. Most likely it is an oversight as the use of petroleum was shown to be 34% in 2019 and would not have been reduced to o% just two years later. The trends in usage are more important.

The three existing forms of energy (natural gas, coal and nuclear all show a decrease in use (natural gas is reduced from 37% in 2021 to 34% in 2050, coal is reduced from 23% in 2021 to 19% in 2050, and nuclear is reduced from 19% in 2021 to 12% in 2050). The use of renewables is projected to be more than doubled (from 21% in 2021 to 44% in 2050).

Wind turbine power is dependent on the wind blowing, and no country including the USA has a constant wind. Therefore, there must be a power storage system to be developed as a back-up mode for storing wind energy collected when the wind was blowing and excess stored energy needs to be used for meeting human demands in periods of the day and the season regardless of winds blowing or not. That is not a new concept, as many towns and cities use windmills to pump underground water into storage tanks or reservoirs at an elevation when the wind is blowing. When the windmills are dormant, the water in the elevated storage is used – capitalizing on the ever-present gravity – to flow with adequate pressure into the small businesses and the residential facilities of the town. Any maintenance on the wind turbine system can be conducted during periods of inoperability. There has been reported instances of the blades when rotating injuring and killing passing birds, and many local residents complaining of the low humming noise being damaging to human aural system. The wind power approach is not adequate for the larger industrial cities which generally has a high population with "creature comfort" needs that are seasonal, as well as constantly high industrial power needs and demands.

Solar power is displaying a very similar set of pros and cons to that with wind power. Solar power is well known to function only when solar irradiation is impinging on the solar collector panels. Because the Earth rotates one rotation every day, approximately half of each day is in darkness and there are no solar rays to be collected. Additionally, if the weather is cloudy or inclement, or if it is raining or snowing, again the solar rays are not being collected. The use of solar radiation is perhaps best suited to small applications such as residential use, but it clearly is not a highly useful source in industrial us because of its unreliability. There also is a maintenance problem keeping the solar panels clean from any dust blowing, especially after a rain shower.

Other more mature technologies have overcome their original "cons" – hydrodynamic dams for flood control and clean generation of electricity, but with the climate changing many dams are not being provided sufficient water because of drought conditions to remain as reliable sources of energy. Nuclear energy facilities – those powered by Uranium - provide clean, reliable, energy independent of the weather or climate and have a place in future use of alternative energy sources, but only if accidents such as that at Three-Mile Island and at Chernobyl in the Ukraine/Russian region are made 100% possible to recur. The use of thorium as an alternative to the more hazardous uranium is emerging as a better technological choice in many countries. Geothermal energy holds the highest potential for energy sources that are clean, reliable, and use a "fuel" that is safe and abundantly available as a regional asset.

A common mistake that is made in the development of a new technology is that it is developed for a current need and is "life-limited". The human population is exponentially growing – at a much faster rate than the improvement of mature technologies (coal burning being the best example of that) or development of new technologies for safety, health, comfort, transportation, employment opportunities of the population both currently and near term (when the population size has doubled or tripled).

To this point in this book, there has been much data presented and much description and discussion regarding interpretation of the mountain of information relative to the environmental science community hypothesis that *"When everything else is eliminated, it is settled science that Man burning fossil fuels and creating "greenhouse gas emissions is the dominant cause for global warming."* An in-depth system engineering review of several databases of data for those "everything else" factors reveals that the environmental science was not justified to summarily eliminate or ignore those databases.

The independent review of the databases has been done by an "outsider" of the environmental science community – as a peer professional and with the aim of offering constructive criticism. Admittedly, this "outsider" is one who has investigated many major accidents with fatalities and for determining root cause to reduce the possibility of an accident being repeated for the same root cause. In the accident investigation there is a basic credo: "To make a mistake is human. To repeat the mistake is unforgiveable."

The science profession explores observations and most often with an aim to develop a formula or a model for being able to predict how that observation made is trending and will reappear in the future. The engineering profession applies those scientific observations and formulae and models in the form of technology with the aim of making that technology become of utility to humans and protect their safety, health, and efficiency. Scientists belong to a segment of specialized interest – environmental scientists have the aim to explore and protect the environment. Engineers have the same aim except to apply their talents to protecting the human population. That should serve to at least partially explain to Readers why the conclusions being presented seem to be in the category of "naysaying" or "denying" by "outsiders" they are not intended to be as such. The conclusions are what the databases indicate factually.

One such "outsider" is the Author, who is a retired system safety engineer and developer of many technologies applied to major weapons systems for the Department of Defense as well as to commercial aircraft in worldwide use. The Author summarizes his experience as Chief System Engineer of the F-16 fighter for the USAF in "The F-16 Fighting Falcon Multinational Weapon System, 1972 to 2019".

An extract of a letter from the Chairman Dr. John L. McLucas (former Secretary of the Air Force) of the President's (Ronald Reagan) Task Force on Aircraft Crew Complement to USAF Headquarters at the conclusion of an intensive six months (January to June 1981 at Department of Transportation facilities in Washington, D.C..) review of a proposed new commercial aircraft with an advanced technology cockpit. This letter and the F-16 book substantiate the Author's qualifications for accomplishing major complex system engineering analysis of large databases:

> **As you may know, on July 2 our Task Force completed its study and submitted its report to the President and to the Secretary of Transportation on questions concerning the safety of operating certain jet transports that are being and will be introduced into commercial service over the next decade.**
>
> **The Aircraft Crew Complement study centered on the ability of a flight crew composed of two crew members instead of the conventional three crew members on the flight deck safely operating the aircraft, with advanced technology instruments and displays, in heavy airport traffic and under harsh environmental conditions. Carrying out our study and preparing our report would have been difficult and perhaps impossible had it not been for the excellent cooperation of the Air Force which made available to us a wide range of talent and expertise.**
>
> **This is to thank you specifically for making available Mr. Herbert A. Hutchinson, Chief, Flight Technology Division, ASD Deputy for Engineering, Wright Patterson AFB. Mr. Hutchinson served as leader of one of our panels that was responsible for studying the Federal Aviation Administration's flight safety certification process for new commercial aircraft. He provided excellent leadership to the Panel staff, and his outstanding analytical talents and extensive system safety training and military aircraft crash accident investigation experience contributed very substantially to the work of the Task Force.**
>
> **My colleagues, Lt. Gen. Howard Leaf of the Air Force and Mr. Fred Drinkwater of NASA, and I thank you and other senior officials of the Air Force for the strong support given to our Task Force and for providing the invaluable system safety engineering services of Mr. Hutchinson.**

(Source: Letter from "The President's Task Force on Aircraft Crew Complement" by Dr. John L. McLucas (former Secretary of the Air Force) to the Vice Chief of Staff General Robert Mathis) at HQ USAF)

Several key observations were formed during the prior "big picture" review and unbiased scientific analysis of the more important databases (shown in the preceding four chapters) on the topics of global warming, climate changing, and ocean levels rising.

The addition of meteorology temperatures by different organizations to that of the environmental scientists tracking of the amount of carbon dioxide in the atmosphere, caused by the environmentalists' hypothesis that Man burning fossil fuels and creating the "greenhouse gas emissions" that are the dominant cause for the global warming" have generated a valid database of global average temperatures for the period from 1850 to 2015 showing the magnitude of those changes:

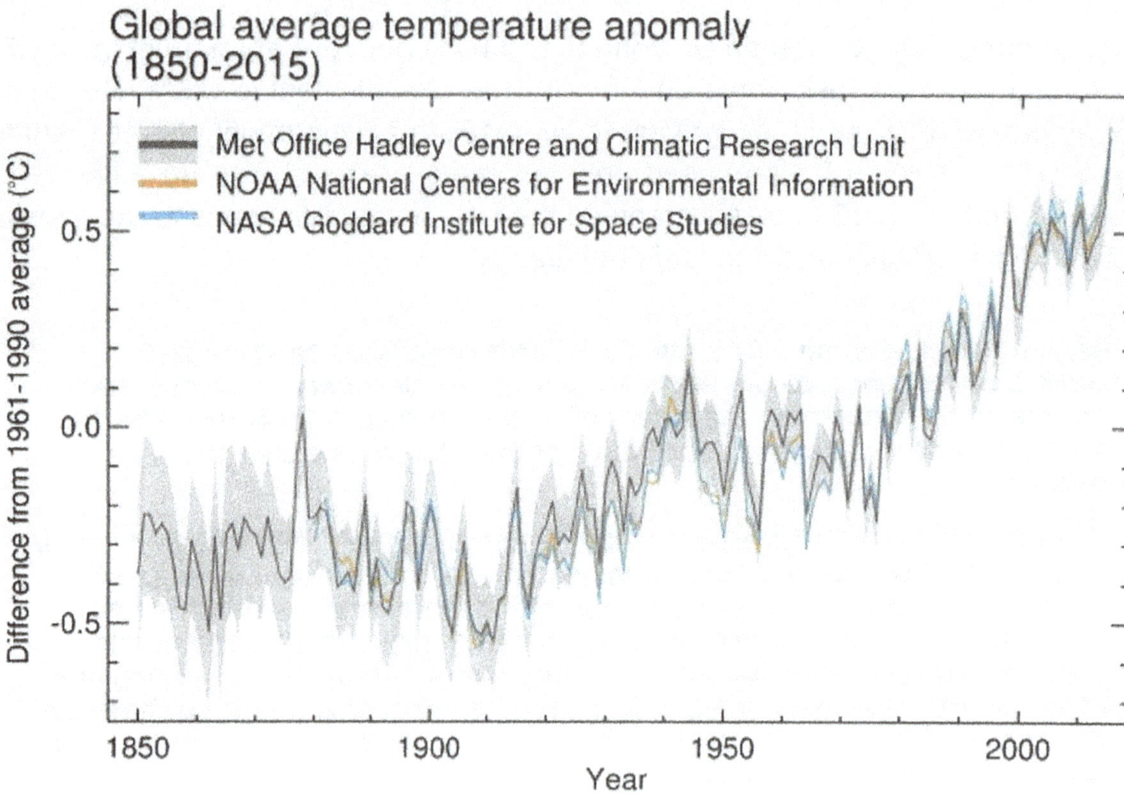

(Sources: NOAA, NASA, Meteorology Office Hadley Center, Climatic Research Unit)

The United States and other developed countries embarked on massive technology development of solar power and wind power added to the existing hundreds of dams to harness hydrodynamic power and reduce the dependency on the fossil fuels – coal, oil, and natural gas. Nuclear power technology facilities were built and because of a few minor accidents – raising the specter of nuclear catastrophe which did not materialize – nuclear power was largely truncated and several nuclear facilities closed rather than to be upgraded for adequate safety. The EPA joined the "crisis" movement that was escalated to be a global crisis, and the United Nations formed the Intergovernmental Panel on Climate Change and the "snowball" effect to reduce Man's dependency on fossil fuels began in the early 1970s.

In 2014, the International Energy Agency (IEA) issued a "report card" as to how the developed countries were progressing with the "combat on climate change" to control the "grim outlook" for the World because of the "greenhouse gas emissions" the public has generated:

Changes in Global Energy Consumption and Other Factors
(Change from 1973 to 2014)

	Technology Used	1973 % of Total	Change %	2014 % of Total
FOSSIL	Coal	24.5	4.1	28.6
	Oil	46.2	-14.9	31.3
	Natural Gas	16	5.2	21.2
	Total Fossil Fuels	86.7	Net -5.6%	81.1
	Hydrodynamic	1.8	0.6	2.4
	Nuclear	0.9	3.9	4.8
	Biofuels & Waste	10.5	0.2	10.3
	Other Solar Wind Geothermal	0.1	1.3	1.4
	Total "Green" Fuels	13.3	Net 6.0%	18.9
	TOTAL - All Fuels	100%	Net 0.4%	100%
	Energy Used - Toe (Tonnes oil equivalent)	6.101×10^9	225%	13.699×10^9
	10^{20} joules	2.56	225%	5.74
	Global Population	3.7×10^9	186%	6.9×10^9
	Use Index - BToe/BPersons	1.649	20.4%	1.985
	Surface Temperature Anomaly (NASA, NOAA, WMO)	-0.01 °C	0.53 °C	0.52 °C

International Energy Agency (IEA), 2014)

(Source: International Energy Agency, IEA)

Many countries reduced their use of coal as an energy source but the larger countries in the Asia Pacific region did not, and the Mideast countries, which were not dependent on coal for most of their power needs, had little objection at UN IPCC conferences to reducing the dependency on coal but refused participation in any drastic movement to eliminate the entire fossil fuel technology.

In the "big picture", the summary above reflects that in the net the global consumption of fossil fuels (shown in the yellow background) was reduced by about 5.6% between 1973 and 2014 but in 2014 remained in consumption as 81.1% of the world energy fuel sources. There were minimal increases in the use of hydrodynamic and nuclear power, a negligible change in the use of biofuels and waste, and a small increase in the use of containing hundreds of volcanoes active and inactive). The San Andreas Fault in

solar, wind, and geothermal power but they continued being a disappointingly small percent of the energy demands/needs of the world population.

All categories of non-fossil fuel technologies (shown in green background) had an increase in consumption between 1973 and 2014 such that the total change in percentages of net increases and net decreases showed a small net increase of 0.4%. In the "big picture" perspective, the world does not accept that the use of fossil fuels creates a global warming and climate change "crisis". That is the same conclusion the Author comes to after a thorough analysis of the other factors that are causing the changes in the planet's surface temperature. There is no question that the surface temperatures are changing but there is the observation made by many others in the STEM professions that cannot accept that those changes which are seen to be extremely small constitutes the "crisis" that the environmental scientists' claim is due entirely to Man burning fossil fuels.

The percentage type analysis of the IEA is overshadowed by the consideration (shown in the red background color) that the total energy used by the world population increased by 225% between 1973 and 2014. That large increase in consumption triggered the Author to research what else changed so largely between 1973 and 2014. Three possibilities were researched (from a system engineering perspective) – one is the growth of the population having increased, second was whether the world population is becoming less efficient in the use of energy and needs more energy at less efficient use, and thirdly had the planet surface temperature increased sufficiently (as a "crisis") to cause the 225% increase in energy consumption.

The largest factor was found to be that the world population had increased 186% between 1973 and 2014, which accounts for most of the 225% increase in energy consumption. The second largest influencing factor was noted to be in the increase shown in the use index of dividing the energy used by the population count. The use index was seen also to increase by 20.4%, but that does not necessarily mean that there is inefficient use by the public but rather that as the world becomes more technology dependent especially in the area of electricity use for many new appliances, tools, recreation and sports gear, and more electrical use in their different forms of transportation – especially the move toward electric vehicle use. That form of transportation technology advances are discussed (i.e., the pros and cons prioritized with respect to their effect on the environment and on the human population) in a later chapter of this book.

The Global Average Temperature chart shown prior to the Changes in Energy Consumption tabular summary reflects a surface temperature anomaly of -0.01 °C in 1973 increasing to a 0.52 °C anomaly in 2014 for a net increase in Global Surface Temperature of 0.53 °C in 41 years. The early chapters of this book conclude that variable (planet average surface temperature) is of scientific interest perhaps but in the "big picture" of astrophysical and seasonal variations in surface temperature varying between + / - 25 °F (14 °C) daily, + / - 40 °F seasonally, varying about 80-100 °C (44 – 55 °F due to latitude, and elevation, plus a variable depending on location of a nearby large body of water (ocean, sea, or gulf) having warm and cool currents far overshadows the importance of a global average anomaly temperature change of 0.053 °C (0.095 °F) in the 41 years between 1973 and 2014. That cause for concern by NOAA being at "crisis" level is believed to be an environmental concern but of insignificance to human, safety, health, comfort, and any minor effect on the infrastructure. The inadvertent revelation of that "grim outlook for aquatic life and anemones on coral reefs bleaching white" was revealed in the following chart:

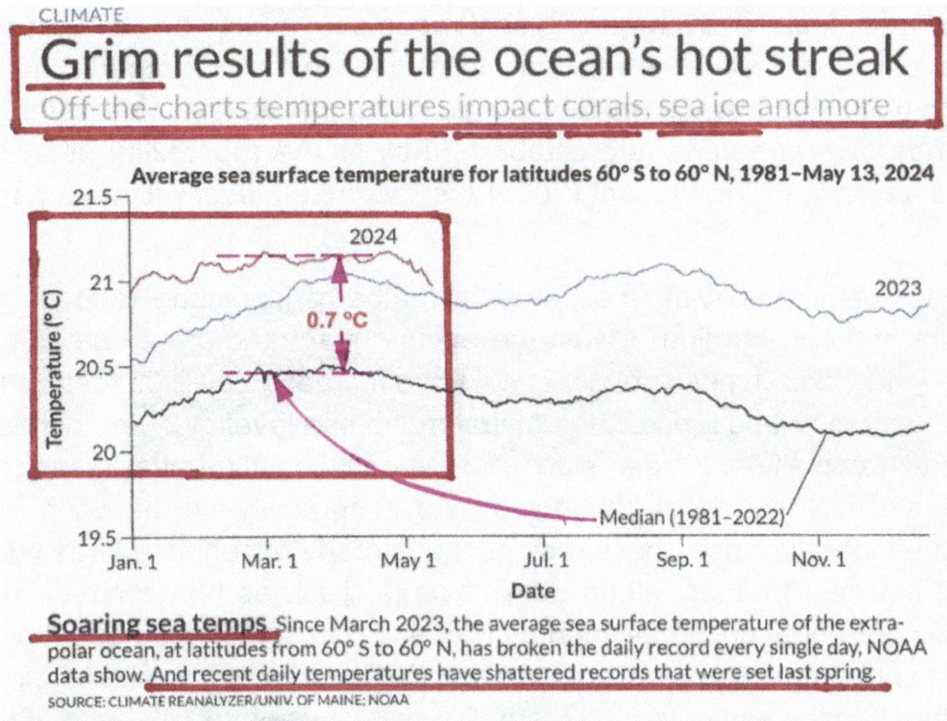

(Source: NOAA)

When a contained volume of water is heated, its contained volume increases. If the average sea surface water temperature is "soaring" (per the NOAA data shown above), there should be an associated increase in the volume of the water contained with the higher temperature. NOAA is concerned that the oceans' levels are rising and pose a flooding hazard to all low-elevation coastal regions of continents and especially to the population on the many islands in the oceans of the planet. NOAA has analyzed the World Meteoroloical Organization database for the global mean sea level surface variation per year since 1993. The following illustration is the graphical database that has been generated for the years from 1993 to 2019:

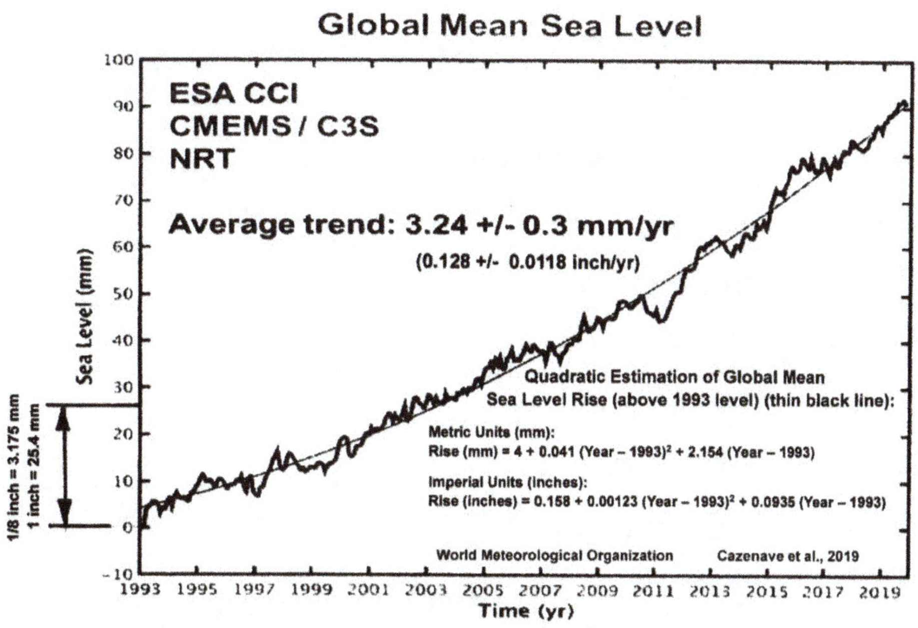

(Source: World Meteorological Organization database analyzed by NOAA and by Author)

NOAA presents the mathematical average of global mean sea level to be trending at the average rate of 3.24 millimeters per year. Using the relationship of 1/8th inches in Imperial units for every 3.178 millimeters in metric units, the average trend of the global mean sea level rise is very close to 1/8th inch per year. At that linear average trend rate established by NOAA (approximately 3.24 millimeters per year), in 2100 – 75 years from the present (2025) the mean sea level will have risen 243 millimeters (9.6 inches)

A more precise trend is estimated with an exponential mathematical model (shown by the Author in the lower right quadrant of the illustration). Using the exponential trend, it can be calculated that by the year 2100, the rise in the global mean sea level will have increased by 2.02 feet above the 1993 level. Although that means there will be a rise in the global mean sea level of 24.24 inches (2.02 feet) or 0.6 meter by 2100, there is ample time to determine what regions (coastline cities, seaports, islands, deltas) may need to have sea walls or dikes constructed. As is suggested to be done by regions/cities to prepare for surface temperature rise, those regions that already experience large tidal sea level cycles by as much as 20 feet and frequent storm surge events (such as hurricanes in the Gulf of Mexico coastal cities or Bay of Bengal countries such as Bangladesh) should increase the height above tidal action and/or storm surge maximums by that two feet. Suffice it to say, ocean levels rising are not at a crisis rate, but neither can the estimated 2.02 feet (0.6 meter) rise by 2100 be totally ignored.

An equally disturbing finding was that of the impact the fallacious emissions controls of the Environmental Protection Agency that were imposed on the largest steel producer in the United States, that of United Steel Corporation. After years of being penalized for emissions violations (the cost of which were passed along to consumers), US Steel divested itself of its capability to produce virgin steel and closed most of its facilities for doing so. Instead, US Steel now produces recycled steel and imports the virgin steel products from other countries – more than half of it from Canada, Mexico, and Brazil. The extra cost for importing the steel rather than producing it is also passed along to consumers.

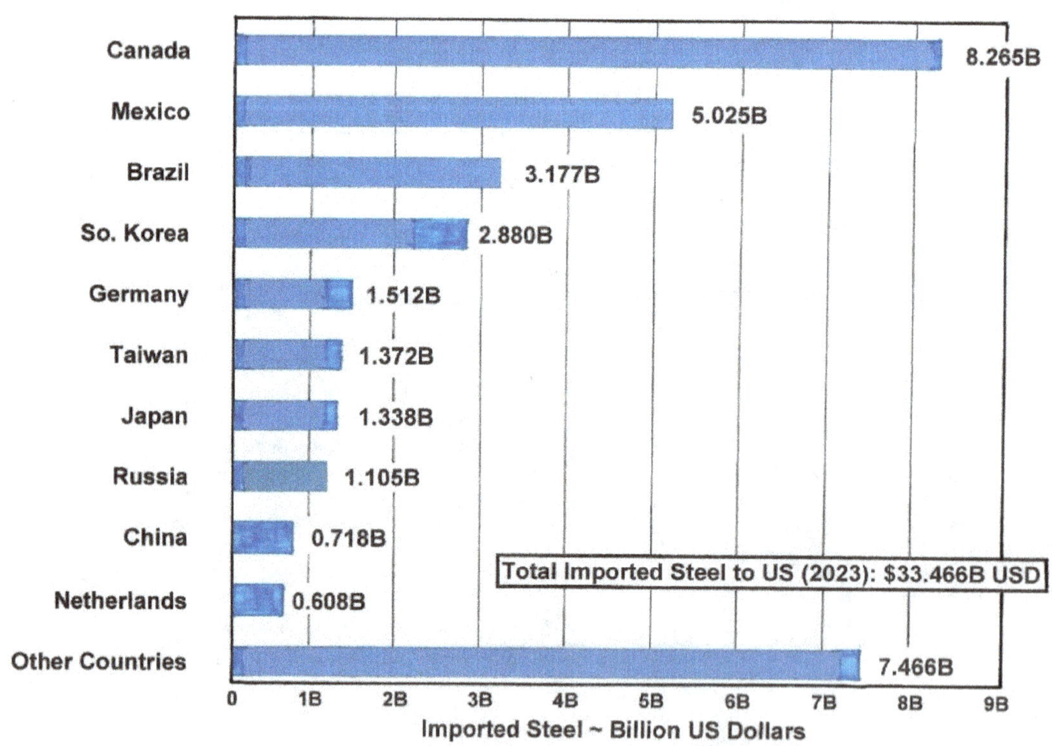

(Source: U.S. Steel Corp.)

The Author is totally in favor of the EPA forcing any corporation – "big" or small - to eliminate emissions generated during the creation of their product that are known to pollute the atmosphere, the land, or the water resources of the United States or its protectorates. There must be scientific measurements made to clearly identify what the limits of those polluting emissions must be that factually are a hazard to the safety and health of the human population as a high priority, and to the environment (land, water, and atmosphere) as a near-equal but secondary priority. However, penalties for "greenhouse gas emissions" levied under the guise of controlling climate change are fallacious and result in the public shouldering the unjustified costs for emission controls penalties or failing to meet any "cap and trade" regulations.

With the revelation that the total concentration of carbon dioxide in the atmosphere is measured to be that as shown in recent valid measurements:

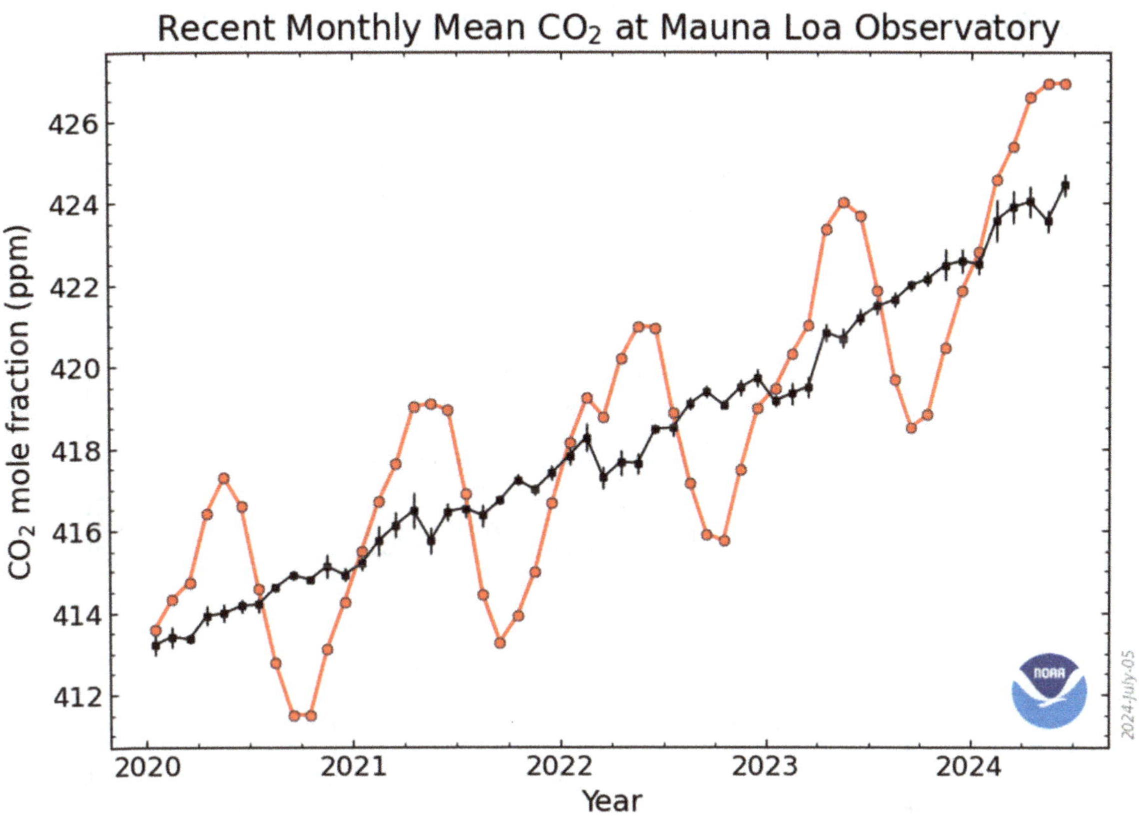

(Source: NOAA)

This book has shown (in previous chapters and earlier in this one) that there is a profound effect of latitude on the surface temperature on the planet, a secondary effect of elevation on surface temperature, and a less definable effect of the proximity of a location on land to an adjacent large body of water (flowing with either a warm or a cold current along the coastal area of that region). Meteorological records from the World Meteorological Organization (WMO) have confirmed temperature changes for 40 cities that the Author has requested the data. Based on the WMO records for those 40 cities (20 in the Northern hemisphere and 20 in the Southern hemisphere, the two graphical summaries for both hemispheres are repeated here side-by-side to show the effect of latitude on average annual temperature (averaging masks the seasonal effect on the temperature to which humans are exposed but is a standard procedure used by scientists studying the longer term effects of climate change).

There can be no refutation by environmental scientists that the latitude effect on surface temperature is far more representative of the real-world circumstances than is the overly simplistic one-size-fits-all "average global temperature" method being used in analyzing the cause(s) of climate change. The red line linearized expression of that latitude effect applicable to both the Northern hemisphere and the Southern hemisphere (shown in graphical form near the start of this chapter) and indexed to the temperature of the Equator can be approximated mathematically (i.e., "modeled") for any latitude between 60 o north or south from the Equator (expressed in decimal format) as:

Imperial Units: Temp (°F) = 90 °F – 0.78 Lat°
Metric Units: Temp (°C) = 32.2 °C – 0.43 Lat°

A refinement of the above latitude effect modeling accounts for the average elevation of the 40 cities selected and the effect of elevation follows that as defined by the International Civil Aircraft Organization (ICAO):

Temperature Lapse with City / Region Elevation
(International Civil Aircraft Organization, ICAO) Standard)

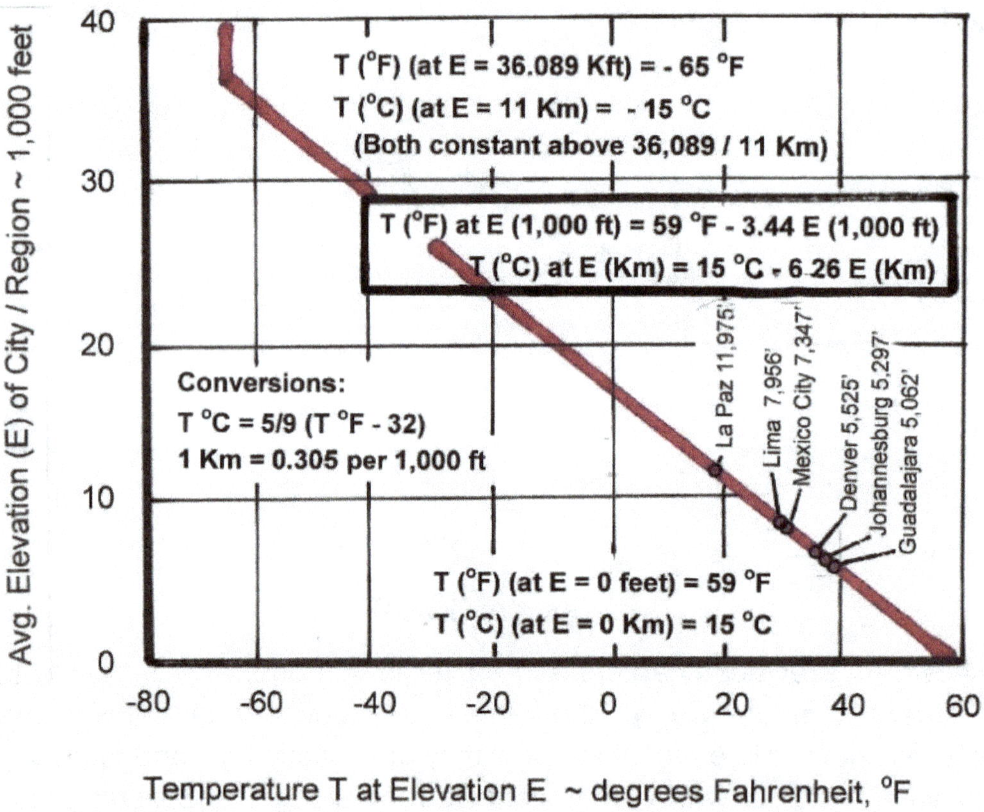

(Source: International Civil Aircraft Organization, ICAO)

The elevations of populated cities are well below the tropopause altitude of 11 kilometers (36,089 feet) and therefore only the lower diagonal-sloped portion of the red line need be considered for the effect of elevation on the average annual temperature of populated cities. The mathematical approximation of the red line in both Imperial Units and Metric Units is shown in the bold outlined box. Mathematically adding that elevation effect to that of the latitude effect on the average annual temperature is a worthwhile improvement in the analysis and modeling of the causes for surface temperature of the cities at any latitude and having any average elevation:

Imperial Units: Temp (°F) = 90 °F - 0.78 Lat° – 3.44 E (Elevation in 1,000 feet)
Metric Units: Temp (°C) = 32.2 °C – 0.43 Lat° – 6.26 E (Elevation in kilometers)

The derivation of the LEO Method described earlier in this book provided a far closer estimation of the average temperature of a region at any latitude and at any elevation above sea level than that used by the environmental scientists – the one-size-fits-all "average global temperature" (approximately 60 °F (26 °C). However, as time permitted, the Author found that while the temperatures estimated by the early LEO method for regions between latitudes 10° and 60° north and south of the Equator, the regions at latitudes between the Equator and 10° north or south gave disturbingly large off-estimates by as much as 8 °F (4.4 °C) from the average annual temperatures measured by the World Meteorological Organization. By setting the temperature between latitude 0 and latitude 10° both north and south of the Equator to be a constant 84 °F (29 °C), there was a much better estimate made for the average annual temperature in that region.

It was also realized that the estimate of average annual temperature at the North and South Poles (90° North and South) the average annual temperatures were found to again be much in error (compared to the meteorological records of the World Meteorological Organization – but by a very different number of °F. The temperature error with the early LEO Method was about 25 °F warmer estimate for the North Pole (90° N) and about 45 °F warmer for the South Pole (90° S).

The Author – humbly - then realized that he had stumbled onto a geophysical observation made by philosopher Copernicus (in the late 1400s) and astronomer Galileo (in the early 1600s) that the tilt angle of the Earth could be causing the average temperature of the planet at the South Pole to be very much colder than the average temperature at the North Pole. It was realized that the World Meteorological Organization database was scientifically correct and was in fact recording temperatures embodying the actual effect of obliquity – the angle solar insolation makes when impinging on the Earth's surface. Obliquity was observed telescopically by Galileo and confirmed to be occurring by the astrophysical mathematician Johanes Kepler mathematically in the early 1600s. The more modern astrophysical mathematician Milutin Milankovic had incorporated Kepler's obliquity angle in the calculations he and his mathematics team had conducted in the mid-1930s.

A further refinement of the LEO method for estimating the effects of elevation and/or ocean proximity on the average annual temperature of a region or city also became necessary after careful comparison of the meteorological database for three cities having different combinations of elevation and ocean proximity locations – and being located either inland or on a coastline. To introduce the refinement for ocean proximity, the following illustration indicates the ocean currents affecting the cities of Cape Town and Durban in South Africa:

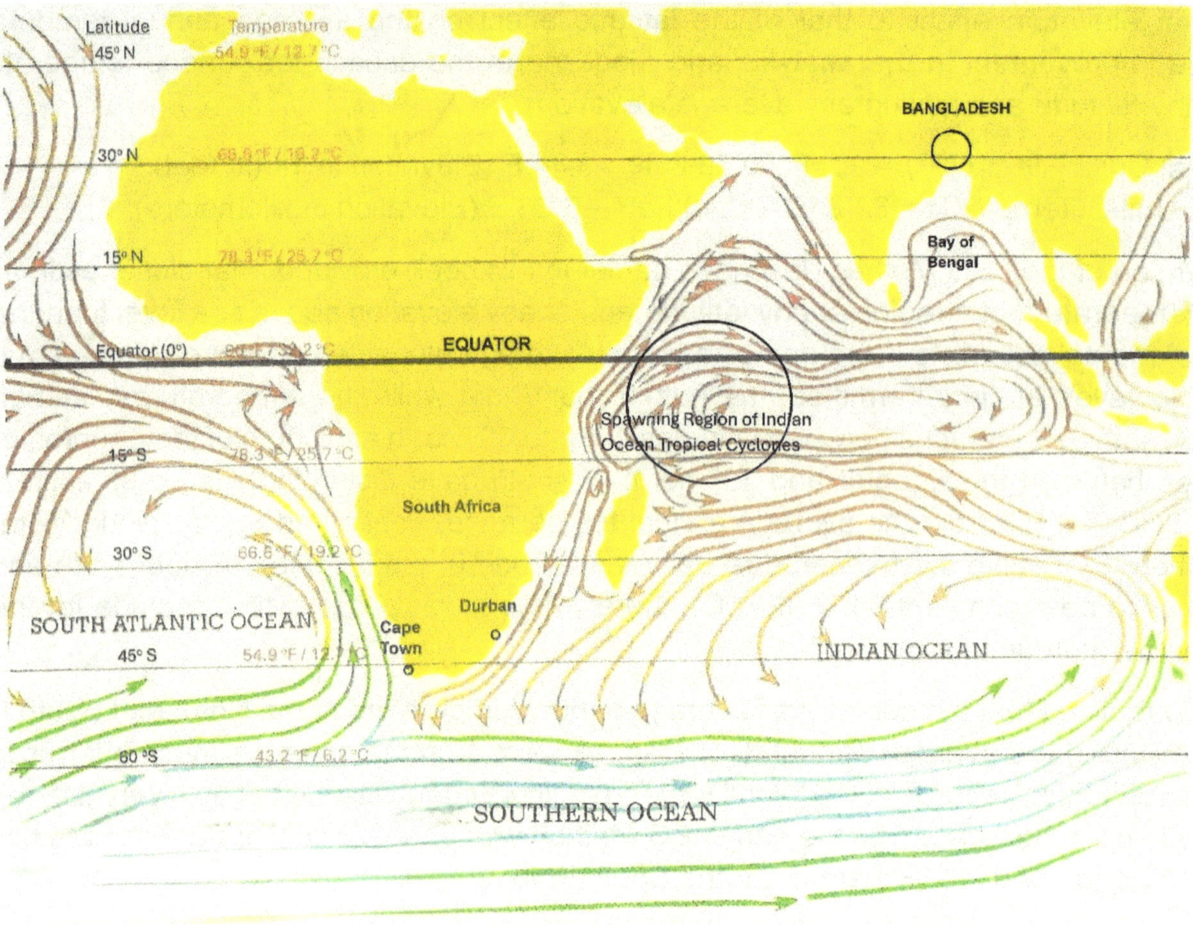

(Source: Atlas of the World, Rand McNally with Author notations)

Shown in the diagram above is that Cape Town – at latitude 33.6° South and having an elevation of 174 feet (53 meters) - located on the southwest tip of South Africa is subjected to the influence of the cold current of the Southern Ocean. Durban – at latitude 29.9° South and having a higher elevation of 5,809 feet (1,771 meters - located on the southeast tip of South Africa is subjected to the warm currents of the Indian Ocean. The road distance between Cape Town and Durban is about 1,015 miles and about 1,000 miles "as the crow flies". The respective average annual temperatures of Cape Town and of Durban might be expected to be only slightly different because the more southern latitude (and cooler effect of latitude) of Cape Town is possibly offset by Durban being at the lower latitude (and warmer) but at more than 5,600 feet higher elevation (and cooler effect of elevation).

Researching the World Meteorological Organization database revealed that the average annual temperature for Cape Town is 62.6 °F (17.0 °C) and for Durban is 75.0 °F (23.9 °C) – reflecting a large, unexpected difference of average annual temperature of more than 12 °F (6.7 °C). The question arose – there are three different effects embodied in the WMO database on temperatures – latitude, elevation, and ocean proximity – and how much has each influenced the factual annual average temperature in the very reliable WMO database? It was decided to estimate the effect of latitude difference with the LEO method and that calculation suggested that at most the effect of small difference of 3.7° latitude was about 3 °F or 1.7 °C. That left about 9 °F (5 °C) to be attributed to the combined effect of proximity to ocean and the effect of elevation (Durban being almost a mile higher in elevation than Cape Town). The average annual temperature for the ocean waters off Cape Town and Durban in close proximity to their respective oceans (Southern Ocean and Indian Ocean) was sought and gleaned from several different organizations (meteorological and travel bureaus) to assist the analysis. A comparison then indicated almost an exact correlation between the average water temperatures and the average annual temperatures for the two cities with their respective meteorological (land-based) data, suggesting that the effect of elevation is not a significant factor relative to the effect of being in close proximity to an ocean (with either warm or cold current).

The following tabulation is presented to make the comparison in number differences easier to interpret:

Comparison of Meteorological Records and LEO Estimation Method for Temperature at Cape Town and Durban, South Africa

(Latitude and Ocean proximity effects on coastal cities)				Average Annual Temperature	
Location	**Latitude**	**Elevation**	**Ocean Temperature (Avg. Annual)**	**Meteorological Database (WMO)**	**LEO Method Estimate** (Latitude only)
Cape Town, S Africa	**33.6° S (-33.6°)**	**174 ft 53 meters**	**62.7 °F / 17.1 °C** (Southern Ocean)	**62.6 °F 17.0 °C**	**65.1 °F 18.4 °C**
Durban, S. Africa	**29.9° S (-29.9°)**	**5,809 ft 1,771 m**	**74.8 °F / 23.8 °C** (Indian Ocean)	**75.0 °F 23.9 °C**	**68.1 °F 20.1 °C**

Note: Road distance between Cape Town and Durban is about 1,015 miles (1,634 kilometers) but both are on the coastal shores of an ocean

(Source: World Meteorological Organization Database, compiled by Author)

To better evaluate the effect of elevation alone without a city being high in elevation but inland and without close proximity to any ocean temperature effect, the latitude and elevation effect portion of the LEO Model for estimating average annual temperature was applied to the city of Denver, Colorado – which is at about 10 degrees different in latitude than Durban (at 29.9° in the southern hemisphere) compared to Denver (at 39.7° in the northern hemisphere). Their respective elevations are nearly the same (Durban at 5,809 ft / 1,771 m and Denver at 5,279 ft / 1,609 m), however Durban is on a coastline of the warm Indian Ocean and Denver is inland in the Rocky Mountains and with no large body of water in proximity. There was uncertainty if the effect of elevation is the same for a city/region in or not in the proximity of a large body of water. It was also considered probable that the large areas within the municipal limits of major cities covered with concrete and/or asphalt (where the solar radiation is absorbed more than that of areas not covered by concrete or asphalt) offsets the cooling of increased atmospheric cooling with elevation.

The results of those comparative calculations revealed that 1) there is a substantial difference shown in the meteorological database (which reflects the natural result of the combined effects of latitude, elevation effect, and ocean proximity) compared to the estimates from the LEO Estimation Method derived earlier; 2) the elevation effects for Durban (5,809 feet above the warm Indian Ocean coastline) appeared to be overcorrected for the amount of elevation effect using the -10 °F (5.6 °C) correction for elevations above 5,000 feet (about 1 kilometer) of the LEO method; and 3) the reduction of 10 °F (5.6 °C) to 5 °F (2.8 °C) for the elevation effect of Denver (at 5,279 feet elevation and inland in the Rocky Mountains – far away from any ocean currents).

Considering the results of those comparative calculations of the LEO Method estimates for average annual temperature with the actual measurements from the World Meteorological Organization database, it was necessary to modify the LEO Estimation Method to eliminate the elevation effect when the city/region is exposed to the ocean proximity effect, and to reduce the elevation effect for cities/regions not exposed to the ocean proximity effect from -10 °F (5.6 °C) to -5 °F (2.8 °C). The following tabulation reflects the average annual temperatures of the three cities as measured by the WMO database compared with the LEO Method revised on June 2, 2025, with the reduced elevation effect:

Comparison of Meteorological Records and LEO Estimation Method
(for Average Annual Temperature of Durban, S Africa, Denver, CO, and Cape Town, S Africa)
(Latitude, Elevation, and Ocean proximity effects on coastal and inland cities)

			Average Annual Temperature				
Location	Latitude	Elevation	Ocean (Closest) (NOAA and Travel Orgs.)	Meteorological Database (WMO)	LEO Method Estimate (Latitude and Elevation)		Ocean Proximity Effect
					Orig LEO	LEO (6/2025)	
Durban, S. Africa	29.9° S (-29.9)	5,809 ft 1,771 m	74.8 °F / 23.8 °C (indian Ocean)	75.0 °F 23.9 °C	58.1 °F 14.5 °C	68.1 °F 20.1 °C	7.0 °F (3.8 °C) warming (avg.)
Denver Colorado	39.7 °N (39.7°)	5,279 ft 1,609 m	No ocean proximity	55.5 °F 13.1 °C	50.2 °F 10.1 °C	55.2 °F 12.9 °C	No ocean proximity effect
Cape Town S. Africa	33.6 °S (-33.6)	174 ft 53 m	62.7 °F / 17.1 °C (Southern Ocean)	62.6 °F 17.0 °C	65.1 °F 18.4 °C	65.1 °F 18.4 °C	2.5 °F (1.4 °C) cooling (avg.)

Note: LEO Method (Revised 6/2/2025) makes no adjustment of average annual temperature for elevation when there is a proximity of a large body of water (either warming or cooling) - as in coastal cities or regions. When there is no proximity of a large body of water, the LEO correction is -1.67 °F (- 0.93 °C) for each 1,000 feet (300 m) of elevation to a maximum of -5 °F (-3 °C) above 3,000 feet (1 kilometer) elevation. The temperature correction for the elevation effect is a constant -5 °F (-3 °C) above that break-point in correction for the elevation effect.
To be observed for latitudes of the southern hemisphere, the latitude is in decimal form with a preceding minus (-) sign.

(Source: WMO Database analyzed with LEO and compiled by Author)

By subtracting the LEO Method average annual temperature estimate (which accounts for the combined effects on temperature of latitude and elevation) from the meteorological database for a coastal city or region, the effect of ocean proximity can be estimated. The previous tabulation indicates (in the last column) the ocean proximity effect for Durban and Cape Town, South Africa. Referring to the ocean currents illustration shown a few pages earlier, Cape Town is rationalized to have about 2.5 °F (1.4 °C) cooling effect of the northeasterly flowing current of the cold Southern Ocean.

In large contrast, the ocean current flowing southwesterly past Durban, South Africa, is deduced to have a much larger effect of ocean proximity – warming of about 7.0 °F (3.8 °C). The westward flowing Equatorial hot current of the Indian Ocean is diverted both north and south by the east coast of Africa and the southern branch of that hot flow passes through the Strait of Madagascar just prior to flowing southerly past Durban while still in a heated condition.

It should be noted for environmental science interest that if there is any effect of "greenhouse gases" in the atmosphere – natural (mostly water vapor) or anthropogenic – the cities of Durban and Cape Town are sufficiently close to each other that any "greenhouse gas" effect will be the same for both cities.

It should also be apparent to Readers that the LEO method for estimating a priori the average annual temperature for any region or city on the planet provides a much better scientific temperature estimation tool for assessing planetary surface temperature changes than does the "one-size-fits-all" average annual global temperature used currently by environmental scientists. That fallacious averaging aimed at representing the global warming phenomenon masks the effects of latitude, elevation, and ocean proximity that affect each city or region differently. The point also being restated so often in this book is that the concentration of anthropogenic (residential and industrial) gases in the atmosphere (0.0168% by volume) as of April 2025 is very low and citing a "crisis" to exist if the average annual global temperature continues to increase and exceeds that fallacious “tipping point” of 60 °F (15.6 °C) by 2050 is unnecessary scientific “fearmongering” by those environmental activists expounding the need to abolish the use of all fossil fuels to control global warming.

The chart that best summarizes graphically and numerically the revised LEO Method is shown as follows: It might be meaningful to envision that the five-segmented green line approximates the surface of the spherically shaper planet, and that a line from the ends of the green line to the points of temperature at 90° north and south latitude approximates the tilt angle of the planet.. The incoming solar radiation energy comes from the Sun at a very distant (several million miles) position in space to the right of the Equatorial line. The mathematical model for the LEO Method is shown in the two outlined boxes at the upper left of the diagram (in Imperial units). A reminder is given that mathematics requires inputting the latitude in decimal form and with a minus sign (-) for latitudes in the Southern hemisphere. Readers wishing to use the LEO Model in metric units of temperature can use the conversion °C = 5 / 9 (°F – 32).

Shown at the lower right of the diagram is the elevation model also corrected to be a constant reduction of 10 °F for cities /regions having an elevation above 3,000 feet (about 1 kilometer). That 10 °F reduction in temperature for elevation effect was reduced to 5 °F (2.8 °C) for cities located inland, and the elevation effect correction does not apply to cities having an ocean proximity effect.

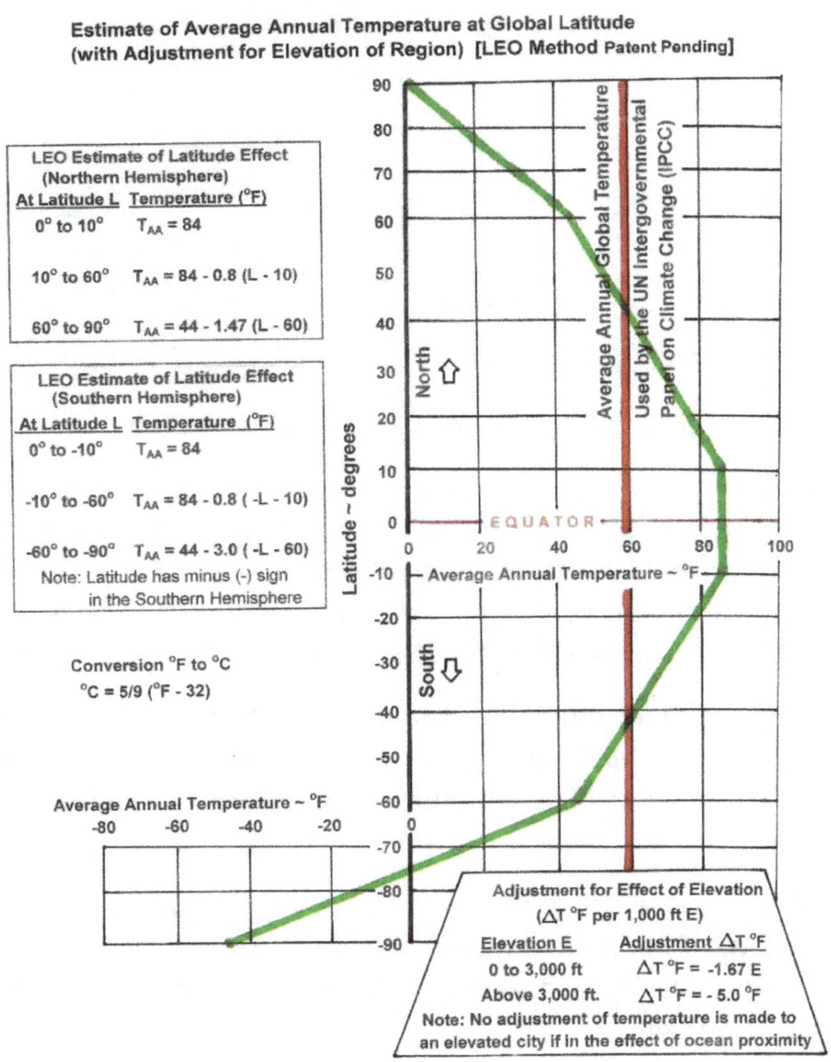

(Source: Author’s analysis of World Meteorological Organization database)

The primary benefit of the LEO Method (green line and mathematical modelling equations) is that it represents the basic average annual temperature as it varies with the latitude of the region anywhere between the North Pole and the South Pole. Because regions are not all at sea level a correction factor for elevation is added to the latitude effect (the elevation factor is included in the lower right of the diagram). The environmental scientists' red line average annual global temperature is seen to represent only the average annual temperature for the entire planet independent of latitude or elevation.

To be also added to the LEO Method is a third temperature that cannot be approximated with a mathematical expression – that of the proximity of a region or a city to a large body of water and which has periods of prevailing wind. The O at the end of LEO stands for that ocean proximity. The following illustration of a portion of the map of the world oceans is shown as follows to assist in the assessment of the ocean currents and prevailing wind currents having an effect on cities or regions:

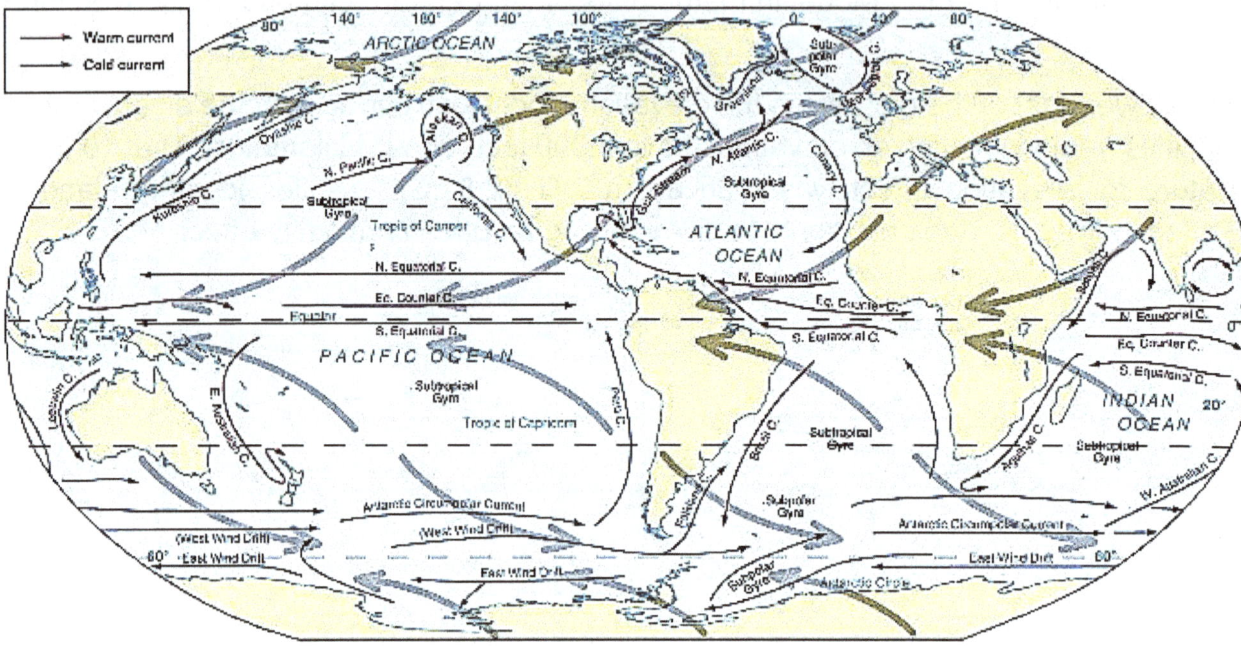

(Source: World Meteorological Organization)

The ocean currents are shown by the smaller width curved lines with an arrowhead portraying their direction. Warm ocean currents are shown in red and cold ocean currents are shown in dark blue. The ocean currents general directions are determined by the shorelines of the various land masses acting as guide vanes. Because of the Coriolis Effect (explained earlier), the circulation of large bodies of water north of the Equator are clockwise. South of the Equator they are counterclockwise.

The wind currents (the heavier width arrowed lines) are not affected significantly by the land masses and are above the surface of the land over which they are shown to influence its temperature by carrying the molecules of warm or cold ocean current water vapor from the proximate large body of water.

Perhaps a more useful illustration which is more localized compared to the global nature of the above illustration is shown as follows: (It is an example explaining how two cities – Cape Town and Durban, South Africa, are at the same latitude and elevation but have different average annual temperatures because of the ocean and winds effect:

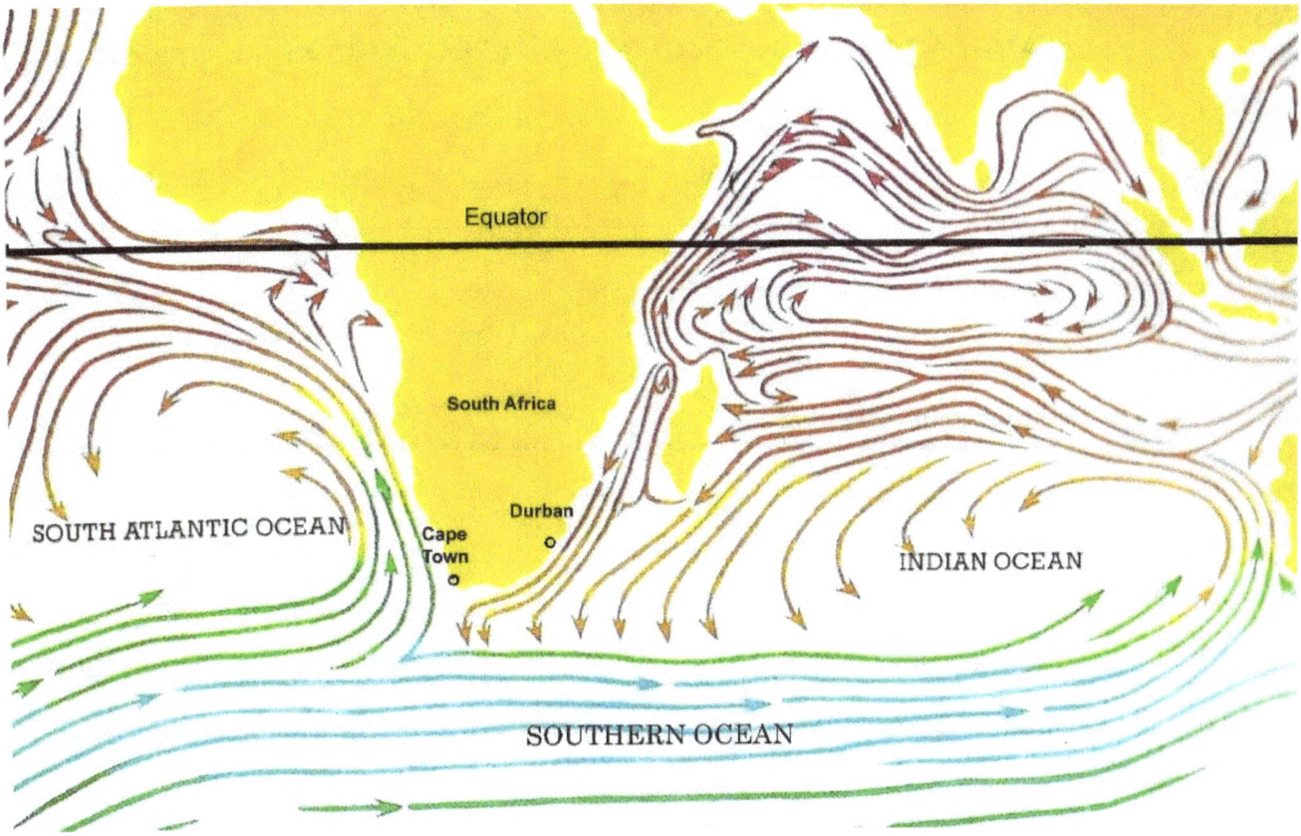

(Source: Atlas of the World, Rand McNally)

A short analysis for the average annual temperature was conducted comparing the results of applying the LEO Method (latitude effect only) with the meteorological records for Cape Town and Durban to evaluate the use of the LEO Method in combination with the World Meteorological Organization database (for the annual average temperature for almost every major and many minor cities of the world. That short analysis was found to be most interesting for the effect of the effect of elevation on cities that are either coastal located with low elevation, coastal located with high elevation, or inland – away from any ocean proximity effect. For the inland city with high elevation, the city of Denver, Colorado, was selected.

By comparing the WMO database with the LEO Method estimates for the average annual temperature in city pairs (Cape Town and Durban, both having ocean different elevations and different proximity to oceans; and Durban and Denver, both having high elevation but Durban having an ocean proximity effect whereas Denver has no ocean proximity effect.

Little if anything varies in Nature in an average, linear pattern of variation. It borders on ludicrous to believe that a 2 °C rise in the average annual temperature (the red line) constitutes a "crisis" when the variation from the actual annual temperature (green line) varies as much as 25 to 40 °F (14 to 22 °C) from the LEO temperature estimate. For each region being analyzed, the LEO Method only provides the annual average temperature for the region. Using the "average" masks the extremes of temperature that realistically occur seasonally (i.e., in four approximately 3 months periods) shown as following for the USA region:

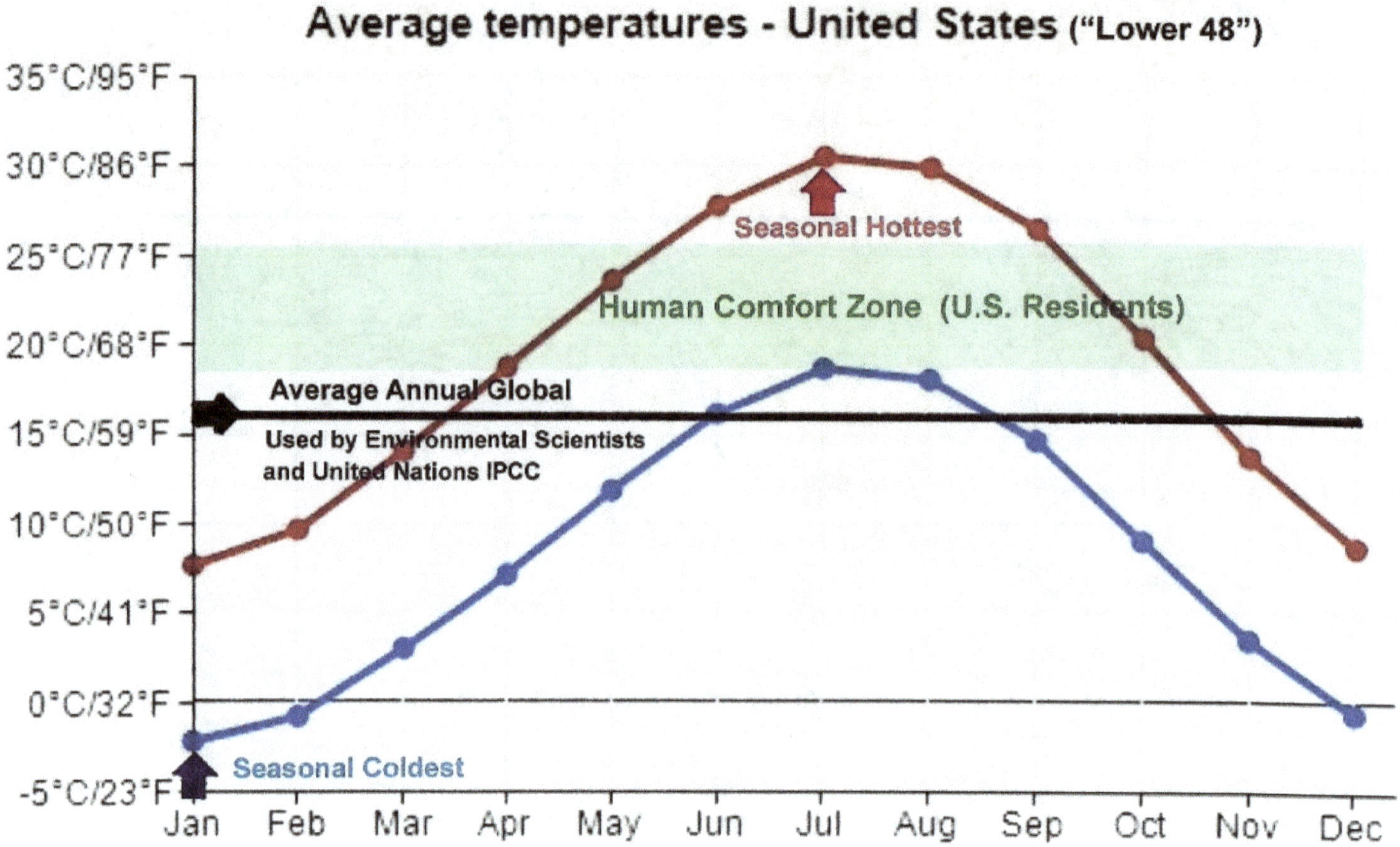

(Source: NOAA)

There are occasions when a combination of natural meteorological events causes a region to have an extremely hot or extremely cold temperature lasting perhaps only a few days; nevertheless, all life forms in the region must survive those extremes well above or below the seasonal extremes shown above and in the LEO method for the annual average temperature. The resulting reality is that every region has a temperature envelope within which everything in that region – animate or inanimate – must survive or perish. Machinery must also operate in the extremes of its environment. The following graph represents the temperature envelope for the United States between the latitudes shown on the graph.

The extremes were also found in this analysis to vary with latitude meaning that other regions should develop the envelope that is derived from the records of the World Meteorological Organization pertinent to the latitudes of that region.

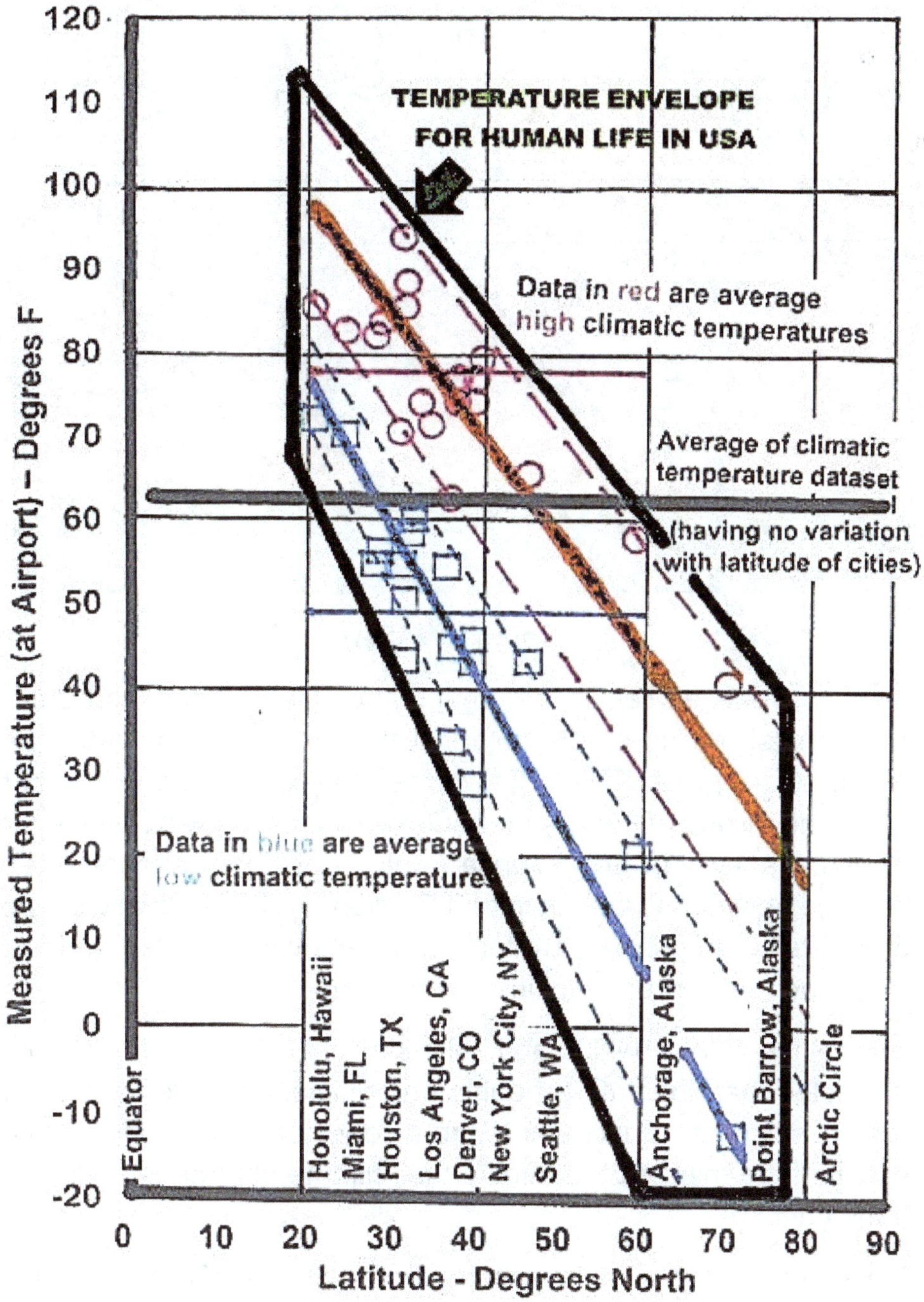

(Source: National Meteorological Service, NOAA, analyzed by Author)

The above temperature envelope was created for the United States region between the latitudes shown on the temperature envelope. The following diagram visually indicates the need for a temperature envelope to be developed by each country/region because the latitude effect on annual average temperature is certainly not a linear variation which the green line on the LEO method graphic has illustrated. The seasonal effect and the occasional onshore and offshore winds affect the ocean proximity effect on coastal cities more than it does the temperature of inland cities. That varies from country to country.

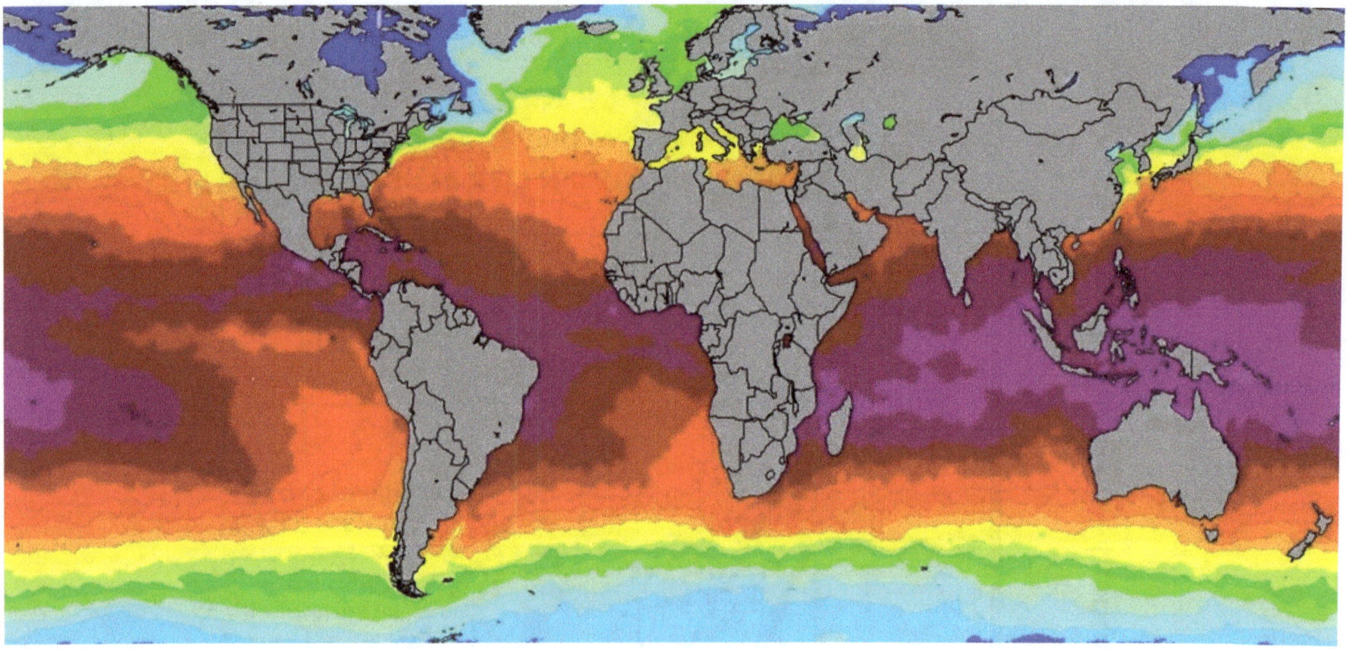

(Source: NASA)

The temperature analysis done earlier in this chapter for Cape Town and Durban in South Africa illustrates the hotter water (shown in purple color) of the Indian Ocean in the Strait of Madagascar causing Durban to be on average 12 °F (6.7 °C) warmer than Cape Town seen to be in the lighter red color (cooler because of the mixing of the Southern Ocean current flowing on the Cape Town coastline. There is no large industrial activity in South Africa to blame for "greenhouse gas" emissions causing the temperature difference between Cape Town and Durban. That is factually the better explanation for global warming and climate change in South Africa – and everywhere else on the planet - not being a dominant result of anthropogenic carbon dioxide emissions.

When it is considered that the safety, health, and efficiency of the humans on the planet is a much more urgent need to be protected than the safety, health, and efficiency of life forms on the ocean floor, the concern of environmental scientists must defer to the safety, health, and efficiency of humans and their technological developments to mitigate the effects that climate change is and will be imposing on the planet and all life forms on it.

The sequel – part 2 in effect – to this book is to be published shortly after this book and is titled "Dealing with the Real Climate Change on Our Planet". There have been 25 technology areas identified as the research for "The Big Picture of Climate Change on Our Planet" was being conducted. Because that research was conducted in a "turn over a stone and more stones were found to be overturned" mode, it was found that there was growing recognition that developing technologies to reduce anthropogenic "greenhouse gases" with the goal of controlling the climate was a futile attempt to change the climate because changes factually happening in the climate have now been identified – in this book and in other books by climate change "naysayers" - to be caused by non-anthropogenic factors.

This Author identified over 35 technologies that had the potential for mitigating the effects of climate change as well as improving the sources of energy to replace the use of the fossil fuel coal. As this book has shown, there is a need (already well underway) to reduce and eventually eliminate the use of coal because its ash residue contains toxins and carcinogens from the coal not being in a pure state of carbon in solid form. The coal ash has been disposed of in landfills and trash dumps but when rainfall washes the landfills and dumps, those toxins and carcinogens are polluting the water aquafers below the landfills and dumps. That polluted water is finding its way into the water supply of the population, the agricultural lands (food supply of humans) and grazing lands of livestock animals. Eliminating the use of coal in the United States will have little or no effect on the climate of the United States. (Other regions of the planet should conduct a similar analysis that was conducted for this book to tailor the technology needs for mitigating climate change and pollution problems unique to their region.

A weighted trade study analyses was conducted for the 25 technologies being developed or strongly considered for development contained technologies mitigating the effects of surface temperatures rising in various regions of the World but more specifically in the United States. Lesser importance was given to avoidance of emissions that can be considered in the "greenhouse gas" heat-trapping category. The methodology used for the weighted trade studies was discussed in an earlier chapter. The laboriously detailed worksheets used to "grade the report card" (i.e., assess the net value of the technology area by weighing the importance of each pro and con of the technology) are contained in the sequel book "Dealing with the Real Climate Change on Our Planet". The methodology for assessing the set of technologies identified in this book is explained more comprehensively in the sequel book.

In the lower grades of education, knowledge is not gained rapidly but it is that base of knowledge – learned by human "twigs" - on which all future gains in knowledge will be built. As older members of humanity – and regardless of language spoken or cultural beliefs – there is an obligation to prepare the youngest members of the region humans reside in for the realities of the world the young people are about to learn if they are to survive. This book hopefully will be "one small step for Man – but a giant leap for Mankind" in the learning process of the youths in the United States.

The second President of the United States John Adams (1735-1826) made a wise statement about teaching children: *"It should be your care, therefore, and mine, to elevate the minds of our children and exalt their courage; to accelerate and animate their industry and activity; to excite in them an ambition to excel in every capacity, faculty, and virtue*." Deficiencies in the education system are not as hazardous as those described in the foregoing sections of this book, but part of the mitigation of climate change is to educate the public about what they can and should do to mitigate the hazardous effects of whatever the climate changes are and will be.

Perhaps the most important finding during this analysis of the causes for the global warming and climate change problem areas was revealed in the "big picture" assessment of the technologies that are underway addressing the general problem of global warming and climate change. The adaption of windmill technology into the generation of electricity from wind power is a logical technology advancement because the wind is a "freebie" form of energy and does not require the combustion of any type fuel to power the generation turbines. Environmental scientists have pursued and championed this technology as a means of obtaining an almost emissions free source of energy to substitute for the reduction in fossil fuels. Wind turbines are not extremely high in efficiency but their "Achilles heel" is that they can only harness wind energy when the wind is blowing – therefore are considered of less value for the industrial needs of reliable energy 24/7/365 regardless of the weather. Wind turbines are also very prone to damage in natural meteorological events such as hurricanes, tornadoes, and floods and mudslides damaging their structural integrity.

A similar finding was with solar energy – harnessing the "freebie" solar radiation from the Sun which is a constant radiation 24/7/365. The principal drawback to solar panels for industrial use is that they are efficient when the solar rays are available – but are dormant when the night hours are cloaking the region in darkness hours or during inclement weather events where clouds and associated precipitation mask the solar panels from the constant solar insolation (incoming solar radiation). For smaller uses, the addition of battery storage added to wind generation and solar generation systems store some of the energy collected during daylight favorable weather time periods. However, in both the wind power and solar power systems there can be much damage done to the systems by natural forces – and electricity for rescue and recovery operations may not be available for long periods in a region for needed recue and repair actions with the need for electrical power in the natural emergency.

Nuclear energy is another older technology that has far more reliability than solar or wind and is attractive because it also does not combust a conventional fuel, thereby avoiding the "greenhouse gas" emission problems inherent with other fuel-powered technologies. There exist many nuclear facilities in almost all developed countries that can afford to have that technology. Lesser developed countries are often unable to develop nuclear facilities and resort to the age-old burning of fossil fuels, especially coal.

There are also a rash of energy generating technologies in the transportation technology area such as all-electric vehicles – with batteries that need to be recharged from recharging stations powered by fuel-powered and solar/wind/nuclear sources. The impetus for all-electric vehicles is the reduction and elimination of greenhouse gases from fuel-powered modes of transportation (on land, on the ocean, and in the atmosphere).

A most promising area for energy generation researched and concluded to not have all the "cons" that are inherent in other energy technologies is that of geothermal energy. The value of geothermal energy beneath Earth's surface is that the systems are not exposed to the often catastrophic and very unreliable "cons" of systems on or above Earth's surface. Geothermal energy has many "pros" and few to none "cons" (other than the initial high costs to drill deep into Earth's surface).

A very enlightening chart was recently issued by the Department of Energy providing the status of the various technologies developed mostly to reduce the use of fossil fuels that are hypothesized to be the cause of global warming and climate change:

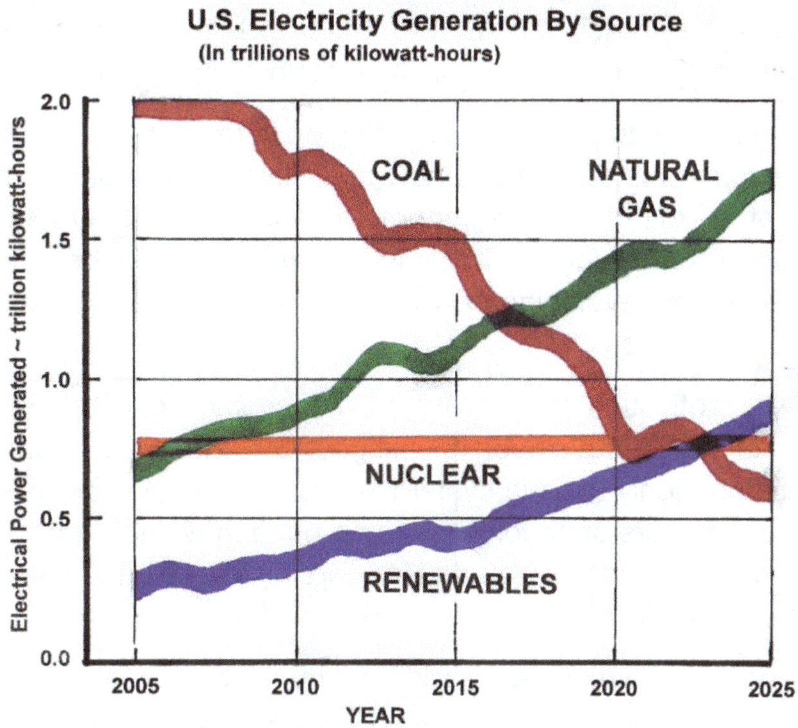

(Source: US Department of Energy)

Regardless of what the environmental scientists are touting as being the technologies that should replace the use of all fossil fuels to protect the environment, the reality is evident that the industries of the United States are using natural gas as the most important replacement for coal. It is somewhat ironic that one fossil fuel (natural gas) is being used to replace another fossil fuel (coal). This trend of greater use of natural gas is not unique to the United States – most of the developed and some lesser developed countries are also increasing their use of natural gas – because it is clean, abundant, and affordable. Evidence of the increased use of liquid natural gas as the preferred distribution mode of natural gas is seen in the following photograph taken at the Panama Canal:

(Credit to Unknown Photographer)

Seen is a behemoth LNG tanker passing through the Panama Canal in the relatively new (2016) Super-Locks constructed to allow the transit of Post Panamax size transport ships on the shortest route between continents in the Pacific Ocean and Atlantic Ocean.

In summary, much has been done to advance technologies in the United States during recent decades but the "big picture" assessment is that all technologies may need to be re-vectored away from their main purpose – raison d'etre – of reducing "greenhouse gases" into the atmosphere to "combat" global warming and climate change. Technologies should be advanced "for the benefit of humanity" – and that is (or should be) the objective of developing new technologies (such as geothermal) and/or improving existing and emerging technologies.

This statement served as motivation for the Author to assess those technologies that need to be revectored or advanced or newly developed to mitigate the effects of global warming and climate change. Using a process commonly used in system engineering known as weighted trade studies, the Author identified the several factors of importance to be used in weighted trade studies and the listing of those Importance Factors is provided to inform Readers how that form of weighting a technology "for its importance to humanity" is as follows:

1. Status of Technology Development (risk, schedule, cost, potential value)
2. Reliability and Sufficiency (time of day/year, weather dependent, regional need dependent)
3. Regional Economy and Resources (mineral, water, labor force to operate/maintain)
4. Cost of Resources (domestic and imported, mode of harvesting / distribution)
5. Importance to Industry, Government, and Human Safety, Health, Comfort
6. Importance and Urgency to Economy and Employment
7. Pros and Cons of By-Products (emissions, pollutants, hazards)
8. a. Effect on Human Population (regional in USA) - Residency.
   b. Effect on Human Population (regional in USA) – Food Supply.
   c. Effect on Human Population (regional in USA) – Safety, Health, Comfort
9. a. Effect on Infrastructure (regional in USA)
   b. Effect on Transportation (regional in USA)
   c. Effect on International Trade
10. a. Effect on Environment (regional in USA) – Atmosphere
    b. Effect on Environment (regional in USA) – Land
    c. Effect on Environment (regional in USA) – Water
11. a. Effect on Animal Life (regional in USA) – Land
    b. Effect on Animal Life (regional in USA) – Aquatic
    c. Effect on Animal Life (regional in USA) – Avian
12. Effect on Forestry and/or Agriculture (regional in USA)

By assessing the pros and cons of each candidate technology for how well the technology satisfies the 12 importance factors, the "report card" for that technology can be given a final grade score.

That process was completed by the Author for 25 candidate technologies – but regrettably is beyond the scope of this book to present the results of that assessment in this book "The Big Picture of Global Warming on Planet Earth". However, the Author has written a sequel to this book – much the same as Leo Tolstoy did in his classic "War and Peace" – and has written "Dealing with the Real Climate Change on Our Planet". In combination, the two books address the two-part question posed at the beginning of this book “What is causing global warming and climate change?” and “What can and must be done to deal with those temperature and climate changes?”

Both books have been written somewhat in a stand-alone manner since the first book is probably of greatest value in the STEM educational arena for how a major problem as complex and of such high importance can be analyzed. The second book is possibly of greater value to government regulators as well as to the public for what areas of effort should the taxpayers tax monies be expended “for the betterment of humanity”.

The next section of this book contains the many conclusions resulting from the “big picture” system analysis the Author has conducted to identify the factual causes of global warming, climate change, and oceans’ levels rising on the planet.

Conclusions (Overall) and Concluding Comments

Extensive research and analysis of a large number of relevant databases - seeking to identify what was or was not causing the phenomenon of global warming and climate change - was conducted on the topic developed over a long span of years by many hundreds of scientists in the many branches of science. During that span of years there has been a flood of factual and fallacious information generated and released to the public and by reputable and not so reputable news reporting agencies and companies. The Author has an extensive background in accident investigation and in technology developments and has done his best to consider only factual information in the investigation being reported on in two books "The Big Picture of Global Warming on Planet Earth" (this book) and in "Dealing with the Real Climate Change".

The open-minded technique used by the Aurhor followed a process of "turning over a stone to see what is under it and finding more stones needing to be overturned" .If the Reader has read the foregoing chapters also with an open mind – and with no bias toward a pre-conceived claim – the conclusions about to be presented and to be read will resemble the evidence used in a jury trial and must prove "beyond reasonable doubt" that the set of conclusions are factually correct or point to additional analysis to come to that conclusion. The topic of global warming, climate change, and oceans levels changing is complex and convoluted because the laws of nature are complex, convoluted, and very interactive. The first many "stones" overturned address the "big picture" of cause of global warming and climate change and are summarized in the numbered conclusions (**C1** – **C25**) contained hereafter. A second set of conclusions address the activities and technologies that need to be developed - "for the betterment of humanity" - and are found in the book "Dealing with the Real Climate Change" by this same Author.

In the Author's "big picture" analysis of the several pertinent databases, the most important scientifically factual causes of global warming and climate change – many of which the environmental science segment chose to ignore or deemphasize – have been surfaced and analyzed. Every effort has been made to evaluate the numbers (valid measurements, accurately recorded, and professionally interpreted) without any preconceived "spin" put on those numbers. The in-depth detailed investigation into the causes for global warming, climate change, and ocean levels rising was conducted with a "big picture" system engineering type approach known to system engineering analysts as a heuristic approach. In lay terms, a heuristic approach is the formal process described in common informal words as "turning over a stone to see what is under it and finding more stones needing to be overturned."

Applying that philosophy to the "big picture" perspective of the planet Earth starts at the top stone – the solar system. So that everyone reading this book would have a common understanding of the solar system, this book showed at the start the solar system in its essential simplicity – but with its unseen complex and dynamic functionality. The "big picture" analysis started with an understanding of the dynamics and geometric relationships which define the geophysics of the planet:

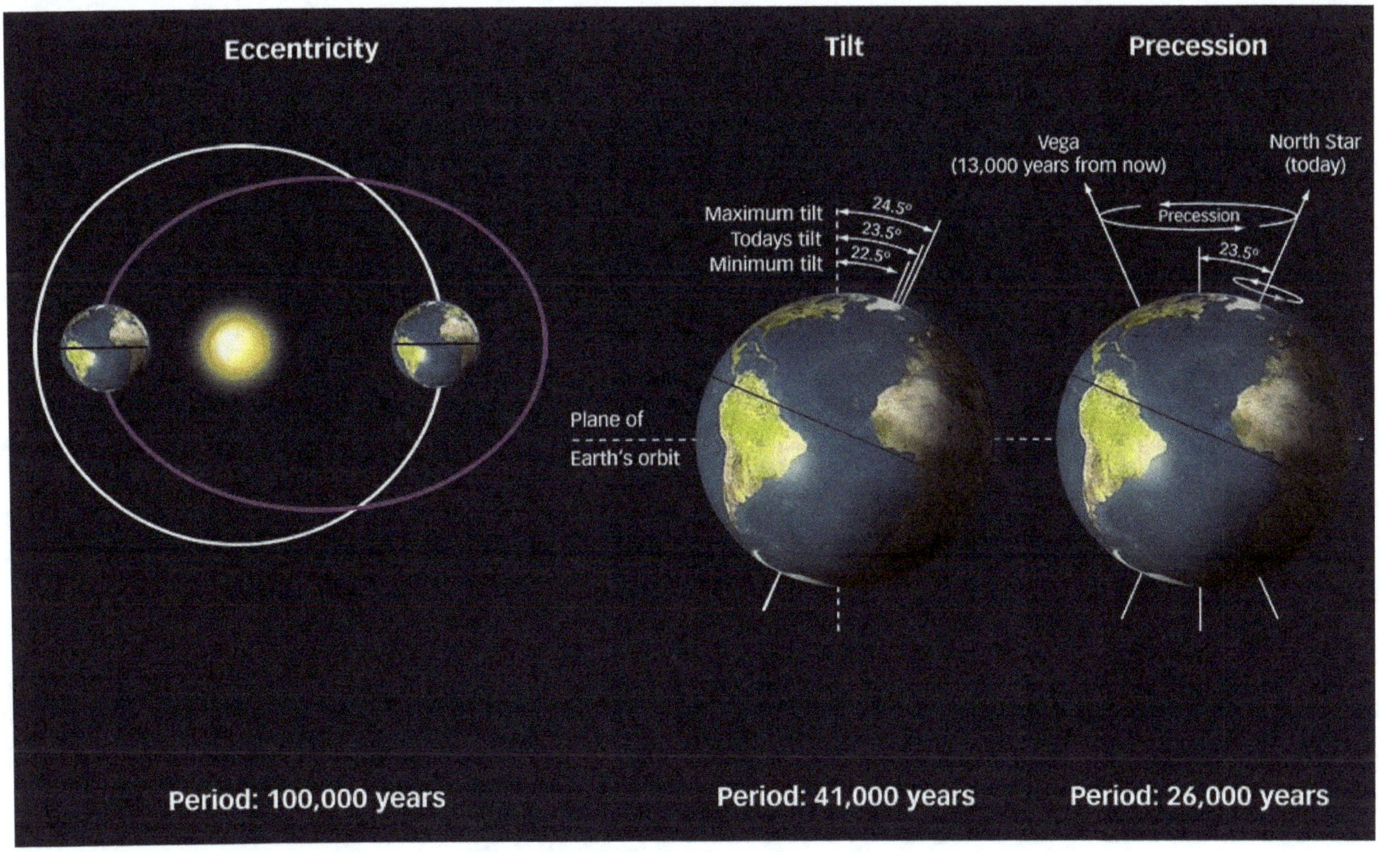

(Source: Smithsonian for Science, modified by Author)

Unquestionably, the geophysical aspects of the Earth – the ellipticity and the eccentricity of the orbit Earth follows in its annual orbit of the Sun, the tilted and wobbling (precessing) axis of rotation about which the Earth rotates once per day, and obliquity – the angle at which the solar rays imping on the Earth at different latitudes - are the primary dynamically changing drivers of the temperature on the surface of the planet. Excluding the known impacts of asteroids with Earth, the dynamic geophysics factors have had the most influence on the temperature of planet Earth for its 4.67 billion years of known existence.

The database that "documents" how the temperature of Earth's surface has varied for the past 420,000 years is that "recorded" (summarized below) at the Vostok Ice Research Station in East Antarctica – at latitude 78.46° South. Shortly after World War 2 ended (1945) and the "Cold War" began, scientists of the Union of Soviet Socialist Republics (USSR – which has now reverted to its original name Russia) began drilling into the ice sheets to collect samples of ice cores at the increasing depths in the ice sheet. In the analysis of the ice core samples, Russian scientists started analyzing, with various types of scientific analyzers, the temperature and gases captured in the ice bubbles and correlated the depth drilled with the approximate year and the measured temperature and gas content in the ice cores to initiate the ice core analysis. A photo of the ice core sample that has been sliced for analysis is shown as follows:

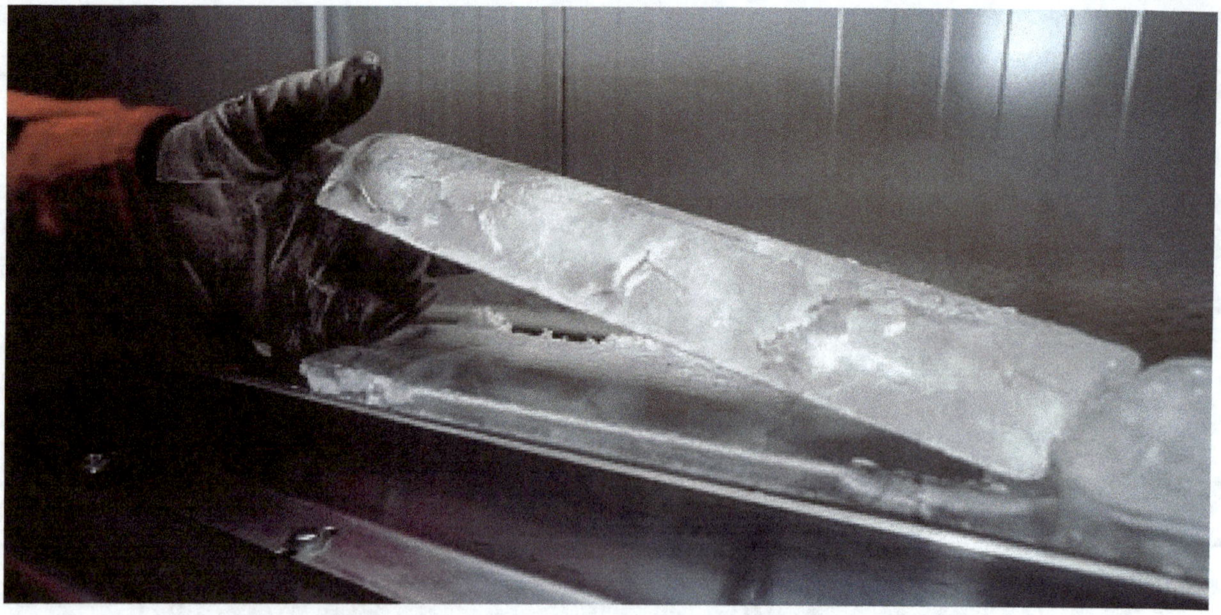

(Source: Smithsonian Institute for Science)

That was explained in detail earlier in the Databases Analyzed chapter of this book. Readers should recognize the following graphical summary of the ice core database and the analysis this Author conducted on the ice core database:

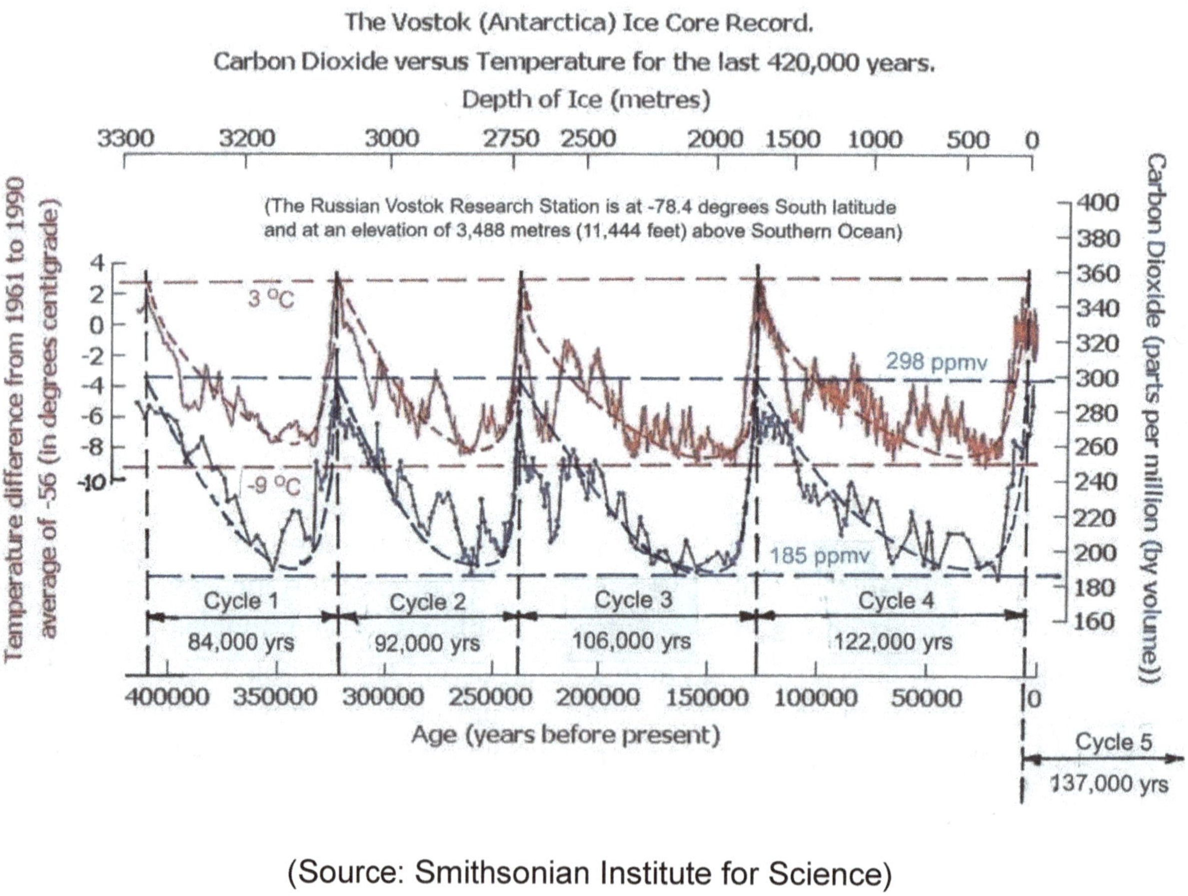

(Source: Smithsonian Institute for Science)

The first 3 conclusions presented in this section of the book (Conclusions and Concluding Comments) – drawn from the ice core sample analysis –

C1 The repetition of four cyclic variations in temperature anomaly at the Vostok Ice Station at latitude 78.46° South in the past 420,000 years has but one major interpretation – with the absence of Man prior to 120,000 years ago ((the last "Little Ice Age"), the four cycles have been caused by astrophysical, geophysical, and natural factors in combination and therefore cannot and do not include any effects of Man or industry on the temperature of the planet's surface or the gases in the planet's atmosphere.

The peak hot and cold temperature and peak high and low carbon dioxide are seen to repeat – at scientifically deduced near-identical levels - for the four cycles captured in the ice core samples. The peak temperature anomalies are 3 °C hotter than normal and 8 °C colder than normal in the four cycles. The peak carbon dioxide concentrations (ppmv) are highest of 295 ppmv and lowest of 185 ppmv. To isolate the dependency of natural carbon dioxide on naturally varied temperature and/or vice versa, an International Ice Research team from several European countries (including Dr. Nicolas Caillon of France) closely analyzed the timing of occurrence of the peak temperatures and the peak carbon dioxide concentrations for their respective occurrence of peaks at the end of the second cycle. Results of that analysis are graphically summarized in the following graph, establishing the timing relationship of the measured temperature and carbon dioxide:

CO_2 lags temperature rises in ice core data

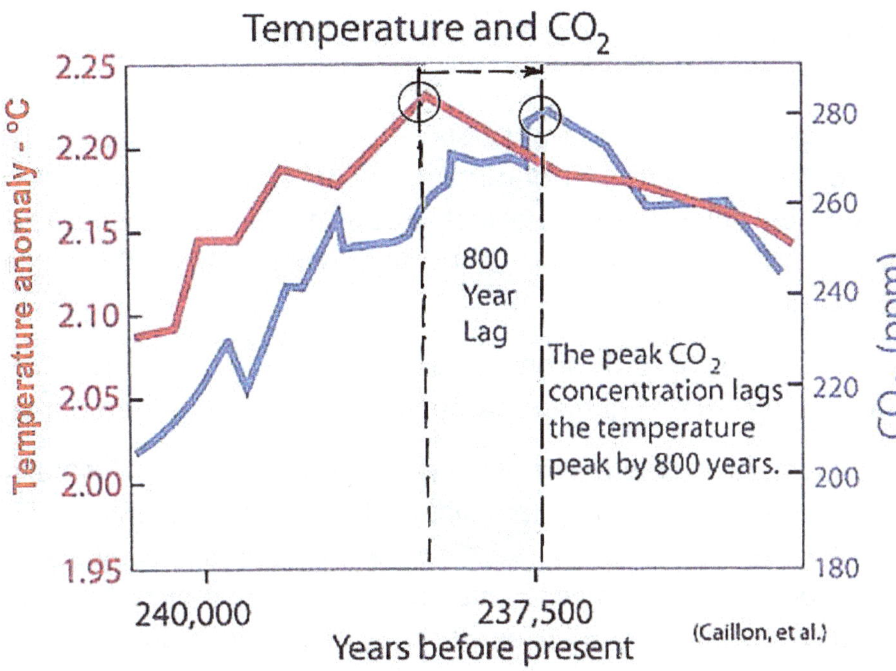

(Source: Smithsonian Institute for Science)

C2 Applying the conclusion shown in **C1** and the timing of temperature and of natural carbon dioxide in the planet's atmosphere (that relationship confirmed by the convincing analysis of an international group of scientists not affiliated with the United Nations Intergovernmental Panel on Climate Change (UN IPCC), the second major conclusion is that the environmental scientists' hypothesis that anthropogenic "greenhouse gas emissions" resulting from the combustion of fossil fuels are the dominant cause of global warming is totally dispelled. Either the temperature or some other unknown forcing factor drives the existence and extent of the carbon dioxide.

C3 Based on Conclusions **C1** and **C2** and revisiting the Author's analysis of the Vostok Ice Core Research team, there can be projected with high probability a repeat occurrence of a fifth cycle of the variation of the planet's surface temperature to occur from now (t=0) to about 137,000 years from now. The planet is seen to be nearing a high temperature at the end of the fourth cycle which coincides with the realization that the global warming problem is factually occurring from astrophysical, geophysical, and natural forcing factors and can be expected to peak within the "window" of time 100 years before now and 200 years after now. The temperature peak at the end of the fourth cycle (seen in the Vostok Ice Core summary shown earlier) is somewhat "burbling" (an erratic variation) and not exhibiting the classical sharp peak characterized by cycles one, two, and three. There were problems early in the drilling for the ice core samples and which may have contaminated the earliest of the ice core samples explaining that "burbling" variation in temperature. The burbling of temperature is not associated with an anthropogenic greenhouse gas effect.

There were over 45 conclusions (including the major three **C1** to **C3** above) formulated during the top-down analysis but several (at least 6) were of lesser importance (for example, the less than professional mathematical technique of averaging of the meteorological records (from sensors at airports which are locations with much ground area covered by heat-trapping concrete and asphalt) and producing a misleading "average annual global temperature" that was in turn used (by environmental scientists in several organizations) as the baseline "average annual global temperature" (59 °F or (15 °C). It was from the oversimplified "average" which the planet's surface temperature was declared to be varying globally and at a rate overexaggerated as being a "crisis". That is a process interpretation fallacy, not a basic science flaw.

A set of 38 conclusions resulting from the analysis of the causes for climate change and rated as significant by this Author are presented in chronological order of their being made, not in order of importance or urgency to be acted upon. The first group of the conclusions are general qualitative conclusions (the initial major 3 conclusions with 15 conclusions **C4** to **C18**) resulting from the "big picture" analyses of the multiple databases.

C4 There were numerous data records from several sources – excluding news media reports - confirming beyond doubt that climate changes are factually occurring on the planet.

C5 There was an abundance of meteorological data (land-based and satellite-based) analyzed indicating that climate changes are occurring to a different extent in different regions and at all latitudes of the planet. In the early twentieth century (1900 to 1950), environmental scientists had only land-based meteorological records from world-wide cities and based their hypothesis of climate change on those land-based records. When NASA satellites were developed for Earth monitoring in the post 1960 period, the database of meteorological variables greatly improved the "big picture" of the full realities of meteorology. Many images similar to the ones below reflect that better definition of factors that are causing regional temperatures, climate change, and anomalistic weather events to be occurring on the planet. The two images shown following reveal the cloud formations that indicate the high concentration of water vapor in the atmosphere and its randomness in location, and also the sensed temperatures as they vary differently in extent and regional location on the planet (and without any "averaging" done to mask important dynamic effects.

(Source: NASA)

In this image cloud formations are seen mostly over the Pacific Ocean and which are not recorded in land-based meteorological records. Environmental scientists basing their global warming and climate change hypothesis on meteorological databases do not include water vapor in their models or estimates. The clouds reveal the higher presence of water vapor in the atmosphere than the anthropogenic greenhouse gas emissions which are focused upon – and overexaggerated - by environmental scientists.

A paradox exists when addressing a problem with either a bottom-up perspective and approach (which environmental scientists employ) or a top-down perspective and approach (i.e., a "big picture" approach used by system engineers) in that most problems are a combination of factors that exist at the intermediate level of the two perspectives. The foregoing satellite from a perspective far above the surface of the Earth shows what appears to be a light "dusting" of snow but in reality is the cloud cover blanketing the planet. From a surface perspective toward the sky of the clouds that may give the appearance of a very tranquil status in the atmosphere, a false impression may be given that the atmosphere is in a quiescent state.

(Credit to unknown photographer)

When the perspective from the window of an airplane at a high altitude is seen, being amid the cloud formation provides a perspective that the amount of water vapor in the air to form such cloud formations must be of high magnitude and when mixed with winds at altitude form some ominous cloud formations:

(Credit to unknown photographer)

Whatever perspective is taken, there can be only one conclusion that water vapor exists in the atmosphere in different amounts of concentration and in different states of dynamic motion. There is very low probability that scientists or statisticians will be able to develop a model that reliably enables the amount and distribution of water vapor in the atmosphere. NASA satellite sensors have occasionally recoded the amount and distribution of water vapor in the atmosphere:

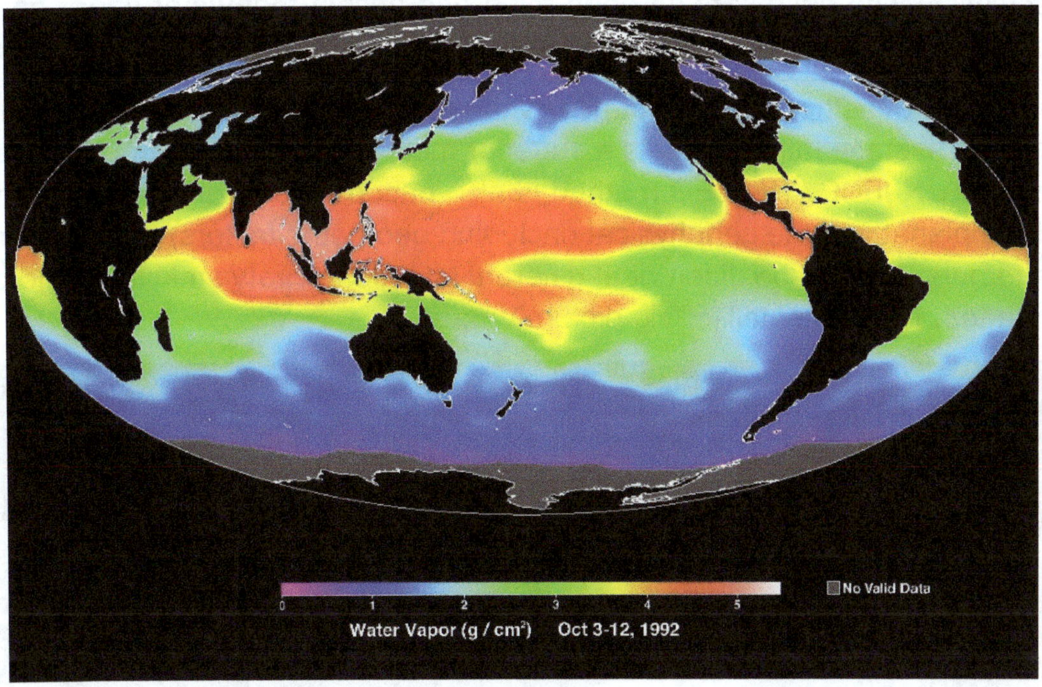

(Source: NASA)

Not surprisingly, there can be made a visual strong correlation obtained when comparing the NASA image of water vapor in the atmosphere with the NOAA image of the temperatures of the land surface and the ocean surface temperatures

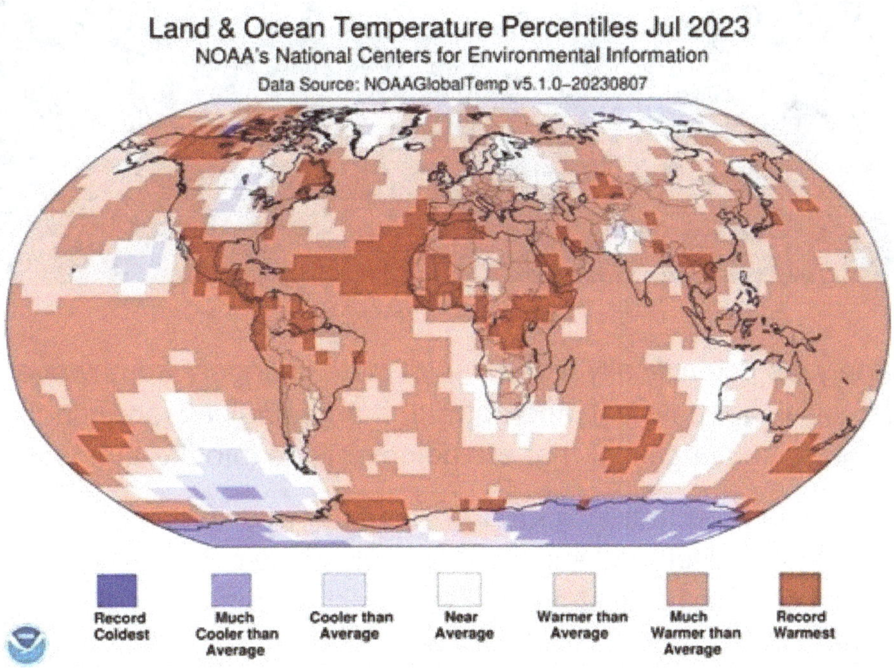

(Source: NOAA)

Even with their own database from satellite sensor measurements indicating the non-average nature of the surface temperatures on the planet, environmental scientists declare there to be a global warming "crisis" – based on the one-size-fits-all "average annual global temperature". This Author has concluded that environmental scientists – as well as the news media, government regulators and about half of the world population have developed the mind-set that Man burning fossil fuels is the cause of global warming and climate change. No amount of valid data appears to be sufficient to alter that fallacious belief. Those scientists who have kept an open mind and accept that water vapor cannot summarily dismissed as an important factor in addressing the problem of global warming and climate change.

This Author developed a method (LEO) that allows the estimation of the average annual temperature for a region using its average latitude that is a superior method than the "average global temperature" method used by the UN IPCC. The LEO method is also based on the "big picture" of the relationship of the Sun and the planet and the geophysical (including geometric factors) that realistically affect the Earth's surface temperatures to be different at different latitudes. A very striking image of that obliquity effect is seen in another NASA satellite "big picture" image:

(Source: NASA)

C6 The databases analyzed scientifically reveal the climate changes are however primarily caused by other factors (astrophysical, geophysical, and natural) than being caused by humans and/or industrial activities generating "greenhouse gases" that are being emitted into the atmosphere

C7 There were no data analyzed that indicated that anthropogenic activities of any type change the frequency or severity of natural meteorological events that cause or contribute to many fatalities and/or much damage. The most convincing illustration of that is the historical record of the paths and severity levels of tropical storms:

Tracks and Intensity of All Tropical Storms

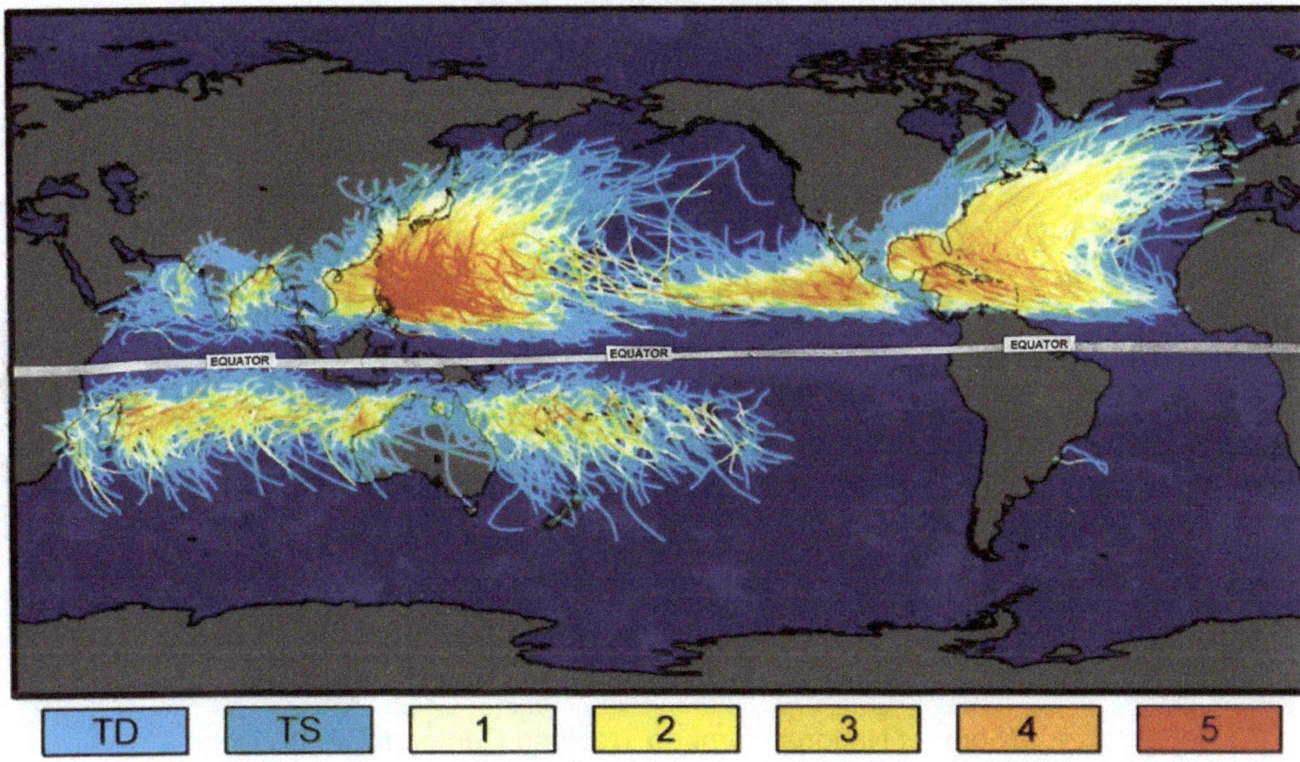

Saffir-Simpson Hurricane Intensity Scale

(Source: World Meteorological Organization)

The origins and ensuing paths of the tropical storms are correlated very closely with the temperature of the oceans and ocean currents below the storms. Environmental science has no basis to indicate that tropical storms are a result of anthropogenic activities. Anthropogenic pollution is a much different problem than an anthropogenic global warming or climate change problem.

A few words of high commendation are very gratefully given to NASA for developing satellites with not only accurate sensors to measure meteorological variables in the big picture but also for developing the capability to track the progress of tropical storms as they progress on their ultimate paths. An example of the value of knowing the size and the path as it changes during the life of a hurricane has allowed the public in the projected path of the approaching storm or hurricane to take actions for protecting the lives of their families and measures to minimize the damage that could occur to their property. The illustration above portrays the composite of those many different paths and on which meteorologists can base prediction models, but each hurricane possesses a "personality" and a behavior unique to itself. In the United States, many hurricanes mature into high Saffir-Simpson Hurricane Intensities in the Gulf of Mexico and move onshore into the coastal regions of the few states having border with the Gulf of Mexico. The Mississippi River Delta is very low elevation and the combination of the hurricane winds and the associated storm surges that are accompanying those high winds annually in "hurricane alley" during "hurricane season" are natural hazards that humans cannot change. A small photograph of a monster hurricane does not adequately describe the hazards of the ball of energy the size of an entire state being unleashed on a region.

(Source: National Tornado Tracking Center, NOAA)

The satellite image above captured the extent of Hurricane Katrina as it was passing over – and causing massive flood damage – in the New Orleans Parish of Louisiana.

C8 Despite the preponderance of water vapor (which has greater heat-trapping properties than carbon dioxide) in the atmosphere – at a ratio of 106 to 1 in concentration of water vapor compared to carbon dioxide – environmental scientists proclaim that human and industrial generated carbon dioxide gas is the primary cause of global warming and climate change. The two charts of comparison of the importance of water vapor being included as a factor of global warming and climate change has apparently had little impact on the environmental science segment:

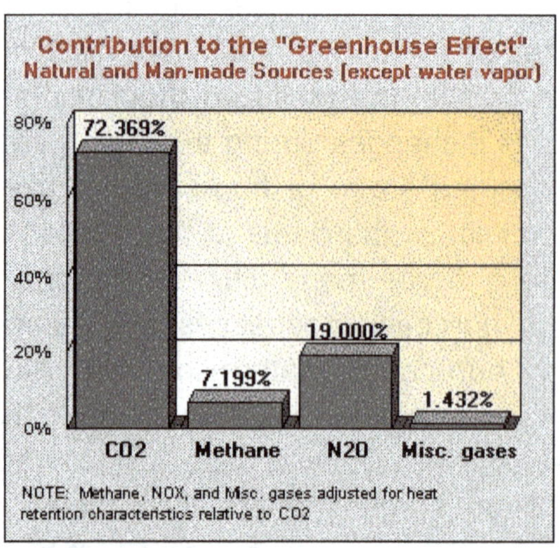

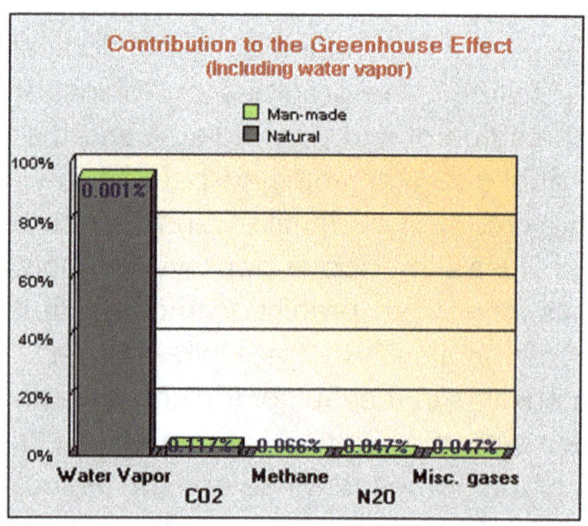

(Source: The Late Dr. S. Fred Singer, Professor Emeritus of Climate Science at the University of Virginia)

An area that environmental scientists also treat much in the same excluding manner as water vapor is that of volcanic eruptions which propel enormous amounts of particulate matter as well as toxic gases into the planet's atmosphere. Admittedly, each volcano that erupts does so with a greatly different "personality" of particulates and gases as well as erupting in an explosive manner or in a lava flow "oozing" manner. The following graph shows the average number of volcanic eruptions that have occurred between 1960 and 2023:

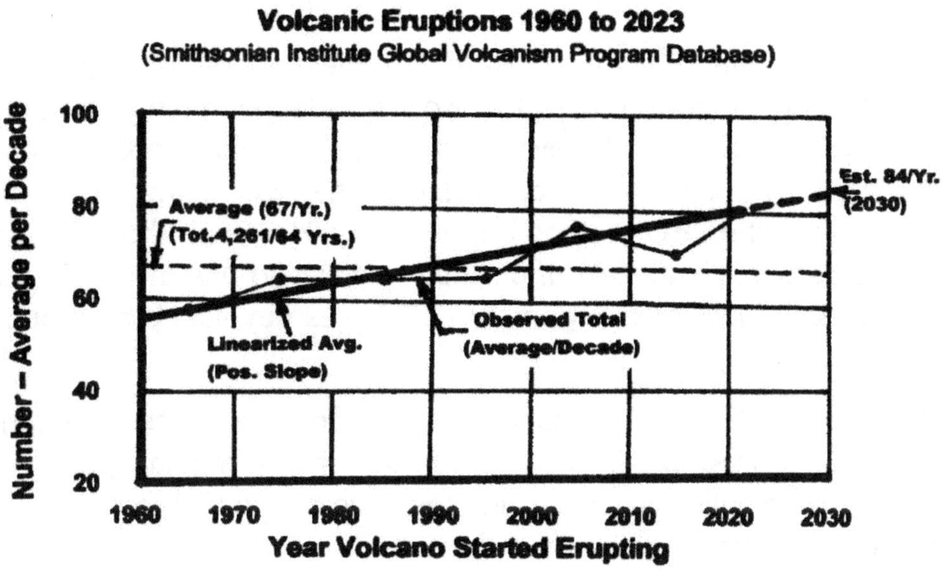

(Source: Smithsonian Institute for Science Global Volcanism Database)

C9 The tactic of suppressing data, such as the data shown above for volcanic eruptions, was found in the "big picture" analysis of global warming and climate change. It is scientifically and professionally reprehensible and is usually accompanied with the fear-mongering tactic to cause a problem of only moderate concern to be a "crisis" is equally reprehensible. The Author made a moderate attempt to review three (AR4, AR5, and AR6) of the most recent UN IPCC Assessment Reports – of which approximately six have been published in the international public domain. All three of the UN IPCC Assessment Reports reviewed revealed that much "cherry-picking" of the available databases – known to exist in different status of understanding for the rapidly changing many factors causing global warming and climate change – practiced by the editors of the UN IPCC Assessment Reports. These two books "The Big Picture..." and "Dealing with . . ." address global warming and climate change with the complete set of pertinent databases being include. A strong recommendation is made that the United States withdraw from the UN IPCC and conduct the activities and programs and technological developments that are "for the betterment of humanity".

C10 The technology of burning coal to be used for industrial energy needs should be discontinued in the United States not for the purpose of reducing "greenhouse gas emissions" causing global warming, but because the ash residue from coal burning is now known to be toxic and carcinogenic. Disposal of hazardous coal ash residue in landfills and trash dumps is polluting the water aquafers and nearby land used for growing food products.

C11 The discontinuance of the fossil fuel coal does not collaterally require the discontinuance of other forms of fossil fuel (oil and/or natural gas) to reduce "greenhouse gas" emissions – in the United States or any other region – as neither oil nor natural gas generates significant toxic or carcinogenic ash residue causing disposal pollution problems.

C12 Pollution (especially from plastics) is a far greater hazard to humans and other forms of animate life than is the "greenhouse gas" emissions hazard affecting the climate.

C13 The databases analyzed indicate that the magnitude of the changes in temperature of the land, of the ocean waters, and of the atmosphere are not increasing equally everywhere and are much slower in rate of change than the public has been led to believe and has been coerced into acceptance of there being a "crisis" as factual. When "fear mongering" claims are surfaced by a scientist the news media adds further exaggeration to the fallacious scientific claim.

C14 The changes in the climate are regional in nature and not of an "average global" nature. Prediction models applicable to different regions (using the Koppen Geiger method and the LEO (Latitude, Elevation, and Ocean proximity) model developed herein) should replace the "average global temperature" model. (A Patent Application has been filed for the LEO Method by the Author; to establish the technological credibility of the LEO Method and its utility. However, the patent holder of the LEO Method encourages any person, organization, or entity wishing to use the LEO model to feel free in using the model without fear of patent infringement.) LEO was developed as a public service as well as a technology advancement.

C15 The fallacious hypothesis of environmental scientists which has caused the unnecessary angst in the public that there is a "doomsday crisis" existing, and/or if a "tipping point" in "average global temperature" is exceeded is scientifically proven in these two books on the factual causes and effects of global warming and climate change to be unwarranted.

C16 The technologies that were developed to reduce "greenhouse gas" emissions – e.g., solar, wind, alternative fuels, carbon removal from the atmosphere, increased use of all-electric vehicles, etc. - are useful but have been found inadequate to meet the increasing need for reliable, clean, affordable electrical energy in the industrial and commercial applications that have mushroomed in recent decades. In the "big picture" perspective, the entire planet is becoming "electrified" and is estimated to become more so as the population grows.

C17 The prospect of the world population growing exponentially creates technological needs in many different categories of technology associated with human and industrial activities (alternative fuels, electric and heat energy storage batteries, desalination of ocean water, non-polluting fertilizers in agriculture, transportation, infrastructure, etc.) The technologies most useful in a region are dependent on the unique resources and industrial needs of that region and on the stage of technological development in that region).

C18 The setting of unrealistic emissions goals based on a sophomoric "average global temperature" and "net zero" goals is a futile attempt to control global warming and climate change. The legislation and judicial actions that penalize any entity for failure to meet those emissions and global temperature goals are not for the benefit of mankind. They are veiled attempts for organizations and government departments to exercise authority in the guise of controlling pollution, and they need to be terminated.

C19 Educators at all levels of education need to instruct students in the factual scientific causes for global warming, climate change, and ocean levels rising, and for the need of the public to be more considerate in reducing or eliminating their out-of-control activities causing pollution problems and/or wasteful expenditures of energy that are not "for the benefit of mankind and to the environment.

There is no question that the planet is experiencing a warming trend in some regions, but it is not the same trend in all regions where it has been shown to be a mixed combination of warming temperature, cooling temperature, or no significant changes in either trend toward warmer or cooler. Correspondingly, there is no question that the climate of the planet is also changing but again it is changing more in some regions than in others in keeping with the pattern of the temperature changes in regions.

The primary causes for the surface temperature and climate changes have scientifically been proven to be from many factors – astrophysical, geophysical, and natural - over which humans have no control. However, the finger of blame has realistically begun shifting from accusing humans and industry of creating "greenhouse gas emissions", which environmental scientists consider as the dominant cause for changes in temperature and climate, to humans being solely to blame for the true crisis of Man polluting his environment and hazarding his own safety and health.

C20 In this book ("The Big Picture of Climate Change on Planet Earth"), an improved method for estimating the temperature envelope of a region has been developed The derivation of the LEO Method is described extensively in the earlier chapters of this book but in summary for these Conclusions, it is a far closer estimation of the average temperature of a region at any latitude and at any elevation above sea level than that used by the environmental scientists – the one-size-fits-all "average global temperature" (approximately 60 °F (26 °C). The chart that best summarizes the LEO Method (revised on June 4, 2025) is shown as follows:

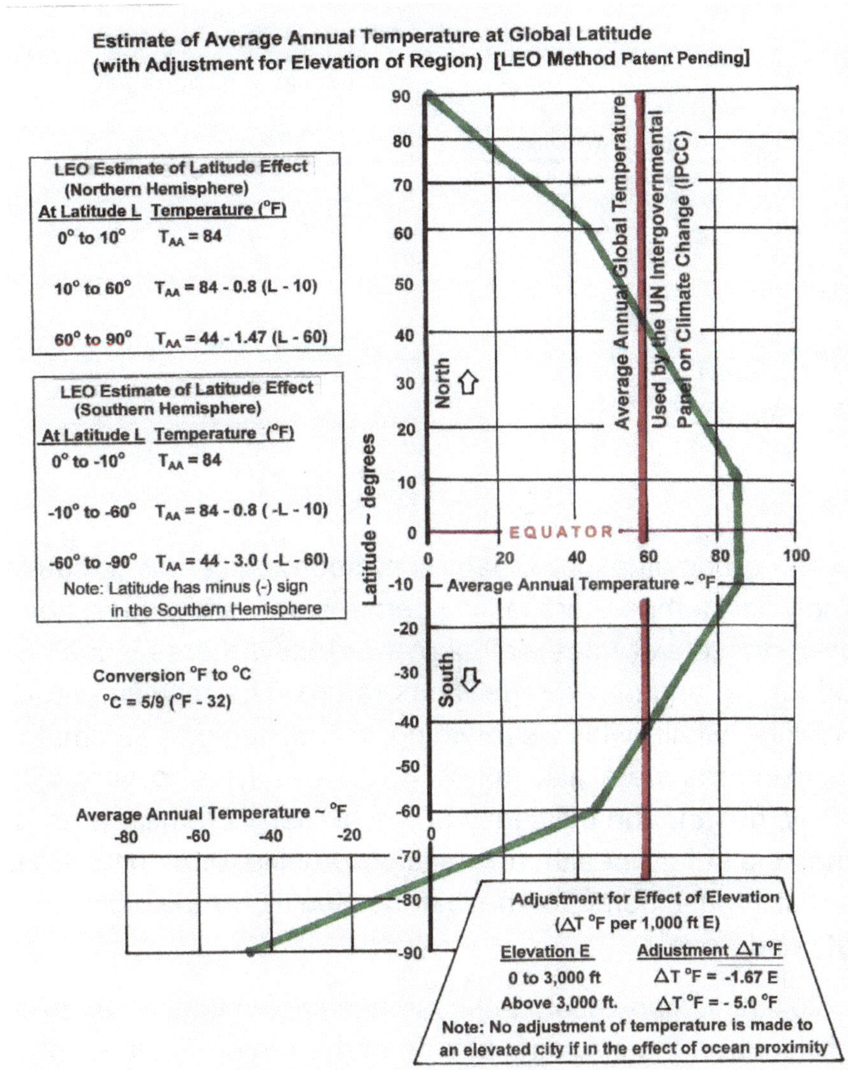

(Source: Author analysis of World Meteorological Organization database)

The primary benefit of the LEO Method (green line and mathematical modelling equations) is that it represents the basic average annual temperature as it varies with the latitude of the region anywhere between the North Pole and the South Pole. Because regions are not all at sea level a correction factor for elevation is added to the latitude effect (the elevation factor is included in the lower right of the diagram). The environmental scientists' red line average annual global temperature is seen to represent only the average annual temperature for the entire planet independent of latitude or elevation. Little if anything varies in Nature in an average, linear pattern of variation. It borders on ludicrous to believe that a 2 °C rise in the average annual temperature (the red line) constitutes a "crisis" when the variation from the actual annual temperature (green line) varies as much as 25 to 40 °F (14 to 22 °C) from the LEO temperature estimate. For each region being analyzed, the LEO Method only provides the annual average temperature for the region.

C21 Using the "average annual global temperature" masks the extremes of temperature that realistically occur seasonally (i.e., in four approximately 3 months periods) shown as following for the USA region:

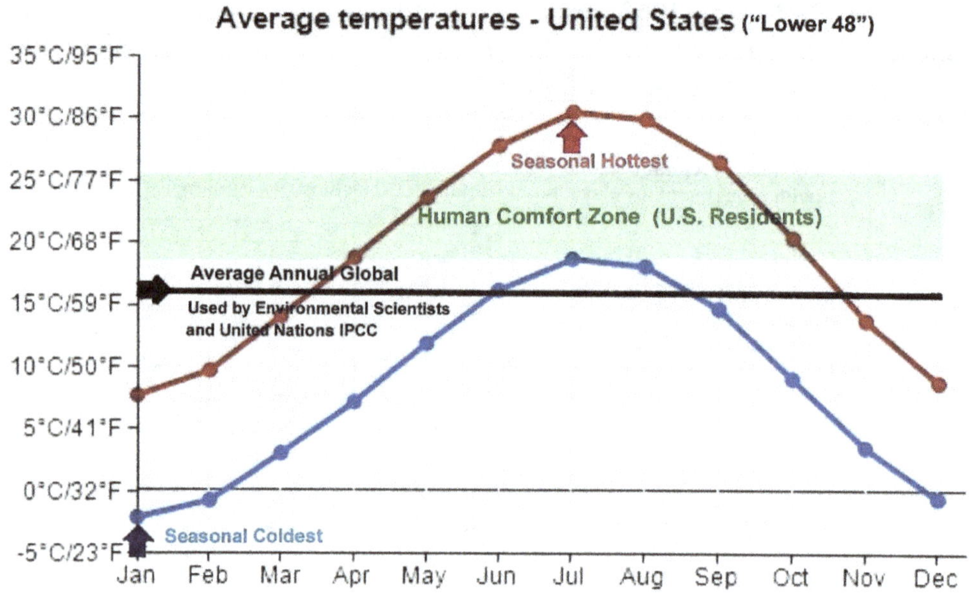

(Source: NOAA)

There are occasions when a combination of natural meteorological evnts cause a region to have an extremely hot or extremely cold temperature lasting perhaps only a few days; nevertheless, all life forms in the region must survive those extremes well above or below the seasonal extremes shown above and in the LEO method for the annual average temperature. The resulting reality is that every region has a temperature envelope within which everything in that region – animate or inanimate – must survive or perish. The extremes were also found in this analysis to vary with latitude. When it is considered that the safety, health, and efficiency of the humans on the planet is a much more urgent need to be protected than the safety, health, and efficiency of life forms on the ocean floor, the concern of environmental scientists must defer to the safety, health, and efficiency of humans and their technological machinery and "toys".

C22 The World Meteorological Organization database has shown that the average annual temperature of a region is very dependent on the average latitude of the region. Accordingly, the seasonal varying temperature – for the hottest temperature and for the coldest temperature – must be ascertained for each region being analyzed for the range of temperatures to which the population of the region will be exposed. For the United States region, the following temperature envelope contains the record extremes of hot and cold temperatures that were recorded in meteorological records (for the cities shown). The bold black horizontal line (at 62 °F) is the average of the temperatures within the temperature envelope and reflects that whether environmental scientists calculate an average annual temperature for a region or for the entire planet it is misleading and has caused much wasted effort and resources in trying to justify their hypothesis that a "crisis" exists because humans and industry are combusting fossil fuels to meet their electrical and heat energy demands (which will continue to increase as the world and the USA population continues to grow exponentially.

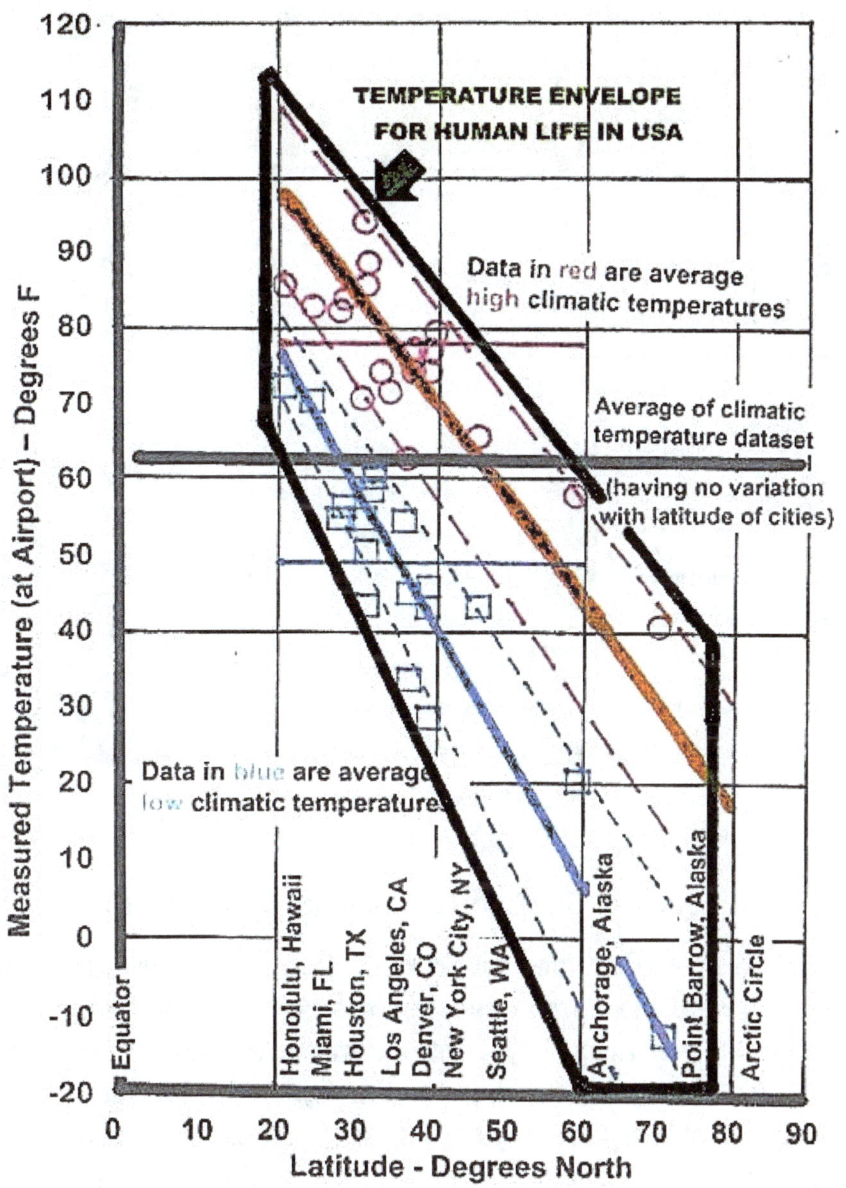

(Source: The meteorological database of the NOAA NWS, compiled by Author)

C23 In addition to the temperature extremes a region experiences, the climate of the region also has a precipitation factor that affects its total climate. For that purpose, the Koppen-Geiger Climate Model is the better model of climate than is the single "average global temperature" that environmental scientists use to determine their version of "climate change". The Köppen-Geiger model climate for the United States is formally contained in climate science as follows:

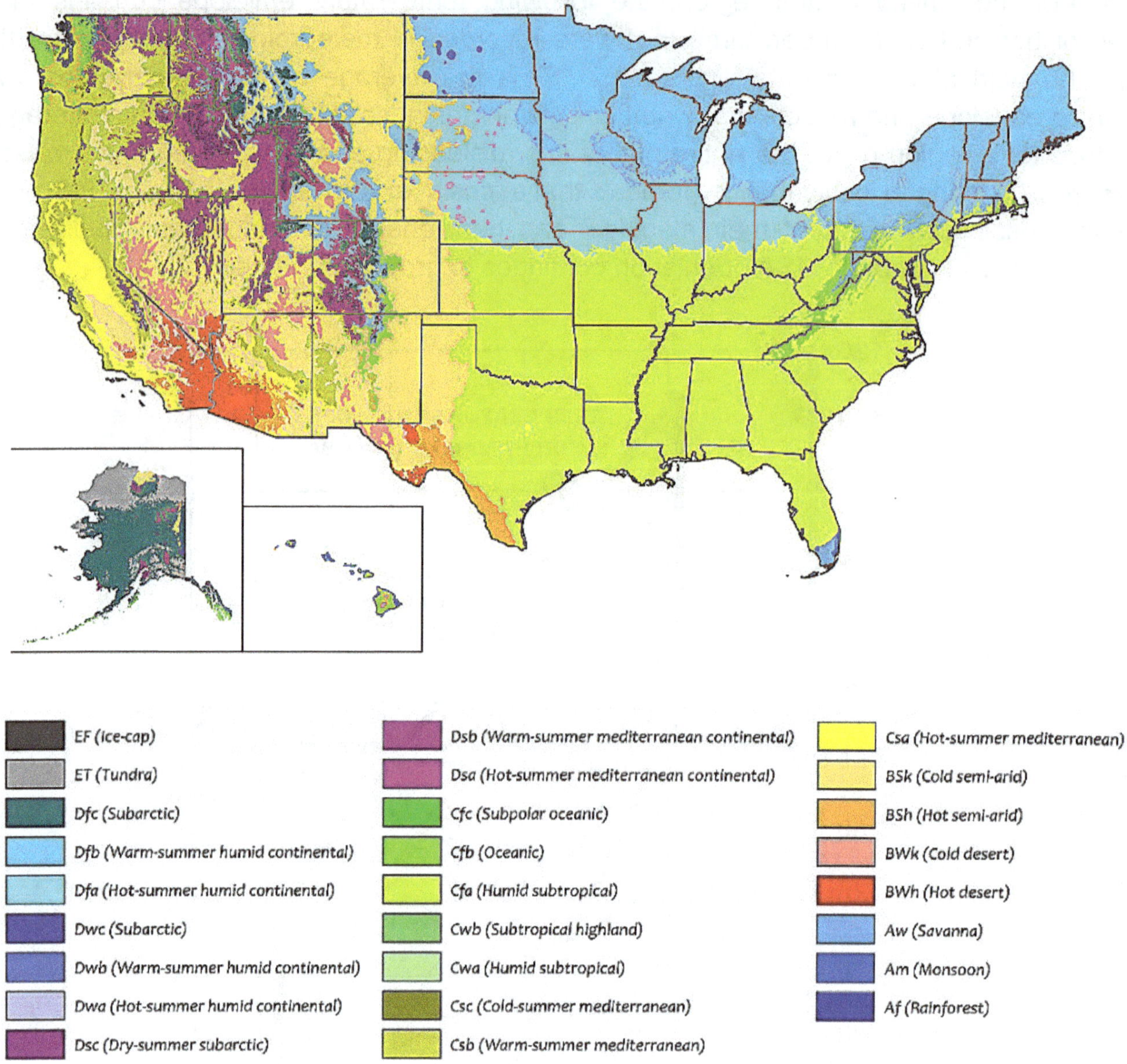

*Isotherm used to distinguish temperate (C) and continental (D) climates is -3°C

(Source: Smithsonian Institute for Science)

A logical combination of the previously described LEO Temperature Model (developed by the Author for this book) to closely estimate the annual average temperature for a region (considering its latitude and elevation and estimate of prevailing adjacent ocean current and prevailing wind current) with the Koppen-Geiger Model would be most meaningful to the agricultural industry of the region. It should be evident that within the climate of a region may be different sub-climates in subregions, as is well-shown above for the United States. The Köppen-Geiger Climate Model has been constructed for all regions of the world and is shown for the entire planet as follows:

Köppen - Geiger Climate Classification (Botanical)

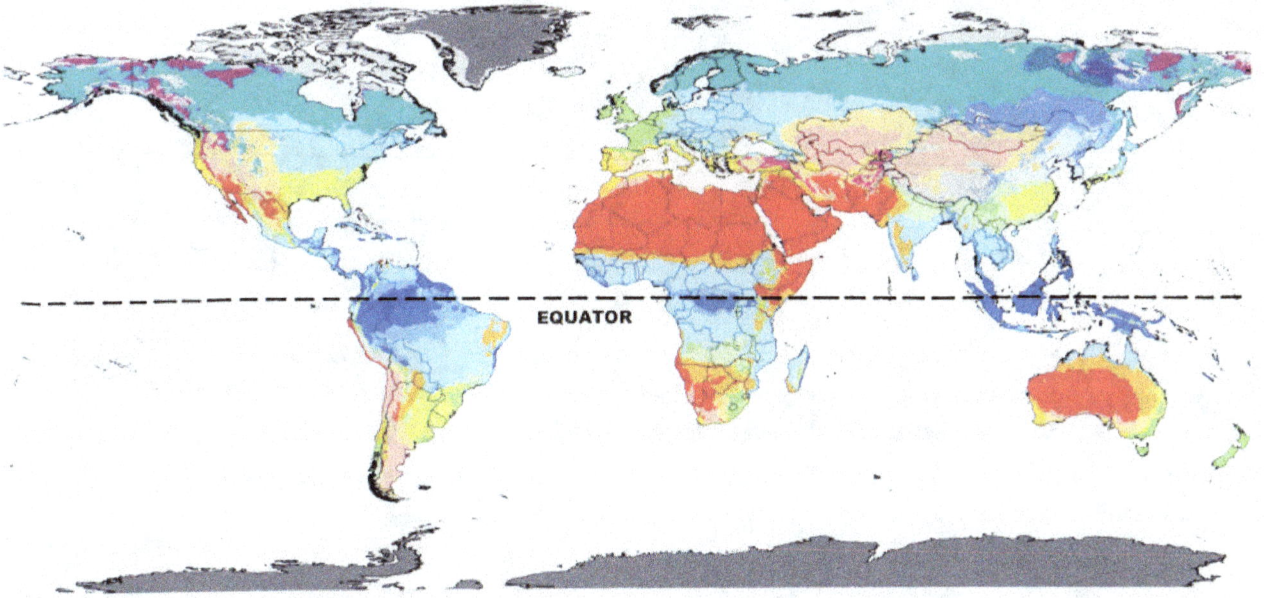

(Source: Smithsonian Institute for Science)

Needless to say, there is no mathematical model that could be developed to describe the Koppen-Geiger Climate Model (although the LEO Method for the temperature consideration of the climate of a region has beem modelled for the latitude and elevation factors).

C24 The amount of effort and amounts of money being expended (inefficiently and wastefully in reducing the "greenhouse gas emissions" as a means of altering the climate is appalling when considering that climate change is an effect not a cause. Environmental scientists are juggling the books in the laws of science to prove the fallacious hypothesis that Man burning fossil fuels is the dominant effect on global warming. A basic principle in physics is that matter having mass cannot be created or destroyed – matter or mass can only be transformed in state. (Matter matters). The thermodynamics of combusting 1 unit mass of coal requires 2.88 mass units of oxygen. If air is taken from the atmosphere to obtain 2.88 mass units of oxygen that requires 25 mass units of air. The 25 units of air are "borrowed" from the atmosphere, such that when environmental scientists proclaim that burning 1 gigaton of coal produces 26 gigatons of carbon dioxide in the atmosphere, the laws of conservation of matter and mass are being juggled fallaciously. Burning 1 gigaton of coal in solid state may increase the total mass of the atmosphere by however many units of gas volume one mass unit of solid-state coal transforms into gas, but the volume of carbon in gaseous state cannot have a mass of carbon greater than the original mass of coal (at near 100% carbon in solid-state). A loan of air borrowed from the atmosphere is much like a financial loan from a bank. The loan amount of air is paid back with some interest added, the coal representing the payback interest.

This Author would be remiss not to give high credit for the Environmental Protection Agency for their diligent efforts to reduce the amount of factual pollutants from the burning of coal being emitted into the atmosphere. A montage of four scenes is shown as follows which visually reflects the accomplishments of the EPA in reducing the pollution of coal burning – by adding capture devices of the carbon-particles from the emissions and by causing the transition from burning coal to that of oil and natural gas.

Coal burning facility with no emissions control

Coal burning facility converted to oil burning

Coal burning facility with EPA emissions control

Coal burning facility converted to natural gas

(Credit to unknown photographers)

Readers should note that the visible emissions (the dreaded "greenhouse gas emissions") in the left two photographs are not seen in the right two photographs. The upper right photograph is of the Consolidated Edison electric and steam generating facilities serving the New York City area, and the lower right photograph is of a facility in Alabama that converted from coal burning to the much- preferred natural gas burning. The environmental scientists who fallaciously claim that the burning of all fossil fuels must cease to avert a climate "crisis" are clearly mistaken.

(Author's note: The Author's father was in the Electric Production Department at the Brooklyn Edison part of the Consolidated Edison Corporation when the decision was made for the Edison Companies to transition from coal-burning to oil-burning. The Author can attest that the decision to make that transition was based on the elimination of the problem of disposal of the coal-ash residue (containing toxins and carcinogens) from the boilers and was not influenced by the coal burning process having any degree of influence on changing the climate. The elimination of the smoke from the four stacks blowing into the Manhattan business district when the winds blew in that direction was also an atmospheric quality improvement consideration but climate change in the 1940s was not a factor entering into the transition from coal to oil burning decision.)

C25 The final conclusion to complete this Conclusions and Concluding Comments section of this book is a recommendation that the young people of the United States be educated with the factual causes and effects of global warming and climate change (as revealed in these two books by the Author). Inasmuch as the purpose of the educational system is to educate (with the factual database of knowledge) the people in how to conduct their lives in the real environment and prepare the students to contend with all the many challenges – and opportunities - that the real world will bring to them. That is the largest reward the Author can hope for with the publication of these two books.

Several more conclusions were drawn as to what the human population of the planet can and should do to mitigate the effects of the temperature, climate, and oceanic changes. This book had the objective to identify the causes of global warming, climate change, and the oceans' surface levels rising and it is felt that that limited objective has been more than adequately met. The technologies that have been developed to date have been developed with the goal of reducing anthropogenic greenhouse gas emissions in hopes that the net result of the technologies achieved will cause an improvement in the surface temperature increases on the planet and in turn improve the climate of all regions of the planet.

The 25 conclusions presented in this book indicate that the technologies developed to date (2025) have had little effect if any on improving the climate, but that does not imply that those developments developed to reduce emissions into the atmosphere have been a waste of time and effort for not improving the climate. To the contrary, a Department of Energy graph (as follows) summarizes very concisely what the industries – and the population – of the United States have technologically achieved in the past twenty years:

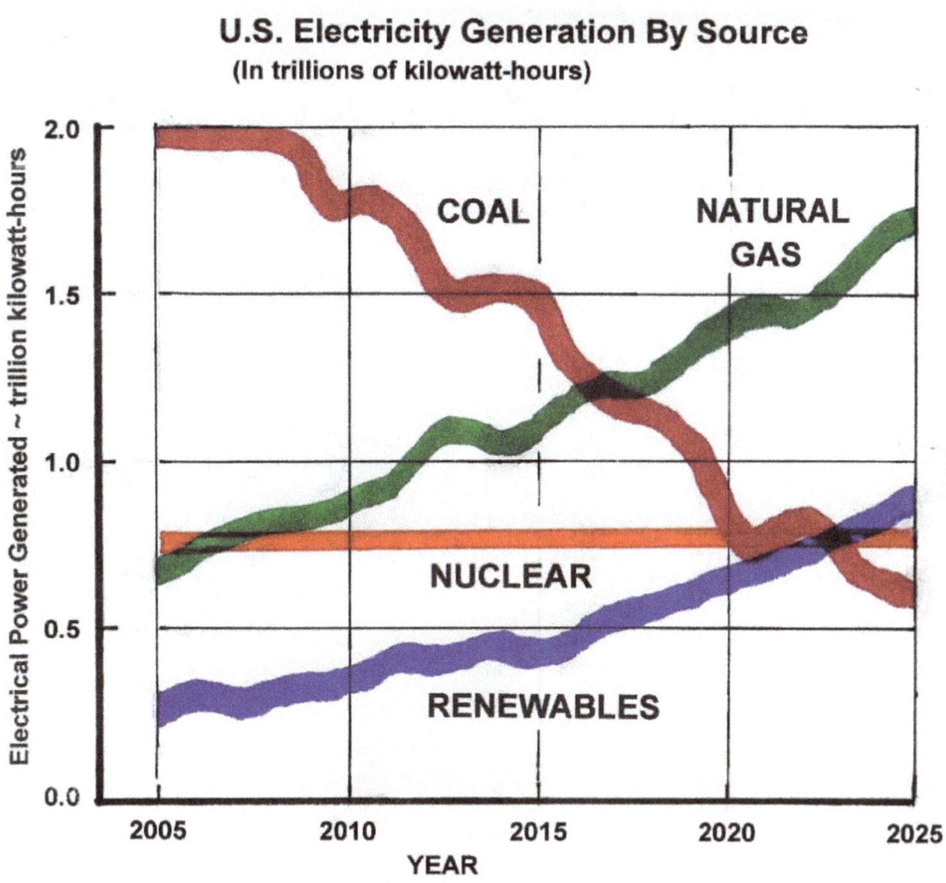

(Source: New York Times, January 2025, colorized by Author)

From a technological perspective, the above diagram must be considered as reflecting an improvement in those technologies which have pros and cons (as all technologies do have) and a reduction in the technology of coal where its technological need has been the major contributor to the development of the United States. The very hazardous con inherent with the use of coal is in the combustion of coal producing an ash residue that contains toxins and carcinogens that are polluting the water and land in the regions where those ash residues are being disposed of. Less polluting technologies (solar, wind, nuclear, geothermal, as well as the other fossil fuels oil and natural gas) have been developed to preclude the further use of a technology having a hazardous con.

The second part of the original question – "What can be done to mitigate the effects of the planet's surface temperature rising (regionally) and mitigate the effects of the associated climate change (also regionally) – is more than adequately addressed in the sequel book by this Author – "Dealing with the Real Climate Change". Readers will find that this second half of the question will be more than adequately answered. As an example of what will be addressed in the "Dealing with the Real Climate Change", the advancement of geothermal energy is the technology of deep drilling to harness the cleanest, most abundant, most economical, non-polluting, and safest form of technology that should be pursued with vigor in the United States. There are various drilling approaches (with or without "fracking" and this Author highly recommends the advanced geothermal system closed loop "pipe-in a-pipe" system functionally summarized in the diagram below:

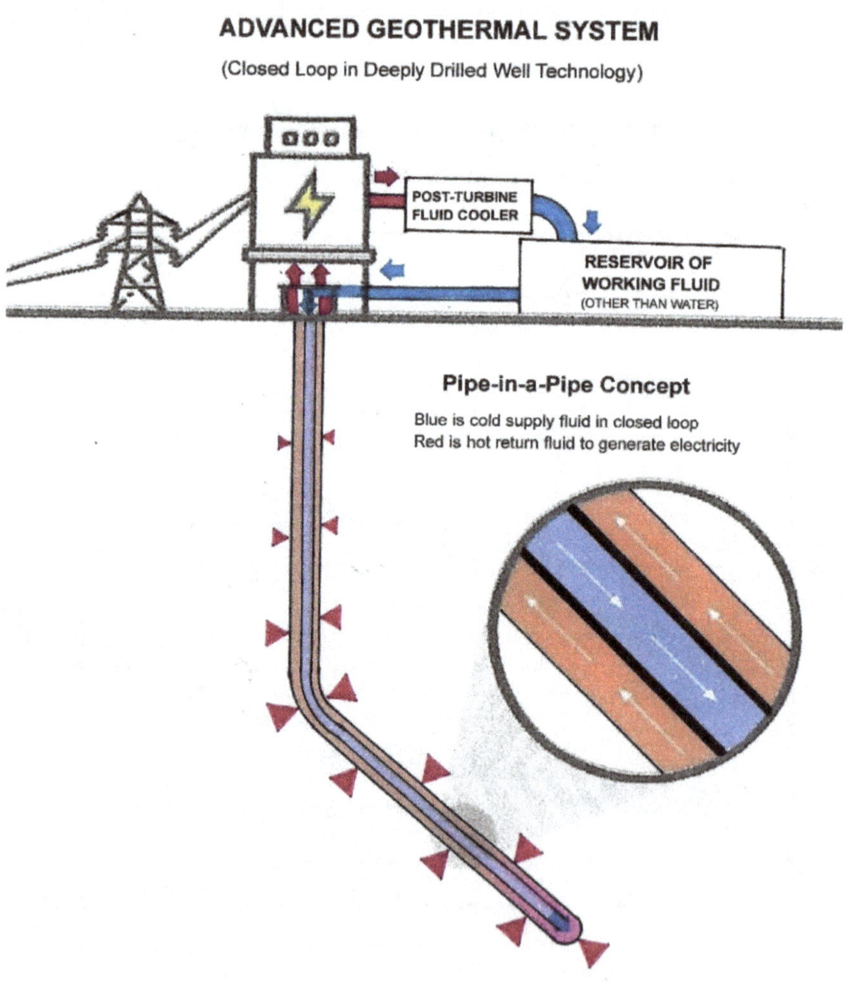

(Source: US Department of Energy with modifications/annotations by Author)

The selection of a technology that appears to be an advancement in the science of that technology does not ensure that the technology will be of use in its real-world region of application. Using a process commonly used in system engineering known as weighted trade studies, the Author identified the several factors of importance to be used in weighted trade studies, and the listing of those Importance Factors is provided to inform Readers how that form of weighting a technology "for its importance to humanity" is as follows:

1. Status of Technology Development (risk, schedule, cost, potential value)
2. Reliability and Sufficiency (time of day/year, weather dependent, regional need dependent)
3. Regional Economy and Resources (mineral, water, labor force to operate/maintain)
4. Cost of Resources (domestic and imported, mode of harvesting / distribution)
5. Importance to Industry, Government, and Human Safety, Health, Comfort
6. Importance and Urgency to Economy and Employment
7. Pros and Cons of By-Products (emissions, pollutants, hazards)
8. a. Effect on Human Population (regional in USA) - Residency.
   b. Effect on Human Population (regional in USA) – Food Supply.
   c. Effect on Human Population (regional in USA) – Safety, Health, Comfort
9. a. Effect on Infrastructure (regional in USA)
   b. Effect on Transportation (regional in USA)
   c. Effect on International Trade
10. a. Effect on Environment (regional in USA) – Atmosphere
    b. Effect on Environment (regional in USA) – Land
    c. Effect on Environment (regional in USA) – Water
11. a. Effect on Animal Life (regional in USA) – Land
    b. Effect on Animal Life (regional in USA) – Aquatic
    c. Effect on Animal Life (regional in USA) – Avian
12. Effect on Forestry and/or Agriculture (regional in USA)

By assessing the pros and cons inherent in each candidate technology for how well the technology satisfies the 12 importance factors identified above, the "report card" for that technology can be mathematically rated and given a final grade score.

That process (evaluating the value of technologies to the United States population) was completed by the Author for 25 candidate technologies that were identified as being of potential use in mitigating the effects of global warming and climate change. Regrettably, it is beyond the scope of this book to present the results of that assessment herein.

However, the Author has written a second part to this book – much the same as Leo Tolstoy did in his classic "War and Peace" – and has written "Dealing with the Real Climate Change". In combination, the two books address the two-part question posed at the beginning of this book "What is causing global warming and climate change?" and "What can and must be done to deal with those temperature and climate changes?" The "Dealing with the Real Climate Change" book completes the "big picture" of what combination of scientific factors are causing global warming and climate change and what can realistically be done by Man to survive the challenges climate change brings with its changes.

Appendix
Flash Flooding in Spain Kills Hundreds and with Much Damage

Eva Manez / Reuters

November 2, 2024

The full horror of the flash flooding in Spain began to emerge on Friday, just as new rainfall lashed southern parts of the country. The storm has killed at least 205 people, with 202 of those in the hardest-hit region of Valencia, emergency services in the region said Friday. It marks Spain's deadliest natural disaster in decades.

The death toll is expected to rise as emergency workers fight to rescue those who are trapped and recover bodies. Authorities warned Friday that roads have collapsed in some areas, with emergency services unable to get access. The Spanish Armed Forces have already rescued 4,607 people, Spain's Minister of Territorial Policies Ángel Víctor Torres Pérez said on Friday. SOS Desaparecidos, an association dedicated to sharing information about those missing through social media, said they have received reports of at least 1,300 people who remain unaccounted for.

The country has experienced significant autumn storms in recent years, but nothing comes close to the devastation wrought in the past few days. More details are emerging of the devastation in the Valencia region, with residents reporting large amounts of damage and horrific encounters with the rapidly rising water. A courthouse was turned into a temporary morgue in the region's capital, Valencia city. In the city's La Torre neighborhood, where the water rose to chest level, volunteers continue to search for more missing people.

Rescue teams discovered the bodies of seven people in an underground parking garage there on Thursday, according to national broadcaster RTVE, citing police.

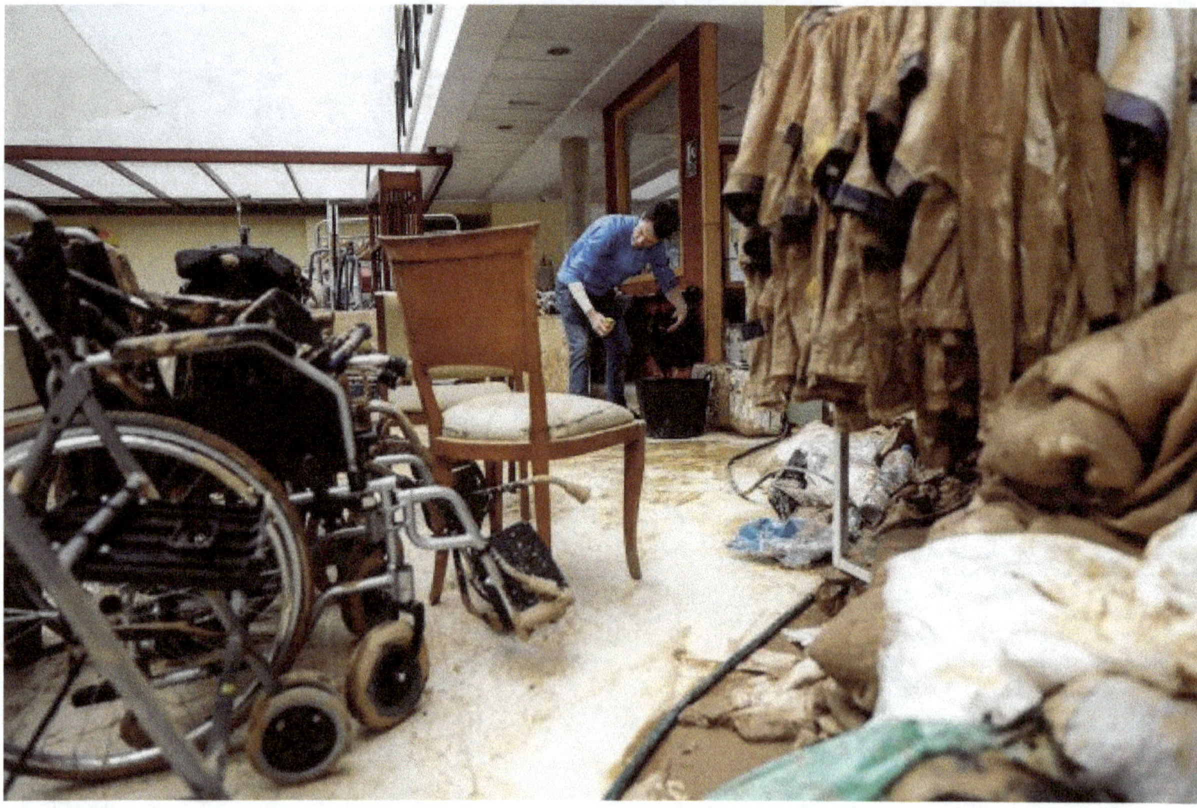

A man cleans mud from inside a home for elderly people in Sedavi on Saturday, November 2.

5454 LCP

DYA

100%

BOMBEROS

MAXAR

SE VENDE
GUARDIA CIVIL
POMPES FÚNEBRES
LA SAFOR

FIATC
INTORERIA
DARKNESS

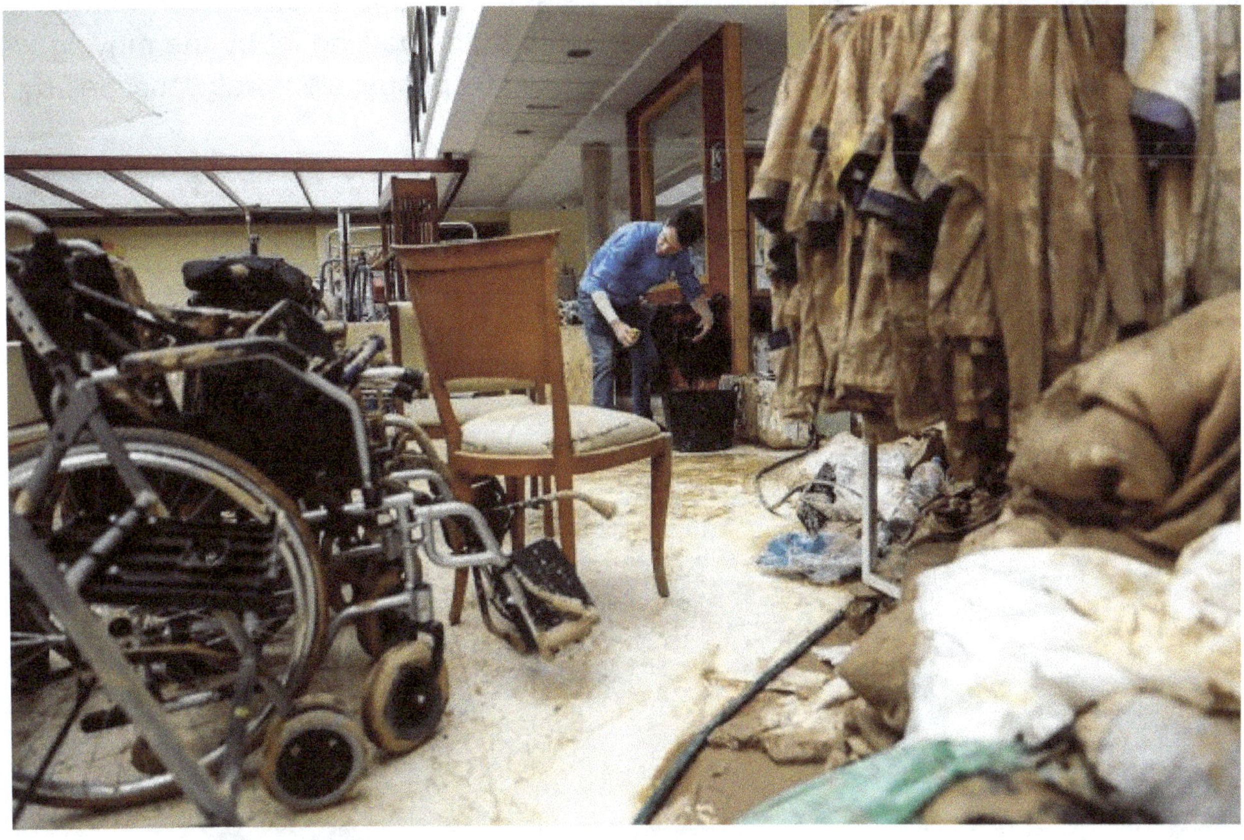

The father of one of those who died in the parking garage, a local policeman, told Spain's *El Mundo* newspaper that residents had rushed to move their cars, but the water rose faster than people were expecting, trapping them. Another woman was dragged into the parking lot by the moving water and died, he said. The town of Paiporta, Valencia, where at least 62 people died, was described by Spanish public broadcaster RTVE as the "ground zero of the tragedy."

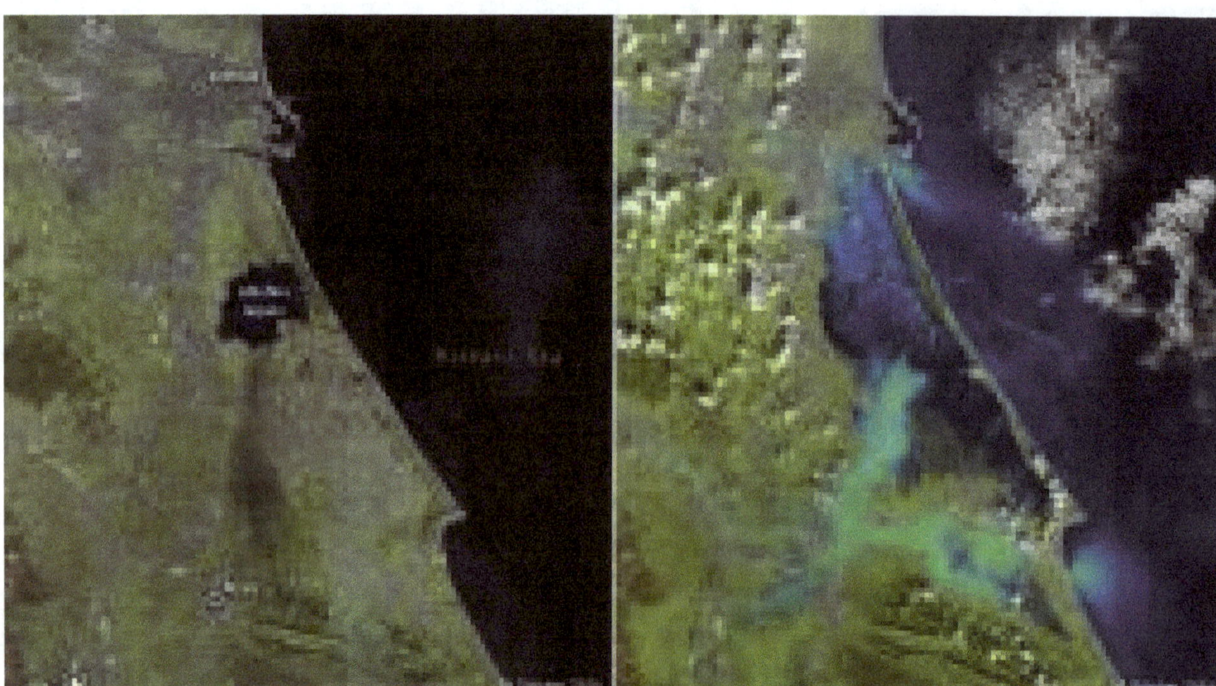

Parts of Spain appear to merge with the sea after historic rain in before-and-after images

A witness who was caught in the flash flood there told RTVE that he saw multiple cars floating past him with people begging for help. Many drivers found themselves caught on a highway and were swept away in their cars, as the road appeared to merge with a nearby river. A bridge also collapsed in the area. At least six people died in a nursing home on the outskirts of the town, Paiporta's mayor told the Spanish national broadcaster. While staff managed to bring most of the elderly people to the first floor, they were unable to save everyone. Mud still fills the streets in many areas, with the mayor of Valencia sharing images of community clean-up efforts on Friday. "Vehicles are being removed, the square is being cleaned and food and water are being collected," Mayor María José Catalá said of La Torre.

The regional government of Valencia said power has been reestablished in 90% of places. Carlos Mazon, the president of the Valencia's regional government, defended his administration amid accusations that authorities failed to alert residents on time. "The forecasts we received initially did not predict this (meteorological) revolution," he said in a video posted on X. Mazon said the regional government sent out "close to hundreds" of red alerts during the day of the storm, "including an SMS alert reserved for the worst possible scenario." Parts of Spain continue to see intense rainfall on Friday, and authorities issued a red warning overnight for the Huelva coast, in Andalusia, which had 140mm (5.5 inches) of precipitation in just 12 hours. Orange and yellow alerts also remain in place in isolated parts of Valencia.

N's Mia Alberti and Sophie Tanno contributed to this report.

About the Author

Herbert A. Hutchinson graduated from the nationally renowned Brooklyn Technical High School (Arista) in New York, and appears as one of a hundred Notables in the BTHS "100 Years of Excellence" who graduated in the first fifty years of Brooklyn Tech. He received a Bachelor of Science degree in Aeronautical Engineering at Georgia Tech in Atlanta, Georgia, in 1953. After serving as a Combat Engineers platoon leader in the U.S. Army Corps of Engineers in the Korean War, he returned to Georgia Tech to earn a Master of Science in Aeronautical Engineering degree (1959).

The initial nine years (1952-1961) of his professional engineering career were as a wind tunnel test engineer (Lockheed C-130) at the Georgia Tech Research Institute, as an aerodynamicist (F-84, F-105B and XF-103) at Republic Aviation in New York, and as a flight test performance engineer (B-58) at General Dynamics in Texas.

The middle half of his career (1961 – 1982) was as the Chief of the Aeromechanics Branch of the Air Force Flight Dynamics Laboratory, and as the Chief System Engineer (GS-15) for several U.S. Air Force weapons systems developments, including the U.S. Air Force Lightweight Fighter program and the U.S. Air Force F-16 Fighting Falcon Multi-National Weapon System, developed by the U.S. Air Force at Wright-Patterson Air Force Base in Dayton, Ohio. He was on the team conducting exploitation testing of foreign flight system technologies at the formerly very highly classified clandestine flight test site now known as Area 51. With extensive applied system engineering experience, he was selected to serve on three Presidential Advisory Committees, participated in numerous aircraft crash investigations (with numerous fatalities), and served three years as Chief of the USAF Flight Technologies Division (Aeronautical Systems Division).

Based on his having been the Chief System Engineer for the USAF YF-16 Lightweight Fighter Prototype Program and the ensuing F-16 Fighting Falcon multinational fighter for the USAF and twenty-eight other countries, former pilots of the advanced technology F-16, used in the several Gulf Wars such as Desert Storm, have awarded him the tactical call name "Stuffer". He has authored two books: "The Inside History of the USAF Lightweight Fighters, 1900 to 1975" and "The F-16 Fighting Falcon Multinational Weapon System, 1972 to 2019", both published by the Xlibris Publishing Company, and which reflect his understanding of the complexities and interactions that exist in applying emerging technologies in "big picture" development of large, complex systems.

The final twelve years (1982 – 1994) of Hutchinson's career were as a Manager of System Engineering at the Northrop Aircraft Division in California managing the system engineering of a new fighter aircraft design for NATO forces in Europe and of the design of stealthy flight vehicles (the Tri-Service Standoff Attack Missile (TSSAM) and the YF-23 Advanced Tactical Fighter (YF-23). He retired from his 41 years professional career at the end of 1994.

After retirement, he applied his system analytical and database research experiences to becoming a world-recognized genealogist and created the extensive family tree histories for his family, of English and British West Indian descendancy (fourteen generations 1595 to 1995), and that of his wife's family, of European descendancy (thirteen generations 1620 to 2000).

In collaboration with the Barbados Museum and Historical Society, he capitalized on the fact his two Barbadian grandfathers were part of the labor force in creating the Panama Canal (in this book as an example of a highly successful early "green" international environmentally-friendly project) to support the Barbados Government and UNESCO in naming the Panama Canal Zone as a Historical World Heritage Site. He invented "An Optronic System for the Measurement of Vehicle Traffic" and holds US Patent 6,750,787, measuring vehicles speeding on neighborhood streets to improve pedestrian safety in neighborhoods. His extensive experience in genealogy and in patenting an invention for improving human safety provided a wealth of knowledge about the human factors involved in everyday living in different regions and migratory habits of humans in different climates.

He is the author of three books on global warming and climate change: this book "The Big Picture of Climate Change on Planet Earth" and "Dealing with the Real Climate Change", and also the contemporary book "Facts and Fallacies of Climate Change on Our Planet", all written as a public service and intended to inform the public worldwide that inevitable climate change is happening but at different rates and with different effects and different state of urgency to be mitigated depending on the latitude of the region on the planet. In his books on global warming and climate change, Hutchinson provides data and a system-level, "big picture" analysis concluding that there is no "one-size-fits-all" solution or goal for mitigating the environmental and human problems brought about by global climate changes occurring in different regions of the planet. His third book assesses a set of twenty-five energy and transportation technologies being developed to mitigate climate change effects.

Hutchinson is an Associate Fellow Emeritus (seventy-four years continuous membership) in the American Institute of Aeronautics and Astronautics (AIAA), is a life member of the Veterans of Foreign Wars, is a member of the Air Force Association, and is a member of Mensa. His name is on the Smithsonian Institute Wall of Honor. He and his wife Arline reside in Durham, North Carolina. Herb continues to be active in the aeronautical engineering profession, passing along his extensive experiences to any group (educational, industrial, or Governmental) pursuing the application of new technologies. A visual indicator of his dedication and passion for describing STEM activities (or fishing with bait at least the length he is indicating?) applied in the "real world" is shown as follows:

Contact the Author:

The Author of this book "The Big Picture of Climate Change On Planet Earth" and the associated "Dealing with the Real Climate Change" is publishing the two-books as a public service to improve the "big picture" understanding of climate change and what can be done by the population (World and United States) to mitigate the global and regional effects. The Author has published a previous book "Facts and Fallacies of Climate Change on Our Planet" which also addresses the topics of global warming and climate change. Two previous books published by the Author regarding "The Inside History of the USAF Lightweight Fighters" and "The F-16 Fighting Falcon Multinational Weapon System" reveal the Author's extensive experience in developing advanced technologies for global "real world" system level applications.

The Author may be contacted with questions or provided with constructive criticism at:
herbineer@aol.com

References for This Book

Date Published	Author	Title
1780	St. Michael Parish Register (RL1/5)	Account of "The Great Hurricane of 1780"; John Orderson, Clerk, St. Michael Parish, Barbados
1982	Mensa Society	"The Constitution of Mensa"; Mensa International Society
1990	Weiner, Jonathan	"The Next One Hundred Years: Shaping The Fate Of Our Living Earth"
1992	Readers Digest	"Atlas of the World" Rand McNally Maps
2001	Lomborg, Bjorn	"The Skeptical Environmentalist"
2007	Aleshire, Peter	"The Extreme Earth: Ocean Ridges and Trenches"
2008	Hutchinson, Herbert A.	"Commemorating the Barbadians who Excavated the Panama Canal"; Journal of the Barbados Museum & Historical Society, Vol. LIV
2008	Singer, S. Fred, and Avery, Dennis T.	"Unstoppable Global Warming" Every 1,500 Years
2009	Gray, Theodore	"The Elements: A Visual Exploration Of Every Known Atom"
2009	Jenkins, Steve	"Down, Down, Down: A Journey to the Bottom of the Sea"
2016	Idso, Craig D., Carter, Robert M., and Singer, S. Fred	"Why Scientists Disagree About Global Warming: The NIPCC Report On Scientific Consensus"; Second Edition
2020	Lomborg, Bjorn	"False Alarm: How Climate Change Fails To Fix The Planet"
2021	Koonin, Steven E.	"Unsettled: What Climate Science Tells Us, What It Doesn't"
2021	Krauss, Lawrence M.	"The Physics Of Climate Change"
2021	Singer, S. Fred, Legates, David R. and Lupo, Anthony R.	"Hot Talk, Cold Science: Global Warming's Unfinished Debate"; 3rd Edition

2022	Corsi, Jerome R.	"The Truth About Energy, Global Warming, and Climate Change: Exposing Climate Lies In An Age Of Disinformation"
2022	Mitchell, Guy K.	"Global Warming: The Great Deception"
2023	Curry, Judith A.	"Climate Uncertainty And Risk: Rethinking Our Response"
2023	Einhorn, Stephen	"Climate Change: What They Rarely Teach in College"
2023	McGee, Steve	"Climate Change: Reasons To Worry Less"
2023	Thunberg, Greta T.	"The Climate Book"
2023	Williams, John R.	"Earth Plight: Global Warming: The Harsh Reality"
2024	Balzer, Lynne	"Exposing the Great Climate Change Lie"
2024	Ineichen, Alexander	"The History of Climatism: Parts 1 through 9"

Vhy this book?

he climate is changing but is factually proven in the "big picture" to be caused by astrophysical, eophysical, and natural regional causes and little by Man-created "greenhouse gas" emissions. nvironmental scientists' use of a one-size-fits-all "average global temperature" (land and oceans urfaces) for the planet is factually disproven in this book as being a valid indicator of the climate ffecting all regions of the planet. Humans must survive seasonal extremes of hot and cold temperature a region, not a benign annual average global temperature.

his book has factually shown that the "average global temperature" and "one-size-fits-all" goals of the nvironmental science groups (especially those in the United Nations Intergovernmental Panel on limate Change (IPCC)) to limit the change in average global temperature to not exceed two degrees entigrade by 2035 and achieve net zero "greenhouse gas" emissions by 2050 are unrealistic. A andidate model (LEO for latitude, elevation, and ocean effect) has been developed to assist the limate scientists in using a more meaningful model than that of the fallacious "average global emperature". From an educational perspective this book illustrates how STEM can be realistically pplied to solve complex, dynamic, multi-variable problems,

ven if fully achieved, the present "combat" of climate change and reduction of "greenhouse gas missions" will bring little if any change in the surface temperature or climate in most regions. The goal hould be to protect human safety, health, comfort and safe transportation, not to preserve the nvironment at any cost to the human population. An assessment (using the weighted importance actor method) for 25 candidate technologies was made to mitigate the effects of the inevitable climate hange on our planet and will be provided in "Dealing with the Real Climate Change" by this same uthor addressing what Man can - and must - do to mitigate the inevitable regional temperature and limate changes and adapt to the ravages brought about by adverse weather catastrophes.

his book was also written to allay the fears of gullible young people who have been misled to believe heir elders are carelessly destroying their future.

here is no "greenhouse gas" global warming or climate change crisis actually occurring on our planet – but there is an urgent need to reduce the Man-made pollutants (toxins, carcinogens, and especially plastics) of the and, water, and atmosphere. Reduction of the use of coal is a timely action as there are several cleaner technologies – such as liquid natural gas – to eplace coal. As the population grows, so does the pollution problem.

Why this book?

www.ingramcontent.com/pod-product-compliance
Lightning Source LLC
LaVergne TN
LVHW082025200926
848481LV00010B/2926
* 9 7 9 8 8 9 3 9 5 1 7 8 3 *